BIOFUEL'S ENGINEERING PROCESS TECHNOLOGY

Edited by **Marco Aurélio Dos Santos Bernardes**

INTECHOPEN.COM

Biofuel's Engineering Process Technology
Edited by Marco Aurélio Dos Santos Bernardes

CBS, Edition 2016

Published by InTech

Janeza Trdine 9, 51000 Rijeka, Croatia

Publishing Process Manager Petra Zobic

Technical Editor Teodora Smiljanic

Cover Designer InTech Design Team

Image Copyright EMJAY SMITH, 2010. Used under license from Shutterstock.com

Additional hard copies can be obtained from orders@intechopen.com

Biofuel's Engineering Process Technology, Edited by Marco Aurélio Dos Santos Bernardes
p. cm.
ISBN 978-953-307-480-1

INTECH
open science | open minds

Contents

Preface

Over the past 20 years, there has been a substantial increase in research and development in the area of biofuels. Many researchers around the world have dealt with environmental, economic, policy and technical subjects aspects relating to these studies. In a way, this book aspires to be a comprehensive summary of current biofuels issues and thereby contribute to the understanding of this important topic. Chapters include digests on: the development efforts on biofuels, their implications for the food industry, current and future biofuels crops, the successful Brazilian ethanol program, insights of the first, second, third and fourth biofuel generations, advanced biofuel production techniques, related waste treatment, emissions and environmental impacts, water consumption, produced allergens and toxins.

Relating theoretical and experimental analyses with many important applied purposes of current relevance will make this book extremely useful for researchers, scientists, engineers and graduate students, who can make use of the experimental and theoretical investigations, assessment and enhancement techniques described in this multidisciplinary field. Additionally, the biofuel policy discussion is expected to be continuing in the foreseeable future and the reading of the biofuels features dealt with in this book, are recommended for anyone interested in understanding this diverse and developing theme

Dr.-Ing. Marco Aurélio dos Santos Bernardes
Environmental Assessment and Management
Postdoctoral Researcher at CRP Henri Tudor
66 rue de Luxembourg

Part 1

Process Control and Dynamics

1

The Effect of Thermal Pretreatment Process on Bio-Fuel Conversion

Aleksander Ryzhkov, Vadim Silin, Tatyana Bogatova, Aleksander Popov and Galina Usova
Ural Federal University named after the first President of Russia B.N.Yeltsin
Russia

1. Introduction

In Russia exploitation of local low-grade fuels assumes the usage of bio fuels for a variety of purposes. The development of advanced combustion and gas-generating facilities operating on low-grade fuel requires the knowledge of burning and gasification processes in moderate low-temperature combustion modes.

As is known, the amount and quality of fuel particle surface open for reaction with oxidizing agent is most important for the rate of thermochemical conversion. Under comparable conditions the fuel having the largest reaction surface will have the highest rate of burn out. In qualitative terms, the properties of internal surface that formed long before it entered the furnace (reactor) and those that are forming directly in the furnace may differ. Gasification of the above fuels in autothermal mode produces gas with high content of complete combustion products (CO_2 and H_2O) and hydrocarbons and low chemical efficiency. To rise the efficiency it is necessary to implement allothermal conditions, to improve heat recirculation. To study marginal allothermal conditions ("ideal gasification") and the ways of their control a number of experiments and calculation-based estimates were made.

2. Experimental procedure

The experiments were performed on "model fuels" that differed greatly in their thermal and kinetic properties. Low-ash high-reactivity bio fuels of medium (wood) and high (dates seeds) density and products of their treatment (charcoal) were used to study the affect of pyrolysis kinetics, material density on formation of coke-ash residue reaction structure. Charcoal as oxydizing pyrolysis product entering the reaction zone of gas generators was used for investigation of gasification modes with different blow conditions. Fuels characteristics are given in Table 1.

Conditions of porous structure formation and porosity during preheating (devolatilization) were studied based on biomass particles with equivalent size $d_p \approx 10$ mm. The particles were heated by two methods: fast heating by placing the particle in muffle furnace preheated up to preset temperature (100, 200, … 800 °C with accuracy ±20 °C) and slow heating simultaneously with muffle heating under conditions of limited oxidizing agent supply. This allowed to simulate real conditions of thermal processes i.e. fast heating (for instance, particle pyrolysis in fluid flow- or fluidized bed-type carbonizer) and slow heating (when the particle enters a cold fluidized bed and gets warmed gradually with the fluidized bed). After cooling the porosity was measured (mercury porometry: volume and sizing the pores with $d > 5.7$ nm) and specific surface area (nitrogen adsorption: surface area of pores with diameter $d > 0.3$ nm).

Parameter	Charcoal	Wood (pine)	Wood pellet	Date seed
Original particle				
Moisture of fuel as received W^{ar}, %	1.4	8	10	4
Ash (dry basis), A^d, %	0.9	1	2	0.97
Volatile content V^{daf}, %	15	88	87	85
Low heat value Q^{daf}, MJ/kg	31.5	18.1	17.5	18.9
Apparent density of fuel as received ρ, kg/m3	380	520	1200	1150
Porosity Π, %	75	65	20	25
Specific surface area S_0, m^2/g	8.6	1.0	2.0	0.01
Coke-ash residue after pyrolysis (fast heating/slow heating)				
Ash content A, %	NA / 1.5	NA / 3	NA / 6	NA / 2
Volatile content V^{daf}, %	NA / 1	NA / 1	NA / 1	NA / 1
Apparent density ρ, kg/m3	280 / 320	230 / 260	NA / 360	200 / 620
Porosity, Π, %	80 / 77	85 / 83	NA / 70	87 / 60
Specific surface area S_0, m^2/g	NA / 29.2	454 / 366	NA / 436	NA / 9.1

NA - not available.

Table 1. Model fuels characteristics

Kinetics of conversion in combustion mode was studied on individual particles with equivalent diameter $d_p = 3–75$ mm. The range of diameters examined corresponds with values showed in (Tillman D.A., 2000) as allowed for individual and co-combustion (gasification) of biofuels. Test sample placed (centered) on thermocouple junction (Ch-A type) was brought into the muffle which was preheated up to preset temperature (100, 200, ... 800 °C with accuracy ±20 °C). The tests were performed with air flow rate 0–3.5 m^3/h (upstream velocity of the flow is 0–0.5 m/s in normal conditions). Average effective burning velocity for coke-ash residue was calculated as loss of coke-ash residue estimated weight per surface unit of equivalent sphere (based on original size) during coke-ash residue burning out: $j = \Delta M / (\Delta\tau_{car} \cdot F)$. Coke-ash residue (CAR) burn-out time ($\Delta\tau_{car}$) was estimated by thermograms (fig. 1.) as time interval between points C and D. The length of A"–B segment was not accounted for.

In some aspects, individual particle burning, combustion in fluidized bed and in flame, may be assessed on the same basis. Both in flame and in fluidized bed the fuel particles are spaced at quite a distance from each other and are usually considered as individual particles. The intensity of heat-mass-exchange of particles burning in FB inert medium is comparatively close to individual particle intensity. Application of experimental data for individual particle burning to calculation and assessment of thermo chemical pretreatment of large-size particles in furnaces with dense bed is justified by the fact that heat-mass-exchange processes in its large size elements are the same as for individual particle, within the statement of the problem. Therefore the experimental data on individual particle burning are usually used in calculations to assess thermo chemical pretreatment of large particles in furnaces of various types.

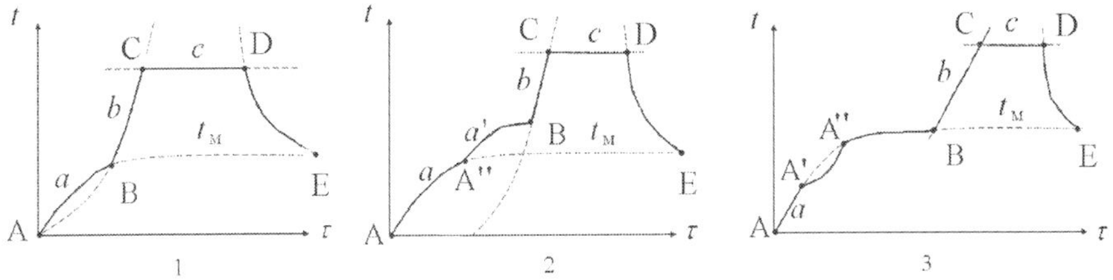

Fig. 1. Schematic view of fuels thermograms at t_m = 400 oC; Moments when: A - the particle enters the furnace, A' - endothermic reaction starts to dominate, A" - process returns back to curve of inert matter heating curve, B - intense oxidation (self ignition) of coke ash residue begins, C - quasi-stationary burning of coke-ash residue begins, D - coke-ash residue has burnt out, E - ash cooled down to muffle temperature. Processes: a - heating by inert matter curve, a' - heating due to pyrolysis gases burning; b - sef-heating of coke-ash residue, c - quasi-stationary process of coke-ash residue burning; temperature in particle center *t*, °C; time since the moment the particle entered the muffle τ, s

Experimental data was compared with other researches' data on thermal pretreatment of low-grade fuel particles in the air showed in Table 2.

No	Material (method)	Particle size, mm	Environment temperature, °C	Speed of blowing, m/s	Re_d
1	Charcoal (IP)	3-80	250-1200	0-0.5	11-43
2	Wood (IP)	3-80	250-1200	0-0.5	11-43
3	Pellet (IP)	13	250-1200	0	0
4	Date seed (IP)	11	100-1200	0	0
5	Pellet (IP)	13	600-1000	0.18	10-24
6	Charcoal (IP)	3-5	280-335	0	0
7	Brown coal (FB)	2.5-5.15	800-950	0.23-0.46	2-11
8	Brown coal (IP)	0.1-1.2	850	0.01	0.004-0.05
9	Brown coal (IP)	0.1-1.0	950-1200	0.02-0.03	0.008-0.13
10	Antracite (IP)	0.1-1.0	950-1200	0.02-0.03	0.008-0.13
11	Antracite (FB)	2-9	750-950	0.54	8-40
12	Antracite (IP)	15	1000-1500	0.27-1.0	18-65
13	Fossil coal (FB)	2-10	800	0.25	4-20
14	Carbon (IP)	5.5-8.5	850-1450	0.01-10	0.5-700
15	Graphite (IP)	15	800	0.6	69
16	Electrode C (IP)	15, 25	1300	0.02	1.3-2.2

Table 2. Experimental data on thermal pretreatment of particles in the air (IP - individual particle, FB - fluidized bed)

Kinetics of conversion in gasification conditions was studied at the plant consisting of quartz retort with inner diameter 37 mm, length 650 mm, located in cylinder-shape muffle furnace (N_{el} = 2.5 kW, T_{max} = 1250°C), air blower (Q_{max} = 6 m³/h, H_{max} = 0.6 m), electric heater, steam generator, rotameter, a set of thermocouples, carbon dioxide cylinder and thermocouple polling and temperature recording system. Combustible gas components (CO, H_2, CH_4) were determined by gas chromatograph, air flow coefficient was determined by effluent gas composition.

The experiments were performed in dense bed which provides for the most strict fulfillment of fuel thermochemical pretreatment as stratified process, stepwise and in compliance with temperature and concentration conditions, without flow disturbances and fluid mechanics problems. Gasification was based on downdraft process. Particles with initial diameter, varying from 3 to 20 mm in different experiments, were placed in retort having a tube welded to its bottom for gas release and sampling for analysis. Fuel bed was heated in muffle furnace up to 600–1000°C (the temperature depended on experiment). Gasifying agent (air, air and water vapor, water vapor, or carbon dioxide) was fed via furnace tuyere inside the bed to a different depth. Blown fluid was heated by electric heater up to 700-750°C. The experiment was considered to be completed at the moment when CO and H_2 content in gas lowered by less than 1% of volume.

3. Experimental results and their analysis

3.1 Fuel structural transformation by pyrolysis

Pyrolysis causes significant changes of physical and chemical properties of fuel particles. Measurements showed a two-fold reduce of bio-fuel particle density in a narrow temperature range. Particle shrinks insignificantly, not more than 30% of its initial size. Since the change of volume does not exceed 20% of initial value, whereas the density decreases greatly, the porosity of particles increases. Due to pore opening the oxygen can reach new surface which was inaccessible earlier (fig. 2).

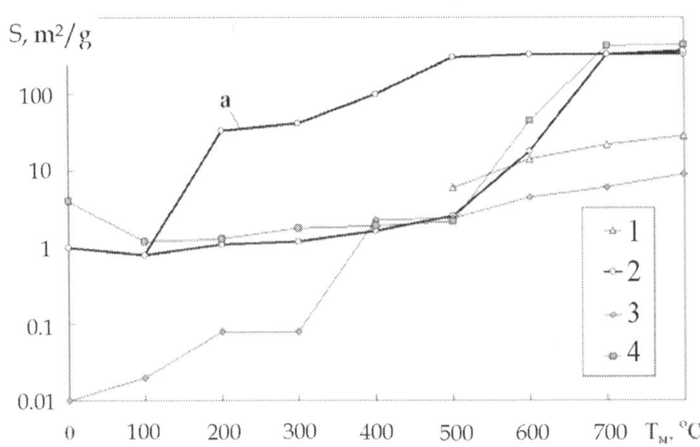

Fig. 2. Specific surface area of fuels during heating. Symbols: 1 - charcoal; 2 - wood chips, 3 - date seeds, 4 - wood pellet; a - fast heating, others - slow heating

In the range from 200 to 700°C the specific surface area of wood chip greatly depends on rate of heating, i.e. in case of fast heating it will be one or two orders higher than area during slow heating. Fuel porosity curves (fig. 3) for test samples revealed three steady peaks at 5–50, 100–3000 и 10000–50000 nm in mezzo- and macro porosity domain (5–50000 nm).

The porosity of the first nanolevel is typical of dense fuel particles. For most coals the pores' average diameter is within 4–10 nm, and more rarely in the range of 30–40 nm. Artificial materials with such pores are highly-scorched activated coals intended for absorption of large molecules such as organic dyes. When a particle enters the furnace the volume of pores of this class will greatly increase due to thermal decomposition of organic compound which is initiated by temperature rise and depressed with pressure increase being a natural regulator of gas formation process in material pores.

Nuclei of pore formation at nanolevel are thinner pores and cracks and it is within their volume that the detachment of gaseous "fragments" of splitting macro molecule of coal material occurs. Initial size of these pores is close to that of gas molecule diameter (0.4-1.0 nm). Cavity degasification process is retarded by molecular repulsion forces hindering the pass through "contracted" points in ultra microcracks and requiring great energy (activation) to overcome them which results in change of the state of dispersion phase. Materials with flexible structure (wood) form swollen-state colloidal systems resistive to both contraction and further expansion. In solid materials (cokes) structures similar to those observed in metals of interstitial compounds can be formed. The speed of gas diffusion from these pores depends on activation energy and temperature level. Gas molecule travel during typical in-furnace process time is compatible with the size of coal macro molecule.

Gas emission from numerous ultra micropores into larger ones acting as collectors continues during the entire particle burning period.

Significant flow resistance due to system porosity results in intra-pore pressure rise (up to saturation pressure) at initial destruction stage and in development of positive flow in the largest pores that hinders external gas inlet into particle pores. Simultaneously mechanical (rupture) stress may develop in the particle. With destruction process transit in its damping stage and intra-pore pressure reduction, pyrolysis gaseous products will be able to react with external oxidizing agent not only on the surface of the particle but inside the latter creating quite favorable conditions for homogeneous intra-pore burning.

As soon as degasification process is completed, free molecule diffusion (Knudsen diffusion) mode is established in nanolevel pores, coupled with convective Stefan's flow. Based on numerous estimates, for particles from 10 to 1000 μm the degree of such porous space (with specific surface area S_p) participation in reaction insignificantly depends on particle size and at 600 °C it is for oxygen within the range of S_p / S_{car} < 0.1 (for fast heating cokes) and S_p / S_{car} < 0.03 (for slow heating cokes).

Pores of the second (medium) peak (d_p = 0.1–3 μm) occur in the domain of transition from Knudsen mode to normal diffusion. They provide a better access for oxidant and can participate in reaction in larger volume. In pores of the third peak (d_p > 10 μm) diffusion runs similar to that in unrestricted space. These pores constitute insignificant part of internal surface and their contribution to burnout rate is known to be negligible. However, their role is quite significant as they can deliver reagent to joined pores of first and second peaks.

The obtained data show that wood particles and pellets have low-porous structure (S_0 < 2 m²/g), charcoal has mesoporous (S_0 < 8.6 m²/g) and seed has dense microporous structure (S_0 < 0.01 m²/g). Specific surfaces vary quite significantly in original state but this difference tends to flatten out for products of their thermal treatment. It increases to the third order for seed (up to 9 m^2/g), to the second order for wood and its products (pellet) (400 m^2/g), and negligibly for charcoal (three times).

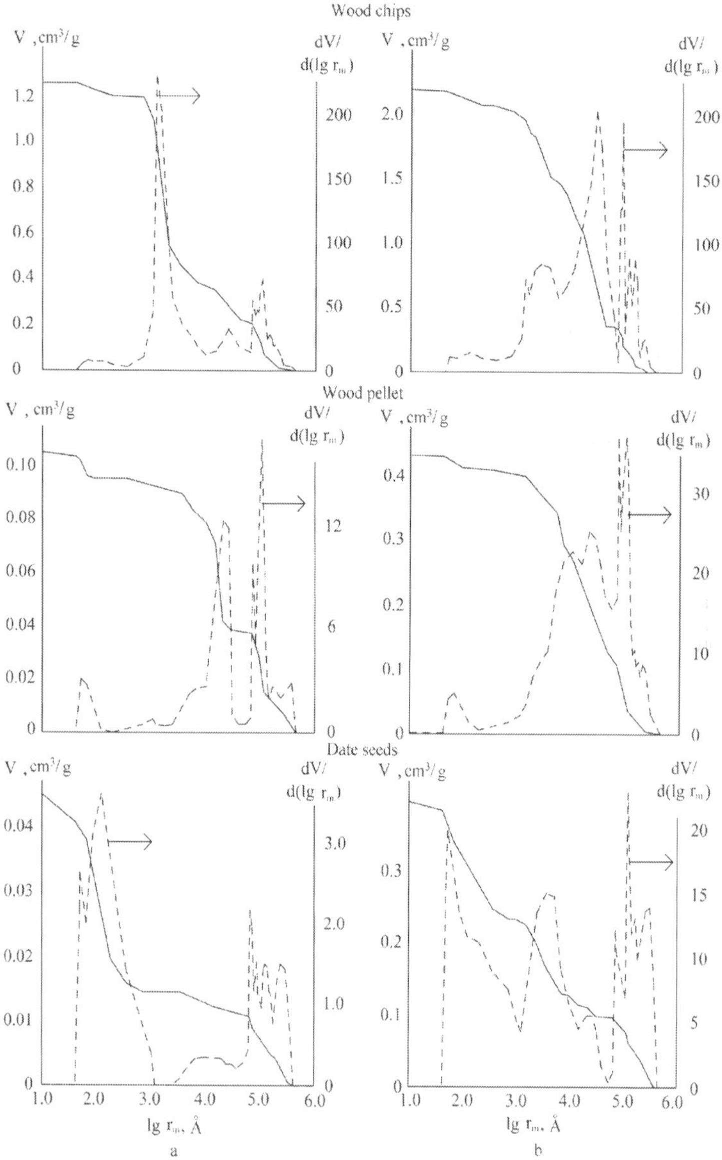

Fig. 3. Porosity curves for test fuels; pyrolysis temperature: a) 20°C, b) 800°C; r_m – medium (average) radius of pores

In seed which relates to bio fuels with the highest natural density and occupies intermediate place between wood and fossil coal the first peak pores dominate (more than 65% by volume). They are followed by the third peak pores (25%). Total volume of seed pores (0.045 cm^3/g) is 2,7 times less than that of pellet (0.12 cm^3/g) and 27 times less than that of the wood (1.22 cm^3/g). Specific surface area of seeds is two orders lower than that of the wood. After thermal treatment the pore volume of the seed increased 10 folds and there appeared a second peak on the background of the first and the third peaks which is compatible with these two peaks, although their heights increased by one order.

In bio fuels with natural density (wood) the pores of second type dominate, whereas the pores of the first type have not been revealed and volume of the third type is insignificant. Thermal treatment of wood results in slight increase of total pore volume (twice), whereas its structure changes to form larger pores. The height of second peak reduced three times and the height of the third peak increased three times.

In pellet the structure of the wood subjected to sever mechanical processing (crushing, pressing) differs greatly from the original one, forming larger pores with drastic reduction of their original total volume (10 fold reduction). Pore distribution in pellet after thermal treatment is qualitatively identical to original one but the total volume increased 4 times and peaks became twice as high.

Comparison of porosity curves for various fuels shows that thermal treatment of bio fuels with different original structure will flatten out the difference with the formation of common transport pore structure for all fuels which may result in similar burn out rates by volume for their coke residues.

3.2 Pyrolysis thermal effects

Thermo-gravimetric analysis (TGA) was used to trace the fine dynamics of conversion (mass transfer) of small-size fuel particle (Bi < 0.1) at controlled temperature in thermally inert medium with hindered oxygen access (which prevent particle overheating and its premature burn-out) and mass transfer correlation having the value and sign of thermal effect.

Experiments were performed at installation Q1500D (Hungary) according to standard procedure in air medium (ground fuel sample weight was 100 mg, inert medium charge - 400 mg, temperature rise at a speed of 0.3 K/s, and final temperature 1000°C). The samples were wood particles, seeds and charcoal, products of their fast and slow thermal treatment by above described procedure and soot from ash box of pilot downdraft gas producer. Thermograms are shown in fig. 4.

Since the samples were actually dry, weight loss was mainly determined by coke-ash residue pyrolysis and oxidation effects. Overheating value and the sign of thermal effect were due to oxidizing exothermic processes in volatiles emitted by coke-ash residue (except the initial stage).

Steady heterogeneous burning of carbon of ground charcoal and coke-ash residue of bio fuels started at medium temperature above 350°C. For charcoal and wood particles this process is distinguished by appearance of specific temperature peak at 500°C. On having passed the peak, the burning of charcoal becomes uniform and finishes with some exposure at T = 1000°C. Overheating curve for wood particles reproduces charcoal curve in shortened variant.

For seeds the pattern differs radically from above cited. In this case there is no overheating in the domain of volatile emission (which is weaker than with wood particles) which means that they behave like chemically inert substances. It is only at T > 370°C the temperature of seed sample begins to exceed the ambient one. However, it exhibits its specific nature in this range too. Burnout curve for seeds has a low and extended (truncated) peak and a bit greater overheating in steady burning domain. Hence, the pyrolysis may be described as time extended process running in parallel with heterogeneous oxidation of coke-ash residue approximately up to 750°C.

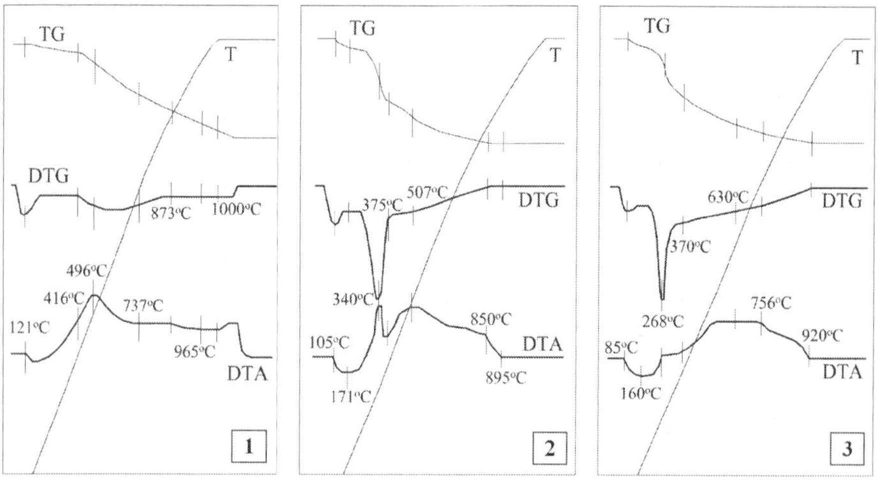

Fig. 4. Fuel thermograms: T - temperature in thermo gravimeter chamber, TG - sample weight, DTG - rate of sample weight loss, DTA - thermal effect; numbers: 1 - charcoal; 2 - wood particle, 3 - date seed

After preliminary thermal treatment according to thermal shock scenario during 15-20 minutes at 400, 600, 800°C the following was found:

Wood after thermal treatment at heating rate of 200 K/20 min retains the peak of volatile emission at 342°C, but the first ("gas") preheating peak disappears, the height of the second peak (coke-ash residue burning) increases but the peak appears with a shift towards higher temperature domain (585°C); conversion process (weight loss) completes earlier (at 706°C instead of 850°C). Hence, the preliminary heating of fossil fuel improves its reactivity.

Charcoal after thermal treatment in thermal shock conditions during 15-20 min at 400, 600, 800°C showed that increase of thermal treatment temperature resulted in the shift of coke-ash residue peak occurrence (at 496°C instead of 462°C), heating value and conversion rate were lower, process time and final temperature were rising and the fuel partially seized to burn.

Thus, in case of thermal treatment at 400 K / 20 min the moment of coke-ash residue burn-out coincide with the moment when maximum temperature is achieved in the plant (1000°C), whereas after thermal treatment at 800 K / 15 min the burning process finishes with incomplete burn out (unburned carbon of 9%) and much later after the furnace has been warmed up to maximum.

Charcoal after thermal treatment according to "heating simultaneously with furnace" scenario is characterized by still lower burn out rate and greater unburned carbon (14%) at the same final temperature values. The behavior of solid-phase volatile decomposition products settling in gas generator ash box (soot) is alike. Inert component content in these products (due to specific sampling conditions) reaches 50%, therefore the burn-out process finishes earlier which corresponds to 800°C.

3.3 Biofuel particle ignition

In low-temperature range (temperature in muffle t_m = 250–500°C) from the moment the particle enters the furnace and up to pyrolysis commencement the particles are heated due to muffle irradiation as inert compound (curve a in fig. 1). Further in the case of charcoal the coke-ash residue will ignite (point B), with temperature upsurge (self-heating) exponentially (curve *b*) and burning in quasi- stationary conditions (curve *c*). Charcoal has the lowest volatile content and preliminary prepared (mostly) reaction surface (75% porosity and specific surface area of 8.6 m^2/g). In the course of pyrolysis charcoal characteristics change insignificantly (porosity increases up to 77–80%, specific surface up to 29.2 m^2/g), self-heating starts at $t_m \approx 250$°C. Quasi-stationary combustion starts earlier than in the case of wood and seeds. Charcoal self-heating starts about 50 s and 400 s earlier than that of wood and seed respectively (time difference between B points).

In the case of wood having porosity close to coal porosity (65%) and specific surface one order lower (1 m^2/g), heating of the particle after the temperature of intense pyrolysis commencement has been reached (particle temperature $t_p \approx 275$°C) follows the curve a' (up to $t_p \approx 420$°C > t_m). The analysis shows that at muffle temperature of 400°C the volatile combustion heat which amounts up to 60% of total heat flux plays a key role. Contribution of bio fuel volatiles combustion to warm-up of their solid residue is of major importance. At muffle temperature of 800°C the main heat flux is muffle irradiation which amounts to approximately 60%.

Intense emission and burning of volatiles hinders the access of oxygen to coke-ash residue, which has been repeatedly described elsewhere, and residue warming is less intense than in the case of charcoal. After the major portion of volatiles has burnt out (point *B*) the coke-ash residue of wood particle gets heated following the curve *b* with its intensity close to that of charcoal self-heating but to a higher temperature, and finally it burns out 1.5-2.0 times quicker. It should be noted that porosity and specific surface of wood coke-ash residue exceed that of the coal.

In the case of wood the sources of warming are heat fluxes from muffle furnace, exothermic reactions of pyrolysis and combustion of volatiles. The amount and ratio of these fluxes depend on muffle and particle temperature.

The time of preheating stage and solid residue combustion at 400°C are compatible for wood (~ 1 : 1). With muffle temperature increase this ratio will change towards increase of relative duration of solid residue burn out and at 800°C it will be 1 : 4. Heat emission intensity (W/m^2) for volatile combustion and coke oxidation is actually 1 : 1. With temperature increase the intensity of heat emission for volatile combustion will increase significantly compared to coke oxidation, i.e. ~ 4 : 1.

Initial characteristics of seed inner surface are much lower than those of other bio fuels (25% porosity, 0.01 m^2/g specific surface). Due to different thermal capacity and under endothermic effects revealed in analysis of the seed, the latter is heated more slowly on segment *A'*-*A''* than "model" inert compound. To the right from *A''* point volatiles release intensively in the form of boiling liquid tar fractions ("cuts") in coke-ash residue pores and on the surface (as is the case with coking coals). During decomposition they partially form soot on porous coke-ash residue surface and partially emit non-ignited as a dense smoke (of fallow color). On segment *A"–B* within approximately 250–300 sec temperature in seed center actually coincides with muffle temperature. Formally this period is a variety of well-known induction period (Pomerantsev, 1973).

Partial overlap of reaction surface by tar fractions and much less initial porosity (see Table 1) result in notable delay of seed self-warming commencement compared to wood and charcoal self-heating. By moment *B* which is characterized by disappearance of liquid phase on the surface, the particle becomes accessible to air and ignition and self heating of coke ash residue commence. In this case the speed of temperature increase is almost 10 times lower than that of wood and charcoal which is apparently due to incomplete pyrolysis at previous stage. By the end of self-heating seed coke-ash residue has the greatest porosity and hence the highest overheating temperature and finally the maximum rate of burning.

For seed the duration of thermal pretreatment and solid residue burning at 400 °C is expressed as 3 : 1 (in this case the preparing stage lasts much longer than that of wood). When muffle temperature reaches 800 °C this ratio changes towards increase of relative solid residue burn out period, similarly to the ratio for wood, and becomes 1 : 4. The relation between heat radiation intensity at stages of volatile combustion and seed coke oxidation is 0.3 : 1 (volatile: coke) at 400 °C and 4:1 at 800 °C.

In high temperature range (at muffle temperature above 500 °C) the distinctions between warming and ignition of different bio fuels are smoothed. Seed warming delay relative to charcoal particle decreases actually to zero and overheating by the end of self-heating and burning rate of coke-ash residue in the main segment come closer. Qualitatively varying pyrolysis scenarios for different bio fuels with close quantitative result for burning intensity are of less importance which correlates well with a well-known high-temperature experiment.

The processes of wood particle and seed ignition are qualitatively different to a large extent and they both differ from charcoal ignition. Visual examination shows that after wood particle is placed in muffle at 800 °C volatiles release is slow with formation of "faint" burning layer close to surface and slightly fluctuating short flame above it (compatible with particle diameter). Volatile release from seeds has "explosive" intensity with continuous burning substance burst out to the distance of up to 3–4 diameters of seed (Fig. 5)

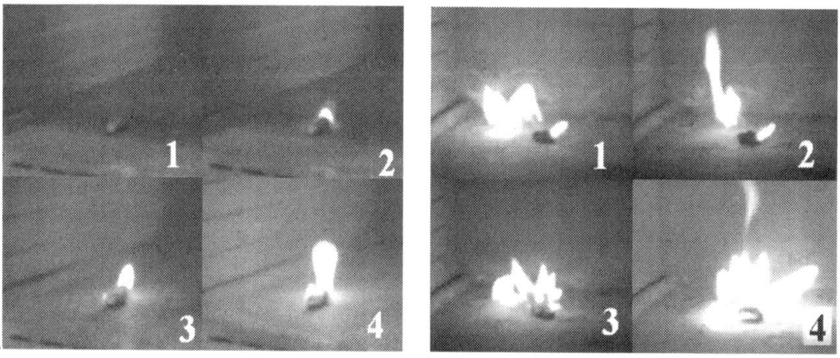

a b

Fig. 5. Process of inflammation: a) wood chip, b) date seed; 1-4 – stages of inflammation from time of placing particle into the muffle (1) separated by 20 s.

3.4 Kinetics of coke-ash residue conversion in burn out conditions

Visual observations of a single particle show that irrespective of low ash content wood chips burn out inside ash enclosure under natural convention conditions: the size of burning carbon nuclear gradually decreases and remaining ash (soft) enclosure retains the original shape of the particle, actually without changing the size. This effect is reached in intense turbulent fluidizing bed too, but when high-ash flotation tailings with rigid mineral enclosure are burnt (Belyaev A.A., 2009).

In low-temperature range the burning of wood and seed coke-ash residues begins after the major portion of volatiles releases at particle temperature $t_p \approx$ 350–400 oC; for charcoal with small volatile content the burning starts at lower temperature $t_p \approx$ 250–300 oC. Coke-ash residue burning at temperature close to above cited values in intra-kinetic mode exhibits significant overheating of particle center relative to environment temperature: ΔT = 360 oC for seed, 250 oC for wood, 300 oC for charcoal (6,a).

In high temperature range with dominating diffusive resistance the overheating of particle center relative to environment is negligible: at 800 °C it is equal to ΔT = 145 °C for pellet and seed, 105 °C for wood, 85 °C for charcoal, at 1200 °C it is 75 °C for wood, 40 °C for charcoal.

Transfer from kinetic to intra-diffusion conditions is most pronounced for charcoal, as its combustion is not aggravated by volatile release in great amounts and actually represents coke residue burn out. Burning rate curve has a "knee" in the range of environment temperature of 400 °C (tp = 600 °C): steep segment corresponds to kinetic mode and flat segment to diffusion mode (fig. 6,a).

In high temperature range the maximum overheating of particle center at t_m = 800 °C is found for seed having maximum porosity after pyrolysis and minimum overheating is exhibited by charcoal which appears to have the least porosity and lowest reactivity by the moment when stationary burning conditions are achieved, compared to any other examined fuel. Overheating of wood particle center is between these two values. Burn out rate ratio for examined fuels are in correlation with the ratio of porosity of examined fuels coke-ash residue porosity relation in point B of thermal curves, similar to overheating relations (table 1, porosity after fast heating).

Fig. (6,b) shows the rate of burning and overheating (fragment) *vs* blow rate in high temperature range. Air speed variations in the range from 0 to 0.5 m/s result in approximately two-fold change of burning rate and overheating of examined fuels. Irrespective of extremely low ash content, the wood particle at zero blow speed burns out inside ash envelope: carbon-including portion shrinks and ash enclosure builds up actually retaining the shape and the size of original particle.

At blow speed equal to 0.1 m/s and more the ash envelope is thrown away by air flow opening the coke-ash residue which is similar to ash enclosure behavior in case of anthracite, charcoal and electrode coal particles combustion. Charcoal burn out rate is lower than that of bio fuels coke-ash residue in the entire range of examined blow speeds.

The mode of fuel thermo chemical conversion which is similar to examined mode without blow is typical of fluidized bed gasifiers and complete combustion furnaces after the blowing has stopped by some reason and low-intense residue burn-out continues for many hours (with fire bed surface intensity $q_R < 0.3$ MW/m^2). Examined modes with blow speed about 0.5 m/s in normal conditions are marginal case of blowing for ordinary FB furnace without bed stability loss and maintain fire bed surface intensity $q_R \approx 4$ MW/m^2. Experimental results were used by the authors for designing the up-to-date gas generator firing the wood fuel. Fig. (6,c) shows the burning rates vs. particle temperature.

Experimental points are located within the "segment" limited by lines of kinetic and diffusion modes. Wood burning rate in high-temperature conditions at t_p = 900 oC and w = 0 m/s is less than estimated speed limit in diffusion mode by approximately 5-15%, charcoal rate by 10-30%.

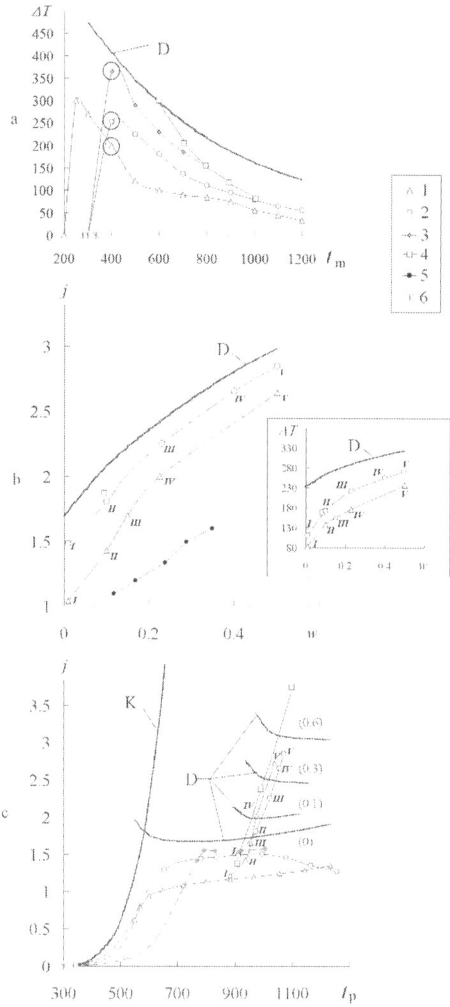

Fig. 6. Curves: a) particle center overheating, °C *vs* muffle temperature, °C, with blow speed w = 0, m/s, b) coke-ash residue burning rate j, g/(m²s) *vs.* blow speed w, m/s, at t_m = 800–900 oC, c) coke-ash residue burning rate j, g/(m2s) vs. particle temperature tp, °C (numbers in brackets indicate blow speed, m/s); designations are per table 2; the fragment shows particle center overheating ΔT, °C vs. blow speed; roman numbers are numbers of experiments; D - diffusive mode, K - kinetic mode

Fig. 7. shows the particle burning rate j, g/(m²s) vs. reverse (1000 / T_p, 1000 / K) and normal (t_p, °C) temperatures of particle.

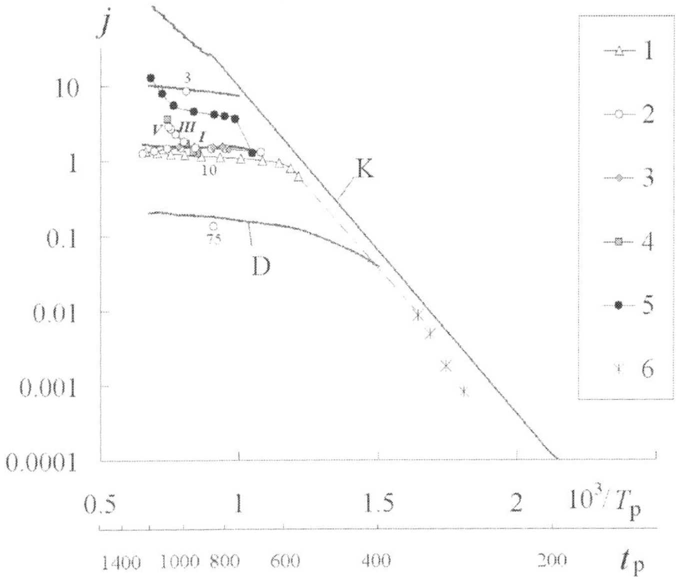

Fig. 7. Particle burning rate j, g/(m2s) vs. reverse (1000 / Tp, 1000 / K) and normal (tp, oC) temperatures of particle; numbers in brackets designate particle diameter, d_p, mm; symbols are per table 2; roman numbers designate burning with blow , the same as in fig. 6

Our data on charcoal at t_m = 300-1200 oC are grouped close to calculated curve for diffusion mode, and charcoal points received at t_m = 280-335oC (Khitrin, 1955) at lower oxygen content in heating medium with burn out without flame-forming self-heating are closer to kinetic curve. Superimposing of experimental points for coal on one curve show that particle temperature shall be chosen as key value in calculations. Choosing the key temperature is important for processing the experimental data in Arrhenius coordinates, because this determines the slope of curve. Choosing the particle temperature as key temperature is important for the range of medium temperatures from 300 to 700°C where significant overheating occurs. Brown coal particles with diameter d_p = 3 mm in FB within temperature range of t_m = 400–900 oC burns out in intra diffusion mode. Due to lower reactivity of brown coal coke-ash residue the rate of its burning in intra diffusion mode is almost half of wood coke-ash residue burning rate with d_p = 3 mm, and transfer to intra kinetic mode occurs at temperature that is higher by 150-200 °C .
Diffusion behavior of bio fuel combustion at environment temperature above 500 °C is observed in the entire examined range of particle diameters, d_p = 3–75 mm. Review of the results obtained in different studies demonstrates that in temperature range from 800 to 950°C fine particles (with sub millimeter diameter) exhibit kinetic mode of burning. Thus, according to fig. 8, the actual overheating of coal dust is below calculated values (line 1). Calculation with actual burning rates (line 2) is close to actual overheating values.

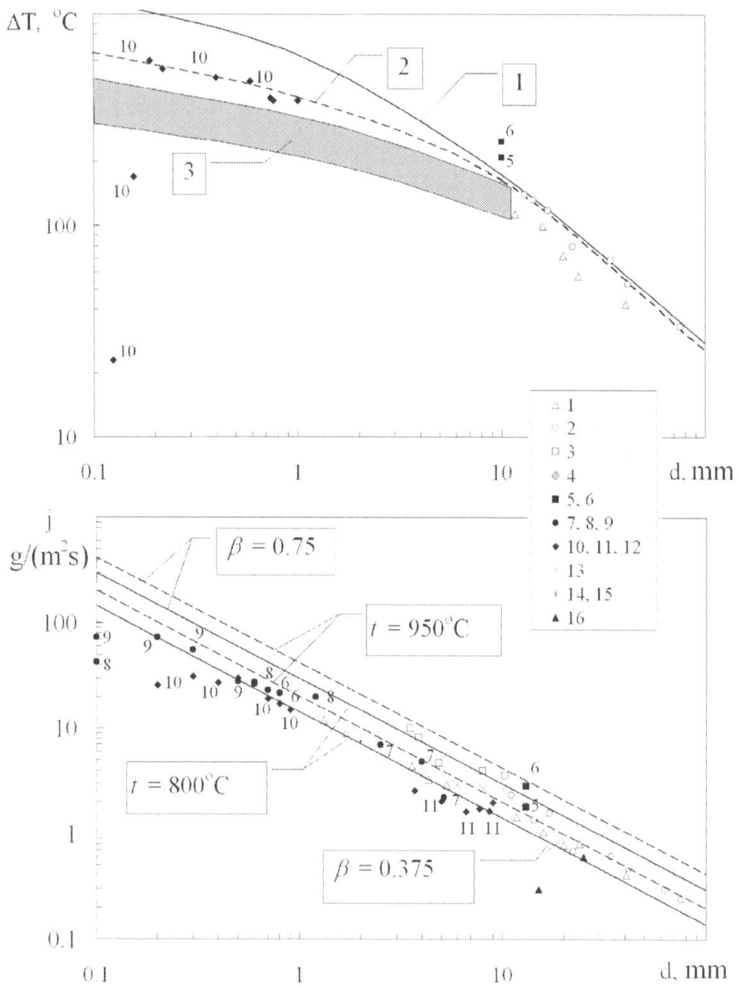

Fig. 8. Overheating and burning rate vs. particle size: 1 - calculated, 2 - ditto for actual burning rates, 3 - calculation for FB with inert material diameter from 0.4 to 1.5 mm. Numbers are per Table 2

With particle size increase its specific surface decreases (per volume unit) and boundary layer becomes larger. This causes decrease of specific heat emission, overheating and burning rate and the particle approaches to isothermal burning. With diameter changing from 10 to 75 mm at $T_m = 800°C$ the overheating lowers from 120-150°C to 30°C, and burning rate from ≈ 2 to ≈ 0.2 g/(m²s). In the range of $d_p = 10–75$ mm calculation with accuracy of ±15% is approximated by dependence $\Delta T = 1130 \cdot d_p^{-0.84}$, °C and $j = 23 \cdot d_p^{-1}$, g/(m²s). Diameter d_p is expressed in mm.

The available results for FB were obtained with fluidization rate which is much higher than the rate of natural convection. The structure of flows in FB is quite specific (Sherwood criterion for inert medium at rest tends to 1.0). At Re = 80 and higher the burning of particle in FB and in inert-free medium occurs with close dimensionless rates, at small Reynolds numbers the burning in FB is less intense than in inert-free medium and occurs with overheating approximately half as much. This is why the estimated set of overheating (field 3 in fig. 8) depending on inert substance particle diameter (from 0.4 to 1.5 mm) is in satisfactory agreement with general dependence for individual fuel particle within a wide range of its diameters but is well below than actual data for dust burning by V.I.Babiy (Babiy, 1986). Overheating of pellet in FB (point 5) is greater than that of wood particle of approximately the same size in natural convection which may be associated with higher density of the pellet (Palchonok, 2002).

Fig. 9 demonstrates generalized relations "dimensionless burning rate vs Reynolds number" at medium temperature 400–1500 °C, with particle diameter d_p = 0.1–80 mm, blow rate w = 0.01–50 m/s. Data by V.I. Babiy for dusts can be generalized by his curve if medium temperature is assumed to be key temperature. Calculations based on particle temperature as key temperature (domain 5 in fig. 9) makes his points lie below his own curve. This means that burn out of dusts at these temperatures occurs in kinetic mode. Experimental data by L.I. Khitrin lie close to his curve.

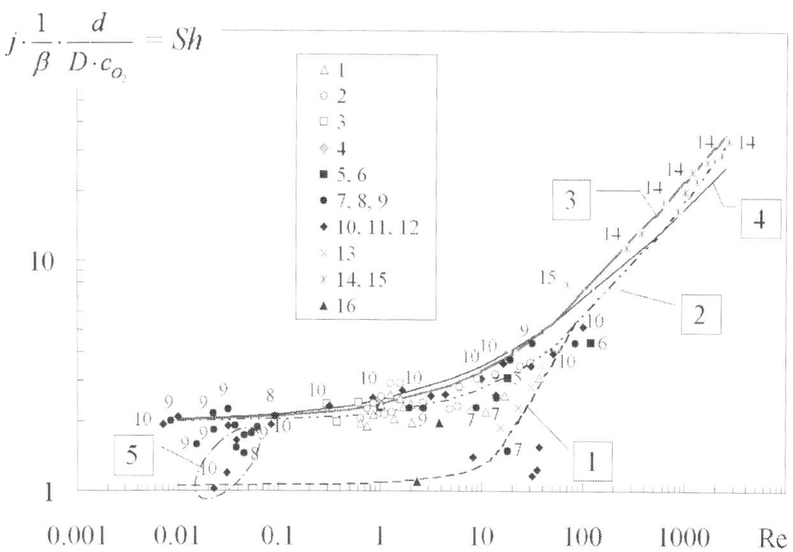

Fig. 9. Dimensionless rate of burning vs. Re number: 1 - calculation for FB, 2, 3, 4 - calculation data; 5 – V.I. Babiy points at particle key temperature; numbers are per Table 2

3.5 Kinetics of coke-ash residue conversion in gasification mode

Gasification process was realized with intermittent charging, gas composition changed with bed burn out: first, combustible content raised, but on having reached (at some definite bed depth) maximum value, it began to decrease (fig. 10,d).

Coal bed depth in retort at which H_2 and CO contents began to decrease was equal to 6 to 7 initial size (average equivalent diameters) of fuel which corresponds to available reference data. At bed depth equal to 1-3 particle size CO content lowered down to 9-10%, and H_2 content approached zero with the accuracy within instrument error range. The depth of 1 to 3 sizes was in compliance with available data on burning zone height.

For two blow types (air and steam-air) charcoal conversion gas composition dependence on air flow rate in maximum concentration point (for bed depth of 6-7 sizes) was studied (fig. 10,a,b). A series of experiments was conducted with sequential blow increase in every next experiment. The greater air flow rate and hence the higher temperature in bed corresponded to higher CO content in gas.

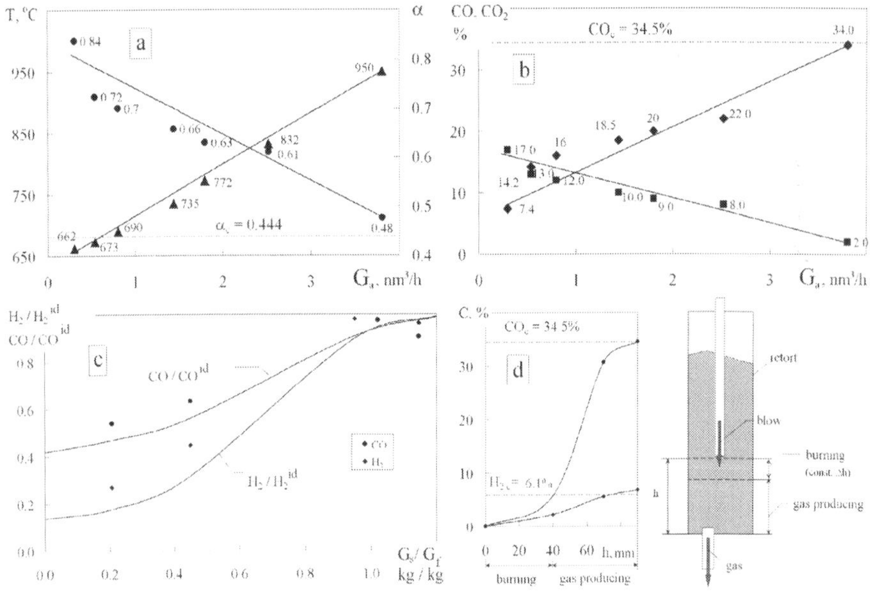

Fig. 10. Relationship between gasification parameters; air flow rate effect on: a) temperature in reduction zone (T) and air flow rate coefficient (α); b) CO and CO_2 content in gas; CO_c and α_c - maximum CO content in gas and blow-limiting factor for air gasification of pure carbon; c) H_2 and CO relative concentration in producer gas depending on steam content in blow: s - steam, f - fuel, id - ideal regime; lines represent calculation of gas equilibrium composition; d) gas composition vs. bed depth in retort

At temperature below 700°C gasification reactions actually did not occur and gas composition corresponded to pyrolysis. CO content tended to estimated limit for pure carbon with air blow (CO_c = 34.5%), and blow coefficient defined by effluent gas composition was approaching α_c. Maximum concentration limits and minimum blow rate coefficient were reached at blow rate ≈ 4 m³/h (speed of blow in bed calculated for empty cross-section is 1.4 m/s) and bed temperature of 950°C.

The results explain the phenomenon of the so called rapid burning. It is believed that the higher is the blow rate in combustion process, the lower is solid-carbon based reduction of CO_2 to CO beyond combustion zone because the time needed for reduction is insufficient. On reaching a certain rate the reduction process stops and CO and CO_2 ratio in gas becomes determined by their ratio that was established in burning zone (primary CO and CO_2). Experiments with activated charcoal showed that at blow rate above 0.3 m/s CO content is 34.5%, and CO_2 is 2%. Study results allow to give a simpler explanation of "rapid gasification" phenomenon, i.e. gasification conditions can be achieved at $\alpha = \alpha_c$. The result confirms the possibility of producing gas that does not contain three-atom gases (CO_2 and H_2O) and methane, and shows one of the ways of its production in dense bed, by providing optimum depth and temperature not less than 900-950°C.

Fig. 10,c demonstrates the results of experiments on allothermal steam-air gasification in dimensionless form: gas concentration is reduced to that in gas of ideal air-steam conversion of char coal (CO^{id} = 43%, H_2^{id} = 26%). Steam content (G_s / G_f, kg steam/kg fuel) was increased in every next experiment with constant air supply corresponding to α = 0.7 for dry blow ("semi-gas" mode). In this case gas was enriched with combustible components by reaction $C + H_2O = CO + H_2$. Steam breakthrough was not observed.

At steam supply in the amount of 1-1.2 kg of steam/kg of fuel (moisture in terms of fuel as received W ≈ 60%) CO and H_2 content in the experiment was close to ideal steam-air concentration of 35-42% and 22-25% respectively, with chemical efficiency for steam-air conversion η_{chem}=1.0.

Fig. 11 demonstrates gas composition change with bed depth for allothermal steam (water) conversion and fig. 12 shows carbon-dioxide conversion of char coal in bed with temperature 950-1000°C and blow at temperature 600-750°C.

Gas composition for steam conversion at temperature T ≈ 1100°C corresponds by 97.5% to stoichiometric one (CO = 42.6%. H_2 = 54.8%), and chemical efficiency of steam conversion η_{chem} =1.26. Ratio of CO content to stoichiometric value for carbon dioxide conversion reached 0.8 and chemical efficiency of carbon dioxide conversion η_{chem}=1.35.

At bed depth less than 70 sizes a linear variation of CO content with bed height was detected for both carbon dioxide or steam blow. Hydrogen content does not depend on bed height and it was increasing during the entire experiment.

Allothermal technique of the experiment allowed to decompose all water steam fed into the bed due to muffle heat. In the experiments allothermal gasification marginal conditions were achieved for steam-water, steam and carbon dioxide blow. In two latter cases homogeneous gas-water shift reaction $CO + H_2O = CO_2 + H_2$ was performed providing for increase of H_2 content above stoichiometric value by heterogeneous reaction and steam conversion of charcoal.

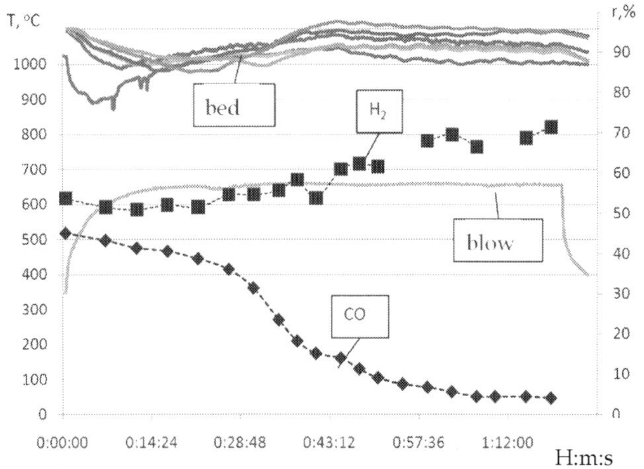

Fig. 11. Steam allothermal conversion

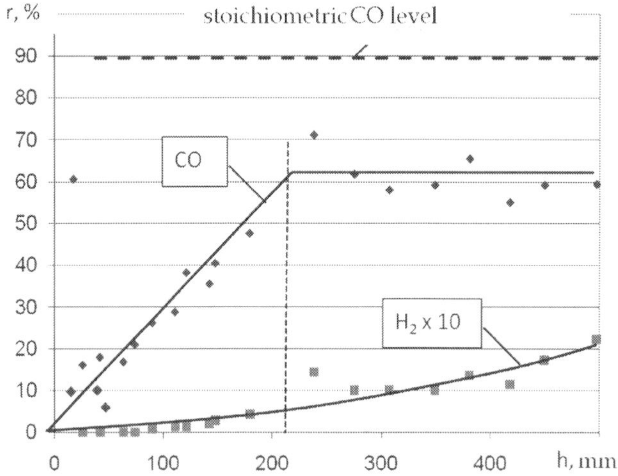

Fig. 12. Carbon dioxide allothermal conversion

3.6 Using heat recovery (recirculation) for advanced CCGT with thermo chemical conversion of low-grade fuels

Temperature limits and ranges of mass concentration of combustible elements and oxidizing agent were determined, in which the process of low-grade fuel gasification is realized with higher conversion efficiency and which include multi-parametric optimum conditions for hypothetic "ideal" tars free conversion into useful products. Fuels are divided into two groups: "A" group includes fuels that are converted into ideal mixture ($CO+H_2$) due to their own internal chemical source of heat (in auto-thermal modes) and "B" group that requires heat power source (allo- and auto-thermal mode) for ideal conversion, moreover in case of auto thermal process the gas is not ideal and has complete combustion products (CO_2 and H_2O). Wood, peat, and some high-moisture brown coals belong to "B" group. In "A" range the ideal gasification is characterized by constant heating value of gas and constant chemical efficiency of fuel conversion into syngas, moreover, high-metamorphized fuels require considerable amount of water vapor to be added whereas the main source of molecular hydrogen for brown coals is fuel hydrogen.

Allothermal conversion procedure allows to obtain maximum effect of chemical efficiency increase (fig. 13, points 8″, 8‴ and 8⁗ are results of allothermal steam-air, steam and carbon dioxide gasification). However, conversion efficiency rise due to external sources is rarely used (p.p. 4′, 7′), which is explained by their "high investments" and low production effectiveness.

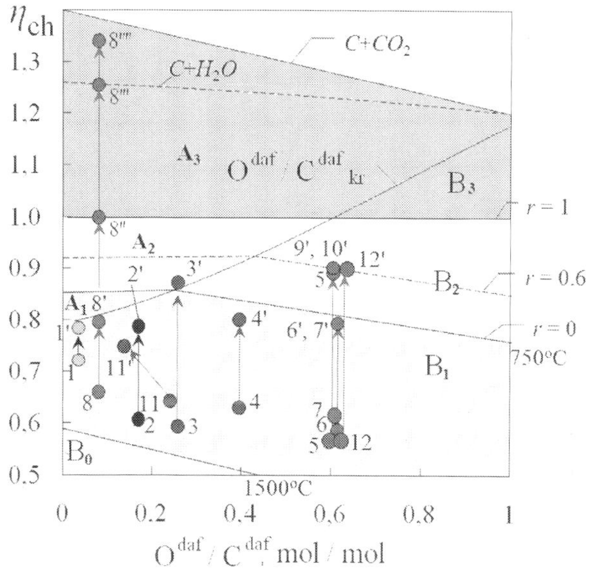

Fig. 13. Chemical efficiency of air thermochemical conversion:

1 - fuel oil gasification without regeneration in flow-type gasifier (by VNII NP); 1′ - ditto, calculation with regeneration; 2 - first-generation dense-bed gasifier operating on black coal; 2′ - tuyere gas producer (JSC VTI) operating on black coal; 3 - gas producer by MHI without regeneration (estimated); 3′ - gas generator MHI (expected upgrading of recirculation system); 4 - first-generation dense bed gas producer operating on peat; 4′ - allothermal pyrolyzer for peat OIVTAN; 5, 6, 7, 8 - first-generation dense bed gas generator; 5′ - first-generation gas producer with superadiabatic heating by IPKhF RAN (estimated); 6′ - upgraded downdraft producer by UrFU with regeneration; 7′ - plasma gas generator, IEE RAN; 8′ - three-zone downdraft producer by UrFU, 8″ - with steam and air blow, 8‴ - with steam blow at 1000oC to the third zone, 8⁗ - with carbon dioxide blow at 1000oC to the third zone; 9′ - mini-CHP gasifier Viking, Denmark (estimated); 10′ - mini-CHP gasifier, UrFU (estimated); 11 - brown coal gasification (estimation); 11′ - joint gasification of brown coal and natural gas, VNII NP; 12, 12′ - downdraft producer by ISEM with different blow temperatures

There is a substantial potential to rise chemical efficiency by improving heat recirculation within gas generator or "gas generator-gas consumer" system. This becomes evident when analyzing the tendencies of development of both industrial large-capacity coal gasifiers (from *Winkler* gasifier to *HTW* plant), and air gasification plants especially those working on low-grade fuels and bio mass. In simple-design (without recirculation) gasifiers operating on biomass with steam and air blow the percent of fuel needed to provide for complete combustion is great (up to 30 %), and temperature is low, therefore gas contains a considerable amount of intermediate gasification products (tars, CO_2 and H_2O) and chemical efficiency is not more than 60 %. Such reactors cannot generate gas with close to ideal characteristics even hypothetically. This is why the development of simple-procedure low-efficiency air reactors does not seem to have any perspective.

4. Conclusion

The kinetics of thermo chemical conversion was studied on low-grade fuels in oxidizing pyrolysis, afterburning and gasification modes in temperature ranges typical of moderate combustion conditions in power reactors. The obtained data were used as the basis for upgrading of effective conversion procedure which will allow to develop gasifying plants with high chemical efficiency in the long term.

The procedure for non renewable resources conversion in fuel gas for gas engines (gas turbine engines, internal combustion engines) and fuel cells has been developed. This procedure provides low tar and hydrocarbon content in gas with high chemical efficiency (80% achieved and 90-95% design efficiency) which is novel in Russia and in prospect it will allow to rise the electrical efficiency of small scale CHP up to 23-25% on bio fuel and up to 50% on fuel cells, and the efficiency of large scale brown-coal CCGT and IGCC up to 50-55% and will make them competitive.

The procedure was implemented at multi-zone dense-bed gasifier with capacity up to 200 kW for bio mass and peat fuels and in a group of pyrolysers having capacity of 10 MW each for pulverized carbon-rich black coal with after-burning of conversion products in coal-fired boiler with capacity of 420 tons of steam per hour.

Small-scale gasification studies are aimed at designing of CHP-ICE with electrical output up 500 kW, using the fluidized-bed gasifier. In the course of works the chemical efficiency of 80% was achieved for gasification process. Gasification of wood fuel having 35% moisture content with air blow resulted in steam-air gas product with heating value of 7.4 MJ/m^3. Investigation results were used at pilot mini-CHP with 200 kW capacity.

5. References

Babiy V.I., Kuvaev Yu.F. (1986). Burning of fuel dust and modeling of coal torch. EnergoAtomIzdat, Russia, Moscow.

Belyaev A.A. (2009). Teploenergetika. Jan. 2009, №1, pp. 9-18.

Khitrin L.I. (1955). Physics of burning and explosion. AS USSR, Russia, Moscow.

Palchonok G. (2002). Combustion characteristics of wood pellets. Proceedings of the First World Pellets Conference. Stockholm, Sept. 2002.

Pomerantsev V.V. (1978). Self-burning and Explosions of Fossil Fuels Dust. Energiya, Russia, Leningrad.

Tillman D.A. (2000). Biomass co-firing: the technology, the experience, the combustion consequences. Biomass and Bioenergy. 2000, №19, pp. 365-384.

2

The Challenge of Bioenergies: An Overview

Nicolas Carels
Fundação Oswaldo Cruz (FIOCRUZ), Instituto Oswaldo Cruz (IOC), Laboratório de Genômica Funcional e Bioinformática, Rio de Janeiro
Brazil

1. Introduction

The rapid rise in the price of crude oil, the decrease in oil reserves, security concerns and greater recognition of the environmental impacts of fossil fuels have generated considerable interest in biofuels as an alternative energy source. The revolution in transportation that occurred at the beginning of the last century created dependence of Western economies on the combustion of hydrocarbon fuels. The invention of the electric light bulb by T.A. Edison led to the installation of the first energy distribution plant in 1883 (in Roselle, New Jersey) and subsequently the electric grid came to the world. Because of the high rate (typically >90%) and ease of alternative to constant voltage conversion or vice versa, as well as mechanical conversion of electricity, the electric grid became the nervous system of our civilization.

During the 20th century, humanity created a foundation in which electric applications are used in all segments of society. This revolution is now creating a system of pervasive computing and is preparing society for the era of nanotechnologies and robotics. However, the electric grid is dangerously dependent on the availability of carbon-based resources such as coal and natural gas (Song, 2006). Human civilization is now challenged with finding renewable and environmentally friendly energy sources for feeding both our electric grid and our economic growth. For the next two decades, fossil fuels will continue to be the most cost-effective energy resource. However, despite their higher costs, alternative energies have begun to be seriously investigated. It is the purpose of the present work to provide an overview of this issue.

Existing fossil fuels are believed to have originated over the course of millions (M) of years (yrs) from biochemical and geochemical transformations of organic substances that were present on the earth's surface. The geological storage of carbon in the form of fossil fuels can be viewed as one alternative route to the reduction/oxidation cycle of carbon that is occurring at the earth's surface. Numerous studies have shown that coal is formed from biochemical degradation and geochemical maturation of higher-plant materials that were originally generated via photosynthesis. Crude oil also shows fingerprint molecules such as phytane that testify to its plant origin (Johnston et al., 2007).

Crude oil contains various hydrocarbons that range from light gases (e.g., C1-C5) to heavy residues (Fialkov et al., 2008). These hydrocarbons are separated via distillation into three main products, namely naphtha, middle distillate and a residual fraction. Naphtha (boiling range 90-190°C) is mainly used for motor gasoline (C3-C12) and comprises approximately 20% of the total crude oil. The middle distillate can be separated into two categories consisting of kerosene-range products (light-end) and diesel-range products (heavy-end). The light-end middle distillates (25-30% of total crude oil, boiling range 150-260°C) are used for the manufacture of solvents, kerosene (C8-C16), commercial and military jet fuels (C3-C10) and light diesel fuel (diesel fuel No. 1, C8-C22). The heavy-end middle distillates (25-30% of total crude oil, boiling range 190-400°C) are processed to produce diesel fuel No. 2 (C10-C25) and heating oils (15-20%) (Kaplan et al., 1997). Lubricating oil (>C18) accounts for approximately 5-7% of the total crude. In addition to the alkane fraction, there is also a fraction of polycyclic aromatic hydrocarbons (PAH) whose relative importance in the composition of distillates increases from kerosene to heating oil. Beside the gas with green house effects (GHG) and their consequences on climate changes, it is the PAH that cause the most important environmental concerns because of oil spills. Actually, PAHs are rather resistant to microbial degradation (bioremediation) under anaerobic conditions, with the consequence that their removal from impacted environments is very slow. PAHs are also a source of concern regarding human health because they may stick to DNA, resulting in deleterious mutations and ultimately cancer.

World oil consumption is approximately 79 M barrels (bl) per day (a barrel is 159 l). The transport sector represents 50% of oil consumption (or 20% of energy consumption) and has an annual average growth rate of >2% per year. Energy demand worldwide is expected to rise by approximately 50%-60% over the next 20 years, reaching 112 M bl/day (Song, 2006). Most of the "cheap oil" that was (relatively) easily removed from the ground (and sold for US$ 20-30/bl) is already gone, and fossil fuel prices have risen over the past decades. The price of crude oil increased from US$ 18/bl in 1990 to US$ >100/bl in 2008 (Tan et al., 2008). Fuel prices will probably continue to increase in the future because of the following factors: (i) fossil fuel resources are limited; (ii) there is a lack of balance between supply and demand; (iii) the demand for fossil fuel is rising rapidly; and (iv) geopolitical instability and international conflicts are increasing (Rout et al., 2008).

With fossil fuel prices at US$ 45/bl, renewable energy from a range of biofuels is becoming economically competitive. Biodiesel fuel usually costs over US$ 0.5/l (as compared with US$ 0.35/l for fossil diesel fuel - see in Zhang et al., 2003) such that the increase in the use of biodiesel has been particularly rapid over the two last decades. It has grown from essentially zero in 1995 to more than 20 billion (G) liters (l) in 2009 (in the US as well as in Brazil). Brazil and the United States (US) have the largest biofuel programs in the world, with the European Union (EU) ranking in the third position.

Biofuels can potentially mitigate greenhouse gases, can provide means of energy independence and may even provide new employment opportunities. Because they are compatible with existing technologies, they will also alleviate natural resistance to change and may act as a medium to allow a "smooth" transition to alternative technologies (Shinnar & Citro, 2006) and regulations. Many countries are now utilizing biofuel products resulting from agriculture and forestry. The most commonly cited advantages of biofuels are as follows: (i) they are available from common biomass sources; (ii) they are integrated into the combustion cycle of carbon dioxide (CO_2); (iii) they are compatible with environmental constraints; and (iv) they contribute to environmental sustainability. For example, biodiesel fuel prepared from vegetable oils or animal fats is compatible with common diesel engines and is therefore a potential alternative to fossil-based diesel fuel. Sulfate emissions resulting from biodiesel combustion are close to zero and the net contribution of CO_2 from biofuels when considering their entire life cycle (i.e., cultivation, oil production and conversion to biodiesel) can be low. The rate of pollutant emission over the life cycle of biofuels is

comparable with that of fossil diesel fuel; however, because it is renewable, biofuel combustion does not result in CO_2 accumulation in the atmosphere (Agarwal, 2007). Actually, plants use solar energy to turn atmospheric CO_2 and water into organic carbon and hydrogen, thereby storing energy. The organic molecules are then broken down as the plant decays and the carbon is returned to the atmosphere as CO_2. When growing a crop for fuel, part of the biomass produced by the plant is used directly to produce energy and the CO_2 that was originally metabolized by the plant is returned to the atmosphere during the combustion process. This CO_2 is therefore "renewable" because it is simply a portion of the total CO_2 that is involved in the natural cycle. However, to produce a biofuel, a certain amount of energy is required. This sophisticated process of production needs a significant energy amount for growing, harvesting and processing the necessary biomass.
The concept of "well-to-wheel" (WTW) is used to characterize the energy consumption that is required to complete the entire process of production and transport of a fuel (Gnansounou et al., 2009). The WTW analysis is often divided into the following five stages: (i) feedstock production, (ii) feedstock transport, (iii) fuel production, (iv) fuel distribution and (v) vehicle use. These stages can be divided even further into "well-to-tank" (WTT) and "tank-to-wheel" (TTW) processes. Fig. 3 in Agarwal (2007) shows that the rates of pollution with the corresponding increase in energy consumption of crude oil, natural gas, biomass, wind power are 1.3, 1.8, 7.1 and 45 times lower, respectively, than that of coal. Nevertheless, biofuels (such as ethanol and biodiesel) fall on a line with a slope equal to crude oil. However, the intercept is lower, which demonstrates that the benefit of biofuels over fossil diesel fuel is due to the recycling of the former through photosynthesis. More advanced technologies (e.g., synthetic fuels based on biomass gasification or wind electricity) use virtually only renewable energy for the conversion processes and result in very low GHG emissions.

1.2 The CO_2 balance

Carbon cycling does not occur with fossil fuel use because the released CO_2 does not return to fossil resources. The amount of CO_2 that has accumulated in the atmosphere is very large and has been calculated to be ~780 billion (G) tons (t) as of 2002 (Song, 2006). CO_2 concentrations are measured by infrared absorption spectroscopy and, because of heterogeneity in the data, estimates may vary from the simple to the double (400 to 780 Gt). However, all measurements agree with a pattern of monotonic increase in CO_2. The concentration is now approximately 390 part per million (ppm), as compared to 310 ppm in 1955 (Song, 2006). Over the same period, the atmospheric temperature has increased by roughly 0.6-0.74 °C and has resulted in numerous effects on the environment, climate (Odling-Smee, 2007; Rout et al., 2008; Tomkiewicz, 2006), ecosystems (Yue et al., 2010) and (consequently) human populations (Simões et al., 2010). For example, growing seasons are becoming longer with the consequence that the amount of carbon accumulating in terrestrial biomass increased by a factor of 7 between the 1980s and the 1990s. In addition to this terrestrial carbon sink (which plays an important role in slowing the rate of CO_2 increase), one must consider the oceanic carbon sink, which sequesters twice as much anthropogenic carbon as does the terrestrial sink (Ridgwell & Zeebe, 2005). Together, these sinks sequester approximately half of the ~24 Gt of anthropogenic carbon emitted each year. Although values of CO_2 emissions should be considered carefully because of the heterogeneity among estimates, one may conclude that phototropic carbon sinks are keys for the mitigation of atmospheric CO_2 concentrations and, consequently, of global temperatures that will increase during the 21st century (Reay, 2007).

However, there is a limit to CO_2 mitigation by phototropic carbon sinks. With the rise in CO_2 concentrations in the atmosphere, falling oceanic pH levels will result in changes in ocean chemistry. This should decrease the efficiency of carbon recycling and, together with the decrease in precipitation, could induce a switch from oceans as a sink to a source of CO_2 by 2050. In fact, warmer soils will release more carbon and forests suffering increased drought will eventually decline. In addition, 40 Gt of CO_2 that is sequestered in 50 Gt of terrestrial biomass could be released because of changes in land use. Deforestation is the single biggest threat to the terrestrial carbon sink because woody biomass is the most efficient biological system for carbon sequestration (Miles & Kapos, 2008).

The world's forests cover ~4 billion (G) hectares (ha) or 30% of the earth's land surface. In 2005, 3.5 Gm^3 of wood were removed from 434 Gm^3 of forests; ~60% of this amount was consumed for industrial purposes and the remainder was used for fuel. Terrestrial forests are either primary (36%) or modified (53%). The area of primary forest has been slowly decreasing at a rate of ~6 Mha/yr since the 1990s. This rate is especially high in Brazil and Indonesia; these two countries are responsible for the loss of 4.9 Mha of forests annually. Forest loss tends to occur in low-income countries (largely in the tropics), whereas higher-income countries have reversed their trend of forest reduction (Kirilenko & Sedjo, 2007).

Converting rainforests, peatlands, savannas, or grasslands to areas used for biofuel production and food crops in Brazil, Southeast Asia, and the United States creates a "carbon debt" by releasing 17 to 420 times more CO_2 than the annual greenhouse gas (GHG) reductions that these biofuels would provide by displacing fossil fuels. In contrast, biofuels from waste biomass or from biomass grown on marginal lands planted with perennial species incur little or no carbon debt (Fargione et al., 2008; Searchinger et al., 2008). Tropical deforestation is estimated to cause approximately one-quarter of anthropogenic carbon emissions, in addition to loss of biodiversity and environmental services. In Brazil, the Amazon forest has been cleared since the 1990s at a gross rate of ~25,000 km^2/yr (for comparison, Belgium is 30,000 km^2). Kindermann et al. (2008) estimated that a program providing a 10% reduction in deforestation from 2005 to 2030 could provide a reduction of 0.3-0.6 Gt of CO_2 emissions per year and would cost approximately US$ 0.4-1.7 billion/year (bn/yr) over 30 years. A 50% reduction in deforestation from 2005 to 2030 could provide a reduction of 1.5-2.7 Gt of CO_2 emissions per year and would cost US$ 17.2-28.0 bn/yr. In comparison, illegal logging was estimated by the World Bank to result in losses of revenue to developing countries of US$ 15 bn/yr (Agrawal et al., 2008). This suggests that better control of illegal logging would compensate for the loss of revenue caused by the reduction in deforestation.

Currently, it is thought that the protection and enhancement of the terrestrial carbon sink is the best way to engineer climate against deleterious evolution (Reay, 2007). Worldwide monitoring of each country's contributions to GHG emissions should therefore be given the highest priority (Yang & Sirianni, 2010). In this spirit, the CarbonTracker is an integrated system for CO_2 measurement that assesses the net CO_2 exchange between the terrestrial biosphere and the atmosphere and that will be useful for the scientific community and for policymakers (Peters et al., 2007). However, to establish effective GHG-reduction, without which CO_2 emissions will be soon unsustainable, governments need consistency and harmonization of policies. For example, it is the objective of the United Nations and other international organizations to raise the GDP/capita in developing countries to the same level as has been achieved in developed countries (e.g., US$ 9,000/capita). However, such politics would result in an increase in yearly global CO_2 emissions by a factor of 5 (Tomkiewicz, 2006) for the following reasons: (i) below the saturation level of US$ 25, 000/capita, there is a positive correlation between energy consumption per capita and purchasing power and (ii) the world average GDP/capita is US$ 5,210 (Tomkiewicz, 2006). Increase in CO_2 emissions will therefore have to be compensated for by one or more of the following: (i) an increase in the global area planted with forests; (ii) a change in the technological paradigm (such as a commitment to convert to solar energy); or (iii) a decrease in human population size. Actually, global warming because of the increased use of fossil fuels matches predator-prey equations (Lonngren & Bai, 2008). The solutions to these equations are periodic in time. If humans continue to adhere to the current trends of fossil fuel consumption (as we have done since the beginning of the industrial revolution), we should observe large fluctuations in human population size in the future. Because earth and fossil fuel reserves are not infinite, the global-warming scenario leads (on a long timescale) to an eventual decrease in both the human population and the fossil fuel reserves, with the ultimate possibility being human extinction. Among catastrophic scenarios that could affect the human population because of climate change, we cite the prediction that one billion people could lack drinking water by 2050 if CO_2 emissions are not drastically reduced (Parry et al., 2008).

1.3 Greenhouse gases and their management

The perturbation in atmospheric trace gases (e.g., SO_2, O_3, CO, CO_2, CH_4, NO_2, and CFCs, among others) is an important factor affecting climate change (Hopkin, 2007). In turn, climate change may promote changes in agricultural conditions that could have deleterious socioeconomic effects (Howden et al., 2007). Atmospheric trace gases have strong absorption bands in the infrared (IR) and interact with IR radiation emitted both by the earth's surface and the atmosphere. This directly influences the thermal structure of the atmospheric environment and contributes to the greenhouse effect. Gases such as NH_3, SO_2 and their derivatives have lifetimes of only a few days, but they can have significant effects on the atmosphere (Begum, 2005). Emissions of N_2O and CH_4 are currently the dominant contributors. Although CO_2 is the main greenhouse gas in terms of volume, others must also be considered. In agricultural practices, the main culprit is nitrous oxide (N_2O), significant quantities of which are released from cultivated fields (particularly with the intensive use of fertilizers) (Snyder et al., 2009; Ceschia et al., 2010; Mander et al., 2010). Because N_2O is >300 times more potent as a GHG than is CO_2, even modest volumes can have significant impacts on the overall balance (Cherubini, 2010).

Harnessing the carbon sequestration capabilities of the terrestrial biosphere has been recognized as a potentially powerful, yet relatively low-cost, tool to offset carbon emissions (Dorian et al., 2006) and models for that purpose have been investigated (Werner et al., 2010). However, terrestrial carbon sequestration has been considered insufficient for meeting more than 25% of the CO_2 emissions reductions that are globally required by 2050. Given that carbon sinks are the best currently available scenario, an *emissions credit system* has been established to provide CO_2 emitters (companies or countries) with a means to satisfy the carbon liability associated with their release of carbon into the atmosphere. The emitter temporarily satisfies his liability by "storing" (for a fee) the equivalent carbon in a terrestrial carbon sink (such as a forest) (Sedjo & Marland, 2003). This concept is the application of the "willing-to-pay" principle within the international economic market. More simply, the right to emit CO_2 (in the form of a *carbon credit*) is compensated for by growing biomass that will sequester an equivalent amount of carbon. The marketing of carbon credits has been organized to allow for rewarding activities that result in the "permanent" immobilization of CO_2 in a nongaseous form. Ultimately, a carbon fee has been proposed that would be paid by industrial countries (in proportion to their emission contributions to GHG) to developing countries; these countries could then invest them in carbon mitigation practices (such as establishing or maintaining forest sinks) (Jones, 2010). The Kyoto Protocol is now legitimating activities such as revegetation, forest management, cropland management, grazing-land management, and also carbon sequestration in deep crustal layers (such as oil fields and deep saline aquifers) for trading with carbon credits (United Nations Framework Convention on Climate Change [UNFCCC], 2002). Principles of justice in proposals and policy approaches to avoided deforestation are also being pursued (Okereke & Dooley, 2010) through negotiations on Reducing Emissions from Deforestation and forest Degradation (REDD).

It has been estimated that the biological sink may attain a cumulative CO_2 sequestration of 100 Gt over the next 50–100 years, with most of it in forest systems. This implies the capture of 10–20% from the anticipated net fossil fuel emissions until 2050 (Sedjo & Marland, 2003). However, carbon sequestered in the terrestrial biosphere may lack permanence. Forests may be harvested for timber that can be used to produce short-lived products or may be cleared for other purposes. Wild fires can release large amounts of sequestered carbon. Farmers may return to agricultural practices that release carbon that was previously captured. In that sense, terrestrial carbon sequestration may simply represent a delay in the flow of fossil fuel carbon to the atmosphere. However, economic incentives for carbon sequestration should increase permanent sequestration. That is, wherever and whenever there are incentives (payments) for carbon-sequestration services, one would expect more sequestration to occur than if no payments were made (Johnston & Holloway, 2007; Tollefson, 2008).

Carbon sequestration in living forests can be performed on lands with low productivity that are not suitable for agriculture or for intensive forestry and that are compatible with goals of biodiversity conservation over large areas. In contrast, to be economically viable, intensive crops for biofuels generally need land that is more productive. Intensive biofuel crops may compete with food production or even with the less-productive lands that are currently sheltering most of the earth's biodiversity (Huston & Marland, 2003; Miles & Kapos, 2008). For example, this phenomena has been observed in Brazil, Indonesia and Malaysia, where cattle, soybean, sugarcane and palm oil may compete with standing forest (Darussalam, 2007; Laurance, 2007; Malhi et al., 2008; Stone, 2007; Venter et al., 2008). In Indonesia, this competition has disastrous consequences for wildlife. To resolve this problem, the Kyoto protocol and subsequent versions should include "wildlife credits" (in addition to carbon credits) to sustain wildlife and its buffering effect on human activities (Lovelock & Margulis, 1974). This strategy would have the advantage of recognizing the fundamental roles played by ecosystem services and to begin to account for them (Mäler et al., 2008). New financial incentives are needed to act as a countervailing force to the economic pressures for deforestation (Jones, 2010). The recent agreement known as the "Bali Roadmap", which aims to extend the Kyoto Protocol beyond 2012, includes directives for providing compensation to rainforest-holding nations in exchange for control of deforestation and environment degradation. Such compensation could be managed either through international carbon markets or through voluntary funds. These directives have the potential to shift the balance of underlying economic market forces that currently favor deforestation by raising billions of dollars to pay for the ecosystem services provided by rainforests. However, to be effective they will require exceptional planning, execution and long-term follow-through. The new proposal also aims to reduce EU CO_2 emissions by 30% by 2020 if a global climate deal is reached in international negotiations (if not, the cut will be 20%) (Schiermeier, 2008). The EU is also planning to protect its economy against carbon "dumpers" by applying leverage that aims to force companies that import goods from polluting countries to buy emissions permits (Barnet, 2008).

Typically, carbon-credit compensation funds are used in developing countries for establishing new long-term plantations (such as rubber trees or oil palm). One difficulty is that the actual goal of carbon sequestration can be negated in cases where the renter first illegally burns the original forest, earns the carbon-credit funds and subsequently establishes a new plantation that will never be as productive, in terms of carbon sequestration, as the original forest. In some regions, environmental crimes are not easily detected and may also not be "significantly" punished. Key recommendations to ensure the environmental sustainability of biofuels through certification (including international approaches and global monitoring) should help to control the process (Scarlat & Dallemand, 2010). Despite these problems, the carbon-credit market was stabilized as of 2007 and is not expected to collapse any further (Haag, 2007). The next few years represent a unique opportunity (perhaps the last) to maintain the resilience of biodiversity and ecosystem services (Malhi et al., 2008; García-Montero et al., 2010; Hagerman et al., 2010).

1.4 Why biofuels?

Compared with an array of solar cells, plants are strikingly poor transducers of the sun's energy. Energy storage by photosynthesis is approximately <2 watts (W) per m^2. The important difference between plants and solar cells is that plants are very cheap. They are able to grow with a moderate supply of water, nutrients and CO_2 that they turn into stable organic compounds. No fuel technology is perfect, but the GHG crisis and concern over oil supplies means that diversifying the range of fuel options makes good sense; at the very least, such diversification places humanity on a healthy learning curve (Haug et al., 2011). Continuous increases in energy needs have encouraged governments to search for new alternatives to fossil fuels. The rationale is to facilitate the transition from a fossil-energy based economy to an economy based on renewable sources of energy. Numerous low-emission scenarios have demonstrated that the goals of the Kyoto Protocol cannot be achieved without providing a large role for biofuels by 2050 in the global energy economy. Among the reasons why biofuels are appropriate for such a transition are the following: (i) their simplicity; (ii) their production via well-known agricultural technologies; (iii) their potential for mitigation of climate warming without complete restructuring of the current working energy system; (iv) the use of existing engines for their transportation (even considering the conventional turbofan used in aviation) (Kleiner, 2007; Rothengatter, 2010); (v) their potential to facilitate worldwide mobilization around a common set of regulations; (vi) their potential as a directly available energy source with good public acceptance; (vii) their more uniform distribution than the distributions of fossil fuel and nuclear resources; and (viii) their potential to create benefits for rural areas, including employment creation.

1.5 What are biofuels?

The term biofuel refers to liquid, gas and solid fuels that are predominantly produced from biomass. The production of biofuels may ignite concerns about security, the environment, trading, and socioeconomic issues related to the rural sector. Biofuels include bioethanol, biomethanol, vegetable oils, biodiesel, biogas, bio-synthetic gas (bio-syngas), bio-oil, bio-char (**charcoal** created by **biomass pyrolysis**), Fischer-Tropsch liquids and biohydrogen. Biogas and bio-oil are primary products, the preliminary processing of which is almost reduced to collecting the raw material. Most traditional biofuels (such as ethanol from corn, wheat, or sugar beets and biodiesel from oil seeds) are produced from classic agricultural food crops that require high-quality agricultural land for growth. The biofuel economy will grow rapidly during the 21st century (Demirbas, 2008a).

1.5.1 Biogas

There are two basic procedures for transforming solid biomass into liquid or gaseous biofuels. The first is to transform it by microbiological fermentation (Gavrilescu & Chisti, 2005) (i.e., to convert the polysaccharides into alcohols such as bioethanol or biobutanol) or to convert raw plant biomass by anaerobic fermentation into biomethane (Demirbas & Balat, 2006). One of the main drawbacks of the anaerobic system is that it is slow (because of the small amount of energy that is available to the organisms). Therefore, the amount of methane that can be produced is limited and this technology is only sustainable under selected scenarios (Asam et al., 2011). However, the introduction of even a small installation for transforming agricultural and human wastes into methane can have an enormous effect on the living standards of small communities (Arthur et al., 2011; Parker, 2002).

The second procedure aims to thermochemically convert the total biomass into a synthesis gas of high calorific value (also called syngas, i.e., H_2 + CO) with subsequent production of various liquid and gaseous fuels (Tijmensen et al., 2002). The production of syngas is a potential area for large-scale CO_2 conversion and utilization. The reforming of CO_2 to CH_4 has been extensively studied and reported on in the literature (Song & Pan, 2004). The catalytic reduction of CO_2 to form methanol (or even CH_4) using renewable energy sources could become a viable alternative to scarce or expensive fossil resources.

Biodiesel from plant oils and bioalcohol from sugar use only a portion of the total biomass. Next-generation processes are being developed to convert biomass to syngas (Baker & Keisler, 2011; Fagernäs et al., 2010) that can then be converted into fuels or chemicals by a synthetic process (the so-called Fischer-Tropsch, or FT, process).

$$\mathrm{CH0.8} + 0.7\mathrm{O_2} \rightarrow \mathrm{CO} + 0.4\mathrm{H_2O} \tag{1}$$

$$\mathrm{CH0.8} + \mathrm{H_2O} \rightarrow \mathrm{CO} + 1.4\mathrm{H_2} \tag{2}$$

$$\mathrm{CO} + \mathrm{H_2O} \rightarrow \mathrm{CO_2} + \mathrm{H_2} \tag{3}$$

$$(2n+1)\mathrm{H_2} + n\mathrm{CO} \rightarrow \mathrm{CnH_{2n+2}} + n\mathrm{H_2O} \tag{4}$$

Considering that coal inputs supply a 0.8 to 1 ratio of H/C, the whole FT process can be briefly written as follows. The partial oxidation of coal by oxygen gives equation (1). The interaction of water with carbon monoxide through "steam reforming" produces equation (2). Subsequently, the H/CO ratio is improved by "shifting" (transferring) the oxygen from the molecular water to CO, producing an additional hydrogen and carbon dioxide following equation (3). After removing the sulfur and carbon dioxide contaminants, the syngas is reacted over a catalyst in the FT reactor to produce high-quality clean fuels following the formula (4) (Greyvenstein et al., 2008).

Biomass is more reactive than coal and (depending on the technology) is usually gasified at temperatures of between 550 °C and 1,500 °C and at pressures varying between 4 and 30 bars (Damartzis & Zabaniotou 2011; Leibold et al., 2008; Steinberg, 2006). Typically, biomass is burned in an electrically heated furnace consisting of several multiple-tube units that can be heated separately up to 1,350 °C (Theis et al., 2006). Alternatively, the conversion of fossil fuel or biomass can be performed in hydrogen plasma. The temperature induced by an electric arc in hydrogen plasma is very high (~1,500 °C); therefore, this technology produces hydrogen and CO gas with a conversion rate of near 100% (Steinberg, 2006). FT synthesis generates intermediate products for synthetic fuels (Liu et al., 2007). The thermal efficiency of producing electricity and hydrogen through hydrogen plasma and carbon fuel cells varies from 87% to 92%, depending on the type of fuel and the biomass feedstock. This is more than twice as efficient as a conventional steam plant that burns coal and generates power with a ~38% efficiency. In addition, coupling hydrogen plasma and carbon fuel-cell technologies allows for a 75% reduction in CO_2 emission per unit of electricity (Steinberg, 2006).

Because FT produces predominantly linear hydrocarbon chains, this process is currently attracting considerable interest. FT has already been applied on a commercial scale by Sasol, Petro S.A. and Shell, mainly to produce transportation fuels and chemicals (the feedstock being coal or natural gas). This fuel option has several notable advantages. First, the FT process can produce hydrocarbons of different lengths (typically <C15, Liu et al., 2007) from any carbonaceous feedstocks; these hydrocarbons can then be refined to easily transportable liquid fuels. Secondly, because of their functional similarities to conventional refinery products, the synthetic fuels (synfuels) produced by the FT process (i) can be handled by existing transportation systems; (ii) can be stored in refueling infrastructure for petroleum products; (ii) are largely compatible with current vehicles; and (iv) can be blended with current petroleum fuels (Tijmensen et al., 2002). Thirdly, synfuels are of high quality (this is especially true for FT diesel), have a very high cetane number and are free of sulfur, nitrogen, aromatics, and other contaminants typically found in petroleum products. The principal drawbacks of the FT process are that the capital cost of FT-conversion plants is relatively high and that the energy efficiency for the production of FT liquids by conventional techniques is lower than the energy efficiency for the production of alternative fuels (Takeshita & Yamaji, 2008).

1.5.2 Bio-oil

Bio-oils are dark red-to-black liquids that are produced by biomass pyrolysis. Biomass is typically obtained from municipal wastes or from agricultural and forestry by-products (Demirbas, 2007). With an efficiency rate as high as ~70%, pyrolysis is among the most efficient processes for biomass conversion. The density of the liquid is approximately 1,200 kg/m^3, which is higher than that of fuel oils and significantly higher than that of the original biomass. The gasification of bio-oil with pure oxygen and the further processing of syngas into synthetic fuel by the FT process, is being investigated; however, this process does not appear to be economically attractive (Demirbas & Balat, 2006).

1.5.3 Plant oils

There has been interest in the use of virgin plant oils to fuel diesel engines. At least 2,000 oleaginous species, growing in almost all climates and latitudes, have been identified. There are more than 350 plant species that produce oil that could be used to power diesel engines (Goering et al., 1982). The plant oils are made up of 98% triglycerides and small amounts of mono- and diglycerides. There are basically two types of vegetable oils: those in which the majority of fatty acids are in C12 (e.g., palms) and those in which the majority of fatty acids are in C18.

The direct use of plant oils (and/or blends of these oils with fossil fuels) has generally been considered to be unsatisfactory or impracticable for both direct and indirect diesel engines. Obvious problems include their high viscosity (Ramadhas et al., 2005), acidic composition, free fatty acid content, tendency to deposit carbon, tendency for lubricating-oil thickening, and gum formation because of oxidation polymerization during storage and combustion. When blending vegetable oils with fossil diesel fuel, the viscosity can be extensively adjusted. Based on EN 14214 recommendations, the maximum blending rate of most vegetable oils is B30 (30% plant oil/fossil diesel, v/v) (Abollé et al., 2008). The oil viscosity (because of the presence large triglycerides) can also be reduced by pyrolysis, which produces an alternative fuel for diesel engines (Lima et al., 2004). Using plant oils in blends also significantly increases their cloud points and thus limits their use to climatically compatible countries.

1.5.4 Bioalcohol

Because of the energy crisis and climate warming, humanity faces the need for a huge short-term supply of biofuels (see below). Bioethanol and biodiesel have been considered the best candidates for satisfying these needs and are what we consider the first generation of biofuels. Ethanol can be produced from a range of crops including sugarcane, sugar beets, maize, barley, potatoes, cassava, and mahua (Baker & Keisler 2011; Kremer & Fachetti 2000). Flexible-fuel motors have been developed that can burn hydrous ethanol/gasoline blends in any combination, including pure ethanol. The automatic adjustment of combustion parameters is controlled electronically in these engines as a function of the oxygen level needed by the fuel in the tank (Marris, 2006). The so-called "gasohol" is a blend of ethanol and gasoline. Ethanol is produced via fermentation of a sugar slurry that is typically prepared from sugar or grain crops. The action of yeast on the sugar produces a solution that contains approximately 12% ethanol. The yeast invertase catalyzes the sucrose hydrolysis into glucose and fructose. Subsequently, yeast zymase converts the glucose and the fructose into ethanol. The alcohol can then be concentrated by distillation to produce up to 96% ethanol (hydrous ethanol).

Ethanol is a polar solvent and its chemistry is very different from that of hydrocarbon fuels (which are non-polar solvents). As a result, blending ethanol into hydrocarbon fuels introduces some specific challenges. These challenges include (i) higher fuel volatility at low rates of ethanol/gasoline blends, (ii) higher octane ratings, (iii) an increase in dissolved-water content in motor gasoline that promotes heterogeneity of fuel blends and resulting engine corrosion and (iv) higher solvency. However, Akzo Nobel Surface Chemistry and the Lubrizol Corporation have developed and produced a low cost additive that makes it possible to blend ethanol with diesel fuel to obtain a stable and clear fuel (Lü et al., 2004). This fuel is called "Dieshol".

Biomethanol can be produced from biomass using bio-syngas obtained from the steam-reforming processing of biomass. Biomethanol is considerably easier to recover from biomass than is bioethanol. However, sustainable methods of methanol production are not currently economically viable. The production of methanol from biomass is a cost-intensive chemical process. Therefore, under current conditions, only waste biomass, such as wood or municipal waste, is used to produce methanol.

1.5.5 Biodiesel

Biodiesel has the advantage that it can be used in any diesel engine without modification. It is produced by the transformation of renewable oils, such as those synthetized by plants, *algae*, bacteria and fungi. First-generation biodiesel is considered to be the result of a two-stage process that involves (i) the crushing of raw material (typically oilseeds) in specialized mills to expel the oils and (ii) the transformation of oil into biodiesel. Free fatty acids (FFA) or triglycerides are converted into alkyl-esters by reaction with short-chain alcohols (such as methanol or ethanol) in the presence of a catalyst. The reaction involved in the conversion of FFA to alkyl-esters is called esterification, whereas that involved in the conversion of triglycerides is called transesterification. Fatty acid methyl-esters are only partly biological, as the methanol involved is generally produced from fossil methane (natural gas). However, biodiesel can also be produced by replacing methanol with ethanol, resulting in fatty acid ethyl-esters. If the ethanol is of biological origin, the product is fully biological. The purpose of the transesterification process is to lower the viscosity of the oil with transesterification being less expensive than the pyrolysis that is used in bio-oil processing. According to the EU standards for alternative diesel fuels, alkyl-esters in biodiesel must be ≥96.5 wt%.

1.5.6 The four generations of biofuels

The first generation of biofuels demonstrated that energy crops are technically feasible, but that no single solution exists to cover every situation (Venturi & Venturi, 2003). In addition, the production of first-generation biofuels is complicated by issues that are contrary to biofuel philosophy, such as the destruction of tropical rainforests (Kleiner, 2008). Tropical rainforests are the most efficient carbon sinks on earth. Therefore, if biofuels contribute to their destruction, this implies that the carbon balance of biofuels is negative. This consideration limits the viability of first-generation biofuels. It also comes with the corollary that raw materials for biofuel production will have to be diversified over the long term. Second-generation bioethanol is precisely an attempt to overcome this challenge.

Second-generation bioethanol will be produced from lignocellulosic biomass, which is a more suitable source of renewable energy (Frondel & Peters, 2007; Tan et al., 2008; Tilman et al., 2007). Lignocellulose is obtained from inexpensive cellulosic biomass that is encountered throughout the world. However, the low-cost transformation of lignocellulose into bioethanol is still challenging. Some possible technologies involve genetic modification of plants, which is a source of concern for society. Whatever the future evolution of the technology, the introduction of energy policies is crucial to ensure that biomass ethanol is effectively developed to become a major source of renewable energy (Tan et al., 2008).

Algae and cyanobacteria are far more efficient than higher plants in capturing solar energy and will surpass first- and second-generation biofuels in terms of energy capture per unit of surface area (Brennan & Owende, 2010). *Algae* are already used in pilot CO_2-sequestration units for emissions cleaning in some conventional power plants running on fossil fuels. This technique is called CO_2 filtration. Unfortunately, *algae* require capital for investing in reactors that can grow them, making CO_2 filtration an excellent opportunity for developing this technology. In that sense, *algae* can be regarded as a third-generation fuel. New methods and technologies for the production of second- (such as synfuels, Baker & Keisler, 2011) and third-generation biodiesel fuels are being developed and will result in the modification of the definition of biofuels that is generally used in government regulations (Lois, 2007).

Finally, one can also envision the exploration of photosynthetic mechanisms for biohydrogen and bioelectricity production. These would constitute fourth-generation biofuels (Gressel, 2008). The development of effective fourth-generation biofuels is not expected before the second half of the 21st century.

2. Plant biofuels

2.1 Bioethanol

The technique of alcohol fermentation has been known for thousands of years. Ethanol distillation has been carried out for decades by industry because it has been part of the process of the regulation of sugar prices on the international market. Ethanol is regularly produced from the isomerose (high-fructose syrup) of grain crops such as maize or wheat and from sugar crops such as sugar beet or sugarcane. In Europe, sugar beet is preferred. This is especially true in countries such as the UK, France, Holland, Belgium and Germany, where it is highly productive, as 1 ha of this crop can produce 5.5 t of ethanol, (1 ha of wheat only produces 2.5 t of ethanol) (Demirbas & Balat, 2006). These numbers must be compared to the ethanol production from sugarcane, which reaches 7.5 t in Brazil (Bourne, 2007).

The USA produces ethanol from corn, whereas India uses sugarcane, China uses sweet potatoes and Canada uses wheat. Countries such as China, Austria, Sweden, New Zealand, and even Ghana are now building their biofuels infrastructure around wood-based feedstocks (Herrera, 2006).

The growing area used for sugarcane production in Brazil accounts for 8 Mha (Brazil is 850 Mha). Sugarcane produces an eight-fold return on the energy that is used to produce it. One ton of sugarcane used for ethanol production represents a net economy in CO_2 emissions equivalent to 220.5 when compared with fossil fuel. Thus, if rain forest is not destroyed to grow the sugarcane, ethanol from Brazilian sugarcane reduces greenhouse gas emissions by the equivalent of 25.8 Mt CO_2/yr (Marris, 2006; Walter et al., 2010). Fortunately, the Amazon, the Pantanal and the Alto Rio Paraguai regions have been prohibited for sugarcane cultivation by government decree since 2009 to preserve these ecosystems. In 2009, ethanol accounted for approximately 47% of transport fuel used in Brazil. The "Flex" car fleet can use 100% of either ethanol or gasoline (Orellana & Neto, 2006). In fact, ethanol gives 20% to 30% fewer kilometers per liter than does gasoline and people adapt the blend in proportion to the best consumption/price ratio (Marris, 2006).

The ethanol export capacity of Brazil is currently ~8 Gl. The export-destination countries are mainly the US, the EU, Japan and Central America. Conservative estimates suggest that the area used for sugarcane production in Brazil should increase from 8 to 11 Mha by 2015. By government decree, the maximum possible area to be used for sugarcane cultivation has been limited to 64 Mha (i.e., 18.5% of national territory). In the short-to-medium term, Brazil is the only country that is able to sustain the emerging international ethanol market. For long-term establishment in the market, other countries, such as Australia, Columbia, Guatemala, India, Mexico and Thailand, will need to increase their exports (Orellana & Neto, 2006).

Brazil began ethanol production in 1973. At that time, it was heavily dependent on foreign crude oil, with nearly 80% of its oil being imported. It launched the program PROALCOOL in 1975 (Goldemberg et al., 2004) and began to offer subsidies and low-interest loans to bioethanol producers to increase existing capacity. A policy of price dumping was maintained by the government to boost the use of gasohol. The ethanol content of common gasoline was originally 5% and is now 25% by law (Pousa et al., 2007).

2.1.1 Bioethanol from lignocellulosic biomass

The most abundant sources of renewable carbon in the biosphere are plant structural polysaccharides. Approximately 1,011 t of these polymers (with an energy content equivalent to 640 Gt of oil) are synthesized annually (Proctor et al., 2005). For example, non-food plant species for bioenergy production include *Sorghum halepense, Arundo donax, Phalaris arundinacea* (Raghu et al., 2006), poplar, switchgrass (*Panicum virgatum*), the hybrid grass *Miscanthus* x *giganteus* and big bluestem. These species are considered to have energetic, economic and environmental advantages over first-generation biofuel crops (Hill et al., 2006; Havlík et al., 2010). Switchgrass, for example, produces a net energy of 60 Giga Joule per hectare and per year (GJ/ha/yr) (Schmer et al., 2008). The potential terrestrial fuel yield from cellulosic biomass production (135 GJ/ha/yr) is somewhat higher than that from corn (85 GJ/ha/yr) or soybean biodiesel (18 GJ/ha/yr). The optimal types of specialized biofuel crops are likely to be perennial and indigenous species that are well adapted for growth on marginal lands.

In tropical and Mediterranean countries, eucalyptus is a fast-growing woody species that is cultivated for biomass production. In wet and temperate countries, high-yielding varieties of willow (*Salix nigra*), *Miscanthus* (a high-yielding rhizomatous grass that yields up to 26 t of dry matter/ha/yr) and poplar are available. These energy crops require relatively low chemical and energy inputs compared with conventional crop production and they are able to grow on marginal lands (thus avoiding the problem of competition with food crops). Considering an Ireland-based scenario, the utilization of *Miscanthus* and willow for heat and electricity generation would allow for savings of as much as 5.2% of 2004 GHG emissions while using only 4.6% of the total agricultural area (Styles & Jones, 2008). It has been estimated that lignocellulosic biomass could contribute 70-100 exajoules (1 exajoule = 1,000,000,000 gigajoules) by 2020 (Gielen et al., 2002).

Poplar is a candidate for short rotations of ~5 years. Poplar disperses its seeds and pollen much farther than do other crops, it does so for many years before harvesting and it has many wild relatives with which it can outcross. In addition, poplar can be multiplied vegetatively, which would allow for the valorization of low-lignin transformants through the multiplication of sterile accessions. The biotechnology of poplar has been dominated for several years and its genome has been sequenced.

Trees not only can achieve a lignocellulosic energy-conversion factor of 16 (compared with 1-1.5 for corn and 8–10 for sugarcane), but they can also be grown on marginal lands, thus reducing competition with food crops.

The world consumption of wood is 3.4 Gm^3/yr and will substantially increase with the production of ethanol from biomass. The development of high-yield plantations is essential to sustain the increased demand for wood (Fenning et al., 2008). Small towns, schools, buses, ski resorts and factories in Sweden and Austria have long relied on the byproducts of the forest industry to produce liquid and solid fuel (Herrera, 2006).

Biotechnology and systems biology can be envisaged for plant breeding. Many plant species used for bioenergy production are wild to semi-domesticated. Molecular approaches can speed up domestication and productivity (Chen & Dixon, 2007).

A number of candidate genes for domestication traits have been identified by comparing the genomes of poplar, rice and Arabidopsis for large-scale gene function and expression. The genes investigated were involved in synthesis of cellulose and hemicellulose, as well as in various morphological growth characteristics (such as height, branch number and stem thickness) (Ragauskas et al., 2006; Chapple et al., 2007; Sticklen, 2008). Transgenic plants that overexpressed mutant alleles or showed RNA interference (RNAi) for silencing endogenous genes have been designed and cell-wall components that were more easily converted to ethanol have been obtained (Chen & Dixon, 2007; Himmel et al., 2007). Examples of these strategies include the complementary decrease of lignin and the increase of cellulose components in cell walls or the directed overexpression of cellulases in plant cells to drastically decrease the cost of cell wall conversion to ethanol (Sticklen, 2008). However, the strategy involving lignin interference must be evaluated carefully in the context of biomass production because it could have side effects such as excessive sensitivity to fungal pathogens.

Because lignin is relatively resistant to enzymatic degradation, low-lignin transgenic trees have been investigated (Herrera, 2006). RNAi-mediated suppression of p-coumaroyl-CoA 3′-hydroxylase in hybrid poplars generally correlated very well with the reduction of lignin content. Up to ~13.5% more cell-wall carbohydrates have been observed in the suppressed lines as compared to wild-type poplars (Coleman et al., 2008).

Currently, lignocellulose pretreatment followed by enzymatic hydrolysis is the key process used for the bioconversion of lignocellulosic biomass (Sanderson, 2006). The type of pretreatment defines the optimal enzyme mixture to be used and the composition of the sugar mixture that is produced. Finally, the sugars are fermented with ethanol-producing microorganisms such as yeasts, *Zymomonas mobilis, Escherichia coli*, or *Pichia stipitis* (Fischer et al., 2008).

2.2 Biodiesel

2.2.1 The process of biodiesel production

The main components of plant oils are the fatty acids and their derivatives the mono-, di- and triacylglycerides. Tri-acylglycerides make up 95% of plant oils. Glycerides are esters formed by fatty acid condensation with tri-alcohol glycerol (propanetriol). Depending on the number of fatty acids fixed on the glycerol molecule, one can have mono-, di- or triacylglycerides. Of course, the fatty acids can be the same or different. As stated in the introduction, biodiesel can be obtained by esterification or transesterification. *Esterification* is the process by which a fatty acid reacts with a mono-alcohol to form an ester. The esterification reaction is catalyzed by acids. Esterification is commonly used as a step in the process of biodiesel fabrication to eliminate FFAs from low-quality oil with high acid content. *Transesterification* (or alcoholysis) is the displacement of alcohol from an ester by another alcohol in a process similar to hydrolysis. This process has been widely used to reduce triglyceride viscosity. The transesterification reaction is represented by the general equation (5).

$$RCOOR' + R''OH \rightarrow RCOOR'' + R'OH \quad (5)$$

This stepwise reaction occurs through the successive formation of di- and monoglycerides as intermediate products (Canakci et al., 2006). Theoretically, transesterification requires three alcohol molecules for one triglyceride molecule; however, an excess of alcohol is necessary because the three intermediate reactions are reversible (Marchetti et al., 2007; Om

Tapanes et al., 2008). After the reaction period, the glycerol-rich phase is separated from the ester layer by decantation or centrifugation. The resulting ester phase (crude biodiesel) contains contaminants such as methanol, glycerides, soaps, catalysts, or glycerol that must be purified to comply with the European Standard EN 14214.

Different technologies can be used for biodiesel production; these include chemical or enzyme catalysis and supercritical alcohol treatment (Demirbas, 2008b). EN 14214 establishes 25 parameters that must be assessed to certify the biodiesel quality.

In conventional transesterification and esterification processes for the production of biodiesel, strong alkalis or acids are used as chemical catalysts. These processes are highly energy consumptive and the poor reaction selectivity that often results from the physicochemical synthesis justifies the ongoing research on enzymatic catalysis. In addition, an extra purification step is required to remove glycerol, water, and other contaminants from alkyl-esters.

The base catalysis is much faster than the acid catalysis. Low cost and favorable kinetics have turned NaOH into the most-used catalyst in the industry. However, soap and emulsion can be formed during the reaction and complicate the purification process.

2.2.2 Non-edible feedstocks for biofuel production

Currently, approximately 84% of the world biodiesel production is met by rapeseed oil. The remaining portions are from sunflower oil (13%), palm oil (1%), soybean oil and others (2%) (Gui et al., 2008). More than 95% of biodiesel is still made from edible oils. To overcome this undesirable situation, biodiesel is increasingly being produced from non-edible oils and waste cooking oil (WCO). Non-edible oils offer the advantage that they do not compete with edible oils on the food market.

Used cooking oil is a waste product, and for that reason, it is cheaper than virgin plant oil. The higher initial investment required by the acid-catalyzed process (stainless-steel reactors and methanol-distillation columns) is compensated for by low feedstock cost (Zhang et al., 2003). Reusing WCO esters provides an elegant form of recycling, given that waste oils are prohibited for use in animal feed, are harmful to the environment, and human health and disrupt normal operations at wastewater treatment plants (increasing the costs of both maintenance and water purification). The production of biodiesel from WCO is still marginal, but it is increasing worldwide. The USA and China are leaders in WCO use, with 10 and 4.5 Mt/yr, respectively. Other countries and regions, such as the EU, Canada, Malaysia, Taiwan and Japan, produce approximately 0.5-1 Mt/yr (Gui et al., 2008). The potential use of WCO as a primary source for biodiesel fuel is important because such use would negate most of the actual concerns regarding the competition of food and biodiesel crops for land (Bindraban et al., 2009; Odling-Smee 2007). By converting edible oils into biodiesel fuel, food resources are actually being converted into automotive fuels. It is believed that large-scale production of biodiesel fuels from edible oils may bring global imbalance to the food supply-and-demand market, even if such a trend has been contested (Ajanovic, 2010). However, nothing prevents the use of edible oils first for cooking and then for biodiesel fuel.

2.2.3 Biofuel feedstocks in the world

Concerned by potential climate change-related damages (including changes to coastlines and the spread of tropical diseases, among others), the US faces the necessity of finding solutions for the 17.7%-reduction of GHG emissions (Lokey, 2007). Because of the fact that the electrical sector accounts for 40% of all GHG emissions, investments in cost-competitive renewable energy sources, such as wind, geothermal and hydroelectricity, have been recommended. Given the ample solar resources that exist in the US, it has a plethora of untapped sources for renewable-energy generation (Flavin et al., 2006). The Biomass Program of the US Department of Energy (launched in 2000) recommended 5% use of biofuels by 2010, 15% by 2017, and 30% by 2050. However, it is predicted that the ethanol market penetration for transportation should attain ~50% of gasoline consumption by 2030 (Szulczyk et al., 2010). Currently, maize and other cereals (such as sorghum) are the primary feedstocks for US ethanol production. At 40 Ml of ethanol per day, maize is still considered a low-efficiency biofuel crop because of its high required input, excessive topsoil erosion (10 times faster than sustainable) and other negative side effects (Donner & Kucharik, 2008; Laurance, 2007; Sanderson, 2006; Scharlemann & Laurance, 2008). By comparison, biodiesel from soybean requires lower inputs. However, neither of these biofuels can displace fossil fuel without impacting food supplies. Even if all US corn and soybean production were dedicated to biofuels, only 12% of the gasoline and 6% of the diesel demand, respectively, would be met (Hill et al., 2006). However, agricultural, municipal, and forest wastes could together sustainably provide 1 Gt of dry matter annually and should complement the other biofuel crops (Vogt et al., 2008). It was proposed that 3.1-21.3 Mha of land should be converted to biomass production (Schmer et al., 2008). Algal biodiesel is also being included in an integrated renewable-energy park (Singh & Gu, 2010; Subhadra, 2010).

Bioethanol from Brazil results in over 90% GHG savings (Hill et al., 2006). In addition to the PROALCOOL program, the Brazilian government created the PRO-ÓLEO program in 1980 and expected a 30% mixture of vegetable oils or derivatives in diesel and full substitution in the long term. Unfortunately, after the price drop of crude oil on the international market in 1986, this program was abandoned and was only reintroduced in 2002. Because of its great biodiversity and diversified climate and soil conditions, Brazil has a variety of plant-oil feedstocks, including mainly soybean, sunflower, coconut, castor bean, cottonseed, oil palm, physic nut and babassu (Nass et al., 2007). Brazil celebrated the inauguration of the Embrapa Agroenergia research center in 2010 to promote the integration of the oil from these feedstocks into the network of biodiesel sources. The National Program of Production and Use of Biodiesel (PNPB) was launched in 2004 with the objective of establishing the economic viability of biodiesel production together with social and regional development. The current diesel consumption in Brazil is approximately 40 Gl/yr and the potential market for biodiesel currently of 800 Ml and that should achieve 2 Gl by 2013. In addition, B5 has been mandatory since 2010. Auction prices have varied between US$ 0.3 and 0.8/l according to the area of production (Barros et al., 2006). Between 1975 and 1999, US$ 5 bn were invested in bioenergy resulting in the creation of 700,000 new jobs and US$ 43 bn saving in gasoline imports (Moreira & Goldemberg, 1999). The rate of job creation related to biodiesel production has been estimated to be 1.16 jobs/Ml of annual production (Johnston & Holloway, 2007). However, the recent trend of business centralization is expected to reduce this rate (Hall et al., 2009). Petrobras is now processing (with a capacity of 425,000 t) a mixture of plant oil and crude oil under the name of "H-Bio". With a tropical climate in the major part of its extention, the country has a potential 90 Mha that could be used for oleaginous crop production and that extends over Mato Grosso (southwest), Goiás, Tocantins, Minas Gerais (center), Bahia Piauí, and Maranhão (northeast).

The EU accounts for 454 million people (i.e., 7% of the world's population and 50% more people than live in the US) (Solomon & Banerjee, 2006). The EU is dedicated to a long-term conversion to a hydrogen economy. Renewable energy sources and eventually advanced nuclear power, are envisioned as the principal hydrogen sources on the horizon for use in 2020-2050 (Adamson, 2004). However, even for the distant future, the EU foresees hydrogen production from fossil fuels with carbon sequestration still playing a major role (together with renewable energy and nuclear power). Because of their renewability, biodiesel and bioethanol in the EU have been calculated to result in 15-70% GHG savings when compared to fossil fuels. Frondel and Peters (2007) found that the energy and GHG balances of rapeseed biodiesel are clearly positive.

Bioethanol from sugar beets or wheat and biodiesel from rapeseed are currently the most important options available to the EU for reaching its target biofuel production. Because of increased land use for biofuel production, biofuel crops are now competing with food crops (Odling-Smee, 2007) and they are expected to have substantial effects on the economy. The European consumption of fossil diesel fuel is estimated to be approximately 210 Gl and that of biodiesel to be 9.6 Gl (Malça & Freire, 2011). The EU produces over ~2 Mha (i.e., ~1 Gl) of rapeseed (0.5 kl/ha) and sunflower (0.6 kl/ha) (Fischer et al., 2010), which shows that it depends heavily on importation of biofuels to approach the recommended target of B5.75. Given the higher energy potential of synfuel from biomass and the constraints on the availability of arable land, second-generation biofuels should soon enter the race for biofuel production (Fischer et al., 2010; Havlík et al., 2010).

The price for biodiesel that meets the EU quality standard (EN 14214) is approximately € 730/t. By subtracting the biodiesel export value from the EU market price, one obtains the profit obtained by selling biodiesel from abroad on that market. The export value includes production and exportation costs. Production costs are made up of the plant oil or animal fat production plus the biodiesel processing minus the value of by-products (glycerol for example). Exportation costs include scaling, insurance, taxes and administrative costs (see the calculations in Johnston & Holloway, 2007). The price of US$ 0.88/l for biodiesel was 45% higher than the price of fossil diesel fuel during the same period (2006). Although this price is a convenient baseline, the biodiesel price on the EU market can change quickly depending on factors such as current domestic production, fossil diesel-fuel prices, agricultural yields, and legislation. The same rules will apply to emerging markets in China. Based on volume and profitability estimated in this manner, the top five countries that have the best combination of high volumes and low production costs are Malaysia, Indonesia, Argentina, the US, and Brazil. Collectively, these countries account for over 80% of the total biodiesel production. Plant oils currently used in biodiesel production account for only approximately 2% of global vegetable-oil production, with the remainder going primarily to food supplies.

Despite the fact that India has not attained the high level of ethanol production seen in Brazil, it is the largest producer of sugar in the world. Indian ethanol is blended at 5% with gasoline in nine Indian states and an additional 500 Ml would be needed for full directive implementation. The total demand for ethanol is approximately 4.6 Gl (Subramanian et al., 2005). The country burns 3 times more fossil diesel fuel than gasoline (i.e., roughly 44 Mt), mainly for transportation purposes.

Because India imports 70% of its fuel (~111 Mt), any source of renewable energy is welcome. Therefore, India has established a market for 10% biodiesel blends (Kumar & Sharma, 2008). Because India is a net importer of edible oils, it emphasizes non-edible oils from plants such as physic nut, karanja, neem, mahua and simarouba. Physic nut and karanja are the two leaders on the Indian plant list for biodiesel production.

Of its 306 Mha of land, 173 Mha are already under cultivation. The remainder is classified as either eroded farmland or non-arable wasteland. Nearly 40% (80-100 Mha) of the land area is degraded because of improper land use and population pressures over a number of years. These wasted areas are considered candidates for restoration with physic nuts (Kumar & Sharma, 2008). Nearly 80,000 of India's 600,000 villages currently have no access to fuel or electricity, in part because there is not enough fuel to warrant a complete distribution network. Physic nuts could bring oil directly into the villages and allow them to develop their local economies (Fairless, 2007). This also applies to developing areas of Brazil and Africa.

In addition to the biodiesel initiative, regular motorcycles with 100 cm^3 internal combustion engines have been converted to run on hydrogen. The efficiency of these motorcycles has been proven to be greater than 50 km/charge. This development has had great significance because 70% of privately owned vehicles in India are motorcycles and scooters. Efforts are also underway to adapt light cars and buses to hydrogen, a move that will likely be helped by the growing number of electric and compressed natural gas (CNG) vehicles in and around New Delhi (Solomon & Banerjee, 2006).

In China, the area of arable land per capita is lower than the world's average. As a result, most edible oils are imported and the demand for edible oils in 2010 is projected to be 13.5 Mt. Because of its large population, China desperately needs sustainable energy sources. Because little arable land is available, China is exploring possibilities for the production of second- and third-generation biofuels (Meng et al., 2008). China is a large developing country that has vast degraded lands and that needs large quantities of renewable energy to meet its rapidly growing economy and accompanying demands for sustainable development. The energy output of biomass grown on degraded soil is nearly equal to that of ethanol from conventional corn grown on fertile soil. Biofuel from biomass is far more economic than conventional biofuels such as corn ethanol or soybean biodiesel. Potential energy production from biomass could reach 6,350,971 terajoules per year (TJ/yr) and an increased value of biomass in China's energy portfolio is considered unavoidable (Zhou et al., 2008).

Taking advantage of seawater availability, biodiesel from micro*algae* could also be conveniently grown along the 18,000 km Chinese coastline (Song et al., 2008). Marine micro*algae* production requires unused desert land, seawater, CO_2 and sunshine. Given the abundant areas of mudflats and saline lands in China, there is great potential to develop biodiesel production from marine micro*algae*.

Sales of electric bicycles and scooters in China have grown dramatically in the last 10 years and now total over 1 million per year. The growth of this demand has been facilitated by bans on gasoline-fueled bicycles and scooters in Beijing and Shanghai (among other large cities) because of increasing concerns about pollution (Solomon & Banerjee, 2006). For this reason, China has become one of the largest potential markets for hydrogen fuel cells in the transportation sector.

Frequent droughts in many Asian countries have made it difficult for them to replicate Brazil's success with sugarcane, which needs an abundant water supply. Thailand and Indonesia are tapping the potential with palm oil.

Because of its need to retain its position as the high-tech superpower for new technologies, Japan has become one of the most important players in the international development of a hydrogen-based economy. Following Japanese estimations, the hydrogen production potential from renewable energy in Japan is 210 GNm^3/yr (Nm^3 is the gas volume in m^3 at 0 °C and one atmosphere), which is 4 times more than what it will actually need in 2030. However, hydrogen based on renewable sources is only expected to contribute approximately 15% of the hydrogen consumed by 2030. It is estimated that on-board reforming of methanol or gasoline for fuel cell propelling would be the most practical technology in the near term, but the long-term goal is to adopt pure hydrogen (Solomon & Banerjee, 2006).

3. Microdiesel

Oleaginous microorganisms are microbes with an oil content that exceeds 20%. Biodiesel production from microbial lipids (known as single-cell oil or microdiesel) has attracted great attention worldwide. Although microorganisms that store oils are found among various microbes (such as micro*algae*, *bacillus*, fungi and yeast), not all microbes are suitable for biodiesel production (Demirbas, 2010).

Most bacteria are generally not good oil producers. Some exceptions are actinomycetes, which are capable of synthesizing remarkably high amounts of fatty acids (up to ~70% of their dry weight) from simple carbon sources such as glucose under growth-restricted conditions and which accumulate these fatty acids intracellularly as triglycerides (Alvarez & Steinbuchel, 2002).

The most efficient oleaginous yeast, *Cryptococcus curvatus,* can accumulate >60% lipids when grown under nitrogen-limiting conditions. These lipids are generally stored as triglycerides with approximately 44% percent saturated fatty acids, which is similar to many plant seed oils. *Rhodotorula glutinis* has been used for the wastewater treatment in monosodium-glutamate manufacturing. Monosodium-glutamate wastewater is as a cheap fermentation broth for the production of biodiesel using lipids from *R. glutinis*. To be efficient, the fermentation process needs a complementary source of glucose to obtain the proper C:N:P ratio (1:2.4:0.005). This process leads to a lipid production corresponding to 20% of the biomass after 72 h of culture and to an oil transesterification rate of 92% (Xue et al., 2008). In addition, *R. glutinis* can use various carbon sources including dextrose, xylose, glycerol, dextrose and xylose, xylose and glycerol, or dextrose and glycerol and can accumulate 16, 12, 25, 10, 21, and 34% triglycerides, respectively. The rate of unsaturated fatty acid accumulation was found to depend on the carbon source, with 25% and 53% accumulation when *R. glutinis* was grown on xylose and glycerol, respectively (Easterling et al., 2008). These results indicate that the use of *R. glutinis* can add value to several by-products, including glycerol. However, the resulting high levels of unsaturated fatty acids may require some additional saturation step to meet biodiesel standards.

Cyanobacteria are gram-negative photoautotrophic prokaryotes that can be cultivated under aqueous conditions ranging from freshwater to extreme salinity. They are able to produce a wide range of fats, oils, sugars and functional bioactive compounds such that their inclusion to wastewater treatment processes has been proposed (Markou & Georgakakis, 2011). Their duplication time is 3.5 h in the log phase of cell multiplication (Chisti, 2007). Using light energy, they are able to convert carbon substrates into oil (with a fatty acid composition that is similar to that of plants) at a rate of 20–40% of dry biomass (Meng et al., 2008).

Although micro*algae* are high-lipid-storing microbes, they require larger areas and longer fermentation times than do bacteria. The micro*algae* market produces approximately 5,000 t of dry biomass/year and generates approximately US$ 1.25 bn/yr (Pulz & Gross, 2004).

Eukaryotic diatoms, green *algae* and brown *algae* isolated from oceans and lakes typically reach dry-mass levels of 20%-50% lipids (Brennan & Owende, 2010). The quantities of lipids found in micro*algae* can be extraordinarily high. In *Botryococcus*, for instance, the concentration of hydrocarbons may exceed 80% of the dry matter. In comparison, dry-biomass plant oil levels are generally around 15-40% lipids (Spolaore et al., 2006).

There are approximately 300 strains of *algae*, among which diatoms (including genera *Amphora*, *Cymbella*, and *Nitzschia*) and green *algae* (particularly genera *Chlorella*) that are the most suitable for biodiesel production. The oil is accumulated in almost all micro*algaes* as triglycerides (>80%) that are rich in C16 and C18 (Meng et al., 2008). Lipid accumulation in oleaginous microorganisms begins with nitrogen exhaust or when carbon is in excess (Ratledge 2002).

Chlorella protothecoides can accumulate lipids at a rate of 55% by heterotrophic growth under CO_2 filtration. Large quantities of microalgal oil have been efficiently recovered from these heterotrophic cells by n-hexane extraction. The microdiesel from heterotrophic microalgal oil obtained by acidic transesterification is comparable to fossil diesel and should be a competitive alternative to conventional biodiesel because of higher photosynthetic efficiency, larger quantities of biomass, and faster growth rates of micro*algae* as compared to those of plants (Song et al., 2008).

As stated above, microalgal oils differ from most plant oils in being quite rich in polyunsaturated fatty acids with four or more double bonds (Belarbi et al., 2000). This makes them susceptible to oxidation during storage and reduces their suitability for commercial biodiesel (Chisti, 2007). However, fatty acids with more than four double bonds can be easily reduced by partial catalytic hydrogenation (Dijkstra, 2006).

Changes in the degree of fatty acid unsaturation and the decrease or increase of fatty acid length are major challenges in modifying the lipid composition of microalgal oils. These features are regulated by enzymes that are mostly bounded to the cell membrane, which complicates their investigation (Certik & Shimizu, 1999). Currently, most of the genetic manipulations that have aimed to optimize metabolic pathways have been carried out on oleaginous microorganisms. This is mainly because of their abilities to accumulate high amounts of intracellular lipids, their relatively fast growth rates and their similarities of oil composition with plants (Kalscheuer et al., 2006a, 2006b).

Micro*algae* are often used for the sequestration and recycling of CO_2 by "CO_2 filtration" (Haag, 2007) and can reduce CO_2 exhaust by 82% on sunny days and by 50% on cloudy days (Vunjak-Novakovic et al., 2005). This process is much more elegant than carbon storage (CCS) in depleted oil fields or in aquifers because the carbon can be recycled via microdiesel. The storage capacity of CCS is estimated to range between 2,000-11,000 Gt CO_2; however, such aquifers are not evenly distributed around the world (Schiermeier et al., 2008). In addition, CCS does not result in any profit from the CO_2 that is stored and is actually an additional cost in the whole process. In contrast, *algae* convert CO_2 into oil. This means that the energy contained in the CO_2 can be re-injected into the power plant after being filtered by the *algae* and transformed into microdiesel.

The stimulation of fish production by increasing phytoplankton biomass through CO_2 injection into specific ocean localities has also been proposed (Markels & Barber, 2001). However, ocean fertilization has been severely challenged because it would eventually destroy the local ecosystem (Bertram, 2010; Glibert et al., 2008).

4. Biohydrogen

The main alternative energy carriers considered for transportation are electricity and hydrogen. With interest in its practical applications dating back almost 200 years, hydrogen energy is hardly a novel idea. Iceland and Brazil are the only nations where renewable-energy feedstocks are envisioned as the major or sole future source of hydrogen (Solomon & Banerjee, 2006). Fuel-cell vehicles (FCVs) powered by hydrogen are seen by many analysts as an urgent need and as the only viable alternative for the future of transportation (Cropper et al., 2004).

Unlike crude oil or natural gas, reserves of molecular H_2 do not exist on earth. Therefore, H_2 must be considered more as an energy carrier (like electricity) than as an energy source (Song, 2006). H_2 can be derived from existing fuels such as natural gas, methanol or gasoline; however, the best long-term solution is to produce H_2 from water by (for example) using heat from solar sources and O_2 from the atmosphere.

Today, hydrogen is mainly manufactured by decarbonizing fossil fuels, but in the future it will be possible to produce hydrogen by alternative methods such as water photolysis using semiconductors (Khaselev & Turner, 1998) or by ocean thermal-energy conversion (Avery, 2002). Such methods are still in the research and development stage and are not yet ready for industrial application.

Hydrogen production from biomass requires multiple reaction steps. The reformation of fuels is followed by two steps in the water-gas shift reaction, a final carbon monoxide purification step and carbon dioxide removal.

Biomass can be thermally processed through gasification or pyrolysis. The main gaseous products resulting from the biomass are expressed by equations (6), (7) and (8) (Kikuchi, 2006).

$$\text{pyrolysis of biomass} \rightarrow H_2 + CO_2 + CO + \text{hydrocarbon gases} \quad (6)$$

$$\text{catalytic steam reforming of biomass} \rightarrow H_2 + CO_2 + CO \quad (7)$$

$$\text{gasification of biomass} \rightarrow H_2 + CO_2 + CO + N_2 \quad (8)$$

Hydrogen from organic wastes has generally been produced through equations (9), (10) and (11).

$$\text{solid waste} \rightarrow CO + H_2 \quad (9)$$

$$\text{biomass} + H_2O + \text{Air} \rightarrow H_2 + CO_2 \quad (10)$$

$$\text{cellulose} + H_2O + \text{Air} \rightarrow H_2 + CO + CH_4 \quad (11)$$

In the long run, the methods used for hydrogen production are expected to be specific to the locality. They are expected to include steam reforming of methane and electrolysis when hydropower is available (such as in Brazil, Canada and Scandinavia) (Gummer & Head, 2003). When hydrogen will become a very common energy source, it will likely be distributed through pipelines. Existing systems, such as the regional H_2-distribution network that has been operated for more than 50 years in Germany and the intercontinental liquid-hydrogen transport chain, demonstrate that leak rates of <0.1% can be achieved in industrial applications (Schultz et al., 2003). However, a major threat associated with the hydrogen paradigm is the fact that it is the smallest atom and that leakage is apparently unavoidable. One has to face the possibility that a significant amount of H_2 will be released into the stratosphere. Hydrogen is expected to react with ozone following the reaction $H_2+O_3 \rightarrow H_2O+O_2$. This mechanism (reviewed by Kikuchi, 2006) is a potentially dangerous promoter of ozone depletion. Alternatively, hydrogen can be produced from another fuel (e.g., ethanol, biodiesel, gasoline, or synfuel) via onboard reformers (hydrogen fuel processors). This is probably the best solution because synfuel can be produced from local feedstocks through the Fischer-Tropsch process, transported and distributed through existing technologies and infrastructures (Agrawal et al., 2007; Takeshita & Yamaji, 2008). This consideration also applies to biofuels. In addition, the feasibility of cars with onboard reformers has already been proven. The importance of synfuel is expected to increase rapidly because growing reserves of natural gas (or "stranded" gas) are available in remote locations and are considered to be too small for liquefied natural gas (LNG) or pipeline projects.

The biological generation of hydrogen (or biohydrogen) provides a wide range of approaches for generating hydrogen, including direct biophotolysis, indirect biophotolysis, photo-fermentation and dark-fermentation (Lin et al., 2010). Biological hydrogen production processes are found to be more environmentally friendly and less energy intensive as compared to thermochemical and electrochemical processes. There are three types of microorganisms that produce hydrogen, namely cyanobacteria, anaerobic bacteria, and fermentative bacteria (Demirbas, 2008a).

Photosynthetic production of H_2 from water is a biological process that can convert sunlight into useful, stored chemical energy. Hydrogen production is a property of many phototrophic organisms and the list of H_2 producers includes several hundred species from different genera of both prokaryotes and eukaryotes. The enzyme-mediating H_2 production seen in green *algae* is effected by a reversible hydrogenase that can catalyze ferredoxin oxidation in the absence of ATP (Beer et al., 2009). The enzyme is sensitive to oxidation; however, tolerant allozymes are being selected (Seibert et al., 2001). Hydrogen production has also been obtained from glucose using NADP+-dependent enzymes, glucose-6 phosphate dehydrogenase (G6PDH), 6-phosphogluconate dehydrogenase (6PGDH) and hydrogenase (Heyer & Woodward, 2001).

Carbon monoxide (CO) can be metabolized by a number of naturally occurring microorganisms along with water to produce H_2 and CO_2 following equation (12), which is the "water-gas shift" reaction, at ambient temperatures.

$$CO + H_2O \leftrightarrow CO_2 + H_2 \qquad (12)$$

The biological water-gas shift reaction has been used in the processing of syngas from biomass with the bacterium *Rubrivivax gelatinosus* (Wolfrum & Watt, 2001).

Nitrogenases can produce hydrogen but require relatively high energy consumption. However, the nitrogenase reaction is essentially irreversible, which allows for hydrogen pressurization. *Rhodopseudomonas palustris* can drive the nitrogenase reaction using light (Wall, 2004).

5. The future of transport technology

5.1 Fuel cells

The fuel cell is the central component of hydrogen cars; it performs the conversion of fuel energy into electricity through proton mobilization. Fuel cells do not have moving parts, they produce only clean water and low-voltage electricity using hydrogen and oxygen, they are not noisy and they are 60% efficient, which is more than internal combustion engines (ICE, 45% efficiency). Laboratory tests indicate that fuel cells have a potential efficiency of 85% or more, which when combined with an 80%-efficient electric motor could make them 2 times more efficient than the direct use of hydrogen in an ICE (Ross, 2006).

Because of the security and cost problems related to infrastructure for hydrogen distribution and storage, ethanol is currently the most convenient alternative for fuel cells. Ethanol can be converted in hydrogen by onboard steam reforming or can be more conveniently used as a proton donor in specific fuel-cell technologies (Lamy et al., 2004). Ethanol-based steam reforming is performed following equation (13) (Velu et al., 2005).

$$C_2H_5OH + 3H_2O \rightarrow 2CO_2 + 6H_2 \quad (13)$$

Deluga et al. (2004) described an onboard system for hydrogen production by auto-thermal reforming from ethanol. Following this system, ethanol and ethanol-water mixtures were converted directly into H_2 by catalytic oxidation with ~100% selectivity and >95% conversion and with a residence time on rhodium catalysts of <10 milliseconds. This process has great potential for low-cost H_2 generation in fuel cells for small portable applications in which liquid-fuel storage is essential and in which systems must be small, simple, and robust.

Another strategy of energy extraction from simple organic molecules is the glycerol biofuel cell (Arechederra et al., 2007). A biofuel cell is similar to a traditional proton exchange membrane (PEM) fuel cell. Rather than using precious metals as catalysts, biofuel cells rely on biological molecules (such as enzymes) to carry out the reactions. Arechederra et al. (2007) were able to immobilize two oxidoreductase enzymes (pyrroloquinoline quinine-dependent alcohol dehydrogenase and pyrroloquinoline quinine-dependent aldehyde dehydrogenase) at the surface of a carbon anode and to undertake a multi-step oxidation of glycerol into mesoxalic acid with 86% use of the glycerol energy. The bioanodes resulted in power densities of up to 1.21 mW/cm^2 using glycerol at concentrations up to 99 %. Because Nafion (the membrane) does not swell under glycerol, the biofuel cell longevity is expected to be higher than the technology used at moment.

Formula 1 has entered the race for optimizing green technologies. From 2009 on, new regulations for Formula 1 have forced the racing teams to recover the energy lost in braking and to use it to propel the car (Trabesinger, 2007). The technology that accomplishes this is called a "kinetic-energy recovery system" (KERS, better known as "regenerative braking"). In a hybrid car with both combustion and electric motors, batteries can be charged either by the ICE or by regenerative braking. The stored electric energy is then used to power the car at low speeds (i.e., in the city traffic) where the ICE efficiency is low because of continuous "stop-and-go" motion.

Fuel cells are still very expensive and currently cost approximately US$ 4,000/kW, which is 100 times more expensive than the cost of ICEs. Fuel-cell stacks must be replaced 4–5 times during the lifetime of current generations of vehicle. It is thus the cost of 4-5 fuel-cell units that must be compared with alternative ICEs (Marcinkoski et al., 2008; Sorensen, 2007).

Therefore, to be competitive with ICEs, the technology must reach the threshold of US$ 30/kW. To address this situation, Honda is selling its first prototype fuel-cell car under a leasing contract in California. BMW has been a pioneer of fuel-cell technology and produced its first hydrogen-car prototype in the 1960s (Hissel et al., 2004). Its current vehicle uses liquid hydrogen with autonomy of up to 386 km. The Ford Motor Company has set a new land-speed record for a fuel-cell powered car (334 km/h).

Despite these pilot experiments, it is likely that urban buses will be among the first large scale commercial applications for fuel cells. This is due to the fact that urban buses are highly visible to the public, contribute significantly to air and noise pollution in urban areas, have few size limitations and are fueled via a centralized infrastructure. Folkesson et al. (2003) reported the following: (i) the net efficiency of a Scania bus powered by a hybrid PEM fuel-cell system was approximately 40%; (ii) the fuel consumption of the hybrid bus was between 42 and 48% lower than that of a standard ICE Scania bus; and (iii) regenerative braking saved up to 28% energy. The bus prototype was equipped with a fuel cell of 50 kW and was fueled with compressed ambient air and compressed hydrogen stored on the roof.

All of the fossil fuel options result in large amounts of GHG emissions. Ethanol and hydrogen have the potential to significantly reduce greenhouse gas emissions. However, their use will be highly dependent on pathways of ethanol and hydrogen production. Some of the hydrogen options result in higher GHG emissions than do ICEs running on gasoline. The vehicle options that will be competitive during the next two decades are those that use improved ICEs (including hybrids burning 'clean' gasoline or diesel). In the present state of the technology, cars running on hydrogen using onboard reforming of carbon fuel are still ecologically less efficient than are gasoline ICEs. The relatively high energy consumption required to produce hydrogen is expected to affect the geographic distribution of hydrogen-powered cars. One can speculate that such cars would be more appropriate in areas where solar (Eugenia Corria et al., 2006), wind or hydro-electricity power sources are abundant.

5.2 Energy storage

A variation on the hybrid vehicle is the 'plug-in hybrid', which can be connected to the electric grid. The savings in energy costs over the whole cycle of charging an onboard battery and then discharging it to run an electric motor in an electric-hybrid (e-hybrid) car is 80%. This figure is approximately 4 times higher than the savings from fuel-cell cars running on hydrogen made using electrolysis and 30% higher than savings from cars running on gasoline (Romm, 2006). These vehicles allow the replacement of a substantial portion of the fuel consumption and tailpipe emissions. If the electricity is produced from CO_2-free sources, then e-hybrids can also have dramatically reduced net greenhouse gas emissions.

The electrical storage system is the key element of the e-hybrid car because its power capacity and lifetime decisively define the costs of the overall system (Bitsche & Gutmann, 2004).

Bio-based energy-management processes are emerging and could make a significant contribution in the medium term. The production of electricity is also possible with whole-microorganism fermentation. Fe(III)-reducing microorganisms in the family *Geobacteraceae* can directly transfer electrons onto electrodes (Bond et al., 2002; Bond & Lovley, 2003). However, the range of electron donors that these organisms can use is limited to simple organic acids. By contrast, *Rhodoferax ferrireducens* is capable of oxidizing glucose and other sugars (such as fructose and xylose) with similar efficiency and of quantitatively transferring electrons to graphite electrodes. The sugar is consumed in the anode chamber. The oxidation of one molecule of glucose produces CO2, H+ and 24 electrons with a ~83% efficiency. The reaction produces a long-term steady current that is sustained after glucose-medium refreshing in the anode chamber. This microbial fuel cell can be recharged by changing the anode medium. It does not show severe capacity fading in the charge/discharge cycling and only presents low-capacity losses under open circuits and prolonged idle conditions (Chaudhuri & Lovley, 2003).

Another bacterium that is able to transfer electrons to solid metal oxides is *Shewanella oneidensis* MR-1. In addition, to their remarkable anaerobic versatility, analyses of the genome sequences of *Shewanellae* species suggest that they can use a broad range of carbon substrates; this creates possibilities for their application in biofuel production (Fredrickson, 2008). Production and storage of electricity are expected to evolve quickly within the new paradigm of emerging bioelectronics (Willner, 2002).

Sol-gels have been demonstrated to be usable for the entrapment of membrane-bound proteins in a physiologically active form and have been proven to be capable of maintaining protein activity over periods of months or more (Luo et al., 2005). Using a membrane-associated F_0F_1-ATP synthase, Luo et al. (2005) showed that the photo-induced proton gradient can be used to 'store' light energy as ATP. This has the advantage of eliminating passive leakage of ions across the membrane. In addition, ATP can be used for direct powering of motor proteins for the conversion of chemical energy to mechanical energy (Browne & Feringa, 2006). Nano power plants based on the rotation of magnetic bead propellers mounted on F_0F_1-ATP synthase rotors that are fed by ATP to induce electric current in microarrays of nanostators are now being designed and are in the research and development stage of construction (Soong et al., 2000; Yasuda et al., 2001).

6. Options for grid contributions

Electricity is the foundation of modern societies, yet more than 1.6 billion people remain without access to the electrical grid. A majority of this population lives in South Asia and sub-Saharan Africa. Despite global economic expansion and advances in energy technologies, roughly 1.4 billion people (or 18% of the world's population) will still be without power by 2030 unless major governmental incentives are put into place (Dorian et al., 2006).

The world average annual electricity consumption is between 2 and 4 TW. The cost of fossil-derived electricity is now in the range of US$ 0.02-0.05/kW/hr, including storage and distribution costs (Lewis & Nocera, 2006). For comparison, the options of non-biological electricity generation are as follows. (i) The light-water reactors that make up most of the world's nuclear capacity produce electricity at costs of US$ 0.025-0.07/kW/; however, there is no consensus as to the solution to the problem of how to deal with the nuclear wastes that have been generated in nuclear power plants over the past 50 years (Schiermeier et al., 2008). (ii) Hydroelectric energy sources have a generating capacity of 800 GW (i.e., 10 times more power than geothermal, solar and wind power sources combined) and currently supply approximately one-fifth of the electricity consumed worldwide. Annual operating costs are US$ 0.03-0.10/kW/h, which makes such sources competitive with coal and gas. Because only approximately 30% of worldwide hydroelectric capacity is currently used, energy from these sources can still be tripled (Schiermeier et al., 2008). (iii) Wind turbines can produce 1,500 kW at US$ 0.05-0.09/kW/h making wind competitive with coal; wind power could provide up to 20% of the electricity in the grid. The EU should be able to meet 25% of its current electricity needs by developing wind power in less than 5% of the North Sea and is heavily investing in that option. (iv) Exploitation and resulting use of the best geothermal sites is estimated to cost approximately US$ 0.05/kW/h. Thus, 70 GW of the global heat flux is seen as exploitable. However, because of the great deal of investment required, exploitation of geothermal power lies outside of current priorities except in regions with significant volcanic activity (Schiermeier et al., 2008). (iv) Commercial photo-voltaic (PV) electricity costs US$ 0.25-0.30/kW/h, which is still 10 times more than the current price of electricity on the grid.

The possibility for use of current PV technology is limited to 31% by theoretical considerations. A conversion efficiency of >31% is possible if photons with high energies are converted to electricity rather than to heat. With use of such technology, the conversion efficiency could be >60% (Lewis, 2007). The absence of a cost-effective storage method for solar electricity is also a major problem. Currently, the cheapest method of solar-energy capture, conversion, and storage is solar thermal technology, which can cost as little as US$ 0.10-0.15/kW/h for electricity production. This requires the focusing of the energy in sunlight for syngas or synfuel synthesis (Lewis & Nocera, 2006) or its thermal capture by heat-transfer fluids that are able to sustain high temperatures (>427 °C) and resulting electricity generation through steam production (see in Shinnar & Citro, 2006). Solar power is among the most promising carbon-free technologies available today (Schiermeier et al., 2008). The earth receives approximately 100,000 TW of solar energy each year. There are areas in the Sahara Desert, the Gobi Desert in central Asia, the Atacama in Peru and the Great Basin in the US that are suitable for the conversion of solar energy to electricity. The total world energy needs could be fed using solar energy captured in less than a tenth of the area of the Sahara. Residential and commercial roof surfaces are already being used in several countries to allow the people to sell their own PV electricity to the grid (and in this way saving substantial annual costs). This elegant strategy could be extended to other systems of energy production.

The capital costs of biomass are similar to those of fossil fuel plants. Power costs can be as little as US$ 0.02/kW/h when biomass is burned with coal in a conventional power plant. Costs increase to US$ 0.04-0.09/kW/h for a co-generation plant, but the recovery and use of the waste heat makes the process much more efficient. The biggest problem for new biomass power plants is finding a reliable and concentrated feedstock that is available locally. Biomass production is limited by land-surface availability, the efficiency of photosynthesis, and the water supply. Biomass potential is estimated at ~5 TW (Schiermeier et al., 2008). Photosynthesis is relatively inefficient if one considers that in switchgrass (one of the fastest-growing crops), energy is stored in biomass at an average rate of <1 $W/m^2/yr$. Given that the average insolation produces 200-300 W/m^2, the average annual energy conversion and storage efficiency of the fastest growing crops is only <0.5% (Lewis 2007; Lewis & Nocera, 2006). However, photosynthetic efficiency can be improved by genetic engineering (Ragauskas et al., 2006). Another potential problem with biomass production is that it could result in an increase of water consumption of two to three orders of magnitude. This is an important consideration because basic human necessities and power generation are increasingly competing for water resources (King et al., 2008).

The potential availability of wind (Pryor & Barthelmie, 2010), solar and biomass energy varies over time and location. This variation is not only caused by the individual characteristics of each resource (e.g., wind and solar regimes, soils), but also by geographic (land use and land cover), techno-economic (scale and labor costs) and institutional (policy regimes and legislation) factors (de Vries et al., 2007). The regional potential in energy units/year must be integrated over the geographical units that belong to a particular region. The model from de Vries et al. (2007) showed the following: (i) electricity from solar energy is typically available from Northern Africa, South Africa, the Middle East, India, and Australia; (ii) wind is concentrated in temperate zones such as Chile, Scandinavia, Canada, and the USA; (iii) biomass can be produced on vast tracts of abandoned agricultural land typically found in the USA, Europe, the Former Soviet Union (FSU), Brazil, China and on grasslands and savannas in other locations. In many areas of India, China, Central America , South Africa and equatorial Africa, these energy sources are available at costs of below US$ 0.1/kW/h and are found in areas where there is already a large demand for electricity (or there will be such demand in the near future). A combination of electricity from wind, biomass and/or solar sources (Eugenia Corria et al., 2006) may yield economies-of-scale in transport and storage systems. Regions with high ratios of solar-wind-biomass potential to current demand for electricity include Canada (mainly wind), African regions (solar-PV and wind), the FSU (wind and biomass), the Middle East (solar-PV) and Oceania (all sources). In other region (such as Southeast Asia and Japan), the solar-wind-biomass supply is significantly lower than the demand for electricity. Ratios of around one are found in Europe and South Asia. The potentials just described depend on many parameters, and their achievement will depend on future land-use policies (de Vries et al., 2007; Miles & Kapos, 2008).

7. Management and sustainability

Adam Smith's notion that by pursuing his own interest a man "frequently promotes that of society more effectively than when he really intends to promote it" and Karl Marx's picture of a society in which "the free development of each is the condition for the free development of all" are both limited by one obvious constraint. The world is finite. This means that when one group of people pursues its own interests, it damages the interests of others (Vertès et al., 2006). The model of Western economies was established using this logic. The theoretical framework of this philosophy is a mathematical model that is based on energy-conservation equations formulated by von Helmholtz in 1847, in which physical variables were arbitrarily substituted by economic ones. The consequences of this model are as follows: (i) the market is a closed circular flux between production and consumption, without inflows or outflows; (ii) natural resources are located in a domain that is separate from that of the closed market system; (iii) the costs of environmental destruction because of economic activities must be considered as unrelated to the closed market system (or at least they cannot be included in the price-formation processes of that system); (iv) the natural resources that are used by the market system are endless and those that are limited in quantity can be substituted by others that are endless; and (v) biophysical limits to the increase of the market system simply do not exist (Nadeau, 2006). This model is obsolete and is based on hypotheses that have no grounding in scientific bases. Sustainable economic solutions to global warming and environmental destruction are impossible to establish under the logic of this model.

As a consequence, the US alone has reached a level of oil consumption in the transportation sector that approaches 14 Mbl/day and corresponds to a release of 0.53 gigatons of carbon per year (Gt C/yr). The current global release of carbon from all fossil fuel usage is estimated to be at 7 Gt C/yr and is expected to rise to ~14 Gt C/yr by 2050 (Agrawal et al.,

2007). It has been estimated that global energy consumption could reach 30-60 TW by 2050. With world population expected to reach 8 billion by 2030, the scale-up in energy use that is needed to maintain economic growth is critical. China, with 1.3 billion people and a fast-growing economy, has overtaken Japan to become the second-largest oil consumer behind the US. The Asian giant is currently the largest producer and consumer of coal (Tollefson, 2008) and has announced the construction of 24-32 new nuclear reactors by 2020 (Dorian et al., 2006). If current trends continue, the world will need to spend an estimated $16 trillion over the next three decades to maintain and expand its energy supply. Generation, transmission, and distribution of electricity will absorb almost two-thirds of this investment, whereas capital expenditures in the oil and gas sectors will amount to almost 20% of global energy investment.

Experts believe that peak of world oil production should not occur before at least 30-40 years from now. To put global oil needs into perspective, demand for oil is projected to rise from nearly 80 Mbl/day today to over 120 Mbl/day by 2030. The OPEC nations are currently operating at near full capacity, which caused oil prices to reach US$ 120/bl in August 2008. Clearly, the world must find more efficient ways to manage energy. Some argue that the supplies of oil needed to satisfy the growing world demand will become available because of a combination of price and technology incentives (Rafaj & Kypreos, 2007). As oil prices continue to rise because of increasing difficulties in reaching remaining oil resources, other energy forms will appear (Herrera, 2006). A transition from oil to renewable energy should occur at some point before the world runs out of oil resources (Dorian et al., 2006). Renewable energy sources, including solar, wind, and geothermal, but excluding biofuels, currently provide only 3% of world energy demand (Dorian et al., 2006). Solutions that use these energy sources should be increased worldwide and should be connected to the electricity grid.

Renewable biodiesel from palm oil and bioethanol from sugarcane are currently the two leaders of plant bioenergy production per hectare. They are being grown in increasing amounts; however, continuous increases in their production are not sustainable and will not resolve the enormously increasing demands for energy. Palm oil yields ~5,000 l/ha. In Brazil, the best bioethanol yields from sugarcane are 7,500 l/ha. Most of the energy needed for growing the sugarcane and converting it to ethanol is gained from burning its wastes (e.g., bagasse). For every unit of fossil energy that is consumed by producing sugarcane ethanol, ~8 units of energy are recovered (Bourne, 2007). The rates of energy recovery from other biofuel crops are usually less than 5. Biofuel crops from the EU are much less productive than palm oil and sugarcane; therefore, B5 enforcement would require that ~13% of the EU25 arable land be dedicated to biofuel production. This is hardly sustainable (the present situation is ~5 times less).

Regarding environmental impact, ethanol from corn (for example) contains costs that stem from the copious amounts of nitrogen fertilizer used and the extensive topsoil erosion associated with cultivation. Every year, pesticides, herbicides and fertilizers run off the corn fields and bleed into groundwater. River contamination promotes eutrophication, algal blooms and 'dead zones'. In addition, ethanol importation by industrialized nations could lead to increased ecological destruction in developing countries as indigenous natural habitats are cleared for energy crops (Gui et al., 2008; Marris, 2006; Thomas 2007).

The general feeling is that first-generation biofuels are already reaching saturation because of the limited availability of arable lands. Brazil has additional lands available for sugarcane and physic nut production, whereas India is promoting physic nut cultivation on its extensive wastelands. However, the development of these fuels has already been a success because they have demonstrated that motor technology running on ethanol or biodiesel is feasible and can (at least) be used to power public transport.

Fortunately, second-generation biofuels from biomass offer additional opportunities. The cost of feedstock is lower for lignocellulose as compared to the agricultural crops that now contribute up to 70% of the total production costs for first-generation bioethanol. Even if they are more expensive now, synfuel from biomass sources (such as poplar, willow, and reed grass) could have higher cost effectiveness in the near future than does fuel from sugar beets, wheat and rapeseed sources (Wesseler, 2007; Styles & Jones, 2008).

Biomass fuels will be another opportunity for the EU to meet its target of energy production from renewable sources. However, this goal has not been met by 2010 as was initially expected (Fischer et al., 2010; Havlík et al., 2010). The European CO_2 emissions-trading system of carbon credits seems to be much more cost effective than its biodiesel program because it allows for the purchase of units of CO_2 sequestration in tropical climates that have much higher rates of fixation than do temperate ones (Frondel & Peters, 2007).

Third-generation biofuels have also entered the race for fuel renewability. In terms of total dry matter, sugarcane typically yields ~75 t biomass per hectare, whereas micro*algae* are able to produce two times more biomass per hectare (Brennan & Owende, 2010; Chisti, 2007, 2008). Considering a productivity of 150 t/ha and an average dry-weight oil content of 30%, the oil yield per hectare would be ~123 m^3 over 90% of the year (i.e., 98.4 m^3/ha). If 0.53 Gm^3 of biodiesel are needed in the US to power transport vehicles, micro*algae* should be grown over an area of ~5.4 Mha (3% of the US). Producing algal biomass in a 100 t/yr facility has been estimated to cost approximately US$ 3,000/ton. The feasibility of oil extraction for microalgal biomass has been demonstrated (Belarbi et al., 2000; Sánchez Mirón et al., 2003) and the majority of algal biomass residues from oil extraction can be recycled by anaerobic digestion to produce biogas.

Impediments to large-scale culture of micro*algae* are mainly economic and are tied to the investment requirements for the *algae* cultivation. One solution would be to increase the oil productivity by genetic and metabolic engineering (León-Bañares et al., 2004; Mathews & Wang, 2009). One may expect the expansion of algal technology via CO_2 filtration because power plants can incorporate this technology immediately into their management systems. This technology is expected to spread slowly with the accumulation of experience.

Nearly half of the world's oil consumption is dedicated to the transportation sector, which also accounts for 32% of GHG emissions. The overall efficiency of energy conversion to work in the transportation segment is lower than it is in large-scale power plants and the goal is to increase it from the current level of 15–35 to 60–80% (Song, 2006).

Unfortunately, advanced transportation technologies (such as hydrogen fuel cell vehicles and alternative fuels including gas-to-liquids, coal-to-liquids, and biodiesels) are not likely to significantly penetrate the conventional transportation fuel market before 2030 (except on a regional basis). The growth in oil consumption for transportation use in the coming decades may be slowed by the adoption of fourth-generation technologies such as hybrids and fuel cell cars. However, the necessary technological breakthroughs will not occur without unprecedented policy actions worldwide to promote the use and inclusion of these technologies in everyday life (Doniger et al., 2006; Haug et al., 2011; Michel 2009). Currently, there are approximately half a million hybrids and 30 million advanced clean-diesel engines globally. The use of hybrid cars is growing in the US and Japan, whereas advanced clean-diesel motors are mostly concentrated in Europe (Dorian et al., 2006).

Actually, auto-mobility is a self-organizing and non-linear system that presupposes and calls into existence an assemblage of cars, drivers, roads, fuel supplies, and other objects and technologies. Modern social life has become interconnected with auto-mobility. However, this mode of mobility is neither socially necessary nor inevitable (Urry, 2008). One billion cars were produced during the last century. World automobile travel is predicted to triple between 1990 and 2050 (Hawken et al., 2002). Today, world citizens move 23 Gkm annually. Auto-mobility forces people to contend with the temporal and spatial constraints that it itself generates (Mills et al., 2010). Fortunately, some 35-year-old projects have begun to be finally implemented (i.e., the integration of car and bicycle rentals into public transportation systems, such as occurs in some European cities). A post-car future will involve changes in lifestyles, city architecture, thinking and social practices. Increased active transport (e.g., walking and bicycling) will help to achieve substantial reductions in emissions while improving public health. Cities require safe and pleasant environments for active transport as well as easy accessibility of public transport. Adverse health effects because of transportation include traffic injuries, physical inactivity (the cost of obesity in the USA is estimated to be around US$ 139 bn/yr), urban air pollution, energy-related conflicts, and environmental degradation. For instance, urban air pollution accounts for 750,000 deaths each year, of which 530,000 are in Asia (Woodcock et al., 2007). Because of limited energy resources, it has been argued that the world will be required to move toward virtual travel (such as internet surfing, virtual sensorial traveling, and video conferences) to replace physical travel as much as possible (Moriarty & Honnery, 2007).

In reality, the situation outlined above is the result of consideration of humanity only within social contexts and without the necessary environmental perspective (Thomas, 2007). The concept of environmental crime barely operational; if it exists at all, it is very recent and is not generally applied. Logical human societies should take into account the amount of land that human beings and wildlife actually need to reasonably sustain themselves. Not doing this will lead to increasing worldwide destruction (Urry, 2008) and will threaten the future of humanity. These considerations led to the formulation of the Gaia principle (Lovelock & Margulis, 1974). This principle states that one should consider the planet Earth as a whole, with the consequence that the destruction of one ecosystem can affect all of the others. Concern for the value of ecosystems is recent (Costanza et al., 1997). Society has only begun to address human integration with the environment because of the threat of global warming and its potentially disastrous effects (Stern, 2006). A discussion of the economic accounting for ecosystem services from the perspective of sustainable development has also been proposed (Mäler et al., 2008).

The concept of "willingness-to-pay" (WTP) has also been recently introduced. This concept allows for the monetary measurement of individual preference to avoid a negative impact. It aims to estimate the need for improved environmental quality. WTP measures how much individuals are ready to pay to improve their quality of life or that of other people. The sum of the WTP of all individuals gives the value that a group of individuals are ready to pay to maintain their environment in an unaffected state. For example, the pathways of polluting substances are followed from their release sources to the points of damage occurrence with associated "external" costs of reparation. Taking external costs into account in the full cost of energy production leads to the estimation of the "real" cost of an activity and supplies an efficient policy instrument for reducing the negative impacts of energy use (Nast et al., 2007). The approach of merging production costs with external costs into a total specific cost serves as a comparative indicator for the evaluation of the economic-environmental performance of energy options and technologies (Rafaj & Kypreos, 2007). The scenarios proposed under this new cost-accounting strategy reveal the possibilities for the diffusion of advanced technologies and fuel switching into the electricity production system. Following this model, renewable energies increase their competitiveness and the dependency of the electricity sector on fossil fuels is decreased considerably. Additionally, emissions of SO_2 and NO_x decrease by 70–85% by 2030. Although the analysis indicates that advanced technologies with emission controls and carbon sequestration will undergo significant cost reduction and will become competitive in the long run, policies supporting these technologies are a prerequisite to their establishment in electricity markets (especially during their initial period of market penetration). This model refers to policy measures for the stimulation of technological progress via investments in research and development that assist carbon-free technologies to progress along their necessary learning curves (Haug et al., 2011; Rafaj & Kypreos, 2007).

8. Conclusions

The time has come for the integration of the technological and social sciences to find a route to environmental and economic sustainability on earth. If such a solution is not reached, economic growth will occur at the cost of the human population size (Urry, 2008). Fortunately, because of the continuous increase in the price of fossil fuel, investigations into sources of renewable energy have become economically viable. It is now clear that technologies for renewable energies have reached a pivotal stage such that there is no turning back. There are at least 5 regional blocks (the USA, the EU, China, Brazil, and India) that are interested in decreasing their dependence on fossil fuels. It does not appear to be in anyone's interest to shut this process down by mean of aggressive oil price cutting and market dumping. In fact, biotechnology is intimately bound to agricultural processes that are also supported by governments because of geostrategic issues. In addition, climate change is becoming obvious and will soon overcome particular interests to become a general concern of humanity.

Biofuels and sources of bioenergy will pass through a rapid succession of technological improvements and developments before they arrive in their final forms. It is expected that bioethanol from sweet crops will be surpassed by bioethanol from biomass. Synfuel from biomass and solar energy should also progressively replace plant biodiesel. Biotechnology is expected to increase its participation in microdiesel fuel production, in genetic engineering of plants and microorganisms and in the contribution of enzymes to nanotechnology.

The integration of renewable energies into the electricity grid is just beginning, but is already progressing rapidly. It is expected to make a significant contribution; however, it should be accompanied by policies of energy management and urbanization to avoid unnecessary energy waste that could negate the benefits of technological breakthroughs and developments. New concepts (such as willingness-to-pay, carbon credits and external costs) are now being taken into account in the calculation of energy life cycles. This toolbox will expand with increasing government regulations and should include fundamental concepts such as "biodiversity credits" and the definition of a "minimal territorial unit" for living entities to warrant sustainability of wildlife and humanity. Biodiversity is a source of nanostructures and nanomachines. It should not be destroyed without consideration when we are aware that it required three billions years to develop and that humanity is just beginning to investigate it.

As a result of energy saving requirements, the cars of the near future will run on combinations of fuel combustion and electricity. Such options can reduce fossil fuel consumption and greenhouse gas emissions by 30 to 50%, with no gross vehicle modifications required. In addition, they will allow for connection to the electricity grid for additional cost saving on electricity consumption. These so-called plug-in hybrids will likely travel three to four times farther per kW/h than other vehicles. Ideally, these advanced hybrids will also be flexible and capable of running on bio/fossil blends and gas (Romm, 2006).

At some point during the first half of this century, a transition from fossil fuels to a non-carbon-based world economy will begin and will seriously affect the type of society experienced by future generations (Dorian et al., 2006).

9. Acknowledgement

We thank the Coordenação de Aperfeiçoamento de Pessoal de Nível Superior (CAPES) and Fundação Oswaldo Cruz (FIOCRUZ) for providing a research fellowship from the Centro de Desenvolvimento Tecnológico em Saúde (CDTS) to N. Carels. This work received financial support from Conselho Nacional de Desenvolvimento Científico e Tecnológico (CNPq), Brazil (no. 471214/2006-0).

10. References

Abollé, A., Kouakou, L. & Planche, H. 2008. The viscosity of diesel oil and mixtures with straight vegetable oils: Palm, cabbage palm, cotton, groundnut, copra and sunflower. *Biomass Bioenergy* 33:1116-1121.

Adamson, K.A. 2004. Hydrogen from renewable resources - the hundred year commitment. *Energ. Policy* 32:1142–1231.

Agarwal, A.K. 2007. Biofuels (alcohols and biodiesel) applications as fuels for internal combustion engines. *Prog. Energ. Combust. Sci.* 33: 233-271.

Agrawal, R., Singh, N.R., Ribeiro, F.H. & Delgass, W.N. 2007. Sustainable fuel for the transportation sector. *Proc. Natl. Acad. Sci. USA* 104:4828-4833.

Agrawal, A., Chhatre, A. & Hardin, R. 2008. Changing Governance of the World's Forests. *Science* 320:1460-1462.

Ajanovic, A. 2011. Biofuels versus food production: Does biofuels production increase food prices? *Energy*, doi:10.1016/j.energy.2010.05.019

Alvarez, H.M. & Steinbuchel, A. 2002. Triacylglycerols in prokaryotic microorganisms. *Appl. Microbiol. Biot.* 60:367-376.

Arechederra, R.L., Treu, B.L. & Minteer, S.D. 2007. Development of glycerol/O_2 biofuel cell. *J. Power Sources* 173:156-161.

Arthur, R., Baidoo, M.F. & Antwi, E. 2011. Biogas as a potential renewable energy source: A Ghanaian case study. *Renew. Energ.* 36:1510-1516.

Asam, Z.-u.-Z. 2011. How can we improve biomethane production per unit of feedstock in biogas plants? *Appl. Energ.*, doi:10.1016/j.apenergy.2010.12.036

Avery, W. 2002. Ocean-thermal energy conversion. In: Encyclopedia of Physical Science and Technology, edited by Meyers, R.A., 123- 160. San Diego: 3rd ed. Academic Press.

Baker, E. & Keisler, J.M. 2011. Cellulosic biofuels: Expert views on prospects for advancement. *Energy* 36:595-605.

Barnet, A. 2008. Europe's 2020 vision. *Nat. Rep. Clim. Change* 2:36.

Barros, G.S.A.C., Silva, A.P., Ponchio, L.A., Alves, L.A., Osaki, M. & Cenamo, M. 2006. Custos de produção de biodiesel no Brasil. *Revista de Política Agrícola* 3:36-50.

Beer, L.L., Boyd, E.S., Peters, J.W. & Posewitz, M.C. 2009. Engineering *algae* for biohydrogen and biofuel production. *Curr. Opin. Biotech.* 20:264–271.

Begum, Z.N. 2005. Modeling of chemical tracer transport in the atmospheric environment and its impact on the global climate. *J. Quant. Spectrosc. Radiat. Transfer* 95:423-427.

Belarbi, E.-H., Molina Grima, E. & Chisti, Y. 2000. A process for high yield and scaleable recovery of high purity eicosapentaenoic acid esters from micro*algae* and fish oil. *Enzyme Microb. Tech.* 26:516-529.

Bertram, C. 2010. Ocean iron fertilization in the context of the Kyoto protocol and the post-Kyoto process. *Energ. Policy* 38:1130–1139

Bindraban, P.S., Bulte, E.H. & Conijn, S.G. 2009. Can large-scale biofuels production be sustainable by 2020? *Agr. Syst.* 101:197–199

Bitsche, O. & Gutmann, G. 2004. Systems for hybrid cars. *J. Power Sources* 127:8–15.

Bond, D.R., Holmes, D.E., Tender, L.M. & Lovley, D.R. 2002. Electrode-reducing microorganisms that harvest energy from marine sediments. *Science* 295:483-485.

Bond, D.R. & Loveley, D.R. 2003. Electricity production by *Geobacter sulfurreducens* attached to electrodes. *Appl. Environ. Microb.* 69:1548-1555.

Bourne, J.K.Jr. 2007 Biofuels: green dreams. *National Geographic Magazine* (October), 41–59.

Brennan, L. & Owende, P. 2010. Biofuels from micro*algae* - A review of technologies for production, processing, and extractions of biofuels and co-products. *Renew. Sustain. Energ. Rev.* 14:557-577

Browne, W.R. & Feringa, B.L. 2006. Making molecular machines work. *Nature Nanotechnology* 1:25-35.

Canakci, M., Erdil, A. & Arcaklioglu, E. 2006. Performance and exhaust emissions of a biodiesel engine. *Appl. Energ.* 83, 594-605.

Certik, M. & Shimizu, S. 1999. Biosynthesis in microorganisms being used for single cell oil production. *Biochimie* 87:1-14.

Ceschia, E. et al. 2010. Management effects on net ecosystem carbon and GHG budgets at European crop sites. *Agricult. Ecosys. Environ.* 139:363–383.

Chapple, C., Ladisch, M. & Meilan, R. 2007. Loosening lignin's grip on biofuel production. *Nat. Biotechnol.* 25:746-748.

Chaudhuri, S.K. & Lovley, D.R. 2003. Electricity generation by direct oxidation of glucose in mediatorless microbial fuel cells. *Nat. Biotechnol.* 21:1229-1232.

Chen, F. & Dixon, R.A. 2007. Lignin modification improves fermentable sugar yields for biofuel production. *Nat. Biotechnol.* 25:759-761.

Cherubini, F. 2010. GHG balances of bioenergy systems - Overview of key steps in the production chain and methodological concerns. *Renew. Energ.* 35:1565–1573.

Chisti, Y. 2007. Biodiesel from micro*algae*. *Biotechnol. Adv.* 25:294-306.

Coleman, H.D., Park, J.-Y., Nair, R., Chapple, C. & Mansfield, S.D. 2008. RNAi-mediated suppression of p-coumaroyl-CoA 3′-hydroxylase in hybrid poplar impacts lignin deposition and soluble secondary metabolism. *Proc. Natl. Acad. Sci. USA* 105:4501-4506.

Costanza, R. et al. 1997. The value of the world's ecosystem services and natural capital. *Nature* 387, 253-260.

Cropper, M., Geiger, S. & Jollie, D. 2004. Fuel cells: a survey of current developments. *J. Power Sources* 131:57–61.

Damartzis, T. & Zabaniotou, A. 2011. Thermochemical conversion of biomass to second generation biofuels through integrated process design-A review. *Renew. Sustain. Energ. Rev.* 15:366–378.

Darussalam, B. 2007. Last-Gasp effort to save Borneo's tropical rainforests. *Science* 317:192.

Deluga, G.A., Salge, J.R., Schmidt, L.D. & Verykios, X.E. 2004. Renewable hydrogen from ethanol by autothermal reforming. *Science* 13:993-997.

Demirbas, M.F. & Balat, M. 2006. Recent advances on the production and utilization trends of bio-fuels: A global perspective. *Energ. Convers. Manage.* 47:2371-2381.

Demirbas, A. 2007. The influence of temperature on the yields of compounds existing in bio-oils obtaining from biomass samples via pyrolysis. *Fuel process. Technol.* 88:591-597.

Demirbas, A. 2008a. Biofuels sources, biofuel policy, biofuel economy and global biofuel projections. *Energ. Convers. Manage.* 49:2106-2116.

Demirbas, A. 2008b. Comparison of transesterification methods for production of biodiesel from vegetable oils and fats. *Energ. Convers. Manage.* 49:125-130.

Demirbas A. 2010. Use of *algae* as biofuel sources. *Energ. Convers. Manage.* 51:2738–2749.

de Vries, B.J.M., van Vuuren, D.P. & Hoogwijk, M.M. 2007. Renewable energy sources: Their global potential for the first-half of the 21st century at a global level: An integrated approach. *Energ. Policy* 35:2590-2610.

Dijkstra, A.J. 2006. Revisiting the formation of trans isomers during partial hydrogenation of triacylglycerol oils. *Eur. J. Lipid Sci. Tech.* 108:249-264.

Doniger, D.D., Herzog, A.V. & Lashof, D.A. 2006. An ambitious, centrist approach to global warming legislation. *Science* 314:764-765.

Donner, S.D. & Kucharik, C.J. 2008. Corn-based ethanol production compromises goal of reducing nitrogen export by the Mississippi River. *Proc. Natl. Acad. Sci.* USA 105:4513-4518.

Dorian, J.P., Franssen, H.T. & Simbeck, D.R. 2006. Global challenges in energy. *Energ. Policy* 34:1984-1991.

Easterling, E.R., French, W.T., Hernandez, R. & Licha, M. 2008. The effect of glycerol as a sole and secondary substrate on the growth and fatty acid composition of *Rhodotorula glutinis*. *Bioresource Technol.* 100:356-61

Eugenia Corria, M., Melian Cobas, V. & Silva Lora, E. 2006. Perspectives of Stirling engines use for distributed generation in Brazil. *Energ. Policy* 34:3402-3408.

Fagernäs, L., Brammer, J., Wilén, C., Lauer, M.& Verhoeff, F. 2010 Drying of biomass for second generation synfuel production. *Biomass Bioenergy* 34:1267-1277.

Fairless, D. 2007. The little shrub that could - maybe. *Nature* 449:652-655.

Fargione, J., Hill, J., Tilman, D., Polasky, S. & Hawthorne, P. 2008. Land clearing and the biofuel carbon debt. *Science* 319:1235-1238.

Fenning, T.M., Walter, C. & Gartland, K.M.A. 2008. Forest biotech and climate change. *Nat. Biotechnol.* 26:615-616.

Fialkov, A.B., Gordin, A. & Amirav, A. 2008. Hydrocarbons and fuels analyses with the supersonic gas chromatography mass spectrometry - The novel concept of isomer abundance analysis. *J. Chromatogr.* A 1195:127-135.

Fischer, C.R., Klein-Marcuschamer, D. & Stephanopoulos, G. 2008. Selection and optimization of microbial hosts for biofuel production. *Metab. Eng.* 10:295–304.

Fischer, G., Prieler, S., van Velthuizen, H., Lensink, S.M., Londo, M. & de Wit, M. 2010. Biofuel production potentials in Europe: Sustainable use of cultivated land and pastures. Part I: Land productivity potentials. *Biomass Bioenergy* 34:159 - 172.

Flavin, C. 2006. American Energy: The Renewable Path to Energy Security. 40 p. Washington DC: Worldwatch Institute.

Folkesson, A., Andersson, C., Alvfors, P., Alaküla, M. & Overgaard, L. 2003. Real life testing of a Hybrid PEM Fuel Cell Bus. *J. Power Sources* 118:349-357.

Fredrickson, J.K. et al. 2008. Towards environmental systems biology of *Shewanella*. *Nat. Rev. Microbiol.* 6:592-603.

Frondel, M. & Peters, J. 2007. Biodiesel: a new Oildorado? *Energ. Policy* 35:1675-84.

García-Montero, L.G., López, E., Monzón, A. & Pastor, I.O. 2010. Environmental screening tools for assessment of infrastructure plans based on biodiversity preservation and global warming (PEIT, Spain). *Environ. Impact Assess. Rev.* 30:58–168.

Gavrilescu, M. & Chisti, Y. 2005. Biotechnology - a sustainable alternative for chemical industry. *Biotechnol. Adv.* 23:471-499.

Gielen, D.J., Fujino, J., Hashimoto, S. & Moriguchi, Y. 2002. Biomass strategies for climate policies? *Climate Policy* 2:319–333.

Glibert, P.M. et al. 2008. Ocean urea fertilization for carbon credits poses high ecological risks. *Mar. Pollut. Bull.* 56:1049-1056.

Gnansounou, E., Dauriat, A., Villegas, J. & Panichelli, L. 2009. Life cycle assessment of biofuels: Energy and greenhouse gas balances. *Bioresource Technol.* 100:4919–4930.

Goering, E., Schwab, W., Daugherty, J., Pryde, H. & Heakin, J. 1982. Fuel properties of eleven vegetable oils. *Transactions ASABE* 25:1472-1483.

Goldemberg, J., Coelho, S.T. & Lucon, O. 2004. How adequate policies can push renewables. *Energ. Policy* 32:1141-1146.

Gressel, J. 2008. Transgenics are imperative for biofuel crops. *Plant Sci.* 174:246-263.

Greyvenstein, R., Correia, M. & Kriel, W. 2008. South Africa's opportunity to maximise the role of nuclear power in a global hydrogen economy. *Nucl. Eng. Des.* 238:3031-3040.

Gui, M.M., Lee, K.T. & Bhatia, S. 2008. Feasibility of edible oil vs. non-edible oil vs. waste edible oil as biodiesel feedstock. *Energy* 33:1646-1653.

Gummer, J.H. & Head, C.R. 2003. Hydrogen, hydropower and world poverty. *Int. J. Hydropower Dams* 5:67-71.

Haag, A.L. 2007. *Algae* bloom again. *Nature* 447:520-521.

Hagerman, S., Dowlatabadi, H., Satterfield, T. & McDaniels, T. 2010. Expert views on biodiversity conservation in an era of climate change. *Global Environ. Change* 20:192–207.

Hall, J., Matos, S., Severino, L. & Beltrão, N. 2009. Brazilian biofuels and social exclusion: established and concentrated ethanol versus emerging and dispersed biodiesel. *J. Clean. Prod.* 17:S77–S85.

Haug, C., Huitema, D. & Wenzler, I. 2011. Learning through games? Evaluating the learning effect of a policy exercise on European climate policy. *Technol. Forecast. Soc. Change*, doi:10.1016/j.techfore.2010.12.001

Havlík, P. et al. 2010. Global land-use implications of first and second generation biofuel targets. *Energ. Policy*, doi:10.1016/j.enpol.2010.03.030

Hawken, P., Lovins, A. & Lovins, L.H. 2002. Natural Capitalism: Creating the Next Industrial Revolution. London: Earthscan. Accessed February 2, 2010. http://www.natcap.org/.

Herrera, S. 2006. Bonkers about biofuels. *Nat. Biotechnol.* 24:755-760.

Heyer, N.I. & Woodward, J. 2001. Efficient hydrogen production using enzymes of the pentose phosphate cycle. Proceedings of the 2001 DOE Hydrogen Program Review, NREL/CP-570-30535. Accessed February 2, 2010. http://www1.eere.energy.gov/hydrogenandfuelcells/pdfs/30535g.pdf.

Hill, J., Nelson, E., Tilman, D., Polasky, S. & Tiffany D. 2006. Environmental, economic, and energetic costs and benefits of biodiesel and ethanol biofuels. *Proc. Natl. Acad. Sci. USA* 103:11206-11210.

Himmel, M.E., Ding, S.-Y., Johnson, D.K., Adney, W.S., Nimlos, M.R., Brady, J.W. & Foust, T.D. 2007. Biomass Recalcitrance: Engineering Plants and Enzymes for Biofuels Production. *Science* 315:804-807.

Hissel, D., Péra, M.C. & Kauffmann, J.M. 2004. Diagnosis of automotive fuel cell power generators. *J. Power Sources* 128:239-246.

Hopkin, M. 2007. Climate sceptics switch focus to economics. *Nature* 445:582-583.

Howden, S.M., Soussana, J.-F., Tubiello, F.N., Chhetri, N., Dunlop, M. & Meinke, H. 2007. Adapting agriculture to climate change. *Proc. Natl. Acad. Sci. USA* 104:19691-19696.

Huston, M.A. & Marland, G. 2003. Carbon management and biodiversity. *J. Environ. Manage.* 67:77-86.

Johnston, M. & Holloway, T. 2007. A global comparison of national biodiesel production potentials. *Environ. Sci. Technol* 41:7967-7973.

Johnston, C.D., Bastow, T.P. & Innes, N.L. 2007. The use of biodegradation signatures and biomarkers to differentiate spills of petroleum hydrocarbon liquids in the subsurface and estimate natural mass loss. *Eur. J. Soil Biol.* 43:328-334.

Jones, D.B. 2010. Proposal for an international carbon fee. *Energy for Sustainable Development* 14:245–250.

Kalscheuer, R., Stoveken, T. & Steinbuchel, A. 2006a. Microdiesel: *Escherichia coli* engineered for fuel production. *Microbiology* 152:2529-2536.

Kalscheuer, R., Stoveken, T., Luftmann, H., Malkus, U., Reichelt, R. & Steinbuchel, A. 2006b. Neutral lipid biosynthesis in engineered *Escherichia coli*: Jojoba oil-like wax esters and fatty acid butyl esters. *Appl. Environ. Microbiol.* 72:1373-1379.

Kaplan, I.R., Galperin, Y., Lu, S.-T. & Lee, R.-P. 1997. Forensic environmental geochemistry: differenciation of fuel-types, their sources and release time. *Org. Geochem.* 27:289-317.

Khaselev, O. & Turner, J. 1998. A monolithic photovolatic–photochemical device for hydrogen production via water splitting. *Science* 280, 425-427.

Kikuchi, R. 2006. Penetration of hydrogen-based energy system and its potential for causing global environmental change: Scoping risk analysis based on life cycle thinking. *Environ. Impact Asses. Rev.* 26:206-218.

Kindermann, G., Obersteiner, M., Sohngen, B., Sathaye, J., Andrasko, K., Rametsteiner, E., Schlamadinger, B., Wunder, S. & Beach, R. 2008. Global cost estimates of reducing carbon emissions through avoided deforestation. *Proc. Natl. Acad. Sci. USA* 105:10302-10307.

King, C.W., Holman, A.S. & Webber, M.E. 2008. Thirst for energy. *Nat. Geosci.* 1:283-286.

Kirilenko, A.P. & Sedjo, R.A. 2007. Climate change impacts on forestry. *Proc. Natl. Acad. Sci. USA* 104:19697-19702.

Kleiner, K. 2007. Civil aviation faces green challenge. *Nature* 448:120-121.

Kleiner, K. 2008. The backlash against biofuels. *Nat. Rep. Clim. Change* 2:9-11, doi:10.1038/climate.2007.71.

Kremer, F.G. & Fachetti, A. 2000 Alcohol as automotive fuel-Brazilian experience. SAE Technical Paper No. 2000-01-1965, doi:10.4271/2000-01-1965

Kumar, A. & Sharma, S. 2008. An evaluation of multipurpose oil seed crop for industrial uses (*Jatropha curcas* L.): A review. *Ind. Crop. Prod.* 28:1-10.

Lamy, C., Rousseau, S., Belgsir, E.M., Coutanceau, C. & Léger, JM. 2004. Recent progress in the direct ethanol fuel cell: development of new platinum–tin electrocatalysts. *Electrochim. Acta* 49:3901-3908.

Laurance, W.F. 2007. Switch to corn promotes amazon deforestation. *Science* 318:1721.

Leibold, H., Hornung, A. & Seifert, H. 2008. HTHP syngas cleaning concept of two stage biomass gasification for FT synthesis. *Powder Technol.* 180:265-270.

León-Bañares, R., González-Ballester, D., Galván, A. & Fernández, E. 2004. Transgenic micro*algae* as green cellfactories. *Trends Biotechnol.* 22:45-52.

Lewis, N.S. & Nocera, D.G. 2006. Powering the planet: Chemical challenges in solar energy utilization. *Proc. Natl. Acad. Sci. USA* 103:15729-15735.

Lewis, N.S. 2007. Toward cost-effective solar energy use. *Science* 315:798-801.

Lima, D.G. et al. 2004. Diesel-like fuel obtained by pyrolysis of vegetable oils. *J. Anal. Appl. Pyrol.* 71:987-996.

Lin, C.-Y. et al. 2010. A pilot-scale high-rate biohydrogen production system with mixed microflora. *Int. J. Hydrogen Energ.*, doi:10.1016/j.ijhydene.2010.07.115.

Liu, X., Li, X., Suehiro, Y. & Fujimoto, K. (2007). Elucidation of reaction network and effective control of carbon number distribution in the three phase Fischer-Tropsch synthesis. *Appl. Catal. A-Gen.* 333:211-218.

Lois, E. 2007. Definition of biodiesel. *Fuel* 86:1212-1213.

Lokey, E. 2007. How the next US president should slow global warming. *Energ. Policy* 35:5399-5402.

Lonngren, K.E. & Bai, E.-W. 2008. On the global warming problem due to carbon dioxide. *Energ. Policy* 36:1567-1568.

Lovelock, J.E. & Margulis, L. 1974. Atmospheric homeostasis by and for the biosphere- The Gaia hypothesis. *Tellus* 26:2-10.

Lü, X-c, L., Yang, J-g, Zhang, W-g, Z. & Huang, Z. 2004. Effect of cetane number improver on heat release rate and emissions of high speed diesel engine fueled with ethanol-diesel blend fuel. *Fuel* 83:2013-2020.

Luo, T.-J.M., Soong, R., Lan, E., Dunn, B. & Montemagno, C. 2005. Photo-induced proton gradients and ATP biosynthesis produced by vesicles encapsulated in a silica matrix. *Nat. Materials* 4:220–224.

Malça, J. & Freire, F. 2011. Life-cycle studies of biodiesel in Europe: A review addressing the variability of results and modeling issues. *Renew. Sustain. Energ. Rev.* 15:338-351.

Mäler, K.-G., Aniyar, S. & Jansson, Å. 2008. Accounting for ecosystem services as a way to understand the requirements for sustainable development. *Proc. Natl. Acad. Sci. USA* 105:9501-9506.

Malhi, Y., Roberts, J.T., Betts, R.A., Killeen, T.J., Li, W. & Nobre, C.A. 2008. Climate change, deforestation, and the fate of the Amazon. *Science* 319:169-172.

Mander, U. et al. 2010. Assessment of methane and nitrous oxide fluxes in rural landscapes. *Landscape Urban Plan.* 98:172–181.

Marchetti, J., Miguel, V. & Errazu, A. 2007. Possible methods for biodiesel production. *Renew. Sustain. Energ. Rev.* 11:1300-1311.

Marcinkoski, J., Kopasz, J.P. & Benjamin, T.G. 2008. Progress in the US DOE fuel cell subprogram efforts in polymer electrolyte fuel cells. *Int. J. Hydrogen Energ.* 33:3894-3902.

Markels, M. & Barber, R.T. 2001. Sequestration of carbon dioxide by ocean fertilization. Preprints - American Chemical Society, Division of Petroleum Chemistry 46:45-48.

Markou, G. & Georgakakis, D. 2011. Cultivation of filamentous cyanobacteria (blue-green *algae*) in agro-industrial wastes and wastewaters: A review. *Appl. Energ.*, doi:10.1016/j.apenergy.2010.12.042

Marris, E. 2006. Drink the best and drive the rest. *Nature* 444:670-672.

Mathews, J. & Wang, G. 2009. Metabolic pathway engineering for enhanced biohydrogen production. *Int. J. Hydrogen Energ.* 34:7404 - 7416.

Meng, X., Yang, J., Xu, X., Zhang, L., Nie, Q. & Xian, M. 2008. Biodiesel production from oleaginous microorganisms. *Renew. Energ.* 34:1-5

Michel, D. 2009. Foxes, hedgehogs, and greenhouse governance: Knowledge, uncertainty, and international policy-making in a warming World. *Appl. Energ.* 86:258–264.

Miles, L. & Kapos, V. 2008. Reducing greenhouse gas emissions from deforestation and forest degradation: global land-use implications. *Science* 320:1454-1455.

Mills, G. et al. 2010. climate information for improved planning and management of mega cities (needs perspective). *Procedia Environmental Sciences* 1:228–246

Moreira, J.R. & Goldemberg, J. 1999. The alcohol program. *Energ. Policy* 27:229-245.

Moriarty, P. & Honnery, D. 2007. The prospects for global green car mobility. *J. Clean. Prod.* 16:1717-1726.

Nadeau, R. 2006. The Environment Endgame: Mainstream Economics, Ecological Disaster, and Human Survival. 214 p. London: Rutgers University Press.

Nass, L.L., Pereira, P.A.A. & Ellis, D. 2007. Biofuels in Brazil: An Overview. *Crop Sci.* 47:2228-2237.

Nast, M., Langniss, O. & Leprich, U. 2007. Instruments to promote renewable energy in the German heat market-Renewable Heat Sources Act. *Renew. Energ.* 32:1127–1135.

Odling-Smee, L. 2007. Biofuels bandwagon hits a rut. *Nature* 446:483.

Okereke, C. & Dooley, K. 2010. Principles of justice in proposals and policy approaches to avoided deforestation: Towards a post-Kyoto climate agreement. *Global Environ. Change* 20:82–95.

Om Tapanes, N.C., Gomes Aranda, D.A., de Mesquita Carneiro, J.W. & Ceva Antunes, O.A. 2008. Transesterification of *Jatropha curcas* oil glycerides: Theoretical and experimental studies of biodiesel reaction. *Fuel* 87:2286-2295.

Orellana, C. & Neto, R.B. 2006. Brazil and Japan give fuel to ethanol market. *Nat. Biotechnol.* 24:232.

Parker, J. 2002. Turning manure into gold. *EMBO rep.* 3:1114-1116.

Parry, M., Palutikof, J., Hanson, C. & Lowe, J. 2008. Squaring up to reality. *Nature Reports Climate Change* 2, 68-70.

Peters, W. et al. 2007. An atmospheric perspective on North American carbon dioxide exchange: CarbonTracker. *Proc. Natl. Acad. Sci. USA* 104:18925-18930.

Pousa, G.P.A.G., Santos, A.L.F. & Suarez, P.A.Z. 2007. History and policy of biodiesel in Brazil. *Energ. Policy* 35:5393-5398.

Proctor, M.R., Taylor, E.J., Nurizzo, D., Turkenburg, J.P., Lloyd, R.M., Vardakou, M., Davies, G.J. & Gilbert, H.J. 2005. Tailored catalysts for plant cell-wall degradation: Redesigning the exo/endo preference of *Cellvibrio japonicus* arabinanase 43A. *Proc. Natl. Acad. Sci. USA* 102:2697-2702.

Pryor, S.C. & Barthelmie, R.J. 2010. Climate change impacts on wind energy: A review. Renew. Sustain. *Energ. Rev.* 14:430–437.

Pulz, O. & Gross, W. 2004. Valuable products from biotechnology of micro*algae*. *Appl. Microbiol. Biot.* 65:635-48.

Rafaj, P. & Kypreos, S. 2007. Internalisation of external cost in the power generation sector: Analysis with global multi-regional MARKAL model. *Energ. Policy* 35:828-843.

Ragauskas, A.J. et al. 2006. The path forward for biofuels and biomaterials. *Science* 311:484-489.

Raghu, S., Anderson, R.C., Daehler, C.C., Davis, A.S., Wiedenmann, R.N., Simberloff, D. & Mack, R.N. 2006. Adding biofuels to the invasive species fire? *Science* 313:1742.

Ramadhas, A.S., Jayaraj, S. & Muraleedharan, C. 2005. Biodiesel production from high FFA rubber seed oil. *Fuel* 84:335-340.

Ratledge, C. 2002. Regulation of lipid accumulation in oleaginous micro-organisms. *Biochem. Soc. Trans.* 30:1047-1050.

Reay, D. 2007. Spring-time for sinks. *Nature* 446:727-728.

Ridgwell, A.T. & Zeebe, R.E. 2005. The role of the global carbonate cycle in the regulation and evolution of the Earth system. *Earth Planet. Sci. Lett.* 234:299-315.

Romm, J. 2006. The car and fuel of the future. *Energ. Policy* 34:2609-2614.

Ross, D.K. 2006. Hydrogen storage: The major technological barrier to the development of hydrogen fuel cell cars. *Vacuum* 80:1084-1089.

Rothengatter, W. 2010. Climate change and the contribution of transport: Basic facts and the role of aviation. *Transportation Research* Part D 15:5–13.

Rout, U.K., Akimoto, K., Sano, F., Oda, J., Homma, T. & Tomoda, T. 2008. Impact assessment of the increase in fossil fuel prices on the global energy system, with and without CO_2 concentration stabilization. *Energ. Policy* 36:3477-3484.

Sánchez Mirón, A. Ceron Garcia, M.C., Contreras Gomez, A., Garcia Camacho, F., Molina Grima, E. & Chisti, Y. 2003. Shear stress tolerance and biochemical characterization of *Phaeodactylum tricornutum* in quasi steady-state continuous culture in outdoor photobioreactors. *Biochem. Eng. J.* 16:287-297.

Sanderson, K. 2006. A field in ferment. *Nature* 444:673-676.

Scarlat, N. & Dallemand, J.-F. 2010. Recent developments of biofuels/bioenergy sustainability certification: A global overview. *Energ. Policy*, doi:10.1016/j.enpol.2010.12.039

Scharlemann, J.P.W. & Laurance, W.F. 2008. How green are biofuels? *Science* 319:43-44.

Schiermeier, Q. 2008. Europe spells out action plan for emissions targets. *Nature* 451:504-505.

Schiermeier, Q., Tollefson, J., Scully, T., Witze, A. & Morton, O. 2008. Electricity without carbon. *Nature* 454:816-823.

Schmer, M.R., Vogel, K.P., Mitchell, R.B. & Perrin, R.K. 2008. Net energy of cellulosic ethanol from switchgrass. *Proc. Natl. Acad. Sci. USA* 105:464-469.

Schultz, M.G., Diehl, T., Brasseur, G.P. & Zittel, W. 2003. Air pollution and climate-forcing impacts of a global hydrogen economy. *Science* 302:624-627.

Searchinger, T. et al. 2008. Use of U.S. croplands for biofuels increases greenhouse gases through emissions from land-use change. *Science* 319 :1238-1240.

Sedjo, R.A. & Marland, G. 2003. Inter-trading permanent emissions credits and rented temporary carbon emissions offsets: some issues and alternatives. *Climate Policy* 3:435-444.

Seibert, M., Forestier, M., Zhang, L. & Ghirardi, M. 2001. Development of an efficient algal H_2-production system. Proceedings of the 2001 DOE Hydrogen Program Review NREL/CP-570-30535, Accessed February 2, 2010. http://www1.eere.energy.gov/hydrogenandfuelcells/pdfs/30535c.pdf.

Shinnar, R. & Citro, F. 2006. A road map to U.S. decarbonization. *Science* 313:1243-1244.

Simões, A.F., Kligerman, D.C., La Rovere, E.L., Maroun, M.R., Barata, M. & Obermaie, M. 2010. Enhancing adaptive capacity to climate change: The case of smallholder farmers in the Brazilian semi-arid region. *Environmental Science & Policy* 13:801-808.

Singh, J. & Gu, S. 2010. Commercialization potential of micro*algae* for biofuels production. *Renew. Sustain. Energ. Rev.* 14:2596–2610.

Snyder, C.S., Bruulsema, T.W., Jensen, T.L. & Fixen, P.E. 2009 Review of greenhouse gas emissions from crop production systems and fertilizer management effects. *Agric. Ecosyst. Environ.* 133:247–266.

Solomon, B.D. & Banerjee, A. 2006. A global survey of hydrogen energy research, development and policy. *Energ. Policy* 34:781-792.

Song, C. 2006. Global challenges and strategies for control, conversion and utilization of CO_2 for sustainable development involving energy, catalysis, adsorption and chemical processing. *Catalysis Today* 115:2-32.

Song, C.S. & Pan, W. 2004. Tri-reforming of methane: a novel concept for catalytic production of industrially useful synthesis gas with desired H/CO ratios. *Catalysis Today* 98:463-484.

Song, D., Fu, J. & Shi, D. 2008. Exploitation of oil-bearing micro*algae* for biodiesel. *Chinese Journal of Biotechnology* 24:341-348.

Soong, R.K., Bachand, G.D., Neves, H.P., Olkhovets, A.G., Craighead, H.G. & Montemagno, C.D. 2000. Powering an inorganic nanodevice with a biomolecular motor. *Science* 290:1555-1558.

Sorensen, B. 2007. Assessing current vehicle performance and simulating the performance of hydrogen and hybrid cars. *Int. J. Hydrogen Energ.* 32:1597-1604.

Spolaore, P., Joannis-Cassan, C., Duran, E., & Isambert, A. (2006). Commercial applications of micro*algae*. *J. Biosci. Bioeng.* 101:87-96.

Steinberg, M. 2006. Conversion of fossil and biomass fuels to electric power and transportation fuels by high efficiency integrated plasma fuel cell (IPFC) energy cycle. *Int. J. Hydrogen Energ.* 31:405-411.

Stern, N. 2006. Review on the economics of climate change. Accessed February 2, 2010. http://www.hm-treasury.gov.uk/independent_reviews/stern_review_economics_climate_change/stern_review_report.cfm.

Sticklen, M.B. 2008. Plant genetic engineering for biofuel production: towards affordable cellulosic ethanol. *Nature Rev. Genet.* 9:433-443.

Stone, R. 2007. Can palm oil plantations come clean? *Science* 317:1491.

Styles, D. & Jones, M.B. 2008. Life-cycle environmental and economic impacts of energy-crop fuel-chains: an integrated assessment of potential GHG avoidance in Ireland. *Environmental Science and Policy* 11:294-306.

Subhadra, BG. 2010 Sustainability of algal biofuel production using integrated renewable energy park (IREP) and algal biorefinery approach. *Energ. Policy* 38:5892–5901.

Subramanian, K,A., Singal, S.K., Saxena, M. & Singhal, S. 2005. Utilization of liquid biofuels in automotive Diesel engines: an Indian perspective. *Biomass Bioenergy* 29:65-72.

Sun, F. & Chen, H. 2008. Enhanced enzymatic hydrolysis of wheat straw by aqueous glycerol pretreatment. *Bioresource Technol.* 99:6156-6161.

Szulczyk, K.R., McCarl, B.A. & Cornforth, G. 2010. Market penetration of ethanol. *Renew. Sustain. Energ. Rev.* 14:394–403.

Takeshita, T. & Yamaji, K. 2008. Important roles of Fischer-Tropsch synfuels in the global energy future. *Energ. Policy* 36:2773-2784.

Tan, K.T., Lee, K.T. & Mohamed, A.R. 2008. Role of energy policy in renewable energy accomplishment: The case of second-generation bioethanol. *Energ. Policy* 36:3360-3365.

Theis, M., Skrifvars, B.-J., Zevenhoven, M., Hupa, M. & Tran, H. 2006. Fouling tendency of ash resulting from burning mixtures of biofuels. Part 2: Deposit chemistry. *Fuel* 85:1992-2001.

Thomas, C.D. 2007. A sixth mass extinction? *Nature* 450:349.

Tijmensen, M.J.A., Faaij, A.P.C., Hamelinck, C.N. & Van Hardeveld, M.R.M. 2002. Exploration of the possibilities for production of Fischer Tropsch liquids and power via biomass gasification. *Biomass Bioenergy* 23:129-152.

Tilman, D., Hill, J. & Lehman, C. 2007. Response to comment on "carbon-negative biofuels from low-input high-diversity grassland biomass". *Science* 316:1567c.

Tollefson, J. 2008. Stoking the fire. *Nature* 454:388-392.

Tomkiewicz, M. 2006. Global warming: science, money and self-preservation. *Comptes rendus Chimie* 9 :172-179.

Trabesinger, A. 2007. Power games. *Nature* 447:900-903.

UNFCCC 2002. Report of the Conference of the Parties on its Seventh Session, Marrakesh, 29 October to 10 November 2001, Document FCCC/CP/2001/13/add.1

Urry, J. 2008. Governance, flows, and the end of the car system? *Global Environ. Change* 18:343-349.

Velu, S., Suzuki, K., Vijayaraj, M., Barman, S. & Gopinath, C. 2005. In situ XPS investigations of Cu1_xNixZnAl-mixed metal oxide catalysts used in the oxidative steam reforming of bio-ethanol. *Applied Catalysis* B: Environmental 55:287-299.

Venter, O., Meijaard, E. & Wilson, K. 2008. Strategies and alliances needed to protect forest from palm-oil industry. *Nature* 451:16.

Venturi, P. & Venturi, G. 2003. Analysis of energy comparison for crops in European agricultural systems. *Biomass Bioenergy* 25:235-255.

Vertès, A.A., Inui, M. & Yukawa, H. 2006. Implementing biofuels on a global scale. *Nat. Biotechnol.* 24:761-764.

Vogt, K.A. et al. 2008. Bio-methanol: How energy choices in the western United States can help mitigate global climate change. *Renew. Energ.* 34:233-241.

Vunjak-Novakovic, D., Kim, Y., Wu, X., Berzin, I. & Merchuk, J.C. 2005. Air-lift bioreactors for algal growth on flue gas: mathematical modeling and pilot-plant studies. *Ind. Eng. Chem. Res.* 44:6154-6163.

Wall, D. 2004. Rain or shine - a phototroph that delivers. *Nat. Biotechnol.* 22:40-41.

Walter, A., Dolzan, P., Quilodrán, O., de Oliveira, J.G., da Silva, C., Piacente, F. & Segerstedt, A. 2010. Sustainability assessment of bio-ethanol production in Brazil considering land use change, GHG emissions and socio-economic aspects. *Energ. Policy*, doi:10.1016/j.enpol.2010.07.043

Werner, F., Taverna, R., Hofer, P., Thürig, E. & Kaufmann, E. 2010. National and global greenhouse gas dynamics of different forest management and wood use scenarios: a model-based assessment. *Environmental Science & policy* 13:72 - 85.

Wesseler, J. 2007. Opportunities ('costs) matter: A comment on Pimentel and Patzek "Ethanol production using corn, switchgrass, and wood; biodiesel production using soybean and sunflower". *Energ. Policy* 35:1414-1416.

Willner, I. 2002. Biomaterials for sensors, fuel cells, and circuitry. *Science* 298:2407-2408.

Wolfrum, E.J. & Watt, A.S. 2001. Bioreactor design studies for a novel hydrogen-producing bacterium. Proceedings of the 2001 DOE Hydrogen Program Review, NREL/CP-570-30535.

Woodcock, J., Banister, D., Edwards, P., Prentice, A.M. & Roberts, I. 2007. Energy and health 3, energy and transport. *Lancet* 370:1078-1088.

Xue, F., Miao, J., Zhang, X., Luo, H. & Tan, T. 2008. Studies on lipid production by *Rhodotorula glutinis* fermentation using monosodium glutamate wastewater as culture medium. *Bioresource Technol.* 99:5923-5927.

Yang, Z. & Sirianni, P. 2010. Balancing contemporary fairness and historical justice: A 'quasi-equitable' proposal for GHG mitigations. *Energ. Econ.* 32:1121-1130.

Yasuda, R., Noji, H., Yoshida, M., Kinosita, K. & Itoh, H. 2001. Resolution of distinct rotational substeps by submillisecond kinetic analysis of F1-ATPase. *Nature* 410:898-904.

Yue, T.-X., Fan, Z.-M., Chen, C.-F., Sun, X.-F.& Li, B.-L. 2010. Surface modeling of global terrestrial ecosystems under three climate change scenarios. *Ecol. Model.*, doi:10.1016/j.ecolmodel.2010.11.026

Zhang, Y., Dubé, M.A., McLean, D.D. & Kates, M. 2003. Biodiesel production from waste cooking oil: Process design and technological assessment. *Bioresource Technol.* 89:1-16.

Zhou, X., Xiao, B., Ochieng, R.M. & Yang, J. 2008. Utilization of carbon-negative biofuels from low-input high-diversity grassland biomass for energy in China. *Renew. Sustain. Energ. Rev.* 13:479-485.

3

Biogas Upgrading by Pressure Swing Adsorption

Carlos A. Grande
SINTEF Materials and Chemistry, Oslo
Norway

1. Introduction

Biogas is a raw gaseous stream produced by anaerobic decomposition of organic matter. The main component of biogas is methane, reason why this stream is considered to be a renewable source of energy and fuel. The most positive aspects of biogas rely on its worldwide decentralized production and on the environmental benefits of avoiding methane emissions to atmosphere while using bio-methane it to replace fossil fuels.

In order to use the energy obtained in biogas, its production should be controlled. The production of biogas from organic matter is a complex process involving many different bacterial groups. In a simple way, the entire biogas conversion from organic matter can be divided into four steps (Gavala et al., 2003; Demirbas et al., 2011):

1. Hydrolisis: complex organic molecules are hydrolyzed into smaller units (sugars, amino-acids, alcohols, fatty acids, etc.
2. Acidogenese: acidogenic bacteria further break down the molecules into volatile fatty acids, NH_3, H_2S and H_2.
3. Acetanogese: the acetanogens transform the molecules into CO_2, H_2 and mainly acetic acid.
4. Metanogese: at the end of the process, the methanogenic archaea transform the H_2 and acetic acid molecules into a mixture of CO_2, CH_4 and water.

These production steps can be controlled in reactors (digesters) or are naturally occurring in landfills that can be optimized for collection of biogas (see www.epa.gov/lmop). The digesters can operate in mesophilic and thermophilic modes, which means that the biogas is generated at 293-313 K and 323-333 K, respectively (Gavala et al., 2003). Biogas generation in landfills mainly operates in psychrophilic conditions (285-290 K) (Monteiro et al., 2011). Biogas main constituents are methane, carbon dioxide, sulphur compounds (H_2S, siloxanes), water and minor contaminants (O_2, N_2, ammonia, chlorine, fluorines, etc) (Wellinger, 2009; Pettersson and Wellinger, 2009, www.epa.gov/lmop). The final composition of biogas is variable and strongly depends on the source of organic matter (Pettersson and Wellinger, 2009). Major sources of biogas production are landfills, waste-water treatment plants, manure fermentation and fermentation of energy crops. The composition of the biogas obtained from these sources is given in Table 1. For comparison, the composition and properties of natural gas are included in Table 1. The methane content vary strongly due to the different kind of molecules processed: i.e, fat has a much higher bio-methane yield than carbohydrates. The biogas yield of cereal residues is also high (around 200 m^3 CH_4/ton)

(Pettersson and Wellinger, 2009), representing an interesting opportunity for fermentation in farms around the world.
Since the content of methane in biogas streams is higher than 50%, its emissions to atmosphere result in a waste of an efficient hydrocarbon which also has a greenhouse warming potential 23 times higher than CO_2. For this reason, adequate collection and utilization of the bio-methane contained in biogas avoid important emissions of methane to atmosphere and also transforms this stream in a valuable source of renewable energy. The bio-methane can be directly burned and converted to electric power. In fact, biogas utilization for production of electric energy is increasing worldwide.

Gas	Biogas	Landfill gas	Natural gas
CH_4 (%)	90-70	65-65	90
Hydrocarbons (%)	0	0	9
H_2 (%)	0	0-3	0
CO_2 (%)	30-40	15-50	1
N_2 (%)	~0.2	5-40	0.3
O_2 (%)	0	0-5	0
H_2S (ppm)	0-4000	0-100	3
NH_3 (ppm)	100	5	0
Heating value, kWh/Nm^3	6.5	4.4	11.0

Table 1. Average composition and properties of natural gas and different biogas streams.

An alternative application of biogas is its upgrading to obtain purified bio-methane that can be either injected in the natural gas grid or be directly used as vehicle fuel. The purification of biogas involves several steps: removal of impurities (sulphur compounds mainly), water removal and the upgrading process which consist in bulk removal of CO_2. Normally the removal of sulphur compounds is the first step in biogas cleaning. The order of separating H_2O and CO_2 depends on the specific technology employed for upgrading. The bio-methane obtained as product should hold certain purity and the maximum quantity of CO_2 allowed is between 2-4% depending on the legislation of each country. From all the purification steps, the bulk removal of CO_2 is the most expensive one and is the one that will be considered in this Chapter.

2. Technologies for removal of carbon dioxide from biogas

The upgrading of biogas (CO_2 removal) takes between 3-6% of the energy of biogas and may cost up to € 10/GJ in small streams. Nevertheless, it has been shown that transporting upgraded biogas may result in better overall energy efficiency than converting it to electricity on-site (Pettersson and Wellinger, 2009). There are several commercial technologies to remove CO_2 from biogas streams. So far, there is not a clear best-technology and the cost of upgrading of all technologies is quite similar decreasing only with plant capacity (Wellinger, 2009). The selection of the proper technology to upgrade biogas depends on gas flowrate, value of utilities, legislation, etc.
The most important biogas upgrading techniques are scrubbing with water or other physical solvent, chemical scrubbing, membranes and Pressure Swing Adsorption (PSA). In this Chapter, only a small description of the phenomena underlying these processes will be given. The main purpose of this Chapter is to explain in detail the PSA technique for biogas

upgrading, some examples of application and also some new trends in research that can be employed to reduce the size of PSA processes to produce bio-methane economically in small scale (farm-oriented).

2.1 Water scrubbing

Carbon dioxide is more soluble in water than methane. This phenomenon is employed to remove CO_2 from biogas in water scrubbing technologies. Biogas is fed to a column where it is "washed" with counter-current water that is sprayed from the top of the column. The column is normally filled with some material to enhance the interface area promoting CO_2 absorption. The CO_2 is dissolved in the water that is then pumped to a "regeneration column" where CO_2 is released. The regeneration of the water scrubbing process can be carried out at higher temperatures or at lower pressures. In this technology, H_2S is removed with CO_2. Also the purified CH_4 stream (with purity up to 98%) should be dried after leaving the scrubber.

The solubility of CO_2 in water strongly increases at lower temperatures. In order to reduce pumping energy, the water should be available at low temperatures. In fact, this technique is been employed in several countries with cold weather (Sweden, Switzerland, Germany, Austria, etc). Cooling down water may still be efficient for large facilities, but not for small applications.

Nowadays, water scrubbing is the most employed technique for upgrading biogas. Plants with processing capabilities from 80 to 10000 m^3/hour are in operation. Main technology developers in this area are: Malmberg (www.malmberg.se), Flotech Inc. (www.flotech.com), Rosroca (www.rosroca.de), DMT (www.dirkse-milieutechniek.com), etc. In some of these webpages there is a video which actually explain graphically in detail how the process works.

2.2 Chemical scrubbing

In a similar manner to water scrubbing, it is possible to use other chemicals to absorb CO_2. The technology is also composed by AN absorption tower where the chemical solvent is flushed to selectively absorb CO_2. The saturated absorbent is then heated in a regeneration tower, releasing CO_2. This technology is widely employed to clean large facilities in the natural gas industry. The selection of the solvent for this process is quite important since the "energy" of CO_2 absorption dictates the final consumption of energy of the system. Chemicals which strongly absorb CO_2 (like amines) are more suitable to upgrade methane with relatively low content of CO_2 to a very high purity. This process may have higher energetic penalties since the CO_2 removal in biogas is a bulk removal process. On the other side, for bulk CO_2 removal to obtain a CH_4 purity in the range 97-98%, physical solvents consume less energy being more energy efficient. Different examples of physical absorbents are: methanol, Selexol, Rectisol, Genosorb, Morphysorb, etc. Plants able to process biogas flowrates of 55 to 13000 m^3/hour are in operation. Several companies provide this technology(Pettersson and Wellinger, 2009).

2.3 Membranes

The use of membranes for gas cleaning is a well established technology in chemical industries. The membrane is a porous material that let some gases permeate through its structure. Employing an adequate material, it is possible to have selectivity between the gases of the mixture to be separated. For this particular application, two different streams are obtained: a permeate gas (mainly CO_2, water and ammonia) and the retentate (concentrated CH_4). The most commonly employed materials are hollow fibres made of different polymers. Several companies provide this technology, being Air Liquide the largest company in this area (Air Liquide, 2011). In their process, the biogas is compressed to 16 bars and then routed to a two-stage membrane process where methane with purity higher than 90% can be obtained. To upgrade CH_4 to a higher purity, a PSA process can be used in series.

2.4 Pressure Swing Adsorption (PSA)

Pressure Swing Adsorption (PSA) is the second most employed techniques for biogas upgrading. Several companies develop and commercialize this technology: Carbotech (www.carbotech.de), Acrona (www.acrona-systems.com), Cirmac (www.cirmac.com), Gasrec (www.gasrec.co.uk), Xebec Inc (www.xebecinc.com), Guild Associates (www.moleculargate.com), etc. Small scale plants (flowrate of 10 m^3/hour of biogas) are in operation, but this technology is also available for much higher flowrates (10000 m^3/hour of biogas).

In PSA processes, biogas is compressed to a pressure between 4-10 bar and is fed to a vessel (column) where is putted in contact with a material (adsorbent) that will selectively retain CO_2. The adsorbent is a porous solid, normally with high surface area. Most of the adsorbents employed in the commercial processes are carbon molecular sieves (CMS) but also activated carbons, zeolites and other materials (titanosilicates) are employed. The purified CH_4 is recovered at the top of the column with a very small pressure drop. After certain time, the adsorbent is saturated with CO_2, and the column needs to be regenerated by reducing the pressure (normally to vacuum for biogas upgrading). The adsorption of H_2S is normally irreversible in the adsorbents and thus a process to eliminate this gas should be placed before the PSA. Alternatively, depending on the choice of the adsorbent, the humidity contained in the biogas stream can be removed together with CO_2 in the same unit. Multi-column arrays are employed to emulate a continuous process. For small applications subjected to discontinuities, a single column with storage tanks may be used. One of the most important properties of the PSA process is that is can be adapted to biogas upgrading in any part of the world since it does not depend on the availability of cold or hot sources. A detailed explanation of this process follows.

3. Principles of Pressure Swing Adsorption for biogas upgrading

In a PSA unit for biogas upgrading, an adsorbent material is subjected to pressure changes to selectively adsorb and desorb CO_2. Adsorption is an exothermic spontaneous process and the loading of CO_2 in the adsorbent depends specifically on the properties of the material employed (surface area and composition, pore size, etc). Once the material is specified, then its regeneration should be realized. Note that since the material is continuously used and regenerated, there comes a point where the process achieves a "cyclic steady state" (CSS). The largest part of engineering a PSA process rely in designing a regeneration protocol for the adsorbent able to spent small amount of energy (reduce energetic penalty) and do it in the fastest way possible (increase productivity).

The operational principle of the PSA process can be observed in Figure 1 where two generic CO_2 isotherms (representing two different materials) are shown. In both materials, the adsorbent may take CO_2 up to the loading established by its partial pressure in the feed step (P_{feed}) which is $q_{feed,1}$ and $q_{feed,2}$ for adsorbents 1 and 2, respectively. After the adsorbent is saturated, it is regenerated to a lower pressure, P_{reg}, where the loading of CO_2 decreases to $q_{reg,1}$ and $q_{reg,2}$. The material 1 has a higher CO_2 capacity than material 2 for the entire pressure range. However, the difference in loading between q_{feed} and q_{reg} (Δq) is higher for material 2, indicating that the "cyclic capacity" will be better for this material. In fact, the conclusion from this image is that it is important to know the shape of the isotherm in order to design the PSA process. Also, ideally for PSA applications, linear or mild non-linear isotherms are better than very steep isotherms with high loading. Furthermore, when the isotherms are steep, it is more difficult to regenerate the adsorbent since the energy required to desorb CO_2 is higher. From Figure 1, it is possible to see that the selection of the regeneration pressure has also an important effect in the cyclic operation of the PSA process. For this reason, in the next sections, the properties of the adsorbents and the cycles used for adsorbent regeneration in PSA for biogas upgrading will be discussed.

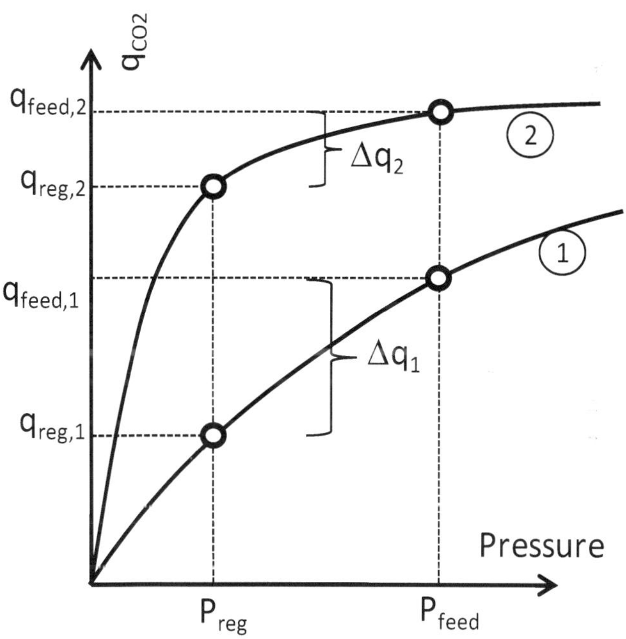

Fig. 1. Generic CO_2 isotherms for two different materials (1) and (2) indicating partial pressure of CO_2 in feed step and regeneration (low) pressure.

3.1 Adsorbents

It can be said that the adsorbent material is the "heart" of the PSA unit. All the properties of the cycle (operating conditions and operating mode) depend on the initial choice of the adsorbent. As mentioned before, several materials can be employed in PSA technology. The material selected should at least satisfy one of two criteria:

i. have a higher selectivity to CO_2: this gas should be more "attached" to the surface of the material than CH_4; in most solids CO_2 can create stronger bonds with surface groups than CH_4. This kind of materials will be termed as equilibrium-based adsorbents since its main selectivity is due to differences of interaction forces between CO_2 and CH_4 with and the surface.
ii. the pores of the adsorbent can be adjusted in such a way that CO_2 (kinetic diameter of 3.4 Å) can easily penetrate into their structure while larger CH_4 molecules (kinetic diameter of 3.8 Å) have size limitations to diffuse through them. These materials will be termed as kinetic adsorbents since its main selectivity is due to diffusion constrains.

Carbon molecular sieves are one of the most employed materials for biogas upgrading. Adsorption equilibrium isotherms of CO_2 and CH_4 in CMS-3K (Takeda Corp., Japan) are shown in Figure 2 (Cavenati et al., 2005). This material has a clear selectivity towards CO_2, but the most important property in CMS-3K is not its equilibrium selectivity, but the kinetic selectivity. In this material, CO_2 adsorbs much faster than CH_4: adsorption equilibrium of CH_4 is achieved only after two days of solid-gas contact. In fact, the pore mouth of CMS-3K is narrowed to dimension closer to the kinetic diameter of CH_4 creating a specific resistance (mass transfer in the micropore mouth) (Srinivasan et al., 1995) that can be seen in the initial moments of CH_4 uptake in Figure 2 (b). Another material that also presents strong resistance to CH_4 diffusion is ETS-4 (titanosilicate-4) modified with alkali-earth metals (Kuznicki, 1990; Marathe et al., 2004; Cavenati et al., 2009). In this material, the pore diameter can be tuned with different heating temperatures resulting in a "molecular gate" effect that actually named the process commercialized by Guild Associates Inc. (USA).

Other normally employed adsorbents are activated carbons and zeolites. In these materials, the diffusion of both gases can be very fast and actually what is exploited is the difference between loadings of CO_2 and CH_4. An example of these equilibrium-based materials is given in Figure 3, where adsorption equilibrium of CO_2 and CH_4 on zeolite 13X (CECA, France) is shown (Cavenati et al., 2004). Note that the loading of CO_2 is much higher than the loading of CH_4 at given P,T conditions. Furthermore, recasting the conclusions taken from Figure 1, the cyclic CO_2 capacity at lower temperatures is smaller than at higher temperatures, which means that if zeolite 13X is employed at 323 K, it will be easier to regenerate than at 298 K.

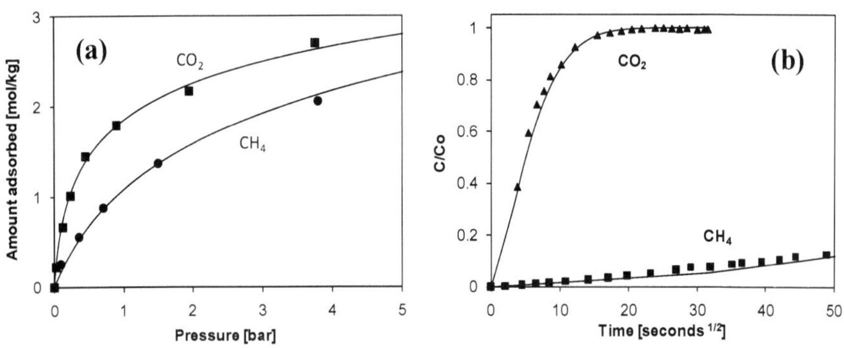

Fig. 2. Adsorption of CO_2 and CH_4 in carbon molecular sieve 3K (Takeda Corp, Japan) at 298 K: (a) adsorption equilibrium; (b) uptake rate curves (data from Cavenati et al., 2005).

Another topic that is important for the selection of materials for the PSA process for biogas upgrading, is the presence of contaminants. Apart from CH_4 and CO_2, other gases present in biogas are H_2S and H_2O. In almost all adsorbents, H_2S is irreversibly adsorbed, reason why it has to be removed before the PSA process. When carbonaceous materials are employed it is possible to remove H_2O in the same vessel as CO_2. However, that is not possible using zeolites since water adsorption is also very steep, resulting in a very difficult desorption.

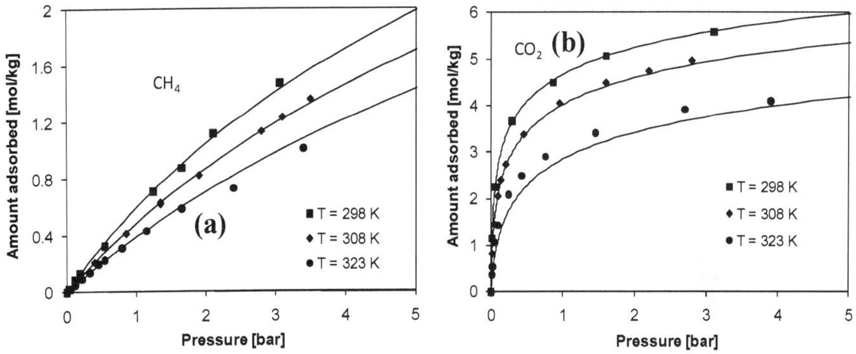

Fig. 3. Adsorption equilibrium of CO_2 (a) and CH_4 (b) on zeolite 13X at 298, 308 and 323 K (Data from Cavenati et al., 2004).

3.2 Packed-bed performance

Adsorption is a spontaneous process and when the gas is putted in contact with the adsorbent, a new equilibrium state will be established, depending on the partial pressure of each of the gases and on the total temperature of the system. After achieving such equilibrium, no more adsorption takes place and the adsorbent should be regenerated. For this reason, a PSA column should be regenerated periodically to be able to absorb CO_2 in different cycles. In order to keep constant feed processing, more than one column are employed in parallel: when biogas is fed for selective removal of CO_2, the other column(s) are being regenerated.

The operation of a PSA process for biogas upgrading can be explained by showing what happens when a mixture of CH_4-CO_2 is fed to a column filled with adsorbent. For simplicity, the column will be considered to be at the same pressure of the biogas stream and filled with an inert gas (helium). An example of such behaviour is normally termed as "breakthrough experiments". An example of a breakthrough curve of CH_4 (55%) - CO_2 (45%) mixture in CMS-3K is shown in Figure 4 (Cavenati et al., 2005). It can be observed that in the initial moments, methane molecules travel across the column filling the gas phase in the inter-particle space, but also in the intra-particle voids (macropores), replacing helium. Due to the very large resistance to diffuse into the micropores, CH_4 adsorption is very difficult, reason why it breaks through the column very fast. On the other side, CO_2 takes a very long time to break through the column since it is being continuously adsorbed. Note that before CO_2 breakthrough, there is a period of time where only methane is obtained at the column product end. In Figure 4(b) also the temperature increase on the different positions of the column is shown. Note that in this experiment, temperature increase is due solely to CO_2 adsorption. This experiment was carried out under non-isothermal and non-adiabatic conditions. In the case of larger adsorbers where adiabatic conditions can be found, temperature increase should be higher having a stronger negative impact in the adsorption of CO_2 (faster breakthrough).

Another important thing that can be observed in Figure 4 is the dispersion of the CO_2 curve. The perturbation in the feed stream was a step increase in CH_4 and CO_2 partial pressure and the breakthrough result indicates that the response to that input after passing through the column is quite spread. The shape of the adsorption breakthrough curves is associated to diverse factors:

1. Slope of the adsorption isotherms: comprise the concentration wave if isotherm is favourable (Langmuir Type) and dispersive if the adsorption equilibrium is unfavourable (desorption for Langmuir-type isotherms). No effect if the isotherm is linear,
2. Axial dispersion of the adsorption column: disperse the concentration wave,
3. Resistance to diffusion within the porous structure of the adsorbent: disperse the concentration wave.
4. Thermal effects: normally in gas separations the thermal wave travels at the same velocity as the concentration wave (Yang, 1987; Ruthven et al., 1994; Basmadjian, 1997) and its effect is to disperse the concentration wave. Thermal effects can control the shape of the breakthrough curve.

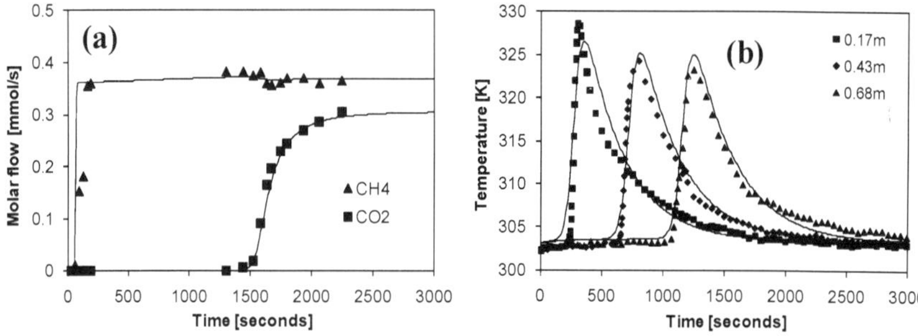

Fig. 4. Binary CH_4 (55%) - CO_2 (45%) breakthrough curve experiment in fixed-bed filled with CMS-3K extrudates. Temperature: 303 K; Pressure: 4 bar (data from Cavenati et al., 2004). (a): molar flow of CH_4 and CO_2; (b) temperature evolution in three different points of the column.

To compare the performance of different adsorbents, the thermal effects associated to adsorption of CO_2 in zeolite 13X extrudates can be observed in Figure 5 where a breakthrough of CO_2 was carried out (Cavenati et al., 2006). The experiment was conducted at 299 K and a total pressure of 3.2 bar. It can be observed that CO_2 breaks through the bed quite sharply due to the strong non-linearity of the CO_2 adsorption isotherm that tends to compress the concentration front. After the initial sharp breakthrough, the shape of the curve gets quite dispersed due to thermal effects. It can be seen in Figure 5(b) that the temperature increase in certain points of the column is quite high, reducing the loading of CO_2 and making breakthrough quite faster than it should be if carried out at isothermal conditions. The opposite effect will take place in desorption of CO_2: the temperature in the bed will drop increasing the steepness of the adsorption isotherm, making desorption more unfavourable.

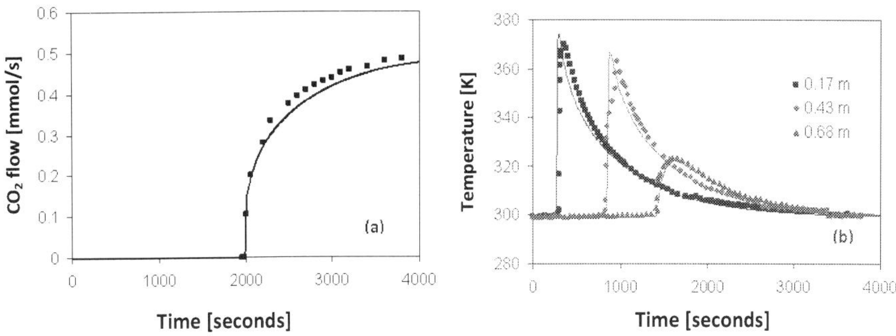

Fig. 5. Breakthrough curve of pure CO_2 in fixed-bed filled with zeolite 13X extrudates. Temperature: 299 K; Pressure: 3.2 bar (data from Cavenati et al., 2006). (a): molar flow of CO_2; (b) temperature evolution in three different points of the column.

Due to the thermal effects and the steepness of the CO_2 isotherm on zeolite 13X, it was concluded that using a similar PSA cycle, if the temperature of the biogas stream is close to ambient temperature, it is better to use the Carbon Molecular Sieve (CMS-3K) than zeolite 13X (Grande and Rodrigues, 2007).

The solid lines shown in Figures 4 and 5, represent the prediction of a mathematical model, based on pure gas adsorption equilibrium and kinetics (Cavenati et al., 2004; Cavenati et al., 2005). The resulting equations for the prediction of the fixed-bed behaviour are (Da Silva, 1999):

i. mass balances in the column, particle and micropores (crystals) of the adsorbent.
ii. Energy balances in the gas and solid phases and column wall
iii. Momentum balance (simplified to the Ergun equation)
iv. Multicomponent adsorption isotherm model.

Note that the mass, energy and momentum balances are partial differential equations linked by a (generally) non-linear equation (isotherm model). The mathematical model was tested under diverse adsorbents and operating conditions for CH_4-CO_2 separation as well as for other gas mixtures. The mathematical model employed is termed as "homogeneous model" since it considers mass and heat transfer in different phases using different equations. Heterogeneous models (single energy balance) and also more simplified mass transfer models can also be employed to predict column behaviour with good accuracy (Ruthven, 1984; Yang, 1987; Ruthven et al., 1994).

3.3 Packed-bed regeneration: basic cycles

Once that the adsorbent is selected to perform a given CH_4-CO_2 separation under specific operating conditions (T, P, y_{CO2}), there are only few actions that can be taken to make the adsorption step more efficient (dealing with energy transfer, for example). When designing the upgrading PSA, the most important task is to make desorption efficiently.

The initial work reporting Pressure Swing Adsorption technology was signed by Charles W. Skarstrom in 1960 (Skarstrom, 1960). A similar cycle was developed by Guerin - Domine in

1964 (Guerin and Domine, 1964). The Skarstrom cycle is normally employed as a reference to establish the feasibility of the PSA application to separate a given mixture.
The Skarstrom cycle is constituted by the following cyclic steps:

1. Feed: the CH_4-CO_2 mixture is fed to the fixed bed where the adsorbent is placed. Selective adsorption of CO_2 takes place obtaining purified CH_4 at the column product end at high pressure.
2. Blowdown: immediately before CO_2 breaks through, the column should be regenerated. This is done by stopping the feed step and reducing the pressure of the column counter-currently to the feed step. Ideally, this step should be carried out until a new equilibrium state is established as shown in Figure 1. However, the blowdown step is stopped when the flowrate of CO_2-rich stream exiting the column is small. With the reduction of pressure, CO_2 is partially desorbed from the adsorbent. In this step, the lowest pressure of the system is achieved.
3. Purge: when the low pressure is achieved, the column will have CO_2 molecules in the adsorbed phase but also in the gas phase. In order to reduce the amount of CO_2 in both phases, a purge step is performed counter-current to feed step. In the purge, some of the purified methane is recycled (light recycle) to displace CO_2 from the CH_4 product end.
4. Pressurization: Since the purge is also performed at low pressure, in order to restart a new cycle, the pressure should be increased. Pressurization can be carried out co-currently with the feed stream of counter-currently with purified CH_4. The selection of the pressurization strategy is not trivial and may lead to very different results (Ahn et al., 1999).

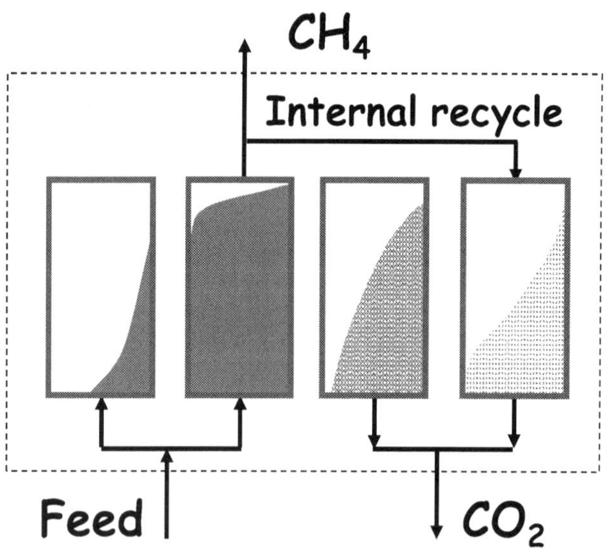

Fig. 6. Schematic representation of the different steps in a Skarstrom cycle. The dotted line represents the external boundary used to calculate performance parameters.

A schematic representation of the different steps of one column in a single cycle is shown in Figure 6. Note that in this image an external boundary was established. This boundary is used to define the performance parameters of the PSA unit: CH_4 purity, CH_4 recovery and unit productivity. They are calculated using the following equations:

$$PURITY = \frac{\int_0^{tfeed} C_{CH4}\, u\big|_{z=L}\, dt}{\left(\int_0^{tfeed} C_{CH4}\, u\big|_{z=L}\, dt + \int_0^{tfeed} C_{CO2}\, u\big|_{z=L}\, dt\right)} \qquad (1)$$

$$RECOVERY = \frac{\int_0^{tfeed} C_{CH4}\, u\big|_{z=L}\, dt - \int_0^{tpurge} C_{CH4}\, u\big|_{z=L}\, dt}{\int_0^{tfeed} C_{CH4}\, u\big|_{z=0}\, dt + \int_0^{tpress} C_{CH4}\, u\big|_{z=L}\, dt} \qquad (2)$$

$$PRODUCTIVITY = \frac{\left(\int_0^{tfeed} C_{CH4}\, u\big|_{z=L}\, dt - \int_0^{tpurge} C_{CH4}\, u\big|_{z=L}\, dt\right).A_{col}}{t_{cycle} w_{ads}} \qquad (3)$$

where C_{CH4} is the concentration of methane, u is the velocity, t_{cycle} is the total cycle time, A_{col} is the column area and w_{ads} is the total adsorbent weight. Note that the calculation of CH_4 recovery and unit productivity involves the molar flowrates of the different steps where some CH_4 is recycled. In the case of changing the cycle configurations, the equations to calculate the process parameters may also be different.

In the cycle developed by Guerin-Domine, a pressure equalization step between different columns take place between feed and blowdown and after the purge and the pressurization. The pressure equalization steps are very advantageous for PSA applications since they help to improve the recovery of the light product, they reduce the amount of gas lost in the blowdown step and as a direct consequence, the purity of the CO_2-rich stream obtained in the blowdown (and purge) steps increases and also less power is consumed if blowdown is carried out under vacuum. It should be mentioned that in the PSA process for biogas upgrading, it is important to perform some pressure equalization steps to reduce the amount of methane that is lost in the blowdown step. The amount of CH_4 lost in the process is termed as CH_4 slip and in PSA processes is around 3-12% (Pettersson and Wellinger, 2009). More advanced cycles for other applications also make extensive use of the equalization steps: up to three pressure equalizations between different columns take place in H_2 purification (Schell et al., 2009; Lopes et al., 2011). As an example, in Figure 7, the pressure history over one cycle is shown for the case of a two-column PSA process using a modified Skarstrom cycle with one pressure equalization step (Santos et al., 2011).

Continuing with the example of CMS-3K as selective adsorbent for biogas upgrading, the cyclic performance of a Skarstrom cycle is shown in Figure 8. In this example, the feed was a stream of CH_4 (55%) - CO_2 (45%) resembling a landfill gas (T = 306 K), with a feed pressure of 3.2 bar. The blowdown pressure was established in 0.1 bar and pressurization step was carried out co-current with feed stream (Cavenati et al., 2005). Figure 8(a) shows the pressure history over one entire cycle while Figure 8(b) shows the molar flowrate of each gas exiting the column. It can be seen that in the feed step, a purified stream of CH_4 is obtained. In this experiment, the purity of CH_4 was 97.1% with a total recovery of 79.4%

(Cavenati et al., 2005). An important feature of the CMS-3K adsorbent is related to the very slow adsorption kinetics of CH_4. In Figure 8(c) the simulated amount of CH_4 adsorbed is shown. It can be observed that after reaching the cyclic steady state (CSS), the loading of CH_4 per cycle is constant: this means that no CH_4 is adsorbed in the column. This is very important since no CH_4 will be adsorbed in the pressurization step, even with a very strong increase in its partial pressure. Unfortunately, the narrow pores also make CO_2 adsorption (and desorption) difficult, reason why only part of the capacity of the bed is employed as shown in Figure 8(d) resulting in small unit productivity.

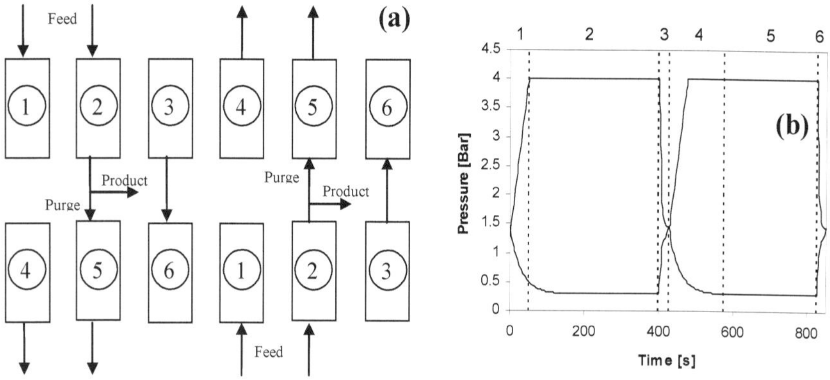

Fig. 7. Scheduling of a Skarstrom cycle in a two column PSA unit: (a) step arrangement: 1. Pressurization; 2. Feed; 3. Depressurization; 4. Blowdown; 5. Purge; 6. Equalization. (b) Pressure history of both columns during one cycle.

As can be seen, an important amount of CH_4 is lost in the blowdown step, since there is no pressure equalization: pressure drops from 3.2 bar to 0.1 bar having at least 55% of CH_4 in the gas phase. The main problem of using the Skarstrom cycle for biogas upgrading is that the CH_4 slip is quite high. Since the Skarstrom cycle is potentially shorter than more complex cycles, the unit productivity is higher. Keeping this in mind, it may be interesting to employ this cycle in the case of combining the production of fuel (bio-CH_4) and heat or electricity where the gas obtained from the blowdown step can be directly burned or blended with raw biogas.

In order to avoid large CH_4 slip, at least, one pressure equalization should be employed to reduce the amount of methane in the gas phase that is lost in the blowdown stream. If such step is performed, it is possible to increase the methane recovery from 79.4% to 86.3% obtaining methane with a similar purity (97.1%). It can be concluded that the increase of number of equalization steps will reduce the methane lost in the blowdown step. Furthermore, if less gas is present in the column when the blowdown step starts, the vacuum pump will consume less power. However, to perform multiple pressure equalizations, the number of columns and the complexity of operation of the unit increase. Furthermore, the time required by the multiple pressure equalization steps will reduce the unit productivity resulting in larger units. A trade-off situation is normally achieved in PSA units with four-columns employing up to two pressure equalization steps before blowdown (Wellinger, 2009).

Another source of CH_4 slip is the exit stream of the purge step: in the purge, part of the purified CH_4 stream is recycled (counter-currently) to clean the remaining CO_2 in the column. Since CH_4 is not adsorbed, after a short time it will break through the column. However, if the purge step is too short, the performance of the PSA cycle is poor. In order to achieve very small CH_4 slip keeping an efficient purge, one possible solution is to recompress and recycle this stream (Dolan and Mitariten, 2003). Furthermore, if this stream is recycled, the flowrate of the purge can be used to control the performance of the PSA cycle when strong variations of the biogas stream take place (CO_2 content or total flowrate).

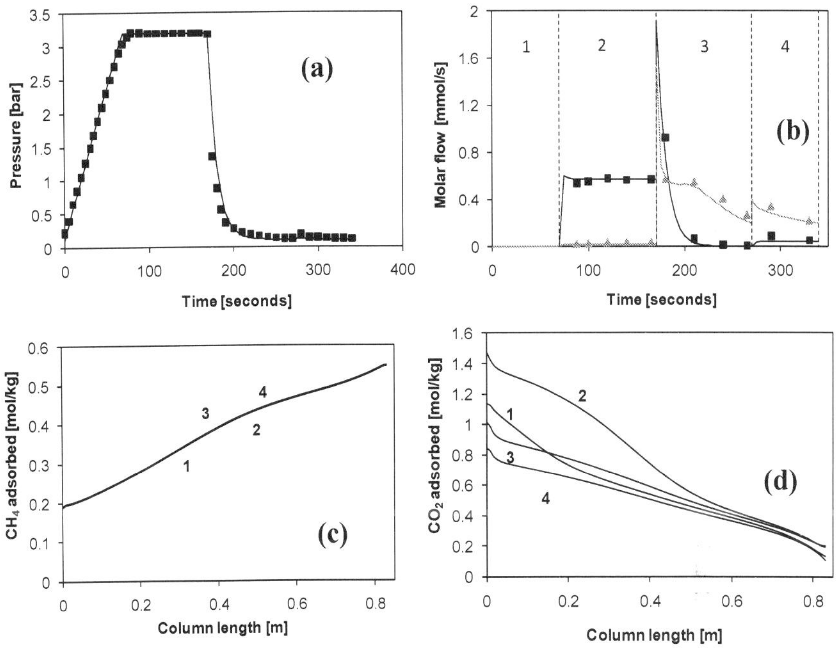

Fig. 8. PSA separation of a mixture of CH_4 (55%) - CO_2 (45%) using a packed bed filled with CMS-3K operating with a Skarstrom cycle (1. Pressurization; 2. Feed; 3. Blowdown; 4. Purge). Feed pressure: 3.2 bar, blowdown pressure: 0.1 bar. (a) Pressure history over one cycle; (b) molar flowrate exiting the column; (c) loading of CH_4 at the end of each step; loading of CO_2 at the end of each step. Data from Cavenati et al., 2005.

4. New markets and improvements of PSA technology

As mentioned before, the biogas market has enormous possibilities to grow. One of the most important sectors that may trigger large growth of PSA development is within small farms. In such cases, the biogas can be employed for heating and to generate electricity, but a portion of the stream (or the exceeding) can be upgraded to fuel. In such applications, besides the specifications of process performance, six characteristics are desired for any upgrading technology:

1. Economic for small streams,
2. Compact,
3. Automated,
4. Minimal attendance (by non-expert person most of the time),
5. Possible to switch on / off quite fast
6. Deliver product specifications even when subjected to strong variations in feed.

The PSA technology can potentially be employed in such applications since it can satisfy most of the criteria established above. As an example it can be mentioned that some plants of the Molecular Gate technology are operated remotely (automated with minimal attendance) transported in trucks (compact) and they are employed for small streams of natural gas (Molecular Gate, 2011). However, the scale of small biogas application is quite small (smaller than 10 m^3/hour). Furthermore, fast switch on/off a PSA unit for several times was not reported in literature and surely require dedicated research as well as PSA design to handle strong variations in feed streams.

The two major areas where research should be conducted to deliver a PSA unit to tackle such applications are: new adsorbents and design engineering.

4.1 New adsorbents

Despite of the explosion in discovery of new materials with a wide range of possibilities, most of the PSA units existing in the market still use the well-known zeolites (4A, 5A and 13X), activated carbons, carbon molecular sieves, silica gel and alumina. Since the adsorbent material is the most important choice for the design of the PSA unit, more efficient materials should be employed to satisfy more market constrains (energy consumption and size). One interesting example of the possibility of application of new materials is the Molecular Gate technology, where the utilization of narrow pore titanosilicates (ETS-4) lead to a successful technology for CH_4 upgrading (Kuznicki, 1990; Dolan and Mitariten, 2003). The ETS-4 materials when partially exchanged with alkali-earth metals present a unique property of pore contraction when increasing the temperature of activation (Marathe et al., 2004; Cavenati et al., 2009). This property is very important since the pores can be adjusted with a very high precision to do separations as complex as CH_4-N_2. Within this kind of inorganic substrates, other interesting material that deserves attention are the aluminophosphates. Even when these materials do not present a very high CO_2 capacity, they have quite linear isotherms (ideal for utilization in PSA applications) and also some of them present Type V isotherms for water adsorption, which means that they have certain tolerance (and regenerability) if traces of water are present (Liu et al., 2011).

In the last years, a new family of materials with extremely high surface area has been discovered (Li et al., 1999; Wang et al., 2002; Millward and Yaghi, 2005; Mueller et al., 2005; Kongshaug et al., 2007). The metal-organic frameworks (MOFs) can actually adsorb extremely large amounts of CO_2 when compared with classical adsorbents. Furthermore, it is possible to adjust the structure in such a way that the steepness of the isotherm is mild and thus regeneration is simpler. An example of this high CO_2 loading on MOFs is given in Figure 9 where the isotherms of CO_2 and CH_4 on Cu-BTC are shown at different temperatures (Cavenati et al., 2008). Comparing these isotherms with the ones presented by zeolite 13X (Figure 3), it can be observed that the steepness of the isotherm is quite mild leading to much higher "cyclic capacity" than zeolite 13X. Several MOFs were studied to separate CH_4-CO_2 mixtures (Schubert et al., 2007; Cavenati et al., 2008; Llewellyn et al., 2008; Dietzel et al., 2009; Boutin et al., 2010). Most of them present excellent properties for CO_2

adsorption, eventually with mild-non-linearity of CO_2 isotherms. Issues to commercialize these materials are related to the correct formulation and final shaping without significantly loosing their surface area.

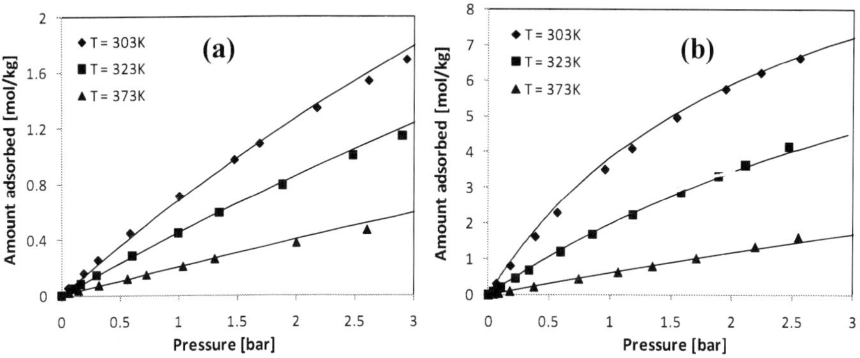

Fig. 9. Adsorption equilibrium of CO_2 (a) and CH_4 (b) on Cu-BTC MOF at 303, 323 and 373 K (data from Cavenati et al., 2008).

The extremely high CO_2 loading of MOFs indicate that the size of the PSA unit can be significantly reduced using this material instead of classical adsorbents. Furthermore, the CO_2 adsorption kinetics in several MOFs is quite fast, thus most of its loading can actually be employed per cycle. One of the main issues with MOFs is that water cannot be present in the system and should be removed in a previous step (which should not be an important problem since water must be removed anyway).

4.2 Alternative PSA design

A possible route to design a new PSA unit involve the selection of the adsorbent, the selection of the PSA cycle that should be used, the sizing of the unit, the definition of operating variables for efficient adsorbent regeneration and finally the arrangement of the multi-column process for continuous operation (Knaebel and Reinhold, 2003). However, in the development of new applications in small scale, other parameters can be considered, particularly the ones related to the design of the unit. One example of the possibility of out-of-the-box process design is the rotary valve employed by Xebec that has allowed the industrial application of rapid-PSA units for biogas upgrading (Toreja et al., 2011). When designing small units, the shape of the columns can be different to the traditional ones and this fact can be used to maximize the ratio of adsorbent employed per unit volume. Furthermore, in some cases of high CO_2 contents, the heat of adsorption may increase the temperature of the adsorbent in such a way that the effective capacity decreases significantly. In such cases, the possibility of effective heat exchange with the surroundings can be an alternative (Bonnissel et al., 2001) as well as increase the heat capacity of the column (Yang, 1987). Other alternative to increase the unit productivity when using kinetic adsorbents (like CMS-3K) is to use a second layer of adsorbent with larger pores (fast adsorption) and with easy regenerability (Grande et al., 2008). By using this layered arrangement, it is possible to "trap" the CO_2 in the final layer for some additional time, which is enough to double the unit productivity of the system (keeping similar CH_4 purity

and recovery). This layering of adsorbents can also be employed to remove water and CO_2 in the same bed as it is being done in other CO_2 applications (Li et al., 2008).
Perhaps the most important engineering challenges of new PSA design are related to the modification of the PSA cycles. Most of the PSA units existing in industry nowadays use the Skarstrom cycle (or small variations of it) with several pressure equalizations to reduce the CH_4 slip. The utilization of different cycles can be adjusted for different applications of the biogas stream: production of extremely high CH_4 purity, small CH_4 slip, combined heat / electricity and/or fuel generation, etc. The possibility of "playing" with the step arrangement in a PSA cycle for a given application is virtually infinite. Extreme variations in PSA cycles can be achieved with PSA units with three or four columns. An example of such possibilities is given in Figure 10 where a different cycle is presented in order to radically improve the unit productivity of kinetic adsorbents (Santos et al., 2011b). This 4-column PSA cycle was designed keeping in mind that the adsorption should be continuous, that at least one equalization step is necessary to reduce CH_4 slip and also to improve the contact time between gas and solid which is particularly important to increase the loading of CO_2 in the adsorbent. To enhance the contact time between the adsorbent and the feed stream, a lead-trim concept is employed (Keller et al., 1987).

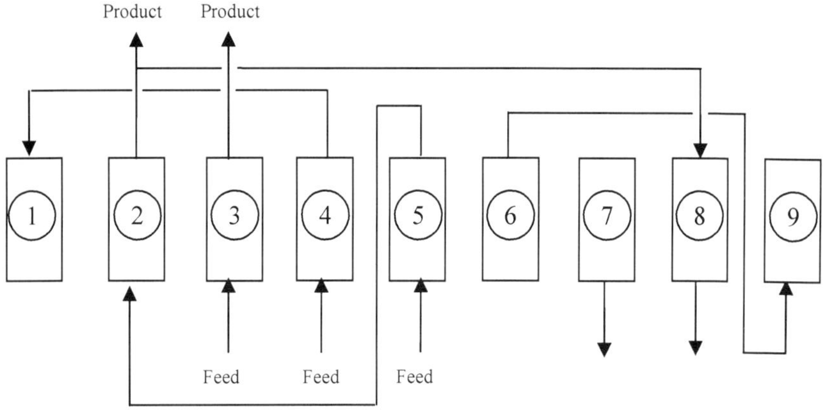

Fig. 10. Scheduling of a column for a PSA cycle for biogas upgrading using lead-trim concept. The steps are: 1. Pressurization; 2. Trim feed; 3-4. Feed; 5. Lead adsorption; 6. Depressurization; 7. Blowdown; 8. Purge; 9. Pressure equalization.

In a kinetic adsorbent, the CO_2 breakthrough happens relatively fast and the mass transfer zone is quite large as shown in Figure 8(d). In order to avoid contamination of the CH_4-rich stream, the feed step is normally stopped, but using the lead-trim cycle arrangement, the gas exiting one column is routed to a second column where this residual CO_2 can be adsorbed, giving the first column extra time to adsorb CO_2. This column arrangement leads to a column with virtually the double of the size (only for some adsorption steps). Also, the column that is ready for regeneration has a higher content of CO_2, which also result in small CH_4 slip. A simulation of the performance of this PSA cycle using CMS-3K is shown in Figure 11. Using this column arrangement, CH_4 purity of 98.3% could be obtained with a total recovery of 88.5% and a unit productivity of 5.5 moles of CH_4 per hour per kilogram of adsorbent (Santos et al., 2011b). From the 11.5% of CH_4 lost in blowdown and purge steps, around 7% is lost in the purge step, which means that if this stream is recycled, the CH_4-slip will drop to values lower than 5%. Note that in Figure 11(b), CO_2 started to break through the column at the end of the feed step. In this case the objective was to produce CH_4 with purity higher than 98%, but this cycle can be regulated if higher purity is required. Furthermore, the cycle is quite efficient and it does not require going to 0.1 bar for regeneration and only 0.3 bar are employed, which significantly reduced the power consumption when compared to classical step arrangements.

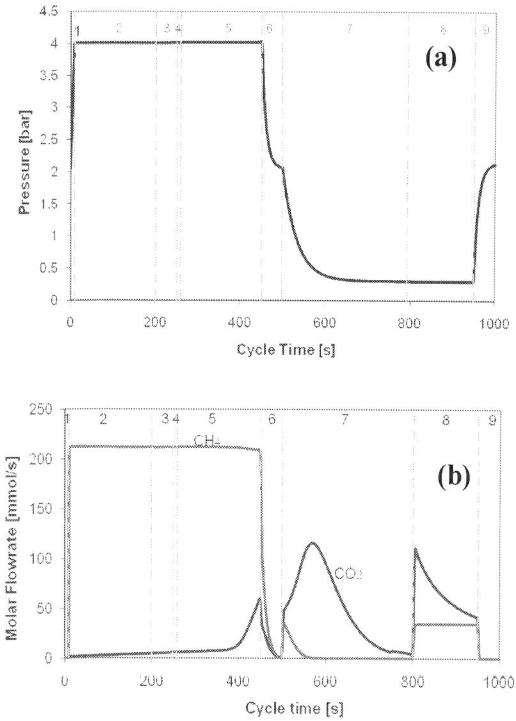

Fig. 11. Simulation of a 4-column PSA process using the lead-trim cycle (see Figure 10) with CMS-3K for separation of a mixture of CH_4 (67%) and CO_2 (33%). (a) pressure history of one cycle; (b) molar flow of CH_4 and CO_2 after cyclic steady state was achieved. Feed pressure: 4 bar; Blowdown pressure: 0.3 bar; Temperature: 323 K.Data from Santos et al., 2011b.

5. Conclusions

Pressure Swing Adsorption (PSA) has already proved that it is an efficient technology for biogas upgrading under different operating conditions. This work presents a summary of the available technologies for biogas upgrading (water and chemical scrubbing and membranes) and gives a special focus to PSA technology. A brief overview of the operating principles of PSA technology is given, with some insights in the adsorbents employed and the design possibilities of the PSA units. A final section shows some of the new range of possibilities to improve its design for new applications, oriented to small biogas flowrates encountered in farms. Certainly, there is still much research required to successfully develop PSA technology for small flowrates applications. Certainly, a strong link between materials science and process engineering can contribute to develop this technology faster. Successful application of PSA in such market should expand the application of biogas utilization as environmentally-friendly and sustainable fuel.

6. Acknowledgments

The author would like to acknowledge Prof. Alirio E. Rodrigues for its constant guidance in adsorption science along several years. The assistance of many former colleagues of the Laboratory of Separation and Reaction Engineering at the University of Porto was also essential in developing most of the research activities reported in this work. Also, I would like to express my gratitude to the support of SINTEF Materials and Chemistry, particularly to Dr. Richard Blom, in writing this Chapter.

7. References

Ahn, H.; Lee, C-H; Seo, B.; Yang, J.; Baek, K (1999). Backfill Cycle of a Layered Bed H2 PSA Process. Adsorption, Vol. 5, No. 4, (October 1999), pp 419-433, ISSN 0929-5607.

Air Liquide (2011). Biogas Recovery System. 02.04.2011. Available at: http://www.dta.airliquide.com/en/our-offer/decentralized-energies/biogas.html

Basmadjian, D. (1997). The Little Adsorption Book: A Practical Guide for Engineers and Scientists; CRC Press, 1997.

Bonnissel, M.P.; Luo, L.; Tondeur, D (2001). Rapid Thermal Swing Adsorption. Ind. Eng. Chem. Res. Vol. 40, No. 10, (April 2001), pp 2322-2334, ISSN 0888-5885.

Boutin, A.; Coudert, F-X.; Springuel-Huet, M-A.; Neimark, A.V.; Ferey, G.; Fuchs, A.H (2010). The Behavior of Flexible MIL-53 (Al) upon CH_4 and CO_2 Adsorption. J. Phys Chem. C. Vol. 114, No. 50, (December 2010), pp 22237-22244, ISSN 1932-7447.

Cavenati, S.; Grande, C.A.; Rodrigues, A.E (2004). Adsorption Equilibrium of Methane, Carbon Dioxide and Nitrogen on Zeolite 13X at High Pressures. J. Chem. Eng. Data, Vol. 49, No. 4, (June 2004), pp 1095-1101, ISSN 0021-9568.

Cavenati, S.; Grande, C.A.; Rodrigues, A.E (2005). Upgrade of Methane from Landfill Gas by Pressure Swing Adsorption. Energy & Fuels, Vol. 19, No. 6, (August 2005), pp 2545-2555, ISSN 0887-0624.

Cavenati, S.; Grande, C.A.; Rodrigues, A.E (2006). Removal of Carbon Dioxide from Natural Gas by Vacuum Pressure Swing Adsorption. Energy & Fuels, Vol. 20, No. 6, (September 2006), pp 2648-2659, ISSN 0887-0624.

Cavenati, S.; Grande, C.A.; Rodrigues, A.E (2008). Metal Organic Framework Adsorbent for Biogas Upgrading. Ind. Eng. Chem. Res. Vol. 47, No. 16, (July 2008), pp 6333-6335, ISSN 0888-5885.

Cavenati, S.; Grande, C.A.; Lopes, F.V.S.; Rodrigues, A.E (2009). Adsorption of Small Molecules on Alkali-Earth Modified Titanosilicates. Microp. Mesop. Mater, Vol. 121, No. 1-3, (May 2009), pp 114-120, ISSN 1387-1811.

Da Silva, F. A. Cyclic Adsorption Processes: Application to Propane/Propylene Separation. Ph.D. Dissertation, University of Porto, Portugal, 1999.

Dietzel, P.D.C.; Besikiotis, B.; Blom, R (2009). Application of Metal-Organic Frameworks with Coordinatively Unsaturated Metal Sites in Storage and Separation of Methane and Carbon Dioxide. J. Mater. Chem. Vol. 19, (August 2009), pp 7362-7370, ISSN 0959-9428.

Demirbas, M.F.; Balat, M.; Balat, H (2011). Biowastes-to-biofuels. Energy Conv. Management, Vol.52, No. 4, (April 2011), pp 1815-1828, ISSN 0196-8904.

Dolan, W.B.; Mitariten, M.J (2003). CO_2 Rejection from Natural Gas. United States Patent US 2003/0047071, 2003.

Gavala, H.N.; Yenal, U.; Skiadas, I.V.; Westermann, P.; Ahring, B.K. (2003). Mesophilic and Thermophilic Anaerobic Digestion of Primary and Secondary Sludge. Effect of Pre-treatment at Elevated Temperature. Water Research. Vol. 37, No. 19, (November 2003), pp 4561-4572, ISSN 0043-1354.

Grande, C.A.; Rodrigues, A.E (2007). Biogas to Fuel by Vacuum Pressure Swing Adsorption I. Behavior of Equilibrium and Kinetic-Based Adsorbents. Ind. Eng. Chem. Res., Vol. 46, No. 13, (May 2007), pp 4595-4605, ISSN 0888-5885.

Grande, C.A.; Cavenati, S.; Rodrigues, A.E (2008). Separation Column and Pressure Swing Adsorption Process for Gas Purification. World Patent Application WO/2008/072215.

Guerin de Montgareuil, P.; Domine, D. Process for Separating a Binary Gaseous Mixture by Adsorption. US Patent 3,155,468, 1964.

Keller, G.E.; Anderson, R.A.; Yon, C.M (1987). Handbook of Separation Process Technology. JohnWiley, New York.

Knaebel, K.S.; Reinhold, H. E (2003). Landfill Gas: From Rubbish to Resource. Adsorption Vol. 9, No. 1, (March 2003), pp 87-94, ISSN 0929-5607.

Kongshaug, K.O.; Heyn, R.H.; Fjellvag, H.; Blom, R (2007). MOF-compounds as Gas Adsorbers. World Patent WO/2007/128994, 2007.

Kuznicki, S.M. (1990). Preparation of Small-Pored Crystalline Titanium Molecular Sieve Zeolites. U.S. Patent 4,938,939. July 1990.

Llewellyn, P.L.; Bourrelly, S.; Serre, C.; Vimont, A.; Daturi, M.; Hamon, L.; De Weireld, G.; Chang, J-S; Hong, D-Y; Hwang, Y.K.; Jhung, S.W.; Ferey, G (2008). High Uptakes of CO2 and CH4 in Mesoporous Metal-Organic Frameworks MIL-100 and MIL-101. Langmuir, Vol. 24, No. 14, (July 2008), pp 7245-7250, ISSN 0743-7463.

Li, G.; Xiao, P.; Webley, P.; Zhang, J.; Singh, R.; Marshall, M. (2008). Capture of CO_2 from High Humidity Flue Gas by Vacuum Swing Adsorption with Zeolite 13 X. Adsorption, Vol. 14, No. 2-3, (June 2008), pp 415-422, ISSN 0929-5607.

Li, H.; Eddaoudi, M. O'Keeffe, O. M. Yaghi (1999). Design and Synthesis of an Exceptionally Stable and Highly Porous Metal-Organic Framework. Nature, Vol. 402, (November 1999), pp 276-279, ISSN 0028-0836.

Liu, Q.; Cheung, N.C.O.; Garcia-Bennet, A.E.; Hedin, N (2011). Aluminophosphates for CO2 Separation. ChemSUSChem, Vol. 4, No. 1, (January 2011), pp 91-97, ISSN 1864-5631.

Lopes, F.V.S.; Grande, C.A.; Rodrigues, A.E. (2011). Activated Carbon for Hydrogen Purification by Pressure Swing Adsorption. Multicomponent Breakthrough Curves and PSA Performance. Chem. Eng. Sci., Vol. 66, No. 3, (February 2011), pp 303-317, ISSN 0009-2509.

Marathe, R.P.; Mantri, K.; Srinivasan, M.P.; Farooq, S. (2004). Effect of Ion Exchange and Dehydration Temperature on the Adsorption and Diffusion of Gases in ETS-4. Ind. Eng. Chem. Res., Vol. 43, No. 17, (July 2004), pp 5281-5290, ISSN 0888-5885.

Millward, A.R.; Yaghi, O.M (2005). Metal-organic Frameworks with Exceptionally High Capacity for Storage of Carbon Dioxide at Room Temperature. J. Am. Chem. Soc. Vol. 127, No. 51, (December 2005), pp 17998-17999, ISSN 0002-7863.

Molecular Gate Adsorption Technology. 02.04.2011. Available at: www.moleculargate.com/molecular-gate-CO2-removal-nitrogen-rejection.html.

Monteiro, E.; Mantha, V.; Rouboa, A. (2011). Prospective Application of Farm Cattle Manure for Bioenergy Production in Portugal. Renewable Energy. Vol. 36, No. 2, (February 2011), pp 627-631, ISSN 0960-1481.

Mueller, U.; Lobree, L.; Hesse, M.; Yaghi, O.; Eddaoudi, M (2005). Shaped Bodies Containing Metal-Organic Frameworks. U. S. Patent 6,893,564, 2005.

Pettersson, A.; Wellinger, A (2009). Biogas Upgrading Technologies - Developments and Innovation, 02.04.2011. Available at: http://www.iea-biogas.net/

Ruthven, D.M. (1984) Principles of Adsorption and Adsorption Processes; John Wiley & Sons: New York.

Ruthven, D. M.; Farooq, S.; Knaebel, K. S. (1994) Pressure Swing Adsorption; Wiley-VCH, New York, 1994. ISBN: 0471188182.

Santos, M.P.S.; Grande, C.A.; Rodrigues, A.E (2011). Pressure Swing Adsorption for Biogas Upgrading. Effect of Recycling Streams in PSA Design. Ind. Eng. Chem. Res., Vol. 50, No. 2, (December 2010), pp 974-985, ISSN 0888-5885.

Santos, M.P.S.; Grande, C.A.; Rodrigues, A.E (2011b).New Cycle Configuration to Enhance Performance of Kinetic PSA Processes. Chem. Eng. Sci., Vol. 66, No. 8, (April 2011), pp 1590-1599, ISSN 0009-2509.

Schell, J.; Casas, N; Mazzotti, M (2009). Pre-Combustion CO_2 Capture for IGCC Plants by an Adsorption Process. Energy Procedia, Vol. 1, No. 1, (February 2009), pp 655-660, ISSN 1876-6102.

Schubert M.; Müller, U.; Hesse, M.; Diehlmann, U (2007). Process for Preparing Porous Metal-Organic Framework Materials. World Patent WO/2007/090809.

Skarstrom, C. W. Method and apparatus for fractionating gaseous mixtures by adsorption. U.S. Patent No. 2 944 627; 1960

Srinivasan, R.; Auvil, S.R.; Schork, J.M (1995) mass Transfer in Carbon Molecular Sieves- an Interpretation of Langmuir Kinetics. Chem. Eng. J. Vol. 57, No. 2, (April 1995), pp 137-144, ISSN 0923-0467.

Toreja, J.; VanNostrand, B.; Chan, N.; Dickinson, J.P (2011). Rotary-Valve, Fast-Cycle Pressure-Swing Adsorption Technology Allows West Coast Platform to Meet Tight California Specifications and Recover Stranded Gas. Laurence Reid Gas Conditioning Conference. 02.04.2011.Available at: www.xebecinc.com/pdf/Rotary-Valve-Fast-Cycle-Pressure-Swing-Adsorption-Paper.pdf

Wang, Q.M.; Shen, D.; Bülow, M.; Lau, M.L.; Deng, S.; Fitch, F.R.; Lemcoff, N.O.; Semanscin, J (2002). Metallo-organic Molecular Sieve for Gas Separation and Purification. Microp. Mesop. Mater, Vol. 55, No. 2, (September 2002), pp 217-230, ISSN 1387-1811.

Wellinger, A (2009). Gas Upgrading Issues. European Biomethane Fuel Conference. Göteborg, Sweden, September 2009. Available at: http://www.biogasmax.eu/european-conference-on-biomethane/download/

Yang, R. T (1987) . Gas Separation by Adsorption Processes Butterworths: Boston, 1987. ISBN: 0409900044.

Use of Rapeseed Straight Vegetable Oil as Fuel Produced in Small-Scale Exploitations

Grau Baquero, Bernat Esteban, Jordi-Roger Riba, Rita Puig and Antoni Rius
Escola d'Enginyeria d'Igualada, Universitat Politècnica de Catalunya
Spain

1. Introduction

The current dependence on oil in most industrial sectors and mainly in the transport sector is unsustainable neither in short nor in long term. This encourages to consider alternatives in most industrial sectors and incentivises to promote renewable energy use. In addition, the EU is promoting or even forcing the use of renewable energies in order to accomplish the commitments under the Kyoto Protocol.

In Europe the most common biofuels in transport are biodiesel and bioethanol. These biofuels are mostly obtained from large-scale plants and its production involves serious environmental and social problems as shown by several authors (Russi, 2008; Galan et al., 2009). In this scenario it is necessary to implement other biofuels currently not present in the Spanish market.

Straight vegetable oil (SVO) is a biofuel that can be small-scale produced from rapeseed planted in dry Mediterranean areas. The small-scale production presents several advantages and is more sustainable than large-scale production as cited by several authors (Baquero et al., 2010).

This chapter presents a method to produce rapeseed and process it to obtain rapeseed oil and rapeseed cake meal from a small-scale point of view. It also shows how rapeseed oil can be used as fuel in diesel engines for agriculture self-consumption. A production, processing and use-as-fuel model for rapeseed oil is also presented, analysing environmentally and economically the use of rapeseed oil as fuel compared to other agricultural production alternatives. The results are evaluated for dry Mediterranean area conditions.

2. Rapeseed production

Rapeseed is an oleaginous plant widely distributed all around the world. It has the capacity to grow and develop under temperate climate. Rapeseed is adapted to many soils, being the fertile and well-drained soils the more advantageous, as it has low tolerance to floods. The best are loamy soils, composed of clay, silt and sand. The desirable pH is from 5.5 to 7, but it also withstands some alkalinity, up to 8.3. It is resistant to periods of drought due to its deep taproot and the fibrous near-surface root system and has a good recovery after the drought (Sattell et al., 1998). An image of the rapeseed flower is shown in Figure 1.

In the studied zone the rapeseed is a dry farming plant. Thanks to its deep roots, rapeseed can gain access to subterranean water resources better than wheat and barley, grains usually grown in the area studied. The recommended field rotation for rapeseed is planting every five years in rotation with wheat (1 year) and barley (3 years). If there were strong price expectations, producers might keep rapeseed in the same field for two or even three years at the risk of the crop developing fungal diseases (Provance et al., 2000).

Fig. 1. Image of the flower and siliqua of rape (Photo J.F. Marti).

In order to select the rapeseed variety better adapted to the area of study (Anoia area in Catalonia, Spain, selected as a dry Mediterranean area) a test has been carried out in an experimental and representative field. The yield and the oil content of 9 rapeseed varieties were studied during the harvest of 2006. The experimental field was divided into 36 rectangular divisions, this is to say, 4 replicas of each one of the 9 studied rapeseed varieties were performed.

This study is still being carried out in order to average the results obtained in various years. Table 1 shows the preliminary results obtained in the harvest of 2006. The results obtained in 2008 were unusable because of the hard drought suffered in the autumn of 2007 and the winter of 2007-2008.

Variety	**Supplier**	**Average oil content (%)**	**Rapeseed yield (kg/ha)**
Bellini	Aceites Borges Pont	41.6	3636
Pacific	Limagrain Iberica	42.6	4645
Madrigal	Koipesol semillas	39.1	4525
Aviso	Aceites Borges Pont	40.5	4348
Sun	Agrusa	41.5	5251
Potomac	Limagrain Iberica	38.7	5251
Bambin	Agrusa	42.0	-
Royal	Koipesol semillas	39.5	5110
Standing	S.A. Marisa	40.3	4722

Table 1. Studied varieties of rapeseed. Average oil content and yield.

The average oil content of the 9 varieties and rapeseed yield are presented in Table 1 with an average content of humidity of 9.0% in the harvest of 2006. The analysis was carried out by applying the method described by EUETII-UPC (2006).

It should be pointed out that edge effects associated to experimental small rectangular divisions results in higher experimental yields than those found in real arable fields. From

Table 1 it seems clear that the rapeseed variety with more oil content is the Pacific, but the varieties with higher yield are Sun and Potomac. Thus, the Sun rapeseed variety maximized the rapeseed oil yield in the study of the harvest of 2006.
As a ground fertilization, the application was 450 kg/ha of a fertilizer of 15% nitrogen, 0 % phosphorus, and 15% potassium oxide. Additionally, 260 kg/ha of ammonium nitrosulphate of 27% nitrogen was spread out as a fertilizer coverage.
Before sowing, an herbicide treatment consistent in Trifluralin (48%, 2.5 l/ha), Glyphosate (36%, 1.0 l/ha) and Metazachlor (50%, 3.5 l/ha) was applied. The insecticide treatment was an application of Deltamethrin 2.5% of 0.4 l/ha.
Rapeseed agricultural production includes the use of different products (fertilizers, pesticides, herbicides, fungicides, rapeseed seed to plant) for its cultivation as long as the agricultural work done (mainly tractor work). Considering the studied region, dry farming conditions for rapeseed are taken into account. The yields in Table 1 are very high because they are obtained from an experimental study, where the edge effect and other variables increase this production value. In this study, the rapeseed yield mean value considered is 2300 kg/ha. The use of 3 kg/ha of fertilizer and 2kg/ha of herbicide are considered. In the area of study, the straw from the collected seeds is usually left in the field as fertilizer, so the straw is considered a co-product used as fertilizer for next year.

3. Rapeseed processing

The processing of the rapeseed to obtain SVO to be used as engine fuel is made through three mechanical steps: cleaning of seed, pressing and purification (see Fig. 2). The first step consists of cleaning the seeds from stones, metal pieces and straw. In this process it is very important to reduce the risk of damaging the press.

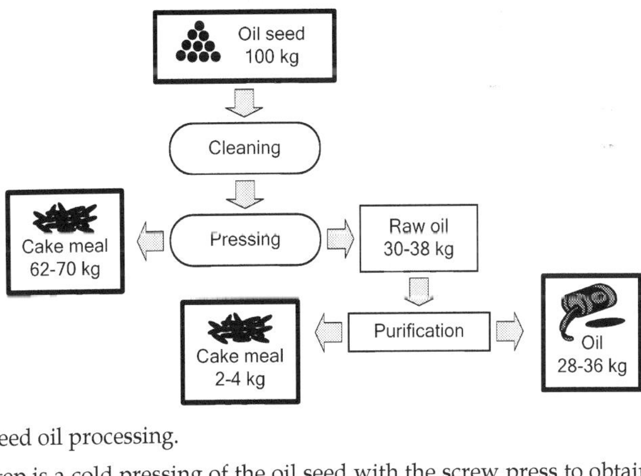

Fig. 2. Rape seed oil processing.

The second step is a cold pressing of the oil seed with the screw press to obtain oil. This step must be done carefully to reduce the incorporation of undesirable materials from the solid by-product (rapeseed cake) The pressing process influences the content of phosphorus, calcium and magnesium as well as the content and dimension of the particles. The variability of those elements depends on the speed and the pressing temperature. A low speed (low throughput) increases the oil yield and the content of particles. A high speed (high throughput), produces the opposite effect, decreasing the oil yield and also the particles. It is possible to find an optimal compromise according to the necessities of production and capacity of filtering. The oil yield should be between 32-36% of rapeseed mass, due to the amount of undesirable particles obtained in the oil if the pressure is too high or if a second pressing is done (Ferchau, 2000).

As a final step, purification of raw oil obtained from the press is needed. It is recommended to use a press filter and to perform a security filtration after a decantation. A general filtration procedure must be done after decantation in order to remove the suspended particles from the oil. Usually a pressure filter is used, either a chamber filter or a vertical one. As a final step, a security filtration of a defined pore size (between 1 and 5 μm) is recommended to remove the finest particles that still remain in the oil. In this step is very important to pass the quality control exposed in section 4.5. After this final step and after complying with the quality control, the oil is prepared for combustion in a modified diesel engine.

The cake meal and the filter cake obtained in the process to obtain SVO both have a high content of protein and are suitable for being incorporated as part of animal fodder

There is a variation of this process to extract more oil from the seed using a solvent. The abovementioned process is the first step. About 70% of oil from the seed is extracted, leaving 30% in cake meal. The next stage is a process of extraction using hexane as solvent. It reaches up to 95% extraction of the seed oil. In this stage, a solvent (hexane) is mixed with rapeseed cake. The solvent dissolves the oil remaining in the rapeseed cake. After its evaporation, the solvent is recovered for its use. The outline of the process is shown in Figure 3. In case of hexane extraction, the cake meal obtained has less protein than when just pressing the seed. Even though, there is no problem to use it as animal food.

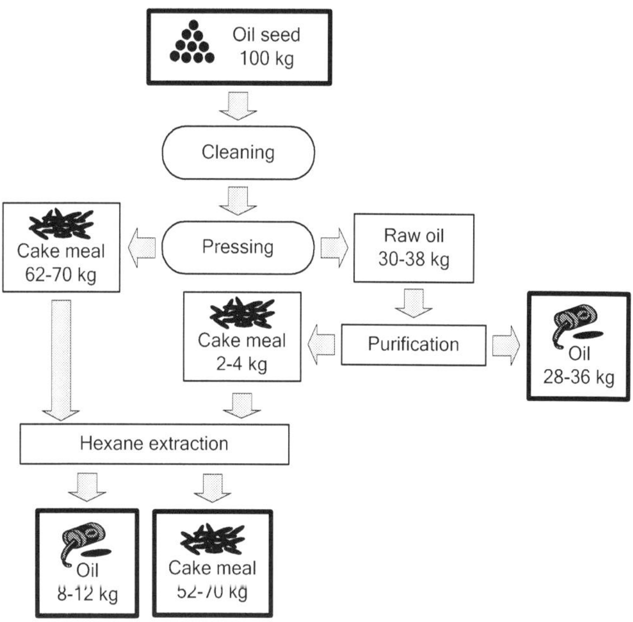

Fig. 3. Rapeseed oil hexane extraction process.

4. Use of rapeseed oil as fuel

4.1 Use of rapeseed SVO in diesel engines

Rapeseed oil can be used as fuel in diesel engines. Other vegetable oils can also be used as SVO to fuel diesel engines because they have similar properties. In Table 2 the properties of different oils are shown. The differences in the oil properties are small. However, to replace diesel fuel, some modifications are required to adjust the physical properties of the oil to be pumped to the engine and pulverized in diesel common injectors.

Fuel type		Diesel fuel	Rapeseed oil	Corn oil	Soybean oil
LHV[a] (MJ/kg)[b]		43.35	37.62	37.83	39.62
Density 20°C (kg/m³)[c]		828	915	920	920
Energy content (MJ/l)[b,c]		35.81	34.42	34.80	36.45
Viscosity (mm²/s)[c]	20°C	4.64	75.27	70.8	64.37
	80°C	1.64	12.27	11.65	11.29
Cetane number[b]		47	37.6	37.6	37.9
Flame point (°C)[b]		58	275-290	270-295	230
Chemical formula[b]		$C_{16}H_{34}$	$C_{57}H_{105}O_6$	$C_{56}H_{103}O_6$	$C_{56}H_{102}O_6$
[a]LHV: Lower Calorific Value; [b](Altin et al., 2001); [c](Riba et al., 2010)					

Table 2. Physical and chemical specifications of some vegetable oil fuels.

The modifications are aimed to heat the rapeseed oil to reduce its viscosity and density. During start-up, the vehicle runs with diesel to avoid the engine working at low temperatures with straight vegetable oil. Once the engine has warmed, it will be able to heat and use SVO. Note that the engine shouldn't be stopped for a long time when using SVO, otherwise it will be complicated to cold start the engine with SVO.

The components that need to be installed in the fuel supply system:

- an additional deposit for the start-up diesel
- a water-oil heat exchanger
- a temperature sensor
- two solenoid valves to select the fuel to be used
- filters for oil and diesel fuels

The use of vegetable oil as fuel started long ago. Rudolf Diesel used peanut oil to run a diesel engine at the World Exhibition in Paris in 1900 (Baquero et al., 2010). He also suggested that vegetable oils could be the future fuel for diesel engines, but diesel fuel from oil substituted vegetable oil due to its abundance and price.

The use of SVO in diesel engines carries also some difficulties, namely:

- difficulties in operating the motor itself because of the different ignition temperatures of the two fuels. These difficulties can be solved just by preheating the vegetable oil.

- problems of engine durability due to deposit formation in the combustion chamber and mix of the vegetable oil with the engine lubricating oil. The first problem is solved by increasing the vegetable oil temperature, so it decreases its viscosity and density, which allows a correct injection and burning of the vegetable oil. The second problem is solved by reducing the life of the engine lubricant, (Agarwal et al., 2008; Vaitilingom et al., 2008).

Despite these difficulties, it is noteworthy that both fuels have very similar energy content: 34.42 MJ/l for rapeseed SVO and 35.81 MJ/l for diesel fuel. This makes the engine performance and consumption very similar for both fuels. If we compare the performance of both fuels in the same engine, experimental results show that the performance of a vehicle running on diesel is optimal at low loads, whereas working with vegetable oil is optimal at high loads.

4.2 Oil as fuel quality control

In order to use rapeseed oil as fuel, some physical and chemical properties of the oil must be met. The description of these properties as well as its effect on the diesel engine should be taken into account. Thus, the German norm DIN 51605 is to be followed.

This norm establishes the maximum and minimum values for the parameters selected to accept a rapeseed vegetable oil as appropriate biofuel to substitute diesel in modified engines. The parameters include some intrinsic rapeseed oil properties and some which are variable and indicate if the oil has been correctly processed. Between these properties, acid value, iodine index and oxidation time are the ones which indicate the vegetable oil degradation.

4.3 Use of SVO as fuel

The authors experience in the use of a car with a modified diesel engine is described in this section. The car which engine was adapted to run with SVO is a VW Caddy 2.0 SDI using the parts described in section 4.1.

Table 3 presents the results of a test performed by the authors of this paper with the modified VW Caddy 2.0 SDI after 45000 km of trial. The consumption of this vehicle using diesel is nearly the same as with SVO, as the calorific value of both fuels are almost the same.

Own average consumption (l/100km)	7.54
Total distance (km)	45000
Fuel consumption (l)	3393
Average rate oil/total (%)	91.36%
SVO consumption (l)	3100

Table 3. SVO consumption as fuel.

From the technical data available from Volkswagen, the urban consumption for this vehicle is 7.5 l/100km, the extra-urban is 5.3 l/100km and the combined consumption is 6.1 l/100km. The test carried out with the above-mentioned 70 HP vehicle shows that maintaining an average speed of 70-80 km/h leads to an average consumption of about 6 l/100km. Driving faster, maintaining 120 km/h during long periods of the ride, leads to a consumption of about 9 l/100km.

5. Use of rapeseed cake for animal feeding

Due to its high content of protein, it is interesting to consider the use of rapeseed cake for animal feeding. The incorporation of cake meal in animal fodder is studied in many works, which support the fact that cake meal is suitable as animal fodder complement.

The introduction of rapeseed cake as part of the fodder has been largely studied. A lot of studies have been carried out and the results show that the introduction of rapeseed cake in

little proportions in the fodder (until 10-15%) entails no significant changes in parameters such as nitrogen, lipid and mineral metabolism and also for the health status of the animals (Gopfert et al., 2006). Even in cow milk, no significant differences were found in fat, protein, casein, solids and non-solids fat content in the milk from cows fed with 15% of rape cake in fodder (Simek et al., 2000). Other studies of rapeseed used in different forms (Brzoska, 2008; Kracht et al., 2004) and (Rinne et al., 1999) show no negative effects on animal neither to their meat nor the milk obtained.
Rapeseed is nowadays used as a component in the fodder of many animals. The limit proportion is not determined by law in Spain, but some recommendations have been given by the Spanish Animal Nutritional Foundation (FEDNA, 2003) for the different species and ages. In Table 4 the mean chemical composition of rapeseed meal is shown (Moss & Givens, 1994).

Crude protein (g/kg DM[a])	397
Crude fiber (g/kg DM[a])	106
Cellulose (g/kg DM[a])	177
Starch (g/kg DM[a])	45
Water-soluble carbohydrates (g/kg DM[a])	115
Gross Energy (MJ/kg DM[a])	19.7
[a]DM: Dry matter	

Table 4. Mean chemical composition of rapeseed meal.

The most representative groups of farm animals in the studied area are cattle, pigs and poultry (IDESCAT, 2008). Using the total number of animals and the characteristic intake of each species, the potential fodder demand is calculated. In Table 5 the values of fodder consumption in the Anoia region are shown for these representative groups. The proportion of cake meal in fodder was calculated using FEDNA (2003) recommendations. The cake meal yield (1500 kg_{cake}/ha) is calculated based on the yield of rapeseed in the regions -2300 kg/ha as detailed in section 2- and the amount of oil extracted through pressing -35% from rapeseed w/w as seen in section 3-.

Animal group	Bovines	Pigs	Poultries	Total
Number of livestock per year	6779	89439	607491	-
Fodder (t/year)	16321.8	62093.2	22958.4	-
Maximum cake meal in fodder (%)	17%	7%	5%	-
Maximal cake meal consumption (t/year)	2774.7	4346.5	1147.9	8269.2
Cake meal yield (kg/ha)	1500	1500	1500	-
Rapeseed land (ha)	1849.8	2897.7	765.3	5512.8

Table 5. Rapeseed land requirement.

The fodder demand in the considered region could absorb completely the amount of rapeseed cake meal produced if a tenth of the arable land (about 3000 ha) was dedicated to rapeseed production. As seen in Table 5, the amount of land requirement for rapeseed cultivation to cover the maximal cake meal consumption of the studied area is about 5500 ha.

6. Proposed cropping model and agricultural exploitation

The previous sections show the rapeseed production, the rapeseed processing to obtain oil and the use of the cake meal obtained from the seed processing. This information can be used to develop a cropping model that comprises the introduction of rapeseed to the current agricultural rotation based on wheat and barley (WBBB, where W stands for wheat and B for barley). The proposed rotation would preserve the 3 years of barley after one year of wheat in each field portion adding on year rapeseed prior to wheat (RWBBB). The introduction of rapeseed increases the two next following crop yields by 10% (wheat) and 3% (barley) for normal weather conditions. Additionally to the introduction of rapeseed to the rotation, the processing of the seed into oil and cake meal would allow its use as straight vegetable oil to fuel the exploitation tractor.

The proposed model for small-scale biofuel self consumption exploitations is graphically represented in Fig. 4, where the basis model, the rapeseed processing and the fate of the different products obtained are shown.

In order to design this model some hypotheses have to be made. First of all, small-scale producers are considered. The mean farmer is supposed to work an arable land of about 100 ha. The proposal involves using approximately 10% of the arable land for self-supply. In the studied area, as a dry Mediterranean zone, irrigated lands are nearly inexistent, being the traditional sowed crops wheat and barley. It is proposed to cultivate rapeseed as a dry crop in order to avoid putting pressure on water resources. Secondly, the system of crop-rotation jointly with direct seeding is going to be applied. Thus, rapeseed can be seeded in the same land one out of five years. Only the seeds are extracted whereas the rest of the plant is crushed while gathering the seed and left on the fields to be rot. Doing so, the soil recovers part of the nutrients contained in the straw from the plant, thus avoiding the use of some amount of fertilizer. Finally, the farmers bring the harvest to the farmer's cooperative, which is located near their lands and where there is an industrial press for extracting the oil of the rapeseed harvest.

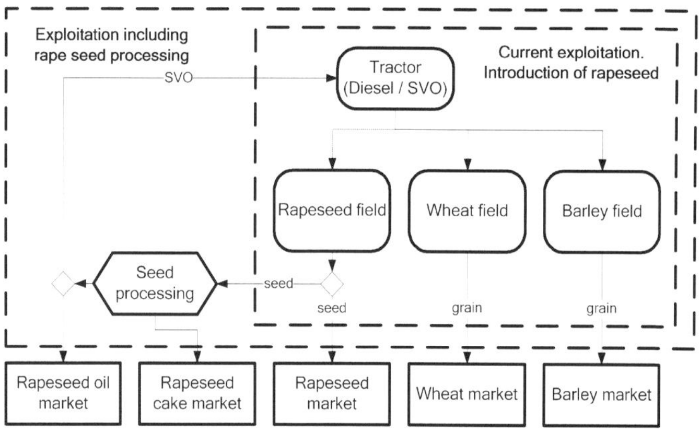

Fig. 4. Exploitation model and products fate.

Important institutions such as the Food and Agriculture Organization (FAO) of the United Nations support good agricultural practices to mitigate negative impacts, in particular on carbon, soil and water resources. Among such practices we find no tillage and direct seeding, retention of soil cover, multiple cropping, appropriate crop choice and crop rotations. There are mainly three systems of harvest namely traditional seeding, minimum cultivation and direct seeding that nowadays coexist in the studied area, being direct seeding the chosen one for its lower impact, better carbon retention in soil and reduced fuel consumption.

General assumptions are made in this model. For example, the press is assumed to extract in average 80-85% of the total oil content from the seeds. This means that after pressing, seeds are converted in a 35%of oil and a 65%of meal cake. Additionally, according to a survey answered by farmers in the Anoia area (EUETII-UPC, 2010), the average yield of the rapeseed harvest in this area is a minimum of about 2300 kg of rapeseed/ha.

Supposing a direct harvesting system of cultivation, the fuel consumption would be about 7000 l per 100 ha. As explained, the production of rapeseed SVO is supposed to be 875 l per ha. Therefore, dedicating 10% of the arable land to cultivating rapeseed is enough for self fuel supply. Also there is a small excess of SVO that could be sold for other needs. Vegetable oils can be also used in the production of additives that are useful for various industrial purposes as pointed out by (Hancsok et al., 2008). The 15000 kg of rapeseed cake per 10 ha would be used to feed the animals in this area as calculated in section 5.

7. Environmental and economic analyses

Life cycle assessment (LCA) is a methodology widely used to evaluate environmentally all kind of processes and products production (Hsu et al., 2010; Huo et al., 2009; Lardon., 2009; Schmidt, 2010). Economic assessment based on LCA methodology is also being used in literature (Lee et al., 2009; Huppes et al., 2010; Ouyang et al., 2009; Nassen., et al 2008).

7.1 Environmental analysis

As FAO indicates (FAO, 2008), a policy objective by many countries entails mitigating climate change by means of bioenergy promotion. Conversely, life-cycle analyses -which measure emissions all over the bioenergy production chain- points toward a wide divergence in carbon balances according to technologies used, locations and production paths. Thus, more research should be carried out in this field. As FAO suggests, important sources of emissions seem to be land conversion, mechanization and fertilizer use at the feedstock production stage, as well as the use of non-renewable energy in processing and transport.

To evaluate the environmental impact of the model suggested in this work, a general analysis of different topics can be done: energy and water requirements, biodegradability, equivalent CO_2 emissions (global warming), tailpipe engine emissions and deforestation. Moreover, LCA methodology (Schmidt, 2010) is used to comparatively evaluate environmental impacts.

Regarding to the use of energy, the proposed method nearly eliminates the impacts related to fuel processing and transport, which allows minimizing energy requirements. Fossil fuels, on the contrary, are transported from remote countries as well as raw materials to produce large-scale first-generation biofuels. Furthermore, both fossil fuels and first-generation biofuels need complex processing, which requires significant amounts of energy. Therefore, the proposed model reduces significantly energy consumption. Additionally, rapeseed cultivation helps crop rotation and direct seeding. This is highly recommended as it reduces the steps of land working, thus minimizing power requirements. This results in less use of fuel for each crop, which is a desirable way to reduce emissions.

As for water requirements, as (FAO., 2008) states, many feedstocks are highly water intensive, meaning that their expansion is likely to create even greater competition for this limited resource, depending upon location and production methods. The method proposed here, moves towards a dry land use, as rapeseed is able to grow in the same conditions as replaced cereals would do. On the other hand, water requirements of SVO production are null whereas as stated by (Pate et al., 2007), water requirements of bioethanol with the current technology are about 4 litres of water per litre of bioethanol produced. Consumptive water use in petroleum refining is about 1.5 l/l and biodiesel refining requires about 1 litre of fresh water per litre of biodiesel produced.

Additionally, concerning biodegradability, commonly used SVOs including rapeseed oil are biodegradable and non-toxic, making them useful for transportation applications in highly sensitive environments, such as marine ecosystems and mining enclosures for example (West et al., 2008). This implies less risk when storing the fuel and less impact to biodiversity if accidentally spoiled.

To compare the CO_2 emissions from both models, their differences have to be considered. As long as use of machinery, fertilizer and herbicide requirements are similar, the main variation between the two systems is the use of SVO instead of petrodiesel.

Emissions associated to transport, production and combustion of 1 litre of petrodiesel are 3.16 kg CO_2/l (Flessa et al., 2002). Approximately a 10% of these emissions result from the extraction, production and transport of the diesel fuel and the remaining 90% are due to its combustion. The fuel consumption for direct seeding and for traditional seeding, according to local farmers, are respectively 70 and 140 l fuel/ha. Thus, the emitted CO_2 due to tractor diesel consumption when using traditional seeding doubles the direct seeding method.

On the other hand, the CO_2 emitted when burning SVO in a diesel engine was absorbed by the crop during growth (CO_2 neutral). Consequently, these emissions are compensated by the photosynthesis absorption. SVO production is very simple and has a low energy requirement, as already seen. Thus, the CO_2 associated emissions of this stage are much lower than the ones from petrodiesel.

According to these results, the proposed system avoids the emission of more than 200 kg CO_2/ha. In future studies, a life-cycle assessment of this model will be carried out in order to take into account all the emissions in the studied area. Life-cycle analyses would measure the emissions throughout all the bioenergy production chain.

Regarding tailpipe engine emissions, diverging results are found (Krahl et al., 2007; Thuneke & Emberger, 2007). As concerns CO, CO_2 or Particulate Matter (PM) emissions, the SVO is clearly better than petrodiesel. Meanwhile, looking at NOx and HC it is not clear if the use of SVO reduces or increases its emissions. Thus, more research is needed to study this field in greater depth.

In relation to deforestation, the high demands of productive soil in large-scale production of biofuels would produce deforestation especially in tropical forests (Russi, 2008). On the other hand, the small scale production plant presented here deals with a small portion of the amount of available land to produce biofuel, thus avoiding the abovementioned impact.

In order to achieve representative results, the general framework for conducting an LCA is followed in this work (ISO 14040, 2006; ISO 14044, 2006). Taking the cropping model presented in section 6, Gabi 4 software (PE International 2010) has been used to carry out the LCA impact assessment.

The use of diesel or straight vegetable oil (SVO) as the tractor fuel is also included to take into account the consumption and the corresponding fuel emissions. Crop types are considered depending on the crop rotation chosen for each scenario. Data on crop works, fertilizing needs and yields were obtained from the Anoia region, a northeastern dry Mediterranean area in Spain.

Different cropping schemes are studied fixing the functional unit in 10^9kcal of energy produced, because this is the energy obtained from approximately 100ha of land. Direct cropping technique is assumed. An energy functional unit is the most suitable to evaluate a system where different outputs are found, namely barley grain, wheat grain, rapeseed seed, cake and oil.

The system boundary includes an agricultural exploitation where different crop types are considered. The fate of the obtained products is not considered, only the energy that each obtained product represents. The boundaries comprehend (i) materials inputs which take into account fertilizers, herbicides, insecticides, fungicides, diesel fuel and planting seeds, (ii) cropping stages including fertilizing, herbicide, insecticide and fungicide treatments, sowing, harvesting and seed/grain transportation to cooperative installations, and (iii) rapeseed processing stage which includes transportation, pressing, filtering and degumming processes.

Three scenarios are considered for this environmental assessment, based on grouping three crop types, namely barley, wheat and rapeseed. Barley, wheat and rapeseed models consist on the production of the grain and seed. Additionally, rapeseed model incorporates the seed processing, to obtain rapeseed oil that can be used as biofuel (SVO) in the exploitation. The use of SVO as fuel is also considered in one scenario. Thus, the first scenario is the current exploitation method (**current scenario**). The second incorporates rapeseed into crop rotation but uses only diesel fuel (**diesel scenario**). The third additionally includes rapeseed processing and SVO fuel use (**SVO scenario**).

Emissions of the considered model are aggregated into impact categories according to an international accepted method in the impact assessment phase. CML method from the Environmental Sciences Institute of Leiden University is the method chosen in this study, because it is the one which generates more international consensus and avoids subjectivity (Guinée et al., 2001; Alvaro-Fuentes et al., 2009). It is a cause-effect method that limits the uncertainty in groups according to impact categories (Dreyer et al., 2003). It calculates the increase of damage and quantifies its effects (Garraín, 2009).

Fig. 5 shows the environmental impact category results using 6 CML non-toxicological impact categories, energy consumption and land use for each scenario taking the current one as a basis. The introduction of rapeseed in the classical rotation and its use to produce SVO for fuel self-consumption slightly lessens some of the environmental impacts considered. Crop energy ratio indicator shows a preference for SVO fuelled scenarios, being the ratio 21.6% superior for SVO scenario compared to the current and the diesel seed one. Adverse environmental impacts to SVO scenario (ODP and POCP) are just 8.5% and 9.8% worse than reference scenario. A slight land requirement increase in both diesel and SVO scenarios is obtained, but not much representative, being lower that 1.7%. Favourable environmental impacts to SVO scenario (ADP and GWP) are down to 21.2% and 6.2% lower than reference scenario. AP and EP are 2.7% and 1.9% lower than reference scenario, not being much representative. On the other hand, diesel scenario impacts compared to reference scenario varies less than 4%, being practically the same. The higher diesel scenario impact is global warming potential (GWP), which increases 3.7% whereas in SVO scenario lessens 6.2%. Sensitivity analysis carried out show practically no variation in tendency, however, the impact of the electrical energy use in SVO scenario can be reduced if renewable electrical energy is used.

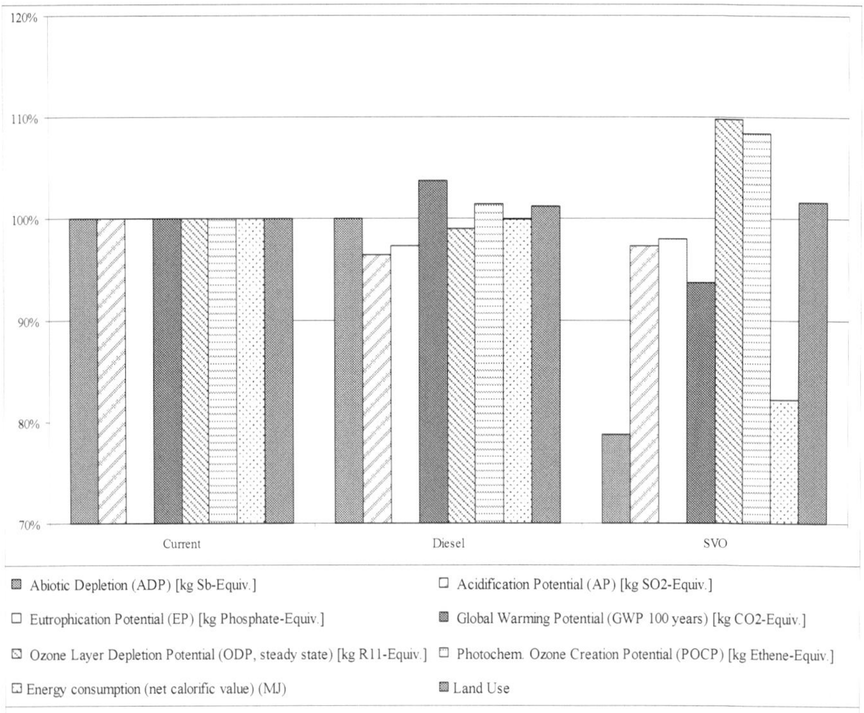

Fig. 5. Environmental impact category results

7.2 Economic analysis

Following the same model as in the environmental analysis, an economic assessment is done to evaluate the different cropping alternatives considered for the region. Thus, a comparative economic result can be achieved. Life cycle costing (LCC) methodology enables the compilation and evaluation of the inputs, outputs and the potential economic benefit of a product system throughout its life cycle (Lee et al., 2009). There are different LCC approaches, depending on their target, the costs involved and the context of the LCC itself. LCC is used in many fields, such as building techniques and rebuilding (Nassen et al., 2008; Ouyang et al., 2009) and also military equipment (Huppes et al., 2004). However as a product oriented method is hardly used for agricultural processes.

Costs include investments linearly distributed during the years of use, as this study pretends to give a mean economic benefit for an exploitation period of a year (a complete season). The benefit is shown as a representative parameter of the viability of each scenario.

Fig. 6 shows a simple diagram of the rapeseed processing and co-products use. This figure also shows schematically the production of rapeseed with the different inputs and its processing. The same input scheme is applied to wheat and barley.

Following the LCA-based LCC methodology, this analysis takes into account all the process stages. The benefit calculation is developed by modelling each crop type as well as the

rapeseed processing stage. Each crop type requires its particular fertilization and crop protection products. Rapeseed processing stage is only considered when transformation of seed is required (SVO scenario). The use of diesel or SVO in the tractor is also considered to take the consumption and the corresponding fuel emissions into account.
Each scenario is obtained by the combination of different crop partitions, each one with its own conditions, as already explained. The different partitions are Barley, Barley-2, Fallow, Rapeseed, Wheat and Wheat-1; where Barley-2 and Wheat-1 stand for barley 2 years after rapeseed and wheat 1 year after rapeseed. The benefits obtained in each scenario are shown in Fig. 7.

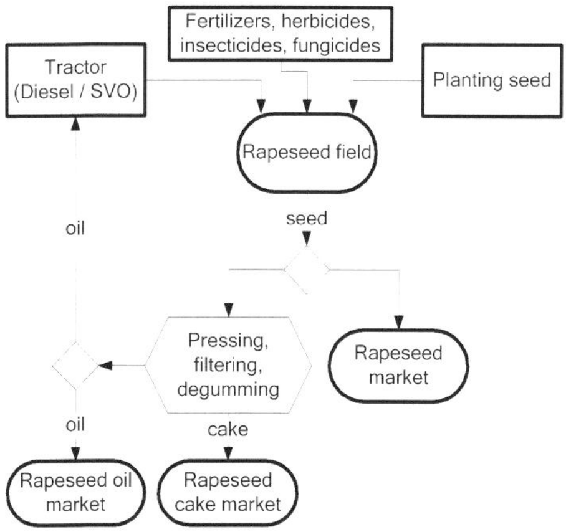

Fig. 6. Rapeseed production scheme

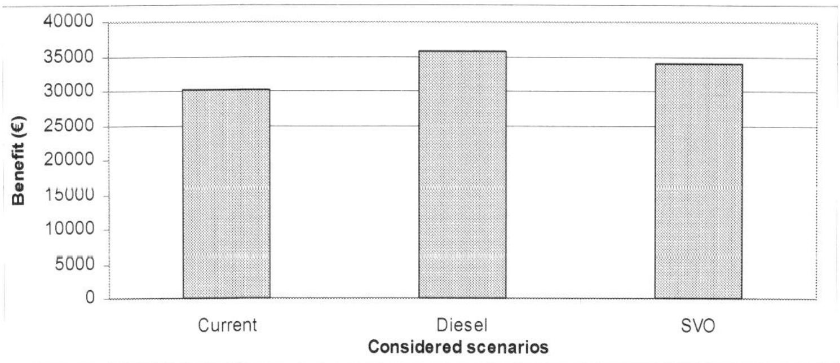

Fig. 7. Economic benefit results for each scenario

Contributions to the benefit are higher or lower according the proportion of each crop partition. Rapeseed -when not processed- gives a higher benefit per ha than the other crop types. It is clear from the results that small scale processing of the rapeseed to sell the oil is not as economically feasible as direct rapeseed sell.
Using diesel or SVO as fuel options have been analyzed, performing a diesel price and taxes sensitivity analysis. Results of these analyses show that the Spanish granted diesel for

agriculture is preventing SVO to be introduced as a fuel for agricultural exploitations. Current policies do not support specifically self-supply fuels for agriculture, thus being unable to compete with already implemented diesel exploitations.

It is clear that in the current economic conditions, applying crop rotation RWBBB with diesel as fuel (diesel seed scenario) is currently the best option. Very close to this option is the SVO seed scenario, which uses the same rotation scheme but destines part of the seed harvested to produce fuel for the agricultural machinery. Thus, it reduces the amount of diesel used in the exploitation. Diesel and SVO scenarios benefit is 15% and 11% respectively higher than reference scenario (current scenario).

8. Conclusions

This chapter explains the production and use of rapeseed oil as self-produced agricultural biofuel and analyzes its use in the study area. It also evaluates from an environmental and economic point of view the presented model.

The first three sections show the small-scale production technology to obtain rapeseed oil and analyse the best rapeseed variety for a specific zone. Similar studies are necessary to analyse the most appropriate variety for each region under study. The fourth section is a summary of the necessary modifications in diesel engines to work with straight vegetable oil, as long as showing real consumption data from an adapted vehicle. It also shows the use of rapeseed cake as animal food. In section 6 an exploitation model is presented, introducing rapeseed to the traditional crop rotation of wheat and barley and incorporating the seed processing and oil use. Section 7 and 8 show the environmental and economic results of the proposed model compared to the traditional rotation and the sole use of diesel in the proposed rotation.

In the proposed exploitation model, all co-products obtained from the rapeseed plant processing (straw, rapeseed cake, oil and seed) have a clear target (field, animal feed, biofuel and seeds market) and a defined market price. Thus, no waste products are generated. Furthermore, the SVO obtaining process is more sustainable than biodiesel production thanks to its lower energy consumption and the avoidance of chemicals use like methanol.

Results for SVO and diesel fuel use in the proposed rotation with rapeseed are compared to current rotation. Life cycle assessment show the environmental impact category results using 6 CML non-toxicological impact categories. The environmental evaluation shows the preference for SVO in most categories, however some others show adverse results. The implementation of this exploitation model should take the latter into account to minimize them. On the other hand, economic feasibility is not clear in the current economic context. However, it might be feasible in future scenarios where the access to fossil fuels was limited. Moreover, small-scale production and consumption of SVO can revitalize rural economies and help them being less dependent on diesel fuel. Furthermore, this model can also be useful in less developed countries, where diesel fuel might be scarce or difficult to obtain. Additionally, research in fields such as crop sustainability, crop emissions and new varieties of plants, diesel engines modifications and new type of lubricants among others is promoted. More research is especially needed in the sustainability assessment of the proposed model along the whole life cycle.

Thus, the use of SVO in diesel engines is a real possibility that can be taken into account when considering small-scale biofuel production.

9. References

Agarwal, D. & Agarwal, A. K. (2007). Performance and emissions characteristics of Jatropha oil (preheated and blends) in a direct injection compression ignition engine. Appl. Therm. Eng., Vol.27, No.13, (September 2007) pp. 2314-2323, ISSN 1359-4311

Altin, R.; Cetinkaya, S. & Yucesu, H. S. (2001). The potential of using vegetable oil fuels as fuel for diesel engines. Energy Conversion and Management, Vol.42, No.5, (March 2001) pp. 529-538, ISSN 0196-8904

Alvaro-Fuentes, J.; Lopez, M. V.; Arrue, J. L.; Moret, D. & Paustian, K. (2009). Tillage and cropping effects on soil organic carbon in Mediterranean semiarid agroecosystems: Testing the Century model. Agriculture Ecosystems & Environment, Vol.134, No.3-4, (2009) pp. 211-217, ISSN 0167-8809

Baquero, G.; Esteban, B.; Rius, A.; Riba, JR. & Puig, R. (2010). Small-scale production of straight vegetable oil from rapeseed and its use as biofuel in the Spanish territory. Energy Polic, Vol.38, No.1, (2009) pp. 189-196, ISSN 0301-4215

Brzoska, F. (2008). Milk production and composition as influenced by soybean meal, rapeseed meal or rapeseed cake in concentrates for dairy cows. Annals of Animal Science, Vol.8, No.2, (March 2008) pp. 133-143, ISSN 1642-3402

Dreyer, L. C.; Niemann, A. L. & Hauschild, M. Z. (2003). Comparison of three different LCIA methods: EDIP97, CML2001 and Eco-indicator 99 -Does it matter which one you choose? International Journal of Life Cycle Assessment, Vol.8, No.4, (2003) pp. 191-200, ISSN 0948-3349

Esteban, B.; Baquero, G.; Puig, R.; Riba J-R. & Rius A. (2011) Is it environmentally advantageous to use vegetable oil directly as biofuel instead of converting it to biodiesel? Biomass & Bioenergy, Vol.35, No.3, (March 2011), pp. 1317-1328, ISSN 0961-9534

EUETII-UPC. PNT GRA 105, Normalised working procedure: Determination of the total amount of oil extractable from rape seed. (2006) Escola d'Enginyeria d'Igualada. Igualada, Spain

EUETII-UPC. Survey conducted between 2009-2010 in the Anoia region, to the farmers of Unio de Pagesos and Coperal agriculture cooperatives. (2010) Escola d'Enginyeria d'Igualada. Igualada, Spain

FAO. Reports on international commodity prices. Food and Agricultures Organization of the United Nations 2.03.2011, Available from: Available from: www.fao.org/es/esc/prices/WeeklyPricesServlet.jsp

FEDNA. (2003) Fundación española para el desarrollo de la nutrición animal, Tables of composition for animal food and quality control, In Fundación Española el Desarrollo de la Nutrición Animal. 2.03.2011, Available from: http://www1.etsia.upm.es/fedna/introtabla.htm

Ferchau, E., (2000) Equipment for decentralised cold pressing of oil seeds, In: Folkecenter for Renewable Energy, 2.03.2011, Available from: http://www.folkecenter.net/mediafiles/folkecenter/pdf/dk/efdcpos_ef.pdf

Flessa, H.; Ruser, R.; Dorsch, P.; Kamp, T.; Jimenez, M. A.; Munch, J. C. & Beese, F. (2002). Integrated evaluation of greenhouse gas emissions (CO2,CH4, N2O) from two farming systems in southern Germany. Agriculture Ecosystems & Environment, Vol.91, No.1-3, (2002) pp. 175-189, ISSN 0167-8809

Galan, M. I.; Bonet, J.; Sire, R.; Reneaume, & J. M.; Plesu, A. E. (2009). From residual to useful oil: Revalorization of glycerine from the biodiesel synthesis. Bioresour. Technol, Vol.100, No.15, (August 2009), pp. 3775-3778, ISS 0960-8524

Garraín, D. (2009) Desarrollo y aplicación de las categorías de impacto ambiental de ruido y de uso de suelo en la metodología de análisis de ciclo de vida. In Universitat Jaume I. 2.03.2011, Available from: http://www.tdr.cesca.es/TESIS_UJI/AVAILABLE/TDX-0630109-01132//garrain.pdf

Gopfert, E.; Trckova, M. & Dvorak, R. (2006). The use of treated rape cake in a calf starter diet, Czech J of Anim Sci, Vol.51, No.11, (August 2006), pp. 491-501, ISSN 1212-1819

Guinée, J. B.; Gorrée, M.; Heijungs, R.; Huppes, G.; Kleijn, R.; Koning, A. d.; Oers, L. v.; Sleeswijk, A. W.; Suh, S.; Haes, H. A. U. d.; Bruijn, H. d.; Duin, R. v. & Huijbregts, M. A. J. (2001) Life cycle assessment: an operational guide to the ISO standards, Center of Enviromental Science- Leiden University 2.03.2011, Available from: http://media.leidenuniv.nl/legacy/new-dutch-lca-guide-part-1.pdf

Hancsok, J.; Bubalik, M.; Beck, A. & Baladincz, J. (2008). Development of multifunctional additives based on vegetable oils for high quality diesel and biodiesel. Chemical Engineering Research & Design, Vol.86, No.7A, (September 2007) pp. 793-799, ISSN 0263-8762

Hsu, D. D.; Inman, D.; Heath, G. A.; Wolfrum, E. J.; Mann, M. K. & Aden, A. (2010) Life Cycle Environmental Impacts of Selected US Ethanol Production and Use Pathways in 2022. Environmental Science & Technology, Vol.44, No.13, (June 2010) pp. 5289-5297, ISSN 0013-936X

Huo, H.; Wang, M.; Bloyd, C. & Putsche, V. (2008). Life-Cycle Assessment of Energy Use and Greenhouse Gas Emissions of Soybean-Derived Biodiesel and Renewable Fuels. Environmental Science &Technology, Vol.43, No.3, (February 2009) pp. 750-756, ISSN 0013-936X

Huppes, G.; Rooijen, M. v.; Kleijn, R.; Heijungs, R.; Koning, A. d. & Oers, L. (2004) ; Life Cycle Costing and the Environment. Report of a project commissioned by the Ministry of VROM-DGM for the RIVM Expertise Centre LCA. 2.03.2011, Available from: http://www.rivm.nl/milieuportaal/images/Report%20LCC%20April%20%202004%20final.pdf

ISO 14040: Environmental management -- Life cycle assessment -- Principles and framework. (2006) International Organization for Standardization Geneve, Switzerland

ISO 14044: Environmental Manangement—Life Cycle Assessment—Requirements and guidelines. (2006) International Organization for Standardization Geneve, Switzerland

IDESCAT database, Agricultural census, 1999. Institut d'Estadística de Catalunya, 2.03.2011, Available from: http://www.idescat.cat/

Kracht, W.; Danicke, S.; Kluge, H.; Keller, K.; Matzke, W.; Hennig, U. & Schumann, W. (2004). Effect of dehulling of rapeseed on feed value and nutrient digestibility of rape products in pigs. Archives of Animal Nutrition, Vol.58, No.5, (August 2006) pp. 389-404, ISSN 0003-942

Krahl, J.; Munack, A. & Bünger, J. (2007) Emissions of Biodiesel and Vegetable Oil, In: UFOP, 2.03.2011, Available from: http://www.ufop.de/downloads/report.pdf

Lardon, L.; Helias, A.; Sialve, B.; Stayer, J. P. & Bernard, O. (2009). Life-Cycle Assessment of Biodiesel Production from Microalgae. Environmental Science & Technology, Vol.43, No.17, (2009) pp. 6475-6481, ISSN 0013-936X

Lee, J. Y.; Yoo, M.; Cha, K.; Lim, T. W. & Hur, T. (2009). Life cycle cost analysis to examine the economical feasibility of hydrogen as an alternative fuel. International Journal of Hydrogen Energy, Vol.34, No.10, (2009) pp.4243-4255, ISSN 0360-3199

Moss, A. R. & Givens, D. I. (1994). The Chemical-Composition, Digestibility, Metabolizable Energy Content and Nitrogen Degradability of Some Protein-Concentrates. Animal Feed Science and Technology, Vol.47, No.3-4, (June 1994) pp. 335-351, ISSN 0377-8401

Nassen, J.; Sprei, F. & Holmberg, J. (2008). Stagnating energy efficiency in the Swedish building sector-Economic and organisational explanations. Energy Policy, Vol.36, No.10, (2008) pp. 3814-3822, ISSN 0301-4215

Ouyang, J. L.; Ge, J. & Hokao, K. (2009). Economic analysis of energy-saving renovation measures for urban existing residential buildings in China based on thermal simulation and site investigation. Energy Policy, Vol.37, No.1, (2009) pp. 140-149, ISSN 0301-4215

Pate, R., M. Hightower, C. Cameron, & W. Einfeld, (2007). Overview of Energy-Water Interdependencies and the Emerging Energy Demands on Water Resources, Report SAND 2007-1349C, 2.03.2011, Available from:http://www.sandia.gov/energy-water/

PE_International (2010) In Institut für Kunststoffprüfung und Kunststoffkunde; GMBH, P. E., Ed. Stuttgart, Echterdingen, 2010.

Provance, P.; Suzanne, M.; Crutchifield, J. & Sandene, P. (March 2000). World Rapeseed Production in 2000/01 Likely to Fall Below Record 1999/2000 Levels, In: Feature commodotity articles, 2.03.2011, Available from:
http://ffas.usda.gov/wap/circular/2000/00-03/wap2.htm

Riba, J.; Esteban, B.; Baquero, G.; Puig, R.; & Rius, A. (2010) Characterization of physical properties of vegetable oils to be used as fuel in diesel engines. Afinidad Vol.67, No.546, (2010), pp. 100-106,

Rinne, M.; Jaakkola, S.; Varvikko, T. & Huhtanen, P. (1999). Effects of type and amount of rapeseed feed on milk production. Acta Agriculturae Scandinavica Section a-Animal Science, Vol.49, No.3, (September 1999), pp. 137-148, ISSN 0906-4702

Russi, D. (2008). An integrated assessment of a large-scale biodiesel production in Italy: Killing several birds with one stone? Energy Policy, Vol.36, No.3, (March 2008), pp. 1169-1180, ISS 0301-4215

Sattell, R.; Dick, R.; Ingham, R.; Karow, R.; Kaufman, D. & McGrath, D. (January 1998) Rapeseed (Brassica campestris/Brassica napus), In: 2.03.2011, Available from: http://extension.oregonstate.edu/catalog/html/em/em8700

Schmidt, J. H. (2010). Comparative life cycle assessment of rapeseed oil and palm oil. International Journal of Life Cycle Assessment, Vol.15, No.2, (2010), pp. 183-197, ISSN 0948-3349

Simek, M.; Sustala, M.; Vrzalova, D. & Trinacty, J. (2000). The effect of rape cakes in feed mixtures on the content parameters of cow milk. Czech Journal of Animal Science, Vol.45, No.4, (April 2000) pp. 161-167, ISSN 1212-1819

Thuneke, K. & Emberger, P. (2007). Exhaust gas emission characteristics of rapeseed oil fuelled tractors - Investigations at a test stand. Conference: Agricultural Engineering 2007, Vol.2001, (2007) pp. 47-52, ISSN 0083-5560

Vaitilingom, G.; Perilhon, C.; Liennard, A. & Gandon, M. (1998). Development of rape seed oil burners for drying and heating. Industrial Crops and Products, Vol.7, No.2-3, (January 1998), pp. 273-279, ISS 0926-6690

West, A. H.; Posarac, D. & Ellis, N. (2008). Assessment of four biodiesel production processes using HYSYS. Plant. Bioresour. Technol., Vol.99, No.14, (2008), pp. 6587-6601, ISSN 0960-8524

5

Nanotech Biofuels and Fuel Additives

Sergio C. Trindade
SE[2]T International, Ltd. and International Fuel Technology, Inc.
USA

1. Introduction

This chapter was inspired by an invited presentation of the author at the Chemindix conference in Bahrain in October 2010. This was the 8th. International Conference & Exhibition on Chemistry in Industry, promoted by the Saudi Chapter of the American Chemical Society and Aramco. The focus is on reviewing the application of nanotechnology to biofuels production and to the utilization of fuel additives, some of which are derived from renewable materials.

To introduce the topic, the broader context of petroleum fuels and biofuels is presented. A smart future of oil refining would be to increasingly utilize margins to finance a transition away from oil towards future alternative providers of mobility, in particular biofuels.

Future scenarios of liquid biofuels involve the market penetration of second and further generations of technologies and the continuous improvement of first generation processes. On the other hand, nanotechnologies are among the candidate technologies for the biofuels of the future. The nanotechnology field is vast and its applications unbound.

This is followed by a brief review of nanotechnology developments, especially as they apply to liquid particles, beyond the more common solid particle applications.

Algae growth, harvesting and conversion are presented and discussed, given the immense potential of their contribution towards an energy future where biofuels play a significant role.

Most of the current effort in second generation conversion to liquid biofuels is based on biomass cellulosics to ethanol and biodiesel. Nano processes are being pursued and will be reviewed in the chapter.

Likewise, the presently used processes to convert oils and animal fat into biodiesel are based on trans-esterification with methanol or ethanol, which inevitably generates glycerol, which must find a market or get disposed properly. Nano processes may be useful in addressing this issue.

Speculative considerations are made about the role of liquid nanoparticles of fuel additives in enhancing the performance of additized biofuel/fuel blends, in connection with surface and combustion effects.

Public concerns over the impacts of nanotechnologies on security, health and the environment are also mentioned and discussed. But, a cautionary optimistic view is presented on the huge benefits of a careful penetration of nanotechnologies in the realm of biofuels and fuel additives, and in many more applications, especially those dealing with human health.

2. The future of oil refining [1] and the oil transition to alternatives [2]

Crucial challenges to the oil industry are evolving, as the demand for energy services (mobility, lighting, rotary movement, heating and cooling) increases, which with the current technology setup, is translated into expanding demand for liquid fossil fuels. Oil producers and refiners face difficulties finding sufficient good quality crude oil in adequate amounts and reasonable costs to meet growing demand over the long run, while users and the public at large are pushing for environmental improvements, such as better air quality in the immediate future. Moreover, concerns about the impacts of climate change caused by increased greenhouse gases emissions from the production and use of liquid hydrocarbons may eventually force a transition to climate friendly energy services providing systems. This offers biofuels a market penetration opportunity in the transition towards a yet undefined new energy future.

Under this background, the oil refining industry in the USA and the European Union has been stagnant. It has been immobilized by environmental obstacles posed by an articulated public, augmented by a "not in my backyard" attitude that makes it difficult to build new refineries. In addition, declining margins for refined products may have led major players to focus more on the upstream.

On the other hand, refining is expanding in other parts of the world, such as India, China, Brazil and the Middle East, as these countries develop and the oil producers attempt to add more value to their resources. An evidence of the shift of refining towards developing economies is the fact that the largest refinery in the world is in India and belongs to Reliance.

But, all over the world, the increase in the long-term marginal cost of oil combined with environmental pressures and stricter government regulations and mandates are likely to lead to the decline of the centrality of oil in the global energy mix in favor of natural gas. This shift in dominance happened to wood and coal over the past two centuries and is now happening to oil. Oil companies are increasingly calling themselves energy companies. Some of them will leverage their current oil production margins to make a smooth transition to alternatives over time. A profitable transition to alternatives in the oil economy would require a gradual transfer of oil profits into green investments and the stretching of current oil supplies.

3. Liquid biofuels issues [3]

The enormous global daily consumption of liquid fuels is of the order of 80 million barrels/day (e equivalent of 12.7 million m^3/day). The sugar cane area required to produce the same volume of ethanol is about 700 million hectares, assuming a yield of 6.5 m^3/ha/year of ethanol. This area is equivalent to 100 times the sugar cane cultivated area in Brazil, the second largest bio-ethanol producer in the world. Biofuels definitely face an issue of scale. In 2010, fuel ethanol and biodiesel combined displaced a mere 3% of oil in the world.

Figure 1[4] below illustrates the scale issue by showing how much land it would take for the USA to grow its own fuel.

It appears that algae require the least area to meet the large scale demand of liquid fuels in the USA, whereas the area required by soybeans is larger than the USA's 48 continental states. The area required by corn is substantial. This suggests that the current biofuels production base of the USA would not be able to meet demand, and imports would be required to meet the colossal American energy appetite.

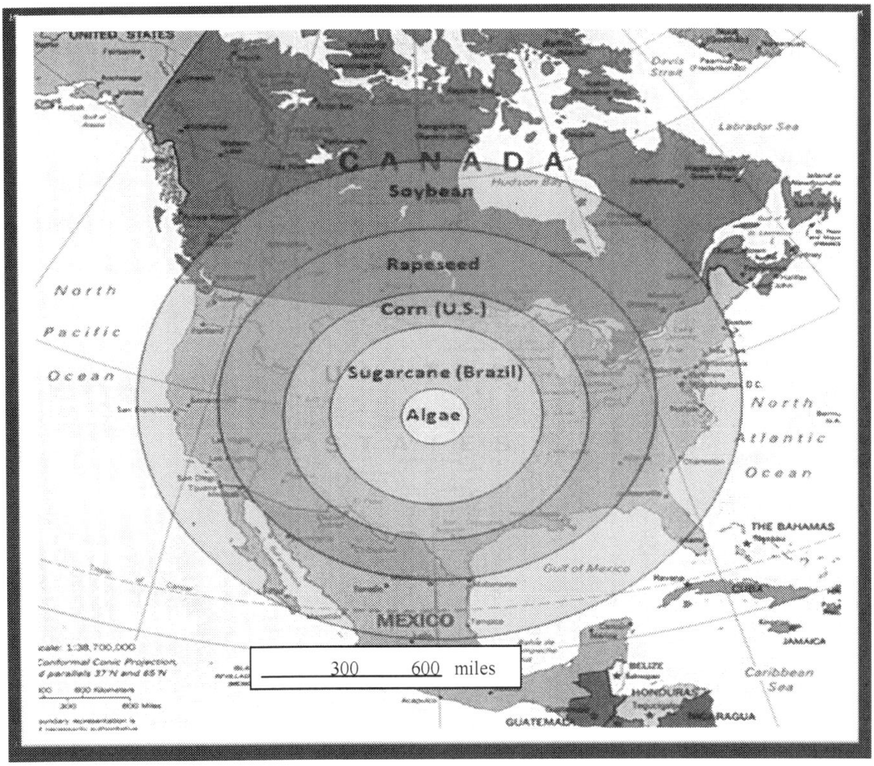

Fig. 1. How much land would it take for the USA to grow its own fuel?

The scale challenge posed to biofuels relates to the labour, management, land, water and sunshine required to produce the biomass and the processing that originates them. These are scarce resources that are also needed to grow food, feed and fibre to ultimately meet various human demands. These are resources that have an opportunity cost from competing markets. To develop biofuels in the scale of commercial liquid fuels require massive financing, a resource that may have alternative uses as well. The mobilization of private capital under a perception of market and other uncertainties is another issue that biofuels have to resolve in order to thrive.

The production of biofuels is accompanied by local environmental issues that need addressing. For instance, in the case of sugar cane ethanol, stillage the liquid residue of distillation, has a high chemical and biological oxygen demand and requires appropriate processing before final disposal. From a global climate change perspective, designed and managed properly, a biofuels production system would add minimally to greenhouse gas emissions. But, in practice, many biofuel production systems in the world are contributing net GHG emissions.

A bone of contention in the development of the biofuels industry is the present competition for feedstocks between the food and fuel industries. In the case of biodiesel, all commercial vegetable oils that are used in preparing food are also convertible to biodiesel. A similar situation exists with respect to fuel ethanol, especially for the starch-based feedstocks (corn and wheat). However, the hike in food prices that happened globally in 2007/8 and is happening in 2010/1 derive mostly from other causes such as droughts and other climate related phenomena, higher oil prices and market speculation.

Since the cost of biofuels is dominated by feedstocks cost, access to feedstocks in the required amounts, timing and at adequate prices is key to the success of the biofuels economy. The combination of the food versus fuel conundrum with the need to have reliable and economic access to feedstocks is shifting the industry towards non-food feedstocks and to the market penetration of second generation technologies to convert cellulosic biomass into liquid biofuels.

Concern in important consuming markets about the sustainability of biofuels producing systems is putting pressure on suppliers to abide by sustainability protocols subject to certification. The sustainability of biofuels is actually linked to freer international trade, which would tend to phase out unsustainably produced biofuels in favour of regions of the world that can meet sustainable production requirements. A valuable discussion on this matter was hosted by the Rockefeller Foundation in 2008 at its Bellagio Centre and produced a sustainable biofuels consensus. The objective was to understand the many drivers for sustainable trade, consumption and production of biofuels, and the comparative advantage of supplying regions combined with demand and technology from consuming regions [5].

However, much remains to be done to achieve free international trade of biofuels. The World Trade Organization Doha rounds have reached an impasse. Currently, biodiesel is considered an industrial product, whereas fuel ethanol is categorized as an agricultural product, which allows more protectionism. What is needed is a unified treatment of biofuels, where they are classified under environmental goods and services. But, irrespective of these drawbacks, a sign pointing to a larger role for biofuels in the future are the new biofuels technology initiatives by large oil companies, such as BP, Chevron, Exxon and Shell. The development of the international trade in biofuels is likely to distribute more evenly the production and consumption of biofuels in the world. For the time being, biofuels production is overwhelmingly concentrated in the USA, Brazil and the European Union, as shown in Fig. 2 below[6].

4. The vastness of nanotechnology

Nanotechnology can be simply defined as the discipline of building machines/devices on the scale of molecules, a few Nanometers (10^{-9}m) wide, way smaller than a cell. Table 1 below show some practical applications of nanotechnologies and confirms the vastness of their domain [7]:

In the practically important area of polymers, nanotechnologies originate nano-structured polymers, where applications can be found in support structures; manufacturing processes; diagnostics and therapy; pharmaceuticals; medical and dental prosthesis; and thin films for surface treatment. The main chemicals involved in nano-structured polymers are: poly-oxides; poly-acrylates; poly-vinylics; poly-saccharides; and poly-ethylenes. The main materials incorporated into polymer nano-matrices are silicon, chromium and carbon[8]

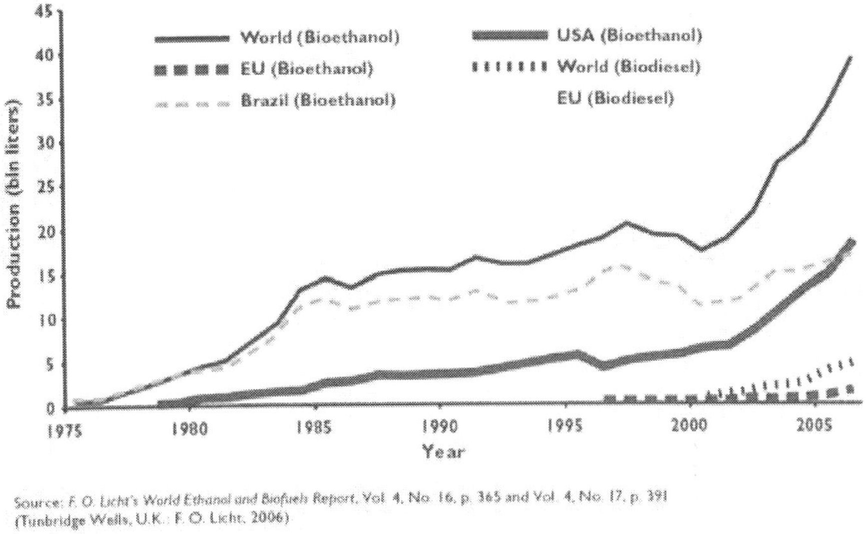

Fig. 2. World's biofuels production is concentrated in the US, Brazil and the EU

Nanocrystalline Drugs	Oxide Nanoparticles
Nanofilm Coatings	Carbon Nanotubes
Cosmetics	Silicon Nanomaterials
Nanocatalysts	Metal nanoparticles
Ceramics	Polymer Nanocomposites
Rubbers	Inorganic Nanocrystals

Table 1. Some applications of nanotechnology

5. Turning algae into biofuels

As shown in Fig. 1, biofuels derived from algae offer a great potential in view of the possible high yields and smaller area requirements. In addition, algae can play a role in carbon mitigation, as one way of growing algae is to feed them carbon-dioxide (CO_2), besides water and sunlight. Algae can be fed other substrates as well, because to grow, cost-effectively, on carbon dioxide there would be a need of concentrated sources of the gas, such as found in combustion off-gases from fossil fueled power plants.

Oil can form up to 50% of the algae mass, in contrast with the best oil-bearing plants - oil palm trees - where less than 20% of the biomass is made out of foil. Algae carbohydrates can also be made into ethanol or gasified into bio-gas, or methane or hydrogen [9].

But, algae development into biofuels must overcome a number of challenges before algae can become significant sources of commercial biofuels. Since algae also need water to grow, expansion of algae production may create a dilemma of water versus fuel, similar to food versus fuel dilemma discussed previously. Another challenge is the low natural carbon dioxide concentration in the atmosphere, hence the consideration of additional sources of carbon for algal growth in a commercial biofuels system. One response to these challenges may include the use of nanotechnology to turn algae into biofuels.

As way of examples, in 2009, the company QuantumSphere received a grant from the California Energy Commission to develop a nano-catalyzed algae biogasification. Also in California, the Salton Sea receives large amounts of agricultural runoff, which sometimes create large algae blooms. These algae and similar biomass have been turned experimentally into methane, hydrogen and other gases [10].

One nanotechnology relevant to algae development is the use of nanoparticles as no-harm harvesters of biofuel oils from algae, as illustrated in Fig. 3 [11]. The nano particles are shown on the left hand side of the photograph before the oil pregnant algae are added. The right hand side shows the contacting between the algae and the nano particles, which results in extracting the oil without harming the algae. Maintaining the algae alive can dramatically reduce production costs and the generation cycle.

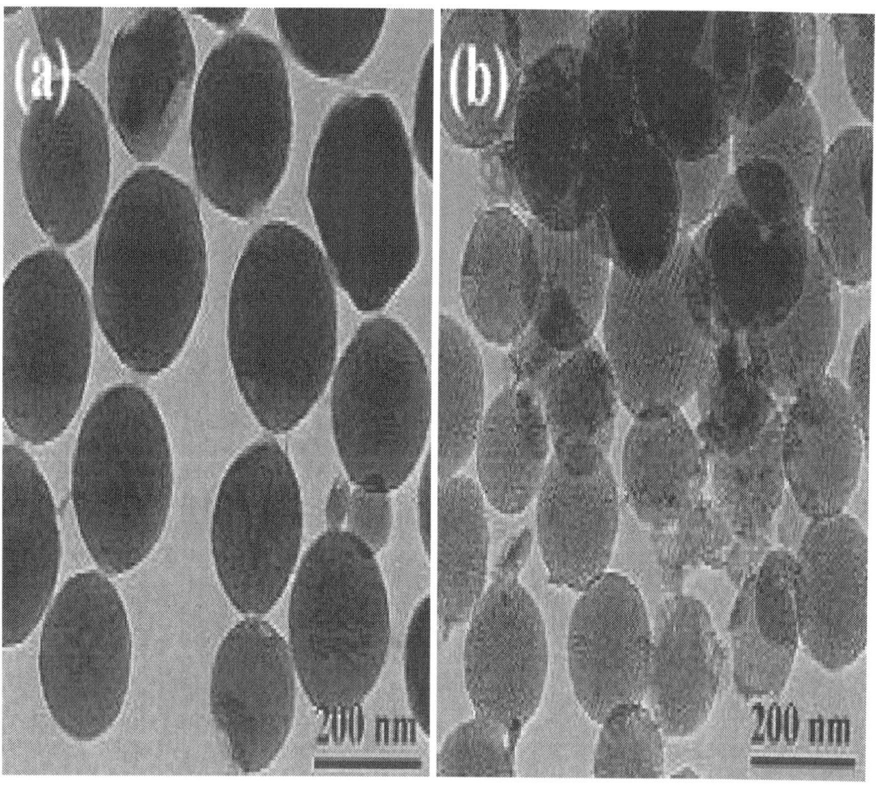

Fig. 3. Nano-particles harvesting oil from algae without harming the organism

One possible downside of the nano-harvesters is the risk that they may be released into the environment, although the spherical nano-particles are made of calcium compounds and sand [12]. The pores of the spheres are lined with chemicals, which extract algal oil without breaking the cell membrane. Nevertheless, prior to commercial market penetration of nano-harvesting, there would be a need to carry out due diligence to ensure the safety of these processes.

6. Nanotechnology applied to landfill facilities[13]

The organic matter in landfills tend to undergo anaerobic fermentation yielding methane and CO_2 [14], which if naturally vented into the atmosphere would add to the greenhouse emissions that warm the climate. And the climate change impact of methane is 25 times larger than that of carbon dioxide for a time horizon of 100 years [15]. Thus, there is a need to sequester the carbon present in landfill methane. Nano-catalysts can crack methane into elemental carbon and hydrogen. The carbon can be produced in high-purity nano-graphite for use in aerospace, automobile, batteries, etc. This approach to handling methane can considerably improve the economics of landfills as well as of anaerobic digester plants that generate electricity from biogas fueled electricity.

7. Nanotechnology to convert biomass into biofuels

Delinking biofuels production from food crops is a necessary condition to expand the scale of the market penetration of biofuels globally. Among the challenges this strategy faces is the inherent resistance of cellulosic feedstocks to conversion to simpler sugars that can be fermented into ethanol. Here, the promise lies in nano-particles used as immobilizing beds for expensive enzymes that can be used over and over again to breakdown the long chain cellulose polymers into simpler fermentable sugars [16].

The Louisiana Tech University is one among many organizations worldwide engaged in this endeavour, through the work of Dr. James Palmer, in collaborating with fellow professors Dr. Yuri Lvov, Dr. Dale Snow, and Dr. Hisham Hegab [17]. The focus is on non-edible cellulosic biomass, such as wood, grass, stalks, etc, to be converted into ethanol. This approach to produce ethanol can reduce GHG emissions by some 86% over fossils fuels.

The broader field of nanotechnology research into converting biomass into biofuels is growing fast. For example, in 2007 the oil company BP has granted a research fund of $500 million to the University of California, at Berkeley, and the University of Illinois, to explore the conversion of corn, plant material, algae and switch grass into fuel [18].

In the past, Berkeley had used nanotechnology in research for cost-effective solar panels [19]. But, the new Energy Biosciences Institute - EBI created at Berkeley will focus on fuel production with minimum environmental impacts and carbon emissions. A three pronged approach is being employed that begins with technologies for better crop production, improved feedstocks processing and development of new biofuels. The application of this approach aims at developing better feedstocks, breaking down plant material into sugars and their conversion to ethanol. Success along this pathway is expected to lead EBI to investigate the use of nanotechnology to develop other alternative fuels, such as butanol and renewable hydrocarbon fuels.

Another relevant application of nanotechnology is the use of nano-catalysts for the trans-esterification of fatty esters from vegetable oils or animal fats into biodiesel and glycerol

[20]. The nano-catalyst spheres replace the commonly used sodium methoxide. The spheres are loaded with acidic catalysts to react with the free fatty acids and basic catalysts to react with the oils. This approach eliminates several production steps of the conventional process, including acid neutralization, water washes and separations. All those steps dissolve the sodium methoxide catalyst so it can't be used again. In contrast, the catalytic nanospheres can be recovered and recycled. The overall result is a cheaper, simpler and leaner process. In summary, the process claims to be economical, recyclable, to react at mild temperatures and pressures, with both low and high FFA (free fatty acid) feedstock, producing cleaner biodiesel and cleaner glycerol, greatly reducing water consumption and environmental contaminants, and can be used in existing facilities.

8. Nanotech liquid additives

All previous presentation and discussion referred to solid nano-particles playing a catalytic role in the obtaining biofuels from algae, landfill methane and biomass. The following segments will examine the practical opportunities that exist for liquid nano-particles or droplets [21]. Consider multifunctional surface active liquid additives, whose lubricity enhancement is achieved via the formation of a monolayer over the surfaces in contact with additized fuel. [22] The treat rate for lubricity is determined by the adsorption saturation concentration. Speculate that the improved detergency and water co-solvency is obtained by the formation of nano emulsions. Also, postulate that the more complete combustion and consequent fuel efficiency increase is the result of the behaviour of nano droplets. These nano droplets result from the surfactant action of the additive in the fuel formulation and the presence of some water in all commercial fuel systems, usually due to evening condensation.
Research by Wulff and colleagues [23] has shown that nano emulsions, which the authors call micro emulsions, with fuel (biofuel included most likely), water and surfactant are:

- Thermodynamically stable and
- Microscopically isotropic, and
- Nano-structured (thus, nano emulsions).

Their research concluded that:

- The use of these nano structures with fuel, water and surfactant is able to break the usual trade off between reduction of soot and NO_x emissions, by achieving them simultaneously, and
- For the same fuel consumption, higher efficiency is obtained.

Strey and collaborators filed patent applications for what they call micro-emulsions used as fuel [24]. The interpretation offered for the behaviour of stable diesel (and most likely biodiesel)-water-surfactant nano emulsions is as follows:

- The surfactant components -oleic acid and nitrogen containing compounds (amines) - dissolve readily in diesel (and possibly in biodiesel) fuel and bind water to it without stirring;
- The water droplets are as small as a nanometer across, helping stabilize the emulsion
- The result is a "liquid sponge", can be stored indefinitely, like ordinary diesel fuel, without risk of phase separation
- This fuel formulation, when burned, results in the near-complete elimination of soot, and a reduction of up to 80% in nitrogen-oxide emissions
- The surfactant in the formulation also burns without creating emissions beyond water, carbon dioxide and nitrogen

9. Public concerns over nanotechnology: security, health and the environment

As with all new technologies, there may be cause to concern about impacts, such as on security, health and the environment. Nanotechnologies have been the subject of many assessments seeking to anticipate possible consequences of their deployment, to humans and to the environment. For instance, the Woodrow Wilson Center carried out a Nanotechnology project [25] from 2005. The project managers said that "manipulating materials at the atomic level can have astronomic repercussions, both positive and negative. The problem is no one really knows exactly what these effects may be." This was the motivation for the Project on Emerging Nanotechnology at the Woodrow Wilson Center.

Another initiative came from the International Risk Governance Council - IRGC's Nanotechnology project [26]. Two expert workshops were held. The first in May 2005 focused on how to frame nanotechnology, its risks and its benefits. A distinction was made between the nanotechnologies of the so-called Frame One (passive or classical technology assessment) and Frame Two (active or the social desirability of innovation). The second, in January 2006, concentrated on identifying gaps in nanotechnology risk governance and developing recommendations for improved risk governance.

A symposium on the subject took place in Zurich in July 2006. A presentation by Ortwin Renn[27] discussed the policy implications of Frame One, referred to in Fig. 4. The fact is that "most people have no clear associations when it comes to nanotech. They expect economic benefits but no revolutionary technological breakthroughs. Risks are often not explicitly mentioned but there is a concern for unforeseen side effects. There is a latent concern about industry, science and politics building a coalition against public interest. And one negative incident could have a major negative impact on public attitudes."

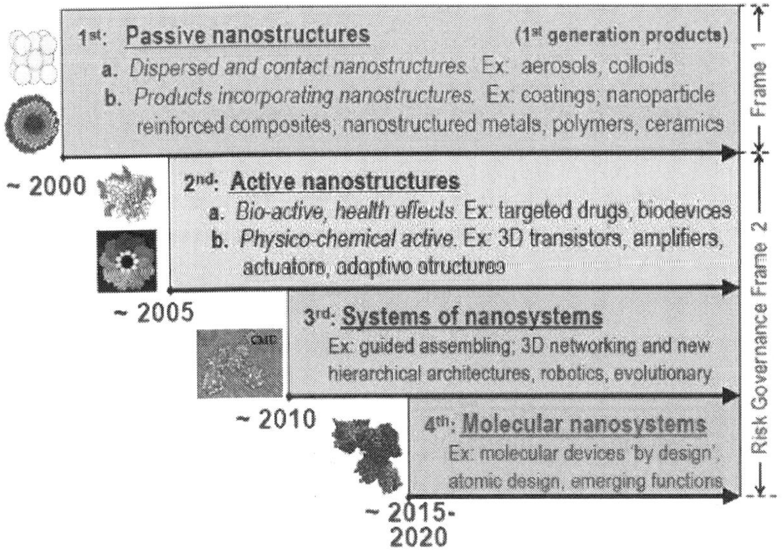

Fig. 4. Frames of reference of nanotechnology generations

The IRGC's Nanotechnology project concluded[28], among other things, that "communication about nanotechnology's benefits and risks should reflect the distinction between passive and active nano-materials and products, stressing that different approaches to managing risks are required for each. Care should also be taken to ensure that potential societal concerns about the possible impacts of Frame Two active nano-materials do not have the effect of unnecessarily increasing anxiety regarding Frame One products using only passive nanostructures." This is further expounded by Renn [29] as follows: "Frame One passive nanostructures are found, for example, in easy-to-clean surfaces, paints or in cosmetics. Frame Two refers to active nanostructures and molecular systems which could be able to interact actively or could be understood as evolutionary biosystems which change their properties in an autonomous process."

In reality, nanotechnologies are already facing challenges. Man-made nano-materials have been banned by the UK Soil Association from all its certified organic products. The 2008 annual report of the Soil Association of the UK contains the following statement [30]: "The Soil Association published the world's first standards banning nanotechnology. The risks of nanotechnology are still largely unknown, untested and unpredictable. Initial scientific studies show negative effects on living organisms, and three years ago scientists warned the Government that the release of nanoparticles should be 'avoided as far as possible'. There are many parallels with GM in the way nanotechnology is developing, particularly because commercial opportunities have run ahead of scientific understanding and regulatory control. What's more, while nano-substances are being rapidly introduced to the market, there is no official assessment process or labeling of the products - which is even worse than GM.

Health and beauty products that use nanoparticles are of concern for their potential toxicity if they get under the skin. Similar concerns exist regarding food and textiles. Definitely, more studies about health and environmental impacts are needed, to alleviate public concerns.

On the other hand, there is so much potential for nanotechnologies to do good, that Frame One and Two assessments should proceed as new applications evolve, including for instance more effective delivery of drugs to fight human and animal disease.

Fig. 5 showing a RNA nano-particle created by Peixuan Guo of Purdue University, illustrates the point. Strands are spliced together from two kinds of RNA - a scaffold and a hunter to find cancer cells. This nano-structure has proven effective against cancer growth in living mice as well as lab-grown human nasopharyngeal carcinoma and breast cancer cells.

10. Conclusions

Increasing demand for energy services in the decades ahead will require an expanding supply of liquid fuels, despite efforts at improving energy efficiency and diversification of energy systems, including growing use of electricity in transportation. Biofuels have a key role to play in this scenario. However, the future supply of biofuels must be of such a scale that non-food feedstocks and new technologies are intensively employed. Nanotechnologies are primary candidates to play a prominent role in this energy future. They will help bring to markets liquid biofuels, including renewable hydrocarbons, from algae, carbohydrates, fatty esters and biogas. Nanotechnologies will also play a role in augmenting the efficiency of using current and future liquid fuels, especially biofuels, by providing improved

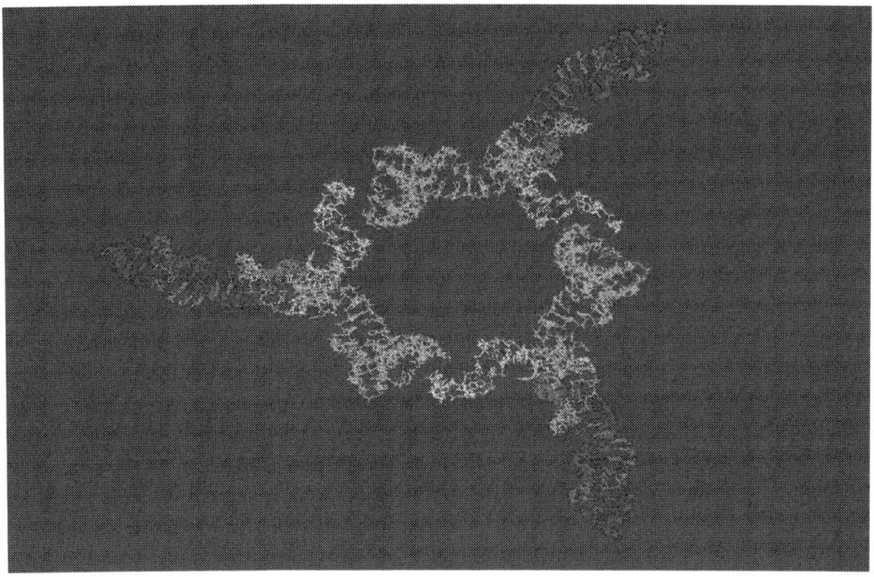

Fig. 5. RNA nano-particle created by Peixuan Guo, Purdue University [31]

combustion of nanodroplets. While there are risks in each and every new technology, the world today is much better equipped to assess risks and act accordingly, that it seems possible to advance nanotechnologies applied to biofuels, without jeopardizing security, public health or the environment. But, the reach of nanotechnologies is vast and goes much beyond biofuels and offer hopes in so many areas, including importantly, human health.

11. References

[1] Trindade, Sergio C. (2010). *Refining will definitely survive.* Pipeline Magazine, 29 August 2010

[2] Trindade, Sergio C. (2010). *Renewable Energy Perspective – a profitable pathway from oil,* Exploration and Processing, Fall 2010 [8-9], Sep.

[3] Trindade, Sergio C. (2010). *International Biofuels Trade: Issues and Options* International Biofuels Conference_São Paulo, 26-28 May.

[4] Santana, G. and S. Quirk (2009). *Growing Green: An In-Depth Look at the Emerging Algae Industry,* Greener Dawn Research, 22 July, 16p.

[5] *A Sustainable Biofuels Consensus (2008).* Statement from a conference hosted by the Rockefeller Foundation Bellagio Study and Conference Center, Bellagio, Italy, 24-28 March 2008

[6] www.defra.gsi.gov.uk (2007), In: F.O. Lichts's *World Ethanol and Biofuels Report,* Vol. 4, No. 16, p.365 and Vol. 4, No. 17, p.391, Turnbridge Wells, U.K.: F.O. Licht, 2006.

[7] Carvalho da Silva, Flávio; Paulo Roberto da Costa Brum and Taís Neno dos Santos (2005). *Nanotechnology/Nanoscience Knowledge Managament emphasizing nanostructured polymers.* Presentation, School of Chemistry, UFRJ, Brazil.

[8] Borschiver, Suzana; Maria José O. C. Guimarães, Taís N. dos Santos, Flávio C. da Silva, Paulo Roberto C. Brum (2005). *Patenteamento em Nanotecnologia: Estudo do Setor de Materiais Poliméricos Nanoestruturados.*, Polímeros: Ciência e Tecnologia, vol. 15, n° 4, p. 245-248.

[9] http://ecolocalizer.com/2009/04/23/nanotechnology-to-aid-the-commercial-viability-of-algal-bio-fuel-production, *April 23, 2009*

[10] http://www.qsinano.com/news/releases/2009_02_24.php

[11] http://www.ameslab.gov/news/news-releases/nanofarming-technology-extracts-biofuel-oil-without-harming-algae

[12] http://www.ameslab.gov/news/news-releases/nanofarming-technology-extracts-biofuel-oil-without-harming-algae

[13] http://biomassmagazine.com/articles/2354/dudek-catalyx-nanotech-to-build-landfill-facilities

[14] http://journalstar.com/news/local/article_6d5b6a34-e86f-11df-ae58-001cc4c002e0.html Nov. 4, 2010

[15] http://en.wikipedia.org/wiki/Global_warming_potential

[16] http://www.sciencedaily.com/releases/2009/10/091008131858.htm

[17] Nanotechnology Used In Biofuel Process to Save Money, *Environment Science Daily (Oct. 10, 2009)*

[18] http://berkeley.edu/news/media/releases/2007/02/01_ebi.shtml

[19] (U.S. Department of Energy. Berkeley Lab Helios Project. (n.d.) *Helios Solar Energy Research Center. Goals and challenges.* Retrieved December 10, 2009 from http://www.lbl.gov/LBL-Programs/heliosserc/html/goals.html

[20] http://www.public.iastate.edu/~nscentral/news/2007/jun/catilin.shtml

[21] *Cleaner diesel engines – pouring water on troubled oils,* The Economist, June 3rd, 2010, p.86 http://www.economist.com/node/16271415

[22] http://www.internationalfuel.com

[23] Wulff, Pascal; Lada Bemert, Sandra Engelskirchen and Reinhard Strey (2008). *Water-biofuel microemulsions.* Institute for Physical Chemistry, University of Cologne. http://strey.unikoeln.de/fileadmin/user_upload/Download/WATER___BIOFUEL_MICROEMULSIONS.pdf http://strey.uni-koeln.de/333.html?&L=1

[24] Strey, R. et al (2007). *Microemulsions and use thereof as a fuel.* US Patent Application 2007/028507 , Feb. 8. http://www.rexresearch.com/strey/strey.htm

[25] http://www.loe.org/shows/segments.htm?programID=05-P13-00050&segmentID=3

[26] http://www.irgc.org/-Nanotechnology-.html

[27]http://www.yasni.ch/ext.php?url=http%3A%2F%2Fwww.irgc.org%2FIMG%2Fpdf%2FOrtwin_Renn_Nanotechnology_Frame_1_Policy_Implications_.pdf&name=Ortwin+Renn&cat=document&showads=1

[28] http://www.irgc.org/Policy-Recommendations,188.html

[29] http://ec.europa.eu/health/ph_risk/documents/ev_20081002_rep_en.pdf, p.14

[30]http://www.soilassociation.org/LinkClick.aspx?fileticket=Moyw3Q7H%2Fp4%3D&taid=303, p.22

[31] http://www.eng.uc.edu/nanomedicine/Papers/1NCI.pdf

6

Bioresources for Third-Generation Biofuels

Rafael Picazo-Espinosa, Jesús González-López and Maximino Manzanera
University of Granada
Spain

1. Introduction

Modern societies' welfare relies greatly on fossil fuels. The current energy model, based on the extensive utilization of fossil fuels, is affected by economic and environmental problems. The United States Department of Energy 2009 report estimates that, within the next two decades, global energy consumption will double (Conti, 2009). On the other hand, the European Commission 2009 report indicates that the management of climate change problems in Europe, since 2000, has been globally unfavourable. Nevertheless, there are some positive signs, such as the 1.4% reduction in 2007 of CO_2 emissions with respect to the figures obtained from 2000 to 2004 in the European Union of Fifteen (E-15). However, considering the 27 European states (E-27), and paying attention to the consumption and production of renewable energy and biofuels, the reduction in emissions has not fulfilled the European Union objectives. Among the motives of this negative evaluation, the fall in the companies' productivity, increased transport and industry emissions and the reduction in research and development areas can be cited (Radermacher, 2009). First- and second-generation biofuels could ameliorate or solve the associated fossil fuel depletion problems, although their recent implantation has raised some doubts. The main problems associated with biofuels are the food vs. fuel controversy; the agricultural and forestry land usage and the actual sustainability of biofuels' production. Third-generation biofuels, based on the microbiological processing of agricultural, urban and industrial residues, could be a possible solution. However, several technical problems must be solved to make them economically viable and easily affordable for the industry (Robles-Medina et al., 2009).

2. First-generation biofuels

The parallel progression in energy demands over depleting oil reserves and rising greenhouse gas emissions entails a high risk of severe impacts on biodiversity, humankind food security and welfare. Thus, a new energy model is needed, based on greener and renewable energy sources, and cleaner as well as more sustainable fuel technology (Fortman et al., 2008; Jegannathan et al., 2009).

2.1 Biogas, syngas, vegetable oils blends and Fischer Tropsch liquids

The first response of heavy industry to the current energy and environmental problems includes some old systems, such as syngas and Fischer Tropsch liquids. Current advances in technology and engineering could bring new opportunities to these classical chemistry and biochemistry solutions, associated with fuel shortage situations such as the Arab oil embargo of the 1970s, or the Second World War. Some of these will be detailed below.

2.1.1 Biogas

Biogas is an attractive source of energy primarily because it is renewable and enables the recycling of organic waste. The production of biogas from manure can help to manage the problems associated with this residue, contributing to the reduction of the greenhouse gas methane. Besides, biomethanation is not only useful for energy production, but also for cleaning up solid waste in urban areas. Compared with bioethanol from wheat and biodiesel from rapeseed, biogas production based on energy crops could generate about twice the net energy yield per hectare per year. Furthermore, biogas could be produced from the by-products generated by the current bioethanol and biodiesel industries (Jegannathan et al., 2009).

Biogas production is based on bacterial methanogenesis in the absence of air of organic matter in a water solution. The process occurs in three steps. The first, hydrolysis, is carried out by strict anaerobes such as *Bacteroides* or *Clostridia,* and facultative anaerobes such as *Streptococci.* It involves the enzymatic transformation of insoluble organic material and higher molecular mass compounds such as lipids, polysaccharides, proteins, nucleic acids etc. into soluble organic materials – energy and cell carbon sources such as monosaccharides and amino acids, among others. In the second step, acidogenesis, other types of microorganisms ferment the mentioned products to acetic acid, hydrogen, carbon dioxide and other lower weight simple volatile organic acids, such as propionic and butyric acid, which are converted to acetic acid. Finally, organic acids, hydrogen and carbon dioxide are converted into a mixture of methane and carbon dioxide by the methanogenic bacteria such as *Methanosarcina spp.* or *Methanothrix spp.* (consuming acetate), as well as microorganisms such as *Methanobacterium sp.* and *Methanococcus sp.,* or others that consume hydrogen and formate to yield methane (Jegannathan et al., 2009).

In spite of its attractions, biogas has only been used in rural areas of developing countries and has received investment from governmental and non-profit organizations. The absence of private investment is due to some technical limitations that hamper its economic viability. The process is relatively slow and unstable, and requires large volumes of digester. The decrease in gas generation during the winter season is a serious problem, and can lead to the clogging of the reactor. Other causes for the reduction in gas production are pH and temperature variations, so the loading rate and solid concentration have to be continuously maintained (Jegannathan et al., 2009).

2.1.2 Syngas, biosyngas and Fischer-Tropsch derivatives

Synthetic gas, known as syngas, is a mixture of H_2, CO and CO_2 in different proportions. Traditionally, syngas was produced through gasification of coal at high temperatures, but it can also be produced by methane reformation (submitting the methane to a high temperature water steam stream, or hydrocracking) or by gasification of biomass. In the latter case, the obtained gas is called biosyngas. Syngas and biosyngas can be used directly as fuel, but they also can serve as precursors for other fuels, such as hydrogen, obtained by the compression of carbon monoxide and dioxide. Also, by Fischer-Tropsch synthesis (FTS), short and long chain hydrocarbons can be obtained from the aforementioned H_2, CO and CO_2 mixture (Srinivas et al., 2007). Fischer-Tropsch synthesis was discovered in the first half of the twentieth century and developed for large-scale production during the Second World War. It is based on the polymerization, through successive stages, of H_2 with CO and CO_2, yielding linear hydrocarbons. Iron, cobalt or ruthenium can be used as catalysts (Huber et al., 2006). FTS can be developed at high or low temperature. The high temperature FTS is performed at 330-350°C yielding mostly short-chain hydrocarbons (gasolines) and light olefins in a fluidized-bed reactor. On the other hand, low temperature FTS develops at 220-250°C in a slurry bubble column reactor, and waxes and long-chain hydrocarbons are obtained (Bludowsky & Agar, 2009). As FTS is an extremely exothermic reaction, it can be coupled with biomass gasification. However, FTS has some drawbacks, such as the fact that complex mixtures of different chain lengths are always obtained. Thus, FTS products have to be separated prior to subsequent processes (Huber et al., 2006).

2.1.3 Vegetable oil blends

The direct usage of crude or filtered vegetable oils for diesel engine fuel is possible by blending them with conventional diesel fuels in a suitable ratio. These blends are easy to obtain and keep stable for short-term use. But vegetable oils present high viscosity, acid contamination and free fatty acids that lead to gum formation by oxidation, polymerization and carbon deposition (Ranganathan, 2008). Thus, the long-term utilization of vegetable oils for fuel leads to filter clogging, nozzle blockage and deposits in the combustion chamber (Sidibé et al., 2010). Alongside the long-term problems in injection systems, filters and combustion chamber, doubts about the sustainability of using crude vegetable oil for fuels have to be considered. Vegetable oils are expensive, and their direct use in engines or as feedstock to produce petro-diesel substitutes would encounter the same economic and environmental problems that affect the conventional biodiesel and bioethanol industries (UNCTAD, 2010).

A more interesting solution is the usage of waste cooking oil (WCO; also called waste frying oil, WFO). Waste cooking oil is widely produced, inedible, and could serve as a low-cost and almost ready-to-use substitute for fossil origin diesel. As crude vegetable oil, waste cooking oil has a high viscosity. Besides, it is enriched with free fatty acids and, hence, can generate clogging problems in unmodified diesel vehicles, especially in temperate climates and during the ignition of the engine. Viscosity problems are usually bypassed by blending WCO with petrol diesel or by using transesterification to produce biodiesel (Pugazhvadivu et al., 2005; Al-Zuhair et al., 2009; Chen et al., 2009). Al-Zuhair et al. studied the production of biodiesel with lipases from *Candida antarctica* and *Burkholderia cepacia*, both free and immobilized in ceramic beads, with or without solvents. They found that clay micro-environments protected immobilized *B. cepacia* lipase from methanol damage (Al-Zuhair et al., 2009). Also, Pugazhvadivu et al. proposed solving the injection and filter-clogging problems by preheating the waste cooking oil (Pugazhvadivu et al., 2005), by comparing the performance of a diesel engine when using conventional diesel and waste frying oil, preheated at different temperatures, as fuel. They found that preheating the waste frying oil to 135°C improved the overall yield of the engine. In particular, the brake specific energy consumption and brake thermal efficiency were improved, and the engine exhaust CO and smoke density were reduced considerably compared to WFO preheated at 75°C. They concluded that WFO could be used as a diesel fuel by preheating it to 135°C.

2.2 Bioethanol and biodiesel

Bioethanol and biodiesel are frequently claimed as the most realistic alternatives to fossil fuels. These renewable fuels can be extensively produced, and both the fossil fuel distribution and engines can be easily adapted to work with blends of ethanol and gasoline, diesel and biodiesel, or even pure ethanol and pure biodiesel (Da Costa et al., 2010). But, in order to play a significant role in fossil fuel substitution, these renewable fuel industries should overcome technical limitations in production process efficiency and feedstock-related issues (UNCTAD, 2010). Decisions about feedstock election, catalysis technology or energy gain along the production process are of paramount importance for proper biodiesel and bioethanol production.

2.2.1 Bioethanol and biodiesel production

Bioethanol is produced from simple sugar-rich raw materials or from starchy feedstock, from which simple sugars can be easily processed and released, which are fermented to produce ethanol. Bioethanol production comprises three steps. Firstly, the complex sugars are hydrolysed to release glucose. Subsequently, the glucose is subjected to a second fermentation step carried out by yeasts such as *Saccharomyces cerevisiae;* for example, yielding ethanol and carbon dioxide. The third step consists of a thermochemical process and is based on the distillation of the diluted ethanol to obtain highly concentrated ethanol. When using lignocellulosic raw materials such as agricultural residues (corn stover, straw, sugar cane bagasse), forestry waste, wastepaper and other cellulosic residues, a chemical or enzymatic hydrolysis pretreatment to degrade the lignin is needed. This additional step reduces the efficiency of the process. Some improvements have been achieved by the engineering of cellulases from the *Trichoderma* genus fungi (Fukuda et al., 2006) and the utilization of microorganisms able to simultaneously express the cellulase and enzymes needed for the ethanol fermentation pathway, such as piruvate descarboxilases and alcohol dehydrogenases (Lu et al., 2006; van Zyl et al., 2007; Jegannathan et al., 2009; Rahman et al., 2009; van Dam et al., 2009). However, these improvements have still not generated an efficient and economically affordable process.

With regard to biodiesel, it consists of a mixture of fatty acid alkyl esters (FAAE) obtained by the transesterification of fatty acids and straight chain alcohols (generally ethanol or methanol), mainly from vegetable oils. When methanol is the alcohol of choice, the term used to refer to the biodiesel is fatty acid methyl esters (FAME), while the ethanol-derived biodiesel is known as fatty acid ethyl esters (FAEE). The properties of the biodiesel obtained from ethanol or methanol are very similar, but methanol is the preferred alcohol in spite of its toxicity and fossil fuel origin because of its low cost and wide availability (Ranganathan et al., 2008; Fjerbaek et al., 2009).

The commercially delivered biodiesel is mainly obtained by the chemical transesterification of the triglycerides contained in sunflower, rapeseed or palm oil. This process can be carried out by acid and alkaline liquid catalysts (Kawahara & Ono, 1979; Jeromin et al., 1987; Aksoy et al., 1988; Fukuda et al., 2001), or heterogeneous solid catalysts such as supported metals, basic oxides or zeolites (Cao et al., 2008). The preferred catalysts are the liquid ones, particularly the alkaline ones, because these catalysts are cheap, very versatile and yield less corrosive fuel than the acid catalysts. Also, liquid catalysts are preferred because the reusable solid catalysts are still withdrawn with mass transfer and reactant diffusion problems. However, the alkaline catalysis has several limitations, especially the futile consumption of the catalyst, problems of viscosity, mass transfer and recovery of biodiesel and by-products owing to the saponification of the catalyst and free fatty acids in the presence of water (Freedman et al., 1984; Zhang et al., 2003; Jaruwat et al., 2010). These problems are bypassed by high temperature reaction conditions, addition of organic solvents to manage the water presence or enhance the interface formation, or increase of the alcohol:catalysts ratio (Kawahara & Ono, 1979; Fukuda et al., 2001). Thus, the process requires high energy inputs to maintain high temperatures conducive to viable transesterification rates, and to separate methanol. Besides, the process generates alkaline waste water that requires treatment prior to its disposal (Jaruwat et al., 2010). Jointly, all these negative factors raise doubts about the sustainability and environmental benefits of the biodiesel industry.

2.2.2 Bioethanol and biodiesel advantages and drawbacks

Extensive bioethanol and biodiesel implantation has been followed by a panoply of economic, sociopolitical and environmental issues (Guerrero-Compeán, 2008). It is worth noting the strong dependency of these biofuels industries on crops used for human nourishment and the feeding of livestock (UNCTAD, 2010). Although a large number of patents have been proposed to solve many technical problems, the sudden peak in demand for biofuels has uncovered serious technical limitations of the currently used production systems. As a consequence, a growing controversy about the real sustainability and environmental friendliness of the actual biofuels industry has been generated (Fortman et al., 2008; Abdullah et al., 2009; Demirbaş, 2009; Yee et al., 2009; Jaruwat et al., 2010).

In addition, the consequences of biofuel production for farming practices or food markets, as well as real greenhouse gases (GHG) emission reduction along the biofuel life cycle, represent an important issue that, frequently, is not clearly treated. Parameters such as the kind of biofuel under study, feedstock, and energy inputs needed to maintain the process of transformation need to be taken into account. Also, the possibility of cogeneration of electricity or the exchange of energy between the biofuel synthesis and the feedstock transformation processes must be added to the model. Thus, wide variations in the net energy gain and consumption of resources can occur owing to the different assumptions made to calculate the overall benefits and drawbacks. Timilsina and collaborators draw a general picture of this issue over the OECD estimations. According to these authors, the most efficient biofuel production scheme is represented by sugarcane-based bioethanol in Brazil, with a 90% GHG reduction with respect to the gasoline equivalent. This high efficiency relies mainly on the high yield of this crop and the usage of sugarcane as an energy source for production plants and the cogeneration of electricity. Second-generation biofuels based on cellulosic feedstocks present a 70–90% GHG reduction relative to gasoline or diesel. Combined with electricity cogeneration, this kind of biofuel could have an even greater effect on GHG reduction, but they are still under development. Ethanol from sugar beet GHG reduction ranges from 40 to 60%, while wheat-based ethanol presents a 30–50% GHG reduction. The corn-based production of bioethanol is the least GHG-reducing biofuel and presents a low efficiency at GHG reductions varying from 0 (even negative in some cases) to 50% compared to gasoline (OECD, 2008; Timilsina & Shrestha, 2010).

3. Second-generation biofuels

Theoretically, biofuel implantation in transport and industry should solve, or at least improve, the ecological and economic problems derived from the unsustainability of the fossil fuel-based energy model.

However, recent field experiences indicate a much more complex scenario. The market economy and unbalanced relations between different sectors of the economy and national markets generate unpredictable dynamics of fuels' raw material prices. In this context, the development of subsequent new commercial and industrial opportunities has altered the already unstable behaviour of the agricultural international markets. The sudden peak in demand for grain, owing to its usage as a raw material for the production of ethanol, has abruptly increased the prices of corn (Fischer et al., 2009). The demand pressure has operated similarly in the palm oil market, generating a palm oil tree and soy culture surface expansion in several regions, with spectacular dimensions in South-East Asia (Abdullah et al., 2009; Jaruwat et al., 2010), where the biofuels fever threatens biodiversity and has a deep social impact because of the proliferation of unregulated, intensive, agricultural practices and the switching of oil usage for traditional human nutrition, housekeeping and livestock feed (Fortman et al., 2008; Guerrero-Compeán, 2008; Demirbaş, 2009; UNCTAD, 2010; Yee et al., 2009).

3.1 Feedstock costs and biofuel competition

Biodiesel usually costs over 0.5 US$/l, compared to 0.35 US$/l for petroleum-based diesel (Demirbaş et al., 2009). It is reported that the high cost of biodiesel is mainly due to the cost of virgin vegetable oil (Krawczyk, 1996; Connemann & Fischer, 1998). For example, the soybean oil price is currently 1.27 $/l while the palm oil price is 1.18 $/l (World-Bank, 2011). Biodiesel from animal fat is currently the cheapest option (0.4–0.5 US$/l), while the traditional transesterification of vegetable oil is, at present, around 0.6–0.8 US$/l (Bender, 1999). Zhang et al. (2007) stated that there is no global market for ethanol. Within the reasons for this, crop types, agricultural practices, land labour costs, production plant sizes, processing technologies and government policies can be cited. The cost of ethanol production in a dry mill plant currently totals 0.44 US$/l. Corn represents 66% of operating costs while energy (electricity and natural gas) to fuel the production plant represents nearly 20% of operating costs. Nevertheless, ethanol from sugar cane, produced mainly in developing countries with warm climates, is generally much cheaper to produce than ethanol from grain or sugar beet (Bender, 1999). For this reason, in countries like Brazil and India, sugar cane-based ethanol is becoming an increasingly cost-effective alternative to petroleum fuels. On the other hand, ethanol derived from cellulosic feedstock using enzymatic hydrolysis requires much greater processing than from starch or sugar-based feedstock, but feedstock costs for grasses and trees are generally lower than for grain and sugar crops. If targeted reductions in conversion costs are achieved, the total cost of producing cellulosic ethanol in EOCD countries could fall below that of grain ethanol.
Estimates show that ethanol in the EU becomes competitive when the oil price reaches 70 US$/barrel, while in the USA it becomes competitive at 50–60 US$/barrel. For Brazil and other efficient sugar producing countries such as Pakistan, Swaziland and Zimbabwe, the competitive ethanol price is much cheaper, between 25–30 US$/barrel. However, anhidrous ethanol, blendable with gasoline, is still more expensive, although prices in India have declined and are approaching the price of gasoline. Although the feedstock costs represent the majority of biofuels' cost, the production plant size can reduce the final cost of the fuel. Thus, the generally larger USA conversion plants produce biofuels, particularly ethanol, at lower cost than plants in Europe. Production costs are much lower in countries with a warm climate such as Brazil, with less than half the costs of Europe. But, in spite of the reduced costs of production, ethanol from Brazil is competitive with gasoline owing to the huge sugar cane production and the cogeneration of electricity (Demirbaş et al., 2009).

3.2 Brazilian and USA models of implementation for the bioethanol industry

Since the Arab oil embargo of the 1970s, Brazil has made an incomparable effort in the reduction of its energy dependency by intensifying and extending sugar cane-based bioethanol production. Although the alternative periods of scarcity and abundance of oil have marked fluctuations in the strength of the Brazilian Alcohol National Programme (Proalcool), the global trend has been an ascending progression in the total production of alcohol, as well as in the yield per hectare of sugar cane, and the implantation of this alcohol as transportation fuel. Today, Brazil is the second largest worldwide ethanol producer. In this way, Brazil has reduced its energy dependency, and has become the first ethanol exporter. According to Brazilian Government data, this milestone has been achieved on the basis of rural employment and welfare improvement. The key aspects of the Proalcool programme are a combination of technological advances, social planification and projection of the bioethanol industry. According to the Brazilian Government (Da Costa et al., 2010), owing to the high productivity of sugar cane, Brazil has expanded ethanol production and use without a significant increment in the fertile land surface used to cultivate sugar cane, or a food vs. fuel competition. However, there are several authors who are not so enthusiastic with the success of the Brazilian model, and point to the sugar cane industry as one of the reasons for the losses in biodiversity and the expansion of agricultural land over doubtfully catalogued marginal land, which is more relevant and dangerous than the Brazilian Government data indicates (Coelho et al., 2008; Gauder et al., 2011).

On the other hand, the American bioethanol industry choice of corn grain as its raw material has been followed by a dramatic rise in the prices of corn derivatives. Although the USA production of bioethanol supersedes the Brazilian one, the production:consumption ratio of the former (1:3) is much smaller than the latter (8:3). Despite its commercial orientation, the global efficiency of the USA model is low compared with the Brazil system and relies on the high importation taxes that protect the American industry from foreign ethanol inputs (Da Costa et al., 2010). Finally, the narrow margin of the USA production:consumption ratio suggests that the model has reached a production glass ceiling that blockades the medium-term implantation of biofuels in American society and hampers their exportation (UNCTAD, 2010; Da Costa et al., 2010).

3.3 Europe and Asia: Chemically catalyzed biodiesel

The European and Asian strategy to improve climate change and fossil fuel depletion problems is based mainly on the chemically catalyzed biodiesel obtained from vegetable oils. There is a variety of feedstocks for the production of this biofuel, from inedible oils, (mainly rapeseed oil in Europe or jatropha oil in Asia), to edible oils (principally sunflower oil in Europe and palm oil or soybean oil in Asia, although corn, peanut, cotton seed or canola oil can also be cited) (Ranganathan et al., 2008; Abdullah et al., 2009). As the elected method for industrial biodiesel production is chemical catalysis, these vegetable oils are preferred to other heterogeneous lipids sources. These other lipids need pretreatment prior to their use (Peterson, 1986; Fortman et al., 2008), and include waste frying oils, waste-activated bleaching earth from the oil refinery industry, and even animal origin lipids such as beef tallow, lard, yellow grease and poultry grease or fat from fat traps, septic tanks, or waste water sludges. The need for economically viable vegetable oils for biodiesel production implies the cultivation of greater areas with oil-producing crops such as sunflowers or palm oil trees. Thus, the previously mentioned rising corn prices, owing to the derivation of huge amounts of grain for the industrial production of bioethanol, is neither an isolated case in developing biofuel industries nor the only aspect of the biofuel industry issue. Like the bioethanol industry, the European and Asian biodiesel industries have the energy and chemical problems associated with the current biofuels model. These limitations can be summarized according to nearly obsolete technology, being strongly dependent on chemical catalysis, non-renewable materials and promotion of non-sustainable market and farming practices (Guerrero-Compeán, 2008; Demirbaş, 2009; UNCTAD, 2010).

3.4 Technical aspects of biodiesel production

The industrial production of biodiesel needs to solve several technical problems in order to obtain this kind of biofuel in an efficient and sustainable way. The physical factors to consider can be summarized by pH, temperature, hydric activity, solvents and supports. Depending on the catalyst used to drive the transesterification reaction, some of the cited factors have different impacts on the global efficiency and feasibility of the process. A non-optimal configuration of the system can reduce significantly the biodiesel yield and compromise the viability of the production plant, especially if the upstream by-products, excess catalyst or auxiliary devices for solvent recovery hinder an easy, clean and rapid downstream processing of the biofuel.

3.4.1 pH, temperature and hydric activity

As mentioned above, the chemical catalysis of the transesterification reaction requires high temperatures to achieve an acceptable reaction rate. In the case of the alkaline catalysis, the minimal temperature to produce conventional biodiesel is 60°C, while in the acid catalysis the temperature ranges from 50 to 80°C (Robles-Medina et al., 2009). Acid catalysis is slower than the alkaline one and generates a more corrosive fuel, so alkaline catalysts are preferred by the industry. It incurs a great energy cost in order to initiate and maintain the reaction (Kawahara & Ono, 1979; Aksoy et al., 1988; Cao et al., 2008). However, the utilization of sodium hydroxide as a catalyst has a serious limitation in the form of saponification of free fatty acids if water is present. This drives the increased consumption of the catalyst and downstream processing problems, such as the separation of glycerol and unreacted precursors. The solutions to manage this problem include using only virgin oils, often edible vegetable oils, instead of oils with high free fatty acids and water content, such as waste cooking oils or animal origin fats, as well as other residual fats. Higher temperatures, up to 120°C, and the addition of organic solvents, or additional steps for free fatty acids esterification with sulphuric acid before performing the alkali-catalyzed transesterification are quite common as well (Jeromin et al., 1987).

When lipases are used as catalysts, it is possible to get over the saponification problems owing to their ability to transesterificate alcohols with both triacylglycerols and free fatty acids. Besides, lipases work as well in the presence of water. In fact, they need a certain hydric activity to maintain their tridimensional structure, so the presence of water is not a problem with this kind of catalyst – although excessive hydric activity affects the transesterification reaction because the substrates are water insoluble (Jaeger & Eggert, 2002; Shah et al., 2004; Gilham & Lehner, 2005; Fjerbaek et al., 2009). Lipases can operate at low or relatively low temperatures in the range of 20 to 70°C, and at even lower temperatures if the enzyme has been obtained from psycrophilic microorganisms (Dabkowska & Szewczyk, 2009). Depending on the chosen lipase and preparation (free, immobilized or *whole cell catalyst*), lower temperatures (below 65°C) can be applied to avoid the thermal denaturation of the enzyme, thus saving in production costs (Fukuda et al., 2008). Within the thermostable lipases, we can cite *Burkholderia cepacia* lipase (Amano PS lipase, from Amano Pharmaceutical Co., Japan), that reaches its highest activity at 60°C (Dabkowska & Szewczyk, 2009), and the lipases obtained from *Thermoanaerobacter thermohydrosulfuricus*

SOL1 and *Caldanaerobacter subterraneous* subsp. tengcongensis, which show their activity maximum at 75°C and tolerate temperatures as high as 95°C (Royter et al., 2009).
On the other side of the spectrum, the lipase from *Bacillus sphaericus* MTCC 7526 presents its optimal temperature at 15°C, keeping stable until 30°C, and the *Microbacterium phyllosphaerae* lipase presents the optimal temperature at 20°C and deactivates when the temperature exceeds 35°C, with the pH value fixed at 8 for both psycrophilic enzymes (Joseph et al., 2006; Srinivas et al., 2009). Therefore, pH plays an important role in the enzymatic production of biodiesel because it influences both the reaction rate and the thermal stability or solvents' susceptibility of the lipases. An adequate pH can facilitate the optimization of the operation temperature and improve the activity of the enzyme. Gutarra and collaborators reported a high stability of the *Penicillium simplicissimum* lipase in the pH range 4.0–6.0, that showed the maximal activity at 50°C and remained stable and active (although with a lower activity) even at 70°C (Gutarra et al., 2009).

3.4.2 Heterogeneous catalysts and immobilized enzymes

An alternative to the chemical transesterification of low quality oils with a relatively high concentration of water or free fatty acids consists of heterogeneous catalysis using acidic cation-exchange resins, supported metals (Zabeti et al., 2009), basic oxides or zeolites (Knezevic et al., 1998; Suppes et al., 2004). Even low cost alternatives such as waste eggshell have been proposed as well (Wei et al., 2009). These kind of catalysts are considered as an intermediate and relatively low-cost solution between the traditional homogeneous catalysts and the lipases. However, the cited heterogeneous catalysts are affected by the slow diffusion of the triglycerydes through their pores and require a higher alcohol:oil ratio to accelerate the reaction, in order to increase the production (Zabeti et al., 2009). Nevertheless, the heterogeneous catalysts can serve to improve the reusability and efficiency of immobilized enzymes and *whole cell catalysts*. Immobilization of enzymes on inert materials such as porous ceramic beads (Iso et al., 2001) or polymeric resins (Dizge et al., 2008; Dizge et al., 2009) can improve their performance. This improvement is owing to the protection that the pores' microenvironment brings to the enzyme, avoiding the inhibition or damage of the enzyme caused by methanol or solvents. Another attractive approach to the immobilization is the so-called protein-coated microcrystals technology (PCMC). PCMC is based on the use of crystalized proteins as a support to the lipases, or in the direct use of crystalized lipases as solid catalysts (Raita et al., 2010). However, the real increase in reaction rate and enzyme stability with these immobilization techniques is usually lower than the theoretically expected. One of the reasons for these lower rate issues responds to the blockage of the pores of the used material as support because of the precipitation of glycerol or the insufficient circulation of substrates around the enzyme (Zabeti et al., 2009). The knowledge generated by the intense research on production and use of supports, resins and porous metallic alloys can be useful for enzymatic production of biodiesel. With this comparative approach, the optimization of the immobilized enzyme technology could be a reality in the short rather than medium term.

3.4.3 Alcohol to oil ratio and solvents

Depending on the kind of catalyst used and the selected operation conditions in the biodiesel production plant, the alcohol to oil molar ratio will present a wide variation. Adding excess alcohol is a common practice, and could serve as reference. However, excess alcohol use implies higher reactant associated costs, especially when the alcohol of choice is ethanol, which is more expensive than methanol. Thus, a more detailed approach to the system optimization in terms of minimal alcohol consumption is needed. Besides, a fine adjustment of the alcohol to oil ratio allows the maximal biodiesel production in the shortest possible time span and with the lowest energy input (Shieh et al., 2003).

The optimization is a relatively simple task when homogeneous catalysts such as sulphuric acid or sodium hydroxide are used to perform the conventional transesterification of vegetable oils with methanol. High yields are achieved with a methanol to oil ratio of 1:1 with an alkaline catalyst (although to improve the yield this proportion rises to 6:1) and a 30:1 ratio when an acid catalyst is used (Zhang et al., 2003).

However, in the case of lipase-catalyzed biodiesel, the situation is more complex and the molar ratio of alcohol to oil varies depending on the type of lipase, the use of an immobilized or free enzyme, and the alcohol used. Similar to the chemical catalysts, an increase of the molar alcohol:oil ratio elevates the efficiency of the reaction, but an excessive alcohol content inhibits and even damages the enzyme, especially when using methanol and free enzymes. Although the lipase-based solvent-free systems are under intensive research, owing to advantages such as the direct saving in solvents and the indirect cost reductions in downstream processes, the utilization of lipases does not necessary mean abandoning the use of a certain amount of solvents. The addition of solvents like *t*-butanol, diesel oil, hexane or dioxane to the precursors of biodiesel usually allows a better mixing of the reactants. Thus, solvents relieve the problems associated with the different water solubility of lipids and alcohols. In addition, solvents provide a more durable interaction between the enzyme and its substrates, and can favour the circulation of reactants through resins and support pores in immobilized enzyme systems. This improved circulation confers some protection to the lipases against inhibition by substrates and damages by excessive alcohols. However, solvents' addition has to be carefully studied, since an excess of solvent or an inadequate amount of solvent can affect the enzyme activity and stability. For example, Shieh et al. studied the optimal operation conditions to transesterificate soybean oil with methanol by *Rhizomucor miehei* lipase immobilized on macroporous weak anionic resin beads. They found that the best transesterification rate was obtained when the methanol:oil molar proportion was 3.4:1 at 36.5°C (Shieh et al., 2003). Raita et al. studied the transesterification of palm oil with ethanol by *Thermomyces lanuginosa* lipase-coated microcrystals in the presence of *t*-butanol. In this case, the optimal conditions were ethanol to fatty acids 4:1 molar ratio and *t*-butanol:tryacylclycerides 1:1 molar ratio, at 45°C (Raita et al., 2010. However, Tongboriboon et al. worked on the solvent-free transesterification of used palm oil with *Thermomyces lanuginosa* and *Candida antarctica* lipases immobilized in porous polypropylene powder, reporting that the best yield was achieved at an ethanol to oil ratio of 3:1, and the yield decreased when the molar ratio was increased to 4:1 at 45°C (Tongboriboon et al., 2010). These authors pointed to the inhibition of the enzymes by an excessive amount of ethanol, although it is worth emphasizing that they worked on a solvent-free system, so the enzyme was relatively vulnerable to alcohol-driven damage. On the other hand, Shah et al used 4:1 ethanol to oil molar ratio as standard reaction settings in their study about the transesterification of jatropha oil with ethanol at 40°C. The experimental design consisted of a solvent-free system and three different lipases (free and immobilized on Celite), namely *Chromobacterium viscosum*, *Candida rugosa* and *Porcine pancreas* lipases, although they did not try different alcohol to oil molar ratios (Shah et al., 2004).

4. Third-generation biofuels

As a response to the problems associated with the recent worldwide implantation of second-generation biofuels, some authors propose focusing on the processes involved in the production of such biofuels. This new approach consists of the utilization of microbial enzymes to achieve the current chemical pretreatment steps of cellulosic or starchy raw materials (Carere et al., 2008). Microorganisms deal with the degradation of lignocellulose, hemicellulose or lipid-rich materials by means of enzyme catalyzed processes at near to room temperature. Therefore, microbial enzymes could be used to make the current biofuels industry cleaner and greener. Furthermore, the production of biofuels would be coupled with the management of woody and oily wastes, converting these residues into suitable and cheap raw materials (Steen et al., 2010).

4.1 Microalgae-based biodiesel production

Another promising lipids source, still not implemented but currently being studied worldwide, is represented by microalgae. Microalgae have a high potential as biodiesel precursors because many of them are very rich in oils, sometimes with oil contents over 80% of their dry weight, although not all species are suitable as biodiesel production oils (Chisti, 2008; Manzanera, 2011). Besides, these microorganisms are able to double their biomass in less than 24 hours, achieving a reduction between 49 and 132 fold in the medium culture time required by a rapeseed or soybean field. Furthermore, microalgae cultures require low maintenance and can grow in wastewaters, non-potable water or water unsuitable for agriculture, as well as in seawater (Mata et al., 2010). The production of microalgae biodiesel could be combined with the CO_2 removal from power generation facilities (Benemann, 1997), the treatment of waste water from which microalgae would remove NH_4^+, NO_3^- and PO_4^{3-} (Aslan & Kapdan, 2006), or the synthesis of several valuable products, from bioethanol or biohydrogen to organic chemicals and food supplements (Banerjee et al., 2002; Chisti, 2007; Rupprecht, 2009; Harun et al., 2010). However, microalgae biomass-based biofuels have several problems ranging from the optimization of high density and large surface units of production to the location of the microalgae production unit. Anyway, the main decisions to take are the adoption of open or closed systems, and the election of batch or continuous operation mode. As will be discussed below, depending on the system and mode of operation choice, there will be different advantages and drawbacks.

4.1.1 Open vs. closed systems

Microalgae can be cultivated in open-culture systems such as lakes or (raceway) ponds, and in closed-culture systems called photobioreactors (PBRs). Open-culture systems are normally cheaper to build and operate, more durable and have a higher production capacity than PBRs. However, open systems are more energy expensive in terms of nutrient distribution owing to mass transfer problems, and have their depth limited to 15 cm, to ensure that the microalgae receive enough light to grow. Moreover, ponds are more sensitive to weather changes, and temperature, evaporation and light intensity controls are not feasible. Furthermore, these open systems require more land area than PBRs, and are more susceptible to contamination, both from bacteria and from microalgae present in the surroundings of culture installations (Manzanera, 2011).

In contrast, PBRs are more flexible and are intensive land-usage systems that can be configured according to the specific physical-chemical requirements of the algae of choice, allowing the cultivation of species unsuited to open ponds. Nutrient homogenization, light distribution, pH, temperature, CO_2 and O_2 control can be achieved in photobioreactors. Thus, closed systems provide more stable and appropriate growing conditions, allowing higher cell densities and minimizing contamination. Nevertheless, PBRs have several technical problems that make them non-competitive in applications that can be achieved in raceway ponds. Such problems are overheating, bio-fouling, shearing stress, oxygen accumulation, scaling-up difficulties and the high costs of building, operation and maintenance (Chen et al., 2011).

Within these problems, it is worth highlighting capital building investment and high operation costs. PBRs biomass production costs may be one order of magnitude higher than in open systems. If the biomass added value is high, PBRs can be competitive. Otherwise, open ponds will be the preferred option. However, the evaluation of performance of open and closed systems is complex and depends on several factors, such as algal species or productivity computation method. Three parameters are commonly used to evaluate productivity in microalgae cultivation installations. Firstly, volumetric productivity (VP), that is, productivity per unit of reactor volume ($g/l \cdot d$). The second parameter is area productivity (AP), defined as productivity per unit of ground area occupied by the reactor ($g/m^2 \cdot d$). The third one is illuminated surface productivity (ISP), namely the productivity per unit of reactor illuminated surface area ($g/m^2 \cdot d$). Nevertheless, the election of closed or open systems relies on more aspects apart from productivity, as will be discussed below (Richmond, 2010).

4.1.2 Continuous vs. batch operation mode

PBRs can be operated in batch or continuous mode. There are several advantages when using them in continuous mode. Firstly, continuous culture provides a higher control than batch mode. Secondly, growth rates can be regulated and keep in a steady state for long periods, and the biomass concentration can be modulated by dilution rate control. In addition, results are more reliable and reproducible owing to the steady state of continuous reactors, and the system yields better quality production (Molina et al., 2001).

However, there are limitations that can make the continuous process unsuitable for some cases. One of these limitations is the difficulty in controlling the production of some non-growth-related products. For instance, the system often requires feed-batch culturing and continuous nutrient supply that can lead to wash-out. Filamentous organisms can be difficult to grow in continuous PBRs because of the viscosity and heterogeneity of the culture medium. Another problem is that the original strain can be lost if it is displaced by a faster-growing contaminant. The contamination risk and loss of reliability of the bioreactor becomes more relevant when long incubation periods are needed, so the potential initial investment in necessary better quality equipment could rise and hamper the economic viability of the production unit (Mata et al., 2010).

The possible coproduction of high value chemicals could lead to the solution of the above problems, but it implies taking multiple parameters and options into consideration. The microalgae production units will suffer drastic changes, both in the operational aspect (temperature, insolation, wind, microalgal and bacterial or fungical contaminations etc.) and in the commercial one (oscillations in value of by-products, improvements in centrifugation or extraction strategies or development of non-algal biofuels, etc). Taking into consideration all the above mentioned parameters, it can be ascertained that any microalgae-based biodiesel project is unique. Hence, such projects must be designed by thinking in terms of a flexible or even multipurpose and adaptable installation (Richmond, 2010).

4.2 Biodiesel production from oily biomass

Microalgae are not the only option to produce biofuels from oily biomass. Multiple prokaryotes and eukaryotes can accumulate high amounts of lipids. But, as occurred with microalgae, not all species are suitable for biodiesel production owing to differences in the kind of storage lipids. Thus, as stated by Waltermann & Steinbüchel (2010), many prokaryotes synthesize polymeric compounds such as poly(3-hydroxybutyrate) (PHB) or other polyhydroxyalkanoates (PHAs), whereas only a few genera show accumulation of triacylglycerols (TAGs) and wax esters (WEs) in the form of intracellular lipid bodies. On the other hand, storage TAGs are often found in eukaryotes, while PHAs are absent, and WE accumulation has only been reported in jojoba (*Simmondsia chinensis*). All these lipids are energy and carbon storage compounds that ensure the metabolism viability during starvation periods. Similar to the formation of PHAs, TAGs and WE, synthesis is promoted by cellular stress and during imbalanced growth; for instance, by nitrogen scarcity alongside the abundance of a carbon source (Kalscheuer et al., 2004).

The most interesting prokaryote genera in terms of accumulation of TAGs are nocardioforms such as *Mycobacterium sp.*, *Nocardia sp.*, *Rhodococcus sp.*, *Micromonospora sp.*, *Dietzia sp.*, and *Gordonia sp*, alongside streptomycetes, which accumulate TAGs in the cells and the mycelia. TAGs storage is also frequently shown by members of the gram-negative genus *Acinetobacter* (although, in this case, WE are the dominant inclusion bodies components) (Waltermann & Steinbüchel, 2010). Within eukaryotes, with the exception of algae, yeasts of the genera *Candida* (non *albicans*) (Amaretti et al., 2010), *Saccharomyces* (Kalscheuer et al., 2004; Waltermann & Steinbüchel, 2010) and *Rhodotorula* (Cheirsilp et al., 2011) are the most interesting ones to produce biodiesel feedstocks.

Steinbüchel and collaborators have worked on the heterologous expression of the non specific acyl transferase WS/DGAT from *Acinetobacter calcoaceticus* ADP1 in *Saccharomyces cerevisiae* H1246 (a mutant strain unable of accumulating TAGs) (Kalscheuer et al., 2004). These authors found that the yeast recovered the ability to accumulate TAGs, as well as fatty acid ethyl esters and fatty isoamyl esters. This finding showed that the *Acinetobacter calcoaceticus* transferase had a high potential for biotechnological production of a large variety of lipids, either in prokaryotic and eukaryotic hosts. From this basis, as will be discussed in detail in Section 4.3, they worked on *Escherichia coli* TOP 10 (Invitrogen) and obtained an engineered strain able to produce fatty acid ethyl esters (biodiesel) directly from oleic acid and glucose (Kalscheuer et al., 2006).

Another possibility is combining the biomass obtained from microalgae and yeast, as recently proposed by Cheirsilp et al. (2011). These authors studied a mixed culture of oleaginous yeast *Rhodotorula glutinis* and microalga *Chlorella vulgaris* in industrial wastes. The used effluents, including both a seafood processing wastewater and molasses from a sugar cane plant. They found a synergistic effect in the mixed culture. *R. glutinis* grew faster and accumulated more lipids in the presence of *C. vulgaris*, that acted as an oxygen generator for yeast, while the microalgae obtained surplus CO_2 from yeast. The optimal conditions for lipid production were 1:1 microalga to yeast ratio initial pH of 5.0, molasses concentration at 1%, 200 rpm shaking, and light intensity at 5.0 klux under 16:8 hours light and dark cycles (Cheirsilp et al., 2011).

4.3 Whole cell catalysts

Pure or immobilized enzymes obtained from microorganisms could reduce the energy costs of industrial ethanol and biodiesel production. Nevertheless, the cellulases used to treat (ligno)cellulosic materials such as forestry residues, waste paper or straw are difficult to purify, like the lipases used for the transesterification of lipids yielding biodiesel. Hence, their price is still too high to make their usage economically viable (Shieh et al., 2003; Ranganathan et al., 2008). Another limiting factor for the use of enzymes is the inactivation and inhibition by reactants and substrates. These drawbacks are the object of an intensive effort to make possible the reutilization of enzymes through protein engineering (Ebrahimpour et al., 2008), in order to increase their stability and activity. Research interest is also targeted on immobilization in different supports or the usage of genetically engineered microorganisms, called whole cell catalysts, which carry the necessary enzymes, avoiding their exposure to inhibiting substrates and operating as microrefineries (Kalscheuer et al., 2006). In the case of biodiesel microbiological production that will be revealed in detail below, the authors proposing and developing this technology refer to this third-generation biofuel as 'Microdiesel'. The microbial production of biodiesel requires the construction of genetically modified microorganisms, able to transesterificate ethanol with lipids and, if possible, able to produce it by themselves to optimize the whole process. Since their 2006 work on microdiesel production on the laboratory scale using an engineered *Escherichia coli* strain, Steinbüchel and collaborators have established the guidelines of microdiesel industry development. Their approach consisted of expressing heterologously in *E. coli* the genes from *Zymomonas mobilis,* encoding for piruvate decarboxylase (*pdc*) and alcohol dehydrogenase (*adhB*), as well as the *Acinetobacter baylyi* non specific acyl transferase ADP1 (*atfA*). The obtained strain was able to carry out the aerobic ethanol fermentation from sugars, as well as the enzymatic transesterification of this alcohol with the fatty acids derived from the lipidic metabolism, yielding FAEE, referred to as 'microdiesel' by the authors (Kalscheuer et al., 2006). Recently, Elbahloul and Steinbüchel have used the aforementioned microdiesel producing *E. coli* at a pilot plant scale, using glycerol and sodium oleate as carbon and fatty acids sources respectively, with promising results (Elbahloul & Steinbüchel, 2010). Nevertheless, their conclusions for both studies indicate that there is still a long way to go to the industrial application of their findings, and that the technique needs to be modified to make the engineered strains adaptable to different lipids rich sources and to lignocellulosic raw materials. These modifications would allow the usage of forestry and agricultural wastes, making the biodiesel production process at least as versatile as chemical transesterification.

4.4 Microdiesel production from residues

Vegetable oils are expensive and require large areas of farmland for their production, so the direct usage of these oils for biodiesel production is expensive and unsustainable. However, there are multiple and as yet unexploited alternative fatty acid sources. Similarly, bioethanol production for its direct use as a biofuel or as a biodiesel precursor requires huge amounts of corn grain or sugar cane. Nevertheless, industrial residues such as the vegetable oil refinery waste, as well as farming, forestry, livestock and domestic solid and liquid waste (Chen et al., 2009; Dizge et al., 2009) are widespread and huge sources of lipids and carbon. Wang et al. proposed the soybean oil deodorizer distillate (SODD), a by-product from the soybean oil refineries that represents 0.3-0.5% of the soybean oil processed, to produce biodiesel. With 45-55% of triglycerides and 25-35% of free fatty acids, these authors estimated that around 80% of the SODD can be transformed into biodiesel in a transesterification with methanol by the *Thermomyces lanuginosa* and *Candida antarctica* lipases in the presence of tertbutanol and 3Å molecular sieve (Wang et al., 2006). Park et al.

used waste-activated bleaching earth (ABE), a residue of the rapeseed or palm oil refinery industry that stores 35–40% of oil and can be used to synthesize multiple bulk chemicals, including biodiesel. As in the Wang example, these authors chose methanol as alcohol, but their solvent choice was fuel oil and kerosene, the catalyst was *Candida cylindracea* lipase and the obtained FAME was extracted with a filter press (Park et al., 2008). Al-Zuhair and colleagues studied the production of biodiesel from simulated waste cooking oil (SWCO) with free- and immobilized- on ceramic beads *Candida antarctica* and *Burkholderia cepacia* lipases, with or without organic solvent. They obtained the best yield when they used *B. cepacia* without organic solvent, and observed that the system worked better when the enzymes were immobilized, probably because the clay structural microenvironments offered the lipases protection against the methanol derived denaturation (Al-Zuhair et al., 2009). Recently, Steen et al., among others, have proposed the direct fermentation of cellulosic biomass to produce biodiesel, fatty alcohols, waxes and other valuable chemicals (Steen et al., 2010). Their approach combines the waste management and the guidelines defined by Steinbüchel et al. with the new trends in synthetic biology and consolidated bioprocesses. This multidisciplinary approach brings a new flexible, easy-to-modify toolbox, composed of genetically modified FAEE synthetic strains, harbouring the enzymatic apparatus needed to produce ethanol from raw (hemi)cellulosic materials, to transesterificate it with fatty acids, or to synthesize both the fatty acids and the ethanol directly from the cellulose (Steen et al., 2010).

4.5 Wastewater sludges-based microdiesel

The microdiesel concept initiated by Steinbüchel et al. can be combined with the management and reutilization of waste waters by the application of microbial lipases to transesterificate the lipids present in the dairy industry or urban wastewater sludges. The lipidic fraction of sludges from urban wastewater treatment represents between 17 and 30% of the dry weight. This lipidic fraction is formed by direct absorption of fats present in the water by the sludge particles and by the phospholipids released from the cell membranes of micro-organisms, as well as from metabolites and cell lysis by-products (Boocock et al., 1992; Shen & Zhang, 2003; Jardé et al., 2005).

Lipid-rich wastewaters require pretreatment in order to reduce the amount of lipids and ease the subsequent conventional treatment. The pretreatment is usually based on physical processes, the most common of which are fat traps, tilted plate separators (TPS), and dissolved air flotation (DAF) units. In addition, centrifuges and electroflotation systems are used occasionally (Willey, 2001). Fat traps are rectangular or circular vessels through which the wastewater passes under laminar-flow conditions, at a rate that allows the lipids to rise to the surface near to the outlet end of the trap. The separation principle is based on Stoke's law, relating rising velocity of a particle to its diameter, so the theoretical separation efficiency is dependent on depth. In practice, fat traps have a depth of 1.5 m, although if the accumulation of a bottom sludge is expected, then an additional 0.5 m would be added to the total liquid depth. Gravity flow is preferred to pumping when feeding the trap, in order to minimize the wastewater emulsification. Fat traps are used in the food industry and in restaurants (Willey, 2001).

Meanwhile, tilted plate separators were developed in the petrochemical industry and are based on the fact that surface area, rather than depth, determines the oil separation. The introduction of tilted plates into a vessel provides many parallel gravity separators with a high surface to volume ratio in a shallow tank. Typically, TPS can occupy less than 10% of the area needed to install a conventional fat trap, although they have some disadvantages. They are susceptible to fouling if solid or semi-solid fat is present in the effluent and a crane is required to remove the plate pack for cleaning. Besides, the pumping systems have to be carefully selected and controlled to avoid surging and liquid depth fluctuations (Zeevalkink & Brunsmann, 1983; Willey, 2001). Finally, dissolved air flotation units are based on the flotation of lipids by means of microbubble clouds (60-70 μm bubble diameter) created by the injection into water of 6 bar pressure air through nozzles. Microbubbles attach to the surface of the fat and oil particles, increasing their rise rate. These systems are used both in the food industry (Willey, 2001) and in the mining wastewater treatment (Tessele, 1998). Once the pre-treatment has finished, wastewater can receive further treatment prior to its disposal or biological treatment. Thus, chemical treatment may be used to reduce the total fatty matter in wastewater. Such treatment uses aluminium sulphate, ferric chloride, or more usually, lime, to break the emulsion and coagulate the fat particles. Subsequently, the fats can be separated by flotation or sedimentation. The rate of sedimentation can be improved by a second-stage flocculation, involving the addition of low levels of polyelectrolyte (0.5-5.0 mg/l) to the wastewater once coagulation has taken place (Willey, 2001).

The use of sludges to produce biofuels is not a new idea itself, but the available literature focuses mainly on the methane production by anaerobic fermentation, currently applied in the majority of waste water treatment plants (WWTP) to provide energy to these installations, or for fermentative biohydrogen production, which is still not industrially available (van Groenestijn et al., 2002; Wang et al., 2003). Several groups have studied the *in situ* transesterification of WWTP sludges, but have focused on the chemical catalysis of the transesterification with methanol (Haas & Foglia, 2003; Mondala et al., 2009). However this method still presents the same limiting factors that affect the chemical transesterification of edible vegetable oils (Freedman et al., 1984). In spite of their chemical approach, these works provide useful information about several aspects of the biodiesel production process, especially at the first stages of the process. Thus, one common problem of chemical and enzymatic biodiesel production is the need for the pretreatment of the feedstock to make its lipids easily available to the catalyst. In the case of wastewater treatment sludge, this pretreatment step usually implies the use of organic or non-polar solvents to release the lipids from the organic matter (Antczak et al., 2009; Siddiquee & Rohani, 2011).

The most extended protocols rely on chloroform:methanol mixtures, as used in the Folch's method (Folch, 1957), in which a 2:1 chloroform:methanol reactant is mixed with the sample, where water acts as ternary component to form an emulsion. After equilibration with a fourth volume in saline solution, the emulsion separates in two phases: the lower one containing chloroform:methanol:water in the proportions 86:14:1 alongside the lipids; and the upper one containing the same solvents in proportions 3:48:47 and carrying the non-lipidic components of the sample. Bligh and Dyer's method is a simplified variant of the former, but requires the re-extraction of the sample residue with chloroform (Bligh & Dyer, 1959). Nevertheless, there are some methods with near to Folch's reagent yielding which use less toxic reagents, such as pure hexane or different combinations of hexane and other solvents, such as the hexane-isopropanol (3:2) blend proposed by Hara and Radin (Hara & Radin, 1978), or the ethyl acetate-ethanol (2:1) mixture used by Lin et al. (Lin et al., 2004). For a detailed revision of the solvents based extraction protocols, see Kuksis, 1994; Murphy, 1994; Kates, 1996.

In spite of being slightly less toxic than chloroform, the cited solvents are hazardous and present enough management risks to consider other extraction strategies. Several authors propose solvent-free methods based on ionic liquids (Ha et al., 2007), boiling the sludge or subjecting it to supercritical gases, mainly *t*-butanol (Wang et al., 2006; Royon et al., 2007), propane (Rosa et al., 2008), syngas (Tirado-Acevedo et al., 2010) and CO_2 (Helwani et al., 2009), or even to extreme pressures and temperatures (cracking) (Saka & Kusdiana, 2001). All of them are costly and not feasible with the current technology (Siddiquee et al., 2011). A more realistic and ready-to-use option is extraction using hot ethanol, which can be used to perform the lipids' extraction without using coadyuvant solvents. This approach to extraction can be illustrated with the works developed by Holser and Akin (2008) or Nielsen and Shukla (2004), among others. Although these authors have focused on the ethanol-based extraction of high value lipids from flax processing wastewater and egg yolk powder, respectively, their findings could be scaled and applied to biodiesel production from wastewater sludges. Nielsen and Shukla found that the use of ethanol at room temperature led to the extraction of nearly all the phospholipids, together with cholesterol and a minor part of the triacylglycerols, without special extraction and filtration devices. On the other hand, Holser and Akin performed a serial extraction of the lipids present in flax wastewater in three steps, under different temperature values (50, 80, 90 and 100°C). They found that the most efficient extraction was achieved when the sample-ethanol mixture was heated to 90°C and the reaction time was 15 minutes (Holser & Akin, 2008).

Considering the above findings, and taking into account the fact that the enzymatic production of biodiesel generally requires high alcohol to oil ratios, to improve the solubilization of the lipids and the formation of water-oil, enzyme-oil and enzyme-alcohol-oil interfaces, we propose that a suitable scheme for the production of biodiesel from WWTP sludge could be as simple as using a pressurized pretreatment tank, where the sludge is soaked in ethanol, kept at 90°C under stirring and refluxed to subject the mixture to three extraction cycles. This is followed by incubation in a reaction tank where the extracted lipids, alongside with part of the ethanol used in the previous step, are added to a reaction mix containing the enzyme (free, immobilized or *whole cell catalyst*) and kept at the optimal temperature and pH conditions, to ensure both enzyme stability and an acceptable microdiesel production rate. Heat exchangers between the two tanks could serve to save energy, using the heat released before entering the second tank to preheat the sludge before entering the first one.

The system could even be autonomous in terms of ethanol requirements if the engineered microorganism used to produce the lipase was able to produce ethanol simultaneously, or if the cited tanks were coupled with a third reactor where bioethanol was produced from sugars present in the non-lipidic products obtained in the pretreatment tank by means of ethanol-producing yeast or bacteria strains. In the case of economic restrictions, some short-term cost reduction could be achieved by replacing the pretreatment pressurized tank with a non-pressurized unit, and keeping the temperature of the extraction mix below 79°C, although it would imply medium-term economic losses because the lower extraction efficiency must be compensated by performing more extraction cycles at a higher reflux rate and a greater ethanol volume in the pretreatment tank, or even by the use of at least two serial pretreatment tanks.

5. Conclusion

As a short-term response to the consequences of greenhouse gas emissions and the unsustainability of the fossil fuel-based energy model, the industry has developed ready-to-use substitutes for traditional fossil fuels, delivered generally and ambiguously under the commercial 'bio' denomination. However, the first- and second-generation of so-called biofuels are neither of completely biological origin nor based on renewable and environmentally friendly feedstocks. In addition, the production techniques rely on high energy inputs, both in feedstock production (as is the case for rapeseed, soybean or palm oil) and in the biofuel synthesis (acid catalyzed biodiesel or corn bioethanol perfectly illustrate the neat energy gain problems). Alongside these problems, new and complex problems have emerged. Firstly, the increase in the prices of grain and vegetable oils used both to produce biofuels and for human nourishment and livestock feeding; and secondly, the expansion of agricultural land to increase production of sugar cane or vegetable oils to satisfy the huge demand for these sugar and lipid sources, generated by the abrupt increase in biofuels production. Thus, the development of cleaner and more sustainable biofuels is required to achieve the challenge to totally replace traditional fossil fuels by third-generation biofuels, independent of non-renewable precursors or inefficient industrial processes, that damage the environment directly and indirectly and threaten biodiversity and food security (UNCTAD, 2010).

A great variety of domestic, agricultural and industrial residues, from lignocellulosic forestry and agriculture waste to fatty acid rich waste waters, generated by the dairy, poultry or vegetable oil refinery industries, as well as the sludges from urban waste waters, can be used as precursors of biofuels. The treatment of these residues could be combined with the production of third-generation biofuels by enzymatic catalysis because the high cost of enzymes could be compensated by the low cost of the residues (or even the presence of incentives for residue reduction and management). But the massive application of these concepts requires a series of technical and biotechnological improvements, such as the optimization of lipids and sugars extraction, feedstock pretreatment processes, biofuels production plant design, heterogeneous catalysts and enzyme immobilization techniques, protein engineering of lipases, alcohol dehydrogenases or hydrolases to increase their activity and reusability, genetic engineering of microbes to facilitate both the pretreatment of precursors, and the synthesis and purification of the biofuels.

6. Acknowledgement

We thank the Junta de Andalucía (Spain) for funding this study through project reference P08-RNM-04180 and the Spanish Ministry of Science and Technology for funding through project reference CTM2009-09270. M. Manzanera received grants from the Programa Ramon y Cajal, (Ministerio de Educacion y Ciencia MEC, Spain, and ERDF, European Union).

7. References

Abdullah, A.Z., Salamatinia, B., Mootabadi, H. & Bathia, S. (2009). Current status and policies on biodiesel industry in Malaysia as the world's leading producer of palm oil. *Energy Policy* 37(12): 5440-5448

Aksoy, H.A., Kahraman, I., Karaosmanoglu, F., & Civelekoglu, H. (1988). Evaluation of Turkish sulphur olive oil as an alternative diesel fuel. *Journal of the American Oil Chemists Society* 65: 936-938.

Alauddin, Z.A.B.Z., Lahijani, P., Mohammadi, M., & Mohamed, A.R. (2010). Gasification of lignocellulosic biomass in fluidized beds for renewable energy development: A review. *Renewable and Sustainable Energy Reviews* 14: 2852-2862.

Al-Zuhair, S., Dowaidar, A. & Kamal, H. (2009). Dynamic modelling of biodiesel production from simulated waste cooking oil using immobilized lipase. *Biochemical Engineering Journa.l* 44: 256-262.

Amaretti, A., Raimondi, S. & Rossi, M. (2010). Production of Single Cell Oils from Glycerol By Oleaginous Yeasts. *Journal of Biotechnology* 150: 389.

Antczak, M.S., Kubiak, A., Antczak, T., & Bielecki, S. (2009). Enzymatic biodiesel synthesis: Key factors affecting efficiency of the process. *Renewable Energy*. 34: 1185-1194.

Aslan, S., & Kapdan, I.K. (2006) Batch kinetics of nitrogen and phosphorus removal from synthetic waste water by algae. *Ecological Engineering*. 28: 64-70.

Banerjee, A., Sharma, R., Chisti, Y., & Banerjee, U.C. (2002). *Botryococcus braunii*: A renewable source of hydrocarbons and other chemicals. *Critical reviews in Biotechnology*. 22: 245-279.

Bender, M. (1999). Economic feasibility review for community-scale farmer cooperatives for biodiesel. *Bioresource Technology*. 70: 81.

Benemann, J.R. (1997). CO_2 mitigation with microalgae systems. *Energy Conversion and Management*. 38: S475-S479.

Bligh, E.G. & Dyer, W.J. (1959). A rapid method of total lipid extraction and purification. *Canadian Journal of Biochemistry and Physiology*. 37: 911-917, 1959.

Bludowsky, T. & Agar, D.W. (2009). Thermally integrated bio-syngas-production for biorefineries. *Chemical Engineering Research and Design*. 87: 1328.

Boocock, D.G.B., Konar, S.K., Leung, A. & Ly, L.D. (1992). Fuels and chemicals from sewage sludge. The solvent extraction and composition of a lipid from a raw sewage sludge. *Fuel*. 71: 1283.

Cao, F., Chen, Y., Zhai, F., Li, J., Wang, J., Wang, X., Wang, S., & Zhu, W. (2008). Biodiesel production from high acid value waste frying oil catalyzed by superacid heteropolyacid. *Biotechnology and Bioengineering*. 101: 93- 100.

Carere, C.R., Sparling, R., Cicek, N. & Levin, D.B. (2008). Third-generation biofuels via direct cellulose fermentation. *International Journal of Molecular Sciences*. 9: 1342-1360.

Cescut, J., Severac, E., Molina-Jouve, C. & Uribelarrea, J.L. (2011). Optimizing pressurized liquid extraction of microbial lipids using the response surface method. *Journal of Chromatography A*. 1218: 373.

Cheirsilp, B., Suwannarat, W. & Niyomdecha, R. (2011). Mixed culture of oleaginous yeast *Rhodotorula glutinis* and microalga *Chlorella vulgaris* for lipid production from industrial wastes and its use as biodiesel feedstock. *New Biotechnology* In Press, Corrected Proof.

Chen, Y., Xiao, B., Chang, J., Fu, Y., Ly, P. & Wang, X. (2009). Synthesis of biodiesel from waste cooking oil using immobilized lipase in fixed bed reactor. *Energy Conversion and Management*. 50: 668-673.

Chen, C-Y., Yeh, K-L., Aisyah, R., Lee, D-J. & Chang, J-S. (2011). Cultivation, photobioreactor design and harvesting of microalgae for biodiesel production: A critical review. *Bioresource Technology*. 102: 71.

Chisti, Y. (2008). Biodiesel from microalgae beats bioethanol. *Trends in Biotechnology*. 26: 126-131.

Chisti, Y. (2007). Biodiesel from microalgae. *Biotechnology Advance.*, 25: 294-306.

Chisti, Y. (1992). Assure bioreactor sterility. *Chemical Engineering Progress*. 88: 80-85.

Choi, K.H., Chisti, Y. & Moo-Young, M. (1995). Split-channel rectangular airlift reactors: Enhancement of performance by geometric modifications. *Chemical Engineering Communications*. 138: 171-181.

Coelho, C.H., Francisco, J.G., Nogueira, R.F.P. & Campos, M.L.A.M. (2008). Dissolved organic carbon in rainwater from areas heavily impacted by sugar cane burning. *Atmospheric Environment*, 42: 7115.

Connemann, J., & Fischer, J. (1998). Biodiesel in Europe 1998: Biodiesel Processing Technologies. *Paper presented at the International liquid biofuels congress*. July 19-22, 1998, Curitiba-Parana, Brazil

Conti, J.J. (2009). Annual Energy Outlook 2009 With Projections to 2030. USDE, EIA. *Energy Information Administration Office of Integrated Analysis and Forecasting US. Department of Energy Washington, DC 20585.*

Da Costa, A.C.A., Junior, N.P. & Aranda, D.A.G. (2010). The situation of biofuels in Brazil: New generation technologies, *Renewable and Sustainable Energy Reviews*. 14: 3041-3049.

Dabkowska, K. & Szewczyk, K.W. (2009). Influence of temperature on the activity and enantioselectivity of *Burkholderia cepacia* lipase in the kinetic resolution of mandelic acid enantiomers. *Biochemical Engineering Journal* 46: 147-153.

Demirbaş, A. (2009). Political, economic and environmental impacts of biofuels: A review. *Applied Energy*, 86: S108-S117.

Dizge, N., Aydiner C., Imer D.Y., Bayramoglu, M., Tanriseven, A. & Keskinler, B. (2009). Biodiesel production from sunflower, soybean, and waste cooking oils by transesterification using immobilized lipase. *Bioresource Technology*, 100: 1983-1991.

Dizge, N. & Keskinler, B. (2008). Enzymatic production of biodiesel from canola oil using immobilized lipase. *Biomass and Bioenergy*, 32: 1274-1278.

Domínguez, A., Fernández, Y., Fidalgo, B., Pis, J.J. & Menéndez, J.A. (2008). Bio-syngas production with low concentrations of CO_2 and CH_4 from microwave-induced pyrolysis of wet and dried sewage sludge. *Chemosphere*, 70: 397.

Ebrahimpour, A., Rahman, R.N.Z.R.A., Ch'ng, D.H.E., Basri, M. & Salleh, A.B. (2008). A modelling study by response surface methodology and artificial neural network on culture parameters optimization for thermostable lipase production from a newly isolated thermophilic *Geobacillus sp.* strain ARM. *BMC Biotechnology*. 8: 96.

Elbahloul, Y. & Steinbüchel, A. (2010). Pilot-scale production of fatty acid ethyl esters by an engineered *Escherichia coli* strain harboring the p(Microdiesel) plasmid. *Applied and environmental microbiology*. 76: 4560-4565.

Fischer, G., Hizsnyik, E., Prieler, S., Shah, M. & van Velthuizen, H. (2009). Biofuels and food security. Prepared by the International Institute for Applied Systems Analysis (IIASA) for OPEC Fund for International Development (OFID). *OFID Pamphlet Series*.

Fjerbaek, L., Christensen, K.V. & Norddahl, B. (2009). A review of the current state of biodiesel production using enzymatic transesterification. *Biotechnology and Bioengineering* 102: 1298-1315.

Folch, J., Lees, M. & Sloane-Stanley, G.H. (1957). A simple method for the isolation and purification of total lipides from animal tissues. *The Journal of Biological Chemistry,* 226: 497-509.

Fortman, J.L., Chhabra, S., Mukhopadhyay, A., Chou, H., Lee, T.S., Steen, E. & Keasling, J.D. (2008). Biofuel alternatives to ethanol: Pumping the microbial well. *Trends in Biotechnology.* 26: 375-381.

Freedman, B., Pryde, E.H. & Mounts, T.L. (1984). Variables affecting the yields of fatty esters from transesterified vegetable oils. *Journal of the American Oil Chemists Society.* 61: 1638-1643.

Fukuda, H., Kondo, A. & Noda, H. (2001). Biodiesel fuel production by transesterification of oils. *Journal of Bioscience and Bioengineering.* 92: 405-416.

Fukuda, H., Hama, S., Tamalampudi, S. & Noda, H. (2008). Whole-cell biocatalysts for biodiesel fuel production. *Trends in Biotechnology.* 26: 668-673.

Fukuda, T., Ishikawa, T., Ogawa, M., Shiraga, S., Kato, M., Suye, S-I. & Ueda, M. (2006) Enhancement of cellulase activity by clones selected from the combinatorial library of the cellulose-binding domain by cell surface engineering. *Biotechnology progress* 22: 933-938.

Gauder, M., Graeff-Hönninger, S. & Claupein, W. (2011). The impact of a growing bioethanol industry on food production in Brazil. *Applied Energy.* 88: 672.

Gilham, D. & Lehner, R. (2005). Techniques to measure lipase and esterase activity in vitro. *Methods.* 36: 139-147.

Grothe, E. & Chisti, Y. (2000). Poly(β-hydroxybutyric acid) thermoplastic production by Alcaligenes latus: Behavior of fed-batch cultures. *Bioprocess Engineering.* 22: 441-449.

Guerrero-Compeán, R. (2008). Regional, economic, and environmental effects of traditional and biotechnologically enhanced ethanol production processes in Brazil. Thesis (M.C.P.) -- Massachusetts Institute of Technology, Dept. of Urban Studies and Planning. Fulltext available at http://hdl.handle.net/1721.1/45426

Gutarra, M.L.E., Godoy, M.G., Maugeri, F., Rodrigues, M.I., Freire, D.M.G. & Castilho, L.R. (2009). Production of an acidic and thermostable lipase of the mesophilic fungus *Penicillium simplicissimum* by solid-state. *Bioresource Technology.* 100: 5249-5254.

Ha, S.H., Lan, M.N., Lee, S.H., Hwang, S.M. & Koo, Y-M. (2007). Lipase-catalyzed biodiesel production from soybean oil in ionic liquids. *Enzyme and Microbial Technology.* 41(4): 480-483.

Haas, M.J. & Foglia, T.A. (2003). In situ production of fatty acid alkyl esters. United States patent and trademark office pre-grant publication, Dec 2003 patno: US20030229237

Hara, A. & Radin, N.S. (1978). Lipid extraction of tissues with a low-toxicity solvent. *Analytical Biochemistry.* 90: 420.

Harun, R., Singh, M., Forde, G.M. & Danquah, M.K. (2010). Bioprocess engineering of microalgae to produce a variety of consumer products. *Renewable and Sustainable Energy Reviews.* 14: 1037-1047.

Helwani, Z., Othman, M.R., Aziz, N., Fernando, W.J.N. & Kim, J. (2009). Technologies for production of biodiesel focusing on green catalytic techniques: A review. *Fuel Processing Technology.* 90: 1502.

Holser R.A. & Akin D.E. (2008). Extraction of lipids from flax processing waste using hot ethanol. *Industrial Crops and Products* 27: 236.

Huber, G.W., Iborra, S. & Corma, A. (2006). Synthesis of transportation fuels from biomass: chemistry, catalysts, and engineering. *Chemical reviews* 106 (9): 4044-4098.

Hung, T-C., Giridhar, R., Chiou, S-H. & Wu, W-T. (2003). Binary immobilization of *Candida rugosa* lipase on chitosan. *Journal of Molecular Catalysis B: Enzymatic.* 26: 69-78.

Iso, M., Chen, B.X., Eguchi, M., Kudo, T. & Shrestha, S. (2001). Production of biodiesel fuel from triglycerides and alcohol using immobilized lipase. *Journal of Molecular Catalysis B-Enzymatic.* 16: 53-58.

Jardé, E., Mansuy, L. & Faure, P. (2005). Organic markers in the lipidic fraction of sewage sludges. *Water research.* 39: 1215.

Jaeger, K.E. & Eggert, T. (2002). Lipases for biotechnology. *Current Opinion in Biotechnology.* 13:390-397.

Jaruwat, P., Kongjao, S. & Hunsom, M. (2010). Management of biodiesel waste water by the combined processes of chemical recovery and electrochemical treatment. *Energy Conversion and Management.* 51: 531-537.

Jegannathan, K.R., Chan, E-S. & Ravindra, P. (2009). Harnessing biofuels: A global Renaissance in energy production? *Renewable and Sustainable Energy Reviews.* 13: 2163-2168.

Jeromin, L., Peukert, E. & Wollmann, G. (1987). Process for the pre-esterification of free fatty acids in fats and oils US Patent-4698186. 6 October, 1987. 327. Henkel Kommanditgesellschaft auf Aktien.

Joseph, B., Ramteke, P.W. & Kumar, P.A. (2006). Studies on the enhanced production of extracellular lipase by *Staphylococcus epidermidis. Journal of General and Applied Microbiology.* 52: 315-320.

Kalscheuer, R., Luftmann, H. & Steinbüchel, A. (2004). Synthesis of novel lipids in *Saccharomyces cerevisiae* by heterologous expression of an unspecific bacterial acyltransferase. *Applied and Environmental Microbiology* 70 (12): 7119-7125.

Kalscheuer, R., Stolting, T. & Steinbuchel, A. (2006). Microdiesel: *Escherichia coli* engineered for fuel production. *Microbiology-Sgm.* 152: 2529-2536.

Kates, M. (1996). Review of "Advances in Lipid Methodology-Three" (Ed. Christie, WW) pp. 1 177, 1996, The Oily Press, Dundee, Scotland. *Chemistry and Physics of Lipids* 83: 87-89. Book review.

Kawahara, Y. & Ono, T. (1979). Process for producing lower alcohol esters of fatty acids. United States Patent and Trademark Office granted patent, Aug 1979 patno:US4164506.

Knezevic, Z., Mojovic, L. & Adnadjevic, B. (1998). Palm oil hydrolysis by lipase from *Candida cylindracea* immobilized on zeolite type Y. *Enzyme and Microbial Technology.* 22: 275-280.

Krawczyk, T. (1996). Biodiesel: Alternative fuel makes inroads but hurdles remain. *INFORM.* 7: 801-829.

Kuksis, A. (1994). Advances in Lipid Methodology-Two Edited by W.W. Christie, The Scottish Crop Research Institute, Innergowrie, Dundee DD22 5DA, Scotland Published by The Oily Press, Dundee, 1993. *Chemistry and Physics of Lipids* 71: 115-116. Book review.

Lin, J-H., Liu, L-Y., Yang, M-H. & Lee, M-H. (2004). Ethyl acetate/ethyl alcohol mixtures as an alternative to Folch reagent for extracting animal lipids. *Journal of Agriculture and Food Chemistry*. 52: 4984- 4986.

Lu, Y.P., Zhang, YHP. & Lynd, L.R. (2006). Enzyme-microbe synergy during cellulose hydrolysis by *Clostridium thermocellum*. *Proceedings of the National Academy of Sciences of the United States of America* 103: 16165-16169.

Manzanera, M. (2011). Biofuels from Oily Biomass, In: *Carbon-neutral fuels and energy carriers*. M. Muradov & N. Veziroghu (Ed.), 635-663, CRC, Taylor & Francis Group, ISBN 978-143-9818-57-2 Orlando,USA.

Mata, T.M., Martins, A.A. & Caetano, N.S. (2010). Microalgae for biodiesel production and other applications: A review. *Renewable and Sustainable Energy Reviews*. 14: 217-232.

Matsumura, M. & Murakami, S. (2002). Method and equipment of refining plant oil and waste vegetable oil into diesel engine fuel. Japan Patent JP2000219886 (A). Also published as EP1026224 (A1) and US6364917 (B1).

Molina-Grima, E., Acién-Fernández, F.G. & Chisti, Y. (2001). Tubular photobioreactor design for algal cultures. *Journal of Biotechnology*. 92: 113-131.

Molina-Grima, E., Acién-Fernández, F.G., García-Camacho, F., Camacho-Rubio, F. & Chisti, Y. (2000) Scale-up of tubular photobioreactors. *Journal of Applied Phycology*. 12: 355-368.

Mondala, A., Liang, K., Toghiani, H., Hernandez, R. & French, T. (2009). Biodiesel production by in situ transesterification of municipal primary and secondary sludges. *Bioresource Technology*. 100: 1203-1210.

Murphy, E.J. (1994). *Advances in Lipid Methodology One. Edited by W.W. Christie, The Oily Press, Ayr, Scotland ISBN 0 9514171 1 8. Chemistry and Physics of Lipids* 69: 179-180. Book review.

Nielsen, H. & Shukla, VKS. (2004). In situ solid phase extraction of lipids from spray-dried egg yolk by ethanol with subsequent removal of triacylglycerols by cold temperature crystallization. *LWT - Food Science and Technology*. 37: 613.

OECD (2008). Biofuel Support Policies: An Economic Assessment. *OECD Publishing*. Also available in French under the title Politiques de soutien des biocarburants: Une évaluation économique. Online publication.
http: / / dx.doi.org / 10.1787 / 9789264050112-en

Park, E.Y., Sato, M. & Kojima, S. (2008). Lipase-catalyzed biodiesel production from waste activated bleaching earth as raw material in a pilot plant. *Bioresource Technology*. 99: 3130-3135.

Peterson, C.L. (1986). Vegetable oil as a diesel fuel: Status and research priorities. *Transactions of the Asae*. 29: 1413-1422.

Pugazhvadivu, M. & Jeyachandran, K. (2005). Investigations on the performance and exhaust emissions of a diesel engine using preheated waste frying oil as fuel. *Renewable Energy*. 30: 2189.

Radermacher, W. & Director-General E. European Commission (2009). Measuring progress towards a more sustainable Europe statistical books 2009 monitoring reportof the EU sustainable development strategy. *Office for Official Publications of the European Communities*. ISBN 978-92-79-12695-6

Rahman, Z., Shida, Y., Furukawa, T., Suzuki, Y., Okada, H., Ogasawara, W. & Morikawa, Y. (2009). Application of *Trichoderma reesei* cellulase and xylanase promoters through homologous recombination for enhanced production of extracellular beta-glucosidase I. *Bioscience, biotechnology, and biochemistry.* 73: 1083-1089.

Raita, M., Champreda, V. & Laosiripojana, N. (2010). Biocatalytic ethanolysis of palm oil for biodiesel production using microcrystalline lipase in tert-butanol system. *Process Biochemistry.* 45: 829.

Rakopoulos, C.D., Antonopoulos, K.A. & Rakopoulos, D.C. (2006). Multi-zone modelling of Diesel engine fuel spray development with vegetable oil, biodiesel or diesel fuels. *Energy Conversion and Management,* 47: 1550.

Ranganathan, S.V., Narasimhan, S.L. & Muthukumar, K. (2008). An overview of enzymaticproduction of biodiesel. *Bioresource Technology.* 99: 3975-3981.

Ribeiro, A.M., Santos, J.C. & Rodrigues, A.E. (2010). PSA design for stoichiometric adjustment of bio-syngas for methanol production and co-capture of carbon dioxide. *Chemical Engineering Journal.* 163: 355.

Richmond, A. (Ed.) (2010). *Handbook of Microalgal Culture,* Blackwell Science. Iowa State Press, a Blackwell Publishing Company, 2121 State Avenue, Ames, Iowa 50014-8300, USA. ISBN 0-632-05953-2 .

Robles-Medina, A., Gonzalez-Moreno, P.A., Esteban-Cerdan, L. & Molina-Grima, E. (2009). Biocatalysis: Towards ever greener biodiesel production. *Biotechnology Advances.* 27: 398-408.

Rosa, C.D., Morandim, M.B., Ninow, J.L., Oliveira, D., Treichel, H. & Oliveira, J.V. (2008). Lipase-catalyzed production of fatty acid ethyl esters from soybean oil in compressed propane. *Journal of Supercritical Fluids.* 47: 49-53.

Royon, D., Daz, M., Ellenrieder, G. & Locatellu, S. (2007). Enzymatic production of biodiesel from cotton seed oil using t-butanol as a solvent. *Bioresource Technology.* 98: 648–653.

Royter, M., Schmidt, M., Elend, C., Höbenreich, H., Schäfer, T., Bornscheuer, U.T. & Antranikian, G. (2009). Thermostable lipases from the extreme thermophilic anaerobic bacteria *Thermoanaerobacter thermohydrosulfuricus* SOL1 and *Caldanaerobacter subterraneus subsp. tengcongensis. Extremophiles.* 13: 769-783.

Rupprecht, J. (2009). From systems biology to fuel-*Chlamydomonas reinhardtii* as a model for a systems biology approach to improve biohydrogen. *Journal of Biotechnology.* 142: 10-20.

Saka, S. & Kusdiana, D. (2001). Biodiesel fuel from rapeseed oil as prepared in supercritical methanol. *Fuel.* 80: 225-231.

Shah, S., Sharma, S. & Gupta, M.N. (2004). Biodiesel preparation by lipase-catalyzed transesterification of Jatropha oil. *Energy & Fuels.* 18: 154-159.

Shen, L. & Zhang, D-K. (2003). An experimental study of oil recovery from sewage sludge by low-temperature pyrolysis in a fluidised-bed. *Fue.l* 82: 465.

Shieh, C.J., Liao, H.F. & Lee, C.C. (2003). Optimization of lipase-catalyzed biodiesel by response surface methodology. *Bioresource Technology.* 88: 103-106.

Sidibé, S.S., Blin, J., Vaitilingom, G. & Azoumah, Y. (2010). Use of crude filtered vegetable oil as a fuel in diesel engines state of the art: Literature review. *Renewable and Sustainable Energy Reviews* 14: 2748-2759.

Siddiquee, M.N. & Rohani, S. (2011). Lipid extraction and biodiesel production from municipal sewage sludges. A review. *Renewable and Sustainable Energy Reviews* 15. 1067–1072.

Silvis, T.W. (1977). Waste vegetable oil as a diesel fuel. Thesis. The Ohio State University. Department of Mechanical Engineering.

Srinivas, S., Malik, R.K. & Mahajani, S.M. (2007). Fischer-Tropsch synthesis using bio-syngas and CO_2. *Energy for Sustainable Development*. 11: 66.

Srinivas, T.N.R., Nageswara Rao, S.S.S., Vishnu-Vardhan-Reddy, P., Pratibha, M.S., Sailaja, B., Kavya, B., Hara-Kishore, K., Begum, Z., Singh, S.M. & Shivaji, S. (2009). Bacterial diversity and bioprospecting for cold-active lipases, amylases and proteases, from culturable bacteria of Kongsfjorden and Ny-Alesund, Svalbard, Arctic. *Current microbiology*. 59: 537-547.

Steen, E.J., Kang, Y., Bokinsky, G., Hu, Z., Schirmer, A., McClure, A., B.-del-Cardayre, S. & Keasling, J.D. (2010). Microbial production of fatty-acid-derived fuels and chemicals from plant biomass. *Nature*. 463: 559-563.

Suppes, G.J., Dasari, M.A., Doskocil, E.J., Mankidy, P.J. & Goff, M.J. (2004). Transesterification of soybean oil with zeolite and metal catalysts. *Applied Catalysis A: General*. 257: 213-223.

Tessele, F., Misra, M. & Rubio, J. (1998). Removal of Hg, As and Se ions from gold cyanide leach solutions by dissolved air flotation. *Minerals Engineering*. 11: 535.

Timilsina, G.R. & Shrestha, A. (2010). How much hope should we have for biofuels? *Energy*. Corrected Proof. doi:10.1016/j.energy.2010.08.023

Tirado-Acevedo, O., Chinn, M.S. & Grunden, A.M. (2010). Production of biofuels from synthesis gas using microbial catalysts. *Advances in Applied Microbiology*. 70: 57.

Tongboriboon, K., Cheirsilp, B. & H-Kittikun, A. (2010). Mixed lipases for efficient enzymatic synthesis of biodiesel from used palm oil and ethanol in a solvent-free system. *Journal of Molecular Catalysis B: Enzymatic*. 67: 52.

UNCTAD, (2010). United Nations Conference on Trade and Development - Harvard - Belfer Center for Science and International Affairs. United Nations. New York and Geneva.

van Dam, J., Faaij, APC., Hilbert, J., Petruzzi, H. & Turkenburg, W.C. (2009). Large-scale bioenergy production from soybeans and switchgrass in Argentina: Part A: Potential and economic feasibility for national and international markets. *Renewable and Sustainable Energy Reviews*. 13: 1710-1733.

van Groenestijn, J.W., Hazewinkel, J.H.O., Nienoord, M. & Bussmann, P.J.T. (2002). Energy aspects of biological hydrogen production in high rate bioreactors operated in the thermophilic temperature. *International Journal of Hydrogen Energy*. 27: 1141-1147.

van Zyl, W.H., Lynd, L.R., den Haan, R. & McBride, J.E. (2007). Consolidated bioprocessing for bioethanol production using *Saccharomyces cereviside*. *Biofuels*. 108: 205-235.

Waltermann, M. & Steinbüchel, A. (2010). Neutral lipid bodies in prokaryotes: Recent insights into structure, formation, and relationship to eukaryotic lipid depots. *Journal of Bacteriology* 187 (11): 3607-3619.

Wang, C.C., Chang, C.W., Chu, C.P., Lee, D.J., Chang, B.V. & Liao, C.S. (2003). Producing hydrogen from waste-water sludge by *Clostridium bifermentans*. *Journal of Biotechnology*. 102: 83-92.

Wang, L., Du, W., Liu, D.H., Li, L.L. & Dai, N.M. (2006). Lipase-catalyzed biodiesel production from soybean oil deodorizer distillate with absorbent present in tert-butanol system. *Journal of Molecular Catalysis B-Enzymatic*. 43: 29-32.

Wei, Z., Xu, C. & Li, B. (2009). Application of waste eggshell as low-cost solid catalyst for biodiesel production. *Bioresource Technology*. 100(11): 2883-2885.

Willey, R. (2001). Fats, oils, and greases: the minimization and treatment of wastewaters generated from oil refining and margarine production. *Ecotoxicology and Environmental Safety* 50: 127-133.

World-Bank (2011). World Bank Commodity Price Data (Pink Datasheet). 1818 H Street, N.W. , MSN MC 2-200 Washington, D.C. 20433 U.S.A.

Yee, K.F., Tan, K.T., Abdullah, A.Z. & Lee, K.T. (2009). Life cycle assessment of palm biodiesel: Revealing facts and benefits for sustainability. *Applied Energy*. 86: S189-S196.

Zabeti, M., Wan Daud, WMA. & Aroua, M.K. (2009). Activity of solid catalysts for biodiesel production: A review. *Fuel Processing Technology*. 90: 770-777.

Zeevalkink, J.A. & Brunsmann, J.J. (1983). Oil removal from water in parallel plate gravity-type separators. *Water research*. 17: 365.

Zhang, Y., Dube, M.A., McLean, D.D. & Kates, M. (2003). Biodiesel production from waste cooking oil: 1. Process design and technological assessment. *Bioresource Technology*. 89: 1-16.

Zhang, Y., Habibi, S. & MacLean, H.L. (2007). Environmental and economic evaluation of bioenergy in Ontario, Canada. *Journal of the Air & Waste Management Association* 57: 919-933.

Zhao, Z.K., Zhao, X., Hu, C., Shen, H. & Wu, S. (2009). Towards industrial lipid production using oleaginous yeasts. *New Biotechnology*. 25: S233.

Overview of Corn-Based Fuel Ethanol Coproducts: Production and Use

Kurt A. Rosentrater
United States Department of Agriculture, Agricultural Research Service
U.S.A.

1. Introduction

Modern societies face many challenges, including growing populations, increased demands for food, clothing, housing, consumer goods, and the raw materials required to produce all of these. Additionally, there is a growing need for energy, which is most easily met by use of fossil fuels (e.g., coal, natural gas, petroleum). For example, in 2008, the overall U.S. demand for energy was 99.3 x 10^{15} Btu (1.05 x 10^{14} MJ); 84% of this was supplied by fossil sources. Transportation fuels accounted for 28% of all energy consumed during this time, and nearly 97% of this came from fossil sources. Domestic production of crude oil was 4.96 million barrels per day, whereas imports were 9.76 million barrels per day (nearly 2/3 of the total U.S. demand) (U.S. EIA, 2011). Many argue that this scenario is not sustainable in the long term, and other alternatives are needed.

Biofuels, which are renewable sources of energy, can help meet some of these increasing needs. They can technically be produced from a variety of materials which contain either carbohydrates or lipids, including cereal grains (such as corn, barley, and wheat), oilseeds (such as soybean, canola, and flax), legumes (such as alfalfa), perennial grasses (such as switchgrass, miscanthus, prairie cord grass, and others), agricultural residues (such as corn stover and wheat stems), algae, food processing wastes, and other biological materials. Indeed, the lignocellulosic ethanol industry is poised to consume large quantities of biomass in the future (Agrawal et al., 2007; Alexander and Hurt, 2007; Cassman, 2007; Cassman et al., 2006; Cassman and Liska, 2007; Dale, 2007; De La Torre Ugarte et al., 2000; Dewulf et al., 2005; Lynd and Wang, 2004). At this point in time, however, the most heavily used feedstock for biofuel production in the U.S. is corn grain. Industrial-scale alcohol production from corn starch is readily accomplished, and at a lower cost (generally between \$1/gallon and \$1.4/gallon), compared to other available biomass substrates in the U.S. The most commonly used process for the production of fuel ethanol from corn is the dry grind process, the primary coproduct of which is distillers dried grains with solubles (DDGS) (Figure 1), which will be discussed subsequently.

Corn-based ethanol has been used as a liquid transportation fuel for more than 150 years, although up until recent times the industry has been quite small. The modern corn-based fuel ethanol industry, however, has reached a scale which can augment the nation's supply of transportation fuels. In 2008, for example, ethanol displaced more than 321 million barrels of oil (Urbanchuk, 2009), which accounted for nearly 5% of all oil imports. Only recently has this industry become truly visible to the average citizen. This has been due, in part, to the growing demand for transportation fuels, escalating prices at the fuel pump, positive economic effects throughout rural America, as well as questions and controversies surrounding the production and use of corn ethanol.

Fig. 1. Corn-based distillers dried grains with solubles (DDGS), which is currently available from most U.S. fuel ethanol plants.

To help meet the increasing demand for transportation fuels, the number of ethanol plants has been rapidly increasing in recent years, as has the quantity of fuel ethanol produced (Figure 2).

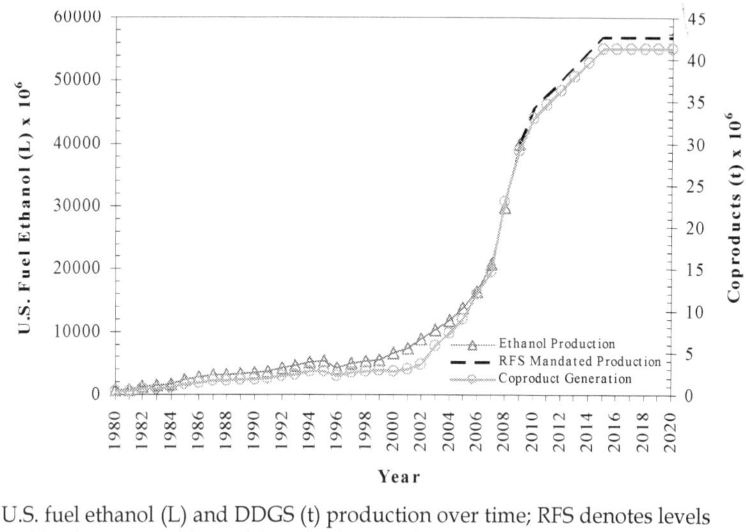

Fig. 2. U.S. fuel ethanol (L) and DDGS (t) production over time; RFS denotes levels mandated by the Renewable Fuel Standard. Inset shows number of U.S. ethanol plants over time (adapted from RFA, 2009a, 2009b, 2011)

In 2005, 87 manufacturing plants in the U.S. had an aggregate production capacity of 13.46 billion L/y (3.56 billion gal/y). At the beginning of 2011, however, that number had risen to 204 plants with a production capacity of nearly 51.1 billion L/y (13.5 billion gal/y), which is an increase of nearly 380% in six years (RFA, 2011). Most new ethanol plants have been dry-grind facilities (Figure 3), which will be discussed subsequently. And, over the next several years, the

Fig. 3. U.S. dry grind corn-to-ethanol manufacturing plants. A. 450 x 10^6 L/y plant. B. 80 x 10^6 L/y plant.

Renewable Fuel Standard (RFS) mandates the use of 15 billion gal/y (56.8 billion L/y) of renewable biofuels (i.e., which will primarily be corn-based ethanol) (RFA, 2009a), although the RFS does mandate the growing use of advanced and cellulosic biofuels as well. Because the industry is dynamic and still evolving, these current production numbers will surely be outdated by the time this book is published. As production volume increases, the processing residues (known collectively as "distillers grains" -will increase in tandem (as shown in Figure 2). It is anticipated that over 40 million metric tonnes (t) of distillers grains (both wet and dry) will eventually be produced by the U.S. fuel ethanol industry as production reaches equilibrium due to the RFS.

It is true that as the industry has grown, the concomitant consumption of corn has grown as well (Figure 4). Since 2008, for example, over 30% of the U.S. corn crop has been used to produce ethanol. When examining these numbers, however, it is important to be aware of several key points: exports have been relatively constant over time, there has been a slight decline in the corn used for animal feed, and the overall quantity of corn which is produced by U.S. farmers has been substantially increasing over time. Thus, it appears that the corn which is used to produce ethanol is actually arising mostly from the growing corn supply. It is also important to note that the corn which is redirected away from animal feed is actually being replaced by DDGS and other ethanol coproducts in these animal feeds. Thus coproducts (especially DDGDS) are key to the sustainability of both the ethanol and livestock industries. In other words, fuel, feed, and food needs can be simultaneously met.

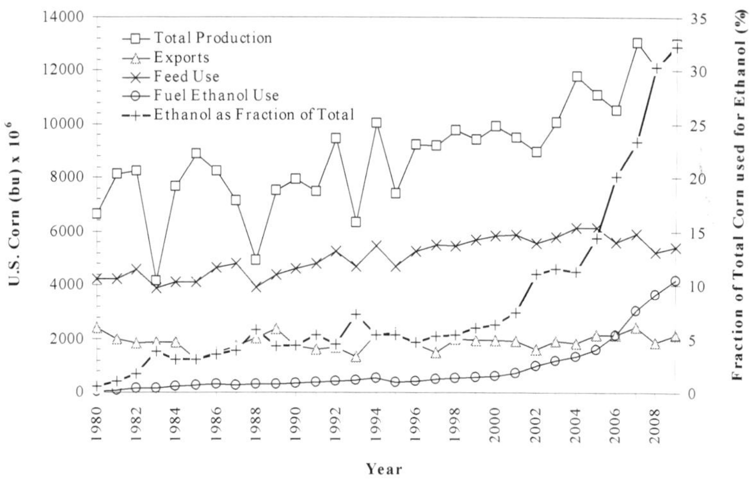

Fig. 4. Historic U.S. corn production (bu) and major categories of use (adapted from ERS, 2011).

2. Objectives

The goals of this chapter are three-fold: 1) to briefly discuss U.S. fuel ethanol and coproduct manufacturing processes; 2) to explain the importance of coproducts to the fuel ethanol and livestock industries; and 3) to describe how coproduct quality is improving and potential uses are expanding as the ethanol industry continues to evolve.

3. Manufacturing processes

Corn can be converted into fuel ethanol by three commercial processes: wet milling, dry milling, and dry grind processing. Over the last decade, many new fuel ethanol plants have been built (Figure 2), and considerable innovations have occurred throughout the industry vis-à-vis production processes used and final products produced, as well as raw materials, water, and energy consumption. Many of these innovations have arisen with the advent of dry grind processing. Due to many advantages, including lower capital and operating costs (including energy inputs), most new ethanol plants are dry grind facilities as opposed to the older style mills. For example, in 2002, 50% of U.S. ethanol plants were dry grind; in 2004 that number had risen to 67%; in 2006 dry grind plants constituted 79% of all facilities; and in 2009 the fraction had grown to over 80% (RFA, 2009a).

The dry grind process (Figure 5) entails several key steps, including grain receiving, distribution, storage, cleaning, grinding, cooking, liquefaction, saccharification, fermentation, distillation, ethanol storage and loadout, centrifugation, coproduct drying, coproduct storage and loadout. Additional systems that play key roles include energy / heat recovery, waste management, grain aeration, CO_2 scrubbing and extraction, dust control, facility sanitation, instrumentation and controls, and sampling and inspection. Figure 5 depicts how all of these pieces fit together in a commercial plant.

Grinding, cooking, and liquefying release and convert the corn starch into glucose, which is consumed during the fermentation process by yeast (Sacchharomyces cerevisiae). After fermentation, the ethanol is separated from the water and nonfermentable residues (which consist of corn kernel proteins, fibers, oils, and minerals) by distillation. Downstream dewatering, separation, evaporation, mixing, and drying are then used to remove water from the solid residues and to produce a variety of coproduct streams (known collectively as distillers grains): wet or dry, with or without the addition of condensed solubles (CDS). Distillers dried grains with solubles (known as DDGS), is the most popular, and is often dried to approximately 10% moisture content (or even less at some plants), to ensure an extended shelf life and good flowability, and then sold to local livestock producers or shipped by truck or rail to various destinations throughout the nation. DDGS is increasingly being exported to overseas markets as well. Distillers wet grains (or DWG) has been gaining popularity with livestock producers near ethanol plants in recent years; in fact, it has been estimated that, nationwide, more than 25% of distillers grains sales are now DWG. But, because the moisture contents are generally greater than 50 to 60%, their shelf life is very limited, especially in summer months, and shipping large quantities of water is expensive. DDGS is still the most prevalent type of distillers grain in the marketplace.

Dry grind ethanol manufacturing results in three main products: ethanol, the primary end product; residual nonfermentable corn kernel components, which are sold as distillers grains; and carbon dioxide. A common rule of thumb is that for each 1 kg of corn processed, approximately 1/3 kg of each of the constituent streams will be produced. Another rule of thumb states that each bushel of corn (~ 56 lb; 25.4 kg) will yield up to 2.9 gal (11.0 L) of ethanol, approximately 18 lb (8.2 kg) of distillers grains, and nearly 18 lb (8.2 kg) of carbon dioxide. Of course, these will vary to some degree over time due to production practices, equipment settings, residence times, concentrations, maintenance schedules, equipment conditions, environmental conditions, the composition and quality of the raw corn itself, the location where the corn was grown, as well as the growing season that produced the corn.

During fermentation, carbon dioxide arises from the metabolic conversion of sugars into ethanol by the yeast. This byproduct stream can be captured and sold to compressed gas markets, such as beverage or dry ice manufacturers. Often, however, it is released to the atmosphere because location and/or logistics make the sales and marketing of this gas economically unfeasible. In the future, however, the release of carbon dioxide may eventually be impacted by greenhouse gas emission constraints and regulations.

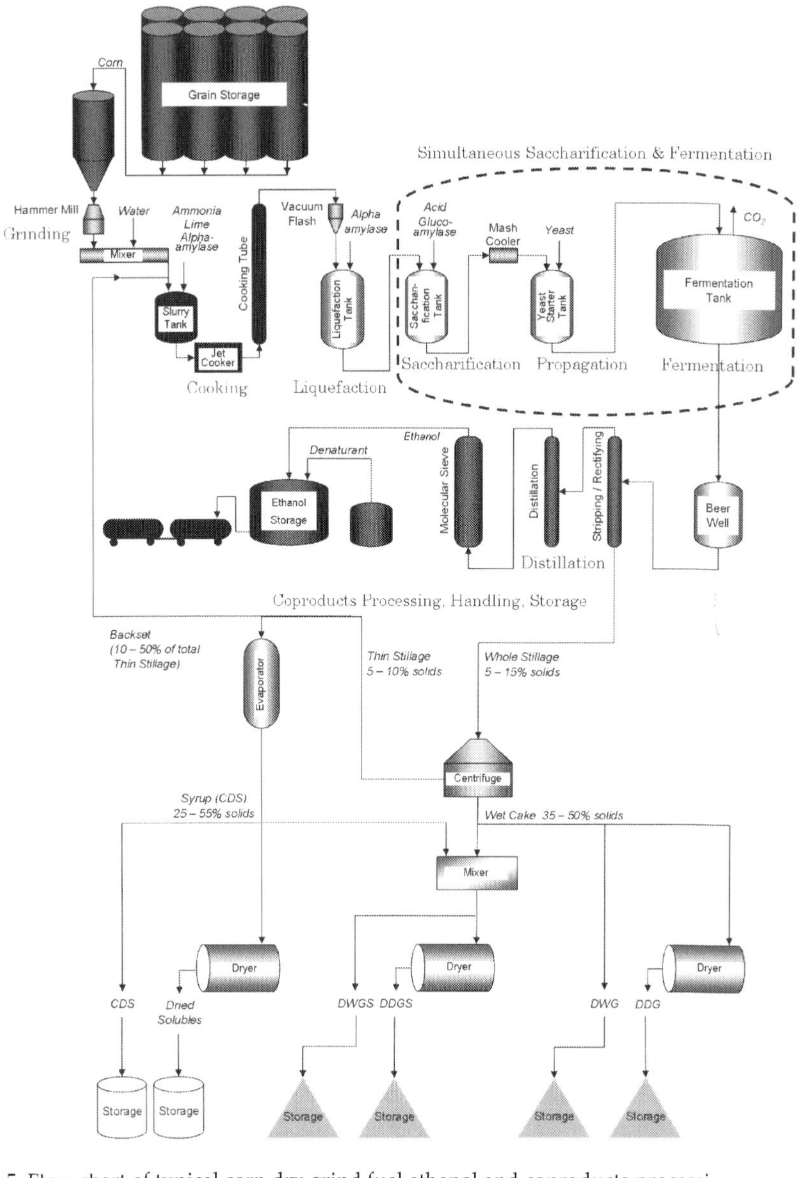

Fig. 5. Flow chart of typical corn dry grind fuel ethanol and coproducts processing.

Additional detailed information on ethanol and DDGS processing steps can be found in Tibelius (1996), Weigel et al. (1997), Dien et al. (2003), Jaques et al. (2003), Bothast and Schlicher (2005), Rausch and Belyea (2006), and Ingledew et al. (2009).

4. Importance of coproducts

DDGS from most modern U.S. fuel ethanol plants typically contains about 30% protein, 10% fat, at least 40% neutral detergent fiber, and up to 12% starch (Rosentrater and Muthukumarappan, 2006). Composition, however, can vary between plants and even within a single plant over time, due to a number of factors. For example, Table 1 summarizes composition of DDGS samples collected from five ethanol plants in South Dakota. On a dry basis, crude protein levels ranged from 28.3 to 31.8%; crude lipid varied between 9.4 and 11.0%; ash ranged from 4.1 to 13.3%. In terms of within-plant variability, the crude protein, crude lipid, and starch content all exhibited relatively low variation, whereas neutral detergent fiber (NDF), acid detergent fiber (ADF), and ash all had substantially higher variability.

Plant	Protein	Lipid	NDF	ADF	Starch	Ash
1	28.33^{b} (1.25)	10.76^{a} (1.00)	31.84^{b} (4.02)	15.56^{a} (2.29)	11.82^{a} (1.20)	13.27^{a} (3.10)
2	30.65^{a} (1.20)	9.75^{a} (1.05)	39.90^{a} (3.95)	15.21^{a} (3.95)	9.81^{a} (1.52)	12.84^{a} (2.56)
3	28.70^{a} (1.32)	10.98^{a} (0.95)	38.46^{a} (4.01)	17.89^{a} (4.01)	11.59^{a} (1.42)	11.52^{a} (3.05)
4	30.65^{a} (1.23)	9.40^{b} (0.16)	36.73^{a} (1.07)	15.28^{a} (0.49)	9.05^{b} (0.33)	4.13^{b} (0.21)
5	31.78^{a} (0.63)	9.50^{b} (0.41)	38.88^{a} (0.86)	17.24^{a} (1.12)	10.05^{a} (0.65)	4.48^{b} (0.22)

Table 1. Composition (% db) of DDGS from five ethanol plants in South Dakota (± 1 standard deviation in parentheses). Statistically significant differences among plants for a given nutrient are denoted by differing letters, α=0.05, LSD (adapted from Bhadra et al., 2009).

Furthermore, DDGS from 49 plants from 12 states were analyzed for proximate composition (Table 2) and amino acid profiles (Table 3) (UMN, 2011). Dry matter content varied from 86.2% to 92.4%, while protein varied from 27.3% to 33%. Crude fat content displayed even higher variability, and ranged from 3.5% to 13.5%; crude fiber ranged from 5.37% to 10.58%; and ash content varied from 2.97% to 9.84%. On average, geographic trends were not readily apparent for any of the nutrient components. In terms of amino acids, lysine ranged from 0.61% to 1.19%, but again, no geographic trends were apparent.

Some plants are beginning to implement various fractionation processes (either pre-fermentation or post-fermentation) in order to produce multiple product streams (RFA, 2009a). These new processes can lead to additional differences in DDGS nutrient levels. For example, various techniques for dry fractionation and wet fractionation have been developed to concentrate protein, fiber, and oil components from the endosperm (which contains the starch). This allows a highly-concentrated starch substrate to be introduced to the fermentation process, and it allows the other components to be used for human food applications. Singh and Johnston (2009) have provided an extensive discussion regarding various pre-fermentation fractionation approaches. On the other hand, post-fermentation fractionation techniques have also been examined. For example, Srinivasan et al. (2005) used a combination of (air classification and sieving to separate fiber particles from DDGS. Processes have also been developed to remove corn oil from thin stillage and CDS; although the resulting corn oil fractions cannot be used as food-grade oil, they can readily be converted into biodiesel. All of these approaches, if implemented commercially, will alter the composition of the resulting DDGS.

State	Plants Sampled	Dry Matter (%)	Crude Protein (%)	Crude Fat (%)	Crude Fiber (%)	Ash (%)
Minnesota	12	89.03	30.70	11.73	6.96	6.63
Illinois	6	89.72	29.98	11.48	7.26	5.60
Indiana	2	90.55	29.40	12.80	8.07	5.86
Iowa	7	88.92	31.23	10.27	7.57	5.76
Kentucky	3	90.57	29.43	9.77	9.28	4.47
Michigan	1	89.60	32.60	11.00	7.37	6.06
Missouri	2	87.90	30.45	10.25	7.17	5.39
Nebraska	4	89.02	30.40	11.35	8.13	4.23
New York	1	88.21	30.00	9.60	7.87	4.55
North Dakota	4	89.21	31.75	11.70	6.89	6.32
South Dakota	4	88.61	31.80	11.53	6.65	4.78
Wisconsin	3	89.68	31.70	11.63	7.59	5.77
Overall Average	49 (Total)	89.25	30.79	11.09	7.57	5.45

Table 2. Composition (% db) of DDGS samples from 49 ethanol plants from 12 states (adapted from UMN, 2011).

State	Plants Sampled	Agrinine (%)	Histidine (%)	Isoleucine (%)	Leucine (%)	Lysine (%)	Methionine (%)
Minnesota	12	1.39	0.84	1.20	3.63	0.99	0.61
Illinois	6	1.37	0.82	1.15	3.45	0.94	0.63
Indiana	2	1.19	0.79	1.08	3.28	0.85	0.60
Iowa	7	1.34	0.86	1.20	3.63	0.95	0.61
Kentucky	3	1.35	0.79	1.09	3.33	0.89	0.66
Michigan	1	1.28	0.86	1.18	3.67	0.87	0.71
Missouri	2	1.35	0.83	1.18	3.68	0.89	0.73
Nebraska	4	1.46	0.88	1.18	3.61	1.05	0.65
New York	1	1.46	0.85	1.21	3.64	1.04	0.61
North Dakota	4	1.37	0.88	1.24	3.76	0.97	0.65
South Dakota	4	1.47	0.87	1.22	3.70	1.08	0.62
Wisconsin	3	1.45	0.86	1.24	3.75	1.07	0.59
Overall Average	49	1.37	0.84	1.18	3.59	0.96	0.64

State	Plants Sampled	Phenylalanine (%)	Threonine (%)	Tryptophan (%)	Valine (%)	Tyrosine (%)
Minnesota	12	1.59	1.17	0.24	1.62	1.20
Illinois	6	1.51	1.11	0.22	1.52	1.22
Indiana	2	1.45	1.04	0.21	1.44	-
Iowa	7	1.57	1.14	0.25	1.60	-
Kentucky	3	1.48	1.09	0.26	1.43	-
Michigan	1	1.52	1.15	0.25	1.57	-
Missouri	2	1.53	1.15	0.24	1.58	-
Nebraska	4	1.58	1.15	0.26	1.58	1.14
New York	1	1.63	1.11	0.20	1.59	1.19
North Dakota	4	1.62	1.19	0.25	1.67	-
South Dakota	4	1.67	1.19	0.23	1.63	1.35
Wisconsin	3	1.65	1.14	0.22	1.64	1.25
Overall Average	49	1.56	1.13	0.24	1.57	1.22

Table 3. Amino acid profiles (% db) of DDGS samples from 49 ethanol plants from 12 states (adapted from UMN, 2011).

The U.S. ethanol industry's primary market for distillers grains has historically been as a commodity livestock feed. Most often this has been in the form of DDGS, and to a lesser degree in the form of DWG; the other coproducts are sold in much lower quantities than either DDGS or DWG and some are not always produced either). Feeding ethanol coproducts to animals is a practical method of utilizing these materials because they contain high nutrient levels, and they are digestible (to varying degrees) by most livestock. And, use of DDGS in animal feeds (instead of corn grain) helps to offset the corn which has been redirected to ethanol production. Over 80% of all distillers grains is used in beef and dairy diets; due to their ability to utilize high levels of fiber, ruminant animals have become the dominant consumers of DDGS. But, as livestock producers and animal nutritionists increase their knowledge, through research and experience, the swine and poultry markets are also increasing their consumption as well (UMN, 2011). Over the years, numerous research studies have been conducted on coproduct use in livestock diets, for both ruminant and monogastric feeds. Table 4 lists some of this research. Depending on the diet composition used, all livestock species have been shown to thrive at 10% DDGS inclusion, and most can tolerate levels up to 20% (or even more).

Species	Citation	Species	Citation
Beef		Dairy	
	Loy et al., 2007		Kleinschmit et al., 2007
	MacDonald et al., 2007		Anderson et al., 2006
	Martin et al., 2007		Kleinschmit et al., 2006
	Roeber et al., 2005		Leonardi et al., 2005
	Al-Suwaiegh et al., 2002		Birkelo et al., 2004
	Peter et al., 2000		McKendrick et al., 2003
	Lodge et al., 1997a		Al-Suwaiegh et al., 2002
	Lodge et al., 1997b		Liu et al., 2000
	Fron et al., 1996		Huang et al., 1999
	Klopfenstein, 1996		Schingoethe et al., 1999
	Ham et al., 1994		Batajoo and Shaver, 1998
	Larson et al., 1993		Nichols et al., 1998
	Donaldson et al., 1991		Clark and Armentano, 1997
	McCann et al., 1991		DePeters et al., 1997
			O'Mara et al., 1997
			Zhu et al., 1997
			Arosemena et al., 1995
			Murphy et al., 1995
			Powers et al., 1995
			Ham et al., 1994
			Clark and Armentano, 1993
Swine		Poultry	
	Stein and Shurson, 2009		Waldroup et al., 2007
	Pedersen et al., 2007		Wang et al., 2007a
	Widmer et al., 2007		Wang et al., 2007b
	Fastinger et al., 2007		Wang et al., 2007c
	Stein et al., 2006		Batal and Dale, 2006
	Whitney et al., 2006a		Fastinger et al., 2006
	Whitney et al., 2006b		Martinez-Amezcua et al., 2006
	Whitney et al., 2006c		Noll, 2006
	Whitney et al., 2006d		Lumpkins and Batal, 2005
	Nyachoti et al., 2005		Lumpkins et al., 2005
	Whitney and Shurson, 2004		Roberson et al., 2005
	Gralapp et al., 2002		Biggs et al., 2004
	Spiehs et al., 2002		Lumpkins et al., 2004
	Nicolai et al., 1999		Martinez Amezcua et al., 2004
	Cromwell et al., 1993		Batal and Dale, 2003
			Roberson, 2003
			Cromwell et al., 1993

Table 4. Summary of livestock research on fuel ethanol coproducts.

DDGS use in livestock diets has continued to increase over the years. Predictions of peak potential for DDGS use in domestic U.S. beef, dairy, swine, and poultry markets have estimated that between 40 and 60 million t could be used in the U.S. each year, depending upon inclusion rates for each species (Staff, 2005; Cooper, 2006; U.S. Grains Council, 2007). Globally, the need for protein-based animal feeds continues to grow. Of the 23 million t of DDGS produced in 2008 (RFA, 2009b), 4.5 million t were exported to international markets (FAS, 2009); this accounted for nearly 20% of the U.S. DDGS production that year (Figure 6). And the potential for global exports is projected to increase for the foreseeable future (U.S. Grains Council, 2007).

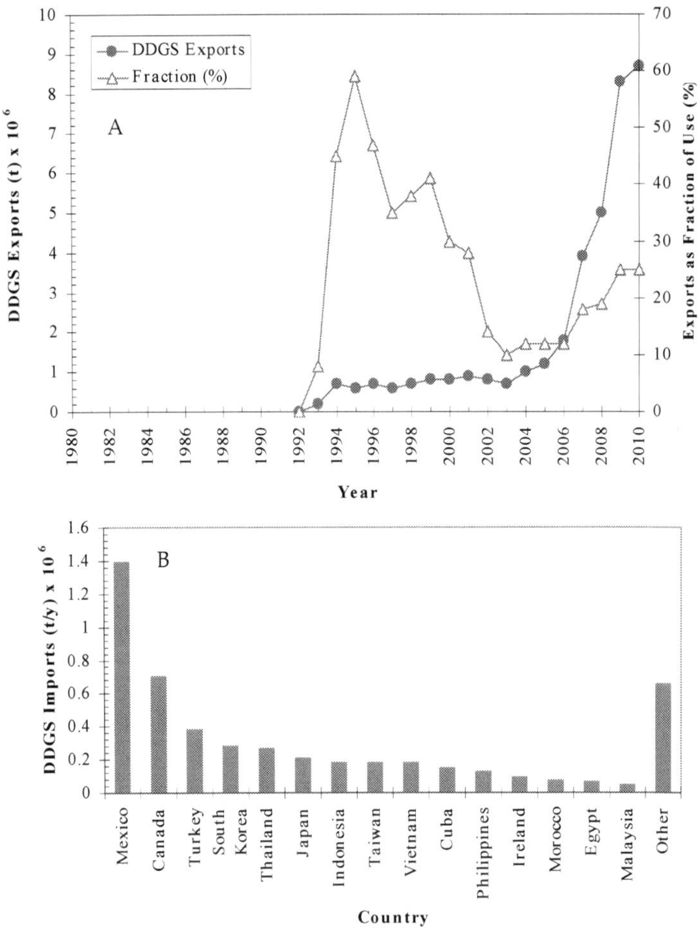

Fig. 6. A. U.S. DDGS exports in 2008. B. Countries who imported DDGS in 2008 (adapted from Hoffman and Baker, 2010).

Not only are coproducts important to the livestock industry as feed ingredients, but they are also essential to the sustainability of the fuel ethanol industry itself. In fact, the sale of distillers grains (all types - dry and wet) contributes substantially to the economic viability of each ethanol plant (sales can generally contribute between 10 and 20% of a plant's total revenue stream (Figure 7), but at times it can be as high as 40%), depending upon the market conditions for corn, ethanol, and distillers grains. This is the reason why these process residues are referred to as "coproducts", instead of "byproducts" or "waste products"; they truly are products in their own right along with the fuel.

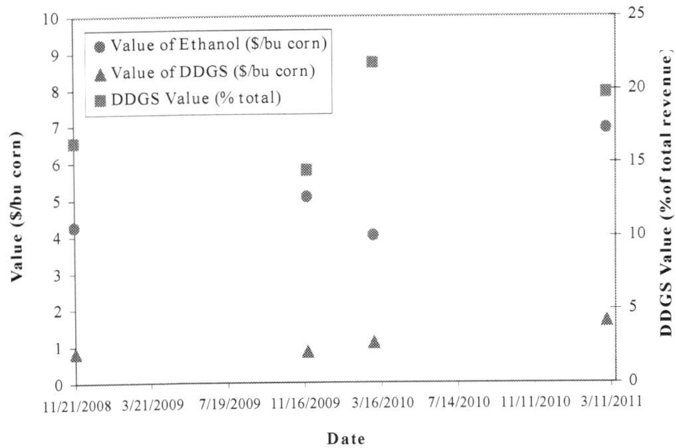

Fig. 7. Some relative comparisons of the value of DDGS and fuel ethanol to ethanol plant profits (adapted from DTN, 2011).

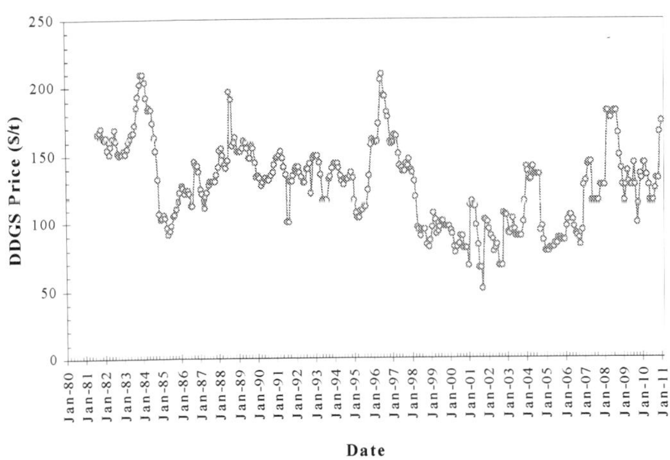

Fig. 8. DDGS sales price over time (monthly averages) (adapted from ERS, 2011).

So the sales price of DDGS is important to ethanol manufacturers and livestock producers alike. Over the last three decades, the price for DDGS has ranged from approximately \$50.71/t up to \$209.44/t (Figure 8). DDGS and corn prices have historically paralleled each other very closely (Figure 9). This relationship has been quite strong over the last several

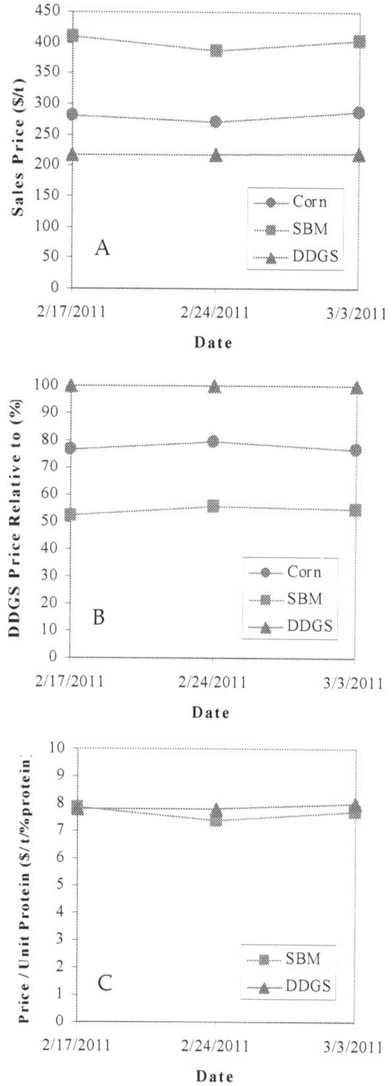

Fig. 9. A. Some comparisons of DDGS, soybean meal (SBM), and corn sales prices. B. Relative price comparisons. C. Cost comparisons on a per unit protein basis (adapted from DTN, 2011).

years. This is not surprising, as DDGS is most often used to replace corn in livestock diet formulations. DDGS has increasingly been used as a replacement for soybean meal as well, primarily as a source of protein. Even so, DDGS has historically been sold at a discounted price vis-à-vis both corn and soybean meal. This has been true on a volumetric unit basis, as well as per unit protein basis (Figure 9).

5. Coproduct evolution

The ethanol industry is dynamic and has been evolving over the years in order to overcome various challenges associated with both fuel and coproduct processing and use (Rosentrater, 2007). A modern dry grind ethanol plant is considerably different from the inefficient, input-intensive Gasohol plants of the 1970s. New developments and technological innovations, to name but a few, include more effective enzymes, higher starch conversions, better fermentations, cold cook technologies, improved drying systems, decreased energy consumption throughout the plant, increased water efficiency and recycling, and decreased emissions. Energy and mass balances are becoming more efficient over time. Many of these improvements can be attributed to the design and operation of the equipment used in modern ethanol plants. A large part is also due to computer-based instrumentation and control systems.

Many formal and informal studies have been devoted to adjusting existing processes in order to improve and optimize the quality of the coproducts which are produced. Ethanol companies have recognized the need to produce more consistent, higher quality DDGS which will better serve the needs of livestock producers. The sale of DDGS and the other coproducts has been one key to the industry's success so far, and will continue to be important to the long-term sustainability of the industry. Although the majority of DDGS is currently consumed by beef and dairy cattle, use in monogastric diets, especially swine and poultry, continues to increase. And use in non-traditional species, such as fish, horses, and pets has been increasing as well.

Additionally, there has been considerable interest in developing improved mechanisms for delivering and feeding DDGS to livestock vis-à-vis pelleting/densification (Figure 10). This is a processing operation that could result in significantly better storage and handling characteristics of the DDGS, and it would drastically lower the cost of rail transportation and logistics (due to increased bulk density and better flowability) (Figure 11). Pelleting could also broaden the use of DDGS domestically (e.g., improved ability to use DDGS for rangeland beef cattle feeding and dairy cattle feeding) as well as globally (e.g., increased bulk density would result in considerable freight savings in bulk vessels and containers).

There are also many new developments underway in terms of evolving coproducts. These will ultimately result in more value streams from the corn kernel (i.e., upstream fractionation) as well as the resulting distillers grains (i.e., downstream fractionation) (Figure 12). Effective fractionation can result in the separation of high-, mid-, and low-value components. Many plants have begun adding capabilities to concentrate nutrient streams such as oil, protein, and fiber into specific fractions, which can then be used for targeted markets and specific uses. These new processes are resulting in new types of distillers grains (Figure 13).

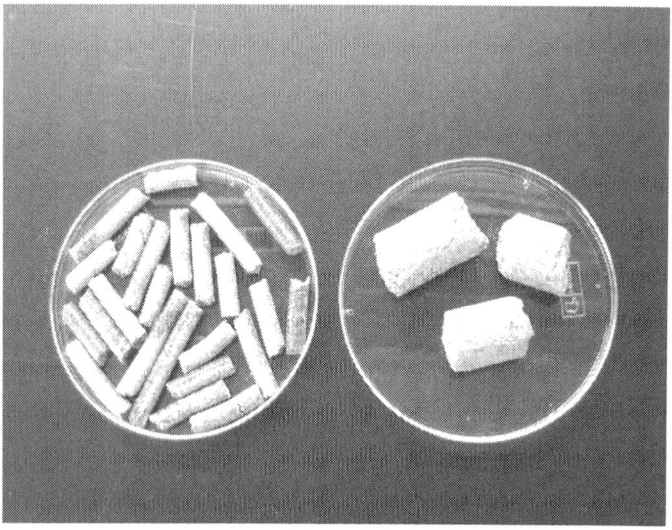

Fig. 10. Pelleting is a unit operation that can improve the utility of DDGS, because it improves storage and handling characteristics, and allows more effective use in dairy cattle feeding and range land settings for beef cattle.

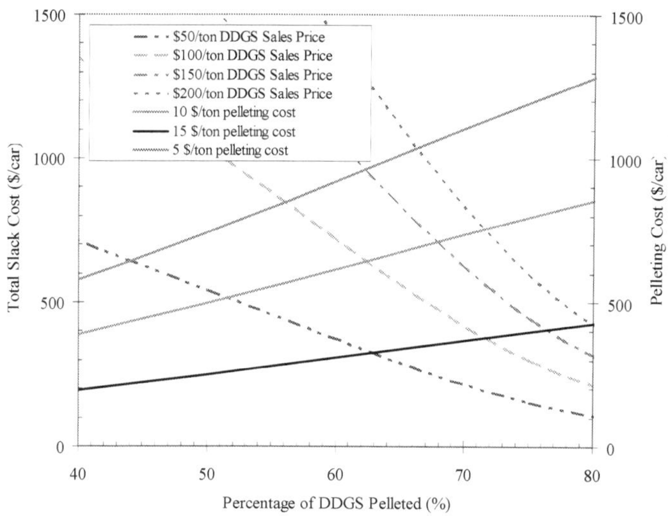

Fig. 11. By pelleting, empty space in rail cars is minimized during shipping. Techno-economic analysis of the resulting slack (i.e., wasted space) costs and costs of pelleting for each rail car due to differing DDGS sales prices and pelleting costs indicates the proportion of DDGS which needs to be pelleted in order to achieve breakeven for this process (adapted from Rosentrater and Kongar, 2009).

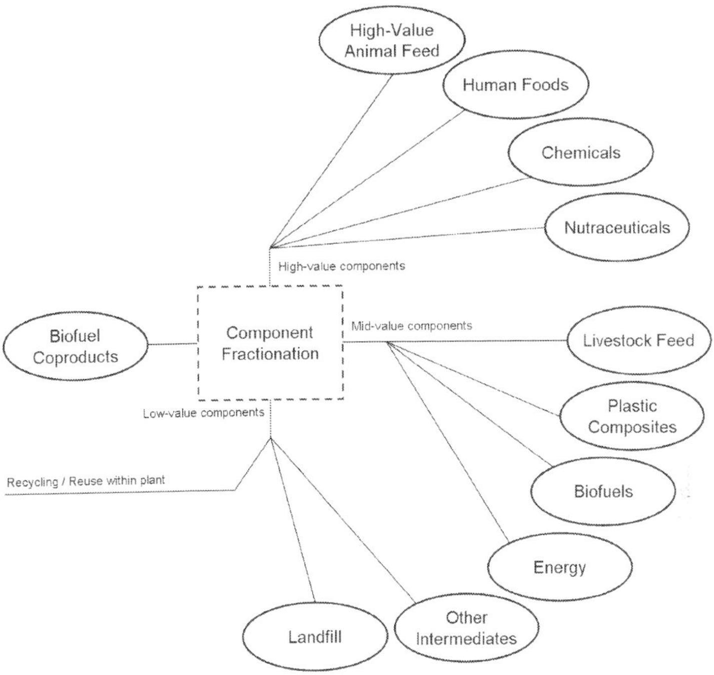

Fig. 12. Fractionation of DDGS into high-, mid-, and low-value components offers the opportunity for new value streams.

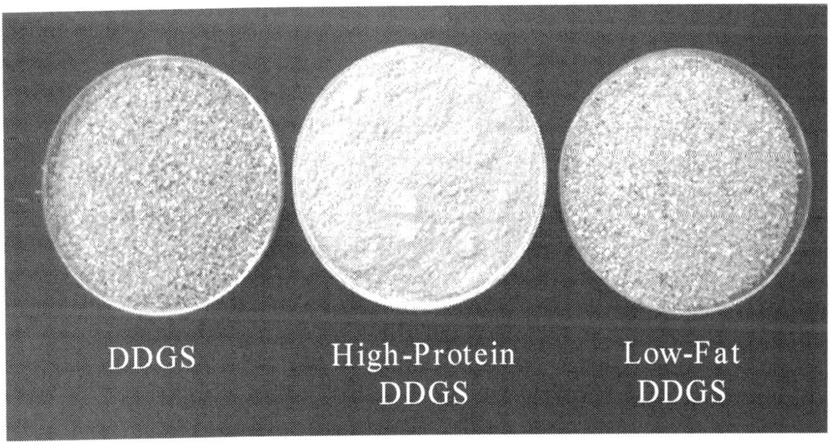

Fig. 13. Examples of traditional, unmodified DDGS and some fractionated products (e.g., high-protein and low-fat DDGS) which are becoming commercially available in the marketplace.

For example, if the lipids are removed from the DDGS (Figure 14), they can readily be converted into biodiesel, although they cannot be used for food grade corn oil, because they are too degraded structurally. Another example is concentrated proteins, which can be used for high-value animal feeds (such as aquaculture or pet foods), or other feed applications which require high protein levels. Additionally, DDGS proteins can be used in human foods (Figure 15). Furthermore, other components, such as amino acids, organic acids, or even nutraceutical compounds (such as phytosterols and tycopherols) can be harvested and used in high-value applications.

Mid-value components, such as fiber, can be used as biofillers for plastic composites (Figure 16), as feedstocks for the production of bioenergy (e.g., heat and electricity at the ethanol plant via thermochemical conversion) (Figure 17), or, after pretreatment to break down the lignocellulosic structures, as substrates for the further production of ethanol or other biofuels.

In terms of potential uses for the low-value components, hopefully mechanisms will be developed to alter their structures and render them useful, so that they will not have to be landfilled. Fertilizers are necessary in order to sustainably maintain the flow of corn grain into the ethanol plant, so land application may be an appropriate venue for the low value components.

As these process modifications are developed, validated, and commercially implemented, improvements in the generated coproducts will be realized and unique materials will be produced. Of course, these new products will require extensive investigation in order to determine how to optimally use them and to quantify their value propositions in the marketplace.

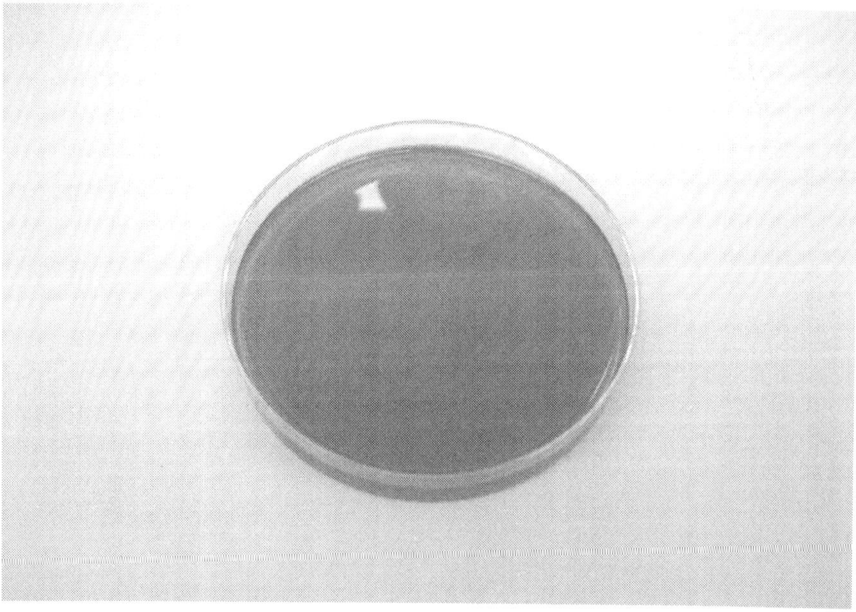

Fig. 14. Corn oil which has been extracted from DDGS can be used to manufacture biodiesel.

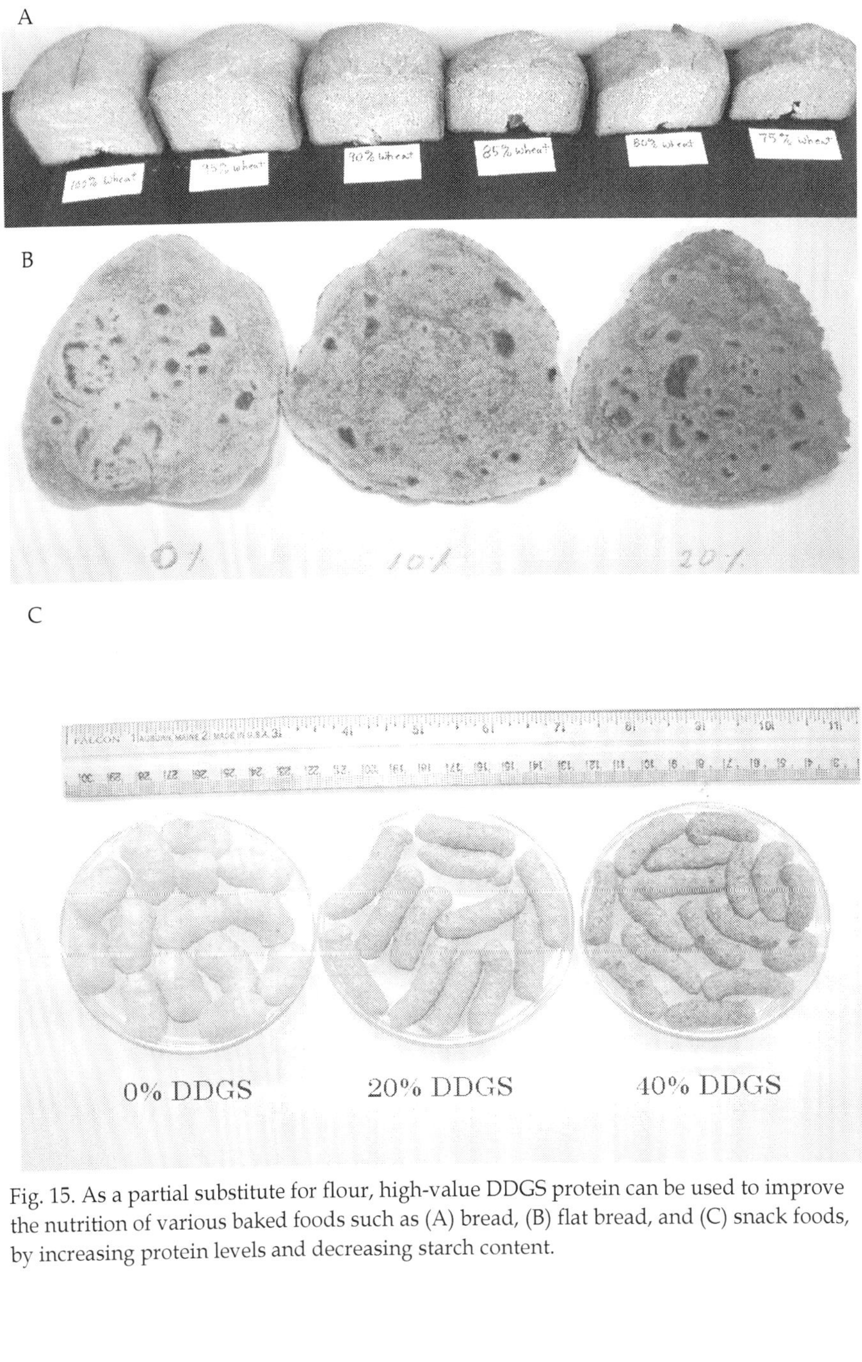

Fig. 15. As a partial substitute for flour, high-value DDGS protein can be used to improve the nutrition of various baked foods such as (A) bread, (B) flat bread, and (C) snack foods, by increasing protein levels and decreasing starch content.

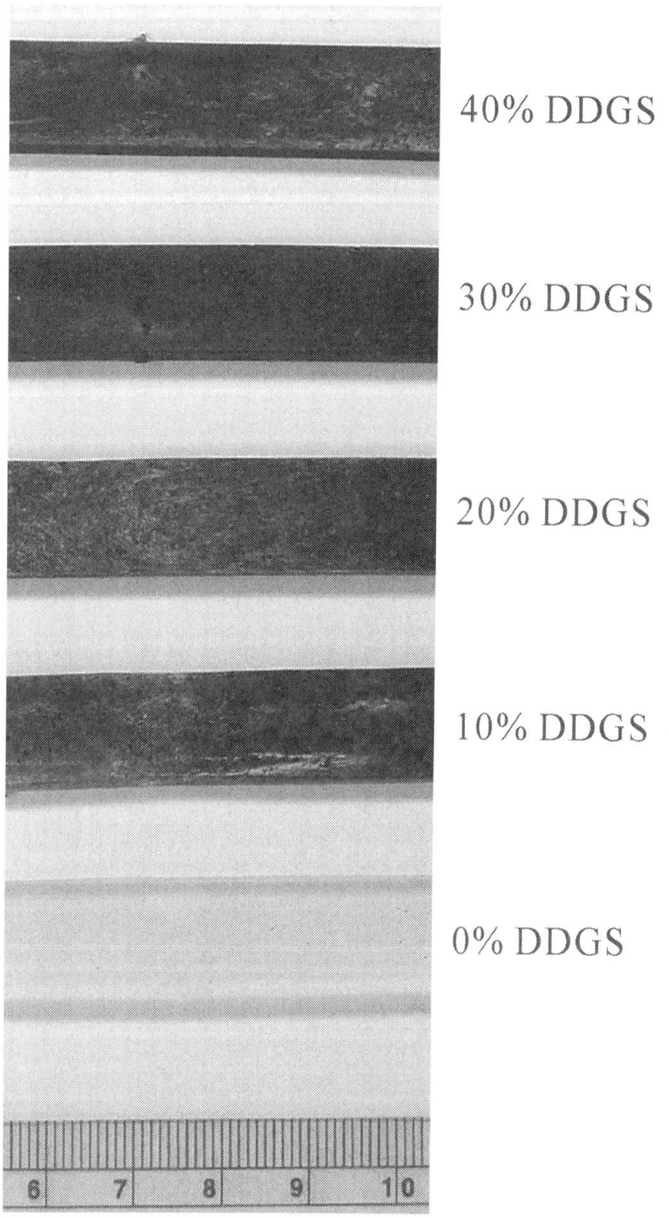

Fig. 16. Mid-value or low-value fractions from DDGS (such as fiber) have been shown to be an effective filler in plastics, replacing petroleum additives and increasing biodegradability. Scale bar indicates mm.

Fig. 17. Mid-value or low-value fractions from DDGS (such as fiber) can be thermochemically converted into biochar, which can subsequently be used to produce energy, fertilizer, or as a precursor to other bio-based materials.

6. Conclusion

The fuel ethanol industry has been rapidly expanding in recent years in response to government mandates, but also due to increased demand for alternative fuels. This has become especially true as the price of gasoline has escalated and fluctuated so drastically, and the consumer has begun to perceive fuel prices as problematic. Corn-based ethanol is not the entire solution to our transportation fuel needs. But it is clearly a key component to the overall goal of energy independence. Corn ethanol will continue to play a leading role in the emerging bioeconomy, as it has proven the effectiveness of industrial-scale biotechnology and bioprocessing for the production of fuel. And it has set the stage for advanced biorefineries and manufacturing techniques that will produce the next several generations of advanced biofuels. As the biofuel industry continues to evolve, coproduct materials (which ultimately may take a variety of forms, from a variety of biomass substrates) will remain a cornerstone to resource and economic sustainability. A promising mechanism to achieve sustainability will entail integrated systems (Figure 18), where material and energy streams cycle and recycle (i.e., upstream outputs become downstream inputs) between various components of a biorefinery, animal feeding operation, energy (i.e., heat, electricity, steam, etc.) production system, feedstock production system, and other systems. By integrating these various components, a diversified portfolio will not only produce fuel, but also fertilizer, feed, food, industrial products, energy, and most importantly, will be self-sustaining.

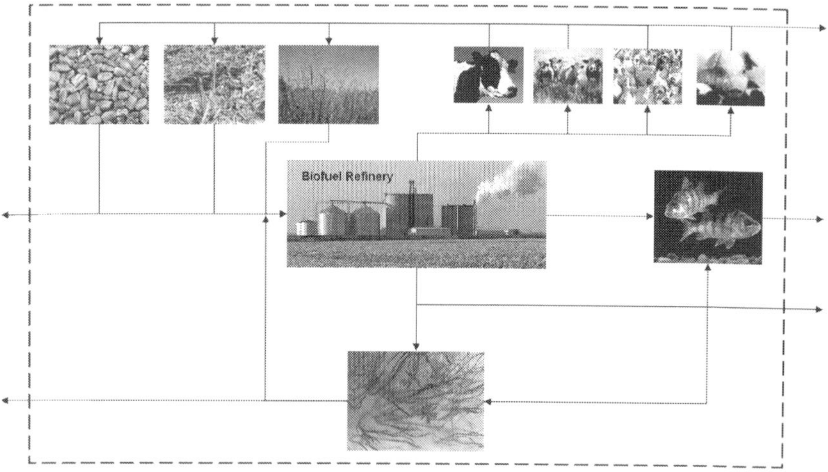

Fig. 18. Coproducts such as DDGS will continue to play a key role as the biofuel industry evolves and becomes more fully integrated. This figure illustrates one such concept.

7. References

Agrawal, R., N. R. Singh, F. H. Ribeiro, and W. N. Delgass. (2007). Sustainable fuel for the transportation sector. Proceedings of the National Academy of Sciences 104(12): 4828-4833.

Alexander, C. and C. Hurt. (2007). Biofuels and their impact on food prices. Bioenergy ID-346-W. Department of Agricultural Economics, Purdue University: West Lafayette, IN.

Al-Suwaiegh, S., K. C. Fanning, R. J. Grant, C. T. Milton, and T. J. Klopfenstein. (2002). Utilization of distillers grains from the fermentation of sorghum or corn in diets for finishing beef and lactating dairy cattle. J. Anim. Sci. 80: 1105-1111.

Anderson, J. L., D. J. Schingoethe, K. F. Kalscheur, and A. R. Hippen. (2006). Evaluation of dried and wet distillers grains included at two concentrations in the diets of lactating dairy cows. J. Dairy Sci. 89: 3133–3142.

Arosemena, A., E. J. DePeters, and J. G. Fadel. (1995). Extent of variability in nutrient composition within selected by-product feedstuffs. Animal Feed Sci. and Technology 54: 103-120.

Batajoo, K. K. and R. D. Shaver. (1998). In situ dry matter, crude protein, and starch degradabilities of selected grains and by-product feeds. Animal Feed Science Technology 71: 165-176.

Batal, A. and N. M. Dale. (2003). Mineral composition of distillers dried grains with solubles. J. Appl. Poult. Res. 12: 400-403.

Batal, A. B. and N. M. Dale. (2006). True metabolizable energy and amino acid digestibility of distillers dried grains with solubles. J. Appl. Poult. Res. 15: 89-93.

Bhadra, R., K. A. Rosentrater, and K. Muthukumarappan. (2009). Cross-sectional staining and the surface properties of DDGS and their influence on flowability. Cereal Chemistry 86(4): 410-420.

Biggs, P. E., M. E. Persia, K. W. Koelkebeck, and C. M. Parsons. (2004). Further evaluation of nonfeed removal methods for molting programs. Poultry Sci. 83(5) 745-752.

Birkelo, C. P., M. J. Brouk, and D. J. Schingoethe. (2004). The energy content of wet corn distillers grins for lactating dairy cows. J. Dairy Sci. 87: 1815-1819.

Bothast, R. and M. Schlicher. (2005). Biotechnological processes for conversion of corn into ethanol. Applied Microbiology and Biotechnology 67(1): 19-25.

Cassman, K. G. (2007). Climate change, biofuels, and global food security. Environmental Research Letters 2(011002): DOI # 10.1088/1748-9326/2/1/011002.

Cassman, K. G. and A. J. Liska. (2007). Food and fuel for all: realistic or foolish? Biofuels, Bioproducts and Biorefining 1(1): 18-23.

Cassman, K. G., V. Eidman, and E. Simpson. (2006). Convergence of agriculture and energy: implications for research policy, QTA2006-3. Council for Agricultural Science and Technology: Ames, IA.

Clark, P. W. and L. E. Armentano. (1993). Effectiveness of neutral detergent fiber in whole cottonseed and dried distillers grains compared with alfalfa haylage. J. Dairy Sci. 76: 2644-2650.

Clark, P. W. and L. E. Armentano. (1997). Replacement of alfalfa neutral detergent fiber with a combination of nonforage fiber sources. J. Dairy Sci. 80: 675-680.

Cooper, G. (2006). A brief, encouraging look at 'theoretical' distillers grains markets. Distillers Grains Quarterly 1(1): 14-17.

Cromwell, G. L., K. L. Herkelman, and T. S. Stahly. (1993). Physical, chemical, and nutritional characteristics of distillers dried grains with solubles for chicks and pigs. J. Anim. Sci. 71: 679-686.

Dale, B. E. (2007). Thinking clearly about biofuels: ending the irrelevant 'net energy' debate and developing better performance metrics for alternative fuels. Biofuels, Bioproducts, and Biorefining 1: 14-17.

De La Torre Ugarte, D. G., M. E. Walsh, H. Shapouri, and S. P. Slinsky. (2000). The economic impacts of bioenergy crop production on U.S. agriculture, Agricultural Economic Report 816. USDA Office of the Chief Economist, U.S. Department of Agriculture: Washington, D.C.

DePeters, E .J., J. G. Fadel, and A. Arosemena. (1997). Digestion kinetics of neutral detergent fiber and chemical composition within some selected by-product feedstuffs. Animal Feed Science Technology 67: 127-140.

Dewulf, J., H. Van Langenhove, and B. Van De Velde. (2005). Energy-based efficiency and renewability assessment of biofuel production. Environmental Science and Technology 39(10): 3878-3882.

Dien, B. S., R. J. Bothast, N. N. Nichols, and M. A. Cotta. (2003). The U.S. corn ethanol industry: an overview of current technology and future prospects. In: The Third International Starch Technology Conference - Coproducts Program Proceedings, eds. M. Tumbleson, V. Singh, and K. Rausch, 2-4 June, 2003, University of Illinois, Urbana, IL, pp. 10-21.

Donaldson, R. S., M. A. McCann, H. E. Amos, and C. S. Hoveland. (1991). Protein and fiber digestion by steers grazing winter annuals and supplemented with ruminal escape protein. J. Anim. Sci. 69: 3067-3071.

DTN. (2011). DTN Weekly Distillers Grains Update. Available online: www.dtnprogressivefarmer.com.

ERS. (2011). Feed Grains Database: Yearbook Tables. Economic Research Service, U.S. Department of Agriculture: Washington, D.C. Available online: www.ers.usda.gov/data/feedgrains/.

FAS. (2009). Foreign Agricultural Service, U. S. Department of Agriculture: Washington, D.C. Available online: www.fas.usda.gov/.

Fastinger, N. D. and D. C. Mahan. (2006). Determination of the ileal amino acid and energy digestibilities of corn distillers dried grains with solubles using grower-finisher pigs. J. Anim. 84: 1722–1728.

Fastinger, N. D., J. D. Latshaw, and D. C. Mahan. (2006). Amino acid availability and true metabolizable energy content of corn distillers dried grains with solubles in adult cecectomized roosters. Poultry Sci. 85: 1212-1216.

Fron, M., H. Madeira, C. Richards, and M. Morrison. (1996). The impact of feeding condensed distillers byproducts on rumen microbiology and metabolism. Animal Feed Sci. Technology 61: 235-245.

Gralapp, A. K., W. J. Powers, M. A. Faust, and D. S. Bundy. (2002). Effects of dietary ingredients on manure characteristics and odorous emissions from swine. J. Anim. Sci. 80: 1512-1519.

Ham, G. A., R. A. Stock, T. J. Klopfenstein, E. M. Larson, D. H. Shain, and R. P. Huffman. (1994). Wet corn distillers byproducts compared with dried corn distillers grains with solubles as a source of protein and energy for ruminants. J. Anim. Sci. 72: 3246-3257.

Hoffman, L. and A. Baker. (2010). Market issues and prospects for U.S. distillers' grains: supply, use, and price relationships. Report FDS-10k-01. United States Department of Agriculture, Economic Research Service: Washington, D.C. Available online: www.ers.usda.gov.

Huang, H. J., P. Wen-Shyg Chiou, C. R. Chen, J. K. Chiang, and B. Yu. (1999). Effects of dried rice distillers' and grain supplementation on the performance of lactating cows. Animal Feed Science and Technology 77: 303-315.

Ingledew, W. M., D. R. Kelsall, G. D. Austin, and C. Kluhspies. (2009). The Alcohol Textbook, 5th Edition. W. M. Ingledew, D. R. Kelsall, G. D. Austin, and C. Kluhspies, ed. Nottingham University Press: Nottingham, UK.

Jaques, K. A., T. P. Lyons, and D. R. Kelsall. (2003). The Alcohol Textbook, 3rd ed. Nottingham University Press: Nottingham, UK.

Kleinschmit, D. H., D. J. Schingoethe, A. R. Hippen, and K. F. Kalscheur. (2007). Dried distillers grains plus solubles with corn silage or alfalfa hay as the primary forage source in dairy cow diets. J. Dairy Sci. 90: 5587-5599.

Kleinschmit, D. H., D. J. Schingoethe, K. F. Kalscheur, and A. R. Hippen. (2006). Evaluation of various sources of corn dried distillers grains plus solubles for lactating dairy cattle. J. Dairy Sci. 89: 4784-4794.

Klopfenstein, T. (1996). Distillers grains as an energy source and effect of drying on protein availability. Animal Feed Science Technology 60: 201-207.

Larson, E. M., R. A. Stock, T. J. Klopfenstein, M. H. Sindt, and R. P. Huffman. (1993). Feeding value of wet distillers byproducts for finishing ruminants. J. Anim. Sci. 71: 2228-2236.

Leonardi, C., S. Bertics, and L. E. Armentano. (2005). Effect of increasing oil from distillers grains or corn oil on lactation performance. J. Dairy Sci. 88: 2820-2827.

Liu, C., D. J. Schingoethe, and G. A. Stegeman. (2000). Corn distillers grains versus a blend of protein supplements with or without ruminally protected amino acids for lactating cows. J. Dairy Sci. 83: 2075–2084.

Lodge, S. L., R. A. Stock, T. J. Klopfenstein, D. H. Shain, and D. W. Herold. (1997a). Evaluation of corn and sorghum distillers byproducts. J. Anim. Sci. 75: 37–43.

Lodge, S. L., R. A. Stock, T. J. Klopfenstein, D. H. Shain, and D. W. Herold. (1997b). Evaluation of wet distillers composite for finishing ruminants. J. Anim. Sci. 75: 44-50.

Loy, T. W., J. C. MacDonald, T. J. Klopfenstein, and G. E. Erickson. (2007). Effect of distillers grains or corn supplementation frequency on forage intake and digestibility. J. Anim. Sci. 85: 2625–2630.

Lumpkins, B. S. and A. B. Batal. (2005). The bioavailability of lysine and phosphorus in distillers dried grains with solubles. Poultry Science 84: 581-586.

Lumpkins, B. S., A. B. Batal, and N. M. Dale. (2004). Evaluation of distillers dried grains with solubles as a feed ingredient for broilers. Poultry Science 83: 1891-1896.

Lumpkins, B., A. Batal, and N. Dale. (2005). Use of distillers dried grains plus solubles in laying hen diets. J. Appl. Poult. Res. 14: 25-31.

Lynd, L. R. and M. Q. Wang. (2004). A product-nonspecific framework for evaluating the potential of biomass-based products to displace fossil fuels. Journal of Industrial Ecology 7(3-4): 17-32.

MacDonald, J. C., T. J. Klopfenstein, G. E. Erickson, and W. A. Griffin. (2007). Effects of dried distillers grains and equivalent undegradable intake protein or ether extract on performance and forage intake of heifers grazing smooth bromegrass pastures. J. Anim. Sci. 85: 2614–2624.

Martin, J. L., A. S. Cupp, R. J. Rasby, Z. C. Hall, and R. N. Funston. (2007). Utilization of dried distillers grains for developing beef heifers. J. Anim. Sci. 85: 2298–2303.

Martinez Amezcua, C., C. M. Parsons, and S. L. Noll. (2004). Content and relative bioavailability of phosphorus in distillers dried grains with solubles in chicks. Poultry Sci. 83: 971-976.

Martinez-Amezcua, C., C. M. Parsons, and D. H. Baker. (2006). Effect of microbial phytase and citric acid on phosphorus bioavailability, apparent metabolizable energy, and amino acid digestibility in distillers dried grains with solubles in chicks. Poultry Sci. 85: 470–475.

McCann, M. A., R. S. Donaldson, H. E. Amos, and C. S. Hoveland. (1991). Ruminal escape protein supplementation and zeranol implantation effects on performance of steers grazing winter annuals. J. Anim. Sci. 69: 3112-3117.

McKendrick, E. J., D. J. Roberts, and N. W. Offer. (2003). The value of malt distillers' grains ensiled with molassed sugar beet pellets as a feed for dairy cows. Grass and Forage Sci. 58: 287-294.

Murphy, J. J., J. F. Connolly, and G. P. McNeill. (1995). Effects on milk fat composition and cow performance of feeding concentrates containing full fat rapeseed and maize distillers grains on grass-silage based diets. Production Science 44: 1-11.

Nichols, J. R., D. J. Schingoethe, H. A. Maiga, M. J. Brouk, and M. S. Piepenbrink. (1998). Evaluation of corn distillers grains and ruminally protected lysine and methionine for lactating dairy cows. J. Dairy Sci. 81: 482-491.

Nicolai, R. E., M. J. Spiehs, G. C. Shurson, and M. H. Whitney. (1999). Deep pit simulator modeling protocol for individual metabolism crates during diet studies. ASAE Paper No. 994135. ASAE, St. Joseph, MI.

Noll, S. (2006). Maximizing alternative protein ingredient use in market turkey diets. Midwest Poultry Research Program Paper. Available online: www.ddgs.umn.edu.

Nyachoti, C. M., J. D. House, B. A. Slominski, and I. R. Seddon. (2005). Energy and nutrient digestibilities in wheat dried distillers' grains with solubles fed to growing pigs. J. Sci. Food Agric. 85: 2581-2586.

O'Mara, F. P., J. J. Murphy, and M. Rath. (1997). The amino acid composition of protein feedstuffs before and after ruminal incubation and after subsequent passage through the intestines of dairy cows. J. Anim. Sci. 75: 1941-1949.

Pedersen, C., M. G. Boersma, and H. H. Stein. (2007). Digestibility of energy and phosphorus in ten samples of distillers dried grains with solubles fed to growing pigs. J. Anim. Sci. 85: 1168–1176.

Peter, C. M., D. B. Faulkner, N. R. Merchen, D. F. Parrett, T. G. Nash, and J. M. Dahlquist. (2000). The effects of corn milling coproducts on growth performance and diet digestibility by beef cattle. J. Anim. Sci. 78: 1–6.

Powers, W. J., H. H. Van Horn, B. Harris, Jr., and C. J. Wilcox. (1995). Effects of variable sources of distillers dried grains plus solubles on milk yield and composition. J. Dairy Sci. 78: 388-396.

Rausch, K.D. and R.L. Belyea. (2006). The future of coproducts from corn processing. Applied Biochemistry and Biotechnology 128: 47-86.

RFA. (2009a). Growing Innovation. 2009 Ethanol Industry Outlook. Renewable Fuels Association. Washington, D.C. Available at: www.ethanolrfa.org.

RFA. (2009b). Industry resources: co-products. Renewable Fuels Association: Washington, D.C. Available online: www.ethanolrfa.org.

RFA. (2011). Biorefinery locations. Renewable Fuels Association: Washington, D.C. Available online: www.ethanolrfa.org.

Roberson, K. D. (2003). Use of dried distillers' grains with solubles in growing-finishing diets of turkey hens. International Journal of Poultry Sci. 2 (6): 389-393.

Roberson, K. D., J. L. Kalbfleisch, W. Pan, and R. A. Charbeneau. (2005). Effect of corn distiller's dried grains with solubles at various levels on performance of laying hens and egg yolk color. International Journal of Poultry Sci. 4 (2): 44-51.

Roeber, D. L., R. K. Gill, and A. DiCostanzo. (2005). Meat quality responses to feeding distiller's grains to finishing Holstein steers. J. Anim. Sci. 83: 2455-2460.

Rosentrater, K. A. (2007). Ethanol processing coproducts - a review of some current constraints and potential directions. International Sugar Journal 109(1307): 1-12.

Rosentrater, K. A. and E. Kongar. (2009). Modeling the effects of pelleting on the logistics of distillers grains shipping. Bioresource Technology 100: 6550-6558.

Rosentrater, K. A. and K. Muthukumarappan. (2006). Corn ethanol coproducts: generation, properties, and future prospects. International Sugar Journal 108(1295): 648-657.

Schingoethe, D. J., M. J. Brouk, and C. P. Birkelo. (1999). Milk production and composition from cows fed wet corn distillers grains. J. Dairy Sci. 82: 574-580.

Singh, V. and D. B. Johnston. (2009). Fractionation technologies for dry-grind corn processing. Pages 193-207 in: The Alcohol Textbook, 5th Ed. M. W. Ingledew, D. R. Kelsall, G. D. Austin, and C. Kluhspies, ed. Nottingham University Press: Nottingham, UK.

Spiehs, M. J., M. H. Whitney, and G. C. Shurson. (2002). Nutrient database for distiller's dried grains with solubles produced from new ethanol plants in Minnesota and South Dakota. J. Anim. Sci. 80: 2639-2645.

Srinivasan, R., R. A. Moreau, K. D. Rausch, R. L. Belyea, M. D. Tumbleson, and V. Singh. (2005). Separation of fiber from distillers dried grains with solubles (DDGS) using sieving and elutriation. Cereal Chemistry 82: 528-533.

Staff, C. H. (2005). Question and answer. Biofuels Journal 3(4): 26-27.

Stein, H. H. and G. C. Shurson. (2009). The use and application of distillers dried grains with solubles in swine diets. J. Anim. Sci. 87: 1292-1303. doi:10.2527/jas.2008-1290.

Stein, H. H., M. L. Gibson, C. Pedersen, and M. G. Boersma. (2006). Amino acid and energy digestibility in ten samples of distillers dried grain with solubles fed to growing pigs. J. Anim. Sci. 84: 853-860.

Tibelius, C. (1996). Coproducts and Near Coproducts of Fuel Ethanol Fermentation from Grain. Agriculture and Agri-Food Canada - Canadian Green Plan Ethanol Program: Starchy Waste Streams Evaluation Project. Available online: http://res2.agr.ca/publications/cfar/index_e.htm.

U.S. EIA. (2011). Annual Energy Review. Energy Information Administration, U.S. Department of Energy: Washington, D.C. Available online: www.eia.doe.gov/emeu/aer/.

U.S. Grains Council. (2007). An Independent Review of US Grains Council Efforts to Promote DDGS Exports. U.S. Grains Council: Washington, D.C. Available online: www.grains.org/ddgs-information.

UMN. (2011). The value and use of distillers grains by-products in livestock and poultry feeds. University of Minnesota: Minneapolis, MN. Available online: www.ddgs.umn.edu/.

Urbanchuk, J. M. (2009). Contribution of the Ethanol Industry to the Economy of the United States. LECG: Wayne, PA.

Waldroup, P. W., Z. Wang, C. Coto, S. Cerrate, and F. Yan. (2007). Development of a standardized nutrient matrix for corn distillers dried grains with solubles. Internatl. Journal of Poultry Sci. 6 (7): 478-483.

Wang, Z., S. Cerrate, C. Coto, F. Yan, and P. W. Waldroup. (2007a). Effect of rapid and multiple changes in level of distillers dried grain with solubles (DDGS) in broiler diets on performance and carcass characteristics. International Journal of Poultry Sci. 6 (10): 725-731.

Wang, Z., S. Cerrate, C. Coto, F. Yan, and P. W. Waldroup. (2007b). Use of constant or increasing levels of distillers dried grains with solubles (DDGS) in broiler diets. International Journal of Poultry Sci. 6 (7): 501-507.

Wang, Z., S. Cerrate, C. Coto, F. Yan, and P. W. Waldroup. (2007c). Utilization of distillers dried grains with solubles (DDGS) in broiler diets using a standardized nutrient matrix. International Journal of Poultry Sci. 6 (7): 470-477.

Weigel, J. C., D. Loy, and L. Kilmer. (1997). Feed Co-Products of the Dry Corn Milling Process. Iowa State University, Iowa Corn Promotion Board, Iowa Department of Agriculture, Renewable Fuels Association, National Corn Growers Association. Available online: www.iowacorn.org/ethanol/ethanol_17.html.

Whitney, M. H. and G. C. Shurson. (2004). Growth performance of nursery pigs fed diets containing increasing levels of corn distiller's dried grains with solubles originating from a modern Midwestern ethanol plant. J. Anim. Sci. 82: 122-128.

Whitney, M. H., G. C. Shurson, and R. C. Guedes. (2006a). Effect of dietary inclusion of distillers dried grains with solubles on the ability of growing pigs to resist a Lawsonia intracellularis challenge. J. Anim. Sci. 84: 1860–1869.

Whitney, M. H., G. C. Shurson, and R. C. Guedes. (2006b). Effect of including distillers dried grains with solubles in the diet, with or without antimicrobial regimen, on the ability of growing pigs to resist a Lawsonia intracellularis challenge. J. Anim. Sci. 84: 1870–1879.

Whitney, M. H., G. C. Shurson, and R. C. Guedes. (2006c). Effect of dietary inclusion of distillers dried grains with solubles, soybean hulls, or a polyclonal antibody product on the ability of growing pigs to resist a Lawsonia intracellularis challenge. J. Anim. Sci. 84: 1880-1889.

Whitney, M. H., G. C. Shurson, L. J. Johnston, D. M. Wulf, and B. C. Shanks. (2006d). Growth performance and carcass characteristics of grower-finisher pigs fed high-quality corn distillers dried grain with solubles originating from a modern Midwestern ethanol plant. J. Anim. Sci. 84: 3356-3363.

Widmer, M. R., L. M. McGinnis, and H. H. Stein. (2007). Energy, phosphorus, and amino acid digestibility of high-protein distillers dried grains and corn germ fed to growing pigs. J. Anim. Sci. 85: 2994-3003.

Zhu, J. S., S. R. Stokes, and M. R. Murphy. (1997). Substitution of neutral detergent fiber from forage with neutral detergent fiber from by-products in the diets of lactating cows. J. Dairy Sci. 80: 2901-2906.

8

Biorefinery Processes for Biomass Conversion to Liquid Fuel

Shuangning Xiu, Bo Zhang and Abolghasem Shahbazi
Biological Engineering Program
School of Agriculture, NC A&T State University
U.S.A

1. Introduction

The development of products derived from biomass is emerging as an important force component for economic development in the world. Rising oil prices and uncertainty over the security of existing fossil reserves, combined with concerns over global climate change, have created the need for new transportation fuels and for the manufacture of bioproducts to substitute for fossil-based materials.

The United States currently consumes more than 140 billion gallons of transportation fuels annually. Conversion of cellulosic biomass to biofuels offers major economic, environmental, and strategic benefits. DOE and USDA predict that the U.S. biomass resources could provide approximately 1.3 billion dry tons of feedstock for biofuels, which would meet about 40% of the annual U.S. fuel demand for transportation (Perlack et al., 2005). More recently, in January 2010, U.S. President Barack Obama delivered a request during his State of the Union speech for Congress to continue to invest in biofuels and renewable energy technology. Against this backdrop, biofuels have emerged as one of the most strategically important sustainable fuels given their potential to increase the security of supply, reduce vehicle emissions and provide a steady income for farmers.

Several biorefinery processes have been developed to produce biofuels and chemicals from the initial biomass feedstock. Of all the various forms energy can take, liquid fuels are among the most convenient in terms of storage and transportation and are conducive to the existing fuel distribution infrastructure. This chapter comprehensively reviews the state of the art, the use and drawbacks of biorefinery processes that are used to produce liquid fuels, specifically bioethanol and bio-oil. It also points out challenges to success with biofuels in the future.

2. Biorefinery concept

2.1 Biorefinery definition

A biorefinery is a facility that integrates biomass conversion processes and equipment to produce fuels, power, heat, and value-added chemicals from biomass. The biorefinery concept is analogous to today's petroleum refinery, which produces multiple fuels and products from petroleum (Smith & Consultancy, 2007).

The IEA Bioenergy Task 42 on Biorefineries has defined biorefining as the "sustainable processing of biomass into a spectrum of bio-based products (food, feed, chemicals, materials) and bioenergy (biofuels, power and/or heat)." The biorefinery is not a single or fixed technology. It is collection of processes that utilize renewable biological or bio-based sources, or feedstocks, to produce an end product, or products, in a manner that is a zero-waste producing, and whereby each component from the process is converted or utilized in a manner to add value, and hence sustainability to the plant. Several different routes from feedstocks to products are being developed and demonstrated, and it is likely that multiple biorefinery designs will emerge in the future.

By producing multiple products, a biorefinery takes advantage of the various components in biomass and their intermediates, thereby maximizing the value derived from the biomass feedstock. A biorefinery could, for example, produce one or several low-volume, but high-value chemical or nutraceutical products and a low-value, but high-volume liquid transportation fuel such as biodiesel or bioethanol (see also alcohol fuel), while also generating electricity and process heat through combined heat and power (CHP) technology for its own use, and perhaps enough for sale of electricity to the local utility. In this scenario, the high-value products increase profitability, the high-volume fuel helps meet energy needs, and the power production helps to lower energy costs and reduce greenhouse gas emissions, as compared to traditional power plant facilities. Although some facilities exist that can be called biorefineries, the technology is not commonplace. Future biorefineries may play a major role in producing chemicals and materials that are traditionally produced from petroleum.

2.1 Two biorefinery platforms

Biomass can be converted to a wide range of useful forms of energy through several processes. As shown in Figure 1, there are two primary biorefinery platforms: sugar and thermochemical. Both platforms can produce chemicals and fuels including methanol, ethanol and polymers. The "sugar platform" is based on the breakdown of biomass into aqueous sugars using chemical and biological means. The fermentable sugars can be further processed to ethanol, aromatic hydrocarbons or liquid alkanes by fermentation, dehydration and aqueous-phase processing, respectively. The residues - mainly lignin - can be used for power generation (co-firing) or may be upgraded to produce other products (e.g., etherified gasoline). In the thermochemical platform, biomass is converted into synthesis gas through gasification, or into bio-oils through pyrolysis and hydrothermal conversion (HTC). Bio-oils can be further upgraded to liquid fuels such as methanol, gasoline and diesel fuel, and other chemicals.

3. Bioethanol production from lignocellulosic biomass

Ethanol is considered the next generation transportation fuel with the most potential, and significant quantities of ethanol are currently being produced from corn and sugar cane via a fermentation process. Utilizing lignocellulosic biomass as a feedstock is seen as the next step towards significantly expanding ethanol production capacity. However, technological barriers - including pretreatment, enzyme hydrolysis, saccharification of the cellulose and hemicellulose matrixes, and simultaneous fermentation of hexoses and pentoses - need to be addressed to efficiently convert lignocellulosic biomass into bioethanol. In addition to substantial technical challenges that still need to be overcome before lignocellulose-to-ethanol becomes commercially viable, any ethanol produced by fermentation has the inherent drawback of needing to be distilled from a mixture which contains 82% to 94% water. This section will review current developments towards resolving these technological challenges.

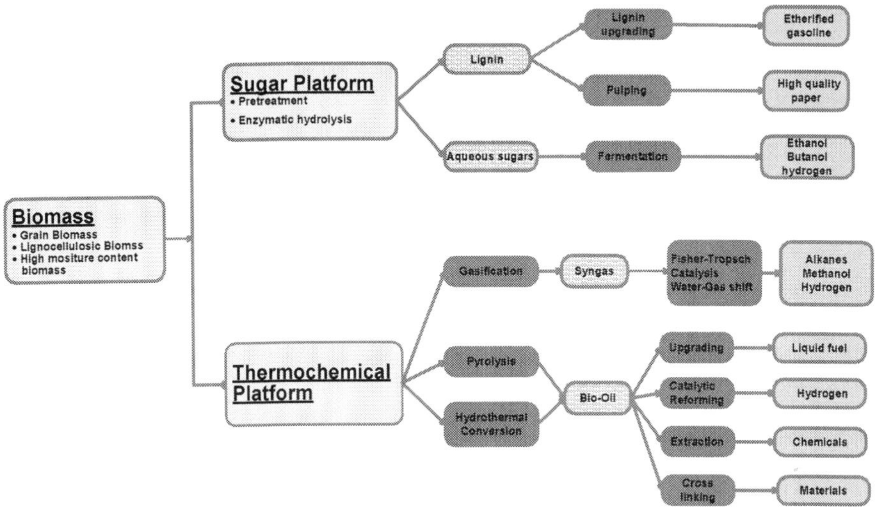

Fig. 1. Primary routes for biofuels conversion

3.1 Pretreatment

Pretreatment is required to break the crystalline structure of cellulosic biomass to make it more accessible to the enzymes, which can then attach to the cellulose and hydrolyze the carbohydrate polymers into fermentable sugars. The goal of pretreatment is to pre-extract hemicellulose, disrupt the lignin seal and liberate the cellulose from the plant cell wall matrix. Pretreatment is considered to be one of the most expensive processing steps in cellulosic ethanol processes, but it also has great potential to be improved and costs lowered through research and development (Lynd et al., 1996; Lee et al., 1994; Mosier et al., 2005).

Many pretreatment techonlogies have been developed and evaluated for various biomass materials. However, each pretreatment method has its own advantages and disadvantages, and one pretreatment approach does not fit all biomass feedstocks. Three widely used pretreatment techologies will be reviewed below.

3.1.1 Alkaline pretreatment

Removing lignin with alkaline chemicals such as dilute sodium hydroxide, aqueous ammonia and lime, has long been known to improve cellulose digestibility (Li et al., 2004). Among these alkaline reagents, sodium hydroxide (NaOH) has been widely used for pretreatment because its alkalinity is much higher than others, but it is also expensive, and the recovery process is complex. The following studies on various feedstocks illustrate this:

Untreated cattails contain 32.0% cellulose, 18.9% hemicellulose and 20.7% lignin. Zhang et al. (2010a) reported that 54.8% of cattail lignin and 43.7% of the hemicellulose were removed with a 4% NaOH solution. The glucose yield from 4% NaOH treated cattails was approximately 80% of the cellulose available.

Adding addtional chemicals along with NaOH could improve pretreatment performance. Applying a NaOH and H_2O_2 solution helps in additional lignin removal through oxidative action on lignin. Maximum overall sugar yield obtained from high lignin hybrid poplar was 80% with 5%NaOH / 5% H_2O_2 at 80°C (Gupta, 2008). Zhao et al. (2008) discovered that a NaOH-urea pretreatment, can slightly remove lignin, hemicelluloses, and cellulose in the lignocellulosic materials, disrupt the connections between hemicelluloses, cellulose, and lignin, and alter the structure of treated biomass to make cellulose more accessible to hydrolysis enzymes. The enzymatic hydrolysis efficiency of spruce also can be remarkably enhanced by a NaOH or NaOH/urea solution treatment. A glucose yield of up to 70% could be obtained at the cold temperature pretreatment of (-15°C) using 7% NaOH/12% urea solution, but only 20% and 24% glucose yields were obtained at temperatures of 238°C and 608°C, respectively.

Two theoretical approaches were used to study the enzyme kinetics of sodium hydroxide pretreated wheat straw, and describe the influence of enzyme concentrations of 6.25-75 g/L on the production of reducing sugars. The first approach used a modified Michaelis-Menten equation to determine the hydrolysis model and kinetic parameters (maximal velocity, Vemax, and half-saturation constant, Ke). The second, the Chrastil approach, was applied to study all the time values from the rate of product formation, taking into account that in a heterogeneous system, these reactions are diffusion limited and the time curves depend strongly on the heterogeneous rate-limiting structures of the enzyme system.

3.1.2 Hot-water pretreatment

Hot water pretreatment is often called autohydrolysis. The major advantages of this method are less expense, lower corrosion to equipment and less xylose degradation and hence fewer byproducts with inhibitory compounds in the extracts (Huang et al. 2008). Hot water under pressure can penetrate the cell structure of biomass, hydrate cellulose, and remove hemicellulose.

Hot water pretreatment could effectively improve the enzymatic digestibility of biomass cellulose. At optimal conditions, 90% of the cellulose from corn stover pretreated in hot-water can be hydrolyzed to glucose (Mosier et al., 2003). When cattails were pretreated at 463K for 15 min, 100% of the hemicellulose was removed and 21.5% of the cellulose was dissolved in the water phase. The process could be further optimized to improve its efficiency (Zhang et al. 2010b).

The pretreatment process of bagasse was studied over a temperature range of 170-203°C, and a time range of 1-46 min. A yield of 80% conversion was achieved, and hydrolysis inhibitors were detected (Laser et al., 2002). Hot water pretreatment also was reported to improve enzymatic digestibility of switchgrass, resulting in 80% glucose yield (Kim et al., 2008). The optimal hot-water pretreatment conditions for hybrid poplar of 15% solids (wt/vol) were 200°C at 10 min, which resulted in the highest fermentable sugar yield of between 54% and 67% (Kim et al., 2008).

3.1.3 Dilute-acid pretreatment

The use of acid hydrolysis for the conversion of cellulose to glucose is a process that has been studied for the last 100 years. Dilute acid (0.5-1.0% sulfuric acid) pretreatment at moderate temperatures (140-190°C) can effectively remove and recover most of the hemicellulose as dissolved sugars. Furthermore lignin is disrupted and partially dissolved, increasing cellulose susceptibility to enzymes (Yang and Wyman, 2004). Under this method, glucose yields from cellulose increase with hemicellulose removal to almost 100% (Knappert et al., 1981). Dilute acid hydrolysis consists of two chemical reactions. One reaction converts cellulosic materials to sugar and the other converts sugars into other chemicals, many of which inhibit the growth of downstream fermentation microbes. The same conditions that cause the first reaction to occur simultaneously cause over-degradation of sugars and lignin, creating inhibitory compounds such as organic acids, furans, and phenols.

Partial cellulose may be degraded as oligomers or monomers during the acid pretreatment process. Sugar (glucose and xylose) yields were often reported for the pretreatment and enzyme hydrolysis stage separately, and as the total for both stages. Lloyd and Wyman (2005) reported that up to 92% of the total sugars originally available in corn stover could be recovered via coupled dilute acid pretreatment and enzymatic hydrolysis. Conditions achieving maximum individual sugar yields were often not the same as those that maximized the total sugar yields, demonstrating the importance of clearly defining pretreatment goals when optimizing the process.

Dilute-sulfuric acid pretreatment of cattails was studied using a Dionex accelerated solvent extractor (ASE) at varying acid concentrations of 0.1 to 1%, treatment temperatures of 140 to 180 °C, and residence times of 5 to 10 min. The yield of extractable products obtained from the pretreatment process increased as the final temperature, treatment time, or acid concentration increased. The highest glucose yield from the pretreatment was 55.4% of the cellulose at 180°C for 15 min with 1% sulfuric acid. The highest glucose yield from the enzyme hydrolysis stage (82.2% of the cellulose) and the highest total glucose yield for both the pretreatment and enzyme hydrolysis stages (97.1% of the cellulose) were reached at a temperature of 180°C, a sulfuric acid concentration of 0.5%, and a time of 5 min.

When switchgrass was pretreated for 60 min with 1.5% acid, the highest glucan conversion yield of 91.8% was obtained (Yang et al. 2009).

3.2 Enzyme hydrolysis

After pretreatment, hydrolysis converts the carbohydrate polymers into monomeric sugars. Although a variety of process configurations have been studied for conversion of cellulosic biomass into ethanol, enzymatic hydrolysis of cellulose provides opportunities to improve the technology so that biomass ethanol is competitive with other liquid fuels(Wyman, 1999). Novozymes (www.novozymes.com) and Genencor (www.genencor.com) are two companies leading research & development for advanced cellulosic ethanol enzymes. In early 2010, Novozymes said its new Cellic® CTec2 enzymes enable the biofuel industry to produce cellulosic ethanol at a price below US$ 2.00 per gallon for the initial commercial-scale plants that are scheduled to be in operation in 2011. This cost is on par with gasoline and conventional ethanol at current US market prices. According to Novozymes, the new enzyme can be used on different types of feedstock including corn cobs and stalks, wheat straw, sugarcane bagasse, and woodchips. The enzyme is designed to break down cellulose in biomass into sugars that can be fermented into ethanol. Genencor, a division of Danisco also introduced its enzyme Accellerase®, which is designed to do the same thing.

The selection of the enzymes needs to match the pretreatment technologies and the feedstock used, as well as the process. For example, if a dilute acid pretreatment is used, most of the hemicellulose is degraded, so hemicellulases is unnecessary. However, if an alkaline or hot-water pretreatment is used, the hemicellulose still needs to be hydrolyzed and hemicellulases will be needed.

The cellulose portion of the biomass is another difficulty. In order to efficiently break it down, a mixture of several enzymes with different activities is required. This mixture includes three basic types of enzymes.

1. Endoglucanases break bonds between adjacent sugar molecules in a cellulose chain, fragmenting the chain into shorter lengths. Endoglucanases act randomly along the cellulose chain, although they prefer amorphous regions where the chains are less crystalline.
2. Cellobiohydrolases attack cellulose chains from the ends of the chain. This exo- or processive action releases mainly cellobiose (glucose dimer). Because endoglucanases create new ends for cellobiohydrolases to act upon, the two classes of enzymes interact synergistically.
3. β-glucosidases break down short glucose chains, such as cellobiose, to release glucose. β-glucosidases are important as they act on cellobiose, which inhibits the action of the other cellulases as it builds up the hydrolysis reactor.

3.3 Fermentation for bioethanol production

Saccharomyces cerevisiae (baker's yeast) has been used for industrial ethanol production from hexoses (C6 sugars) for a thousand years. However, a significant amount of pentoses (C5 sugars) derived from the hemicellulose portion of the lignocellulosic biomass is present in the hydrolysate from the pretreatment process. Modern biotechnologies enable fermenting microorganisms to use both C5 and C6 sugars available from the hydrolysate. This further increases the economic competitiveness of ethanol production and other bio-products from cellulosic biomass.

Recently, microorganisms for cellulosic ethanol production, such as *Saccharomyces cerevisiae, Zymomonas mobilis and Escherichia coli,* have been genetically engineered using metabolic engineering approaches. Lau et al. (2010) compared the fermentation performance of *Escherichia coli* KO11, *Saccharomyces cerevisiae* 424A(LNH-ST) and *Zymomonas mobilis* AX101 for cellulosic ethanol production. Three microorganisms resulted in a metabolic yield, final concentration and rate greater than 0.42 g/g consumed sugars, 40 g/L and 0.7 g/L/h (0-48 h), respectively. They concluded that *Saccharomyces cerevisiae* 424A(LNH-ST) is the most promising strain for industrial production because of its ability to ferment both glucose and xylose.

Vasan et al (2011) introduced an *Enterobacter cloacae* cellulase gene into *Zymomonas mobilis* strain and 0.134 filter paper activity unit (FPU)/ml units of cellulase activity was observed with the recombinant bacterium. When using carboxymethyl cellulose and 4% NaOH pretreated bagasse as substrates, the recombinant strain produced 5.5% and 4% (V/V) ethanol respectively, which was three times higher than the amount obtained with the original E. cloacae isolate.

In 2010, Purde University scientists improved a strain of yeast that can produce more biofuel from cellulosic plant material by fermenting all five types of the plant's sugars: galactose, manose, glucose, xylose and arabinose. Arabinose makes up about 10 percent of the sugars contained in cellulosic biomass (Casey et al., 2010).

3.4 Closing remarks

Ethanol provides the first model for biofuel commercialization. However, in order to make the cellulosic ethanol process economically viable, both government subsidies and scientific

R&D are still required. And it is generally accepted that ethanol alone is not going to provide a long-term solution to meet society's energy needs (Hill et al., 2006). It suffers from a somewhat low energy density, inability to be transported through pipelines and fairly high cost for extraction from fermentation broths. This is opening the door to developing many other molecules as replacements for ethanol and thus, discovering new fuel molecules to be produced via microbial biotechnology.

4. Bio-oil production from lignocellulosic biomass and high moisture content biomass

Bioethanol is only one of the products that may be extracted from lignocellulosic feedstocks. Other forms of energy and a full range of value-added bioproducts may be produced from biomass by thermochemical means. Thermochemical conversion processes include pyrolysis, hydrothermal conversion and gasification. The major product of pyrolysis and hydrothermal conversion, known as "bio-oil" or "biocrude", can be used as a boiler fuel or as fuel in combustion engines. Alternatively, the bio-oil can serve as a raw material for the production of chemicals and biomaterials. One of the major technical obstacles to large scale thermochemical conversion of biomass into bio-oil is its poor oil quality and low biofuel production rate. This section intensively reviewed current technologies used to produce bio-oil and technologies development towards improving the bio-oil yield and quality.

4.1 Current processes for conversion of biomass to bio-oils

Two main types of processes for production of bio-oils from biomass are flash pyrolysis and hydrothermal conversion, as shown in Fig.1. Both of the processes belong to the thermochemical platform in which feedstock organic compounds are converted into liquid products. An advantage of the thermochemical process is that it is relatively simple, usually requiring only one reactor, thus having a low capital cost. However, this process is non-selective, producing a wide range of products including a large amount of char (Huber & Dumesic, 2006).

The characteristic and technique feasibility of the two thermochemical processes for bio-oil production are compared in table 1. Flash pyrolysis is characterized by a short gas residence time (~1s), atmospheric pressure, a relatively high temperature (450-500 °C). Furthermore, feedstock drying is necessary. Hydrothermal processing (also referred to in the literature as liquefaction, hydrothermal pyrolysis, depolymerisation, solvolysis and direct liquefaction), is usually performed at lower temperatures (300-400 °C), longer residence times (0.2-1.0 hr.), and relatively high operating pressure (5-20 Mpa). Contrary to flash pyrolysis and gasification processes, drying the feedstock is not needed in the hydrothermal process, which makes it especially suitable for naturally wet biomass. However, a reducing gas and/or a catalyst is often included in the process in order to increase the oil yield and quality.

The reaction mechanisms of the two processes are different, which have been studied by many investigators (Demirbaş, 2000a; Minowa et al., 1998). The hydrothermal process occurred in aqueous medium which involves complex sequences of reactions including solvolysis, dehydration, decarboxylation, and hydrogenation of functional groups, etc. (Chornet and Overend, 1985). The decomposition of cellulose was studied by Minowa et al. (1998). The effects of adding a sodium carbonate catalyst, a reduced nickel catalyst, and no catalyst addition in the decomposition of cellulose in hot-compressed water were investigated. They found that hydrolysis can play an important role in forming glucose/oligomer, which can quickly decompose into non-glucose aqueous products, oil, char and gases (Fig. 2). Without a catalyst, char and gases were produced through oil as intermediates. However, in the presence of an alkali catalyst, char production was inhibited because the oil intermediates were stabilized, resulting in oil production. Reduced nickel was found to catalyze the steam reforming reaction of aqueous products as intermediates and the machination reaction. Typical yields of liquid products for hydrothermal conversion processes were in the range of 20-60%, depending on many factors including sustrate type, temperature, pressure, residence time, type of solvents, and catalysts employed (Xu and Etcheverry, 2008).

Methods	Treatment condition/ requirement	Reaction mechanism /process description	Technique Feasibility	
			Pros.	Cons.
Flash/Fast Pyrolysis	Relatively high temperature (450-500 °C); a short residence time (~1s); atmosphere pressure; drying necessary	The light small molecules are converted to oily products through homogeneous reactions in the gas phase	High oil yield up to 80% on dry feed; lower capital cost; Commercialized already	Poor quality of fuels obtained
Hydrothermal Processing (HTU)/ liquefaction /hydrothermal pyrolysis	Lower temperature (300-400 °C); longer residence time (0.2-1.0 hr.); High pressure (5-20 Mpa); drying unnecessary	Occurs in aqueous medium which involves complex sequences of reactions	Better quality of fuels obtained (High PTU, low moisture content)	Relatively low oil yield (20-60%); Need high pressure equip, thus higher capital cost

Table 1. Comparison of two typical thermochemical processes for bio-oil production

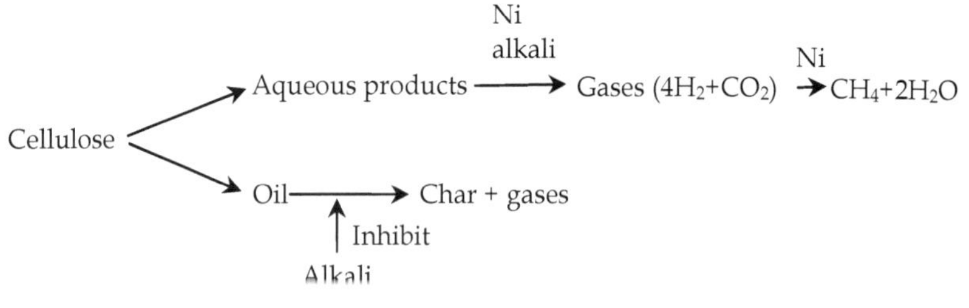

Fig. 2. Reaction pathway for the hydrothermal processing of cellulose

With flash pyrolysis, the light small molecules are converted to oily products through homogeneous reactions in the gas phase. The principle of the biomass flash pyrolysis process is shown in Fig.3. Biomass is rapidly heated in the absence of air, vaporizes, and quickly condenses to bio-oil. The main product, bio-oil, is obtained in yields of up to 80% wt on dry feed, together with the by-product char and gas (Bridgewater and Peacocke, 2000).

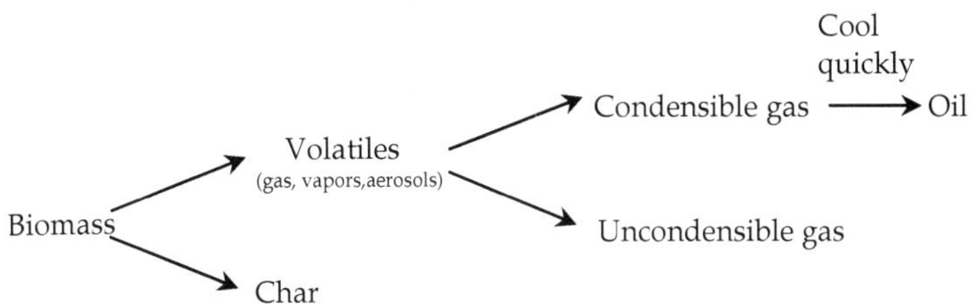

Fig. 3. Reaction pathway for the biomass flash pyrolysis process

4.2 Related research development of flash pyrolysis and hydrothermal process

Flash pyrolysis for the production of liquids has developed considerably since the first experiments in the late 1970s. Several pyrolysis reactors and processes have been investigated and developed to the point where fast pyrolysis is now an accepted, feasible and viable route to renewable liquid fuels, chemicals and derived products. Since the 1990s, several research organizations have successfully established large-scale fast pyrolysis plants. Bridgwater and Peacocke (2000) have intensively reviewed the key features of fast pyrolysis and the resultant liquid product, and described the major reaction systems and processes that have been developed over the last 20 years.

Unlike flash pyrolysis, technological developments in the area of hydrothermal conversion present new ways to turn wastes to fuel. Hydrothermal processing was initially developed for turning coal into liquid fuels, but recently, the technique has been applied to a number of feedstocks, including woody biomass, agricultural residues, and organic wastes (e.g., animal wastes, municipal solid wastes (MSW), and sewage sludge). Table 2 summarizes representative literature data of hydrothermal processing of common types of biomass and the most influential operating parameters. As can be seen from Table 2, organic waste materials are more favourable than woody biomass and agricultural residues for hydrothermal processing, owing to their higher oil yield and the higher heating value of their bio-oil products.

This earlier work was very promising, showing that hydrothermal technology can be used as an efficient method to treat different types of biomass and produce a liquid biofuel. In particular, hydrothermal conversion processes present a unique approach to mitigate the environmental and economic problems related to disposing of large volumes of organic wastes. It not only reduces the pollutants, but also produces useful energy in the form of liquid fuel. Compared with flash pyrolysis, hydrothermal conversion is at an early developmental stage, and the reaction mechanisms and kinetics are not yet fully understood.

Raw Materials	Reactor Capacity	Temp. (°C)	Pressure (Mpa)	Time (min)	Oil Yield (%)	Heating Value(MJ/kg)	Reference
a) Woods							
Beech	--	277-377	--	25	13.8-28.4	27.6-31.3	Demirbaş, et al., 2005a
Spruce	--	277-377	--	25	13.8-25.8	28.3-33.9	Demirbaş, et al., 2005b
Sawdust	0.2 L	280	N/A		7.2	--	Karagöz et al., 2005
b) Agricultural residues							
Corn stalk	0.3 L	300	10 Mpa	30	28.3 on organic basis	29.7	Minowa et al., 1998
Rice husk	0.3 L	300	10 Mpa	30	28.8 on organic basis	30.8	Minowa et al., 1998
Rice straw	1.0 L	260-350	6-18 Mpa	3-5	13.0-38.35	27.6-35.8	Yuan et al., 2007
c) organic wastes							
Swine manure	1-L autoclave	260-340		0-90	14.9-24.2	36.1	Xiu et al., 2010a
Swine manure	Continuous mode	285-305	9-12	40-80	2.8-53.3	25.2-33.1	Ocfemia et al., 2006
Dairy manure	Batch/ continuous mode	250-380	10-34	--	50	--	Appell, et al., 1980
Sewage sludge	5 t/d	300	10	--	48	37-39	Itoh, et al., 1994
Garbage	0.3 L autoclave	250-340	6-18	6-120	27.6	36	Minowa, et al., 1995
Sewage sludge	0.3 L autoclave	150-300	--	0-180	44.5	35.7	Suzuki, et al., 1990
Sewage sludge	4.2L microwave	250-350	8-20	--	30.7	36.4	Bohlmann, et al., 1999
MSW	autoclave	260-340	13-34	--	32	46	Gharieb, et al., 1995
MSW	autoclave	295-450	--	20-90	35-63.3	--	Kranich et al., 1984
Sewage sludge	20 kg/hr.	300-360	10-18	5-20	--	30-35	Goudriaan et al., 2000

Table 2. Overview of literature on hydrothermal processing of common types of biomass

4.3 Properties of bio-oils

The differences in processing conditions result in significant differences in the product yield and product composition of bio-oils. Recently, Lu et al. (2009) intensively reviewed the fuel properties fast pyrolysis oils and discusses how these properties affect the utilization of bio-oils. In general, bio-oils are complex mixtures of volatile hydrocarbons, alcohols, organic acids, aldehydes, ketones, ethers, furans, phenols and other non-volatile compounds. The unstable fragments in bio-oil could rearrange through condensation, cyclization, and polymerization to form new compounds, such as aromatics. Table 3 describes selected properties of bio-oils produced from hydrothermal liquefaction of swine manure and pyrolysis of wood. For comparison purposes, the characteristics of heavy petroleum fuel oil were also presented in Table 3.

Properties		Liquefied bio-oil from swine manure(xiu et al., 2010a)	Pyrolysis bio-oil from wood pyrolysis (Zhang et al., 2007)	Heavy petroleum fuel oil (Oasmaa et al., 1999)
Moisture content (wt%)		2.37	15-30	0.1
PH		--	2.5	-
Specific gravity		1	1.2	0.94
Elemental composition (wt%)	C	72.58	54-58	85
	H	9.76	5.5-7.0	11
	O	13.19	35-40	1.0
	N	4.47	0-0.2	0.3
Ash		0.78	0-0.2	0.1
HHV(MJ/kg)		36.05	16-19	40
Viscosity(at 50 0C)(cP)		843	40-100	180
Solids (wt%)		--	0.2-1	1
Distillation residue (wt%)		63	Up to 50	1

Table 3. Comparison of selected properties of bio-oils produced by hydrothermal liquefaction of swine manure and pyrolysis of wood and heavy fuel oil

As shown in Table 3, liquefied oils have much lower oxygen and moisture contents, and consequently much higher energy value, as compared to oils from fast pyrolysis. The corresponding HHV of liquefied oil from swine manure is 36.05 MJ/kg, which about 90% of that of heavy fuel oil (40 MJ/kg). The properties of bio-oil from both processes are significantly different from heavy petroleum fuel oil. Compared with heavy petroleum fuel oil, the bio-oils have the following undesired properties for fuel applications: high viscosity, high water and ash contents, high oxygen content and low heating value.

Pyrolysis oil is acidic in nature, polar and not miscible with conventional crude oil. In addition, it is unstable, as some (re)polymerization of organic matter in the oil causes an increase in viscosity over time. Overall, bio-oils can not be directly used as transportation fuels due to their high viscosity, high water and ash contents, low heating value, instability and high corrosiveness. Therefore, bio-oil has to be upgraded before it can be used as an engine fuel.

4.4 Typical bio-oil upgrading technologies and their limitation

Considering the above discussion on the properties, it is obvious that the fuel quality of bio-oils is inferior to that of petroleum-based fuels. There have been intensive studies on bio-oil upgrading research and various technologies have been developed for bio-oil upgrading. Table 4 summarizes current techniques in bio-oil upgrading. The characteristics, as well as recent progress, advantages, and disadvantages of each technique are also described as follows:

Upgrading methods	Treatment condition/ requirement	Reaction mechanism /process description	Technique Feasibility	
			Pros.	Cons.
Hydrotreating /hydrofining	Mild conditions, (~500°C /low pressure) chemical needed: H_2/CO, catalyst (e.g., CoMo, HDS, NiMo, HZSM-5)	Hydrogenation without simultaneous cracking (eliminating N, O and S as NH3, H_2O and H_2S)	Cheaper route, Commercialized already	high coking (8-25%) and poor quality of fuels obtained
Hydro-cracking /hydrogenolysis /catalytic cracking	Severe conditions, (>350 °C, 100~2000 Psi), chemical needed: H_2/CO or H_2 donor solvents, catalyst (e.g., Ni/Al_2O_3-TiO_2)	Hydrogenation with simultaneous cracking Destructive(resulting in low molecular product)	Makes large quantities of light products	Need complicated equipment, excess cost, catalyst deactivation, reactor clogging
Supercritical fluid	Mild conditions, organic solvents needed such as alcohol, acetone, ethyl acetate, glycerol	Promotes the reaction by its unique transport properties: gas-like diffusivity and liquid-like density, thus dissolved materials not soluble in either liquid or gaseous phase of solvent	Higher oil yield, better fuel quality (lower oxygen content, lower viscosity)	Solvent is expensive
Solvent addition (direct add solvent or esterification of thethe oil with alcohol and acid catalysts	Mild conditions, polar solvents needed such as water, methanol, ethanol, and furfural	Reduces oil viscosity by three mechanisms: (1) physical dilution (2) molecular dilution or by changing the oil microstructure; (3) chemical reactions like esterification and acetalization	The most practical approach (simplicity, the low cost of some solvents and their beneficial effects on the oil properties)	Mechanisms involved in adding solvent are not quite understand yet
Emulsification /Emulsions	Mild conditions, need surfactant (e.g. CANMET)	Combines with diesel directly. Bio-oil is miscible with diesel fuels with the aid of surfactants	Simple, less corrosive	Requires high energy for production
Steam Reforming	High temperature(800-900 °C), need catalyst (e.g. Ni)	Catalytic steam reforming + water-gas shift	Produces H_2 as a clean energy resource	Complicated, requires steady, dependable, fully developed reactors
Chemical extracted from the bio-oils	Mild conditions	Solvent extraction, distillation, or chemical modification	Extract valuable chemicals	Low cost separation and refining techniques still needed

Table 4. Brief description, treatment condition, and technical feasibility of the current techniques used for upgrading bio-oil

4.4.1 Hydrotreating

It is generally recognized that the higher the hydrogen content of a petroleum product, especially the fuel products, the better the quality. This knowledge has stimulated the use of a hydrogen-adding process in the refinery, which is called hydrogenation. Currently, the most widely used hydrogenation processes for the conversion of petroleum and petroleum products is hydrotreating.

Hydrotreating (HDT) is a nondestructive, or simple hydrogenation process that is used for the purpose of improving product quality without appreciable alteration of the boiling range. It has become the most common process in modern petroleum refineries. Bio-crude may also be processed by a conventional refinery and potentially augmented with petroleum crude. The oxygen in bio-oils can be removed via hydrotreating. The catalysts commonly used for hydrotreating are sulphide CoMo/Al_2O_3, NiMo/ Al_2O_3 systems. (Nava et al., 2009).

Hydrotreating requires mild conditions, while the yield of bio-oil is relatively low. The process also produces a large amount of char, coke, and tar, which will result in catalyst deactivation and reactor clogging.

4.4.2 Hydro-cracking

Hydro-cracking is less popular than the hydrotreating in the petroleum industry.

Hydro-cracking is a thermal process (>350 °C, >660°F) in which hydrogenation accompanies cracking. Relatively high pressure (100 to 2000 Psi) is employed, and the overall result is usually a change in the character or quality of the end products (Ancheyta and Speight, 2007). This process is performed by dual-function catalysts, in which silica-alumina (or zeolite) catalysts provide the cracking function, and platinum and tungsten oxide catalyze the reactions, or nickel provides the hydrogenation function. Alumina is by far the most widely used support

Hydro-cracking is an effective way to make a large amount of light product, but it requires more severe conditions such as higher temperature and hydrogen pressure to deal with acids, which is not economical and energy efficient.

4.4.3 Supercritical fluids (SCFs)

A fluid is considered supercritical when its temperature and pressure go above its critical point. SCFs possess unique transport properties. They can effuse through solids like a gas and dissolve materials like a liquid. In particular, SCFs have the ability to dissolve materials not normally soluble in either liquid or gaseous phase of the solvent, and hence to promote the gasification/liquefaction reactions (Xu and Etcheverry, 2008).

SCFs have been recently used to improve oil yield and quality and have demonstrated a great potential for producing bio-oil or bio-crude with much higher caloric values and lower viscosity. Water is the cheapest and most commonly used supercritical fluid in hydrothermal processing, but utilizing water as the solvent for liquefaction of biomass has the following drawbacks: 1) lower yields of the water-insoluble oil product; 2) it yields a bio-oil that is very viscous, with a high oxygen content. To enhance the oil yields and qualities the utilization of organic solvents such as ethanol (Xu and Etcheverry, 2008; Xiu, et al., 2010b), ethyl acetate (Demirbas, 2000a), acetone (Heitz et al., 1994; Liu and Zhang, 2008), 2-propanol (Ogi et al., 1994), 1,4-dioxane (Bao et al., 2008; Mazaheri et al., 2010; Cemek and Kucuk, 2001), methanol (Minami and Ska, 2003,2005; Yang et al., 2009) and butanol (Ogi et al., 1993) has been adopted. All these solvents have shown a significant effect on bio-oil yield and quality. Minami and Ska (2003, 2005) have reported that 90% of beech wood was successfully decomposed in supercritical methanol. The above motioned supercritical organic solvent fluids are predominately used in hydrothermal treatment processing to improve the bio-oil yield and quality. However, it was also used to upgrade pyrolysis bio-oil. For example, Tang et al. (2009) reported that supercritical ethanol (T = 243.1°C, Pc = 6.37 MPa) can upgrade lignin-derived oligomers in pyrolysis oil, and thus reduce the tar or coke. Although SCFs can be produced at relatively lower temperature and the process is environmentally friendly, these organic solvents are too expensive to make it economically feasible on a large scale. Recently, researchers have been trying to test less expensive organic solvent as a substitute for SCFs. Crude glycerol, the low-value by-product of biodiesel production, has shown very promising results for being using as an SCF solvent. Glycerol has been used as an organic solvent for biomass delignification (Demirbaş, 2008; Demirbaş and Celik, 2005a; Kücük, 2005), bio-oil separation (Li et al., 2009) and to significantly improve the performance of liquefaction in the conversion of biomass into bio-oil (Demirbaş, 2000b; Xiu et al., 2010c; Gan et al., 2010). Xiu et al. found that cross-reactions between swine manure and crude glycerol significantly affected the hydrothermal process, and that the use of crude glycerol dramatically increased bio-oil yield from 23.9% to 70.92% (Xiu et al., 2010c, Fig.4). In addition, they discovered that the free fatty acid in the crude glycerol is the key component that leads to enhancement of the oil yield (Fig.5). Moreover, the oil quality was also improved, having a lower density and viscosity.

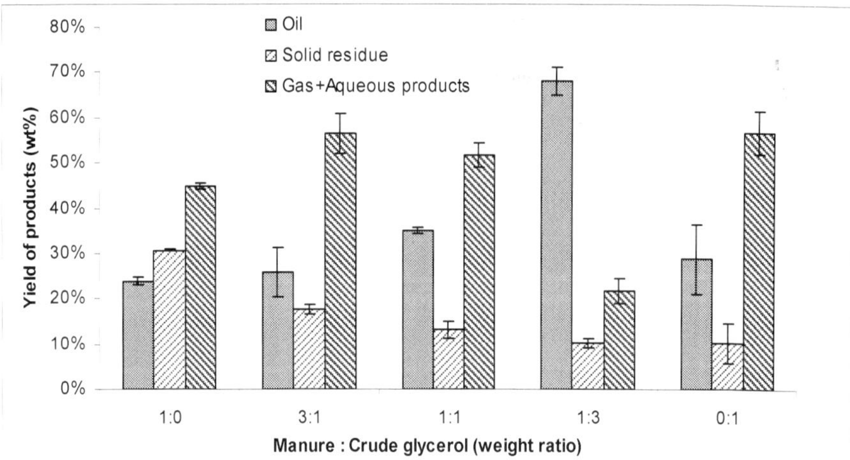

Fig. 4. Effect of swine manure to crude glycerol ratio (weight ratio) on the products yield in hydrothermal pyrolysis of swine manure. (Xiu et al., 2010c)

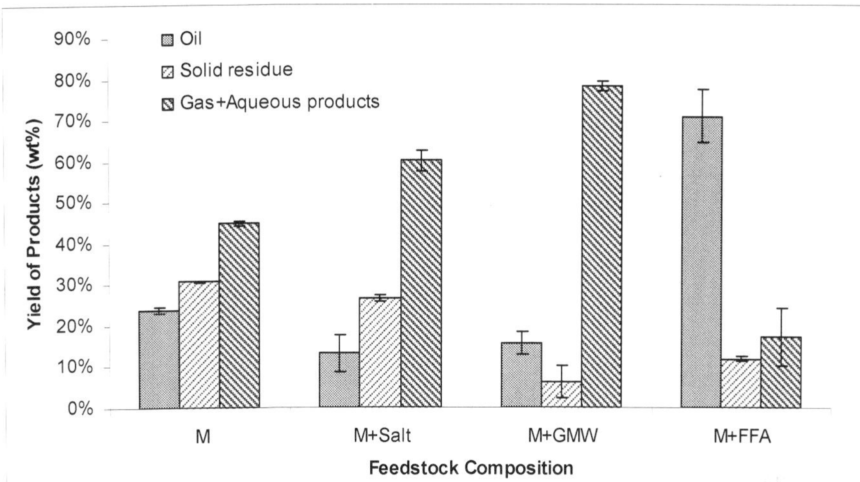

Fig. 5. Effect of crude glycerol fractionated components on the products yield in hydrothermal pyrolysis of swine manure.(Xiu et al., 2010c). M-manure; GMW-glycerol,methnol and water; FFA-free fatty acids.

4.4.4 Solvent addition / etherification

Polar solvents such as methanol, ethanol, and furfural have been used for many years to homogenize and to reduce viscosity of biomass oils (Radlein et al., 1996; Diebold and Czernik, 1997; Oasmma, 2004; Boucher et al., 2000). The immediate effects of adding these polar solvents are decreased viscosity and increased heating value. The increase in heating value for bio-oils mixed with solvents occurs because the solvent has a higher heating value than that of most bio-oils. The solvent addition reduces the oil viscosity due to the following three mechanisms: (1) physical dilution without affecting the chemical reaction rates; (2) reducing the reaction rate by molecular dilution or by changing the oil microstructure; (3) chemical reactions between the solvent and the oil components that prevent further chain growth (Oasmaa and Czernik, 1999).

Most studies have directly added solvents after pyrolysis, which works well to decrease the viscosity and increase stability and heating value. However, several recent studies showed that reacting the oil with alcohol (e.g., ethanol) and acid catalysts (e.g., acetic acid) at mild conditions by using reactive distillation, resulted in a better bio-oil quality (Mahfud et al., 2007; Xu et al., 2008; Tang et al., 2008; Oasmma, et al., 2004; Xu and Etcheverry, 2008). This process is referred to as catalytic etherification or etherification treatment in the literature (Xiong et al., 2009; Wang et al., 2010; Hilten et al., 2010; Yu et al., 2009).

The chemical reactions that can occur between the bio-oil and methanol or ethanol are esterification and acetalization (Fig.6). In such a case, the reactive molecules of bio-oil like organic acids and aldehydes are converted by the reactions with alcohols to esters and acetals, respectively. Thus, in addition to the decrease in viscosity and in the aging rate, they also lead to other desirable changes, such as reduced acidity, improved volatility and heating value, and better miscibility with diesel fuels.

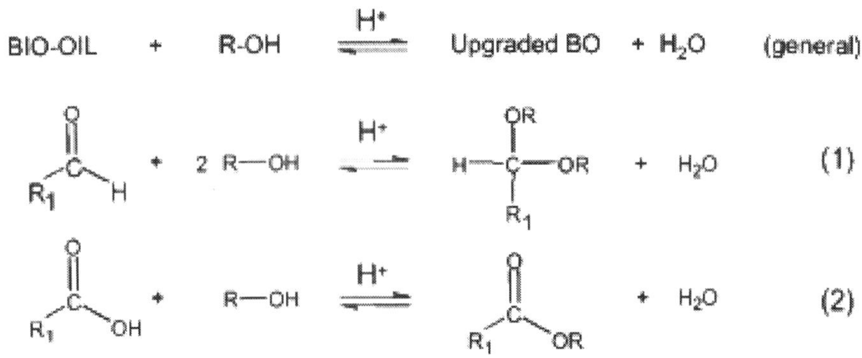

Fig. 6. Reactions involved in bio-oil alcoholysis: (1) acetalization, (2) esterification. (Mahfud et al., 2007)

Most environmental catalysts applied in bio-oil upgrading are heterogeneous catalysts. Solid acid catalysts, solid base catalysts (Zhang et al., 2006), ionic liquid catalysts (Xiong et al., 2009), HZSM-5, and aluminum silicate catalysts were investigated for esterification of bio-oils (Peng et al., 2008, 2009).

Considering the simplicity, the low cost of some solvents such as methanol and their beneficial effects on the oil, this method seems to be the most practical approach for bio-oil quality upgrading.

4.4.5 Emulsification (emulsions)

One of the methods in using bio-oil as a combustion fuel in transportation or boilers is to produce an emulsion with other fuel sources. Pyrolysis oils are not miscible with hydrocarbon fuels, but with the aid of surfactants they can be emulsified with diesel oil. Upgrading of bio-oil through emulsification with diesel oil has been investigated by many researchers (Chiaramonti et al., 2003a, b; Ikura et al., 2003; Jiang & Ellis, 2010; Garcia-Perez et al., 2010).

A process for producing stable microemulsions, with 5-30% of bio-oil in diesel has been developed at CANMET Energy Technology Centre (Oasmaa & Czernik, 1999; Ikura, et al., 1998). Those emulsions are less corrosive and show promising ignition characteristics.

Jiang and Ellis (2010) investigated the bio-oil emulsification with biodiesel while leaving the pyrolytic lignin phase behind. A stable bio-oil/biodiesel emulsion was produced using octanol as an emulsifier. The effects of several process variables on the mixture stability were also examined. They found that the optimal conditions for obtaining a stable mixture between bio-oil and biodiesel are with an octanol surfactant dosage of 4% by volume, an initial bio-oil/biodiesel ratio of 4:6 by volume, a stirring intensity of 1200 rpm, a mixing time of 15 min, and an emulsifying temperature of 30 °C. Various properties of the emulsion have shown more desirable values in acid number, viscosity, and water content compared to the original bio-oil. The reduction in viscosity and corrosively of the emulsion was also reported by Ikura et al (1998).

Chiaramonti et al. (2003b) tested the emulsions from biomass pyrolysis liquid and diesel in engines. Their results suggest that corrosion accelerated by the high velocity turbulent flow in the spray channels is the dominant problem. A stainless steel nozzle has been built and successfully tested. Long term validation however, is still needed.

More recently, He et al. (2010) used a novel high-pressure homogenization (HPH) technique to improve the physicochemical properties and storage stability of switchgrass bio-oil. Compared with the conventional emulsification method, which consists of mixing bio-oil with diesel oil, the HPH technique improved the original properties of bio-oil by decreasing the viscosity and improving its stability in storage. However, the heating value, water content, density, PH value, or ash content did not change.

Overall, upgrading of bio-oil through emulsification with diesel oil is relatively simple. It provides a short-term approach to the use of bio-oil in diesel engines. The emulsions showed promising ignition characteristics, but fuel properties such as heating value, cetane and corrosivity were still unsatisfied. Moreover, this process required high energy for production. Design, production and testing of injectors and fuel pumps made from stainless steel or other materials) are required.

4.4.6 Steam reforming

The term "reforming" was originally used to describe the thermal conversion of petroleum fractions to more volatile products with higher octane numbers, and represented the total effect of many simultaneous reactions such as cracking, dehydrogenation and isomerisation (Yaman, 2004). Reforming also refers to the conversion of hydrocarbon gases and vaporized organic compounds to hydrogen containing gases such as synthesis gas, which is a mixture of carbon monoxide and hydrogen. Synthesis gas can be produced from natural gas, for example, by such processes as reforming in the presence of steam (steam reforming) (Klass, 1998).

Fast pyrolysis of biomass followed by catalytic steam reforming and shift conversion of specific fractions to obtain H_2 from bio-oil was presented as an effective way to upgrade biomass pyrolysis oils. Production of hydrogen from reforming bio-oil was investigated by NREL extensively, including the reactions in a fixed bed and a fluidized bed (Wang et al., 1997,1998; Czernik et al., 2007). Commercial nickel catalysts showed good activity in processing biomass derived liquids (Ekaterini & Lemonidou, 2008).

4.4.7 Chemicals extracted from the bio-oils

There are many substances that can be extracted from bio-oil, such as phenols used in the resins industry, volatile organic acids, nitrogen herocycles and n-alkanes (Ross et al., 2010; Gallivan and Matschei, 1980). Most recently, Cao et al. (2010) extracted triacetonamine (TAA) in a bio-oil from fast pyrolysis of sewage sludge with a high yield (27.9%) and high purity (80.4%) using acetone as the absorption solvent. Hydrothermal bio-oil contains up to 50 wt % asphalt, which makes it a good candidate for the asphalt industry. Recently, Fini and her colleagues fractionated and chemically modified the bio-oil into an effective alphalt bio-binder, which has remarkable potential to replace or augment petroleum-based road asphalt (Fini et al, 2010).

The only current commercially important application of bio-oil chemicals is that of wood flavor or liquid smoke (Mohan et al., 2006).Commercialization of special chemicals from bio-oils requires more devotion to developing reliable low cost separation and refining techniques.

4.5 Closing remarks

Flash pyrolysis processes are so far the only commercially practiced technology for production of bio-oil or bio-crude from biomass. However, pyrolysis oils consist of high oxygen/water contents and hence only about half the caloric value of petroleum (20-25 MJ/kg. In addition, they are strongly acidic and corrosive. Hydrothermal liquefaction with a suitable solvent (water or organics) is superior to pyrolysis. It can potentially produce liquid oils with much higher caloric values. In particular, liquefaction to produce bio-oil from organic wastes is a promising way to not only create value, but also reduce pollutants associated with sludge. There are intensive studies on bio-oil upgrading and several techniques have been developed.

Solvent addition (esterification) appears to be the most practical approach due to simplicity, the low cost of some solvents and their beneficial effects on the oil properties. However, none of these bio-oil upgrading techniques has been commercialized due to low biofuel efficiency and their limitations. Therefore, novel refinery processes are needed to systematically upgrade bio-oils into transportation fuels that have desirable qualities, while producing other value-added co-products to make the economics work.

5. References

Ancheyta, J. & Speight, J.G. (2007). Hyroprocessing of Heavy Oils and Residua. CRC press, ISBN -13: 978-0-8493-7419-7, FL, U.S.A

Appell, H.R., Fu Y.C., Friedman S., Yavorsky P.M., and Wender I. 1980. Converting of organic wastes to oil: a Replenishable Energy Source. U.S. Bureau of Mines. Report of Investigations 7560. Washington, DC.

Bao,G.R. ; Saka, S. & Wang, H. (2008). Cellulose Decomposition Behavior in Hot-Compressed Arotic Solvents. Sci. China Ser B-Chem. Vol.51, No.5, pp.479-486.

Bridgwater, A.V.; Peacocke, G.V.C. (2000). Fast pyrolysis processes for biomass. Renewable and Sustainable Energy Reviews, Vol.4, No.1, pp.1-73.

Bohlmann, B.J.T., Lorth C.M., Drews A., and Buchholz, R. 1999. Microwave high pressure thermochemical conversion of sewage sludge as an alternative to incineration. Chem.Eng.Technol. 21(5): 404-409.

Boucher, M.E.; Chaala, A. & Roy, C. (2000). Bio-oils obtained by vacuum pyrolysis of softwood bark as a liquid fuel for gas turbines. Part I: Properties of bio-oil and its blends with methanol and a pyrolytic aqueous phase. Biomass & Bioenergy, Vol.19, No.5, pp. 337–350.

Cao, J.P.; Zhao, X.; Morishita, K.; Li, L.; Xiao, X.; Obara,R.; Wei, X.& Takarada, T. (2010). Triacetonamine formation in a bio-oil from fast pyrolysis of sewage sludge using acetone as the absorption solvent. Bioresoure Technology, Vol. 101, pp. 4242-4245.

Carrillo, F.; Lis, M.J.; Colom, X.; López-Mesas, M. & Valldeperas, J. (2005). Effect of alkali pretreatment on cellulase hydrolysis of wheat straw: kinetic study. Process Biochem. 40:3360–3364.

Casey, E.; Sedlak, M.; Ho, W.Y. N. & Mosier, S. N. (2010). Effect of acetic acid and pH on the cofermentation of glucose and xylose to ethanol by a genetically engineered strain of Saccharomyces cerevisiae. FEMS Yeast Research.; 10 (4): 385

Cemek, M. & Kucuk, M.M. (2001). Liquid products from verbascum stalk by supercritical fluid extraction. Energy Conver. Manage., Vol.42, pp.125-130.

Chiaramonti, D.; Bonini, M.; Fratini, E.; Tondi, G.; Gartner, K.; Bridgwater, A. V.; Grimm, H P.; Soldaini, I.; Webster, A.; Baglioni, P. (2003a). Development of emulsions from biomass pyrolysis liquid and diesel and their use in engines—Part 1 : emulsion production. Biomass Bioenergy, Vol. 1, pp. 85- 99

Chiaramonti, D.; Bonini, M.; Fratini, E.; Tondi, G.; Gartner, K.; Bridgwater, A. V.; Grimm, H. P.; Soldaini, I.; Webster, A.; Baglioni, P. (2003b). Development of emulsions from biomass pyrolysis liquid and diesel and their use in engines—Part 2: tests in diesel engines. Biomass Bioenergy, Vol. 1, pp.101- 111.

Chornet, E., and R. P. Overend. (1985). Biomass liquefaction: an overview. In Fundamentals of Thermochemical Biomass Conversion, eds. R. P. Overend, T. A. Milne, and L. K. Mudge, 967-1002. New York: Elsevier Applied Science.

Czernik, S.; Evansa, R. & Frencha, R. (2007). Hydrogen from biomass-production by steam reforming of biomass pyrolysis oil. Catalysis Today, Vol.129, No. 3-4, pp. 265-268

Demirbaş, A. (2000a). Mechanisms of liquefaction and pyrolysis reactions of biomass. Energy Conversion & Management, Vol.41, pp.633-646.

Demirbaş, A. 2000b. Effect of lignin content on aqueous liquefaction products of biomass. Energy Conversion and Mgmt, Vol. 41, No. 15, pp. 1601-1607.

Demirbaş, A. & Celik, A. (2005a). Degradation of poplar and spruce wood chips usingalkaline glycerol. Energy Sources, Part A, Vol. 27, No.11, pp. 1073-1084.

Demirbaş, A.; Balat, M. & Bozbas, K. (2005b). Direct and catalytic liquefaction of woodspecies in aqueous solution. Energy Sources Vol.27, No.11, pp. 271-277.

Demirbaş, A. (2008). Liquefaction of biomass using glycerol. Energy Sources, Part A, Vol.30, No. 12, pp.1120-1126.

Diebold, J.P.; Czernik, S. (1997). Additives to lower and stabilize the viscosity of pyrolysis oils during storage. Energy & Fuels, Vol.11, pp. 1081-1091.

Ekaterini C.V. and Lemonidou, A.A. (2008). Hydrogen production via steam reforming of bio-oil components over calcium aluminate supported nickel and noble metal catalysts. Applied Catalysis A: General , Vol.351, No.1, pp. 111-121.

Elliott, D.C. ; Sealock, L.J. & Butner, R.S. (1988). Product Analysis from Direct Liquefaction of Several High-moisture Biomass Feedstocks, In Pyrolysis Oils from Biomass: Producing, Analyzing, and Upgrading, pp.179-188. J. Soltes and T. A. Milne, eds. Denver, Colo.: Am. Chem. Soc.

Fini, E.H. ; Yang S. ; Xiu, S. & Shahbazi, A. (2010). Characterization and Application of Manure-Based Bio-binder in Asphalt Industry. Transportation Research Board, 89th Annual Meeting January 13-17, 2010, Washington, D.C.pp.14

Gallivan, R.M. & Matschei, P.K. (1980). Fraction of oil obtained by pyrolysis of lignocellulosic materials to recover a phenolic fraction for use in making phenolic fraction for use in making phenol-formaldehyde resins. US Patent No. 42009647, American Can Co.

Gan, J., Yuan, W.; Nelson, N.O.& Agudelo, S. C. (2010). Hydrothermal Conversion of Corn Cobs and Crude Glycerol. Biological Engineering, Vol. 2, No. 4, pp. 197-210.

Garcia-Perez, M.; Shen, J.; Wang, X. & Li, C. (2010). Production and fuel properties of fast pyrolysis oil/bio-diesel blends. Fuel Processing Technology, Vol. 91, No. 3, pp. 296-305

Gharieb, H.K.; Faramawy, S. & Zaki, N.N. (1995). Liquefaction of cellulosic waste V. Water formation and evaluation of pyrolytic char as a byproduct of pyrolysis reaction. Fuel Sci. Technol. Int. 13 (7) (1995), pp. 895–909.

Goudriaan, F., Beld B.van de, Boerefijn F.R., Bos G.M., Naber J.E., Wal S.van der, and

Zeevalkink, J.A. (2000). Thermal efficiency of the HTU process for biomass liquefaction. In Proceedings of the Progress in Thermochemical Biomass Conversion Conference, pp.1312-1325.

Gupta, R. (2008). Alkaline Pretreatment OF Biomass For Ethanol Production And Understanding The Factors Influencing The Cellulose Hydrolysis. Ph.D. Dissertation, Auburn University

He, R.; Ye, X.P.; Harte, F.& English, B. (2009). Effects of high-pressure homogenization on physicochemical properties and storage stability of switchgrass bio-oil. Fuel Processing Technology, Vol.90, pp. 415-421.

Heitz, M.; Brown, A. & Chornet, E. (1994). Solvent effects on liquefaction: Solubilizationprofiles of a Canadian prototype wood, Populus deltoides, in the presence of differentsolvents. Canadian J. Chem. Eng, Vol. 72, No. 6, pp. 1021-1027.

Hilten,R.N.; Bibens, B.P.; Kastner, J.R. & Das, K.C. (2010). In-Line Esterification of Pyrolysis Vapor with Ethanol Improves Bio-oil Quality. Energy Fuels, Vol. 24, pp. 673–682

Hill, J.; Nelson, E.; Tilman, D.; Polasky, S. & Tiffany, D. (2006). Environmental, economic and energetic costs and benefits of biodiesel and ethanol biofuels. Proc Natl Acad Sci USA. 103: 11206–11210

Huang, H.; Ramaswamy, S.; Tschirnera, U.W. & Ramaraob, B.V. (2008). A review of separation technologies in current and future biorefineries. Separation and Purification Technology. 62: 1-21

Huber G. W. & Dumesic, J. A. (2006). An overview of aqueous-phase catalytic processes for production of hydrogen and alkanes in a biorefinery. Catalysis Today, Vol. 111, No. 1-2, pp. 119-132.

Ikura, M.; Stanciulescu, M.; Hogan, E. (2003). Emulsification of pyrolysis derived bio-oil in diesel fuel Biomass Bioenergy, Vol. 3, pp. 221- 232.

Ikura, M.; Slamak, M.; Sawatzky, H. (1998). Pyrolysis liquid-in-diesel Oil Microemulsions. U.S.Pat. 5,820,640.

Itoh S., A. Suzuki, T. Nakamura, and S.Yokoyama. (1994). Production of heavy oil from sewage sludge by direct thermochemical liquefaction. Desalination. 98:127-133.

Jiang, X., and Ellis, N. (2010). Upgrading Bio-oil through Emulsification with Biodesel : Mixture Production. Energy & Fuels, Vol.24, pp.1358-1364.

Karagöz, S.; Bhaskar, T.; Muto, A.; Sakata, Y.; Oshiki, T. & Kishimoto, T. (2005). Low temperature catalytic hydrothermal treatment of wood biomass: analysis of liquid products. Chemical Engineering Journal, Vol. 108, pp.127-137.

Kim, Y.; Mosier, N.S. & Ladisch, M.R. (2008). Effect of Liquid Hot Water Pretreatment on Switchgrass Hydrolysis. The 2008 AIChE Annual Meeting. Philadelphia PA

Klass, D.L. (1998). Biomass for renewable energy, fuels, and chemicals. San Diego, CA: Academic Press.

Knappert, D.R.; Grethlein, H.E. & Converse, A.O. (1981). Partial acid hydrolysis of poplar wood as a pretreatment for enzymatic hydrolysis. Biotechnology Bioengineering. 11:67-77

Kücük, M. M. (2005). Delignification of biomass using alkaline glycerol. Energy Sources, Part AVol., 27, No. 13, pp. 1245-1255.

Kranich, W.L. (1984). Conversion of sewage sludge to oil by hydroliquefaction. U.S. Enviromental Protection Agency. EPA-600/2-84-010. Cincinnati,OH

Laser, M.; Schulman, D.; Allen, S.G.; Lichwa, J.; Antal, Jr. M.J. & Lynd, L.R. (2002). A comparison of liquid hot water and steam pretreatments of sugar cane bagasse for bioconversion to ethanol. Bioresource Technology. 81(1): 33-44

Lau, M. W.; Gunawan, C.; Balan, V. & Dale, B. E. (2010). Comparing the fermentation performance of Escherichia coli KO11, Saccharomyces cerevisiae 424A(LNH-ST) and Zymomonas mobilis AX101 for cellulosic ethanol production. Biotechnology for Biofuels. 3:11

Lee, D.; Yu, A.H.C.; Wong, K.K.Y. & Saddler, J.R. (1994). Evaluation of the enzymatic susceptibility of cellulosic substrates using specific hydrolysis rates and enzyme adsorption. Applied Biochemistry and Biotechnology. 45/45: 407–415

Li, D.; Zhao, Y.; Yao, F. & Guo, Q. (2009). Green Solvent for Flash Pyrolysis Oil SeparationEnergy & Fuels, Vol.23, pp. 3337 - 3338

Li, Y.; Ruan, R.; Chen, P.L.; Liu, Z.; Pan, X.; Lin, X.; Liu, Y.; Mok, C.K. & Yang, T. (2004). Enzymatic hydrolysis of corn stover pretreated by combined dilute alkaline treatment and homogenization. Transactions of the ASAE, 47: 821-825

Liu, Z. & Zhang, F. (2008). Effects of Various Solvents on the Liquefaction of Biomass to Produce Fuels and Chemical Feedstocks. Energy Conversion and Management, Vol.49, pp.3498-3504.

Lima, P.; Bonarini, A. & Mataric, M. (2004). Application of Machine Learning, InTech, ISBN 978-953-7619-34-3, Vienna, Austria

Lloyd T.A. & Wyman C.E. (2005). Combined sugar yields for dilute sulfuric acid pretreatment of corn stover followed by enzymatic hydrolysis of the remaining solids. Bioresour Technol. 96(18):1967-77

Lu, Q.; Li, W.Z. & Zhu, X.F. (2009). Overview of fuel properties of biomass fast pyrolysis oils. Energy Conversion and Management, Vol.50, No.5, pp.1376-1383.

Lynd, L.R.; Elander, R.T & Wyman, C.E. (1996). Likely features and costs of mature biomass ethanol technology. Applied Biochemistry and Biotechnology. 57/58: 741–761

Mazaheri, H. ; Lee, K.T. ; Bhatia, S. & Mohamed A.R. (2010). Sub/supercritical Liquefaction of Oil Palm Fruit Press Fiber for the Production of Bio-oil: Effect of solvents. Bioresource Technology. Vol.101, No.19, pp.7641-7647.

Mahfud, F.; Meli_an-Cabrera, I.; Manurung, R.; Heeres, H. (2007). Biomass to fuels: Upgrading of flash pyrolysis oil by reactive distillation using a high boiling alcohol and acid catalysts. Process Saf. Environ. Prot. Vol. 85, No.5, 466–472.

Minowa, T.; Kondo, T. & Sudirjo, S. T. (1998). Thermochemical liquefaction oflndonesian biomass residues. Biomass and Bioenergy, Vol. 14, No.5-6, pp. 517-524.

Minami, E., and Saka S. (2003). Comparison of the Decomposition Behaviors of Hard Wood and Soft Wood in Supercritical Methanol. J Wood Sci. Vol. 49, pp. 73–78.

Minami, E., and Saka S. (2005). Decomposition Behavior of Woody Biomass in Water-added Supercritical Methanol. J Wood Sci. Vol. 51, pp. 395–400.

Minowa T., Murakami M., Dote Y., Ogi T., and Yokoyama S. (1995). Oil production from garbage by thermochemical liquefaction. Biomass and Bioenergy. 8(2):117-120.

Mohan,D; Pittman, C.U. & Steele, P.H. (2006). Pyrolysis of Wood/Biomass for Bio-oil: A Critical Review. Energy & Fuels, Vol. 20, 848-889.

Mosier, N.; Hendricksona, R.; Ho, N.; Sedlaka, M. & Ladisch, M.R. (2005). Optimization of pH controlled liquid hot water pretreatment of corn stover. Bioresource Technology 96: 1986-1993

Mosier, N.; Wyman, C.; Dale, B.; Elander, R,; Lee, Y.Y.; Holtzapple, M. & Ladisch, M. (2005). Features of promising technologies for pretreatment of lignocellulosic biomass. Bioresour Technol. 96(6):673-86

Nava, R. ; Pawelec, B. ; Castañoc, P. ; Álvarez-Galvánb, M.C. ; Loricerab, C.V. & Fierrob, J.L.G. (2009). Upgrading of bio-liquids on different mesoporous silica-supported CoMo catalysts. Applied Catalysis B : Enviromental, Vol. 92, pp. 154-167.

Oasmaa, A. & Czernik, S. (1999). Fuel Oil Quality of Biomass Pyrolysis Oils - State of the Art for the End Users. Energy & Fuels, Vol.13, pp.914-921.

Oasmaa, A.; Kuoppala, E.; Selin, J.F.; Gust, S. & Solantausta, Y. (2004). Fast pyrolysis of forestry residue and pine. 4. Improvement of the product quality by solvent addition. Energy & Fuels, Vol. 18 No.5, pp. 1578–1583.

Ocfmia, K. S.; Zhang, Y. & Funk, T. (2006). Hydrothermal processing of swine manure into oil using a continuous reaction system: Development and testing. Transaction of the ASABE, Vol.49, No. 2, pp. 533-541.

Ogi, T.; Minowa, T.; Dote, Y. & Yokoyama, S. Y. (1994). Characterization of oil produced byhe direct liquefaction of Japanese oak in an aqueous 2-propanol solvent system. Biomassand Bioenergy, Vol.7, No. 1-6: 193-199.

Perlack R.D.; Wright, L.L.; Turhollow, A.F.; Graham, R.L., Stokes B. J. & Erbach, D.C. (2005). Biomass as feedstock for a bioenergy and bioproducts industry: The technical feasibility of a billion-ton annual supply. Sponsored by USDOE and USDA2005.

Peng, J.; Chen, P.; Lou, H.; Zheng, X. (2009). Catalytic upgrading of biooil by HZSM-5 in sub- and super-critical ethanol. Bioresour. Technol. Vol.100, 3415–3418.

Peng, J.; Chen, P.; Lou, H.; Zheng, X. M. (2008). Upgrading of bio-oil over aluminum silicate in supercritical ethanol. Energy & Fuels, Vol.22, pp. 3489–3492.

Radlein, D.J.; Piskorz, J. & Majerski, P. (1996). Method of Upgrading Biomass Pyrolysis Liquids for use as Fuels and as Sources of Chemicals Reaction with Alcohol, in Patent, E. (ed.), Patent: CA2165858.

Ross, A.B.; Biller, P.; Kubacki, M.L.; Lin, H.; Lea-Langton, A. & Jones, J.M.. (2010). Hydrothermal processing of microalgae using alkali and organic acids. Fuel, Vol. 89, pp. 2234–2243.

Smith, W.J. & Consultancy, T. (2007). Mapping the Development of UK Biorefinery Complexes, NNFCC.

Suzuki A., Nakamura T., and Yokoyama S. (1990). Effect of operating parameters on thermochemical liquefaction of sewage sludge. Journal of Chemical Engineering of Japan. 23(1): 6-11.

Tang, Z.; Lu, Q.; Zhang, Y.; Zhu, X.F. & Guo, Q.X. (2009). One step bio-oil upgrading through hydrotreatment, esterification and cracking. Ind. Eng. Chem. Res., Vol. 48, pp. 6923-6929.

Tang, Y.; Yu, W.; Mo, L.; Lou, H. & Zheng, X. (2008). One-step hydrogenation-esterification of aldehyde and acid to ester over bifunctional Pt catalysts:A model reaction as novel route for catalytic upgrading of fast pyrolysis bio-oil. Energy Fuels, Vo. 22, No.5, pp.3484–3488.

Vasan, P. T.; Piriya, P. S.; Prabhu D. I. & Vennison, S. J. (2011). Cellulosic ethanol production by Zymomonas mobilis harboring an endoglucanase gene from Enterobacter cloacae. Bioresource Technolog. 102: 85–2589

Wang, D.; Czernik, S. & Montané, D. (1997). Biomass to hydrogen via pyrolysis and catalytic steam reforming of the pyrolysis oil and its fractions. Ind Eng Chem Res, Vol.36, pp. 1507–1518.

Wang, D.; Czermik, S.& Chornet, E. (1998). Production of hydrogen from biomass by catalytic steam reforming of fast pyrolytic oils. Energy Fuels, Vol. 12, pp. 19–24.

Wang, J.; Chang, J. and Fan, J. (2010). Upgrading of Bio-oil by Catalytic Esterification and Determination of Acid Number for Evaluating Esterification Degree. Energy & Fuels, Vol. 24, pp. 3251–3255.

Wyman C. E. (1999). BIOMASS ETHANOL: Technical Progress, Opportunities, and Commercial Challenges, Annual Review of Energy and the Environment. 24: 189-226.

Xiong, W. M.; Zhu, M. Z.; Deng, L.; Fu, Y.; Guo, Q. X. (2009). Esterification of organic acid in bio-oil using acidic ionic liquid catalysts.Energy & Fuels, Vol. 23, pp. 2278–2283.

Xiu, S.; Shahbazi, A.; Shirley, V. & Cheng, D. (2010a). Hydrothermal pyrolysis of swine manure to bio-oil: Effects of operating parameters on products yield and characterization of bio-oil. Journal of Analytical and Applied Pyrolysis, Vol. 88, No.1, pp.73-79.

Xiu, S.; Shahbazi, A.; Wang, L. & Wallace, C. (2010b). Supercritical Ethanol Liquefaction of Swine Manure for Bio-Oils Production. American J. of Engineering and Applied Sciences, Vol.3, No. 2, pp. 494-500.

Xiu, S.; Shabazi, A.; Shirley, V. B.; Mims, M. R. & Wallace, C. W. (2010c). Effectiveness andMechanisms of crude glycerol on the biofuels production from swine manure through hydrothermal pyrolysis. J. Analytical and Applied Pyrolysis 87(2): 194-198.

Xu, C., and Etcheverry, T. (2008). Hydro-liquefaction of woody biomass in Sub- and Super-critical ethanol with iron-based catalysts. Fuel, Vol.87, pp.335-345.

Xu, J.; Jiang, J.; Sun, Y. & Lu, Y. (2008). Bio-oil upgrading by means of ethyl ester production in reactive distillation to remove water and to improve storage and fuel characteristics. Biomas s & Bioenergy, Vol.32, No.11, pp. 1056-1061.

Yang, B. & Wyman, C.E. (2004). Effect of xylan and lignin removal by batch and flow through pretreatment on the enzymatic digestibility of corn stover cellulose. Biotechnology and Bioengineering. 86(1): 88–95

Yang, Y. ; Gilbert, A. & Xu, C. (2009). Production of Bio-crude from Forestry Waste by Hydro-Liquefaction in Sub-/Super-Critical Methanol. AIChE Journal, Vol.55, No.3, pp.807-819.

Yang, Y.; Sharma-Shivappa, R., Burns, J. C. & Cheng, J. J. (2009). Dilute Acid Pretreatment of Oven-dried Switchgrass Germplasms for Bioethanol Production. Energy Fuels. 23 (7): 3759-3766.

Yaman, S. (2004). Pyrolysis of biomass to produce fuels and chemical feedstocks. Energy Conversion and Management,Vol. 45, Issue 5, March 2004, Pages 651-671

Yu, W.; Mo, L.; Lou, H.& Zheng, X. (2009). One-Step Hydrogenation-Esterification of Aldehyde and

Acid to Ester over Bifunctional Pt Catalysts: A Model Reaction as Novel Route for Catalytic Upgrading of Fast Pyrolysis Bio-Oil.

Yuan X. Z., H. Li, G. M. Zeng, J. Y. Tong, W. Xie. (2007). Sub- and supercriticalliquefaction of rice straw in the presence of ethanol-water and 2-propanol-water mixture. Energy 32(11):2081-2088.

Zhang, B.; Shahbazi, A.;Wang, L.; Diallo, O. & Whitmore, A. (2010A). Alkali Pretreatment and Enzymatic Hydrolysis of Cattails from Constructed Wetlands. American Journal of Engineering and Applied Sciences. 3(2): 328-332

Zhang,B.; Shahbazi,A.; Wang, L.; Diallo O. & Whitmore A. (2010B). Hot-Water Pretreatment of Cattails for Extraction of Cellulose. Journal of Industrial Microbiology and Biotechnology. DOI: 10.1007/s10295-010-0847-x

Zhang, B.; Shahbazi, A.; Wang, L.; Diallo, O. & Whitmore, A. (2011). Dilute-sulfuric acid pretreatment and cellulose conversion of cattails. ASABE Annual Meeting, Louisville, Kentucky

Zhang, Q.; Chang, J.; T. Wang & Xu, Y. (2007). Review of biomass pyrolysis oil properties and upgrading research. Energy Conversion & Management, Vol.48, pp.87-92.

Zhao, Y.; Wang, Y.; Zhu, J.Y.; Ragauskas, A.& Deng, Y. (2008). Enhanced enzymatic hydrolysis of spruce by alkaline pretreatment at low temperature. Biotechnology and bioengineering. 99(6): 1320-1328

9

Utilisation of Waste from Digesters for Biogas Production

Ladislav Kolář, Stanislav Kužel, Jiří Peterka and Jana Borová-Batt
Agricultural Faculty of the University of South Bohemia in České Budějovice
Czech Republic

1. Introduction

1.1 Is the waste from digesters (digestate) an excellent organic fertilizer?

A prevailing opinion of bio-power engineers as well as in literature is that wastes from digesters in biogas production are an excellent fertiliser and that anaerobic digestion is to some extent an improvement process in relation to the fertilising value of organic materials used for biogas production. These opinions are apparently based on the fact that in anaerobic stabilisation of sludge the ratio of organic to mineral matters in dry matter is approximately 2:1 and after methanisation it drops to 1:1. Because there is a loss of a part of organic dry matter of sludge in the process of anaerobic digestion, the weight of its original dry matter will decrease by 40%, which will increase the concentration of originally present nutrients. In reality, anaerobic digestion will significantly release only ammonium nitrogen from the original material, which will enrich mainly the liquid phase due to its solubility; the process will not factually influence the content of other nutrients (Straka 2006).

The opinion that waste from anaerobic digestion is an excellent fertiliser is also due to the observation of fertilised lands. The growths are rich green and juicy. They have a fresh appearance - this is a typical sign of mineral nitrogen, including larger quantities of water retention by plants due to the nitrogen. However, the content of dry matter is changed negligibly, which shows evidence that the fertilisation is inefficient.

If organic matter is to be designated as organic fertiliser, it has to satisfy the basic condition: it has to be easily degradable microbially so that it will release necessary energy for soil microorganisms.

1.2 Mineralisation of organic matter in soil

This microbial transformation of organic matter in soil is mineralisation when organic carbon of organic substances is transformed to CO_2 and from mineralised organic matter those mineral nutrients are released that were already contained in organic matter in mineral (ionic) form and those that were in it in organic form. CO_2 is an important fertiliser in agriculture; it is the basic component for photosynthetic assimilation, for the formation of new organic matter produced by plants. As plants can take up only nutrients in mineral form (K^+, NH_4^+, NO_3^-, Ca^{2+}, Mg^{2+}, $H_2PO_4^-$, HPO_4^{2-}, SO_4^{2-} etc.) and nutrients in organic form (e.g. protein nitrogen, phosphorus of various organophosphates), it is not accessible to plants, and besides its main function - energy production for the soil microedaphon - the mineralization of organic matter in soil is an important source of mineral nutrients for plants. It is applicable solely on condition that organic matter in soil is easily mineralisable, i.e. degradable by soil microorganisms.

1.3 Gain from mineralising organic fertiliser for farmers: energy for soil microorganisms and release of mineral nutrients for plant nutrition

What we appreciated more for organic fertilisers? Gain of energy and enhancement of the microbial activity of soil or savings that are obtained by the supply of mineral nutrients? Unfortunately, simplified economic opinions cause each superficial evaluator to prefer the gain of mineral nutrients released from organic matter. Such a gain is also easy to calculate. The calculation of the gain from an increased microbial activity of soil is difficult and highly inaccurate. Nevertheless, a good manager will unambiguously prefer such a gain. It is to note that the microbial activity of soil is one of the main pillars of soil productivity, it influences physical properties of soil, air and water content in soil, retention of nutrients in soil for plant nutrition and their losses through elution from soil to groundwater. A biological factor is one of the five main factors of the soil-forming process; without this process the soil would not be a soil, it would be only a parent rock or perhaps a soil-forming substrate or an earth at best.

Hence, it is to state that the release of mineral nutrients for their utilisation by plants during mineralisation of organic fertiliser in the soil produces an economically favourable effect but it is not the primary function of organic fertiliser, its only function is the support of microedaphon. The effect of mineral nutrients is replaceable by mineral fertilisers, the energetic effect for the microbial activity of soil is irreplaceable.

1.4 What influences the quality of digestate as a fertiliser?

The digestate, the waste from digesters during biogas production, is composed of solid phase and liquid phase (fugate). We have demonstrated that the solid phase of the digestate is not an organic fertilizer because its organic matter is very stable and so it cannot be a relatively expeditious source of energy for the soil microedaphon (Kolář et al. 2008). Neither is it a mineral fertilizer because available nutrients of the original raw material and also nutrients released from it during anaerobic digestion passed to the liquid phase - fugate. The digestate, and naturally the fugate, have a low content of dry matter (fugate 0.8 - 3% by weight) and this is the reason why analytical data on the ones to tens of weight % of available nutrients given in dry matter foster an erroneous opinion in practice that these wastes are excellent fertilizers. In fact, fugates are mostly highly diluted solutions in which the content of the nutrients that are represented at the highest amount, mineral nitrogen, is only 0.04 - 0.4% by weight.

The surplus of water during fertilization with this waste increases the elution of this nutrient in pervious soils while in less pervious soils the balance between water and air in the soil is impaired, which will have negative consequences.

The quality of the digestate as an organic fertiliser (labile, not organic material that is hard to decompose) substantially influences not only the microbial decomposability of the input material but also the level of anaerobic digestion in the digester. In the past when the sludge digestion was carried out in municipal waste treatment plants in digesters at temperatures of 18°C-22°C (psychrophilic regime), the decomposability of the substrate after fermentation was still good, therefore the digested sludge was a good organic fertiliser. These days we work with less decomposable substrates in mesophilic ranges (around 40°C) or even in thermophilic conditions. The degree of decomposition of organic matter during fermentation is consequently high and the digestate as organic fertiliser is practically worthless.

1.5 A hopeful prospect – IFBB process

It would be ideal to realize biogas production from the liquid phase only - it would be possible to introduce high performance UASB (Upflow Anaerobic Sludge Blanket) digesters and to achieve the large saving of technological volumes but the concentration of substances in the liquid phase should have to be increased. The solid phase of substrates, which cannot be applied as an organic fertilizer after the fermentation process, would be used as biomass for the production of solid biofuels in the form of pellets or briquettes. But it would be necessary to reduce its chlorine content to avoid the generation of noxious dioxins and dibenzofurans during the burning of biofuel pellets or briquettes at low burning temperatures of household boilers and other low-capacity heating units. Wachendorf et al. (2007, 2009) were interested in this idea and tried to solve this problem in a complex way by the hot-water extraction of the raw material (at temperatures of 5°C, 60°C and 80°C) followed by the separation of the solid and liquid phase by means of mechanical dehydration when a screw press was used. This procedure is designated by the abbreviation IFBB (Integrated Generation of Solid Fuel and Biogas from Biomass). These researchers successfully reached the transfer ratio of crude fibre from original material (grass silage) to liquid phase only 0.18, which is desirable for biogas production, but for more easily available organic substances influencing biogas production, e.g. nitrogen-free extract, the ratio is 0.31. The transfer of potassium, magnesium and phosphorus to the liquid phase ranged from 0.52 to 0.85 of the amount in fresh matter, calcium transformation was lower, at the transfer ratio 0.44 - 0.48 (Wachendorf et al. 2009). Transformation to the liquid phase was highest in chlorine, 0.86 of the amount in original fresh matter, already at a low temperature (5°C). The transfer of mineral nitrogen to the liquid phase before the process of anaerobic digestion is very low because there is a minute amount of mineral N in plant biomass and the major part of organic matter nitrogen is bound to low-soluble proteins of the cell walls. Nitrogen from these structures toughened up by lignin and polysaccharides is released just in the process of anaerobic digestion. Because in the IFBB process also organic nitrogen compounds (crude protein - nitrogen of acid detergent fibre ADF) are transferred to the liquid phase approximately at a ratio 0.40, the liquid phase, subjected to anaerobic digestion, is enriched with mineral nitrogen.
Like Wachendorf et al. (2009), we proceeded in the same way applying the IFBB system for the parallel production of biogas and solid biofuels from crops grown on arable land. The IFBB technological procedure is based on a high degree of cell wall maceration as a result of the axial pressure and abrasion induced with a screw press.

2. Crucial problems

2.1 The first problem: organic matter of digestate is poorly degradable in soil, its labile fractions were already utilised in a digester

The point is that the digestate is not an organic fertiliser because its organic substance is poorly degradable. But its liquid fraction contains a small amount of mineral nutrients, mainly of nitrogen. The fugate (and also the digestate) can be considered as a very dilute mineral fertiliser, nitrogenous fertiliser. However, the agriculture sector is exposed worldwide to an enormous pressure on economic effectiveness while the costs of machinery, fuels and agricultural labour force are very high in relation to the price of agricultural products. Therefore the chemical industry helps farmers to save on transportation and application costs incurred by fertilisation when highly concentrated

mineral fertilisers are produced. Even though they are substantially more expensive, from the aspect of cost accounting their use will finally pay off. Before the manufacture of town gas from coal using the ammonia water ended, farmers took the waste containing 1% of ammonia nitrogen only exceptionally even though it was practically free of charge.
With the current output of a biogas plant 526 kW (Chotýčany, South Bohemia) and daily dose of a substrate to the digester 46 t and practically identical production of digestate the daily production of mineral nitrogen is approximately 40 kg, which amounts to a relatively high value per year, almost 15 t of mineral nitrogen, but the dilution is unacceptable.

2.2 The second problem: the digestate contains much water and therefore the solution with plant nutrients is very dilute.

If this waste is applied as a fertiliser, the water surplus increases the elution of this nutrient into the bottom soil in pervious soils. In impervious soils and in less pervious soils the imbalance between water and air in the soil is deteriorated with all adverse consequences: aerobiosis restriction, reduction in the count of soil microorganisms, denitrification and escape of valuable nitrogen in the form of N_2 or N-oxides into the atmosphere. Soil acidification takes place because organic substances are not mineralised under soil anaerobiosis and they putrefy at the simultaneous production of lower fatty acids. These soil processes result in a decrease in soil productivity. Currently, its probability is increasingly higher for these reasons:

1. As a consequence of global acidification the frequency of abundant precipitation is higher in Europe throughout the year.
2. As a result of rising prices of fuels, depreciation on farm machinery and human labour force farmers apply digestates or fugates in the closest proximity of a biogas plant. It causes the overirrigation of fertilised fields even though the supplied rate of nitrogen does not deviate from the required average.

The problem of an excessively high irrigation amount has generally been known since long: it occurred in Berlin and Wroclaw irrigation fields after irrigation with municipal waste water in the 19th and 20th century, in the former socialist countries after the application of agricultural and industrial waste waters and of slurry from litterless operations of animal production. Even though nobody surely casts doubt on the fertilising value of pig slurry or starch-factory effluents, total devastation of irrigated fields and almost complete loss of their potential soil productivity were quite normal phenomena (Stehlík 1988).

2.3 Fundamental issues to solve

A further part of this study should help solve these crucial problems:

1. What is the rational utilisation of digestate and/or fugate and separated solid fraction of digestate in the agriculture sector that are generated by current biogas plants if we know that their utilisation as fertilisers is rather problematic?
2. What are the prospects of utilisation of wastes from biogas production and what modifications in the technology of biogas production from agricultural wastes should be introduced?
3. What problems should be solved by researchers so that the promising utilisation of wastes from biogas production could be realised?
4. What is the optimum form of utilisation of wastes from biogas plants and why?

3. Information

3.1 Current optimum utilisation of digestate from biogas plants in the agriculture sector

3.1.1 Biodegradability (lability) and stability of organic matter

How many labile components of organic matter are lost during anaerobic digestion in a biogas plant can be demonstrated by determination of the degree of organic matter lability. For this purpose a number of methods can be used that are mostly based on resistance to oxidation or on resistance to hydrolysis. Oxidation methods are based on oxidation with chemical oxidants, e.g. with a solution of $K_2Cr_2O_7$ in sulphuric acid at various concentrations - 6 M + 9 M + 12 M (Walkley 1947, Chan et al. 2001) or with a neutral solution of $KMnO_4$ at various concentrations (Blair et al. 1995, Tirol-Padre, Ladha 2004). The degree of organic matter lability is evaluated from the amount of oxidizable carbon in per cent of its total amount in particular variously aggressive oxidation environments or the reaction kinetics of the observed oxidation reaction is examined while its characteristic is the rate constant of the oxidation process.

In 2003 was proposed and tested the method to evaluate the kinetics of mineralisation of the degradable part of soil organic matter by the vacuum measurement of biochemical oxygen demand (BOD) of soil suspensions using an Oxi Top Control system of the WTW Merck Company, designed for the hydrochemical analysis of organically contaminated waters (Kolář et al. 2003). BOD on the particular days of incubation is obtained by these measurements whereas total limit BOD_t can be determined from these data, and it is possible to calculate the rate constant K of biochemical oxidation of soil organic substances per 24 hours as the rate of stability of these substances. A dilution method is the conventional technique of measuring BOD and also rate constants. It was applied to determine the stability of soil organic substances but it was a time- and labour-consuming procedure. The Oxi Top Control method was used with vacuum measurement in vessels equipped with measuring heads with infrared interface indicator communicating with OC 100 or OC 110 controller while documentation is provided by the ACHAT OC programme communicating with the PC, and previously with the TD 100 thermal printer. Measuring heads will store in their memory up to 360 data sentences that can be represented graphically by the controller while it is also possible to measure through the glass or plastic door of the vessel thermostat directly on stirring platforms. The rate of biochemical oxidation of organic substances as the first-order reaction is proportionate to the residual concentration of yet unoxidised substances:

$$dy/dt = K\,(L - y) = KL_z \tag{1}$$

where:
L = total BOD
y = BOD at time t
L_z = residual BOD
k, K = rate constants

By integrating from 0 to t of the above relation the following equation is obtained:

$$L_z = L\,.\,e^{-Kt} = L\,.\,10^{-kt} \tag{2}$$

In general it applies for BOD at time t:

$$y = L\ (1 - 10\text{-}kt) \tag{3}$$

where:
y = BOD at time t
L = BOD_{total}
k = rate constant [24 hrs^{-1}]
Used procedure is identical with the method of measurement recommended by the manufacturer in accordance with the Proposal for German Uniform Procedures DEV 46th Bulletin 2000 - H 55, also published in the instructions for BOD (on CD-ROM) of WTW Merck Company.
The decomposition of organic matter is the first-order reaction. In these reactions the reaction rate at any instant is proportionate to the concentration of a reactant (see the basic equation dy/dt). Constant k is the specific reaction rate or rate constant and indicates the instantaneous reaction rate at the unit concentration of a reactant. The actual reaction rate is continually variable and equals the product of the rate constant and the instantaneous concentration. The relation of the reaction product expressed by BOD at time t (y) to t is the same as the relation of the reactant (L - y) at time t and therefore the equations

$$(L - y) = L\ .\ e\text{-}kt \tag{4}$$

and

$$y = L\ (1 - e\text{-}kt) \tag{5}$$

are analogical.
If in the graph the residual concentration of carbon is plotted on the y-axis in a logarithmic scale *log (L - y)* and the time in days from the beginning of experiment is plotted on the x-axis, we will obtain a straight line, the slope of which corresponds to the value -k/2.303.
The quantity of the labile fraction of organic matter can also be assessed by determination of soluble carbon compounds in hot water (Körschens et al. 1990, Schulz 1990) and their quality by determination of the rate constant of their biochemical oxidation (Kolář et al. 2003, 2005a, b).
Hydrolytic methods are based on resistance of the organic matter different aggressive ways of hydrolysis that is realised at different temperature, time of action and concentration of hydrolytic agent, which is usually sulphuric acid. Among many variants of these methods the hydrolytic method according to Rovira et Vallejo (2000, 2002, 2007) in Shirato et Yokozawa (2006) modification was found to be the best. This method yields three fractions: labile LP1, semi-labile LP2 and stable LP3. The per cent ratio of these three fractions, the sum of which is total carbon of the sample C_{tot}, provides a very reliable picture of the degree of organic matter lability.
Of course, there are a lot of methods based on the study of organic matter biodegradability in anaerobic conditions. First of all, it is the international standard ISO C_D 11734: Water quality - evaluation of the "ultimate" anaerobic biodegradability of organic compounds in digested sludge - Method by measurement of the biogas production, and particularly a very important paper using the Oxi Top Control measuring system manufactured by the German company MERCK for this purpose (Süssmuth et al. 1999).
Tests of methanogenic activity (Straka et al. 2003) and tests examining the activity of a microbial system (Zábranská et al. 1985a, b, 1987) are methods that can describe the degree of organic matter lability in its ultimate effect. Our long-time work experiences in the

evaluation of a huge amount of various analyses for the study of organic matter lability have brought about this substantial knowledge:

1. The study of the ratio of organic matter labile fractions, i.e. of their quantity, is always incomplete. A more authentic picture of the situation can be obtained only if information on the quality of this labile fraction is added to quantitative data. Such a qualitative characteristic is acquired in the easiest way by the study of reaction kinetics of the oxidation process of this fraction. The process of biochemical oxidation and the calculation of its rate constant K_{Bio} are always more accurate that the calculation of its rate constant of oxidation by chemical oxidants K_{CHEM} (Kolář et al. 2009a).
2. It applies to current substrates for biogas production in biogas plants that with some scarce exceptions the degree of organic matter lability is very similar in both aerobic and anaerobic conditions. In other words: organic matter is or is not easily degradable regardless of the conditions concerned (Kolář et al. 2006).
3. A comparison of various methods for determination of organic matter lability and its degradability in the anaerobic environment of biogas plant digesters and also for determination of digestate degradability after its application to the soil showed that hydrolytic methods are the best techniques. They are relatively expeditious, cheap, sample homogenisation and weighing are easy, and the results correlate very closely with methods determining the biodegradability of organic matter directly. E.g. with the exception of difficult weighing of a very small sample and mainly its homogenisation the Oxi Top Control Merck system is absolutely perfect and highly productive - it allows to measure in a comfortable way simultaneously up to 360 experimental treatments and to assess the results continually using the measuring heads of bottles with infrared transmitters, receiving controller and special ACHAT OC programme for processing on the PC including the graph construction. But its price is high, in the CR about 4 million Kč for the complex equipment. Hydrolytic methods require only a small amount of these costs and are quite satisfactory for practical operations (Kolář et al. 2008). However, for scientific purposes we should prefer the methods that determine anaerobic degradability of organic matter, designated by D_C.

The substrate production of methane V_{CH4S} [the volume of produced methane (V_{CH4c}) after the subtraction of endogenous production of methane (V_{CH4e}) by the inocula] was determined by an Oxi Top Control Merck measuring system.

The calculation is based on this equation of state:

$$n = p \times V/RT \tag{6}$$

where:
n = number of gas moles
V = volume [ml]
P = pressure [hPa]
T = temperature [°K]
R = gas constant 8.134 J/mol °K
and the number of CO_2 and CH_4 moles in the gaseous phase of fermentation vessels is calculated:

$$n_{CO2\ g\ CH4} = (\Delta p \times Vg/RT) \times 10^{-4} \tag{7}$$

$$\Delta p = p_1 - p_0 \tag{8}$$

where: p_0 = initial pressure

Fermentation at 35° C and continuous agitation of vessels in a thermostat lasts for 60 days, the pressure range of measuring heads is 500 – 1 350 kPa and the time interval of measuring pressure changes is 4.5 min. Anaerobic fermentation is terminated by the injection of 1 ml of 19% HCl with a syringe through the rubber closure of the vessel to the substrate. As a result of acidification CO_2 is displaced from the liquid phase of the fermentation vessel. The process is terminated after 4 hours. The number of CO_2 moles is calculated from the liquid phase:

$$nCO2\,l = \{[p2\,(Vg - VHCl) - p1 \times Vg]/RT\} \times 10^{-4} \tag{9}$$

The injection of 1 ml of 30% KOH into the rubber container in the second tube of the fermentation vessel follows. The sorption of CO_2 from the gaseous phase of the vessel is terminated after 24 hours and the total number of CO_2 moles in gaseous and liquid phases is calculated from a drop in the pressure in the vessel:

$$n_{CO2\,l,\,CO2\,g} = \{[p_3\,(V_g - V_{HCl} - V_{KOH}) - p_2\,(V_g - V_{HCl})]/RT\} \times 10^{-4} \tag{10}$$

where:

Δp = difference in pressures [hPa]

V_g = the volume of the gas space of the fermentation vessel [ml]

p_1 = gas pressure before HCl application [hPa]

p_2 = gas pressure before KOH application [hPa]

p_3 = gas pressure after KOH application [hPa]

R = gas constant = 8.134 J/mol °K

T = absolute temperature = 273.15 + X °C

V_{HCl} = the volume of added HCl [ml]

V_{KOH} = the volume of added KOH [ml]

Based on the results, it is easy to calculate the number of CO_2 moles in the gaseous phase and by the subtraction from $n_{CO2\,g\,CH4}$ the number of moles of produced methane:

$$n_{CH4} = (n_{CO2\,g\,CH4} + n_{CO2\,l}) - n_{CO2\,l\,CO2\,g} \tag{11}$$

The total number of moles of the gases of transported carbon:

$$n_{CO2\,g\,CH4} + n_{CO2\,l} = n_{total} \tag{12}$$

Baumann's solution A + B in deionised water of pH = 7.0 is used as a liquid medium (Süssmuth et al. 1999).

The standard addition of the inoculum corresponds roughly to an amount of 0.3% by volume (aqueous sludge from the anaerobic tank of the digester). Instead of Baumann's solution it is possible to use a ready-made nutrient salt of the MERCK Company for this system.

The operation of the Oxi Top Control measuring system was described in detail by Süssmuth et al. (1999).

Methane yield was calculated from the substrate production of methane V_{CH4S} by division by the initial quantity of the added substrate:

$$Y_{CH4g} = \frac{(V_{CH4C} - V_{CH4e})}{S} = \frac{V_{CH4S}}{S}[l/g] \tag{13}$$

where:
V_{CH4C} = methane yield of C-source
V_{CH4e} = methane yield of the added inoculum
S = substrate quantity at the beginning [g]
Lord's test and other methods suitable for few-element sets and based on the R range of parallel determinations were used for the mathematical and statistical evaluation of analytical results including the computation of the interval of reliability.
Anaerobic degradability is given by the equation:

$$D_c = \frac{C_g}{C_s}.100 \tag{14}$$

where:
C_s = total C content in the sample
C_g = C content in methane released during the measurement of anaerobic degradability
The value of C_g is computed from the substrate production of methane V_{CH4S}:

$$C_g = \frac{12\, p\, V_{CH4S}}{RT} \tag{15}$$

(because 1 mol CH_4 contains 12 g C)
where:
K = temperature (°K)
R = gas constant
P = pressure
V_{CH4S} = the volume of produced methane after the subtraction of endogenous production by the inoculum from total production
This method, which determines organic matter lability in anaerobic conditions, is so exact that it allows to investigate e.g. the digestive tract of ruminants as an enzymatic bioreactor and to acquire information on its activity, on feed utilisation or digestibility and on the influence of various external factors on the digestion of these animals (Kolář et al. 2010a) or to determine the share of particular animal species in the production of greenhouse gasses (Kolář et al. 2009b).
At the end of this subchapter dealing with the degree of organic matter lability and its changes after fermentation in a biogas plant these experimental data are presented:
A mixture of pig slurry and primary (raw) sludge from the sedimentation stage of a municipal waste water treatment plant at a 1 . 1 volume ratio was treated in an experimental unit of anaerobic digestion operating as a simple periodically filled BATCH-system with mechanical agitation, heating tubes with circulating heated medium at a mesophilic temperature (40°C) and low organic load of the digester (2.2 kg org. dry matter/m^3) and 28-day fermentation.
Acid hydrolysis of sludge, slurry and their mixture was done before and after anaerobic fermentation. The hydrolysis of samples was performed with the dry matter of examined sludge and its mixture with pig slurry including the liquid fraction after screening the material through a 250-µm mesh sieve. The method of hydrolysis according to Rovira and Vallejo (2000, 2002) as modified by Shirato and Yokozawa (2006): 300 mg of homogenised sample is hydrolysed with 20 ml of 2.5 M H_2SO_4 for 30 min at 105°C in a pyrex tube. The

hydrolysate is centrifuged and decanted, the residues are washed with 25 ml water and the wash water is added to the hydrolysate. This hydrolysate is used to determine Labile Pool I (LP I).
The washed residue is dried at 60°C and hydrolysed with 2 ml of 13 M H_2SO_4 overnight at room temperature and continuous shaking. Such an amount of water is added that the concentration of the acid will be 1 M, and the sample is hydrolysed for 3 hours at 105°C at intermittent shaking. The hydrolysate is isolated by centrifugation and decantation, the residue is washed again with 25 ml of water and the wash water is added to the hydrolysate. This hydrolysate is used for the determination of Labile Pool II (LP II). The residue from this hydrolysis is dried at 60°C and Recalcitrant Pool (RP) is determined from this fraction.
C_{tot} is determined in all three fractions.
Degradability of organic matter of the test materials was studied by modified methods of Leblanc et al. (2006) used to examine the decomposition of green mulch from *Inga samanensis* and *Inga edulis* leaves. These authors conducted their study in outdoor conditions (average annual temperature 25.1°C) and we had to modify their method in the cold climate of this country. At first, the liquid phase of sludge, slurry and mixture was separated by centrifugation; the solid phase was washed with hot water several times and separated from the solid phase again. By this procedure we tried to separate the solid phase from the liquid one, which contains water-soluble organic substances and mineral nutrients. Solid phases of tested organic materials were mixed with sandy-loamy Cambisol at a 3:1 weight ratio to provide for inoculation with soil microorganisms and volume ventilation of samples with air. After wetting to 50% of water retention capacity the mixtures at an amount of 50 g were put onto flat PE dishes 25 x 25 cm in size. The material was spread across the surface of the dish. Cultivation was run in a wet thermostat at 25°C, and in the period of 2 - 20 weeks dishes were sampled in 14-day intervals as subsamples from each of the four experimental treatments. The agrochemical analysis of the used topsoil proved that the content of available nutrients P, K, Ca and Mg according to MEHLICH III is in the category "high" and pK_{KCl} = 6.3. After drying at 60°C for 72 hours the content of lipids, crude protein, hemicelluloses, cellulose, lignin, total nitrogen and hot-water-insoluble dry matter was determined in the dish contents.
After twenty weeks of incubation organic substances were determined in the dish contents by fractionation into 4 degrees of lability according to Chan et al. (2001).
The content of hemicelluloses was calculated from a difference between the values of neutral detergent fibre (NDF) and acid detergent fibre (ADF), lignin was calculated from ADF by subtracting the result after lignin oxidation with $KMnO_4$. Because ADF contains lignin, cellulose and mineral fraction, it was possible to determine the cellulose content by ashing the residue in a muffle furnace and by determination of mineral fraction. These methods were described by Van Soest (1963), modifications used by Columbian authors (Leblanc et al. 2006) were reported by López et al. (1992).
Ion exchange capacity [mmol.chem.eq./kg] was determined in dry matter of the examined materials according to Gillman (1979), buffering capacity was determined in samples induced into the H^+-cycle with HCl diluted with water at 1 : 1 and washed with water until the reaction to Cl^- disappears. In the medium of 0.2 M KCl the samples were titrated to pH = 7 with 0.1 M NaOH and buffering capacity was calculated from its consumption.
Tab. 1 shows the analyses of a mixture of pig slurry and primary sludge used in the experiment. Obviously, compared to the values reported in literature our experimental materials had a somewhat lower content of organic substances in dry matter, and perhaps

this is the reason why anaerobic fermentation reduced the content of organic substances by 39% only although the usual reduction by 45 - 65% for primary sludge was expected as reported in literature (Pitter 1981) and by 40 - 50% for pig slurry (Stehlík 1988). As a result of the organic dry matter reduction the content of nutrients in sludge after anaerobic fermentation is higher, nitrogen content is lower by about 20%. In this process organic nitrogen is converted to $(NH_4)_2CO_3$, which partly decomposes into $NH_3 + H_2O + CO_2$ and partly passes into the sludge liquor. Roschke (2003) reported that up to 70% of total nitrogen might pass to the ammonium form at 54% degradation of organic substances of dry matter. Even though concentrations of the other nutrients in dry matter of the aerobically stabilised sludge increased as a result of the organic dry matter reduction, their content in the sludge liquor also increased (Tab. 2).

		Pig slurry	Primary sludge	Mixture of slurry and sludge before methanisation	Mixture of slurry and sludge after methanisation
Organic substances		65.1 ± 2.6	62.7 ± 2.4	64.1 ± 2.4	36.9 ± 1.5
Total nutrients	N	6.2 ± 0.2	2.6 ± 0.1	3.9 ± 0.2	3.1 ± 0.1
	P	1.6 ± 0.1	0.7 ± 0.0	1.1 ± 0.0	1.3 ± 0.1
	K	2.3 ± 0.1	0.2 ± 0.0	1.2 ± 0.0	1.2 ± 0.0
	Ca	2.8 ± 0.1	2.6 ± 0.1	2.5 ± 0.1	2.8 ± 0.1

Table 1. The analysis of experimental pig slurry and primary sludge, mixture of pig slurry and primary sludge before methanisation in a digester and after methanisation in % of dry matter (pig slurry and primary sludge were mixed for anaerobic digestion at a 1:1 volume ratio). (Sample size n = 6, interval of reliability of the mean for a significance level $\alpha = 0.05$)

	[%] A	[%] B	[mg/l] Before fermentation	[mg/l] After fermentation
Total N	8.40	55.20	246.2 ± 14.7	994.7 ± 59.6
Ammonia N	52.60	90.80	153.7 ± 8.4	907.2 ± 48.2
Total P	12.20	25.30	134.5 ± 8.7	176.3 ± 11.6
Total K	19.90	28.10	172.9 ± 10.4	184.1 ± 11.0

Table 2. The analysis of the liquid fraction (sludge liquor) of a mixture of pig slurry and primary sludge from a waste water treatment plant (1 : 1) before fermentation and after fermentation in mg/l. The values A and B express % in the liquid phase of the total amount of sludge before and after fermentation (Sample size n = 5, interval of reliability of the mean for a significance level $\alpha = 0.05$)

Taking into account that the amount of water-soluble nutrients in the sludge liquor and organic forms of N and P dispersed in the sludge liquor in the form of colloid sol (but it is a very low amount) is related not only to the composition of the substrate but also to technological conditions of anaerobic digestion, digester load and operating temperature, it is evident that the liquid fraction of anaerobically stabilised sludge contains a certain amount of mineral nutrients, approximately 1 kg N/m^3, besides the others, although differences in the concentration of P and K in the liquid fraction before and after fermentation are generally negligible. It is a very low amount, and there arises a question whether the influence of the liquid fraction on vegetation is given by the effect of nutrients or water itself, particularly in drier conditions.

After anaerobic digestion the solid phase of sludge still contains a high amount of proteins and other sources of organic nitrogen that could be a potential pool of mineral nitrogen if the degradation of sludge after fermentation in soil is satisfactory.

Material	Proportion		
	LP I	LP II	RP
Primary sewage sludge	68 ± 5	23 ± 2	9 ± 1
Pig slurry	59 ± 5	15 ± 2	26 ± 2
Mixture of primary sludge and pig slurry at a 1:1 volume ratio	63 ± 5	20 ± 2	17 ± 1
Mixture of primary sludge and pig slurry at a 1:1 volume ratio after methanisation	18 ± 2	16 ± 1	66 ± 5

Table 3. Proportions of the three pools of carbon in experimental materials, as determined by the acid hydrolysis method of Rovira and Vallejo (2002),
(Sample size n = 4, interval of reliability of the mean for a significance level α = 0.05),
(Materials including the liquid fraction were used)

The results of hydrolysis in Tab. 3 prove that pig slurry has 59% of its total carbon in LP I, which indicates great lability, corresponding to the hydrolysability of cereals and grasses according to Shirato and Yokozawa (2006). Primary sewage sludge is still better from this aspect, having almost 70% C in LP I. The degree of lability of the sludge and slurry mixture is relatively high and corresponds to the component ratio. After methanisation carbon content in LP I of the sludge and slurry mixture decreases to less than a third of the original amount and carbon of non-hydrolysable matters increases even almost four times in the RP fraction. The sum of LP I and LP II, i.e. the labile, degradable fraction of carbon compounds of the sludge and pig slurry mixture, was reduced by anaerobic digestion from 83% to 34%, that means approximately by 50%. These are enormous differences and they prove that mainly very labile organic substances are heavily destroyed by the anaerobic process even though a reduction in the content of organic substances during anaerobic fermentation is lower (by 39% in our experiment).

Tab. 4 shows the analysis of raw materials (sludge and pig slurry) and their mixture before and after anaerobic fermentation while Tab. 5 shows the analysis of their liquid fraction. The same results (Tab. 4) are provided by the incubation of the solid phase of sludge, pig slurry and their mixture before and after anaerobic fermentation when incubated with soil at 25°C and by the contents of lipids, crude protein, hemicelluloses, cellulose, lignin, total nitrogen and hot-water-insoluble dry matter; the same explicit conclusion can be drawn from the results of the fractionation of organic matter lability of the experimental treatments after 20-week incubation with soil according to Chan et al. (2001) shown in Tab. 5. A comparison of the results in Tab. 3 and 5 indicates that as a result of the activity of microorganisms of the added soil in incubation hardly hydrolysable organic matter was also degraded - differences between the most stable fractions F 3 and F 4 in Tab. 5 are larger by about 73% after anaerobic fermentation while in the course of acid chemical hydrolysis the content of non-hydrolysable fraction was worsened by anaerobic fermentation because it increased by about 290%. But it is a matter of fact that the soil microorganisms are not able to stimulate the anaerobically fermented sludge to degradation as proved by more than ¾ of total carbon in fraction 4.

	I Before incubation (25° C)				II After incubation (25°C, 20 weeks)			
	A	B	C	D	A	B	C	D
Lipids (petroleum ether extractable compounds) [%]	8.60 ± 0.69	14.27 ± 1.14	10.82 ± 0.86	2.01 ± 0.15	7.97 ± 0.65	13.50 ± 1.09	10.39 ± 0,85	2.08 ± 0,17
Proteins (Berstein) [%]	13.43 ± 1.30	17.95 ± 1.62	15.31 ± 1.60	8.50 ± 0.93	11.81 ± 1.20	16.10 ± 1.53	13.89 ± 1.42	8.50 ± 0.98
Hemicelluloses [%]	1.82 ± 0.19	5.03 ± 0.73	3.32 ± 0.61	0.70 ± 0.60	1.43 ± 0.11	4.23 ± 0.51	2.89 ± 0.30	0.69 ± 0.10
Cellulose [%]	7.45 ± 0.92	11.18 ± 1.33	9.61 ± 1.05	6.03 ± 0.95	5.42 ± 0.82	9.27 ± 0.98	7.96 ± 0.94	6.05 ± 0.83
Lignins [%]	4.84 ± 0.62	5.16 ± 0.84	4.99 ± 0.75	5.18 ± 0.92	4.83 ± 0.91	5.18 ± 1.07	4.98 ± 0.84	5.20 ± 0.91
Total N [%]	1.59 ± 0.06	2.70 ± 0.11	2.29 ± 0.10	1.07 ± 0.04	1.51 ± 0.06	2.50 ± 0.11	2.14 ± 0.09	1.08 ± 0.05
Hot-water insoluble dry matter [%]	98.25 ± 2.94	98.26 ± 2.95	98.25 ± 2.95	98.23 ± 2.92	89.05 ± 2.67	85.17 ± 2.60	87.26 ± 2.58	98.20 ± 2.93
Ion exchange capacity [mmol chem. eq./kg]	48 ± 3	55 ± 3	53 ± 3	145 ± 9	50 ± 3	61 ± 4	55 ± 4	168 ±10
Buffering capacity [mmol chem. eq./kg]	62 ± 4	69 ± 4	65 ± 4	157 ± 9	65 ± 4	72 ± 4	70 ± 4	179 ± 11

Table 4. The content of selected organic substances (%) and ion exchange and buffering capacity of the solid phase of primary sludge (A), pig slurry (B), sludge and pig slurry mixture at a 1:1 ratio before fermentation (C) and after fermentation (D) before and after 20 weeks of incubation with sandy-loamy Cambisol topsoil at a 3:1 ratio at 25°C in dry matter (Sample size n = 4 /hot-water-soluble dry matter n = 7/, interval of reliability of the mean for a significance level α = 0.05)

	Unfermented primary sludge	Unfermented pig slurry	Mixture A	Mixture B	Soil only
Fraction 1 (12 N H_2SO_4)	59.84 ± 7.18 (32.00)	55.38 ± 6.52 (28.40)	54.09 ± 6.50 (30.05)	2.65 ± 0.30 (2.60)	1.30 ± 0.17 (7.22)
Fraction 2 (18 N - 12 N H_2SO_4)	42.45 ± 5.13 (22.70)	35.76 ± 4.26 (18.34)	34.22 ± 4.10 (19.01)	9.28 ± 1.10 (9.07)	0.80 ± 0.09 (4.44)
Fraction 3 (24 N - 18 N H_2SO_4)	27.34 ± 3.28 (14.62)	20.18 ± 2.53 (10.35)	20.30 ± 2.42 (11.28)	11.13 ± 1.33 (10.91)	3.70 ± 0.44 (20.56)
Fraction 4 (TOC = 24 N H_2SO_4)	57.37 ± 6.85 (30.68)	83.67 ± 10.01 (42.91)	71.39 ± 8.55 (39.66)	78.97 ± 9.40 (77.42)	1.22 ± 1.42 (67.78)

Table 5. The fractionation of organic carbon (g/kg) of primary sludge, pig slurry, and sludge and slurry mixture at a 1:1 ratio before fermentation (A) and after fermentation (B) in a mixture with sandy-loamy Cambisol (3 : 1) in dry matter after 20 weeks of incubation at 25°C by the modified Walkley-Black method according to Chan et al. (2001) with a change in H_2SO_4 concentration. (The values given in brackets are % of the C fraction in total dry matter carbon) (Sample size n = 5, interval of reliability of the mean for a significance level $\alpha = 0.05$)

The table results document that 20-week incubation decreased more or less the per cent content of examined organic substances except lignin (total N 5 - 8%, cellulose 17 - 25%, hemicellulose 13 - 22%, proteins 9 - 12%, lipids 4 - 7%, and the content of hot-water-insoluble dry matter by 10 - 15%) factually in all experimental treatments except the treatment of the anaerobically fermented mixture of primary sludge and pig slurry where a reduction in these matters is low or nil. Hence, primary sludge, pig slurry and their mixture can be considered as organic fertilisers but only before anaerobic fermentation. We recorded a substantially lower degree of degradation of selected organic substances in sludge, pig slurry and their mixture during incubation with 25% of sandy-loamy soil (5 - 25%) than did Leblanc et al. (2006) with phytomass of *Inga samanensis* and *Inga edulis* leaves, who reported about 50% degradation of total mass, hemicelluloses and nitrogen in mass. We are convinced that it is caused by a very different content of hemicelluloses in our materials compared to the materials used by the above-mentioned authors. No easily degradable hemicelluloses are present in sewage sludge or in pig slurry any longer, and obviously, only more stable forms pass through the digestive tracts of animals and humans. It is also interesting that after anaerobic fermentation and after 20-week aerobic cultivation at 25°C only the compounds (lipids + proteins + hemicelluloses in mixture II D account roughly for 11%) that could be considered as labile remained in the mixture of slurry and sludge. These are apparently their more stable forms as confirmed by the results in Tab. 5 which illustrate that to approximately 11% of organic carbon compounds it is necessary to add the % proportions of the first and second fraction on the basis of oxidisability according to Chan et al. (2001). Literary sources report that the sum of lipids, proteins and hemicelluloses in the anaerobically stabilised sludge from municipal waste water treatment plants amounts to 13% - 39.6% of dry matter, so it is quite a general phenomenon.

The ion exchange capacity of sludge, pig slurry and their mixture before fermentation, before incubation and after incubation is very low and does not reach the values that are typical of sandy soil. It is increased by anaerobic fermentation along with incubation markedly but practically little significantly to the level typical of medium-textured soils. The same relations were observed for buffering capacity, which is not surprising. The results document that degradability of the organic part of anaerobically stabilised sludge worsened substantially and that it cannot be improved very markedly by the use of soil microorganisms and soil.

We have to draw a surprising conclusion that sludge as a waste from the processes of anaerobic digestion is a mineral rather than organic fertiliser and that from the aspect of its use as organic fertiliser it is a material of much lower quality than the original materials. We cannot speak about any improvement of the organic material by anaerobic digestion at all. Their liquid phase, rather than the solid one, can be considered as a fertiliser. If it is taken as a fertiliser in general terms, we do not protest because besides the slightly higher content of mineral nutrients available to plants (mostly nitrogen) it has the higher ion exchange capacity and higher buffering capacity than the material before anaerobic fermentation, but this increase is practically little significant.

3.1.2 Digestate composting

3.1.2.1 What is compost?

Similarly like in the evaluation of digestate when the daily practice has simplified the problem very much because the main functions of mineral and organic fertilisers are not distinguished from each other, the simplification of the problem of composting and application of composts has also led to an absurd situation. In many countries the compost is understood to be a more or less decomposed organic material, mostly from biodegradable waste, which contains a certain small amount of mineral nutrients and water. The main requirement, mostly defined by a standard, is prescribed nutrient content, minimum amount of dry matter, absence of hazardous elements and the fact that the particles of original organic material are so decomposed that the origin of such material cannot be identified. Such 'pseudo' composts are often offered to farmers at a very low cost because the costs of their production are usually paid by producers of biodegradable waste who want to dispose of difficult waste.

The producers of such composts often wonder why farmers do not intend to buy these composts in spite of the relatively low cost. It is so because the yield effect of fertilisation with these composts is minimal, due to a low content of nutrients it is necessary to apply tens of tons per 1 ha (10 000 m^2), which increases transportation and handling costs. In comparison with so called "green manure", i.e. ploughing down green fresh matter of clover, lucerne, stubble catch crops and crops designed for green manure, e.g. mustard, some rape varieties, etc., the fertilisation with these false composts does not have any advantage. The highly efficient decomposing activity of soil microorganisms, supported by equalising the C : N ratio to the value 15 - 25 : 1, works in the soil similarly like the composting process in a compost pile where the disposal of biodegradable material is preferred at the cost of a benefit to farmers.

What should the real compost be like? It is evident from the definition: the compost is a decomposed, partly humified organomineral material in which a part of its organic component is stabilised by the mineral colloid fraction. It is characterised by high ion-exchange capacity, high buffering capacity and is resistant to fast mineralisation. The reader of this text has surely noticed that the nutrients have not been mentioned here at all. Of course, they are present in the compost, their amount may be higher or a lower, but it is not important. It is crucial that the compost will maintain nutrients in the soil by its ion-exchange reactions and that it will protect them against elution from topsoil and subsoil layers to bottom soil or even to groundwater, no matter whether these plant nutrients originate from the compost itself or from mineral fertilisers or from a natural source - the soil-forming substrate in the soil-forming process. In the production of such "genuine" compost it is necessary to ensure that organic matter of the original composted mixture will be transformed not only by decomposing mineralisation, exothermic oxidation processes but also partly by an endothermic humification process that is not a decomposing one, but on the contrary, it is a synthetic process producing high-molecular, polycondensed and polymeric compounds, humic acids, fulvic acids and humins, i.e. the components of soil humus. It is to note that we should not confound the terms "humus" and "primary soil organic matter"; these are completely different mixtures of compounds, of quite different properties! Humus is characterised by high ion-exchange capacity and very slow mineralisation (the half-time of mineralisation of humic acids in soil conditions is 3 000 - 6 000 years!) while primary organic matter, though completely decomposed but not humified, has just opposite properties. Sometimes it may have a high sorption capacity but not an ion-exchange capacity.

The high ion-exchange capacity of humified organic matter is a cause of other two very important phenomena: huge surface forces of humus colloids in soil lead to a reaction with similarly active mineral colloids, which are all mineral soil particles of silicate nature that are smaller than 0.001 mm in size. These particles are called "physical clay" in pedology. The smaller the particles, the larger their specific surface, which implies their higher surface activity. Clay-humus aggregates are formed, which are adsorption complexes, elementary units of well-aerated, mechanically stable and elastic soil microaggregates that may further aggregate to macroaggregates and to form the structured well-aerated soil that has a sufficient amount of capillary, semi-capillary and non-capillary pores and so it handles precipitation water very well: in drought capillary pores draw water upward from the bottom soil while in a rainy period non-capillary pores conduct water in an opposite direction. The basic requirement for soil productivity is met in this way. It is often much more important than the concentration of nutrients in the soil solution (and hence in the soil).

The other important phenomenon related to ion-exchange properties of compost or soil is buffering capacity, the capacity of resisting to a change in pH. Soils generally undergo acidification, not only through acid rains as orthodox ecologists often frighten us but also mainly by electrolytic dissociation of physiologically acid fertilisers and intensive uptake of nutrients from the soil solution by plants. By the uptake of nutrient cations plants balance electroneutrality by the H^+ ion, which is produced by water dissociation, so that the total electric charge does not change. If it were not so, each plant would be electrically charged like an electrical capacitor. The humus or clay or clay-humus ion exchanger in compost or in soil, similarly like any other ion exchanger, behaves in the same way as the plant during nutrient uptake: when any ion is in excess in the environment, e.g. H^+ in an acidifying soil, the plant binds this H^+ and exchanges it for another cation that was bound by it before. The

H^+ ion is blocked in this way and the pH of soil does not change. High buffering capacity is a very favourable soil property and is typical of soils with a high content of mineral or organic colloid fraction, i.e. of heavy-textured soils and of organic soils with a high degree of humification of soil organic matter.

As described above, it is quite obvious what soils should be fertilised with real genuine composts preferentially: these are mainly light sandy and sandy-loam soils in which mineralisation processes are so fast due to high aeration that the organic matter of potentially applied organic fertilisers factually "burns". Mineral nutrients are released from an organic fertiliser but very soon there is a lack of necessary organic matter in such a soil. Energy for the soil microedaphon is not sufficient, ion-exchange capacity is low because decomposed organic matter fails to undergo humification. Such a soil does not hold water while rainfall quickly leaches nutrients from the soil. Only the application of genuine composts can markedly improve the productivity of these soils. Their clay-stabilised organic matter resists the attack of oxygen excess and remains decomposable, so it is able to maintain the required microbial activity of soil.

3.1.2.2 How is "genuine" compost produced?

Modern production of industrial composts is based on an idea that the compost is a substrate for plants with nutrient content. This is the reason why attention is mainly paid to the mechanical treatment of organic material - grinding, crushing and homogenisation. A homogenised blend, enriched with nutrients, applied water and/or compost additives, is subjected to fast fermentation. It is turned at the same time and homogenised again. The turning ensures a new supply of oxygen and if the compost has a sufficient amount of easily degradable organic matter, the temperature during composting increases up to 50 - 60°C, which allows a desirable breakdown of particles of the original organic material. The product acquires a dark colour, it is loose, often has a pleasant earthy smell while the odour of the original organic material is not perceptible any more. Farm sludge is often added to the compost formula as a nitrogen source or the improper C to N ratio is adjusted by the addition of mineral nitrogenous fertilisers. Slurry and liquid manure are used as an N and water source and sometimes limestone is added to prevent acidification. The aeration of the fermented pile of materials is provided by the addition of inert coarse-grained materials, mainly of wood chips, crushed straw, rubble, undecomposable organic waste and other materials available from local sources, whereas the use of horizontal and vertical ventilation systems is less frequent. It is often the type of "aeration" additive which explicitly shows that the compost producer prefers waste processing to the interest of future users of their products, farmers and productivity of their soils. The ion-exchange capacity of these composts is about 40 - 80 mmol chem. eq. 1000 g^{-1} and it is very low. It characterises a light, little fertile sandy soil.

How is the real "genuine" compost produced? The following principles should be observed:

1. Organic material of the compost formula should have a high degree of lability. If the compost producer does not have a sufficient amount of such very easily degradable organic material, its lability should be enhanced by saccharidic waste.
2. The C : N ratio should be adjusted to the value 10 - 15 : 1, not to total C and total N, but to the value of C_{hws} and N_{hws} (hot water extractable carbon and nitrogen). Obviously, it is not worth adding to the compost a nitrogen source e.g. in waste polyamide because this nitrogen is not accessible. It is a flagrant example but we have detected many times that the C : N ratios are completely different from those the compost producers suppose them to be.

3. The compost formula should have a high proportion of buffering agent. It should always be ground limestone or dolomite, it should never be burnt or slaked lime. Do not economize on this additive very much. It will be utilised excellently after the application of this compost to soils.
4. Stabilisation of organic matter should be ensured by a sufficient amount of the clay mineral fraction. It must not be applied in lumps, but in the form of clay slurry, clay water suspension, used also for the watering of the blend of compost materials. Concrete mixers are ideal equipment for the preparation of clay slurry.
5. The compost blend should be inoculated by healthy fertile topsoil. Soil microorganisms are adapted in a different way than the microorganisms of the intestinal tract of animals. Therefore slurry and liquid manure are sources of water and nitrogen but they are not a suitable inoculant even though they are often recommended in literature for this purpose.
6. The basic requirement is to reach a high temperature (55 - 60°C) during composting and to maintain the second phase of temperature (40 - 50°C) for a sufficiently long time. This process will be successful only at a sufficiently high amount of highly labile organic matter in the compost formula, at a correct C : N ratio, at a correct water to air ratio in the pile (the moisture during fermentation should be maintained in the range of 50 - 60% of water-retention capacity) and at a reduction in heat losses. Heat losses of the compost into the atmosphere through the pile surface are relatively small. The highest quantity of heat is lost by conducting the heat through the concrete or the frozen ground of the compost pile, and mainly by an aerating system if it is installed.
7. Humification processes, formation of humus acids and humins or their precursors at least, occur rather in later stages of fermentation and so we should accept that the good compost cannot be produced by short-term fermentation. Old gardeners fermented composts for 10 - 12 years, but their composts reached the ion-exchange capacity of 300 - 400 mmol chem. eq. 1000 g^{-1}.

3.1.2.3 How is the digestate used in compost production?

If besides decomposing exothermic processes synthetic endothermic processes are also to take place in compost when high-molecular humus substances (fulvic acids, humic acids and humins) are formed, these conditions must be fulfilled: very favourable conditions for the microflora development must exist in compost, and minimum losses and the highest production of heat must be ensured. For this purpose it is necessary to use a high admixture of buffering additive (limestone) in the compost formula, sufficient amount of very labile organic matter, thermal insulation of the base of fermented material because the heat transfer coefficient does not have the highest value for transfer from the composted pile into the atmosphere but mainly into solid especially moist materials, i.e. into concrete, moist or frozen earth, clay, bricks, etc. At a sufficient amount of labile fractions of organic matter the maximum heat production can be achieved only by a sufficient supply of air oxygen. Beware of this! The ventilation through vertical and horizontal pipes provides sufficient air for aerobic processes in the fermented material but at the same time the ventilation is so efficient that a considerable portion of reaction heat is removed, the material is cooled down and the onset of synthetic reactions with the formation of humus substances does not occur at all.

When sufficiently frequently turning the fermented material, the safest method of compost aeration and ventilation is the addition of coarse-grained material while inert material such as wood chips, chaff and similar materials can be used. It is however problematic because inert material in the fermented blend naturally decreases the concentration of the labile fraction of organic matter, which slows down the reaction rate of aerobic biochemical reactions and also the depth of fermentation is reduced in this way. It mainly has an impact on the synthetic part of reactions and on the formation of humus substances while the influence on decomposing reactions is smaller.

It would be ideal if during compost fermentation in a microbially highly active environment the inert aeration material were able not only to allow the access of air oxygen into the fermented material but also to decompose itself at least partly and to provide additional energy to biochemical processes in the pile in this way.

These requirements are excellently met by the solid fraction of digestate from biogas plants. It aerates the compost and although it lost labile fractions of organic matter in biogas plant digesters, it is capable of further decomposition in a microbially active environment. It releases not only energy but also other mineral nutrients. So this waste is perfectly utilised in this way. The average microbial activity of even very fertile, microbially active soils is not efficient enough for the decomposition of this stable organic material when the solid phase of digestate is used as an organic fertiliser. The decomposition rate is slow, especially in subsequent years, and therefore the resultant effect of the solid fraction of digestate as an organic fertiliser is hardly noticeable. The combination of anaerobic decomposition in the biogas plant digester and aerobic decomposition in compost could seem paradoxical, and some agrochemists do think so. The preceding exposition has shown that it is not nonsense.

Now let us answer the question: what dose of the solid fraction of digestate should be used in the compost formula? It depends on many factors: on the amount of the labile fraction of organic component and mainly on the degree of its lability (which can be determined in a reliable way by the above-mentioned method according to Rovira and Vallejo 2002, 2007, Shirato and Yokozawa 2006), on the aeration and porosity of materials used in the compost formula, on the number of turnings, on prevailing outdoor temperature, water content, degree of homogenisation and on other technological parameters.

In general: the higher the amount of the labile component of organic matter and the higher its lability (e.g. the content of saccharides and other easily degradable substances), the higher the portion of the solid fraction of digestate that can be used.

Now short evidence from authors own research is presented:

The basic compost blend was composed of 65% fresh clover-grass matter from mechanically mown lawns, 10% ground dolomite, 2% clay in the form of clay suspension, 20% solid phase of digestate (obtained by centrifugation with fugate separation) or 20% crushed wood chips and 3% PK fertilisers. The C : N ratio in the form of C_{hws} : N_{hws} (hot-water-soluble forms) was 15 : 1, nitrogen was applied in NH_4NO_3 in sprinkling water that was used at the beginning of fermentation at an amount of 70% of the beforehand determined water-retention capacity of the bulk compost blend. Inoculation was done by a suspension of healthy topsoil in sprinkling water. Fermentation was run in a composter in the months of April - November, and the perfectly homogenised material was turned six times in total. Water loss was checked once a fortnight and water was replenished according to the increasing water-retention capacity to 60%. The formation, amount and quality of formed humus substances were determined not only by their isolation and measurement but also by their specific manifestation, which is the ion-exchange capacity of the material. The original particles of composted materials were not noticeable in either compost (with the solid part of digestate and with wood chips), in both cases the dark coloured loose material with pleasant earthy smell was produced. Tab. 6 shows the analyses of composted materials and composts. The digestate was from a biogas plant where a mix of cattle slurry, maize silage and grass haylage is processed as a substrate. The material in which the aeration additive was polystyrene beads was used as compost for comparison.

	Solid phase of digestate	Wood chips	Compost		
			PS	Wood chips	Solid phase of digestate
C_{FA} [mg.kg^{-1}]	0	0	38	84	178
C_{HA} [mg.kg^{-1}]	0	0	15	20	62
$C_{HA} : C_{FA}$	-	-	0,39	0,24	0,35
Ion-exchange capacity T [mmol chem. eq.kg^{-1}]	51	12	72	64	224

Table 6. The content of fulvic acid carbon (C_{FA}), humic acid carbon (C_{HA}), their ratio and ion-exchange capacity T of the solid phase of digestate and wood chips at the beginning of fermentation and of composts with polystyrene (PS), wood chips and solid phase of digestate

The results document that the ion-exchange capacity, and hence the capacity of retaining nutrients in soil and protecting them from elution after the application of such compost, increased very significantly only in the digestate-containing compost. The ion-exchange capacity of this compost corresponds to the ion-exchange capacity of heavier-textured humus soil, of very good quality from the aspect of soil sorption. The compost with wood chips produced in the same way does not practically differ from the compost with polystyrene but it does not have any humic acids and the ion-exchange capacity of these composts is on the level of light sandy soil with minimum sorption and ion-exchange properties. However, the total content of humus acids in the compost with the solid phase of digestate is very small and does not correspond to the reached value of the ion-exchange capacity of this compost. Obviously, precursors of humus acids that were formed during the fermentation of this compost already participate in the ion exchange. Humus acids would probably be formed from them in a subsequent longer time period of their microbial transformation. If only humus acids were present in composting products, at the detected low concentration of C_{FA} + C_{HA} the T value of the compost with the solid phase of digestate would be higher only by 1 - 1.2 mmol.kg^{-1} than in the compost with polystyrene or wood chips. Because it is more than a triple, other substances obviously participate in the ion exchange.

3.1.3 Use of digestate for improvement of heavy-textured soils

Optimum values of reduced bulk density O_r for soils are around 1.2 g.cm^{-3}, but more important is the minimum value of bulk density for the restriction of root growth which is about 1.7 - 1.8 g.cm^{-3} for light soils and only 1.40 - 1.45 g.cm^{-3} for heavy-textured clay soils. Bulk density O_r is a crucial parameter for the assessment of the soil compaction rate as an important negative factor of soil productivity. Bulk density of topsoil in the range of 0.95 - 1.15 g.cm^{-3} shows loose topsoil while the value > 1.25 g.cm^{-3} indicates heavily compacted topsoil.

Another important value of soil is soil aeration V_Z. It is expressed in volume % as the difference between porosity P_o and momentous soil moisture W_{obj}.

$$V_z = P_o - W_{obj.} \tag{16}$$

Optimum aeration e.g. for grasslands is 10% by volume, for soils for barley growing it is already as much as 24% by volume. Soil porosity P_o is the sum of all pores in volume per cent, in topsoils it is around 55%, in subsoil it decreases to 45 - 35%. Sandy soils have on average P = 42% by vol., out of this 30% are large pores and 5% are fine pores while heavy-textured clay soils have the average porosity of 48% by vol., out of this only 8% are large pores and 30% are fine pores. Fine pores are capillary and large pores are non-capillary ones. Cereals should be grown in soils with 60 - 70% of capillary pores out of total porosity and 30 - 40% of non-capillary pores. Forage crops and vegetables require the soils with 75 - 85% of capillary pores and only 15 - 25% of non-capillary pores out of total porosity.

Ploughing resistance P is also significant. It is a specific resistance that must be overcome during cutting into and turning over the soil layer. It is expressed by the drawbar pull measured dynamometrically on the coupling hook of a tractor. It is related to the texture and moisture of soil, to its content of organic substances and ploughing depth. Ploughing resistance for light soils is 2 - 4 $t.m^{-2}$, for heavy-textured soils it is 6 - 8 $t.m^{-2}$. The units $kp.dm^{-2}$ are also used. For sandy soils the ploughing resistance of 25 - 28 $kp.dm^{-2}$ is usual, for clay soils it is 70 $kp.dm^{-2}$.

Hence heavy-textured soils are more responsive to the higher reduced bulk density of soil when roots develop poorly, they need more non-capillary pores to allow for the better infiltration of precipitation water, they also need higher aeration because they are mostly too moist and many aerobic processes including the microbial activity take place with difficulty. Of course, the high ploughing resistance is not desirable either for the economics of soil cultivation or for the production process of any crop. Therefore it is necessary to improve heavy-textured soils and the question is how. Organic fertilisers are not sufficient; peat was used previously but now it is banned to use it for the reason of the peat bog conservation, and synthetic soil amendments (Krilium, Flotal etc.) are currently too costly for the agriculture sector. An excellent material for the improvement of heavy-textured soils is the solid phase of digestate if ploughed down at higher doses than those used for the application of farmyard manure or compost, i.e. 100 - 150 $t.ha^{-1}$. Even though we cannot expect any great release of mineral nutrients from organic matter of the solid phase of digestate due to high stability of this material, the improvement and aeration of heavy-textured soil with better conditions for the microbial activity of soil and undisturbed root growth often bring about a higher yield effect than is the yield effect of nutrients from high-quality organic fertilisers as shown by the results of this field trial:

When we still believed that the solid phase of digestate was an organic fertiliser, we laid out an exact field trial on a heavier-textured, loamy-clay soil with medium to good reserve of available nutrients. The trial had two treatments: the one treatment was fertilisation with the solid phase of digestate only (after fugate centrifugation) and the other treatment was the application of only mineral fertilisers in the form of pure salts at such a dose that the level of these easily available nutrients to plants was the same as the amount of unavailable or little available nutrients in the treatment fertilised with digestate. We wanted to find out from the yield of the grown crop what amount of mineral nutrients would be released from the digestate in comparison with completely available nutrients in the first year and in subsequent years of the crop rotation: early potatoes - winter barley - red clover - oats. We intended to compare the digestate with other organic fertilisers, e.g. farmyard manure which in the first year mineralises about a half of its nutrients bound in organic matter. But the result we obtained was surprising: in the first year the yield of early potatoes was higher by 12% in the digestate treatment although nobody could doubt that this treatment had a lower amount of nutrients than the variant fertilised with pure salts. The only explanation is that the higher yield effect in the digestate treatment was not caused by the higher input of nutrients but by the improvement in physical properties of heavy-textured soil that surely occurred as seen in Tab. 7. The favourable effect of the heavy-textured soil improvement on yield was positively reflected in subsequent years also in other crops of the crop rotation that were fertilised in both treatments in the same way, i.e. mineral fertilisers were applied. We drew a conclusion that in practice the yield effect is often ascribed to digestate nutrients although it is caused by better soil aeration and better root growth due to soil loosening after the application of digestate.

		Clay-loamy soils	
		initial	improved by digestate
Reduced bulk density O_r	[g.cm^{-3}]	1.43	1.38
Soil aeration V_z	[% by vol.]	18.5	22.4
Total porosity P_o	[% by vol.]	43.9	43.8
Proportion of large pores in total porosity	[%]	22.7	28.1
Ploughing resistance P	[kp.dm^{-2}]	63	50

Table 7. Bulk density O_r, aeration V_z, total porosity P_o, proportion of large pores in total porosity and ploughing resistance P in a heavy-textured clay-loamy soil and after its improvement with the dose of 150 t-ha^{-1} of the digestate solid phase

3.2 Perspective utilisation of digestate with a modification of conventional technology of biogas production

Perspective utilisation of digestate is connected with envisaged modifications of the technology of biogas production in agricultural biogas plants. These plants have digesters for the solid phase only or the most frequent are liquid (suspension) digesters. These are digesters without partition wall where the biomass of microorganisms is carried by the processed substrate. In reactor systems for the technological processing of waste from chemical and food technologies and from the technology of municipal and industrial waste water treatment those digesters are preferred where the biomass of functional microorganisms is fixed onto a solid carrier or onto partition walls of apparatuses. It is often granulated and is maintained in the digester as a suspended sludge cloud. These reactors may be affected by short-circuiting and therefore they are sensitive to the particle size of the processed substrate but they withstand a much higher organic load than the digesters without partition wall. Of course, the reactor is smaller, cheaper and more efficient.

Hence a perspective modification of the biogas production technology in agricultural biogas plants is gradual transition to the procedures of anaerobic digestion that are currently used in industrial plants for the treatment of organic waste water. The promising utilisation of digestate from such digesters is mainly the manufacture of solid fuels in the form of pellets that are prepared from the solid phase of agricultural waste before the proper aerobic digestion of the material for a biogas plant. The first proposal of this type is the IFBB procedure, the principle of which was explained in Chapter 1.4. The liquid phase from the preparation of processed material, which is destined for anaerobic fermentation in digesters with partition wall, could be used as a liquid or suspension fertiliser but researchers would have to solve the cheap method of nutrient concentration in this waste. The current price of Diesel fuel, machinery and human labour and low purchase prices of agricultural products do not allow the application of highly diluted fertilisers and in fact handling of water.

The problem is that a small biogas plant is only scarcely profitable. Hence economic reasons favour large-capacity plants with the volume of digesters 5 000 - 10 000 m^3. In such large plants the reactors with partition wall would be unjustifiably expensive and therefore in these large-capacity facilities for biogas production it is necessary to use reactors without partition wall. The utilisation of their digestate should be based on this scheme: separation of digestate - concentration of fugate and its utilisation as a liquid mineral nitrogenous fertiliser. The solid phase of digestate should be used as an inert aeration component in compost production and as a material for the improvement and aeration of heavy-textured soils.

In any case, researchers must resolve a cheap method of nutrient concentration in fugate.

A number of different reactors are available for small to medium-sized biogas plants with the treatment of material according to IFBB that were developed on a research basis mainly in the sixties to the nineties of the twentieth century. At first, these were reactors with suspension biomass, e.g. mixing contact anaerobic reactor (ACR - AG), its innovation was a membrane anaerobic reactor system (MARS) and sequencing batch reactors (SBR). Then reactors with immobilised biomass were developed that are divided into reactors with biomass on the surface of inert material and reactors with aggregated (granulated) biomass. The former group is divided into upflow reactors and downflow reactors. Reactors with a mobile filling are the third variant.

The latter group is divided into reactors with the internal separator of biogas and biomass, reactors with the external separator of biomass and reactors with partitions.

Further development brought about biofilm reactors where the biomass of microorganisms is fixed onto a solid carrier. These reactors are considered as facilities with the highest operating stability, very resistant to the fluctuation of operating conditions. But they do not usually allow for such a high load as reactors with suspension biomass. The oldest reactor of this series was an upflow anaerobic filter (UAF) reactor from 1967, then a downflow stationary fixed film reactor (DSFF) and downflow reactor with filling in bulk followed. Great progress was made by designing an anaerobic rotating biological contactor (ARBC) and fluidized bed reactor (FBR) in the eighties of the last century. A similar type of reactor, expanded bed reactor (EBR), also designated by AAFEB (anaerobic attached film expanded bed), is suitable to be operated at low temperatures. The detention time is only several hours and the portion of residual organic impurities is practically the same as in modern aerobic systems for the treatment of organically contaminated waters.

Further advance was the development of reactors with aggregated biomass. The most important representative of this group of digesters is an upflow anaerobic sludge blanket (UASB) reactor. It is a reactor with sludge bed and internal separator of microorganism biomass. The biggest reactor of this type (5 000 m^3) processes waste water from the manufacture of starch in the Netherlands, it withstands the load of 12.7 kg chemical oxygen demand (COD) per 1 m^3/day, 74% of organic matter is degraded and the detention time is

33 hours only. Besides the UASB reactor these reactors belong to this group: hybrid upflow bed filter (UBF) reactor, anaerobic baffled reactor (ABR), expanded granular sludge bed (EGSB) reactor, internal circulation (IC) reactor and upflow staged sludge bed reactor (USBB), often also called biogas tower reactor (BTR), and other design models of the UASB reactor.

At the end of this chapter it is to note that modern anaerobic reactors have almost amazing outputs - unfortunately, the more perfect the reactor, the more expensive, and also their advantage over huge digesters without partition wall we have got accustomed to in biogas plants is gradually disappearing. The selection of modern anaerobic reactors is also more difficult than the selection of conventional technology of reactors without partition wall, because they are mostly rather specific to the substrate to be processed. They also have higher demands on processing, attendance and checks.

The perspective possibility of using modern anaerobic reactors for biogas production in smaller plants and the simultaneous solution to the use of the digestate solid phase as a raw material for the production of solid pelleted biofuels initiated our study of the IFBB procedure (Chap. 1.4.) for the substrate commonly used in biogas plants in the CR. The results of our experimental work are presented below:

The IFBB technological procedure is based on a high degree of cell wall maceration as a result of the axial pressure and abrasion induced with a screw press. Reulein et al. (2007) used this procedure for dehydration of various field crops; it is also known from the technologies of processing rapeseed, sugar beet and leguminous crops for the production of protein concentrates (Telek and Graham 1983, Rass 2001) and in biorefineries for the extraction of lactic acid and amino acids (Mandl et al. 2006).

The basic substrate contained 37.5% by weight of cattle slurry and 62.5% by weight of solid substrates, i.e. a mixture of chopped maize silage and grass haylage of particle size max. 40 mm mixed at a 4.75 : 1 ratio, i.e. 51.6% of silage and 10% of haylage. In total, the substrate accounted for 19.3% of dry matter. This substrate at 15°C is designated by A. A portion of this substrate was mixed with water at a weight ratio of 1 : 5, put into a thermostat with a propeller stirrer at 15°C and intensively stirred for 15 minutes. Analogically, the other portion was also mixed with water at a substrate to water ratio of 1: 5 and put into a thermostat at a temperature of 60°C with 15-minute intensive stirring again. The sample of the substrate with water 15°C was designated by B, the sample with water 60°C was designated by C. The liquid phase from substrate A was separated by centrifugation while the liquid phases from substrate B and C were separated in a laboratory screw press for the pressing of fruits and vegetables. The separated liquid phases of substrates A, B and C were diluted with water to obtain a unit volume and the analytical results were recalculated to a transfer ratio in the liquid phase in relation to the content of particular nutrients in dry matter of the original substrate mixture.

The experiments conducted in an experimental unit of anaerobic digestion and in an equipment for IFBB made it possible to determine the content of mineral nutrients in substrate A after 42-day anaerobic digestion in mesophilic conditions (40°C), in the liquid phase of substrate A after anaerobic digestion, in the liquid phase of substrate B and C after recalculation to the dry matter content and concentration corresponding to substrate A, also after the process of anaerobic digestion under the same conditions (42 days, 40°C).

The above recalculations enable to clearly show the advantages of the IFBB process in nutrient transfer from solid to liquid phase when substrate A and 5 times diluted substrates B and C are compared, but they may unfortunately evoke a distorted idea about the real concentration of nutrients in liquid phases. It is to recall that IFBB increases the mass flow and transfer to the liquid phase but with regard to the 5-fold dilution the nutrient concentration in liquid waste for fertilization continues to decrease. This is the reason why the table below shows the original, not recalculated concentrations in the fugate of fermented substrate A and in the fermented liquid phases of the same substrate in IFBB conditions designated by B and C, which document considerable dilution of these potential mineral fertilizers.

The solid phases of substrates A, B and C after anaerobic digestion were subjected to determination of organic matter hydrolysability in sulphur acid solutions according to Rovira and Vallejo (2000, 2002) as modified by Shirata and Yokozawa (2006); we already used this method to evaluate the degradability of a substrate composed of pig slurry and sludge from a municipal waste water treatment plant (Kolář et al. 2008).

	Cattle slurry	Maize silage	Grass haylage	Substrate	Transfer ratio to liquid phase		
					A	B	C
Dry matter	6.4	28.9	18.7	19.3	0.06 ± 0.01	0.18 ± 0.04	0.20 ± 0.03
N-compounds (N x 6.25)	25.6	11.5	7.4	16.3	0.05 ± 0.01	0.20 ± 0.04	0.26 ± 0.05
Digestible nitrogen compounds	-	6.2	3.8	7.3	-	-	-
Nitrogen-free extract	-	52.8	48.6	49.9	0.30 ± 0.03	0.45 ± 0.05	0.48 ± 0.05
Crude fibre	-	25.7	29.8	18.0	0.01 ± 0.00	0.10 ± 0.00	0.10 ± 0.00
Fat	-	4.8	1.5	2.8	-	-	-
Organic substances	76.4	94.8	87.3	87.0	-	-	-
Mineral N ($N - NH_4^+$, NO_3^-)	2.4	< 0.1	> 0.1	1.0	0.74 ± 0.05	0.89 ± 0.06	0.95 ± 0.06
P	1.3	0.2	0.3	0.6	0.40 ± 0.05	0.52 ± 0.07	0.65 ± 0.08
K	5.3	1.4	1.7	2.9	0.57 ± 0.04	0.60 ± 0.04	0.79 ± 0.05
Ca	1.3	0.4	0.6	0.8	0.31 ± 0.06	0.38 ± 0.08	0.46 ± 0.08
Mg	0.5	0.2	0.3	0.3	0.38 ± 0.07	0.43 ± 0.08	0.55 ± 0.07
Na	0.1	< 0.1	<< 0.1	< 0.1	0.70 ± 0.08	0.77 ± 0.04	0.80 ± 0.08
Cl	0.3	0.2	0.2	0.2	0.77 ± 0.06	0.85 ± 0.05	0.85 ± 0.06

Table 8. Dry matter content in the fresh mass of used materials and their chemical composition in % dry matter. The transfer ratio of mass flow to the liquid phase from the fresh mass of substrate not diluted with water at 15°C (A), diluted with water at a 1:5 ratio at 15°C (B) and diluted with water at a 1:5 ratio at 60°C (C). Liquid phase A was separated by centrifugation, liquid phases B and C with a screw press.
(Sample size n = 5, reliability interval of the mean for a significance level $\alpha = 0.05$)

Table 8 documents that the IFBB procedure proposed by German authors for grass haylage is applicable to the typical substrate of Czech biogas plants, to a mixture of cattle slurry, maize silage and grass haylage. In agreement with German experience the observed transfer ratios are markedly higher at 60°C compared to 15°C of hydrothermal conditions but the value of transfer ratios to the liquid phase is generally lower in our experiments. We ascribe this fact to the properties of the material and also to the achieved axial force of the used press that was apparently lower even though the same perforation size of the conical part of the press (1.5 mm) and slope of the body (1 : 7.5) were used.

The results in Table 8 illustrate that the separation of the liquid and solid phase of the substrate that has not been subjected to anaerobic digestion yet by means of centrifugation only is rather imperfect from the aspect of the mass flow of components. The IFBB system (water dilution, intensive stirring at a temperature of 60°C and subsequent separation of the liquid and solid phase with a screw press) increases the transfer of organic and mineral substances into the liquid phase by about 15 - 20%, and it is also true of the saccharidic nitrogen-free extract and organic nitrogen compounds. This fact documents that the liquid phase has a higher amount of active, well-degradable organic material for anaerobic digestion, and so it is possible to expect not only the higher production of biogas but also more mineral nitrogen in the liquid after anaerobic digestion.

The high mass flow of alkaline metals and chlorine into the liquid phase, and on the contrary, the low transfer of calcium confirm the opinion of German researchers (Wachendorf et al. 2007, 2009) that the IFBB procedure largely increases the quality of biomass solid phase as a material for the production of solid fuels: the production of polychlorinated dioxins and dibenzofurans is reduced, waste gases will be less corrosive and the temperature of ash fusion will be higher.

Nitrogen compounds of the substrate dry matter account for 16.3%, i.e. these nitrogen organic compounds contain 2.6% of nitrogen in dry matter (Table 8). The content of mineral nitrogen in the substrate before anaerobic digestion was 1%. If the digestate contains 2.26% of mineral nitrogen in the same dry matter after anaerobic digestion, it is to state that during anaerobic digestion about a half of organic nitrogen mineralized and enriched the original 1% content of the substrate before the fermentation process. But the dry matter content decreased in the course of fermentation, and therefore the concentration of all nutrients in digestate apparently increased contrary to the original substrate. In our experiment the concentration of substrate dry matter decreased from 19.3% to 13.3% by weight during anaerobic digestion. The content of mineral nitrogen amounting to 3.28% at this dry matter content corresponds to 2.26% of min. N at the original dry matter content of 19.3% by weight (Table 9). The contents of the nutrients P, K, Ca and Mg in digestate dry matter after anaerobic digestion (Table 9) are apparently substantially higher than before fermentation. However, anaerobic digestion did not actually bring about any increase in the content of these nutrients, and the increased concentrations completely correspond to a reduction in dry matter content, 19.3 : 13.3 = 1.45.

It is not a new fact, but Table 9 shows how this mineral nitrogen is transferred to the liquid phase of substrates B and C compared to substrate A. Obviously, the liquid phase of substrate B has a higher amount of mineral nitrogen than that of substrate A, and so the effects of the screw press, which already before anaerobic digestion enriched the liquid substrate B with colloidal solutions (sols) of nitrogen organic compounds from the crushed cell walls of the plant material that provided further mineral nitrogen during fermentation, were significantly positive at the same temperature. It was still more evident in the liquid phase of substrate C while a conclusion can be drawn that a higher temperature contributes to a higher extraction of insoluble or partly soluble nitrogen organic compounds from which further mineral nitrogen is released after subsequent fermentation.

	Substrate A	Liquid phase of substrate A	Liquid phase of substrate after recalculation to dry matter content and concentration of substrate A	
			B	C
N	3.28	2.43	2.92	3.11
P	0.87	0.35	0.45	0.56
K	4.20	2.39	2.52	3.32
Ca	1.16	0.25	0.30	0.36
Mg	0.43	0.11	0.13	0.16

Table 9. Contents of mineral nutrients after anaerobic digestion (42 days, 40°C) in digestate (substrate A), in its liquid phase and in fermented liquids from IFBB in % dry matter by weight

Table 10 documents the original (before their recalculation) concentrations of mineral nutrients in the liquid phase of substrates A, B and C. These results indicate that liquid phase A can be considered as a highly diluted mineral fertilizer. Even though the IFBB process increases the concentration of nutrients (nitrogen) in the liquid phase before and after fermentation (liquid phase B and C), the dilution is very high. The recommended dilution with water, used by Wachendorf et al. (2009) and also in our experiments, produces liquid wastes diluted to such an extent after anaerobic digestion that they are practically hardly utilizable as a solution of mineral nutrients. Fugates are still rather problematic as mineral fertilizers, especially for applications in humid years and to soils with low microbial activity and consequently slow immobilization of mineral nitrogen, and naturally they are hardly applicable to pervious soils.

	Liquid phase		
	A	B	C
N	0.32 ± 0.03	0.09 ± 0.01	0.10 ± 0.01
P	0.05 ± 0.00	0.01 ± 0.00	0.02 ± 0.10
K	0.31 ± 0.04	0.08 ± 0.01	0.11 ± 0.15
Ca	0.03 ± 0.00	0.01 ± 0.00	0.01 ± 0.00
Mg	0.01 ± 0.00	0.00 ± 0.00	0.00 ± 0.00

Note: Statistical evaluation of this recalculation table is based on original data in Table 10.

Table 10. Contents of mineral nutrients after anaerobic digestion (42 days, 40°C) in the liquid phase of digestate A and in liquid phases B, C with the application of IFBB in % by weight of solutions that should be used for mineral fertilization
(Sample size n = 4, reliability interval of the mean for a significance level $\alpha = 0.05$)

Table 11 shows the results of hydrolytic experiments with solid phases of substrates A, B and C. They confirm the previously observed fact in the work with the substrate consisting of a mixture of pig slurry and primary sludge from a municipal waste water treatment plant that the solid phases of wastes from anaerobic digestion cannot be efficient as mineral fertilizers because of their very low degradability (Kolář et al. 2008). The IFBB process, which enriches the liquid phases with organic, easily degradable substances and improves biogas yields during the anaerobic degradation of only liquid phases, further depletes of these substances the solid phases of substrates and impairs their quality as organic fertilizers, even though it is not the case of an increase in the resistant component but only in worse hydrolysable LP II.

Solid phase of substrate	Proportion		
	LP I	LP II	RP
A_1	43 ± 8	41 ± 7	16 ± 2
A_2	22 ± 4	20 ± 3	58 ± 8
B	39 ± 6	44 ± 6	17 ± 3
C	31 ± 6	47 ± 8	16 ± 2

Note: Description of fractions according to the method of Rovira, Vallejo 2002:
LP I = (labile pool I) = the reserve of very labile, easily hydrolysable organic substances expressed as % of the total amount of organic matter in a sample
LP II = (labile pool II) = the reserve of intermediately labile, less easily hydrolysable organic substances in %
RP = (recalcitrant pool) = the reserve of hydrolysis resistant, very hardly degradable organic substances in %

Table 11. Proportions of the three pools of carbon in the solid phase of substrate A before anaerobic digestion (A_1), after anaerobic digestion (A_2) and in the solid phase of substrate A_1 after IFBB procedure before anaerobic digestion at 15°C (B) and at 60°C (C) as determined by the acid hydrolysis method of Rovira and Vallejo (2002).
(Sample size n = 4, reliability interval of the mean for a significance level $\alpha = 0.05$).

Hence it is to state:
We tested the Integrated Generation of Solid Fuel and Biogas from Biomass (IFBB) procedure proposed for ensiled grass matter from the aspect of suitability of its use for a typical substrate of new Czech biogas plants, a mixture of cattle slurry, maize silage and grass haylage. The agrochemical value of the liquid phase from a biodigester was also evaluated. We concluded that this procedure is suitable for the tested substrate and improves the agrochemical value of a fugate from biogas production. By chlorine transfer to the liquid phase it makes it possible to use the solid phase as a material for the production of solid biofuels with a reduced threat of the generation of polychlorinated dioxins and dibenzofurans during combustion. However, the concentration of mineral nutrients in the liquid phase during IFBB procedure is extremely low after anaerobic digestion as a result of dilution with water, and so its volume value is negligible.
Here research must go on.

4. References

Blair G. J., Lefroy R. D. B., Lisle L. (1995): Soil carbon fractions based on their degree of oxidation and the development of a carbon management index for agricultural systems. *Australian Journal of Agricultural Research*, 46, 1459-1466.

Gillman G. P. (1979): A proposed method for the measurement of exchange properties of highly weathered soils. *Australian Journal of Soil Research*, 17, 129-139.

Chan K. Y., Bowman A., Oates A., 2001: Oxidizible organic carbon fractions and soil quality changes in an oxic Paleustalf under different pasture leys. *Soil Science*, 166, (1): 61-67.

Kolář L., Klimeš F., Ledvina R., Kužel S. (2003): A method to determine mineralization kinetics of a decomposable part of soil organic matter in the soil. *Plant, Soil and Environment*, 49, (1): 8-11.

Kolář L., Kužel S., Hanušová A., Gergel J., Ledvina R., Šindelářová M., Silovská Š., Štindl P. (2005a): The use of Spectroquant Merck BOD photometric test to evaluate the stability of organic matters in soil. *Plant, Soil and Environment*, 51, (1): 46-50.

Kolář L., Klimeš F., Gergel J., Kužel S., Kobes M., Ledvina R., Šindelářová M. (2005b): Methods to evaluate substrate degradability in anaerobic digestion and biogas production. *Plant, Soil and Environment*, 51, (4): 173-178.

Kolář L., Ledvina R., Kužel S., Klimeš F., Štindl S. (2006): Soil Organic Matter and its Stability in Aerobic and Anaerobic Conditions. *Soil & Water Research*, 1, (2): 57-64.

Kolář L., Kužel S., Peterka J., Štindl P., Plát V. (2008): Agrochemical value of organic matter of fermenter wastes in biogas production. *Plant, Soil and Environment*, 54, (8): 321-328.

Kolář L., Kužel S., Horáček J., Čechová V., Borová-Batt J., Peterka J. (2009a): Labile fractions of soil organic matter, their quantity and quality. *Plant, Soil and Environment*, 55, (6): 245–251.

Kolář L., Maršálek M., Frelich J., Kužel S., Zedníková J., Švecová M. (2009b): Changes in methane release from organic matter passing through the digestive tract of horses. *Czech Journal of Animal Science*, 54, 112–120.

Kolář L., Frelich J., Brouček J., Kužel S., Borová-Batt J., Peterka J., Šlachta M., Volfová K., Pezlarová J., Hřebečková J., Čechová V. (2010a): Anaerobic degradability of organic matter of cattle faeces and a possibility of its utilization. *Czech Journal of Animal Science*, 55, (12): 538-547.

Kolář L., Kužel S., Peterka J., Borová-Batt J. (2010b): Agrochemical value of the liquid phase of wastes from fermenters during biogas production. *Plant, Soil and Environment*, 56, (1): 23-27.

Körschens M., Schulz E., Böhm R. (1990): Heisswasserlöslicher C und N im Boden als Kriterium für das N-Nachlieferungsvermögen. *Zentralblatt für Mikrobiologie*, Jena, 145, (4): 305-311.

Leblanc H. A., Nygren P., McGraw R. L. (2006): Green mulch decomposition and nitrogen release from leaves of two Inga spp. in an organic alley-cropping practice in the humid tropics. *Soil Biology & Biochemistry*, 38, (2): 349-358.

López F., Rodríguez G., Kass M. (1992): Manual de métodos rutinarios. Laboratorio de nutrición animal. *CATIE*, Turrialba, Costa Rica, 52 pp.

Mandl M., Graf N., Thaller A., Böchzelt H., Schnitzer H., Steinwender M., Wachlhoffer R., Fink R., Kromus S., Ringhofer J., Leitner E., Zentek J., Novalin S., Mihalyi B., Marini I., Neureiter M., Narodoslawsky M. (2006): Green bio-refinery - primary processing and utilization of fibres from green biomass. Series: *Reports from energy and environmental research*. Vol. 67, BMVIT, Vienna.

Pitter P. (1981): Hydrochemie. *SNTL* Praha, 373 pp.

Rass M. (2001): Rheology of biogenic solids under compression using peeled rapeseed as an example. [*Ph.D. thesis*.] University of Essen, Germany.

Reulein J., Scheffer K., Stülpnagel R., Bühle L., Zerr W., Wachendorf M. (2007): Efficient utilization of biomass through mechanical dehydration of silages. In: *Proceedings of the 15th European Biomass Conference Exhibition*, Berlin, Germany, 1770-1774, Florence, Italy: ETA - Renewable Energies.

Roschke M. (2003): Verwertung der Faulsubstrate, in: Biogas in der Landwirtschaft-Leitfaden für Landwirte und Investoren im Land Brandenburg. *Ministerium für Landwirtschaft, Umweltschutz und Raumordnung des Landes Brandenburg*, Postdam, 29-33.

Rovira P., Vallejo V. R. (2000): Examination of thermal and acid hydrolysis procedures in characterization of soil organic matter. *Communications in Soil Science and Plant Analysis*, 31: 81-100.

Rovira P., Vallejo V. R. (2002): Labile and recalcitrant pools of carbon and nitrogen in organic matter decomposing at different depths in soil: an acid hydrolysis approach. *Geoderma*, 107: 109-141.

Rovira P., Vallejo V. R. (2007): Labile, recalcitrant and inert organic matter in Mediterranean forest soils. *Soil Biology & Biochemistry*, 39, (1): 202-215.

Shirato Y., Yokozawa M. (2006): Acid hydrolysis to partition plant material into decomposable and resistant fractions for use in the Rothamsted carbon model. *Soil Biology & Biochemistry*, 38: 812-816.

Schulz E. (1990): Die heisswasserextrahierbaze C-Fraktion als Kenngrösse zur Einschätzung des Versorgungszustands mit organischer Substanz. *Tagung-Berichte, Akad. Landwirtsch. Wiss.*, Berlin, 295, 269-275.

Stehlík K. (1988): Use of waste water for irrigation. Praha, MZVž ČR, 437 pp. (in Czech).

Straka F., Dohányos M., Zábranská J., Dědek J., Malijevský A., Novák J., Oldřich J. (2003): Biogas. *GAS* Říčany, 1, 1-517 pp. (in Czech).

Straka F., Dohányos M., Zábranská J., Dědek J., Malijevský A., Novák J., Oldřich J. (2006): Biogas. *GAS* Říčany, 2, 1-706 pp. (in Czech).

Süssmuth R., Doser Ch., Lueders T. (1999): Determination of the biological biodegradability of organic substances under anaerobic conditions using the Oxi Top Control measuring system. *Universität Hohenheim, Institut für Mikrobiologie, Wissenschaftlich-Technische Werkstäten GmbH & Co.*, KG Weilheim, Germany.

Telek I., Graham H. D. (1983): Leaf protein concentrates. *Westport, CT: AVI Publishing Company*.

Tirol-Padre A., Ladha J. K. (2004): Assessing the reliability of permanganate-oxidizable carbon as an index of soil labile carbon. *Soil Science Society of America Journal*, 68, (3): 969-978.

Van Soest P. J. (1963): Use of detergents in the analysis of fibrous feeds. II. A rapid method for the determination of fiber and lignin. *Journal of the Association of Official Analytical Chemists*, 46, 829-835.

Wachendorf M., Fricke T., Grass R., Stülpnagel R. (2007): Ein neues Konzept für die bioenergetische Nutzung von Grünlandbiomasse. In: Wrage N., Isselstein J.: Mitteilungen der Arbeitsgemeinschaft Grünland und Futterbau, 8, 165-168.

Wachendorf M., Richter F., Fricke T., Grass R., Neff R. (2009): Utilization of semi-natural grassland through integrated generation of solid fuel and biogas from biomass. *Grass and Forage Science*, 64, 132-143.

Walkley A. (1947): A critical examination of a rapid method for determining organic carbon in soils. *Soil Science*, 63, 251-263.

Zábranská J., Dohányos M., Gran P. (1985a): Direct measurement of the Hydrogenase activity of Anaerobic Microorganisms. *Watter Science Technology*, 17, 303-306.

Zábranská J., Schneiderová K., Dohányos M. (1985b): Relation of coenzyme F 420 to the activity of methanogenic microorganisms. *Biotechnology Letters*, 7, 547-552.

Zábranská J., Dohányos M. (1987): Methods of Investigation of the Metabolism of Anaerobic Microorganisms. *Acta hydrochimica hydrobiologica*, 15, 43-55.

10

Biodiesel Production and Quality

Roseli Ap. Ferrari[1], Anna Leticia M. Turtelli Pighinelli[2] and Kil Jin Park[2]
[1]*Food Technology Institute*
[2]*Campinas State University*
Brazil

1. Introduction

Fossil fuels are, nowadays, the most used worldwide but that are some problems involving their utilization. First of all, the price, which is growing often and makes petroleum no longer economically sustainable. Second, during the burning of petrochemical sources there is emission of very dangerous pollutants for human health, like carbon dioxide that is the main reason of the global warming. In addition, fossil fuels are non-renewable resources, so they will last for a limited period of time. For all these reasons, vegetable oils are emerging as a great alternative fuel, because of their renewable nature and environmental benefits (Ferella et al., 2010).

Despite all the advantages, the use of vegetable oils as fuel has some disadvantages. The direct use in internal combustion engines is problematic because vegetables oils have high viscosity than diesel fuel and low volatility, so they do not burn completely and form deposits in the fuel injectors of diesel engine.

According to specialized literature there are five ways to reduce the problems mentioned above: blending of vegetable oil and diesel, thermal cracking (pyrolysis), microemulsions, esterification and transesterification (Ma & Hanna, 1999). Esterification and transesterification reactions are currently the most favored reaction pathways to produce biodiesel (Janaun & Ellis, 2010).

Biodiesel, defined as the simple alkyl monoesters of long chain fatty acids derived from renewable feedstocks, is the most suitable substitute to diesel. For this reason the research on this biofuel are steadily growing all over the planet. In Brazil, the focus of research is the production of biodiesel using ethanol, since this alcohol is produced on a large scale in the country. Ethanolysis produces a biodiesel less damage to the environment than that produced by methyl alcohol, since ethanol is derived from sugar cane or corn. In the rest of the world, the production takes place mostly in the methyl route and with use of heterogeneous catalysts (Pighinelli, 2010).

Biodiesel is highly biodegradable in fresh water as well as in soil and great part of it is mineralized in until 28 days under aerobic or anaerobic conditions (Makareviciene & Janulis, 2003; Pasqualino et al., 2006; Zhang et al., 1998). It is also a carbon-free fuel, as the plants that serve as raw material for its production absorb more carbon than that which is released during the burning of this biofuel (Antolin et al., 2002; Lang et al., 2001; Sharma et al., 2008; Vicente & Martìnez, 2004).

Moreover, when biodiesel is burned in diesel engines the emissions of hydrocarbons, carbon monoxide, particulate matter and sulphur dioxide are reduced with the exception of

nitrogen oxides, that emission increases due to the oxygen content of biodiesel (Canakci et al., 2006; Labeckas & Slavinskas, 2006; Turrio-Baldassarry, 2004).
Biodiesel sold today is still considered expensive, since the production costs involved are influenced by the main raw material, which are vegetable or animal fat and oils. It is estimated that approximately 80% of the total cost of biodiesel production is related to the acquisition of triglycerols source (Pighinelli, 2010). Another problem that has been discussed frequently is the competition between "food production" and "energy production". Some researchers argue that there will be food shortages if the available land is used for oilseed cultivation.
In order to reduce the production costs and to make it competitive with petroleum diesel, biodiesel producers should choose a raw material longer available in their territory, as soybean in Brazil, but also, search for alternatives crops, such as non-edible oils, as *Crambe Abyssinica, Jatropha Curcas* and others, and also waste frying oils (Marchetti et al., 2007).
The production of biodiesel is considered a current topic of great relevance worldwide. Thus, this chapter will be discussed: how biodiesel is produced, which are the main parameters affecting the chemical reactions and the most important issues for assuring biodiesel quality related to its production as well as some post-production parameters.

2. Biodiesel production

Transesterification is the technological route more used for biodiesel production, and can be applied on a small scale, as in laboratories, or in industry, producing millions of gallons of biofuel. Although the esterification also results in biodiesel and is recommended when the raw material is composed of oils rich in free fatty acids, this technique is applied commercially in few industries. That´s why we decide to mention in this chapter only the transesterification process. In this section, mechanisms of the transesterification reaction are going to be explained, identifying all process variables affecting the biodiesel yield and finally some important optimization studies will be presented.

2.1 Transesterification

Transesterification, also known as alcoholysis is the reaction of oil or fat with an alcohol to form esters and glycerol. To complete a transesterification reaction, stoichiometrically, a 3:1 molar ratio of alcohol to triglyceride is needed. In practice, to have a maximum ester yield, this ratio needs to be higher than the stoichiometric ratio. A catalyst is usually used to improve the reaction rate and yield. Because the reaction is reversible, excess alcohol is used to shift the equilibrium to the products side (Ma & Hanna, 1999). The reaction is shown in Fig. 1.

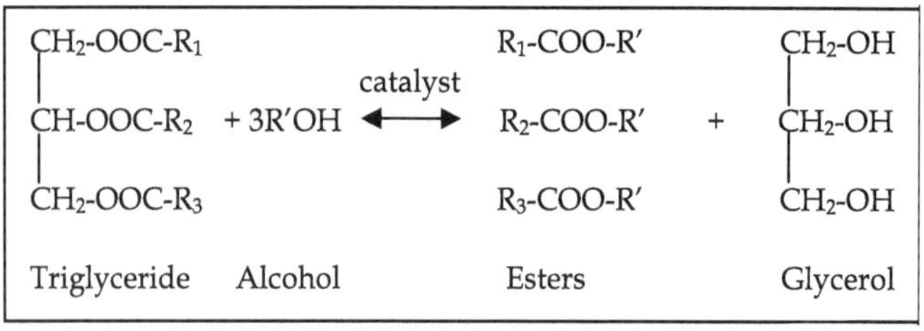

Fig. 1. Transesterification of triglycerides with alcohols.

The conventional technology of biodiesel production employs a basic homogeneous catalyst such as sodium or potassium hydroxides but when oil rich in free fatty acids (FFA) is used the basic catalyst and the FFA will interact to produce soap. This makes the amount of available catalyst for the transesterification reaction to be reduced and also complicates the down streaming separation and the biodiesel purification. Alternative processes for fatty acid ethyl ester (FAEE) production have been under development in order to employ different catalysts as heterogeneous ones (Marchetti & Errazu, 2011) such as metal oxides, metal complexes, active metals loaded on supports, zeolite, resins, membranes, and lipases (Kansedo et al., 2009).
Transesterification consists of a sequence of three consecutive reversible reactions, as can be seen on Figure 2. The first step is the conversion of triglycerides to diglycerides, followed by the conversion of diglycerides to monoglycerides, and finally monoglycerides to glycerol, yielding one ester molecule for each glyceride at each step.

1. Triglyceride (TG) + R'OH $\underset{K_4}{\overset{K_1}{\longleftrightarrow}}$ Diglyceride (DG) + $R'COOR_1$
2. Diglyceride (DG) + R'OH $\underset{K_5}{\overset{K_2}{\longleftrightarrow}}$ Monoglyceride (MG) + $R'COOR_2$
3. Monoglyceride (MG) + R'OH $\underset{K_6}{\overset{K_3}{\longleftrightarrow}}$ Glycerol (GL) + $R'COOR_3$

Fig. 2. The transesterification reactions of vegetable oil with alcohol to esters and glycerol (Freedman et al., 1986).

The main parameters affecting the transesterification reaction are molar ratio of vegetable oil to alcohol, catalyst type and amount, reaction time and temperature, the contents of free fatty acids (FFAs) and water in substrate oil (Freedman et al., 1984) and also the intensity of mixing during the chemical reaction.

2.1.1 The effects of moisture and free fatty acids

The starting materials used for alkali-catalyzed transesterification of glycerides must meet certain specifications. The presence of water during alkali catalyzed transesterification causes a partial reaction change to saponification, which produces soap. For that reason, the glycerides and alcohol must be substantially anhydrous (Wright et al., 1944). A small amount of soap favors the consumption of catalyst and reduces the catalytic efficiency, as well as causing an increase in viscosity, the formation of gels, and difficulty in achieving separation of glycerol. Ma et al. (1998) suggested that the free fatty acid content of the refined oil should be as low as possible, below 0.5%, and Feuge & Grose (1949) also stressed the importance of oils being dry and free of free fatty acids.
The use of alkali catalysts in the transesterification of used cooking oil is somewhat limited because the FFA in used cooking oil reacts with the most common alkaline catalysts (NaOH, KOH, and CH_3ONa) and forms soap. Because water makes the reaction partially change to

saponification, the alkali catalyst is consumed in producing soap and reduces catalyst efficiency. The soap causes an increase in viscosity, formation of gels which reduces ester yield and makes the separation of glycerol difficult. These two problems not withstanding, literature is replete with studies on the transesterification of waste cooking oil using alkaline catalyst (Marchetti & Errazu, 2010).

2.1.2 The effect of molar ratio

One of the most important variables affecting the yield of ester is the molar ratio of alcohol to triglyceride. The stoichiometric ratio for transesterification requires three moles of alcohol and one mole of glyceride to yield three moles of fatty acid ester and one mole of glycerol. The molar ratio is associated with the type of catalyst used. For example, a reaction conduced with an acid catalyzed needed a 30:1 ratio of BuOH to soybean oil, while a alkali-catalyzed reaction required only a 6:1 ratio to achieve the same ester yield for a given reaction time (Freedman et al., 1986). Higher molar ratios result in greater ester conversion in a shorter time. During the ethanolysis of peanut oil with a molar ratio alcohol:oil of 6:1 the amount of glycerin liberated was more than did a 3:1 molar ratio (Feuge and Grose, 1949). In this point is important to consider the type of alcohol that is been used. This is because during ethanolysis, as this alcohol has chemical affinity for both glycerine and ester, the higher the molar ratio is more difficult to separate the both phases.

2.1.3 The effect of catalyst

The catalysts for biodiesel can be separated into two major groups: homogeneous e heterogeneous. Homogeneous type forms a single phase mixture when added to oil and alcohol while the heterogeneous do not mix in the reaction medium. The group of homogeneous catalysts is divided into acid and basic and heterogeneous into metal oxides, metal complexes, active metals loaded on supports, zeolite, resins, membranes, and lipases The criterion for choosing which type of catalyst use should take into account firstly, the quality of raw material, but also the type of alcohol, the costs of the catalysts and technological route to be used for biodiesel production.

2.1.4 The effect of reaction time

The conversion rate of vegetable oils into biodiesel increases with reaction time. Freedman et al. (1984) transesterified peanut, cottonseed, sunflower and soybean oils under the condition of methanol to oil ratio of 6:1, 0.5% sodium methoxide catalyst and 60°C. After 1 minute, a yield of 80% of biodiesel was observed for soybean and sunflower oils, and after 60 minutes, the conversions were almost the same for all four oils.

During the transesterification of beef tallow with methanol the reaction was very slow during the first minute due to the mixing and dispersion of methanol into beef tallow. In the next five minutes, the reaction proceeded very fast (Ma et al., 1998). The production of beef tallow slowed down and reached the maximum value at about 15 min. The di- and monoglycerides increased at the beginning and then decreased. At the end, the amount of monoglycerides was higher than that of diglycerides.

2.1.5 The effect of reaction temperature

Transesterification can occur at different temperatures depending on the vegetable oil used, taking care not to exceed the boiling point of the alcohols used. In methanolysis of castor oil to methyl ricinoleate, with a molar ratio of 6:1-12:1 and 0.005-0.35% (by weight of oil) of

NaOH catalyst, the reaction proceeded most satisfactorily at 20-35°C (Smith, 1949). For the transesterification of refined soybean oil with methanol, molar ratio alcohol:oil of 6:1 and 1% NaOH of catalyst, three different temperatures were used (Freedman et al.,1984). After 0.1 h, ester yields were 94, 87 and 64% for 60, 45 and 32°C, respectively. After 1 h, ester formation was identical for the 60 and 45°C runs and only slightly lower for the 32°C run.

2.2 Optimization of biodiesel production

In the world economic context, the optimization of industrial processes is a tool that has been applied frequently in order to reduce the consumption of raw materials, with the main objective of reducing production costs. Response surface methodology (RSM) is a powerful tool for the optimization of chemical reactions and/or industrial processes. The main advantages of this method include: (1) an understanding of how the test variables (process variables) affect the selected process response; (2) the determination of any possible interrelationship among the test variables; and (3) the characterization of the combined effect that all test variables may have on the process response (Myers & Montgomery, 1995). There are only few examples in the literature involving optimization of transesterification of vegetable oils to produce biodiesel, even knowing that the costs of biofuel production are high (Domingos et al., 2008).

Pinzi et al. (2011) evaluated together with the process variables, how raw material fatty acid composition affects the biodiesel production. The authors applied a factorial design to determine how the operation conditions affected the transesterification process (reaction temperature, initial catalyst concentration by oil mass and methanol concentration), while the reaction yield was considered as the response variable. The vegetable oils studied were maize (MME), sunflower (SFME), olive (OOME), coconut (CME), linseed (LME) and palm (PME). A different range of temperature (from 40 °C to 60 °C) was selected for coconut oil and palm oil, since these oils are rich in unsaturated fatty acids. Molar ratio methanol:oil ranged from 4.2:1 to 5.4:1 and catalyst concentration from 0.8 to 2.1% Optimized conditions for each of the raw materials are shown in Table 1.

A complementary study aimed to define the optimum time of reaction based on a study of reaction kinetics. During the transesterification reaction carried out at the optimized condition, samples were taken at 30s, 1, 2, 5, 10, 35, 60, 90 and 120 min. Kinetic curves showed that every transesterification reaction presents similar performance curve, with the exception of coconut that exhibits the lowest yield of fatty acid methyl esters (FAME). After 20-min reaction oils with longer fatty-acid chains (olive, corn, linseed and sunflower) achieve the optimal yield of FAME, but on the other hand palm oil and coconut oil show the best performance after 40-min reaction. The authors observed that the effect of catalyst concentration was influenced by fatty-acid composition. Vegetable oils composed by unsaturated fatty acids show a directly proportional dependence between the concentration of catalyst and yield, up to a maximum. Whereas, considering vegetable oils with (mono) saturated fatty acids, amounts of catalyst greater than the optimal value lead to soap production. Fatty acids chain length also seems to influence biodiesel conversion. Oils with longer fatty-acid chains need half of the reaction time requested by oils comprising shorter fatty-acid chains to achieve maximum yield.

Silva et al. (2011) discussed in their paper the production process optimization for biodiesel by transesterification of soybean oil with ethanol, where several parameters, including catalyst, alcohol/vegetal oil molar ratio, and temperature would influence the transesterification. The authors' main objective was to study how the process variables (ethanol-to-oil ratio, catalyst concentration, reaction time and temperature) affecting the yield of biodiesel and then optimize this process. The levels of process variables studied were: ethanol/oil ratio (i.e., 3:1, 6:1, 9:1, 12:1 and 15:1), catalyst concentration (0.1%, 0.5%, 0.9%, 1.3% and 1.7% w/v of NaOH), reaction time (40, 60, 80, 100, and 120 min) and temperature (40, 50, 60, 70 and, 80 °C). Optimum values of the process parameter for maximum efficiency (95% of ethyl esters) were molar ratio ethanol: soybean oil 9:1, catalyst concentration 1.3% w/v, temperature 40 °C and reaction time 80 minutes. The analysis of the effects of process variables on yield in ethyl esters showed that the molar ratio, catalyst concentration and reaction time had a positive effect while the temperature had a negative effect. A positive effect means that the larger the values of process variables, the greater the yield of biodiesel.

Factor	Optimum value
MME	Yield: 98.67 (wt%)
Reaction temperature (°C)	47.53
Catalyst concentration (wt%)	1.92
Methanol/oil (molar ratio)	5.4
SFME	Yield: 99.70 (wt%)
Reaction temperature (°C)	59.82
Catalyst concentration (wt%)	1.81
Methanol/oil (molar ratio)	5.4
OOME	Yield: 98.02 (wt%)
Reaction temperature (°C)	45
Catalyst concentration (wt%)	1.6
Methanol/oil (molar ratio)	6.03
CME	Yield: 90.01 (wt%)
Reaction temperature (°C)	60
Catalyst concentration (wt%)	1.7
Methanol/oil (molar ratio)	6.6
LME	Yield: 97.71 (wt%)
Reaction temperature (°C)	53
Catalyst concentration (wt%)	1.8
Methanol/oil (molar ratio)	6.02
PME	Yield: 98.91 (wt%)
Reaction temperature (°C)	65
Catalyst concentration (wt%)	1.81
Methanol/oil (molar ratio)	6.15

Table 1. Optimization results according to Pinzi et al. (2011)

Another optimization study used *Raphanus sativus* (L. Var) crude oil in ethanolysis with sodium hydroxide as catalyst. Three process variables were used to develop the experimental design: the ethanol:oil molar ratio (MR of 6:1 and 12:1), the catalyst concentration in relation to oil mass (C of 0.4 and 0.8 wt% NaOH) and the alcoholysis temperature (T of 45 and 65 °C). This yield was expressed in relation to the oil mass used for ethanolysis, reason why some values were greater than 100%. Reaction temperature had no statistical significance over biodiesel yield. The highest biodiesel yield was 101.7% obtained at 65 °C with a MR of 12:1 and 0.4 wt% of C. Nevertheless, when the alcoholysis temperature was decreased to 45 °C, phase separation improved and lower levels of soap accumulation were obtained in the ethyl ester phase. The authors recommend the following procedure for the ethanolysis of *Raphanus sativus* crude oil: MR of 11.7:1, NaOH concentration of 0.4 wt%, 45 °C and vigorous agitation for 60 min as the first reaction stage, followed by a second stage in which MR and NaOH concentration can be reduced to 6:1 and 0.03 wt%, respectively (Domingos et al., 2008).

Heterogeneous catalysis optimization was studied by Marchetti & Errazu (2011). The reaction temperature's effects (30, 45 and 55 °C), the initial amount of free fatty acid (2.8%, 9.9% and 19.5% w/w), the molar ratio of alcohol/oil (4.2:1, 5.01:1 and 6.1:1) and the type of catalyst (homogeneous - sulfuric acid or heterogeneous - Dowex monosphere 550A) over the main reaction are analyzed and their effects compared. Temperature and molar ratio had a positive effect over biodiesel production: when the temperature and molar ratio increase the final conversion increases as well. When the initial amount of free fatty acid was varied, experimental results show that the final conversion increases as the initial amount of free fatty acid increases. Therefore, this effect could also be seen on the total FAEE production since the final amount of biofuel will be produced from the triglycerides as well as from the fatty acids present in the reaction mixture. The last part of this paper, a comparative study was made between the production of esters using sulfuric acid and a base solid resin with ethanol anhydrous under similar operational conditions, such as T=55 °C, initial amount of FFA=9.9% w/w, 2.2% w/w of each catalyst, and a molar ratio of alcohol/mixture of 6.1:1. Sulfuric acid reaches its final conversion in about 3 days time, while base solid resin reaches almost 100% in 70 hours.

As can be seen by the results showed earlier, biodiesel production is influenced by several process variables. The ideal combination of these variables will result in a higher yield in esters as well as a final product of higher quality. In addition, production costs could be reduced, including the industrial level. It was observed that depending on the feedstock, the type of alcohol and catalyst, the optimum conditions change. Another relevant point is related to the diversification of oil sources, which also help to reduce costs and to produce a quantity of fuel that meets global demand. In this scenario come acid oils, such as waste frying, and the heterogeneous catalysts, which are being actively researched.

3. Biodiesel quality

Generally, the biodiesel quality can be influenced by several factors:
The quality of the feedstock.
The fatty acid composition of the parent vegetable oil or animal fat.
The production process and the other materials used in this process.
Post-production parameters.

3.1 Production process factors

3.1.1 Reaction

The most important issue during biodiesel production is the completeness of the transesterification reaction. The triglycerides are converted to diglycerides, which in turn are converted to monoglycerides, and then to glycerol. Each step produces a molecule of a methyl or ethyl ester of a fatty acid. If the reaction is incomplete, then there will be triglycerides, diglycerides, and monoglycerides left in the reaction mixture. Each of these compounds still contains a glycerol molecule that has not been released. The glycerol portion of these compounds is referred to as bound glycerol. When the bound glycerol is added to the free glycerol, the sum is known as the total glycerol.

3.1.2 Free glycerol

Free glycerol refers to the amount of glycerol that is left in the finished biodiesel. Glycerol is essentially insoluble in biodiesel so almost all of the glycerol is easily removed by settling or centrifugation. Free glycerol may remain either as suspended droplets or as the very small amount that is dissolved in the biodiesel. Alcohol can act as co-solvent to increase the solubility of glycerol in the biodiesel. Most of this glycerol should be removed during the purification process. Water-washed fuel is generally very low in free glycerol, especially if hot water is used for washing. Distilled biodiesel tends to have a greater problem with free glycerol due to glycerol carry-over during distillation. Fuel with excessive free glycerol will usually have a problem with glycerol settling out in storage tanks, creating a very viscous mixture that can plug fuel filters and cause combustion problems in the engine.

3.1.3 Residual alcohol and catalyst

Since methanol or ethanol and the alkaline catalysts are more soluble in the polar glycerol phase, most will be removed when the glycerol is separated from the biodiesel.
However, the biodiesel typically contains 2-4% methanol after the separation, which may constitute as much as 40% of the excess methanol from the reaction. Most processors will recover this methanol using a vacuum stripping process. Any methanol remaining after this stripping process should be removed by the water washing process. Therefore, the residual alcohol level in the biodiesel should be very low. A specific value for the allowable alcohol level is specified in European biodiesel standards (0.2% in EN 14214), but is not included in the ASTM standard. Tests have shown that as little as 1% methanol in the biodiesel can lower the flashpoint of the biodiesel from 170°C to less than 40°C. Therefore, by including a flashpoint specification of 130°C, the ASTM standard limits the amount of alcohol to a very low level (<0.1%). Residual alcohol left in the biodiesel will generally be too small to negatively impact the fuel's performance. However, lowering the flashpoint presents a potential safety hazard as the fuel may need to be treated more like gasoline, which also has a low flashpoint, than diesel fuel.
Most of the residual catalyst is removed with the glycerol phase. Like the alcohol, remaining catalyst should be removed during the water washing. Although a value for residual catalyst is not included in the ASTM standard, it will be limited by the specification on sulfated ash. Excessive ash in the fuel can lead to engine deposits and high abrasive wear levels. The European standard EN 14214 places limits on calcium and magnesium as well as the alkali metals sodium and potassium.

3.2 Post-production factors

3.2.1 Water and sediment

Water and sediment contamination are basically housekeeping issues for biodiesel. Water can be present in two forms, either as dissolved water or as suspended water droplets. While biodiesel is generally considered to be insoluble in water, it actually takes up considerably more water than diesel fuel. Biodiesel can contain as much as 1500 ppm of dissolved water while diesel fuel usually only takes up about 50 ppm (Van Gerpen et al.,

1997). The standards for diesel fuel (ASTM D 975) and biodiesel (ASTM D 6751) both limit the amount of water to 500 ppm. For petroleum-based diesel fuel, this actually allows a small amount of suspended water. However, biodiesel must be kept dry. This is a challenge because many diesel storage tanks have water on the bottom due to condensation. Suspended water is a problem in fuel injection equipment because it contributes to the corrosion of the closely fitting parts in the fuel injection system.

Water can also contribute to microbial growth in the fuel. This problem can occur in both biodiesel and conventional diesel fuel and can result in acidic fuel and sludges that will plug fuel filters. Sediment may consist of suspended rust and dirt particles or it may originate from the fuel as insoluble compounds formed during fuel oxidation. Some biodiesel users have noted that switching from petroleum-based diesel fuel to biodiesel causes an increase in sediment that comes from deposits on the walls of fuel tanks that had previously contained diesel fuel. Because its solvent properties are different from diesel fuel, biodiesel may loosen these sediments and cause fuel filter plugging during the transition period.

3.2.2 Storage stability

Storage stability refers to the ability of the fuel to resist chemical changes during long term storage. These changes usually consist of oxidation due to contact with oxygen from the air. The fatty acid composition of the biodiesel fuel is an important factor in determining stability towards air. Generally, the polyunsaturated fatty acids (C18:2, linoleic acid; C18:3 linolenic acid) are most susceptible to oxidation. The changes can be catalyzed by the presence of certain metals (including those making up the storage container) and light. If water is present, hydrolysis can also occur. The chemical changes in the fuel associated with oxidation usually produce hydroperoxides that can, in turn, produce short chain fatty acids, aldehydes, and ketones.

Under the right conditions, the hydroperoxides can also polymerize. Therefore, oxidation is usually denoted by an increase in the acid value and viscosity of the fuel. Often these changes are accompanied by a darkening of the biodiesel color from yellow to brown and the development of a "paint" smell. When water is present, the esters can hydrolyze to long chain free fatty acids which also cause the acid value to increase.

There is currently no generally accepted method for measuring the stability of biodiesel. The techniques generally used for petroleum-based fuels, such as ASTM D 2274, have been shown to be incompatible with biodiesel. Other procedures, such as the Oil Stability Index or the Rancimat apparatus, which are widely used in the fats and oils industry, seem to be more appropriate for use with biodiesel. However, the engine industry has no experience with these tests and acceptable values are not known. Also, the validity of accelerated testing methods has not been established or correlated to actual engine problems. If biodiesel's acid number, viscosity, or sediment content increase to the point where they exceed biodiesel's ASTM limits, the fuel should not be used as a transportation fuel.

Additives such as BHT (butylated hydroxytoluene) and TBHQ (t-butylhydroquinone) are common in the food industry and have been found to enhance the storage stability of biodiesel. Biodiesel produced from soybean oil naturally contain some antioxidants (tocopherols, i.e., vitamin E), providing some protection against oxidation (some tocopherol is lost during refining of the oil prior to biodiesel production). Any fuel that will be stored for more than 6 months, whether it is diesel fuel or biodiesel, should be treated with an antioxidant additive.

3.3 Quality control

All biodiesel production facilities should be equipped with a laboratory so that the quality of the final biodiesel product can be monitored. It is also important to monitor the quality of the feedstocks. One strategy used by many producers is to draw a sample of the oil (or alcohol) from each delivery and use that sample to produce biodiesel in the laboratory. This test can be fairly rapid (1 or 2 hours) and can indicate whether serious problems are likely in the plant. Measuring feedstock quality can usually be limited to acid value and water content. These are not too expensive and can be operated by less experienced technicians.

To monitor the completeness of the reaction according to the total glycerol level specified in ASTM D 6751 requires the use of a gas chromatograph and a skilled operator. Large producers will find that having this equipment on-site is necessary. Commercial laboratories are available that can analyze the samples but there are costs and the time required may be several days. Smaller producers will need to use a more robust production process involving extra methanol or ethanol and probably multiple reaction steps. Then the product quality can be monitored through periodic testing by an outside laboratory.

Other possibilities for monitoring the transesterification reaction and assessing fuel quality are methods based on spectroscopy (such as near- infrared spectroscopy) or physical properties (such as viscometry). These methods are usually faster and easier to use than gas chromatography. However, some of them require extensive calibration. They also cannot accurately quantify glycerol at the low levels called for in the ASTM standard. To circumvent this, comparison to a reaction and product known to meet ASTM standards is needed.

4. Biodiesel standards

The primary criterion for biodiesel quality is adherence to the appropriate standard. The technical specifications for biodiesel depend on the country or the region where the fuel was produced. Biodiesel has a number of standards for its quality including European standard EN 14214 (Table 2), ASTM D6751 (Table 3), and others.

The European standard for Fatty Acid Methyl Esters (FAME) used as automotive fuel was set in 2003 by the Comité Européen de Normalisation (CEN) and is known under the standard number EN 14214. This standard sets limits and measurement methods for FAME, known as biodiesel that may be used either as a stand alone fuel or as a blending component in diesel fuel. The CEN standard for diesel fuel, EN 590, requires that all biodiesel blended in the fuel must conform to the standard EN 14214. At present, the European diesel fuel allows biodiesel to be blended at up to and including 5% by volume. Some national standards in EU countries allow biodiesel to be distributed as a stand-alone fuel, notably in Germany, for specially adapted vehicles. The CEN is presently studying a revised EN 590 specification for diesel fuel that will permit up to and including 7% of biodiesel blend. Simultaneously CEN is studying a revision of the biodiesel standard EN 14214 with a view to widening the range of feedstock oils that may be used, without compromising the security of vehicles using this product either in blends or as a stand-alone fuel. At the same time the European Commission has mandated CEN to revise the EN 590 specification for diesel fuel up to 10% of biodiesel blend.

The United States of America has chosen to use the specifications developed by ASTM International for both conventional diesel fuel and biodiesel. Specification efforts for biodiesel in the United States of America began in 1993 in Committee D02 on 24 Petroleum

Products and Lubricants. While the initial proposal for the biodiesel specifications at ASTM was for B100 (pure biodiesel) as a stand alone fuel, experience of the fuel in-use with blends above B20 (20% biodiesel with 80% conventional diesel) was insufficient to provide the technical data needed to secure approval from the ASTM members. Based on this, after 1994 biodiesel efforts within ASTM were focused on defining the properties for pure biodiesel which would provide a 'fit for purpose' fuel for use in existing diesel engines at the B20 level or lower. A provisional specification for B100 as a blend stock was approved by ASTM in 1999, and the first full specification was approved in 2001 and released for use in 2002 as "ASTM D6751 Standard Specification for Biodiesel Fuel Blend Stock (B100) for Middle Distillate Fuels".

Property	Test method	Limits		Unit
		min	max	
Ester content	EN 14103	96.5		% (m/m)
Density, 15°C	EN ISO 3675 EN ISO 12185	860	900	kg/m^3
Viscosity, 40°C	EN ISO 3104 EN ISO 3105	3.5	5.0	mm^2/s
Flash point	EN ISO 3679	120		°C
Sulfur content	EN ISO 20846 EN ISO 20884		10.0	mg/kg
Carbon residue (10% dist. residue)	EN ISO 10370		0.30	% (m/m)
Cetane number	EN ISO 5165	51		
Sulfated ash	ISO 3987		0.02	% (m/m)
Water content	EN ISO 12937		500	mg/kg
Total contamination	EN 12662		24	mg/kg
Copper strip corrosion (3hr, 50°C)	EN ISO 2160	1		
Oxidative stability, 110°C	EN 14112	6.0		hr
Acid value	EN 14111		0.50	mg KOH/g
Iodine value	EN 14111		120	g iodine/100g
Linolenic acid content	EN 14103		12	% (m/m)
Content of FAME with ≥4 double bonds			1	% (m/m)
Methanol content	EN 14110		0.20	% (m/m)
Monoglyceride content	EN 14105		0.80	% (m/m)
Diglyceride content	EN 14105		0.20	% (m/m)
Triglyceride content	EN 14105		0.20	% (m/m)
Free glycerol	EN 14105, EN 14106		0.02	% (m/m)
Total glycerol	EN 14105		0.25	% (m/m)
Alkali metals (Na + K)	EN 14108, EN 14109		5.0	mg/kg
Earth alkali metal (Ca + Mg)	prEN 14538		5.0	mg/kg
Phosphorus content	EN 14107		10.0	mg/kg

Table 2. Biodiesel Standard EN 14214 (Europe)

The philosophy used to approve D6751 was the same as that used for the No. 1 and No. 2 grades of fuels within the conventional specification, ASTM D975: If the parent fuels meet their respective specifications then the two can be blended in any percentage and used in conventional diesel engines. No separate set of properties was needed for the finished blends of No. 1 and No. 2, if the parent fuels met their respective specifications. These same conditions hold true for biodiesel; if biodiesel meets D6751 and conventional diesel meets D975 the two can be blended and used in conventional engines with the restriction of the upper limit of 20% biodiesel content in the finished fuel.

Property	Test method	Limits	Unit
Flash point (closed cup)	D 93	130.0 min.	°C
Water and sediment	D 2709	0.050 max.	% vol.
Kinematic viscosity, 40°C	D 445	1.9-6.0	mm^2/s
Sulfated ash	D 874	0.020 max.	% mass
Sulfur	D 5453	0.0015 max or 0.05 max[a]	% mass
Copper strip corrosion	D 130	No. 3 max	
Cetane number	D 613	47 min	
Cloud point	D 2500	Report	°C
Carbon residue (100% sample)	D 4530	0.050 max	% mass
Acid number	D 664	0.80 max	mg KOH/g
Free glycerin	D 6584	0.020 max	% mass
Total glycerin	D 6584	0.240 max	% mass
Phosphorus content	D 4951	0.001 max	% mass
Distillation temperature, atmospheric equivalent temperature, 90% recovered	D 1160	360 max	°C

[a]The limits are for Grade S15 and Grade S500 biodiesel, respectively. S15 and S500 refer to maximum sulfur specifications (ppm).

Table 3. Biodiesel Standard ASTM D6751 (United States)

While this mode of operation has served the US market well, there has been substantial effort since 2003 to develop and formally approve specifications for the finished blend of biodiesel and conventional diesel fuel. In addition, several improvements and changes to D6751 were also undertaken, some as a result of changes needed to secure approval of the finished blended biodiesel specifications. At the time of this report ballots to allow the formal acceptance of up to 5% biodiesel (B5) into the conventional diesel specifications for on/off road diesel fuel (ASTM D975) and fuel oil burning equipment (ASTM D396) and a new stand alone specification covering biodiesel blends between 6% and 20% have been approved through the Subcommittee level of Committee D02. In addition, a ballot to implement a new parameter in D6751 to control the potential for filter clogging above the cloud point in B20 blends and lower has also passed the subcommittee and is on track for a June 2008 vote. Efforts to approve B100 and B99 as stand alone fuels have been discussed at ASTM, but have been put on hold in order to focus on the B5 and B6 to B20 blended fuel specification efforts.

This section describes the parameters of the specifications normally used in the biodiesel standards:

4.1 Ester content

This parameter is an important tool, like distillation temperature, for determining the presence of other substances and in some cases meeting the legal definition of biodiesel (i.e. mono-alkyl esters). Low values of pure biodiesel samples may originate from inappropriate reaction conditions or from various minor components within the original fat or oil source. A high concentration of unsaponifiable matter such as sterols, residual alcohols, partial glycerides and unseparated glycerol can lead to values below the limit.

As most of these compounds are removed during distillation of the final product, distilled methyl esters generally display higher ester content than undistilled ones (Mittelbach and Enzelsberger, 1999).

4.2 Density

The densities of biodiesels are generally higher than those of fossil diesel fuel. The values depend on their fatty acid composition as well as on their purity. Density increases with decreasing chain length and increasing number of double bonds, or can be decreased by the presence of low density contaminants such as methanol.

4.3 Viscosity

The kinematic viscosity of biodiesel is higher than that of fossil diesel, and in some cases at low temperatures becomes very viscous or even solid. High viscosity affects the volume flow and injection spray characteristics in the engine, and at low temperatures may compromise the mechanical integrity of injection pump drive systems (when used as stand alone B100 diesel fuel).

4.4 Flash point

Flash point is a measure of flammability of fuels and thus an important safety criterion in transport and storage. The flash point of petrol diesel fuel is only about half the value of those for biodiesels, which therefore represents an important safety asset for biodiesel.

The flash point of pure biodiesels is considerably higher than the prescribed limits, but can decrease rapidly with increasing amount of residual alcohol. As these two aspects are strictly correlated, the flash point can be used as an indicator of the presence of methanol in the biodiesel. Flash point is used as a regulation for categorizing the transport and storage of fuels, with different thresholds from region to region, so aligning the standards would possibly require a corresponding alignment of regulations.

4.5 Sulfur

Fuels with high sulfur contents have been associated with negative impacts on human health and on the environment, which is the reason for current tightening of national limits. Low sulfur fuels are an important enabler for the introduction of advanced emissions control systems. Engines operated on high sulfur fuels produce more sulfur dioxide and particulate matter, and their emissions are ascribed a higher mutagenic potential. Moreover, fuels rich in sulfur cause engine wear and reduce the efficiency and life-span of catalytic systems. Biodiesel fuels have traditionally been praised as virtually sulfur-free. The national standards for biodiesel reflect the regulatory requirements for maximum sulfur content in fossil diesel for the region in question.

4.6 Carbon residue

Carbon residue is defined as the amount of carbonaceous matter left after evaporation and pyrolysis of a fuel sample under specific conditions. Although this residue is not solely composed of carbon, the term carbon residue is found in all three standards because it has long been commonly used. The parameter serves as a measure for the tendency of a fuel sample to produce deposits on injector tips and inside the combustion chamber when used as automotive fuel. It is considered as one of the most important biodiesel quality criteria, as it is linked with many other parameters. So for biodiesel, carbon residue correlates with the respective amounts of glycerides, free fatty acids, soaps and remaining catalyst or contaminants (Mittelbach 1996). Moreover, the parameter is influenced by high concentrations of polyunsaturated FAME and polymers (Mittelbach and Enzelsberger 1999). For these reasons, carbon residue is limited in the biodiesel specifications.

4.7 Cetane number

The cetane number of a fuel describes its propensity to combust under certain conditions of pressure and temperature. High cetane number is associated with rapid engine starting and smooth combustion. Low cetane causes deterioration in this behaviour and causes higher exhaust gas emissions of hydrocarbons and particulate. In general, biodiesel has slightly higher cetane numbers than fossil diesel. Cetane number increases with increasing length of both fatty acid chain and ester groups, while it is inversely related to the number of double bonds. The cetane number of diesel fuel in the EU is regulated at ≥51. The cetane number of diesel fuel in the USA is specified at ≥40. The cetane number of diesel fuel in Brazil is regulated and specified at ≥42.

4.8 Sulfated ash

Ash content describes the amount of inorganic contaminants such as abrasive solids and catalyst residues, and the concentration of soluble metal soaps contained in the fuel. These compounds are oxidized during the combustion process to form ash, which is connected with engine deposits and filter plugging (Mittelbach, 1996). For these reasons sulfated ash is limited in the fuel specifications.

4.9 Water content and sediment

The Brazilian and American standards combine water content and sediment in a single parameter, whereas the European standard treats water as a separate parameter with the sediment being treated by the Total Contamination property. Water is introduced into biodiesel during the final washing step of the production process and has to be reduced by drying. However, even very low water contents achieved directly after production do not guarantee that biodiesel fuels will still meet the specifications during combustion. As biodiesel is hygroscopic, it can absorb water in a concentration of up to 1000 ppm during storage. Once the solubility limit is exceeded (at about 1500 ppm of water in fuels containing 0.2% of methanol), water separates inside the storage tank and collects at the bottom (Mittelbach 1996). Free water promotes biological growth, so that sludge and slime formation thus induced may cause blockage of fuel filters and fuel lines. Moreover, high water contents are also associated with hydrolysis reactions, partly converting biodiesel to free fatty acids, also linked to fuel filter blocking. Finally, corrosion of chromium and zinc parts within the engine and injection systems have been reported (Kosmehl and Heinrich,

1997). Lower water concentrations, which pose no difficulties in pure biodiesel fuels, may become problematic in blends with fossil diesel, as here phase separation is likely to occur. For these reasons, maximum water content is contained in the standard specifications.

4.10 Total contamination

Total contamination is defined as the quota of insoluble material retained after filtration of a fuel sample under standardized conditions. It is limited to ≤ 24 mg/kg in the European specification for both biodiesel and fossil diesel fuels. The Brazilian and American biodiesel standards do not contain this parameter, as it is argued that fuels meeting the specifications regarding ash content will show sufficiently low values of total contamination as well. The total contamination has turned out to be an important quality criterion, as biodiesel with high concentration of insoluble impurities tend to cause blockage of fuel filters and injection pumps. High concentrations of soaps and sediments are mainly associated with these phenomena (Mittelbach, 2000).

4.11 Copper corrosion

This parameter characterizes the tendency of a fuel to cause corrosion to copper, zinc and bronze parts of the engine and the storage tank. A copper strip is heated to 50°C in a fuel bath for three hours, and then compared to standard strips to determine the degree of corrosion. This corrosion resulting from biodiesel might be induced by some sulfur compounds and by acids, so this parameter is correlated with acid number. Some experts consider that this parameter does not provide a useful description of the quality of the fuel, as the results are unlikely to give ratings higher than class 1.

4.12 Oxidation stability

Due to their chemical composition, biodiesel fuels are more sensitive to oxidative degradation than fossil diesel fuel. This is especially true for fuels with a high content of di - and higher unsaturated esters, as the methylene groups adjacent to double bonds have turned out to be particularly susceptible to radical attack as the first step of fuel oxidation (Dijkstra et al. 1995). The hydroperoxides so formed may polymerize with other free radicals to form insoluble sediments and gums, which are associated with fuel filter plugging and deposits within the injection system and the combustion chamber (Mittelbach & Gangl, 2001). Where the oxidative stability of biodiesel is considered insufficient, antioxidant additives might have to be added to ensure the fuel will still meet the specification.

4.13 Acid value

Acid value or neutralization number is a measure of free fatty acids contained in a fresh fuel sample and of free fatty acids and acids from degradation in aged samples. If mineral acids are used in the production process, their presence as acids in the finished fuels is also measured with the acid number. It is expressed in mg KOH required to neutralize 1g of biodiesel. It is influenced on the one hand by the type of feedstock used for fuel production and its degree of refinement. Acidity can on the other hand be generated during the production process. The parameter characterises the degree of fuel ageing during storage, as it increases gradually due to degradation of biodiesel. High fuel acidity has been discussed in the context of corrosion and the formation of deposits within the engine which is why it is limited in the biodiesel specifications of the three regions. It has been shown that free fatty acids as weak carboxylic acids pose far lower risks than strong mineral acids (Cvengros, 1998)

4.14 Iodine value, linolenic acid ester content and polyunsaturated

Iodine number is a measure of the total unsaturation within a mixture of fatty acids, and is expressed in grams of iodine which react with 100 grams of biodiesel. Engine manufacturers have argued that fuels with higher iodine number tend to polymerize and form deposits on injector nozzles, piston rings and piston ring grooves when heated (Kosmehl and Heinrich 1997). Moreover, unsaturated esters introduced into the engine oil are suspected of forming high-molecular compounds which negatively affect the lubricating quality, resulting in engine damage (Schaefer et al 1997). However, the results of various engine tests indicate that polymerization reactions appear to a significant extent only in fatty acid esters containing three or more double bonds (Worgetter et al. 1998, Prankl and Worgetter 1996, Prankl et al 1999).Three or more-fold unsaturated esters only constitute a minor share in the fatty acid pattern of various promising seed oils, which are excluded as feedstock according to some regional standards due to their high iodine value. Some biodiesel experts have suggested limiting the content of linolenic acid methyl esters and polyunsaturated biodiesel rather than the total degree of unsaturation as it is expressed by the iodine value.

4.15 Methanol or ethanol

Methanol (MeOH) or ethanol (EtOH) can cause fuel system corrosion, low lubricity, adverse affects on injectors due to its high volatility, and is harmful to some materials in fuel distribution and vehicle fuel systems. Both methanol and ethanol affect the flash point of esters. For these reasons, methanol and ethanol are controlled in the specification.

4.16 Mono, di and triglyceride

The EU standard specifies individual limit values for mono-, di- and triglyceride as well as a maximum value for total glycerol. The standards for Brazil and the USA do not provide explicit limits for the contents of partial acylglycerides. In common with the concentration of free glycerol, the amount of glycerides depends on the production process. Fuels out of specification with respect to these parameters are prone to deposit formation on injection nozzles, pistons and valves (Mittelbach et al. 1983).

4.17 Free glycerol

The content of free glycerol in biodiesel is dependent on the production process, and high values may stem from insufficient separation or washing of the ester product. The glycerol may separate in storage once its solvent methanol has evaporated. Free glycerol separates from the biodiesel and falls to the bottom of the storage or vehicle fuel tank, attracting other polar components such as water, monoglycerides and soaps. These can lodge in the vehicle fuel filter and can result in damage to the vehicle fuel injection system (Mittelbach 1996). High free glycerol levels can also cause injector coking. For these reasons free glycerol is limited in the specifications.

4.18 Total glycerol

Total glycerol is the sum of the concentrations of free glycerol and glycerol bound in the form of mono-, di- and triglycerides. The concentration depends on the production process.

Fuels out of specifications with respect to these parameters are prone to coking and may thus cause the formation of deposits on injector nozzles, pistons and valves (Mittelbach et al. 1983). For this reason total glycerol is limited in the specifications of the three regions.

4.19 Metals (Na+K) and (Ca+Mg)

Metal ions are introduced into the biodiesel fuel during the production process. Whereas alkali metals stem from catalyst residues, alkaline-earth metals may originate from hard washing water. Sodium and potassium are associated with the formation of ash within the engine, calcium soaps are responsible for injection pump sticking (Mittelbach 2000).These compounds are partially limited by the sulphated ash, however tighter controls are needed for vehicles with particulate traps. For this reason these substances are limited in the fuel specifications.

4.20 Phosphorus

Phosphorus in biodiesel stems from phospholipids (animal and vegetable material) and inorganic salts (used frying oil) contained in the feedstock. Phosphorus has a strongly negative impact on the long term activity of exhaust emission catalytic systems and for this reason its presence in biodiesel is limited by specification.

4.21 Distillation

This parameter is an important tool, like ester content, for determining the presence of other substances and in some cases meeting the legal definition of biodiesel (i.e. monoalkyl esters).

4.22 Cold climate operability

The behaviour of automotive diesel fuel at low ambient temperatures is an important quality criterion, as partial or full solidification of the fuel may cause blockage of the fuel lines and filters, leading to fuel starvation and problems of starting, driving and engine damage due to inadequate lubrication. The melting point of biodiesel products depend on chain length and degrees of unsaturation, with long chain saturated fatty acid esters displaying particularly unfavourable cold temperature behaviour.

5. Conclusion

Biodiesel is an important new alternative biofuel. It can be produced from many vegetable oil or animal fat feedstocks. Conventional processing involves an alkali catalyzed process but this is unsatisfactory for lower cost high free fatty acid feedstocks due to soap formation. Pretreatment processes using strong acid catalysts have been shown to provide good conversion yields and high quality final products. These techniques have even been extended to allow biodiesel production from feedstocks like soapstock that are often considered to be waste. Adherence to a quality standard is essential for proper performance of the fuel in the engine and will be necessary for widespread use of biodiesel.

6. Acknowledgment

We acknowledge the Faculty of Agricultural Engineering (FEAGRI/UNICAMP)), the Food Technology Institute (ITAL), the State of São Paulo Research Foundation (FAPESP) and the

National Council for Scientific and Technological Development (CNPq) for their financial and technical support.

7. References

Antolin, G.; Tinaut, F.V.; Briceno, Y.; Castano, V.; Perez, C. & Ramirez, A.L. Optimization of biodiesel production by sunflower oil transesterification. *Bioresource Technology*, Vol. 83 (2002), pp. 111–114, ISSN 0960-8524.

Canakci, M.; Erdil, A. & Arcaklioglu, E. Performance and exhaust emission of a biodiesel engine. *Applied Energy*, Vol. 83 (2006), pp. 594-605, ISSN 0306-2619.

Cvengros, J. Acidity and corrosiveness of methyl esters of vegetable oils. *Fett/Lipid*, Vol. 100, No 2 (1998), pp. 41-44, ISSN 1521-4133.

Dijkstra, A. J.; Maes, P. J.; Meert, D. & Meeussen, W. Interpreting the oxygen stability index. *Oils-Fats-Lipids 1995 Proceedings of the World Congress of the International Society for Fat Research*, pp. 629-637, The Hague, Netherland: P. J. Barnes & Associates, 1995.

Domingos, A. K.; Saad, E. B.; Wilhelm, H. M. & Ramos, L. P. Optimization of the ethanolysis of *Raphanus sativus* (L. Var.) crude oil applying the response surface methodology. *Bioresource Technology*, Vol. 99 (2008), pp. 1837–1845, ISSN 0960-8524.

Ferella, F.; Di Celso, G. M.; De Michelis, I.; Stanisci, V. & Vegliò, F. Optimization of the transesterification reaction in biodiesel production. *Fuel*, Vol. 89 (2010), pp. 36 - 42, ISSN 0016-2361.

Feuge, R. O. & Grose, T.: Modification of vegetable oils. VII. Alkali catalyzed interesterilication of peanut oil with ethanol. *Journal of the American Oil Chemists Society*., Vol. 26 (1949), pp. 97-102, ISSN 1558-9331.

Freedman, B.; Butterfield, R.O. & Pryde, E.H. Transesterification kinetics of soybean oil. *Journal of the American Oil Chemists Society*, Vol. 63 (1986), pp. 1375–1380, ISSN 1558-9331.

Freedman, B.; Pryde, E.H. & Mounts, T.L. Variables affecting the yield of fatty esters from transesterified vegetable oils. *Journal of the American Oil Chemists Society*, Vol. 61 (1984), pp. 1638–1643, ISSN 1558-9331.

Janaun, J. & Ellis, N. Perspectives on biodiesel as a sustainable fuel. *Renewable and Sustainable Energy Reviews*, Vol. 14 (2010), pp. 1312 - 1320, ISSN 1364-0321.

Kansedo, J.; Lee, K. T. & Bhatia, S. Biodiesel production from palm oil via heterogeneous transesterification. *Biomass & Bioenergy*, Vol. 33 (2 0 09), pp. 271–276, ISSN 0961-9534.

Kosmehl, S. O. & Heinrich, H. (1997). The Automotive Industry's View on the Standards for Plant Oil-Based Fuels, In: *Plant Oils as Fuels. Present State of Science and Future Developments*, N. Martini & J. Schell (Eds.), 18-28, ISBN 3540647546 9783540647546, Berlin, Germany.

Labeckas, G.; Slavinskas S. The effect of rapeseed oil methyl ester on direct injection diesel engine performance and exhaust emissions. *Energy Conversion Manage*, Vol. 47 (2006), pp. 1954–1967, ISSN 0196-8904.

Lang, X.; Dalai, A.K.; Bakhshi, N.N.; Reaney, M.J. & Hertz, P.B. Preparation and characterization of bio-diesels from various bio-oils. *Bioresource Technology*, Vol. 80 (2001), pp. 53–62, ISSN 0960-8524.

Ma, F. & Hanna, M. A. Biodiesel production: a review. *Bioresource Technology*, Vol. 70 (1999), pp. 1–15, ISSN 0960-8524.

Ma, F., Clements, L. D. & Hanna, M. A.: Biodiesel fuel from animal fat. Ancillary studies on transesterification of beef tallow. *Industrial Engineering Chemistry Research*, Vol. 37 (1998), pp. 3768-3771, ISSN 1520-5045.

Makareviciene, V. & Janulis, P. Environmental effect of rapeseed oil ethyl ester. *Renewable Energy*, Vol. 28 (2003), pp. 2395–2403, ISSN 0960-1481.

Marchetti, J.M. & Errazu, A.F. Biodiesel production from acid oils and ethanol using a solid basic resin as catalyst. *Biomass & Bioenergy*, Vol. 34 (2 0 1 0), pp. 272–277, ISSN 0961-9534.

Marchetti , J.M., Miguel, V. U. & Errazu, A. F. Heterogeneous esterification of oil with high amount of free fatty acids. *Fuel*, Vol. 86 (2007), pp. 906–910, ISSN 0016-2361.

Mittelbach, M. Chemische und motortechnische Untersuchungen der Ursachen der Einspritzpumpenverklebung bei Biodieselbetrieb; Bund-Bundeslander-kooperations-projekt (2000).

Mittelbach, M. & Gangl, S. Long storage stability of biodiesel made from rapeseed and used frying oil. *Journal of the American Oil Chemists Society*, Vol. 78, No 6 (2001), pp. 573-577, ISSN 1558-9331.

Mittelbach, M., & Enzelsberger, H. Transesterification of heated rapeseed oil for extending diesel fuel, *Journal of the American Oil Chemists Society*, Vol. 76 (1999), pp. 545–550, ISSN 1558-9331.

Mittelbach, M. Diesel fuel derived from vegetable oils, VI: specifications and quality control of biodiesel, *Bioresource Technology*, Vol. 56 (1996), pp. 7-11, ISSN 0960-8524.

Mittelbach, M., Worgetter, M.; Pernkopf, J. & Junek, H. Diesel fuel derived from vegetable oils: preparation and use of rape oil methyl ester. *Energy in Agriculture*, Vol. 2 (1983), pp. 369-384, ISSN 0167-5826.

Myers, R.H. & Montgomery, D.C. (1995). *Response surface methodology: process and product optimization using designed experiments*, John Wiley, Canada.

Pasqualino, J.C.; Montane, D. & Salvado, J. Synergic effects of biodiesel in the biodegradability of fossil-derived fuels. *Biomass & Bioenergy*, Vol. 30 (2006), pp. 874–879, ISSN 0961-9534.

Pighinelli, A.L.M.T. Study of mechanical expeller and ethanolic transesterification of vegetable oils. PhD Thesis, School of Agricultural Engineering, State University of Campinas (UNICAMP), Campinas, 2010.

Pinzi, S.; Mata-Granados, J.M.; Lopez-Gimenez, F.J.; Luque de Castro, M.D. & Dorado, M.P. Influence of vegetable oils fatty acid composition on biodiesel optimization. *Bioresource Technology*, Vol. 102 (2011), pp. 1059–1065, ISSN 0960-8524.

Prankl, H. & Worgetter, M.: Influence of the iodine number of biodiesel to the engine performance, *Liquid Fuels and Industrial Products from Renewable Resources. Proceedings of the 3rd Liquid Fuel Conference*, pp. 191-196, Nashville, Tennessee, USA, September 15-17, 1996.

Prankl, H.; Worgetter, M. & Rathbauer, J. Technical performance of vegetable oil methyl esters with a high iodine number. *4th Biomass Conference of the Americas*. Oakland, California, USA, 29 August to 2 September, 1999.

Schafer, A., D. Naber and M. Gairing. Biodiesel als alternativer Kraftstoff für Mercedes-Benz-Dieselmotoren. *Mineralöltechnik*, Vol. 43 (1997), pp. 1-32, ISSN: 0307-6490.

Sharma, Y.C.; Singh, B. & Upadhyay, S.N. Advancements in development and characterization of biodiesel: a review. *Fuel*, Vol. 87 (2008), pp. 2355-2373, ISSN: 0016-2361.

Silva, G. F.; Camargo, F.L. & Ferreira, A.L.O. Application of response surface methodology for optimization of biodiesel production by transesterification of soybean oil with ethanol. *Fuel Processing Technology*, Vol. 92 (2011), pp. 407-413, ISSN: 0378-3820.

Smith, M.K. Process of producing esters. US Patent 2 (1949), pp. 444-486.

Turrio-Baldassarry, L. Emissions comparison of urban bus engine fuelled with diesel oil and biodiesel blend. *Science of the Total Environment*. Vol. 327 (2004), pp. 147-162, ISSN: 0048-9697.

Van Gerpen, J.H.; Hammond, E.H.; Yu, L. & Monyem, A. Determining the influence of contaminants on biodiesel properties. *Society of Automotive Engineers Technical Paper Series, Paper No. 971685, SAE*, Warrendale, PA, USA, 1997.

Vicente, G.; Martìnez, M. & Aracil, J. Integrated biodiesel production: a comparison of different homogeneous catalysts systems. *Bioresource Technology*, Vol. 92 (2004), pp. 297-305, ISSN: 0960-8524.

Worgetter, M.; Prankl, H. & Rathbauer, J. Eigenschaften von Biodiesel. *Landbauforschung Völkenrode. Sonderheft* 190 (Biodiesel- Optimierungspotentiale und Umwelteffekte) (1998), 31-43.

Wright, H. J., Segur, J. B., Clark, H. V., Coburn, S. K., Langdon, E. E. & DuPuis, R.N. A report on ester interchange. *Oil Soap*, Vol. 21(1944), pp. 145-148.

Zhang, X.; Peterson, C.; Reece, D.; Moller, G. & Haws, R. Biodegradability of biodiesel in aquatic environment. *Transactions of The American Society of Agricultural Engineers*, Vol. 41 (1998), pp. 1423-1430, ISSN 0001-2351.

Part 2

Process Modeling and Simulation

11

Perspectives of Biobutanol Production and Use

Petra Patakova, Daniel Maxa, Mojmir Rychtera, Michaela Linhova, Petr Fribert, Zlata Muzikova, Jakub Lipovsky, Leona Paulova, Milan Pospisil, Gustav Sebor and Karel Melzoch
Institute of Chemical Technology Prague
Czech Republic

1. Introduction

Nowadays, with increasing hunger for liquid fuels usable in transportation, alternatives to crude oil derived fuels are being searched very intensively. In addition to bioethanol and ethyl or methyl esters of rapeseed oil that are currently used as bio-components of transportation fuels in Europe, other options are investigated and one of them is biobutanol, which can be, if produced from waste biomass or non-food agricultural products, classified as the biofuel of the second generation. Although its biotechnological production is far more complicated than bioethanol production, its advantages over bioethanol from fuel preparation point of view i.e. higher energy content, lower miscibility with water, lower vapour pressure and lower corrosivity together with an ability of the producer, *Clostridium* bacteria, to ferment almost all available substrates might outweigh the balance in its favour. The main intention of this chapter is to summarize briefly industrial biobutanol production history, to introduce the problematic of butanol formation by clostridia including short description of various options of fermentation arrangement and most of all to provide with complex fermentation data using little known butanol producers *Clostridium pasteurianum* NRRL B-592 and *Clostridium beijerinckii* CCM 6182. A short overview follows concerning the use of biobutanol as a fuel for internal combustion engines with regard to properties of biobutanol and its mixtures with petroleum derived fuels as well as their emission characteristics, which are illustrated based on emission measurement results obtained for three types of passenger cars.

2.Theoretical background

2.1 History of industrial biobutanol production

The initiation of the industrial acetone-butanol-ethanol (ABE) production by *Clostridium* fermentation is connected with the chemist Chaim Weizmann, working at the University of Manchester UK, who wished to make synthetic rubber containing butadiene or isoprene units from butanol or isoamyl alcohol and concentrated his effort on the isolation of microbial producers of butanol. Further, the development of acetone-butanol process was accelerated by World War I when acetone produced by ABE fermentation from corn in Dorset, UK was used for cordite production. However in 1916, the German blockade hampered the supply of grain and the production was transferred to Canada and later with the entry of the United States to the war, two distilleries in Terre Haute were adapted to acetone production. After the war, the group of American businessmen bought Terre Haute plant and restored the production in 1920; at that time butanol was appreciated as solvent for automobile lacquers. Subsequently, with decreasing price of molasses new solventogenic strains were isolated and first plant using this feedstock was built at Bromborough in England near the port, in 1935. In 1936 the Weizmann patent expired and new acetone-butanol plants were erected in U.S.A., Japan, India, Australia and South Africa using usually molasses as the substrate. The Second World War again accelerated the process development and acetone became the most required product; the plant at Bromborough was expanded and semi continuous way of fermentation which cut the fermentation time to 30-32h was accomplished here together with continuous distillation. At the end of the war, two thirds of butanol in U.S.A. was gained by fermentation but rise of petrochemical industry together with increasing price of molasses that started to be used for cattle feeding caused gradual decline of industrial acetone-butanol fermentation. Most of the plants in Western countries were closed by 1960 with the exception of Germiston factory in South Africa where cheap molasses and coal enabled to keep the process till 1983 (Jones & Woods, 1986). In addition to Western countries, the production of acetone and butanol was also supported in the Soviet Union. Here, in Dukshukino plant, in 1980s, the process was operated as semi continuous in multi-stage arrangement with possibility to combine both saccharidic and starchy substrates together with small portion (up to 10%) of lignocellulosic hydrolyzate and continuous distillation (Zverlov et al., 2006). In China, industrial fermentative acetone and butanol production began around 1960 and in 1980s there was the great expansion of the process. Originally, batch fermentation was changed to semi continuous 4-stage process in which the fermentation cycle was reduced to 20 h, the yield was about 35-37% from starch and the productivity was 2.3 times higher in comparison with batch process (Chiao & Sun, 2007). At the end of 20th century the most of Chinese plants were probably closed (Chiao & Sun, 2007) but now hundred thousands of tons of acetone and butanol per year are produced by fermentation in China (Ni & Sun, 2009).

Industrial production of ABE in the former Czechoslovakia started with a slight delay comparing with other already mentioned countries. Bacterial cultures were isolated, selected and tested for many years by professor J. Dyr, head of the Department of Fermentation Technology of the Institute of Chemical Technology in Prague who lead a small research team and preparatory works for the plant design (Dyr & Protiva, 1958). Acetone - butanol plant was fully in operation from 1952 till 1965. The main raw materials were firstly potatoes which were later changed for rye. Various bacteria cultures (all were classified as *Clostridium acetobutylicum)* were prepared for several main crops (potatoes, rye, molasses) which increased flexibility of the production. Annual production of solvents increased from year to year but did not exceed 1000 tons. Concentration of total solvents in the broth varied around 17–18 g.L^{-1}. Process itself was run as batch, pH was never controlled, propagation ratio in large fermentation section was 1 : 35. The whole fermentation time was on average 36–38 h. Critical point for each fermentation was "break" in acidity after which started a strong evolution of gases and solvents. In case of potatoes and rye there were no nutrients supplied to the fermentation broth. The only process necessary for the pre-treatment of the raw materials of starch origin was their steaming under pressure in Henze cooker. Initial concentration of starch ranges from 4.5 to 5% wt. In spite of keeping all sanitary precaution (similarly today´s GMP) two types of unexpected failures occurred. Firstly it was contamination by bacteriophage (not possible to analyze it in those times) which appeared approx. three times during the lifetime and always was followed by a total sanitation and complete change of the producing strain. Secondly there appeared another unexpected event, i.e. a final turn to a complete acidification without initiation of solvent production indicated by a spore creation. This situation appeared in the range from 1 to 4% of the total number of batches.

2.2 Principle of acetone-butanol-ethanol (ABE) fermentation

The butanol production through acetone-butanol-ethanol (ABE) fermentation is an unique feature of some species of the genus *Clostridium*; the most famous of them are strains of *C.acetobutylicum*, *C.beijerinckii* and *C.saccharoperbutylacetonicum* but others with the same ability exist, too. Together with all *Clostridium* bacteria, solvent producers share some common characteristics like rod-shaped morphology, anaerobic metabolism, formation of heat resistant endospores, incapability of reduction of sulphate as a final electron acceptor and G^+ type of bacterial cell wall (Rainey et al., 2009).

ABE fermentation consists of two distinct phases, acidogenesis and solventogenesis. While the first one is coupled with growth of cells and production of butyric and acetic acids as main products the second one, started by medium acidification, can be characterized by initiation of sporulation and metabolic switch when usually part of formed acids together with sugar carbon source are metabolized to 1-butanol and acetone. The biphasic character of ABE fermentation coupled with alternation of symmetric and asymmetric cell division, first mentioned by Clarke et al., (1988), is shown in Fig. 1. In the batch cultivation, first acidogenic phase is connected with internal energy generation and accumulation and also cells growth while second solventogenic phase is bound with energy consumption and sporulation. The tight connection of sporulation and solvents production was proved by finding a gene *spo0A* responsible for both sporulation and solvent production initiation (Ravagnani et al., 2000).

Metabolic pathway leading to solvents production and originating in Embden-Mayerhof-Parnas (EMP) glycolysis is shown in Fig.1, too. Pentoses unlike hexoses are converted to fructose-6-phosphate and glyceraldehyde-3-phosphate prior to their entrance to EMP metabolic pathway. Major products of the acidogenic phase - acetate, butyrate, CO_2 and H_2 are usually accompanied by small amounts of acetoin and lactate (not shown in Fig.1). The onset of solvents production is stimulated by accumulation of acids in cultivation medium together with pH drop. Butanol and acetone are formed partially from sugar source and partially by reutilization of the formed acids; and simultaneously a hydrogen production is reduced to a half in comparison with the acidogenic phase (Jones & Woods, 1986; Lipovsky et al., 2009). Functioning of all enzymes involved in the butanol formation has been reviewed, recently (Gheslaghi et al., 2009). Unfortunately, butanol is highly toxic to the clostridia and its stress effect causes complex response of the bacteria in which more than 200 genes regulating membrane composition, cell transport, sugar metabolism, ATP formation and other functionalities are involved and complicate any effort to increase butanol resistance (Tomas et al., 2004).

Solventogenic clostridia are known for their capabilities to utilize various mono-, di-, oligo- and polysaccharides like glucose, fructose, xylose, arabinose, lactose, saccharose, starch, pectin, inulin and others but usually the specific strain is not able to utilize efficiently all of named substrates. Although all genes of cellulosome were identified in *C.acetobutylicum* ATCC 824 genome, the whole cellulosome is not functional what results in incapability of cellulose utilization (Lopez-Contreras et al., 2004). At first, starchy substrates like corn and potatoes were used for ABE fermentation but later blackstrap molasses became the preferential feedstock. Nowadays, a lot of researchers aim to use lignocellulosic hydrolyzates which, if available at a reasonable price and quality (no inhibitors), would be ideal feedstock for this process because clostridia can utilize diluted solutions of various hexoses, pentoses, disaccharides and oligosacharides efficiently.

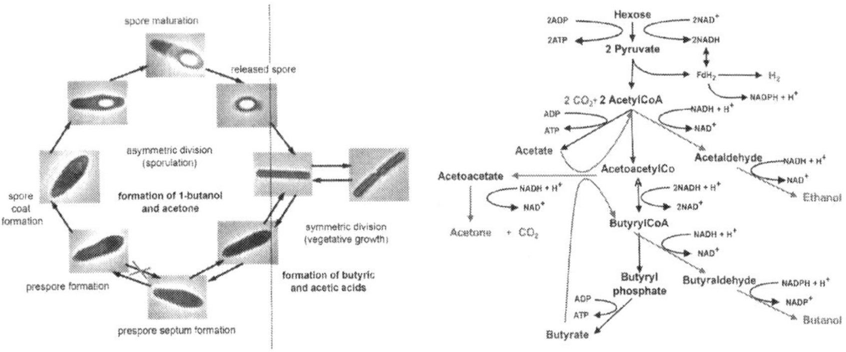

Fig. 1. Life cycle of solventogenic clostridia and simplified metabolic scheme

2.2.1 Challenges of butanol production

Production of biobutanol by clostridia is not straightforward process and 1-butanol is neither a typical primary metabolite, the formation of which is connected with cells growth, nor a typical secondary metabolite like antibiotics or pigments. The metabolic switch from acido- to solventogenesis, regulation of which is usually connected with sporulation initiation, does not need to happen necessarily during the fermentation. Actually, when cells are well nourished and their growth rate approaches its maximum then cells reproduce and form only acids; this state has been many times observed in continuous cultivations (Ezeji et al., 2005) but sometimes it can occur even in batch cultivation as so-called "acid crash" (Maddox et al., 2000; Rychtera et al., 2010) which was generally ascribed to fast acetic and butyric acids formation. The proposed acid crash prevention was careful pH control or metabolism slowdown by lowering cultivation temperature (Maddox et al., 2000). However, very recently the novel possible explanation of this phenomenon has been revealed in intracellular accumulation of formic acid by *C.acetobutylicum* DSM 1731 (Wang et al., 2011). If acid crash is the phenomenon that usually happens at random in the particular fermentation, so-called strain degeneration is a more serious problem when the production culture loses either transiently or permanently its ability to undergo the metabolic shift and to produce solvents. The reliable prevention of the degeneration is maintaining the culture in the form of spore suspension (Kashket & Cao, 1995). A cause of degeneration was investigated in many laboratories using various clostridial strains and therefore also with different results. The degeneration of *C.acetobutylicum* ATCC 824 is probably caused by loss of its mega plasmid containing genes for both sporulation and solvents production (Cornillot et al., 1997) but mechanism and reason of this degeneration were not offered by this study. Actually the authors (Cornillot et al., 1997) compared wild-type strain *C.acetobutylicum* ATCC 824 with isolated degenerated mutants. It is questionable how often or under which conditions the degeneration of *C.acetobutylicum* ATCC 824 happens because in the past, it was reported 218 passages of vegetative *C.acetobutylicum* ATCC 824 cells did not almost influence their solvents formation (Hartmanis et al., 1986). The cells of *C.saccharoperbutylacetonicum* N1-4 degenerated when quorum sensing mechanism in the population was impaired (Kosaka et al., 2007). The very detailed study of *C.beijerinckii* NCIMB 8052 degeneration disclosed two different degeneration causes: involvement of global regulatory gene and defect in NADH generation (Kashket & Cao, 1995). It seems probable that degeneration has no single reason and if other strains were studied different reasons would be found.

ABE industrial fermentation was probably the first process that had to cope with bacteriophage infection of producing microorganism. The first severe bacteriophage attack was reported from Terre Haute plant in the U.S.A. in 1923 and the problems occurred at fermentation of corn by *Clostridium acetobutylicum* (the solvents yield was decreased by half for a year). From that time, *Clostridium* strains used for either starch or saccharose fermentations were attacked by various both lysogenic and lytic bacteriophages what was documented in the literature. The ABE plant in Germiston in South Africa faced to confirmed bacteriophage infection 4- times in its 46-year history (plus two unconfirmed cases). Till now, the best solution in battle against *Clostridium* bacteriophages seems to be the prevention i.e. good process hygiene, sterilization, decontamination and disinfection (Jones et al., 2000).

Lactic acid bacteria represent the most common type of contamination having very similar requests for cultivation conditions (temperature, pH, anaerobiosis, composition of cultivation media) as clostridia and grow faster. These bacteria can cause not only losses in solvents yield but also can hamper the metabolic switch of clostridia because formed lactic acid over-acidifies the medium and poisons the clostridia in higher concentration. Other contaminants like *Bacillus* bacteria or yeast are encountered only scarcely (Beesch, 1953).

2.3 Novel approaches toward biobutanol production

In the past industrial applications, batch fermentation was a usual way how to produce biobutanol due to arrangement simplicity and attaining maximum biobutanol concentration, given by the used strain and cultivation medium, at the end of fermentation. Fed-batch fermentation can be regarded as modification of the batch process offering slight productivity increase by reduction of lag growth phase. However, taking into account possible industrial scale of the process, the preferential process arrangement is continuous ABE fermentation due to a lack of so called "dead" operation times. Nevertheless, its accomplishment in single bioreactor e.g. as chemostat is not usually easy because of biphasic process character when butanol production is not connected with growth directly (see Fig. 1). Theoretically, clostridial culture behaviour under chemostat cultivation conditions should follow an oscillation curve when acidogenesis is coupled with cell multiplication and decrease of substrate concentration. On the contrary, solventogenesis is coupled with decrease of specific growth rate due to sporulation what leads to cells wash-out and increase of substrate concentration in the medium. These two states should cycle regularly (Clarke et al., 1988) but in practice, irregular cycling with various depths of individual amplitudes is more probable as demonstrated several times (S.M. Lee et al., 2008). Moreover, chemostat cultivation conditions induce selection pressure on the microbial culture favouring non-sporulating, quickly multiplying cells what may cause culture degeneration i.e. the loss of the culture ability to produce solvents (Ezeji et al., 2005).

However, there are other options, tested in laboratory scale, how to arrange continuous ABE fermentation like multi-stage process splitting clostridial life cycle into at least two vessels, where first smaller bioreactor serves mainly for cells multiplication under higher dilution rate and in the second bigger bioreactor, actual solventogenesis takes place (Bahl et al.,

1982). In addition, battery of bioreactors working in batch, fed-batch or semi-continuous regime ensuring continuous butanol output can also be considered continuous fermentation (Ni & Sun, 2009; Zverlov et al., 2006).
ABE fermentation in any regime can be combined with cells immobilization performed by different methods - entrapment in alginate (Largier et al., 1985), use of membrane bioreactor (Pierrot et al., 1986) or cells adsorption on porous material (S.Y. Lee et al., 2008; Napoli et al., 2010). Recently, final report of the US DOE grant (Ramey & Yang, 2004) has revealed a novel approach toward ABE fermentation. The principle of this solution is two step butanol production employing two microorganisms; at first *Clostridium tyrobutyricum* produces mainly butyric acid which is consumed by second microorganism *Clostridium acetobutylicum* and utilized for butanol production. The authors claimed they reached 50% yield of butyric acid in the first phase and 84% yield of butanol from butyrate. However, a pilot and a production plant planned for year 2005 have not been realized, yet. Nevertheless, this way of butanol production is still under research in U.S.A. (Hanno et al., 2010), focusing mainly on solventogenic clostridia that are capable of butyrate utilization for butanol production.
One of the main constraints of biotechnological butanol production is its low final concentration in fermented cultivation media caused by its severe toxicity toward producing cells. Average butanol concentration, stable reached in Germiston plant in South Africa, was 13 $g.L^{-1}$ (Westhuizen et al., 1982). Although higher butanol concentration (about 20 $g.L^{-1}$) can be attained using e.g. mutant strain *C.beijerinckii* BA101 (Qureshi & Blaschek, 2001a) cost of distillation separation is still high. Therefore efficient preconcentration methods applied either after the fermentation or more often during the fermentation are being searched now. Moreover, if such separation method is integrated with fermentation process it will increase amount of utilized substrate by alleviating product toxicity. Preferential separation methods in this context seem to be gas stripping (Ezeji et al., 2003), adsorption on zeolites or pervaporation (Oudshoorn et al., 2009).

3. Experience with biobutanol fermentation in ICT Prague

Most of work was performed with the strain *Clostridium pasteurianum* NRRL B-592 which differed from usually employed solvent producing clostridia significantly, especially in sooner onset of solvents production i.e. during exponential growth phase. The strain was also chosen because of its properties i.e. stable growth and solvents production, robustness regarding minor changes in cultivation conditions and resistance toward so-called strain degeneration. Nevertheless in some cases, other, more typical solventogenic strains, *C.acetobutylicum* DSM 1731 and *C.beijerinckii* CCM 6182 were used, too.
Compositions of cultivation media, strains maintenance, description of cultivation, used analytical methods and expressions describing calculation of fermentation parameters i.e. yield and productivity for batch, fed-batch and continuous fermentations are given in Patakova et al., (2009 and 2011a).

3.1 Methods of ABE study

Despite complex process character, fermentation control, which is of key importance, relies only on few on-line measurable values like pH or redox potential of the medium and off-line determined concentrations of substrate(s), biomass and metabolites. In order to understand the process better and to improve fermentation control, fluorescence labelling of selected traits together with microscopy and flow cytometry was applied. Flow cytometry, as high-throughput, multi-parametric technique capable of analysis of heterogenic populations at the level of individual cells, has recently been used for description of clostridial butanol fermentations for the first time, but in totally different context (Tracy et al., 2008).

3.1.1 Use of fluorescent alternative of Gram staining for discrimination of acidogenic and solventogenic clostridial cells

The detailed description of the method development, particular application conditions and its use were published by Linhova et al., (2010a). The main idea of the staining is based on fact that clostridia are usually stained according to Gram as G^+ after germination from spores (motile, juvenile cells) and as G^- when the cells started to sporulate. The change in Gram staining response corresponds to metabolic switch from acids to solvents formation and also with an alteration in a cell membrane composition i.e. thinning of peptidoglycan layer (Beveridge, 1990). Therefore the cells of *C.pasteurianum* were labelled with a combination of fluorescent probes, hexidium iodide (HI) and SYTO 13 that can be considered a fluorescent alternative of Gram staining. Cells of *C.pasteurianum* forming mainly acids fluoresced bright orange-red as G^+ bacteria and the solvent producing, sporulating cells exhibited green-yellow fluorescence as G^- bacteria (see Fig.2). The red colour of labelled young cells was a result of a fact that green fluorescence of SYTO13 was quenched by that of HI while bright green-yellow colour of sporulating and/or old cells was caused by staining only by SYTO13 when HI did not permeate across the cell wall. Jones et al., (2008) used different combination of dyes (propidium iodide and SYTO 9) for labelling *C.acetobutylicum* ATCC 824 during time course of batch cultivation but attained the same conclusion.

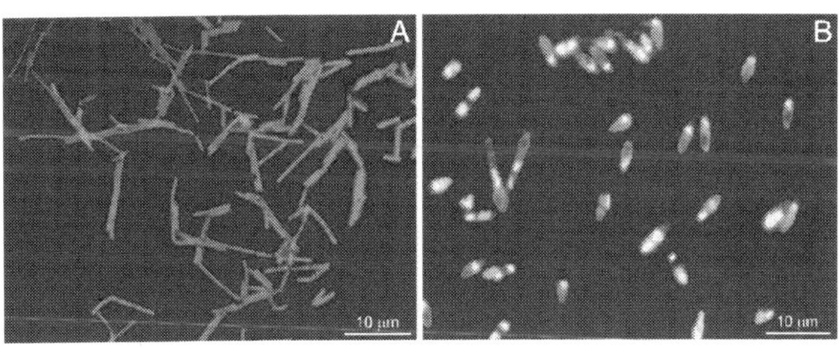

Fig. 2. *C.pasteurianum* cells stained with hexidium iodide and SYTO 13 in acidogenic (A) and solventogenic (B) metabolic phases

Then, flow cytometry enabling quantification of fluorescent intensities of labelled clostridial populations was used for monitoring of physiological changes during fed-batch cultivation (Linhova et al., 2010a). For flow cytometry measurement, the cells were stained only by HI and the signal of fluorescent intensity acquired in a channel FL3 (red colour) was related to forward scatter signal (FSC) which corresponded to cell size in order to gain data independent on cell size. The data measured for *C.pasteurianum* were compared with those for typical G^+ and G^- bacteria i.e. for *Bacillus megatherium* and *Escherichia coli* and there was a striking difference between the values of FL3/FSC for *C.pasteurianum* on one hand and those for *B.megatherium* and *E.coli* on the other hand. While the values for *B.megatherium* (G^+) and *E.coli* (G^-) oscillated ±0.1 and ±0.2, respectively, in time course of 32 h in which they were sampled, the values for *C.pasteurianum* dropped from 3.1 to 0.8 during the cultivation. It was also evident that acidogenic phase had a very short duration and both metabolic phases overlapped. Further experiments are necessary to assess unambiguously the acquired data, however it is tempting to hypothesize that *C.pasteurianum* NRRL B-598 has a different pattern of acids and solvents formation when solvents production is connected rather with exponential growth phase than the well-known solventogenic strain *C.acetobutylicum* ATCC 824 in which solvents production is generally assembled with stationary growth phase.

3.1.2 Use of flow cytometry for viability determination of clostridia

As to perform ABE fermentation means to handle clostridial population in different stages of the life cycle (see Fig. 1), determination of share of metabolically active i.e. vital cells in the population, is very important. Based on testing of various fluorescent viability probes with different principles of functioning, bisoxonol (BOX) was chosen as a convenient dye for *C.pasteurianum* viability determination (Linhova et al., 2010b). BOX stains depolarized cells with destroyed membrane potential i.e. nonviable cells. When the cells were fixed by 5 min boiling, whole population was labelled (Fig.3b) but in case of growing population (Fig.3a) most of cells remained non-stained. After optimization of staining conditions, flow cytometry was used for determination of culture viability (see Fig.4).

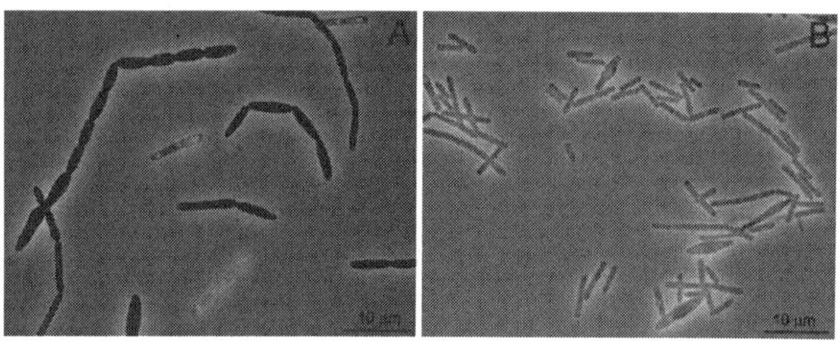

Fig. 3. BOX stained viable (A) and fixed i.e. nonviable (B) cells of *C.pasteurianum*

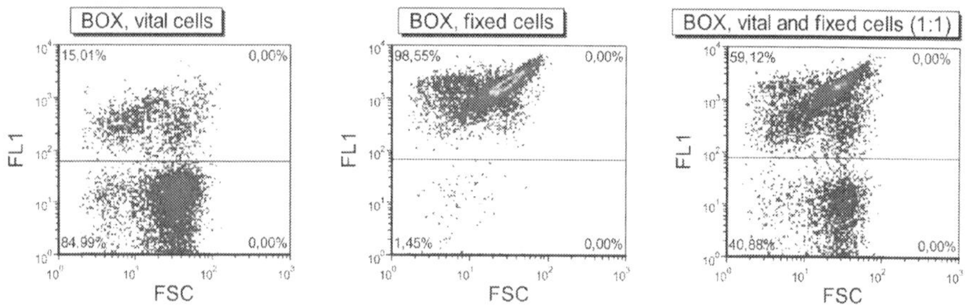

Population of viable cells in the left dot-plot diagram can be seen under the gate (in lower half of the diagram). In upper half of the left diagram, there are rests of cells after spores germination and sporulating cells, the share of which does not exceed 15%.

Fig. 4. Dot-plot diagrams after BOX labelling of *C.pasteurianum* populations of live (1), fixed (2) and mixture of live and fixed cells (3)

Then the method was used for viability determination during batch cultivation (see Fig.5). Bioreactor was inoculated with spore suspension after heat shock that induced spores to grow and killed present vegetative cells. After the heat shock, the viability at the beginning of the fermentation was very low (Fig. 5B). In the exponential growth phase viability increased to ~78%, as expected. With glucose depletion (Fig. 5B) and reaching the highest concentration of 1-butanol (7.5 g.L^{-1} see Fig.5A), the viability began to decrease. Relatively rapid viability decline at nutrient depletion conditions has already been observed by Novo *et al.*, (1999) and Jepras *et al.*, (1995) for *S. aureus*, *E. coli* and *P. aeruginosa*. They observed membrane potential decreased within a few minutes after removal of energy resources. Moreover, in our case, elevated 1-butanol concentration contributed to viability decline, too

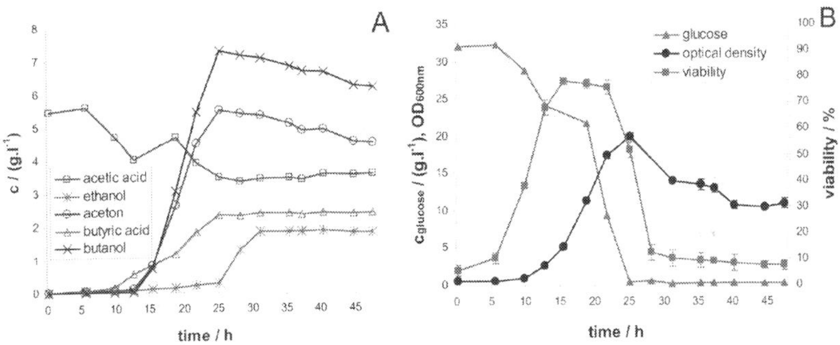

Fig. 5. Comparison of viability determination with fermentation data during batch cultivation of *C.pasteurianum*

3.2 Feedstock comparison

Based on screening in flask fermentations performed in an anaerobic chamber, every feedstock (sugar beet, corn and glucose) was matched with appropriate *Clostridium* strain regarding to yield and productivity values. Sugar beet is a crop grown in the Czech Republic for the last 160 years which provides high yields and can be used in the non food field for the biofuel production. In fact, non-food utilization of sugar beet is already running in CR but only bioethanol is produced in this way in Agroethanol TTD. Regarding corn, its main portion is grown for cattle feeding in CR but at the same time, the size of cattle herds diminishes every year. As the important goals of the biofuels production are, beside others, also the support of farmers and maintenance of arable land areas, corn can be seen as an energetic crop, too. Glucose was taken as feedstock on assumption glucose cultivation medium can be seen as a very simple model of lignocellulosic material hydrolyzate the use of which is supposed in future.

A comparison of butanol production using corn, sugar beet juice and glucose together with relevant strains is provided in Table 1 and sugar beet seems to be the preferable option according to the presented parameters. It is also noteworthy to look at fermentation courses in all compared cases. Fermentation of corn by *C.acetobutylicum* was running with textbook-like biphasic behaviour, when at first acids were formed and in the second solventogenic phase coupled with sporulation a reutilization of acids occurred. However, both fermentation of sugar beet juice by *C.beijerinckii* and fermentation of glucose by

C.pasteurianum differed from this "typical" course by start of butanol formation during exponential growth phase (both cases) and almost no reutilization of acids (*C.pasteurianum*).

species	substrate	B ($g.L^{-1}$)	ABE ($g.L^{-1}$)	$Y_{ABE/S}$ (%)	$Y_{B/S}$ (%)	P_{ABE} ($g.L^{-1}.h^{-1}$)
C.beijerinckii	sugar beet juice	11.6	16.2	37	26	0.40
C.acetobutylicum	corn	9.6	14.4	27	18	0.20
C.pasteurianum	glucose	7.3	11.8	35	18	0.23

Abbreviations B, ABE, YABE/S, YB/S, PABE stand for butanol, total solvents amount, yield of total solvents, yield of butanol and productivity of solvents formation.

Table 1. Comparison of bioreactor cultivations using different substrates and strains

Overall balances of mentioned fermentation courses can be expressed in form of equations (1-3). Similar expression of products in numbers has already been published (see Equation 4) by Jones & Woods (1986) where this equation reflected average results achieved with *C.acetobutylicum* and *C.beijerinckii* strains published in literature till 1986. In the equations (1-4), $C_{12}H_{22}O_{11}$, $C_6H_{12}O_6$, $C_4H_{10}O$, C_3H_6O, C_2H_6O, $C_2H_4O_2$, $C_4H_8O_2$ stand for saccharose, glucose, butanol, acetone, ethanol, butyric acid and acetic acid, respectively.
Butanol production from saccharose by *C.beijerinckii*:

$$1.00C_{12}H_{22}O_{11} \rightarrow 1.22\,C_4H_{10}O + 0.60C_3H_6O + 0.04C_2H_6O + 0.25C_2H_4O_2 + 0.20C_4H_8O_2 + 1.60CO_2 + 0.80H_2 \quad (1)$$

Butanol production from corn (expressed as glucose) by *C.acetobutylicum:*

$$1.00\,C_6H_{12}O_6 \rightarrow 0.42\,C_4H_{10}O + 0.21\,C_3H_6O + 0.04\,C_2H_6O + 0.12C_2H_4O_2 + 0.06C_4H_8O_2 + 0.58CO_2 + 0.36H_2 \quad (2)$$

Butanol production from glucose by *C.pasteurianum*:

$$1.00\,C_6H_{12}O_6 \rightarrow 0.54\,C_4H_{10}O + 0.40\,C_3H_6O + 0.02\,C_2H_6O + 0.19C_2H_4O_2 + 0.06C_4H_8O_2 + 6.77CO_2 + 3.98\,H_2 \quad (3)$$

Butanol production (Jones & Woods 1986):

$$1.00\,C_6H_{12}O_6 \rightarrow 0.56\,C_4H_{10}O + 0.22\,C_3H_6O + 0.07\,C_2H_6O + 0.14\,C_2H_4O_2 + 0.04\,C_4H_8O_2 + 2.21\,CO_2 + 1.35\,H_2 \quad (4)$$

Ratio of 1-butanol per unit of sugar (hexose) was the highest for saccharose (0.61) and the lowest for starch (0.42) but it can be stated the results were similar as presented by Jones and Woods (1986). The only exception was case of *C.pasteurianum*, in which remarkable amounts of carbon dioxide and hydrogen were produced not only in acidogenesis but throughout the whole fermentation period. Other experiences with the mentioned raw materials and also possible alternation of expensive but usual cultivation medium supplements, yeast extract or yeast autolysate, with cheap waste product of milk industry, whey protein concentrate, is presented in Patakova et al., (2009). Detailed description of the use of sugar beet juice as fermentation substrate for biobutanol production has been published, recently (Patakova et al., 2011b).

3.3 Influence of fermentation arrangement on ABE fermentation

An overview of batch, fed-batch and two variants of continuous bioreactor fermentation experiments using glucose cultivation medium and the strain *C.pasteurianum* NRRL B-598 is presented in Table 2. Both batch and fed-batch cultivations were operated about 50h and a ratio of produced solvents (B:A:E) was about 2:1:0.1 in all cases. Batch cultivations were performed in media with initial glucose concentration 40 $g.L^{-1}$ and if usual total solvents yields referred in literature are about 30% (Ezeji et al., 2005; Shaheen et al., 2000) then similar solvents concentrations like those shown in Table 2 were usually obtained. Therefore, higher initial glucose concentrations (60 and 80 $g.L^{-1}$) were tested in flasks cultivations, however solvents concentrations remained either at the same level (for 60 $g.L^{-1}$ glucose) or they were lower (for 80 $g.L^{-1}$ glucose) in comparison with use of glucose concentration 40 $g.L^{-1}$ and significant portion of glucose stayed in media unconsumed what might indicate a phenomenon of substrate inhibition.

Consequently, fed-batch cultivations were employed (see Table 2) in which butanol and total ABE concentrations were moderately increased (about 10%) and lag growth phase was reduced to 50% i.e. 3 h (data not shown). Nevertheless, yield and productivity for both 1-butanol and total solvents remained almost the same as in case of batch cultivations. The reached maximal butanol concentration (8.3 $g.L^{-1}$) is probably near the highest value tolerated by the used strain and a substantial improvement in an overall amount of produced butanol could be attained only by an integration of the cultivation with some on-line separation step.

cultivation	B ($g.L^{-1}$)	ABE ($g.L^{-1}$)	$Y_{ABE/S}$ (%)	B/A ratio	P_{ABE} ($g.L^{-1}.h^{-1}$)	D (h^{-1})
batch	7.3	11.8	35	2.0	0.23	-
fed-batch	8.3	12.3	23	2.2	0.25	-
continuous[a]	4.4	6.2	24	3.4	0.15	0.03
continuous[b]	4.0	5.9	20	1.8	0.20	0.07

Abbreviations B, ABE, $Y_{ABE/S}$, $Y_{B/S}$, P_{ABE} and D stand for butanol, total solvents amount, yield of total solvents, yield of butanol, productivity of solvents formation and dilution rate. Continuous cultivation proceeded as glucose-limited[a] or glucose non-limited[b] experiments; values of yield and productivity were calculated in pseudo steady state. For detailed conditions of continuous fermentations see Patakova et al., 2011a.

Table 2. Parameters of batch, fed-batch and continuous fermentations using *C.pasteurianum*

Surprisingly, glucose-limited fermentation experiment showed superior results in comparison with glucose non-limited fermentation (see Table 2). The only exception was solvents productivity that was higher at the expense of unused substrate. In glucose non-limited continuous experiment, mutually adverse oscillations of butanol and glucose concentrations occurred unlike butanol concentration near constant value (pseudo steady state) achieved in glucose-limited fermentation. The glucose limitation is also believed to support long-term stability and to reduce strain degeneration (Fick et al., 1985). Fermentation courses in both cases were presented in Patakova et al., 2011a.

3.4 Strains comparison

Course of fermentations carried out with *C.acetobutylicum* DSM 1731 and milled corn as substrate was similar to that referred for *C.acetobutylicum* ATCC 824 (Lee S.Y. et al., 2008) i.e. it was characterized by distinct metabolic phases, reutilization of acids during solventogenesis and development of hydrogen that peaked during acidogenesis. According to Johnson et al., (1997), *C.acetobutylicum* DSM 1731 showed 96% DNA sequence similarity with *C.acetobutylicum* ATCC 824. The so-called acid crash i.e. the state when the fermentation finished in acidogenic step was sometimes observed from unclear reason, using this strain and milled corn as substrate (Rychtera et al., 2010). Unfortunately, intracellular level of formic acid was not determined and therefore it was not proved or disproved whether acid crash in these cases was also caused by formic acid (Wang et al., 2011).

The strain *C.beijerinckii* CCM 6218 should be identical with the strain *C.beijerinckii* ATCC 17795 according to data of Czech Collection of Microorganisms. Surprisingly, if the strain *C.beijerinckii* ATCC 17795 was tested for butanol production using molasses cultivation medium (Shaheen et al., 2000), both yield and maximum butanol production was low, 10% and 6.1 g.L^{-1}, respectively. In addition this strain together with *C.pasteurianum* NRRL B-598 showed different fermentation pattern in comparison with *C.pasteurianum* NRRL B-598 and butanol production initiation started during exponential growth phase. The strain also metabolized substrate, saccharose, faster than both other tested strains what was reflected in higher productivity of butanol.

The strain *Clostridium pasteurianum* NRRL B-598 used in this study differed significantly in some physiological traits from both the species characteristics published in Bergey`s Manual of Systematic Bacteriology (Rainey et al., 2009). Although strains of the species *C. pasteurianum* are known rather as acetic and butyric acids or hydrogen producers (Rainey et al., 2009; Heyndrickx et al., 1991), the strain *C.pasteurianum* NRRL B-598 was cited in US Patent No 4539293 as butanol producing when used in mixture with further acidogenic strain e.g. *C.butylicum*. Unfortunately precise cultivation conditions, yields, solvents concentrations and other data are not available in the mentioned patent.

3.5 Separation of biobutanol from fermentation medium by gas stripping

Gas stripping by nitrogen as a method potentially enabling both butanol preconcentration before final distillation and a way how to mitigate butanol toxicity during fermentation was studied separately from fermentation and stripping coefficient β, defined by equation (5) was chosen as main criterion for stripping efficiency:

$$\beta = (-1/P_L)(dP_L/dt) \qquad (5)$$

where P_L is butanol concentration in a liquid phase.

Gas stripping of solvents from fermentation media is however only the first step towards isolation/concentration of products (ABE), further steps consist in product change of state from the gas into liquid phases. There are several ways how to carry out this change of state but there are scarcely discussed. Two of them i.e. application of low temperature (- 4 °C in condenser) and adsorption on charcoal followed by desorption by steam were tested. If low temperature was used for butanol conversion from the gas to liquid, average achieved preconcentration lay in the interval from 7 to 9. However, when the method of freezing was used then only 60% of solvents were captured in one gas cycle (probably due to insufficiency of freezing unit capacity) while at charcoal adsorption, 90% of solvents was captured. This affected stripping efficiency which was lowering gradually at freezing. On the contrary, main disadvantage of charcoal use was a gain of more diluted butanol solution (preconcentration from 2 to 4) after its displacement from charcoal by steam. Energy balance must be done for this process but it needs measurement in pilot scale.

	Model solution ABE				Medium after fermentation			
	Initial conc. (g.L^{-1})	Final conc. (g.L^{-1})	Mean rate of stripping (g.L^{-1}.h^{-1}) (for 24 h)	Stripping coefficient (h^{-1}) (for 24 h)	Initial conc. (g.L^{-1})	Final conc. (g.L^{-1})	Mean rate of stripping (g.L^{-1}.h^{-1}) (for 24 h)	Stripping coefficient (h^{-1}) (for 24 h)
A	3.9	2.6	0.05	0.017	4.8	2.6	0.09	0.025
B	9.2	3.2	0.25	0.044	10.2	2.9	0.30	0.052
E	1.4	0.7	0.03	0.029	0.7	0.5	NA	NA

The profound influence of solution composition on stripping efficiency is shown in Table 3, where comparison of model (water) solution of solvents with medium after fermentation is provided. In this case, the stripping was carried out directly in the bioreactor (liquid volume 3L) using aeration ring as nitrogen distributor (flow rate 2 VVM). Schemes of stripping arrangements are provided in (Fribert et al., 2010).

Table 3. Comparison of butanol stripping from model solution and cultivation medium after fermentation

Nevertheless, if summarized it can be stated that the mean rate of stripping for butanol and butanol preconcentrations achieved after application of freezing corresponded with already published values (Ezeji et al., 2003; Ezeji et al., 2005; Qureshi & Blaschek, 2001b). The mean butanol stripping rate exceeded the butanol productivity what indicated a potential successful integration of gas stripping with fermentation into one process.

4. The use of biobutanol in road transport

4.1 Perspectives of biobutanol use in road transport

The preferred use of biobutanol is the production of motor fuels for spark ignition engines by mixing with conventional gasoline; therefore biobutanol could become an option to bioethanol due to better potential in terms of its physico-chemical properties. Biobutanol concentration in fuel can reach up to 30% v/v without the need for engine modification. Since the butanol fuel contains oxygen atoms, the stoichiometric air/fuel ratio is smaller than for gasoline and more fuel could be injected to increase the engine power for the same amount of air induced. The oxygen content is supposed to improve combustion, therefore lower CO and HC emissions can be expected. Biobutanol and its mixtures can be used directly in the current gasoline supply system, such as transportation tanks and re-fuelling infrastructure. Biobutanol can be blended with gasoline without additional large-scale supply infrastructure, which is a big benefit as opposed to the bioethanol use. Finally biobutanol is non-poisonous and non-corrosive and it is easily biodegradable and does not cause risk of soil and water pollution.

4.2 Physico-chemical properties of biobutanol-gasoline blends

If compared to ethanol, biobutanol exhibits important advantages upon blending with gasoline. The mixtures have better phase stability in presence of water, low-temperature properties, oxidation stability during long-term storage, distillation characteristics and volatility with respect to possible air pollution.

The overview of the selected properties of butanol as compared to bioethanol and automotive gasoline meeting requirements of EN 228 is given in Table 4.

Parameter	Bioethanol	Biobutanol	Gasoline 95
Boiling Point (°C)	78.3	117.7	30-215
Density ($kg.m^{-3}$)	794	809	720-750
Kinematic Viscosity ($mm^2.s^{-1}$)	1.5	3.6	0.4-0.8
Lower Heating Value ($MJ.kg^{-1}$)	28.9	33.1	44.4
Heat of vaporisation ($MJ.kg^{-1}$)	0.92	0.71	0.32
Research Octane Number RON	106-130	94	95
Motor Octane Number MON	89-103	80	85
Reid Vapour Pressure (kPa)	17	2.3	45-90
Stoichiometric air/fuel ratio	9	11.1	14.8
Oxygen Content (% w/w)	34.7	21.6	max 2.7

Table 4. Physico-chemical properties of alcohols and automotive gasoline (Wolf, 2007)

Due to the fact that oxygen content in biobutanol is lower than in ethanol, biobutanol can be added to the gasoline in higher concentrations with respect to EN 228 limit for the oxygen content in gasoline. Higher biobutanol content in gasoline does not require engine modification. The heating value (energy density) of biobutanol is close to that of gasoline, which has a positive effect on the fuel consumption.

Biobutanol has a slightly higher density compared to gasoline but the increase in density of biobutanol/gasoline mixtures is so small that it does not cause problems with fulfilling limits for automotive gasoline containing up to 30% v/v biobutanol. Viscosity of biobutanol is significantly higher compared to the gasoline, which may affect engine fuel injection system at lower temperatures due to higher resistance to flow. However, this impact could be negligible for the blends with gasoline containing up to 30% v/v biobutanol.

4.2.1 Volatility of biobutanol-gasoline blends

The vapour pressure of biobutanol and bioethanol is very low compared to gasoline. A disadvantage of the bioethanol use is a formation of volatile azeotropic mixtures of ethanol and hydrocarbons present in the gasoline which causes the increase in the vapour pressure of gasoline in the range of 6 - 8 kPa (Mužíková et al., 2009). The formation of azeotropes occurs in the concentration up to 10% v/v of biobutanol in gasoline but the highest increase in the vapour pressure is as low as 0.5 kPa at 5% v/v of biobutanol in gasoline. At higher biobutanol concentrations, another volatile and/or oxygen compound has to be added to compensate vapour pressure decrease and to keep good engine startability. The formation of azeotropes is also associated with decrease of the boiling points of the blends. While the addition of bioethanol influences negatively the distillation curve profile, biobutanol has minor effect on the distillation curve.

Because of the use of different gasolines in several European Union countries, the mixing of different oxygen compounds in the vehicle tank can occur, causing the simultaneous presence of ethers like MTBE and ETBE and other alcohols, especially ethanol, in combination with biobutanol in the gasoline. Ethers do not cause problems since their properties are close to hydrocarbons. They influence the vapour pressure of the butanol-gasoline blend proportionally according to the initial vapour pressure of pure components. On the contrary, bioethanol forms azeotropes, which can unpredictably change the vapour pressure of the mixture. The increase of vapour pressure depends on the final ethanol concentration in the mixture.

Biobutanol has significantly higher heat of vaporization than gasoline, which reduces the temperature of the air/fuel mixture and results in higher engine volumetric efficiency. At the same time it leads to lower compression temperature and longer ignition delay, which in turn may decrease the engine performance. The low vapour pressure and higher heat of vaporization is experienced to have a negative effect on the startability and cold start engine performance because of difficult fuel vaporization at low ambient temperatures (Xiaolong et al., 2009).

4.2.2 Phase stability in the presence of water

The water - fuel miscibility is very important factor for distribution of fuel blends. The content of small amounts of free water in the fuel is connected with the risk of corrosion problems, whereas larger amounts of water can impair fuel supply to the engine. Hydrocarbons in gasoline are very slightly miscible with water as opposed to alcohols. The solubility of water in petroleum gasoline is only 100 $mg.kg^{-1}$, while bioethanol is completely miscible with water and solubility of water in biobutanol is 19.7% w/w. Bioethanol is very hygroscopic and its blends with gasoline are partially miscible with water depending on temperature and the ethanol concentration in the blend. Phase separation of water with bioethanol can occur at lower temperatures, which causes formation of the heterogeneous system composed of the hydrocarbon phase and water-ethanol phase. This fact is the reason why the bioethanol-gasoline blends cannot be distributed via common pipelines but only separately using tankers. Contrary to bioethanol, the ability of biobutanol to absorb significant amount of water is very low (Peng et al., 1996). Biobutanol has high affinity to hydrocarbons and the risk of potential phase separation is therefore minimized. Moreover, biobutanol remains in the hydrocarbon phase if the phase separation occurs. Biobutanol is not hygroscopic that is an important factor for the long-term storage of fuels. Accordingly, high stability of gasoline-biobutanol mixtures in the presence of water comparing with bioethanol was reported.

Ethers MTBE and/or ETBE can be added to the gasoline on purpose for increasing the octane number and oxygen content or they can be accidentally mixed in the fuel tank due to another type of gasoline fuelling. The presence of MTBE and ETBE slightly increases the miscibility of butanol-gasoline blend with water and decreases the temperature of the phase separation. Ethanol has the same behaviour, nevertheless due to the bioethanol ability to absorb humidity its presence in the blends is rather unfavourable.

4.2.3 Material compatibility of biobutanol and its mixtures

Biobutanol is not as aggressive as the bioethanol with regard to the engine construction materials, sealants, and plastics. The fuel with 20% v/v of biobutanol has similar properties to the hydrocarbons in terms of swelling of polymers (Wolf, 2007).

The oxidation stability of biobutanol-gasoline blends may be compromised by potential impurities from biobutanol production (acetic and butyric acid, acetaldehyde and lower alcohols). The impurities in concentrations of 0.1% v/v in 10% v/v biobutanol-gasoline blends (which corresponds to 1% of impurities in biobutanol) can decrease the fuel oxidation stability by about 15%, therefore the purification with regard to the removal of fermentation by-products is very important step in biobutanol production.

The high boiling point of butanol may negatively influence its evaporation from engine oil after oil contamination caused by frequent cold starts. This phenomenon can occur especially at low ambient temperatures, when the fuel leaks into the engine oil through piston rings. In a normal engine operation biobutanol evaporates after the engine warm up and the motor oil additives are re-solved. However, the solubility of oil additives may be at risk in case of frequent cold starts and short routes in cold winter conditions.

4.3 Emission characteristics of butanol/gasoline blends in spark ignition engines

Besides the renewability of raw materials used for their production, alcohol fuels are reported to be advantageous over petroleum derived ones thanks to their better environmental characteristics. The oxygen contained in alcohol molecules is supposed to affect combustion process and cause soot and particulate reduction; some studies show that there is the potential for reduction of NO_x emissions. While there was much information collected about the use and combustion behaviour of lower-molecular weight alcohols, such as methanol and ethanol, substantially less effort was yet put to the research of the properties of butanol (especially n-butanol as a product of fermentation during ABE process) upon their use in internal combustion engines.

For the evaluation of emission characteristics, it is very important to study combustion processes at different air/fuel ratio and thermodynamic conditions. The combustion of neat butanol as well as its mixtures with other fuels or chemicals was studied (Agathou & Kyritis, 2011; Broustail et al., 2011; Dagaut & Togbé, 2008; Sarathy et al., 2009) to obtain combustion velocities and kinetic data for modelling processes of butanol oxidation at the conditions of engine cylinder. However, it must be noted that real-world emissions level is affected by the interaction between fuel itself and the engine used, mainly its fuel injection system and engine control unit together with emission control systems - catalytic converters, particulate filters, exhaust recirculation etc.

Although butanol properties (boiling point, viscosity, octane number) predetermine it for the use in spark ignition engines as a partial substitute for conventional gasoline, a number of studies were carried out using butanol/diesel fuel mixtures in compression ignition engines. The addition of butanol (or other alcohols) significantly increases volatility and decrease lubricity of diesel fuel, which requires additional measurements for their use in today's diesel engines. Yao (Yao et al., 2010) studied emission characteristics of CI engine using diesel fuel containing 0 % to 15 % v/v n-butanol. By varying exhaust gas recirculation rates, they kept NO_x emissions constant, while CO and PM (particulate matter) emissions significantly decreased with the concentration of n-butanol in the fuel. Rakopoulos et al. (Rakopoulos et al., 2010a) compared conventional diesel fuel, diesel fuel with 30% biodiesel (FAME), and biodiesel with 25 % n-butanol in turbo-charged CI truck engine; the experiments were focused on transient regimes causing temporary increase of pollutant emissions. Both FAME and butanol helped to improve the particulate emissions in the transient engine regimes, but in both cases the emissions of NO_x increased. In stationary regimes at different engine speed and load, the authors (Rakopoulos et al., 2010b) determined emissions of all regulated pollutants. In all cases, the positive effect of butanol in diesel fuel was found on the emissions of particulates, NO_x, and carbon dioxide, whereas hydrocarbon emissions slightly increased.

Much greater potential and possibility of utilization without necessity to solve technical problems has butanol used as a partial substitute of motor gasoline. The total miscibility with hydrocarbons, boiling point, flash point and other properties allow mixing butanol with gasoline in wide range of concentrations and combustion in common spark ignition engines. In comparison to other alcohols in the range of C_1 to C_5 mixed to gasoline in concentrations matching fuel oxygen content, butanol does not differ significantly in its effect on the emissions of regulated pollutants (Yacoub et al., 1998). The emissions of total hydrocarbons decrease, while significant increase takes place in the emissions of aldehydes, whose main constituent was formaldehyde.

One of the substantial drawbacks connected with the use of alcohols in SI engines is the problem of cold starts especially in winter conditions. Difficulties caused mainly by high heat of vaporization have to be eliminated by greater enrichment of air/fuel mixture in the period in which the engine heats up. This, on the other hand, can bring an increase in emissions of unburned or partially burned fuel due to near zero efficiency of catalytic converter in the early period after engine start. Irimescu (2010) modelled the situation for gasoline/butanol mixtures at different ambient temperatures and successfully verified the results with those obtained in experiments with a port injection engine.

The effect of butanol (or other alcohols) use in spark ignition engines depends also on the technique of fuel injection before its ignition in engine cylinder. Conventional way to prepare air/fuel mixture is the injection of fuel into the engine intake manifold, where it evaporates and the mixture is drawn to the cylinder in the suction cycle. Some engine manufacturers offer engines equipped with direct injection of fuel into the cylinder. Such engines allow the use of advanced techniques of emission control, such as lean (stratified) mixture combustion connected with the use of sequential injection. The direct injection engine was used by Cooney (Cooney et al., 2006), who investigated the effect of ethanol and butanol in blends with gasoline used in a series of engine tests conducted at varied loads. They reported the increase in engine efficiency at higher engine loads by a 4% with either 85 % n-butanol or 85 % ethanol. The efficiency is reported to be affected by lower octane number of n-butanol, even though knock combustion was not observed, and, on the other hand, by the higher flame speed of alcohols. Faster combustion can increase the efficiency if combustion timing was adjusted, while lower octane number should decrease it.

In contrast to modern engines of current passenger cars, there are still applications where carburetted engines or engines with open-loop control of fuel injection are used, without the ability to compensate for air-fuel ratio of specific fuels. In such cases, butanol blends result in approximately 50% enleanment connected with oxygen content in fuel, compared to ethanol. The authors evaluated the effect of the use of butanol-gasoline mixtures on pollutants emission of four different passenger cars equipped with spark ignition engines - from older Euro 2 vehicle to modern multipoint injection turbocharged one. As a baseline, unleaded gasoline with addition of 4 % ethanol was used. Mixtures containing butanol were prepared by addition of 10 %, 20 %, and 30 % pure synthetic n-butanol to the same gasoline. The properties of the mixtures were modified with small amounts of isooctane, toluene, and petroleum ether to keep their octane number and vapour pressure, which deteriorated by the addition of butanol. Four test vehicles manufactured by Skoda were used with different engine displacement, power, and technology level (see Table 5).

The emission tests were performed on a vehicle dynamometer according to ECE 83 emission test with the determination of CO, HC, and NO_x emissions during two driving cycles. In addition to the measurement of regulated emissions, samples were taken during both phases of ECE 83 test for determination of individual hydrocarbons and aldehydes. Basic engine parameters were monitored during the tests using an engine diagnostic unit to detect possible abnormal operation states of engine control unit.

Vehicle type	Year of manufacture	Engine displacement [cm^3]	Maximum power [kW]	Engine characteristics
Felicia Euro 2	1999	1289	50	Multi-point injection, four-cylinder
Fabia Euro4	2004	1198	47	Multi-point injection, three-cylinder
Octavia Euro4	2004	1781	110	Multi-point injection, 20V, five-cylinder, turbocharged

Table 5. Characteristics of vehicles used for emission tests

The addition of butanol to the fuels used caused only little change in regulated emissions (Fig. 6) measured in ECE 83 test. Although more significant changes were found in emission levels determined in individual ECE 83 test phases, with regard to regulated pollutants, total values show only the increase in NO_x emissions for all three vehicles. As expected, the use of butanol caused also small increase in emissions of aldehydes, whose main constituent was formaldehyde.

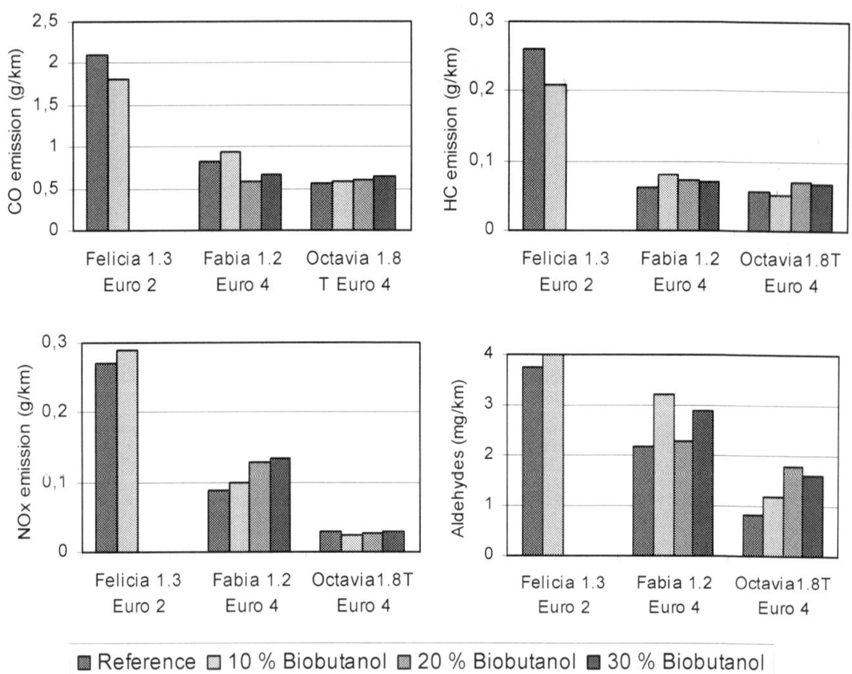

Fig. 6. Effect of butanol in gasoline fuel on emissions of regulated pollutants (CO, HC, NO_x) and aldehydes in ECE 83.03 emission test

5. Conclusion

The significance of the presented fermentation data lies in several fields:

- methodologically - fluorescence staining and flow cytometry proved to be very useful tools for nearly on-line evaluation of physiological state of clostridial population during the fermentation. Both method of discrimination of acidogenic/solventogenic status of individual cells based on fluorescence alternative to Gram staining and vitality staining by bisoxonol were never applied on bacteria of the genus *Clostridium*.
- the greatest attention was concentrated on the strain *C.pasteurianum* NRRL B-598 which was never studied before in such detail. The comparison of three types of fermentation arrangements, batch, fed-batch and continuous represents the unique set of data not usually available for the tested butanol producers. As the strain had somewhat distinct physiology from type *C.pasteurianum* strains and flow cytometry analysis displayed very short acidogenic metabolic phase and presumable overlapping of acidogenic and solventogenic phases, the strain itself and its behaviour is worth further investigation. Moreover, the strain can also be regarded the very promising hydrogen producer
- the best fermentation parameters, yield of ABE 37% and ABE productivity 0.40 $g.L^{-1}.h^{-1}$, were achieved using sugar beet juice as the feedstock and *C.beijerinckii* CCM 6182 as the microbial agent. In Europe and especially in the Czech Republic, the sugar beet has a potential to become significant source of sugar utilizable for non-food purposes. The abilities and the fermentation characteristics of the strain *C.beijerinckii* CCM 6182 (and neither its analog *C.beijerinckii* ATCC 17795) has not been studied intensively although the strain behaved like *C.pasteurianum* NRRL B-598 i.e. favourable butanol production kinetics consisting in onset of butanol formation during exponential growth phase was its typical feature.
- the preliminary experiments dealing with gas stripping as potential concentration and/or separation method for solvents from the fermented media confirmed feasibility of this solution under certain assumptions. The gas stripping must not affect adversely the fermentation and cost of the solvents transition from the gas into liquid phases must be minimized. However further ideally pilot experiments are necessary for full evaluation of gas stripping role in the butanol production.

With reference to the use of biobutanol as a fuel for transportation purposes, it can be concluded:

- in comparison with other bio-components used for blending automobile fuels, especially bioethanol, biobutanol exhibits very attractive properties - high energy content, low water solubility, total miscibility with gasoline hydrocarbons, and appropriate boiling point and vapour pressure
- the use of gasoline containing high concentrations (10 % to 30 % v/v) of butanol did not negatively affect operational parameters of common spark ignition engines used in passenger cars representing current European vehicle fleet. Only slightly increased emissions of NO_x emissions and production of aldehydes was found out during standard ECE 83 emission tests

6. Acknowledgement

This research could be performed thanks to financial support of projects No. QH81323/2008 of the Ministry of Agriculture of the Czech Republic, TIP No. FR-TI1/218 of the Ministry of

Industry and Trade of the Czech Republic, No. MSM6046137305 and No. MSM 6046137304 of the Ministry of Education, Youth and Sport of the Czech Republic.

7. References

Agathou, M.S. & and Kyritis, D.C. (2011) An experimental comparison of non-premixed bio-butanol flames with the corresponding flames of ethanol and methane, *Fuel*, Vol. 90, No. 1, pp. 255-262, ISSN 0016-2361

Bahl, H.; Andersch, W. & Gottschalk, G. (1982) Continuous production of acetone and butanol by *Clostridium acetobutylicum* in two stage phosphate limited chemostat, *European Journal of Applied Microbiology and Biotechnology*, Vol.15, No.4, pp. 201-205, ISSN 0171-1741

Beesch, S.C. (1953) Acetone-butanol fermentation of starches, *Applied Microbiology*, Vol. 1, No. 2, pp. 85-96, ISSN 0099-2240

Beveridge, T.J. (1990) Mechanism of Gram variability in select bacteria, *Journal of Bacteriology*, Vol. 172, No. 3, pp. 1609-1620, ISSN 0021-9193

Broustail, G.; Seers, P.; Halter, F.; Moreac, G. & Mounaim-Rousselle, C. (2011) Experimental determination of laminar burning velocity for butanol and ethanol iso-octane blends, *Fuel*, Vol. 90, No. 1, pp. 1-6, ISSN 0016-2361

Chiao, J-S. & Sun, Z-H. (2007) History of acetone-butanol-ethanol fermentation industry in China: Development of continuous production technology, *Journal of Molecular Microbiology and Biotechnology*, Vol. 13, No. 1-3, pp. 12-14, ISSN 1464-1801

Clarke, K.G.; Hansford, G.S. & Jones, D.T. (1988) Nature and significance of oscillatory behavior during solvent production by *Clostridium acetobutylicum* in continuous culture, *Biotechnology & Bioengineering*, Vol. 32, No. 4, pp. 538-544, ISSN 0006-3592

Cooney, C.; Wallner, T; McConell, C.; Gillen, J.C.; Abell, C. & Miers, C.A. (2006) Effect of blending gasoline with ethanol and butanol on engine efficiency and emissions using a direct-injection, spark-ignition engine, In: *Proceedings of the ASME internal combustion engine division 2009 spring technical conference ICES2009-76155*, Milwaukee, Wisconsin, USA, 2006

Cornillot, E.; Nair, R.; Papoutsakis, E.T. & Soucaille, P. (1997) The genes for butanol and acetone formation in *Clostridium acetobutylicum* ATCC 824 reside on a large plasmid whose loss leads to degeneration of the strain, *Journal of Bacteriology*, Vol. 179, No. 17, pp. 5442-5447, ISSN 0021-9193

Dagaut, P. & Togbé, C. (2008) Oxidation kinetics of butanol-gasoline surrogate mixtures in a jet-stirred reactor: Experimental and modeling study, *Fuel*, Vol. 87, No. 15-16, pp. 3313-3321, ISSN 0016-2361

Dyr, J. & Protiva, J. (1958) Formation of neutral solvents in continuous fermentation by means of *Clostridium acetobutylicum*. In: *Continuous Cultivation of Microorganisms A symposium*, Málek, I., pp. 210-226, Publishing House of the Czechoslovak Academy of Sciences, Prague 1958.

Ezeji, T.C.; Qureshi, N. & Blaschek, H.P (2003) Production of acetone, butanol and ethanol by *Clostridium beijerinckii* BA101 and *in situ* recovery by gas stripping, *World Journal of Microbiology and Biotechnology*, Vol. 19, No. 6, pp. 595-603., ISSN 0959-3993

Ezeji, T.C.; Qureshi, N. & Blaschek, H.P. (2005) Industrially relevant fermentations. In *Handbook on Clostridia*, Dürre, P., pp. 797-812, CRC Press, ISBN 0-8493-1618-9, Boca Raton.

Fick, M.; Pierrot, P. & Engasser, J.M. (1985) Optimal conditions for long-term stability of acetone-butanol production by continuous cultures of *Clostridium acetobutylicum*, *Biotechnology Letters*, Vol. 7, No. 7, pp. 503-508, ISSN 0141-5492

Fribert, P.; Patakova, P.; Jahodova, L.; Lipovsky, J.; Linhova, M.; Rychtera, M. & Melzoch, K. (2010) Recovery of acetone, butanol and ethanol from fermentation broth by gas stripping, *Proceedings of 19th International Congress of Chemical and Process Engineering*, ISBN 978-80-02-02210-7, Prague, Czech Republic, September 2010

Gheslaghi, R.; Scharer, J.M.; Moo-Young, M. & Chou, C.P. (2009) Metabolic pathway of clostridia for producing butanol, *Biotechnology Advances*, Vol. 27, No. 6, pp. 764-781, ISSN 0734-9750

Hanno, R.; Qureshi, N.; Cotta, M. & Largus A. (2010) Mixed community bioreactors to convert lignocellulosic feedstock into the liquid biofuel butanol, February 27, 2011, Available from:http://www.ars.usda.gov/research/publications/publications.htm?seq_no_115=248960

Hartmanis, M.G.; Ahlman, H. & Gatenbeck, S. (1986) Stability of solvent formation in *Clostridium acetobutylicum*, *Applied Microbiology and Biotechnology*, Vol. 23, No. 5, pp. 369-371, ISSN 0175-7598

Heyndrickx, M.; De Vos, P. & De Ley, J. (1991) Fermentation characteristics of *Clostridium pasteurianum* LMG 3285 grown on glucose and mannitol, *Journal of Applied Bacteriology*, Vol. 70, No. 1, pp. 52-58, ISSN 1364-5072

Irimescu, A. (2010) Study of cold start air-fuel mixture parameters for spark ignition engines fueled with gasoline-isobutanol blends, *International Communications in Heat and Mass Transfer*, Vol. 37, No. 9, pp. 1203-1207, ISSN 0735-1933

Jepras, R.I.; Carter, J.; Pearson, S.C.; Paul, F.E. & Wilkinson, M.J. (1995) Development of a robust flow cytometric assay for determining numbers of viable bacteria, *Applied and Environmental Microbiology*, Vol. 61, No. 7, pp. 2696-2701, ISSN 0099-2240

Johnson, J.L., Toth, J., Santiwatanakul, S. & Chen, J.-S. (1997) Culture of *Clostridium acetobutylicum* from various collections comprise *Clostridium acetobutylicum*, *Clostridium beijerinckii* and two other distinct types based on DNA-DNA reassociation, *International Journal of Systematic Bacteriology*, Vol. 47, No. 2, pp. 420-424, ISSN 0020-7713

Jones, D.T.& Woods, D.R. (1986). Acetone-butanol fermentation revisited, *Microbiological Rewievs*, Vol. 50, No. 4, pp. 484-524, ISSN 0146-0749

Jones, D.T.; Shirley, M.; Wu, X. & Keis, S. (2000) Bacteriophage infections in the industrial acetone butanol (AB) fermentation process, *Journal of Molecular Microbiology and Biotechnology*, Vol. 2, No. 1, pp. 21-26, ISSN 1464-1801

Jones, S. W.; Paredes, C. J.; Tracy, B.; Cheng, N.; Sillers, R.; Sanger, R. S.& Papoutsakis, E. T. (2008) The transcriptional program underlying the physiology of clostridial sporulation, *Genome Biology*, Vol. 9, No. 7, pp. R114-R114.21, ISSN 1465-6914

Kashket, E.R. & Cao, Z.-Y. (1995) Clostridial strain degeneration, *FEMS Microbiology Reviews*, Vol. 17, No. 3, pp. 307-315, ISSN 0168-6445

Kosaka, T.; Nakayama, S.; Nakaya, K.; Yoshino, S. & Furukawa, S. (2007) Characterization of the sol operon in butanol-hyperproducing *Clostridium saccharoperbutylacetonicum* strain N1-4 and its degeneration mechanism, *Bioscience, Biotechnology and Biochemistry*, Vol. 71, No. 1, pp. 58-68, ISSN 0916-8451

Largier, S.T.; Long, S.; Santangelo, J.D.; Jones, D.T. & Woods, D.R. (1985) Immobilized *Clostridium acetobutylicum* P262 mutants for solvent production, *Applied and Environmental Microbiology*, Vol. 50, No. 2, pp. 477-481, ISSN 0099-2240

Lee, S.M.; Cho, M.O.; Park, C.H.; Chung, Y-C.; Kim, J.H.; Sang, B-I. & Um, Y. (2008) Continuous butanol production using suspended and immobilized *Clostridium beijerinckii* NCIMB 8052 with supplementary butyrate, *Energy & Fuels*, Vol. 22, No. 5, pp. 3459-3464, ISSN 0887-0624

Lee, S.Y.; Park, J.H.; Jang, S.H.; Nielsen, L.K.; Kim, J. & Jung, K.S. (2008) Fermentative butanol production by clostridia, *Biotechnology and Bioengineering*, Vol. 101, No. 2, pp. 209-228, ISSN 0006-3592

Linhova, M.; Patakova, P.; Lipovsky, J.; Fribert, P.; Paulova, L.; Rychtera, M. & Melzoch, K. (2010a) Development of flow cytometry technique for detection of thinning of peptidoglycan layer as a result of solvent production by *Clostridium pasteurianum*, *Folia Microbiologica*, Vol. 55, No. 4, pp. 340-344, ISSN 1874-9356

Linhova, M.; Branska, B.; Patakova, P.; Lipovsky, J.; Fribert, P.; Rychtera, M. & Melzoch, K. (2010b) Viability determination by flow cytometry during ABE fermentation, *Proceedings of 19th International Congress of Chemical and Process Engineering*, ISBN 978-80-02-02210-7, Prague, Czech Republic, September 2010

Lipovsky, J.; Patakova, P.; Rychtera, M.; Cizkova, H. & Melzoch, K. (2009) Prospects of butanol production from starch and cellulose materials, *Chemicke Listy*, Vol. 103, No. 6, pp. 479-483, ISSN 1213-7103

Lopez-Contreras, A.; Gabor, K.; Martens, A A.; Renckens, B. A. M.; Claassen, P. A. M.; van der Oost, J. & de Vos, W. M. (2004) Substrate induced production and secretion of cellulases by *Clostridium acetobutylicum*, *Applied and Environmental Microbiology*, Vol. 70, No. 9, pp. 5238-43, ISSN 0099-2240

Maddox, I.S.; Steiner, E.; Hirsch, S.; Wessner, S.; Gutierrez, N.A. ; Gapes, J.R. & Schuster, K.C. (2000) The cause of "acid crash" and "acidogenic fermentations" during the batch acetone-butanol-ethanol (ABE-) fermentation process, *Journal of Molecular Microbiology and Biotechnology*, Vol. 2, No. 1, pp. 95-100, ISSN 1464-1801

Mužíková, Z.; Pospíšil, M. & Sebor, G. (2009) Volatility and phase stability of petrol blends with ethanol, *Fuel*, Vol.88, No.8, pp. 1351-1356, ISSN: 0016-2361

Napoli, F.; Olivieri, G.; Russo, M.E.; Marzocchella, A. & Salatino, P. (2010) Butanol production by *Clostridium acetobutylicum* in a continuous packed bed bioreactor, *Journal of Industrial Microbiology and Biotechnology*, Vol. 37, No. 6, pp. 603-608, ISSN 1367-5435

Ni, Y. & Sun, Z. (2009) Recent progress on industrial fermentative production of acetone-butanol-ethanol by *Clostridium acetobutylicum* in China, *Applied Microbiology and Biotechnology*, Vol. 83, No. 3, pp. 415-423, ISSN 0175-7598

Novo, D.; Perlmutter, N.G.; Hunt, R.H. & Shapiro, H.M. (1999) Accurate flow cytometric membrane potential measurement in bacteria using diethyloxacarbocyanine and a ratiometric technique, *Cytometry*, Vol. 5, No. 1, pp. 55-63, ISSN 1552-4922

Oudshoorn, A.; Van der Wielen, L.A.M. & Straathof, A.J.J. (2009) Assessment of options for selective 1-butanol recovery from aqueous solution, *Industrial & Engineering Chemical Research*, Vol. 48, No. 15, pp. 7325-7336, ISSN 0888-5885

Patakova, P.; Lipovsky, J.; Cizkova, H.; Fortova, J.; Rychtera, M. & Melzoch, K. (2009) Exploitation of food feedstock and waste for production of biobutanol. *Czech Journal of Food Sciences*, Vol. 27, No. 4, pp. 276-283, ISSN 1212-1800

Patakova, P.; Lipovsky, J.; Paulova, L.; Linhova, M.; Fribert, P.; Rychtera, M. & Melzoch, K. (2011a) Continuous production of butanol by bacteria of genus *Clostridium*, *Journal of Chemistry and Chemical Engineering*, Vol. 5, No. 2, pp. 124-131, ISSN 1934-7375

Patakova, P.; Toure,S.S.M.; Simacek, P.; Cizkova, H.; Lipovsky, J.; Linhova, M.; Fribert, P., Rychtera, M. & Melzoch, K. (2011b) Exploitation of sugar beet for biobutanol production, *Czech Sugar and Sugar Beet Journal*, Vol. 127, No. 2, pp. 46-49, ISSN 1210-3306

Peng, Ch.; Lewis, C.K.; Stein, F.P. (1996) Water Solubilities in Blends of Gasoline and Oxygenates, Fluid Phase Equilibria, Vol.116 (1996), pp. 437-444, ISSN: 0016-2361

Pierrot, P.; Fick, M. & Engasser, J.M. (1986) Continuous acetone-butanol fermentation with high productivity by cell ultrafiltration and recycling, *Biotechnology Letters*, Vol. 8, No. 4, pp. 253-256, ISSN 0141-5492

Qureshi, N. & Blaschek, H.P. (2001a) Recent advances in ABE fermentation: hyper-butanol producing Clostridium beijerinckii BA101, *Journal of Industrial Microbiology and Biotechnology*, Vol. 27, No. 5, pp. 287-291., ISSN 1367-5435

Qureshi, N. & Blaschek, H.P. (2001b) Recovery of butanol from fermentation broth by gas stripping, *Renewable Energy*, Vol. 22, No. 4, pp. 557-564, ISSN 0960-1481

Rainey, F.A.; Hollen, B.J. & Small, A. (2009) Genus I. *Clostridium* Prazmowski 1880, 23AL. In *Bergey`s manual of Systematic Bacteriology, The Firmicutes*, Second Edition, Volume Three, De Vos, P.; Garrity, G.M.; Jones, D.; Krieg, N.R.; Ludwig, W.; Rainey, F.A.; Schleifer, K.-H. & Whitman, W.B, pp. 738-828, Springer, ISBN 978-0-387-95041-9, New York.

Rakopoulos, C.D.; Dimaratos, A.M.; Giakoumis, E.G. & Rakopoulos D.C. (2010a) Investigating the emissions during acceleration of a turbocharged diesel engine operating with bio-diesel or n-butanol diesel fuel blends, *Energy*, Volume 35, No. 12, pp. 5173-5184, ISSN 0360-5442

Rakopoulos, D.C.; Rakopoulos, C.D.; Hountalas, D.T.; Kakaras, E.C.; Giakoumis, E.G. & Papagiannakis, R.G. (2010b) Investigation of the performance and emissions of bus engine operating on butanol/diesel fuel blends, *Fuel*, Vol. 89, No. 10, pp. 2781-2790, ISSN 0016-2361

Ramey, D. & Yang, S.-T. (2004) Production of butyric acid and butanol from biomass, In: Final report to the US Department of Energy DE-F-G02-00ER86106, February 27, 2011, Available from:
http://74.125.155.132/scholar?q=cache:BqatEryJKOwJ:scholar.google.com/+author:%22Ramey%22+intitle:%22Production+of+butyric+acid+and+butanol+from+biomass%22+&hl=cs&as_sdt=2000

Ravagnani, A.; Jennert, K.C.B.; Steiner, E.; Grünberg, R.; Jefferies, J.R.; Wilkinson, S.R.; Young, D.I.; Tidswell, E.C.; Brown, D.P.; Youngman, P.; Morris, J.G.& Young, M. (2000) *Spo0A* directly controls the switch from acid to solvent production in solvent-forming clostridia, *Molecular Microbiology*, Vol. 37, No. 5, pp. 1172-1185, ISSN 0950-382X

Rychtera, M.; Patakova, P.; Fribert, P.; Lipovsky, J.; Linhova, M.; Paulova, L.; Melzoch, K.& Klefenz, H. (2010) Bottlenecks in bacterial production of fuel butanol, *Proceedings of 19th International Congress of Chemical and Process Engineering*, ISBN 978-80-02-02210-7, Prague, Czech Republic, September 2010

Sarathy, S.M.; Thomson, M.J.; Togbé, C.; Dagaut, P.; Halter, F. & Mounaim-Rousselle, C. (2009) An experimental and kinetic modeling study of n-butanol combustion, *Combustion and Flame*, Vol. 156, No. 4, pp. 852-864, ISSN 0010-2180

Shaheen, R.; Shirley, M. & Jones, D.T. (2000) Comparative fermentation studies of industrial strains belonging to four species of solvent-producing clostridia, *Journal of Molecular Microbiology and Biotechnology*, Vol. 2, No. 1, pp. 115-124, ISSN 1464-1801

Tomas, C.A.; Beamish, J.& Papoutsakis, E.T. (2004) Transcriptional analysis of butanol stress and tolerance in *Clostridium acetobutylicum*, *Journal of Bacteriology*, Vol. 186, No. 7, pp. 2006-2018, ISSN 0021-9193

Tracy, B.P.; Gaida, S.M.& Papoutsakis, E.T. (2008) Development and application of flow-cytometric techniques for analyzing and sorting endospore-forming clostridia, *Applied and Environmental Microbiology*, Vol. 74, No. 24, pp. 7497-7506, ISSN 0099-2240

United States Patent No.4539293 (1985) Production of butanol by fermentation in the presence of cocultures of clostridium, Application 1983-10-05

Wang, S.; Zhang, Y.; Dong, H.; Mao, S.; Zhu, Y.; Wang, R.; Luan, G. & Li, Y. (2011) Formic acid triggers the acid crash of acetone-butanol-ethanol fermentation by *Clostridium acetobutylicum, Applied and Environmental Microbiology,* Vol. 77, No. 5, pp. 1674-1680, ISSN 0099-2240

Westhuizen van der, A.; Jones, D.T. & Woods, D.R. (1982) Autolytic activity and butanol tolerance of *Clostridium acetobutylicum, Applied and Environmental Microbiology,* Vol. 44, No.6, pp. 1277-1281, ISSN 0099-2240

Wolf, L. (2007) 1-Butanol as a Gasoline Blending Biocomponent, *Mobile Sources Technical Review Subcommittee* - March 28, 2007 Meeting, Aarlington, USA, available from http://www.epa.gov/oar/caaac/mstrs/March2007/Wolf.pdf

Yacoub, Y.; Bata, B. & Gautam, M. (1998) The performance and emission characteristics of C_1-C_5 alcohol-gasoline blends with matched oxygen content in a single-cylinder spark ignition engine, *Proceedings of the Institution of Mechanical Engineers. Part A. Journal of Power and Energy,* Vol. 212, No. 5, pp. 363-379, ISSN 0957-6509

Yao, M.; Wang H.; Zheng Z. & Yue, Y. (2010) Experimental study of n-butanol additive and multi-injection on HD diesel engine performance and emissions, *Fuel,* Vol. 89, No. 9, pp. 2191-2201, ISSN 0016-2361

Xiaolong, Y; Jing, Y.; Tieping, L. (2009) The Effect of an SI Engine Using Butanol-Gasoline Blended Fuel on Performance and Enviroment, ICEET '09 Proceedings of the 2009 International Conference on Energy and Environment Technology - Volume 01, pp. 402-405, ISBN: 978-0-7695-3819-8 IEEE Computer Society Washington, DC, USA, 2009

Zverlov, V.V.; Berezina, O.; Velikodvorskaya, G.A. & Schwartz, W.H. (2006) Bacterial acetone and butanol production by industrial fermentation in the Soviet Union: use of hydrolyzed agricultural waste for biorefinery, *Applied Microbiology and Biotechnology,* Vol. 71, No. 5, pp. 587-597, ISSN 0175-7598

12

Paving the Road to Algal Biofuels with the Development of a Genetic Infrastructure

Julian N. Rosenberg, Michael J. Betenbaugh and George A. Oyler
Johns Hopkins University
United States

1. Introduction

It is anticipated that the global demand for energy will double within the next forty years (Hoffert *et al.*, 1998). This leaves a relatively short period of time for a momentous shift in the fundamental sources of global energy. Nonetheless, satisfying our energy requirements with alternative sources can be achieved while allowing for continued technological progress, economic growth, and political stability over this period. The need for clean, sustainable energy sources is even more urgent when considered in light of the environmental consequences related to the liberation of carbon dioxide from fossil fuels.

For instance, increased production of electricity will undoubtedly necessitate a rapid expansion of nuclear, wind, solar, and hydro- power generation. Even if these sources of energy are aggressively developed, few alternatives appear to be available for the continued expansion of coal-based electricity extending to mid-century; thus, there is a pressing need for mechanisms of carbon dioxide (CO_2) abatement. Energy derived from biomass presents a means of both capturing CO_2 and reducing the need for a fossil fuel-based infrastructure. As such, bioenergy has the advantage of being carbon neutral and will prove to be an important asset in our repertoire of renewable energy solutions.

In addition to producing energy from sustainable sources that maintain carbon neutrality, the obligation to use energy efficiently has never been more important – not only in our daily lives, but also in the mechanisms through which we will generate energy at large scales in the future. In biological systems, the utilization of energy is accomplished by a cascade of biochemical reactions mediated by tightly regulated metabolic networks, which are substantially more efficient than the internal combustion engine. One of the most important and impressive molecular mechanisms for harvesting energy is the photosynthetic process. While photovoltaic technology has improved considerably in recent decades, the plants and algae that have been refined over billions of years of evolution represent a fully developed living framework for solar energy collection.

1.1 A focus on biofuels

In the most general of terms, biofuels are biologically derived forms of chemical energy, such as hydrocarbons, that are compatible with the existing infrastructure. In a sense, petroleum is a biofuel because it is the oil-enriched remains of ancient biomass that has been unearthed. In a modern context, however, by actively converting photosynthetic biomass into liquid biofuel, solar energy can be readily stored and utilized to replace the use of petroleum as a transportation fuel. This is not a new concept; in fact, Rudolph Diesel envisioned his piston-driven engine to run on peanut oil so that farmers could grow their own fuel (Nitschke and Wilson, 1965). Currently, biofuels are gaining attention as a valuable piece of the renewable energy puzzle because they fit so seamlessly into the carbon cycle.

According to recent data, liquid transportation fuel use in the United States is between 130 and 140 billion gallons each year. This usage is anticipated to continue increasing to approximately 150 to 160 billion gallons per year, peaking around 2015 (U.S. Energy Information Administration, 2009). Fuel consumption appears likely to remain stable or decrease slowly as we approach the mid-century. Even in the United States, where a more rapid transition to hybrid or fuel-efficient vehicles is predicted, there will be a continued high demand for liquid transportation fuels.

In contrast to the high energy density of liquid fuels, hydrogen-based transportation seems to be a far-reaching goal and may never fully make sense for transportation. With current practices, hydrogen production is an energy intensive process and still requires a large investment in distribution infrastructure. Alternatively, plug-in electric vehicles have cost and environmental limitations related to batteries and would require a significant amount of time to replace the currently fleet of cars. For these reasons, an intensified focus on liquid biofuels is warranted. Moreover, liquid transportation fuels are currently the major use of petroleum throughout the world. In certain cases, the reliance on liquid fuels we have developed cannot easily be substituted (e.g. aviation, trucking, and construction industries). Furthermore, emerging economies will likely opt for the lowest cost vehicle solutions (i.e. internal combustion engines) and will still have a petroleum requirement.

1.1.1 Ethanol: A first-generation biofuel

Ethanol, by nature, has a lower energy density than other liquid fuels and is not entirely compatible with the current distribution infrastructure because it is hygroscopic and can contribute to rust formation. In the United States, ethanol produced in the Midwest also requires costly transportation to sites of consumption, primarily the East and West Coasts. In other parts of the world, particularly Brazil, where ethanol production may be more economically produced from sugar cane, there are still limitations including the high energy input for distillation from the dilute solutions produced biologically.

Biofuels made from food crops may impose stresses on the existing world economy. The most serious of these are the possible effect on food prices due to subsidized farming, substantial input of fossil fuels for production, degradation of agricultural land resources, and potential alterations in ecosystems from expanded land use. Even emissions of non-CO_2 greenhouse gases may increase with first-generation biofuel production. The use of nitrogen fertilizer on biofuel crops also has the potential to increase the release of nitrous oxide, a greenhouse gas more potent than carbon dioxide, as well as the consumption of natural gas for fertilizer production. While first-generation ethanol has initiated the biofuel revolution, advanced biofuels are clearly needed.

1.1.2 The next generation of biofuels

With the limitations of ethanol as a biofuel and the need to expand beyond food crop-based biofuel production, there is a pressing need for second- and third-generation biofuels. Since there appears to be little consensus on the meaning of second, third, and further generations of biofuels, we will adopt the term advanced or *next*-generation biofuels. Next-generation biodiesel is a triglyceride derived fatty acid methyl ester (FAME), which does not originate from food crop sources. Additionally, there are clear limitations in some non-food crop terrestrial plant sources of triglycerides for biodiesel, such as palm oil and jatropha. Palm oil is rapidly becoming a major source of biodiesel to fulfill the European Union mandate of 10% liquid transportation biofuels by 2010 (European Parliament, 2009). Europe is a much larger consumer of diesel fuel for passenger vehicles with approximately 50% of the passenger cars sold in the EU currently having diesel engines (Smolinska, 2008). Unfortunately, the net result of the substantial increase in demand for palm oil has been an accelerated rate of palm plantation development and deforestation in tropical ecosystems.

Another promising group of next-generation biofuels are the alcohols with longer chains than ethanol, such as butanol and branched-chain alcohols. These fuels have a higher energy density than ethanol, do not absorb water as ethanol does, and possess very favorable combustion characteristics, such as high octane ratings. These alcohols are somewhat more unusual and rare in nature, but one example of a microorganism that is adept at producing these compounds is *Clostridium acetobutylicum*. Regrettably, there are limitations on the use of this slow growing anaerobic organism for biofuel production. A search for other means for producing these promising longer and branched-chain alcohol biofuels in photosynthetic organisms is currently underway (Fortman *et al.*, 2008).

While ethanol produced from cellulosic biomass is commonly touted as a promising advanced biofuel solution, the end product is still ethanol, which has all of the limitations stated above. There is strong motivation to move beyond ethanol, but what other means are available? Conversion of the sugars released by deconstruction of cellulosic biomass could easily be directed more usefully to one of the more desirable next-generation biofuels, such as microorganism-derived biodiesel or branched-chain alcohols. In a throwback to biofuels efforts of World War I, recently there has been renewed interest in the ability of *Clostridial* species to produce butanol and possibly other longer chain alcohol biofuels (Sillers *et al.*, 2008). With the availability of genomes for these anaerobic bacteria, means to genetically enhance their productive capacities may be at hand. However, there are still a number of significant barriers to overcome for anaerobic fermentation to be a truly viable means of biofuel production. Alternatively, microbes such as bacteria and microalgae show promise as a renewable feedstock for a biofuels ranging from ethanol to biodiesel. The capacity of photosynthesis to capture solar energy is particularly attractive for producing renewable fuels because no intermediate chemical feedstock is required.

2. Microalgal biomass for biofuel production

Algae are a diverse group of aquatic, photosynthetic organisms generally categorized as either macroalgae (i.e. seaweed) or microalgae, which are typically unicellular. Although the emerging field of algal biofuels remains in its infancy, microalgae have great potential to bring the promise of clean, sustainable fuel production before we must face the reality of fossil fuel depletion and exacerbated climate change. Algae are perhaps the most effective photosynthetic organisms for generating chemical energy from sunlight – the most abundant and renewable global energy source. It is believed that a large percentage of today's fossil fuels, particularly petroleum, originated as prehistoric algal blooms. As single-celled organisms, microalgae are capable of producing a large portion of their biomass as small molecule biofuel precursors since they lack macromolecular structural and vascular components needed to support and nourish terrestrial plants. As such, algae provide one of the most direct routes for the photosynthetic conversion of carbon dioxide and other organic substrates to biofuel. Moreover, the large surface area to volume ratio of these aquatic microorganisms is advantageous for absorption of nutrients and sunlight, which is reflected in the rapid growth rates observed in many species.

As aquatic organisms, microalgae offer many advantages over the terrestrial bioenergy crops with which they contend. Some of the most serious drawbacks of allocating portions of existing food crops to produce biofuels, particularly ethanol from corn and biodiesel from soy or rapeseed, are the obvious competition with food production and encouragement of subsidized operations. Both outcomes are coupled with severe economic ramifications. While cellulosic ethanol may avoid the food versus fuel controversy, this technology has yet to fully mature and will likely remain at the developmental stage for a number of years. In general, terrestrial crops have relatively long growing seasons and require arable land, oftentimes supplemented with costly fertilizers that can have harmful effects on the surrounding ecosystems. Additionally, there are greenhouse gases released in the process of generating fertilizer and harvesting terrestrial biomass. Furthermore, constant irrigation of these crops is yet another impediment, as this can be taxing on natural freshwater resources. While great strides are being made toward the optimization of cellulases for enzymatic degradation of lignocellulose, a significant amount of energy is still required to harvest and pre-treat (thermochemically breakdown) the cellulosic biomass, which constitutes an additional input of fossil fuel-derived energy.

Unlike terrestrial bioenergy crops, microalgae do not require fertile land or extensive irrigation and can be harvested continuously. Several species of algae provide an alternative to freshwater use by growing in brackish, sea, and even hypersaline water. Additionally, since algae consume carbon dioxide through the process of photosynthesis, large-scale cultivation can be used to remediate the CO_2 emissions from fossil fuel combustion (Benemann and Oswald, 1996) (Figure 1). Algal biomass also possesses secondary co-products such as antioxidant pigments, edible proteins, and nutraceutical oils that other alternative fuel crops lack (Spolaore *et al.*, 2006). Lastly, since nearly all microalgae have a simple unicellular structure, algal biomass is devoid of lignocellulose. This strong structural polymer has proven to be a significant obstacle to releasing the energy trapped in terrestrial biomass. Not only do microalgae fully address each of the disadvantages of land-based biofuel crops, but they also are amenable to genetic engineering for the enhanced biosynthesis of a wide range of advanced biofuels and high-value added products.

Currently, three fundamental objectives remain critical to the implementation of economically- and technologically-feasible algal biofuel production: [1] increase of biological productivity through species selection and genetic engineering as well as optimization of culture conditions; [2] development of low-cost vessels for cultivation, whether they be closed photobioreactors or open pond systems; and [3] improvement of inexpensive downstream processing techniques for algal biomass, including harvesting, dewatering, and extraction of biofuel metabolites (Hejazi *et al.*, 2004a; Shelef *et al.*, 1984; Danquah *et al.*, 2009). As with many novel sources of bioenergy, the complexity of the microalgal biofuel production process calls for a multidisciplinary approach in which biotechnological progress will be accompanied by advances in process engineering.

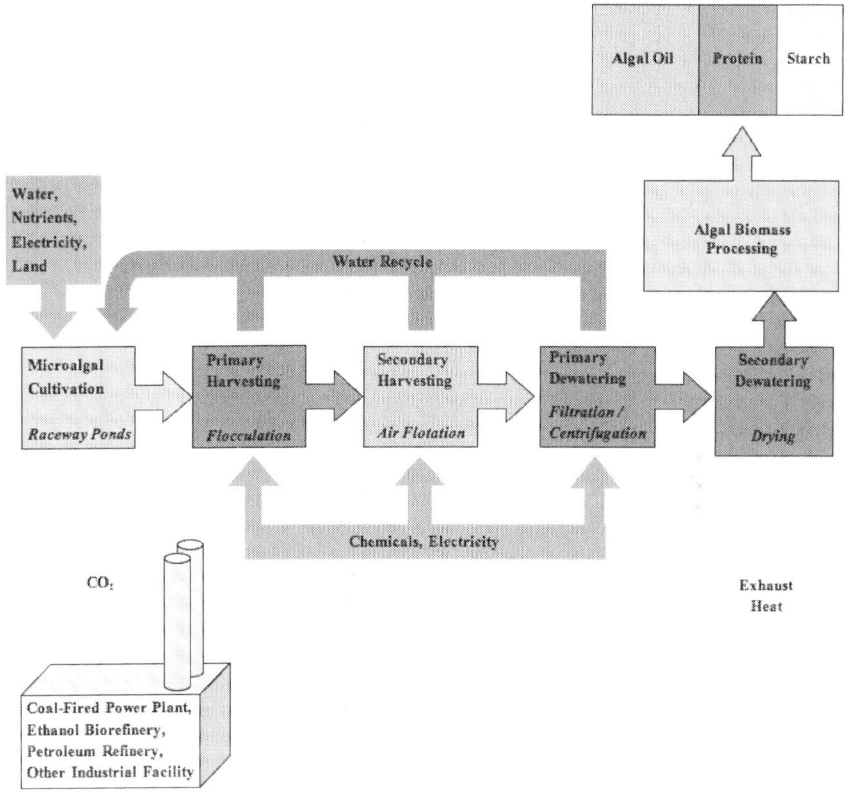

Fig. 1. Algal process flow diagram with integrated industrial CO_2 sequestration.

2.1 Commercial applicability of microalgal biofuels

As the name 'microalgae' suggests, the relative size of these organisms may seem unsuitable for generating massive quantities of biofuel on a global scale; nonetheless, microalgae offer many advantageous qualities for biofuel production, especially when compared to terrestrial bioenergy crops. The basic principle of generating biofuel from microalgae is to exploit these cells as biological factories, whose lipid output can be as much as 70% of their total dry biomass, under optimal conditions. While many species of algae exhibit the natural capacity to produce abundant amounts of oil for conversion to biofuel, a major obstacle in the commercialization of such a process lies in the scalability. Many exciting breakthroughs in algal biotechnology have occurred on the lab bench, but mass cultivation of algae is still associated with some of the most challenging problems. A few areas of intense research focus include highly productive growth systems, temperature control, photooxidative stress tolerance, light intensity regulation, harvesting, and downstream processing. Whether it is through the manipulation of culture conditions or the application of mutagenesis and genetic engineering, the biological networks of these unicellular creatures can potentially be optimized to synthesize and/or secrete biofuel metabolites, particularly in the form of lipids and other hydrocarbons, in order to overcome some of the aforementioned stumbling blocks to large-scale cultivation.

It is difficult to convey a concise list of the ideal species for biofuel production because the organism must be paired not only with the climate in which it will be cultivated, but also the specific mechanism of cultivation and desired end products. Also, the lipid content can vary considerably depending on culture conditions. For example, nitrogen and silicon deprivation has shown to augment lipid accumulation in green algae and diatoms, respectively. As investigated by NREL's Aquatic Species Program, nutrient deprivation experiments and species collection and characterization efforts gave rise to an extensive list of microalgae with particularly high lipid contents (Sheehan *et al.*, 1998).

There is, however, a serious caveat to high lipid accumulation in algae: the energy collected by the cell is partitioned into storage and, thus, made unavailable for immediate use. As a result, oleaginous species exhibit significantly slower growth rates than their more lean relatives. The fatty acid composition and growth characteristics of some of the more promising species are illustrated in Figure 2, where a clear balance can be seen between growth rate and lipid content. In this particular study, the algal species were cultivated first in airlift bioreactors, then in aerated polyethylene bags, and finally in outdoor raceway ponds for a period of four months. Subsequent biochemical evaluation found the neutral lipids to be predominantly C_{16} and C_{18}, which are ideal chain lengths for the composition of biodiesel (Gouveia *et al.*, 2009).

Astonishingly, of the thousands of different species of algae, a mere fifteen organisms are commonly used for commercial applications (Raja *et al.*, 2008) and only eight species' genomes have been sequenced (Hallmann, 2007), but future bioprospecting endeavors paired with high-throughput screening are likely to discover more exemplary candidates for algal biofuel production.

2.2 Opportunities for genetic engineering of algae

While microalgae are an abundant source of naturally oil-rich biomass, these cells can also act as biological factories designed to produce of a variety of promising biofuel precursors. By elucidating the complex metabolic networks involved in carbon utilization, there lies great opportunity for genetic and metabolic engineering of these organisms. Some of the major obstacles to metabolic engineering of algae stem from the lack of basic biological knowledge of these diverse creatures, including sparse genomic information and somewhat primitive methods of genetic transformation. As a result, the introduction of nuclear transgenes to microalgal cells relies on random chromosomal integration, which is highly susceptible to gene silencing; the subsequent recovery of stable transformants is limited to only a handful of species and is oftentimes irreproducible. Overcoming these biotechnological barriers, however, will present enormous opportunities to develop microalgae as versatile platform for biofuel production.

A number of improvements in the productivity of green algae and diatoms would significantly enhance their capabilities as biofuel producers. Photoautotrophic algal growth rates and cell densities at commercial are low compared to microbial fermentation. Enhancement of growth through metabolic engineering with control of cell cycle would be a breakthrough for algal biofuels. Increasing biofuel feedstock production by improving the synthesis of biofuel precursors is imperative. Metabolic engineering of secretion pathways or developing means to readily strip hydrocarbons would allow the organism to survive while producing biofuel metabolites on a continue basis. This would reduce the amount of biomass that is produced per unit of biofuel, thus focusing resources on the primary product. Metabolic engineering might also increase the range of biofuel metabolites and other high-value added materials that can be synthesized. Furthermore, improved performance in a variety of photobioreactors and conditions is necessary. The development of organisms that can survive in environments that exclude invasive species and other contaminant microorganisms is another desirable attribute for open cultivation systems. Finally, the goal of designing suicide genes to prevent the unintended release of genetically modified organisms (GMOs) is an important consideration.

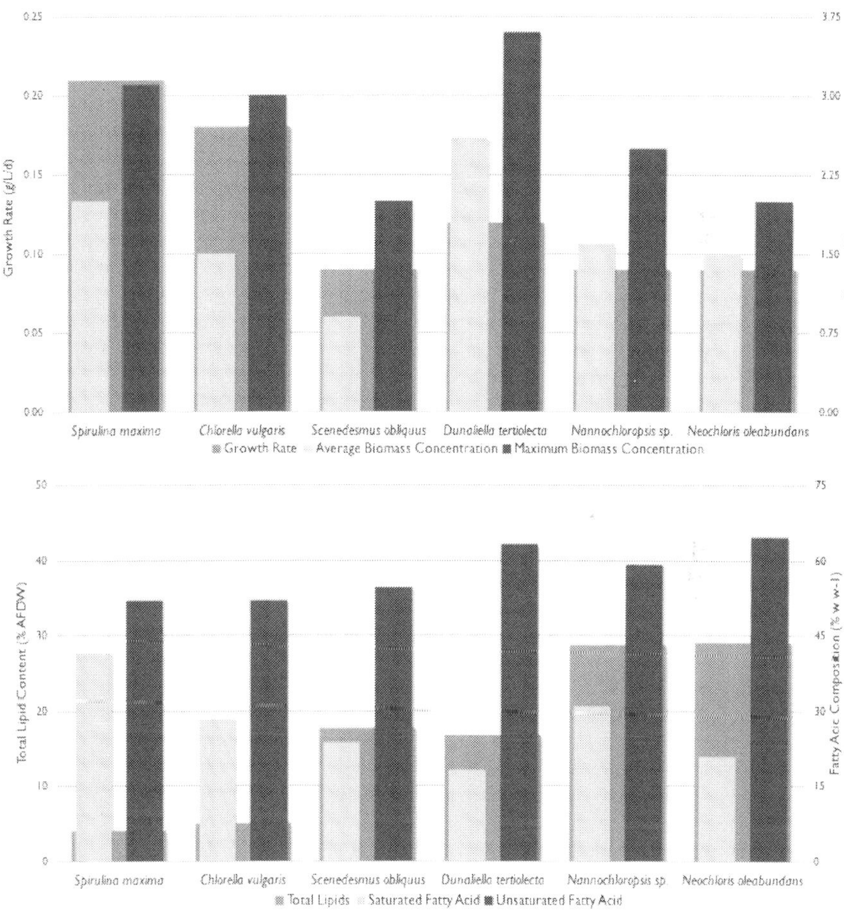

Fig. 2. Growth characteristics and lipid profiles of commercially attractive microalgae.

From a genetic engineering standpoint, there exist certain superior characteristics of eukaryotic algae as compared to other photosynthetic sources of biofuels. Focusing on green algae and diatoms will allow the transfer of well-established metabolic engineering approaches used in other eukaryotic systems. With eukaryotic algae there is the ability to perform both chloroplast and nuclear transformation, possibly increasing the complexity and range of metabolic engineering. Because eukaryotes have evolved a membrane bound secretory pathway, it is conceivable that eukaryotic algae could be genetically engineered to secrete lipid bodies into the media (Benning, 2008). For example, the alga *Botryococcus braunii* has naturally evolved this mechanism for secreting hydrocarbons. If this biological feat could be achieved in other microalgal species, it would greatly simplify downstream processing by eliminating cell harvesting and lysis; thus, reducing the entire procedure to merely skimming the lipids from the top of the medium.

Many accomplishments have already been made in the field of microalgal bioengineering (Leon-Banares *et al.*, 2004; Walker *et al.*, 2005), the most relevant to biofuel production being increased photosynthetic efficiency and light penetration in *C. reinhardtii* (Mussgnug *et al.*, 2007), but augmented lipid production through genetic alterations has yet to be achieved. Currently, *C. reinhardtii* remains the workhorse of algal genetic engineering for its history as a model photosynthetic organism (Harris, 2001). The recently completed genome sequence of *Dunaliella salina* may be a good starting point for genetic research of algal biofuel production; however, a single species cannot be expected to serve every application. With the limited availability of genomic data for microalgae (Hallmann, 2007), exploration of transgenic algae for bioenergy demands a genome project for a model biofuel production strain. Future efforts to probe the metabolic pathways of microalgae will likely employ technologies beyond genomic analysis, such as transcriptomics, metabolomics, proteomics, and lipidomics, to examine the broader biological landscape of algal metabolism (Jamers *et al.*, 2009; Vemuri *et al.*, 2005).

2.3 Mass cultivation of microalgae

The cultivation macro- and microalgae is a well-established practice, providing ample biomass for human nutrition, commercially important biopolymers, and specialty chemicals, that dates back nearly 2,000 years (Spolaore *et al.*, 2006). As an example, growing the gelatinous cyanobacteria *Nostoc* in rice patties enabled much of the Chinese population to survive famine in 200 AD (Qiu *et al.*, 2002). Since that time, the mass cultivation of microalgae has been commercialized for the production of either whole-cell algal nutritional supplements or nutraceutical extracts, such as β-carotene, astaxanthin, and polyunsaturated fatty acids (e.g. DHA, omega-3). In the international market, China, Japan, Australia, India, Israel, and the United States are leaders in algal production.

2.3.1 Constraints on photoautotrophic algal biomass production

In addition to certain biological limitations, several obstacles related to cultivation must be overcome to allow economical industrial scale-up of algal biofuel production. The conversion efficiency of solar energy to biomass by microalgae is governed, in part, by the inherent biological efficiency of photosynthesis, and largely by the effectiveness of light-transfer in liquid cultures. Some species of algae grown heterotrophically (i.e. supplemented with carbon sources other than CO_2, such as sugars) can accumulate a greater amount of lipids (Wu *et al.*, 2006); however, the costs associated with such cultivation may limit its applicability to biofuel production. The approach of heterotrophic algal biofuel production is the model for a number of algal biofuels start-up companies.

On the other hand, generating algal biomass for biofuels with energy directly from the sun rather than a chemical intermediate has its advantages. Microalgae essentially act as biological solar panels directly connected to biorefineries. Photoautotrophic cultivation has the added benefit of CO_2 sequestration from a point source. Although current commercial raceway ponds operate with areal productivities of 2-20 g m^{-2} d^{-1}, there remains much speculation regarding the maximum achievable algal biomass productivity. While heterotrophic modes of cultivation can yield very dense algal cell cultures, photoautotrophic cultures are not expected to exceed 60 g m^{-2} d^{-1}. Figure 3 shows a compilation of realistic areal productivities and theoretical projections for photoautotrophic algal cultivation in open ponds (Chisti, 2007; Weyer *et al.*, 2008; Schenk *et al.*, 2008; Wallace, 2008).

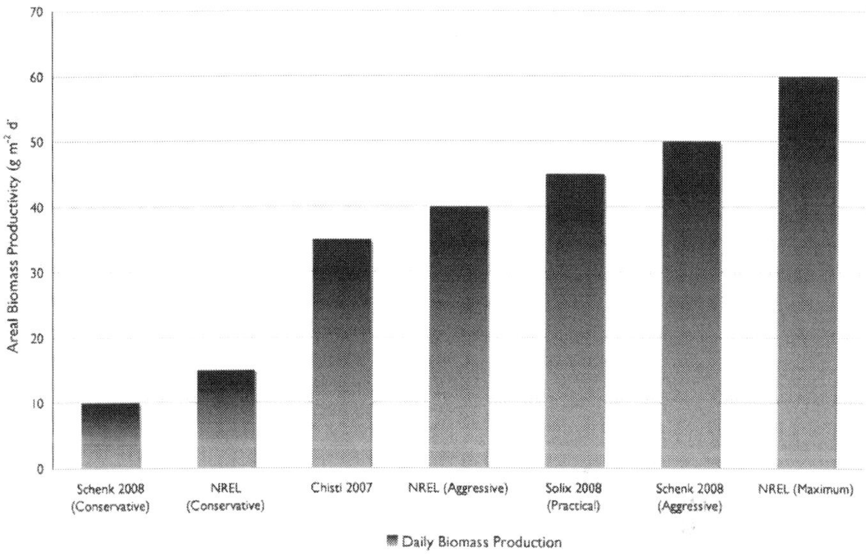

Fig. 3. Projected algal biomass productivities for raceway ponds.

While a wide range of predictions has been made for the maximum attainable productivity of algal cultures, a straightforward analysis of energy transfer in algal biomass production reveals a few key bottlenecks. By following a single photon from its origin at the sun to the desired end product of algal oil, there are a number of unavoidable losses imposed on this conversion of sunlight by the inherent bioenergetics of cellular processes. Additional diversions of solar energy can be attributed to the algal growth system and can be minimized with proper design parameters.

The first impediment that solar radiation faces when traveling to the Earth's surface is the local weather. As we know from our daily observations, cloudy skies can dramatically reduce the amount of light that reaches the ground. Additionally, while the equator receives high-intensity light year-round, solar irradiance diminishes as one travels away from the equator in latitude, thus near-equatorial zones are ideal for algal biomass production. Accordingly, Asia, Australia, and the United States are common sites for algal growth facilities. Figure 4 presents a map of solar data collected from 1990-2004 where black dots represent locations for which detailed weather analysis is available for algal production facilities (Weyer *et al.*, 2008). In geographies that receive more exposure to sunlight, and accompanying high temperatures, evaporative water loss and cooling mechanisms become more important considerations.

Since there is little one can do to change the weather beyond choosing an adequate site for algal cultivation, the next constraint on solar energy collection comes from the limited spectrum of light that plants have adapted to utilize, deemed photosynthetically active radiation (PAR: λ = 400-700 nm), which accounts for only 45% of the total energy in the visible light spectrum (Weyer *et al.*, 2008).

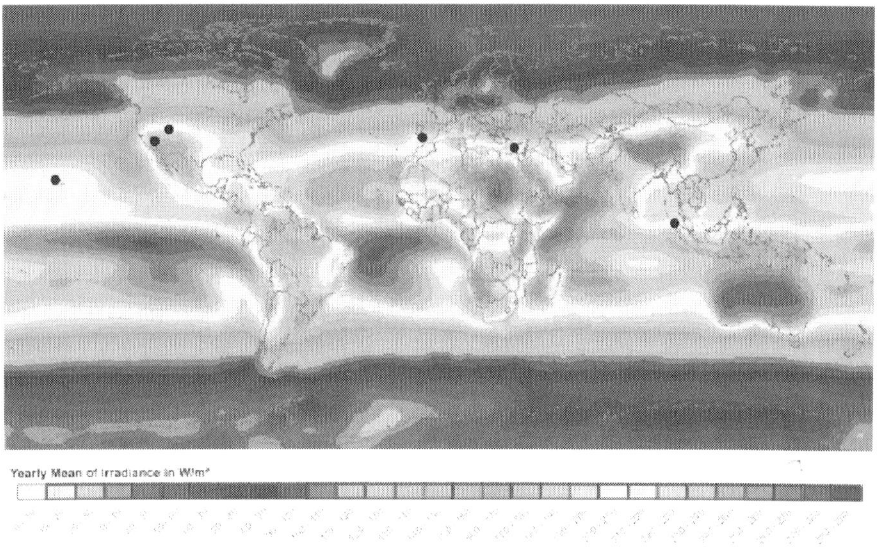

Fig. 4. Global map of average annual solar radiation (Reprinted with permission from SoDa Services, Copyright Mines ParisTech / Armines 2006).

In conventional raceway ponds and photobioreactors, incident sunlight encounters billions of algal cells as it travels through the liquid culture – each cell absorbing some of the available energy. Thus, the transmission of light is severely inhibited by cell shading in these dense solutions. For example, the leaves of a tree have evolved to be essentially two-dimensional structures with only millimeter thicknesses; an algal culture volume can be thought of in a similar manner. In open ponds, only the cells on the surface are exposed to maximum sunlight, and those on the bottom of the 10-30 cm deep trough receive very little of this incoming energy. The advantage, though, of a liquid culture is that the shaded cells in the submerged regions can be recirculated to the surface periodically so that a large volume of biomass can be maintained. Additionally, advanced photobioreactor design can encourage optimal mixing patterns (Tredici and Zittelli, 1998).

The next hurdle that usable photons must surmount is absorption by the molecular photosynthetic apparatus. As microalgae have adapted to survive in conditions of low light, they have evolved biochemical mechanisms that are incredibly adept at collecting light energy.

Light-harvesting complexes, which surround the photosystems in the chloroplast membrane, act as antennae to gather and shuttle photons to the photosynthetic reaction center. One limitation to complete utilization of incident PAR lies in the maximum capacity of energy that each photosystem can handle. In fact, high-intensity light can quickly inundate the photosynthetic machinery, leading to the generation of free oxygen radicals, in a process called photoinhibition (Long *et al.*, 1994). This obviously has a negative effect on the productivity of algal cultures by leading to cellular damage and premature population demise. Unlike the natural environmental conditions to which microalgae are accustomed, industrial cultivation strives for maximum collection of solar energy. In this case, algae can be exposed to high-intensity light for only a short period of time, on the scale of microseconds, allowing each photosystem to absorb light before becoming overwhelmed. Also at the biochemical level, there are inherent limitations to photosynthesis related to electron transfer. Estimates for the efficiency of photosynthesis correlate to roughly 25% energy conversion (Weyer *et al.*, 2008). Optimization of photosynthesis is an ambitious genetic engineering goal and is likely to remain an intrinsic parameter of algal biomass production processes.

After the available sunlight has been harnessed by photosynthesis, there exist various levels of inefficiency related to the biological conversion of this energy to biomass. Principally, nearly 40% of the total energy is required to sustain basic cellular function and growth, leaving an estimated 60% of the total photosynthetically captured energy for biomass accumulation (Weyer *et al.*, 2008). According recent analysis utilizing actual solar irradiance and weather data from various global locations (Figure 4), and taking each of the aforementioned assumptions into account, the projected range of algal biomass production is between 38 and 47 g m^{-2} d^{-1} (Weyer *et al.*, 2008), which is in agreement other predictions (Figure 3). Current values for commercial biomass production in open ponds are typically 2-20 g m^{-2} d^{-1}, which provide sufficient profit margins for high-value products such as β-carotene, but are anticipated to meet the demands of cost-effective biofuel production in the near future. A new partnership between Seambiotic and the Israeli Electric Company has plans to produce algal biomass inexpensively for use as biofuel, with operating costs and profit margins listed in Table 1.

Annual Expenses (USD yr^{-1})	**Nature Beta Technologies**	**Seambiotic/Israeli Electric**
10-ha Raceway Farm	Commercial Plant	Pilot Plant
Manpower	$500,000 (20 Employees)	$120,00 (8 Employees)
Electricity	$180,000	$30,000
Nutrients	$36,000	$36,000
Land	$50,000	$10,000
Carbon Dioxide	$150,000	$5,000
Sea Water	$200,000	$5,000
Fresh Water	$20,000	$10,000
Miscellaneous Expenditures	$30,000	$20,000
Total	$1,166,000	$236,000
Revenue		
Biomass Production (yr^{-1})	70 tons (at 2 g m^{-2} d^{-1})	700 tons (at 20 g m^{-2} d^{-1})
Biomass Cost (USD kg^{-1})	$17.00	$0.34
Market Price (USD kg^{-1})	$4,000 (*D. salina* β-carotene)	< $0.50 (Potential Biofuels)

Table 1. Cost analysis of microalgal biomass production facilities in Israel (Reprinted with permission from Dr. Ami Ben-Amotz).

2.3.2 Raceway pond systems and photobioreactors

Although primitive in design, open raceway ponds are still the predominant algaculture system for high-value added products and have been used for decades (Benemann and Oswald, 1996; Sheehan *et al.*, 1998). The fact that raceway ponds are uncomplicated makes them less productive and offers little control over the culture parameters; however, the low cost of this low-tech cultivation system allows it to compete with complex photobioreactors (Gordon *et al.*, 2007). The surface-to-volume ratio and corresponding light penetration in open ponds are not ideal. As a result, ponds can only support low culture densities; however, the ease of scaling production to industrial proportions (> 1 million liters per acre) justifies their seemingly low efficiency. The use of raceway ponds is also vindicated by their uncomplicated design, which makes them readily available for implementation and relatively simple to clean and maintain. While open ponds are cost effective, they do have a large footprint and contamination by local algal species and threat of algal grazers pose serious risks. Invasion by indigenous microorganisms may be protected against by the use of greenhouse enclosures or by growing algae that can withstand hypersaline environments, such as *Dunaliella salina*. For the purpose of biofuel production, however, the low areal productivities of ponds alone may not be able to provide the necessary biomass feedstock economically. Commercial scale microalgal biofuel facilities will likely rely on integrated systems of high efficiency photobioreactors to provide a dense inoculum for readily scalable raceway ponds (Huntely *et al.*, 2007).

While genetic engineering approaches may improve the photosynthetic and biosynthetic capabilities of microalgae, many innovative methods exist for optimizing photoautotrophic culture conditions to accomplish the same goal of increased yield (Muller-Feuga, 2004). The breadth of chemical engineering knowledge being applied to photobioreactor (PBR) design in order to enhance light and nutrient availability represents an important advance in the field (Tredici and Zittelli, 1998; Miron *et al.*, 1999), particularly for modular and scalable reactors (Janssen *et al.*, 2003; Hu *et al.*, 1996); however, the cost of these technologies remains the ultimate constraint on feasibility.

Photobioreactors aim to optimize many, if not all, of the culture parameters crucial to microalgal growth. One condition achieved in PBRs, but not raceway ponds, is turbulent flow to produce enhanced mixing patterns. By simply increasing the Reynolds number of these systems, the resulting fluid dynamics have a positive effect on nutrient mass transfer, light absorption, and temperature control (Ugwu *et al.*, 2008). However, highly turbulent flow within complex geometries comes at the expense of the cells' sensitivity to shear stress, which is one limitation of photobioreactor design. Another important consideration is the time scale over which the microalgae are transferred from the periphery of the bioreactor to the interior shaded region, as there exists a limit to the amount of light the cells can process before photoinhibiton occurs.

One of the most beneficial aspects of photobioreactors is the extended surface area achievable with tubular and flat-panel designs. This route to increased productivity takes advantage of allowing more algae to have contact with sunlight than an area of land would regularly allow with more basic cultivation systems; however, the additional cost of maintaining ideal temperatures and protecting these sometimes delicate devices from inclement weather pose some concern. Many of these nascent technologies depend entirely on the site of deployment and, as a result, require a great deal of customization (Janssen *et al.*, 2003; Miron *et al.*, 1999).

Some additional drawbacks of photobioreactors include the chemical gradients that can develop, particularly along tubular reactors. As a result of photosynthesis, a significant amount of oxygen can accumulate in these tubes and must be purged periodically. High concentrations of oxygen can both inhibit photosynthesis and, when combined with high irradiance, result in the formation of reactive oxygen species (ROS) (Tredici and Zittelli, 1998). These added difficulties contribute to the higher complexity and cost of photobioreactor operation. Some low-cost alternatives include vertical-column or hanging-bag bioreactors, which still provide a closed system for monocultures, with less control over culture parameters. These systems often rely on sparged air to provide both CO_2 and mixing force, as in airlift bioreactors. The ability to strike a balance between cost and productivity is the major challenge of microalgal cultivation, especially for applications that require closed cultivation, as is the case with genetically modified microalgae.

3. Toward the development of selectable markers for *Dunaliella salina*

In the academic community, green microalgae serve as model organisms for photosynthetic research; *Chlamydomonas reinhardtii* is the most reputable species for this work (Harris, 2001). *Volvox carteri*, a multicellular microalga, is another well-established model species for the elucidation of the genetic basis of cellular differentiation (Miller, 2002). In recent years, the green alga *Dunaliella salina* has been utilized to complement the study of photosynthesis, osmoregulation, carotenogenesis, and glycerol production (Jin *et al.*, 2001; Liska *et al.*, 2004; Thompson, 2005; Shaish *et al.*, 1992; Chitlaru *et al.*, 1991). *Dunaliella salina* is an attractive platform for both commercial and academic pursuits owing to its intriguing and advantageous abilities to survive in conditions of extreme salinity and produce significant amounts of β-carotene. Currently, the aspiration to genetically and metabolically engineer this organism in order to probe its biological networks and eventually enhance its productivity is an ambitious goal. Until recently, the stable expression of transgenes by this organism has been limited due to inexperience with genetic transformation and insufficient knowledge of the species' genome. While molecular methods of manipulation make *C. reinhardtii* and *V. carteri* experimentally tractable at many levels, there is a pressing need for the same tools to be developed for *D. salina*.

This section discusses attempts to genetically engineer *D. salina* through the development of selectable marker systems. The investigation includes detailed characterization of the growth response of *D. salina* to a number of antibiotics and herbicides commonly used for selection of microalgae, such as bleomycin, paromomycin, and phosphinothricin (PPT). Based on reported genetic sequence information for *D. salina*, promoter and 3′-UTR regions of highly active genes were selected as targets for genomic PCR, with the hopes of creating *D. salina*-specific plasmid transformation vectors. Although these efforts did not yield the intended results, this work establishes a foundation for genetic engineering of *D. salina*, which is expected to continue now that the sequenced genome has been made available (Smith *et al.* 2010).

3.1 *D. salina* as a platform organism for biotechnological development

For decades, *D. salina* has been cultivated for its natural ability to produce β-carotene. This valuable bioproduct allows for large-scale cultivation and processing of the biomass to be very profitable, as *D. salina* is the predominant source of natural β-carotene (Ye *et al.* 2008). This green alga is ideal for growth in outdoor ponds due to its ability to grow in high salinity waters - as much as eight times the salt concentration of seawater - greatly reducing the threat of contamination by local microbes and eliminating the need for large quantities of freshwater. *Dunaliella* spp. are similar to *Chlamydomonas* spp. in that they exist as single, flagellated, elongated cells in the size range of 10 microns; however, *D. salina* and *D. tertiolecta,* are capable of osmoregulation by a complex network of ion channels, a flexible cell membrane uninhibited by a cell wall, and glycerol biosynthesis to offset osmotic pressure (Goyal, 2007). As such, the lack of a rigid cell wall makes the algal biomass relatively simple to lyse for the purpose of downstream processing. Furthermore, the technique of "milking" microbial cells for certain metabolites has improved substantially in recent years. In this process, the cells are contacted with a biocompatible organic solvent in order to promote preferential transfer of desired compounds to the solvent phase, leaving the cells viable for continued bioproduction. This process has been successfully demonstrated with *D. salina* for the extraction of β-carotene in a two-phase system (Hejazi *et al.*, 2004b).

While the demand for natural β-carotene dictates the high market price of this compound and continued use of *D. salina*, an increasing desire for biofuel production draws an inquisitive eye to the carotenogenesis pathways of *D. salina* (Lamers *et al.*, 2004). Since all carotenoid compounds are composed of long-chain branched hydrocarbons, it is conceivable that the biosynthetic pathways of *D. salina* could be altered to produce hydrocarbons that are ideal for use as gasoline-like biofuel. With some molecular biology tools already developed for *Dunaliella* spp. (Polle *et al.*, 2009), the sequencing and annotation of its 610 Mbp nuclear genome will now allow for more extensive genetic engineering endeavors with this organism. At the time of these experiments, only the chloroplast and mitochondrial genomes of *D. salina* CCAP 19/18 (GenBank GQ250046, GQ250045) were released. In light of its unique biotechnological application and long history of mass production, *D. salina* is an ideal organism for future development as a biofuel producing microalgae.

3.2 Genetic engineering of *D. salina*

Owing to the attractiveness of *D. salina* for biotechnology, there is a renewed interest in engineering this organism. Publications have reported the genetic transformation of *D. salina* by both microparticle bombardment and electroporation (Geng *et al.*, 2003; Tan *et al.*, 2005). Some of the most impressive progress in the field has come from the Xue group at Zhengzhou University in China. With research covering optimization of transformation techniques, gene characterization, and enhanced gene expressing utilizing matrix attachment regions, their work provides important information and an exemplary research path to follow toward genetic engineering of *D. salina* (Wang *et al.*, 2009; Lu *et al.*, 2009; Jia *et al.*, 2009a; Feng *et al.*, 2009; Wang *et al.*, 2007; Jia *et al.*, 2009b; Liu *et al.*, 2005; Jiang *et al.*, 2003). The down-regulation of specific genes using RNAi in *D. salina* has also been reported (Jia *et al.*, 2009a; Sun *et al.*, 2008). These advances, however, are not readily reproducible and represent solitary accomplishments with an alga that has otherwise been difficult to transform.

3.2.1 Selective agents and genes conferring antibiotic resistance

The bleomycin family of glycopeptide antibiotics is toxic to a wide range of organisms with as intercalator functionality able to cleave DNA. Bleomycin-resistance, attributed to the *ble* gene, is an ideal selectable marker as the BLE protein acts in stoichiometric equivalent. Occurring as a dimer, each protein has a strong affinity for binding and inactivating two molecules of bleomycin. Therefore, the level of expression of this exogenous gene can be directly correlated with antibiotic tolerance observed phenotypically. As such, it has been developed as a selectable marker system for nuclear transformation of *C. reinhardtii* and *V. carteri* and is likely to have similar applicability in other algae (Lumbreras *et al.*, 1998; Hallmann *et al.*, 1999). Since the protein must enter the nucleus, it requires high levels of expression to be active; thus, inherently selecting for clonal isolates with high-level expression. Commercial forms of bleomycin include bleocin™ and zeocin™, which are known to affect mammalian, insect, yeast, bacterial, and plant cells.

Paromomycin has been used extensively as a selective agent for the genetic engineering of microorganisms as well as an antibiotic for the treatment of bacterial infections in humans. Paromomycin acts by binding to the 16S ribosomal RNA of microbes, effectively interfering with protein synthesis. The aminoglycoside 3′-phosphotransferase gene *aph* from *Streptomyces rimosus* already exists in a codon bias similar to that of *Chlamydomonas* and confers resistance to paromomycin, kanamycin, and neomycin. Paromomycin is highly toxic to green algae and has been used as an effective selective agent in both *C. reinhardtii* and *V. carteri* (Sizova *et al.*, 2001; Jakobiak *et al.*, 2004).

Phosphinothricin (PPT), also known as glufosinate-ammonium (GLA) and bialaphos, is a natural amino acid that competitively inhibits glutamine synthesis in plants and animals (Hoerlein, 1994). For this reason, it is used in a number of herbicides, including Basta® and RoundUp®. A number of agricultural crops have been engineered by to resist PPT through introduction of the *bar* (bialaphos resistance) gene from *Streptomyces hygroscopicus*, which encodes a phosphinothricin acetyl transferase (Thompson *et al.*, 1987). In the academic community, PPT is used as a selective agent for work with *Arabidopsis* and the *bar* gene is carried on pGR117 for *Agrobacterium*-mediated transformation of *A. thaliana* (Akama *et al.*, 1995).

3.2.2 Endogenous *D. salina* genetic regulatory elements

Prior to the availability a fully sequenced genome, the genetics of *D. salina* were explored for useful elements such as regulatory sequences of highly expressed genes. Highly active endogenous promoter and 3′-untranslated region (UTR) pairs are of particular interest and significance to expressing transgenes in *Dunaliella*. Recent publications describing the use of the *actin*, *rbcS*, carbonic anhydrase, and ubiquitin promoters (Jiang *et al.*, 2005; Walker *et al.*, 2004; Chen *et al.*, 2009) and nitA 3′-UTR (Li *et al.*, 2007; Xie *et al.*, 2007) are the basis for many of the pioneering attempts to genetically engineer *D. salina*, including the work presented hereafter.

3.3 Materials and methods

3.3.1 Microalgal cell culture

D. salina strains CCAP 19/18 and UTEX 1644 were obtained from the Culture Collection of Algae and Protozoa (UK) and the Culture Collection of Algae at University of Texas at Austin, respectively, and maintained on sterile agar plates (1.5% w/w) containing 1 M NaCl *Dunaliella* medium (Weldy *et al.*, 2007). Cells were cultivated photoautotrophically in 1-L glass Fernbach flasks at 27° C (± 1) using 1 M NaCl *Dunaliella* medium (DM). Each axenic batch culture was inoculated with 10 ml of exponentially growing cells (1×10^6 cells ml^{-1}), constantly stirred, bubbled with sterile air, and illuminated with cool-white fluorescent bulbs at an intensity of 80 µE m^{-2} s^{-1}.

3.3.2 Generation of dosage response curves

In order to test the efficacy of the antibiotics bleocin™ (EMD Biosciences), paromomycin (MP Biomedicals), and the PPT-containing (200 g L^{-1}) commercial herbicide Basta® (Bayer CropSciences) on *D. salina* CCAP 19/18, algal cells were grown in the presence of various concentrations of each selective agent (Table 3) using 100-ml stirrer flasks under the same conditions mentioned previously, with the exception of aeration. Each culture was inoculated with 10 ml of exponentially growing cells (1×10^5 cells ml^{-1}) and the cell density was measured with a Zeiss Axiovert 100 inverted light microscope using a hemocytometer over a one week period. Additionally, cells were plated on agar to analyze viability on solid medium. Measurements were taken in duplicate and experiments were repeated two times.

3.3.3 Molecular cloning and construction of vectors

All plasmids were prepared and isolated from bacterial hosts using the QIAPrep Kit (Qiagen) following the supplier's protocol. *E. coli* DH5α cells (Invitrogen) were grown axenically either on LB agar or in LB medium containing pertinent antibiotic selective pressure (ampicillin or kanamycin) at 37° C. Genomic DNA was extracted from *D. salina* using standard CTAB protocol adapted from *Volvox carteri*. PCR was performed using the MJ Mini (Bio-Rad) to amplify the actin (GenBank: AF541875) and rubisco small subunit (GenBank: AY960592) promoters and nitA 3′-UTR (GenBank: EF156403) from *D. salina* UTEX 1644 genomic DNA. Likewise, the genes *ble* and *bar* were amplified from the plasmids pSP124 and pGR117, respectively. All primers employed in this study are listed below in Table 2. Each PCR product was subsequently ligated into the subcloning plasmid pGEM®-T Easy (Promega) using T4 DNA Ligase and its corresponding buffer (Invitrogen). Sequencing of the genetic fragments derived from PCR was performed at the UMBC Biological Sciences Dept. DNA sequencing facility using BigDye® (Applied Biosystems).

Target Sequence	Primers (including NotI, SmaI, HindIII, and XhoI for subcloning)
actin promoter	5′-AATAATAGCGGCCGCCACGGCTCACCATCTTGTTT-3′ 5′-AATAATACCCGGGTTGATCTCTCTGTCACCCCT-3′
rbcS2 promoter	5′-AATAATAAGCGGCCGCAGACATGAACCTATA-3′ 5′-AATAATAACCCGGGAGGTCTTGGCAATGA-3′
bar	5′-AATAATACCCGGGATGAGCCCAGAACGACGCCC-3′ 5′-AATAATAAAGCTTTCAGATTTCGGTGACGGGCA-3′
ble	5′-AATAATACCCGGGATGTTCTTTACTTTTTTACA-3′ 5′-AATAATAAAGCTTCTAGAGTGGGTCGACGTCGG-3′
aphVIII	5′-AATAATACCCGGGCGAAGCATGGACGATGCGTT-3′ 5′-AATAATAAAGCTTTCAGAAGAACTCGTCCAACA-3′
nitA 3′-UTR	5′-AATAATAAAGCTTGCGGGGTCAGCAGGAGCGAC-3′ 5′-AATAATACTCGAGTCGATCAGCCTTTGCAATCC-3′

Table 2. Primers used to amplify genes and promoters for vector development.

3.3.4 Genetic transformation of *D. salina*

Electroporation

A population of *D. salina* CCAP 19/18 cells was harvested from a 250-ml culture in its exponential growth phase (approximately 1×10^6 cells ml^{-1}), washed repeatedly with pre-electroporation buffer (0.2 M mannitol, 0.2 M sorbitol) to remove residual salts, and resuspended in electroporation buffer (0.08 M KCl, 0.005 M $CaCl_2$, 0.01 M HEPES, 0.2 M mannitol, 0.2 M sorbitol) to achieve a final density of 8×10^7 cells ml^{-1} (Sun *et al.*, 2005). Next, 500 µl electroporation samples were prepared in 0.4 cm electrode gap cuvettes (Bio-Rad), which contained either 4×10^7 or 1×10^6 cells ml^{-1}, 20 mg ml^{-1} of pSP124, and 2 mg ml^{-1} of herring or salmon sperm DNA (Sigma). After a 10 minute incubation on ice, electroporation was executed using the Gene Pulser® II with Capacitance Extender Plus (Bio-Rad). Two electroporation conditions were tested, which correspond to the following respective parameters: [1] capacitance of 500 µF and voltage of 400 V (1 kV cm^{-1}) and [2] capacitance of 25 µF and voltage of 1.6 kV (4 kV cm^{-1}); the resistance of each sample was 50 Ω. Following the electric pulse, the cells were immediately supplemented with DM and allowed to recover in the dark for 12 hours at room temperature. For subsequent selection of potential transformants, the cells were plated on 1.5% agar DM plates containing 4 mg bleocin L^{-1} and monitored for a one week period.

Microparticle Bombardment

Spherical gold particles of less than 10 µm in diameter (Aldrich) were prepared by repeatedly washing with and resuspending in sterile dH_2O to achieve a concentration of 50 mg ml^{-1}. For twenty shots from the microparticle gun, approximately 20 µg of DNA was ethanol-precipitated onto 12.5 mg of gold particles (250 µl) for one hour at -80° C. After briefly spinning the gold solution at 14,000 RPM, the pellet was washed with 70% ethanol, spun again, and finally resuspended in 78% ethanol and kept on ice for use with the transformation gun.

Just before transformation, a 1-L *D. salina* CCAP 19/18 culture in exponential phase (approximately 1×10^6 cells ml^{-1}) was collected by centrifugation (Sorvall® RC-5B Refrigerated Superspeed Centrifuge, Du Pont Instruments) at 5,000 RPM for 10 minutes at 25° C (± 5) and resuspended in 5 ml of fresh DM. Cell bombardment and recovery were as described previously for transformation of *Volvox* (Schiedlmeier *et al.*, 1994). Each filter paper was then submerged in 25 ml of DM and allowed to recover without antibiotic selective pressure for two days using the culture conditions mentioned previously, with the exception of aeration, and transformed colonies were subsequently selected for using the minimum inhibitory concentration (M.I.C.) of the appropriate agent on solid medium.

3.4 Results

3.4.1 Determination of minimum inhibitory concentrations

In liquid culture, the growth rate of *D. salina* CCAP 19/18 was significantly retarded by a bleocin concentration of 0.25 mg L^{-1}. In the presence of 0.5 and 1.0 mg bleocin L^{-1}, however, growth was completely inhibited; yet, cells were able to survive for over a week. In accordance with this restricted cell division, these cells exhibited physical manifestations of stress including minimal chlorophyll pigmentation and reduced motility. Of most relevance to the current study, bleocin concentrations of 2.0 mg L^{-1} and higher proved to eventually eradicate the algal cultures, although some viable cells persisted for three to five days. Similar results were observed when *D. salina* was cultivated on solid medium containing comparable concentrations of bleocin; however, the rate of growth on agar plates was expectedly much slower. An appreciable number of *D. salina* colonies were able to survive on solid medium containing 2.0 mg bleocin L^{-1} for nearly one month, while concentrations of 4.0 mg bleocin L^{-1} and higher killed the cells within one week. Growth curves for liquid cultures that exhibited prolonged viability (0.25, 0.5, 1 mg L^{-1}) are depicted in Figure 5. Based on these findings, the M.I.C. of bleocin for this microalgal strain were found to be 2.0 mg L^{-1} in liquid culture and 4.0 mg L^{-1} on solid medium. Both conditions of selection require at least one week of exposure to the respective M.I.C. of bleocin.

The commercial herbicide Basta®, which employs PPT as its active ingredient, proved to be the most potent and fastest-acting selective agent tested. PPT concentrations of 1 mg L^{-1} and higher were able to kill *D. salina* cells within a matter of hours. In the presence of 0.5 mg PPT L^{-1}, the growth rate was essentially negligible and these cultures were not sustainable for more than three days. Lastly, although 0.25 mg PPT L^{-1} reduced the rate of cell division, cultures remained viable for over a week. This concentration of PPT also induced noticeable signs of toxicity such as inhibited motility and increased carotenoid pigmentation in the algae. On solid medium, *D. salina* was significantly more tolerant to PPT than in liquid culture. Cells survived on agar plates containing 0.25, 0.5 and 1.0 mg L^{-1} for over one month. Qualitatively, an inhibition of cell growth was reflected both in the relative number of surviving cells, which naturally was inversely proportional to PPT concentration, as well as the prolonged viability of these slower growing colonies due to their more gradual exhaustion of available nutrients. After one month, the cells were rich in β-carotene and noticeably orange in color. Figure 6 shows a side-by-side comparison of *D. salina* spot tested on 1 M NaCl medium with no selective agent added (left) and 0.25 mg PPT L^{-1} (right). The threshold for complete cell death on solid medium was clearly crossed at a PPT concentration of 2.0 mg L^{-1} after one week.

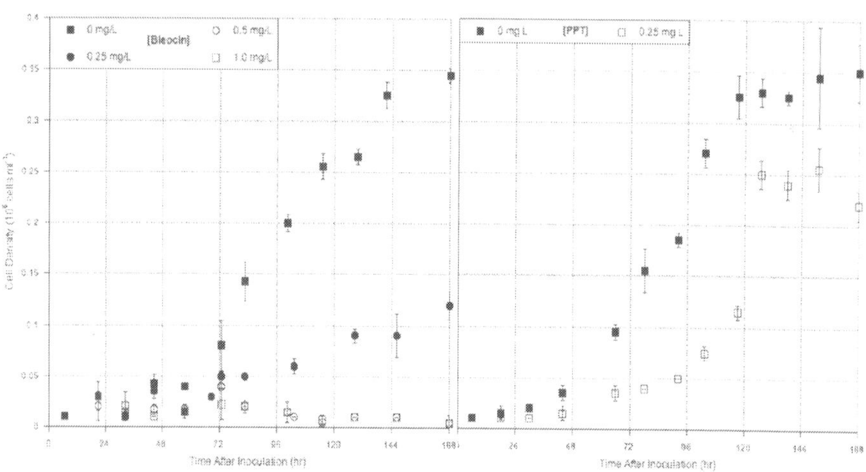

Fig. 5. Dosage response curves for *D. salina* CCAP 19/18.

While paromomycin is toxic to related microalgae, *D. salina* CCAP 19/18 was found to be remarkably insensitive to this antibiotic. In the presence of concentrations as high as 400 mg paromomycin L^{-1}, the algal cells were able to proliferate vigorously. Consequently, paromomycin was deemed unsuitable as a selective agent for the genetic transformation of *D. salina*.

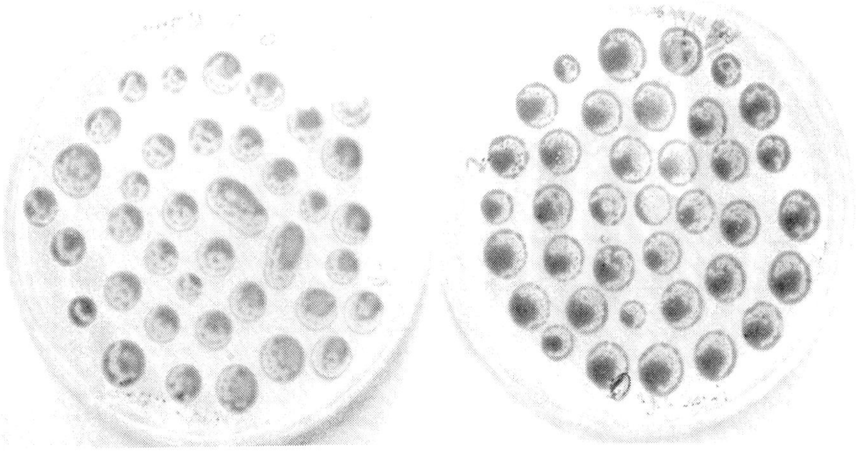

Fig. 6. Stress-induced accumulation of β–carotene by *D. salina*.

[Bleocin] mg L^{-1}	Growth Rate (Liquid)	Cell Viability (Solid)
4.0	×	×
3.0	×	–
2.0	×	+
1.0	0	+
0.5	0	+
0.25	–	+
0	+	+
[Phosphinothricin] mg L^{-1}	**Growth Rate (Liquid)**	**Cell Viability (Solid)**
4.0	×	×
3.0	×	×
2.0	×	×
1.0	×	+
0.5	×	+
0.25	–	+
0	+	+
[Paromomycin] mg L^{-1}	**Growth Rate (Liquid)**	**Cell Viability (Solid)**
400	+	+
300	+	+
200	+	+
100	+	+
50	+	+
25	+	+
0	+	+

Table 3. *D. salina* CCAP 19/18 growth response to antibiotic and herbicide exposure characterized as either normal (+), inhibited (-), or negligible (0). Concentrations that engendered death of the entire algal population within one week are denoted with an ×.

3.4.2 Probing the *D. salina* genome for constitutively active promoters

Despite others' success with genomic PCR of *Dunaliella*-specific nucleotide sequences, our efforts to obtain the *actin* and *rbcS2* promoters and *nitA* 3′-UTR were unsuccessful. Although the products from these PCR attempts appeared to be the correct fragment length, upon sequencing, it became clear that these were not the targets that we set out to amplify. BLAST analysis of some sequences recovered showed relevant homology to genes from *Dunaliella viridis* and *Arabidopsis lyrata*, but not the targeted promoters specific to *D. salina*.

3.5 Discussion

The inefficacy *D. salina* promoter and 3′-UTR amplification greatly inhibited our efforts to develop and test genetic transformation techniques with this alga. The failed attempts to amplify the *actin*, *rbcS2*, and *nitA* regulatory elements raises some concern for the accuracy of the sequences deposited in GenBank. Another possible cause for this lack of success might come from the strain of *D. salina* used as a source for these sequences. All of the prior work with these promoters and 3′-UTR has been done using the UTEX 1644 *D. salina* strain. It is conceivable that the UTEX 1664 sample supplied to us was either misidentified or contaminated. Additionally, the published sequence information might not have actually come from UTEX 1644, as claimed. From our observations, the CCAP 19/18 strain was consistently able to produce β-carotene when cells accumulated and dried on the inner surface of the culture flasks, unlike UTEX 1644. Although, the only way to know the identity of each strain for certain would be to perform genomic analysis of the 18S rRNA. This technique has been established for many species of *Dunaliella*, including both the UTEX 1664 and CCAP 19/18 strains (Olmos *et al.*, 2000; Polle *et al.*, 2008).

Due to our inability to construct *Dunaliella*-specific expression vectors, the attempts to genetically transform *D. salina* were limited to the use of the *C. reinhardtii* bleomycin-resistance plasmid, pSP124. There is evidence that genetic regulatory sequences from *D. salina* demonstrate activity in *C. reinhardtii* (Walker *et al.*, 2004); thus, it is possible that the same is true of *C. reinhardtii* promoters for use in *D. salina*. Unfortunately, after numerous trials of electroporation and microparticle bombardment, no viable transformants were recovered after selection on bleocin plates.

It was surmised that the force of impact imposed by gold microparticles would be too much for *D. salina*, which lacks a cell wall; however, electroporation should have been more accommodating. Testing both high and low voltage (4 and 1 kV cm^{-1}) electroporation conditions as well as high and low cell densities (4×10^7 or 1×10^6 cells ml^{-1}) for transformation proved unacceptable for even transient expression of the *ble* gene. We did find that, with both methods of transformation, control samples remained viable after the procedure, so at least the electrical pulse itself was not killing the cells. Without endogenous *D. salina* promoters, we are unable to determine whether the absence of transgene expression was a result of improper transformation conditions or inactive promoters.

Notwithstanding the pitfalls encountered during the molecular work with *D. salina*, experiments pertaining to antibiotic and herbicidal tolerance yielded results that will complement the microbiological understanding of *D. salina* and aid with future genetic manipulation of the organism.

3.6 Conclusions

Our approach to genetic transformation of a *Dunaliella salina* will hopefully set the stage for future efforts toward genetic engineering of this organism and, perhaps, act as a template for genetic bioprospecting with other novel algal species. Based on our dosage response experiments, we were able to narrow down the already short list of selective agents applicable to *D. salina* to the antibiotic bleomycin and the herbicide phosphinothricin and quantify the minimum inhibitory concentrations in both solid and liquid medium.

Limited by insufficient sequence information, we were unable to construct the proposed *D. salina* transformation vectors and transformation with existing *Chlamydomonas* vectors proved to be unsuccessful. *Dunaliella salina* is known to be a delicate organism due to its lack of a cell wall; thus, established transformation techniques may be too forceful for this organism. It is also reasonable to believe that gene silencing is an issue in *D. salina* and, in addition to optimizing transformation protocols suitable for this alga, molecular methods for promoting stable transgene integration and expression are of great interest to continued work in with *Dunaliella* species.

Although some accomplishments have been made in the area of *D. salina* molecular biology, genetic work with this alga warrants additional investigation. In addition to the chloroplast and mitochondrial genomes of *D. salina* CCAP 19/18, it is anticipated that the recent release of the nuclear genome will greatly encourage further genetic and metabolic engineering of this organism.

In the same way that computers are the coupling of software and hardware, microalgal cultivation systems rely on both the algal organism being grown and the vessel used to amass the cells. While one piece of software can potentially be run on various hardware devices, the two are often developed together and designed accordingly; the same is true with algal culture systems. Whether the growth environment is a raceway pond or a photobioreactor, there exist innumerable prospective algal species that could be cultivated. As the field of microalgal biotechnology moves more toward engineered algae and high-performance PBRs, the unique qualities of the organism will be paired with bioreactor design considerations. Just as the computing power of microchips is always increasing and new versions of operating systems are ever more frequently available, it is expected that algal species that are selected or engineered for high productivity will constantly demand more of their cultivation systems and vice versa.

4. References

Akama K., Puchta H., & Hohn B. 1995. Efficient *Agrobacterium*-mediated transformation of *Arabidopsis thaliana* using the bar gene as selectable marker. *Plant Cell Rep* 14:450-545.

Benemann J.R. & Oswald W.J. 1996. Algal mass culture systems. In *Systems and Economic Analysis of Microalgae Ponds for Conversion of CO2 to Biomass*. US Department of Energy, Pittsburgh, PA, pp. 42-65.

Benning C. 2008. A role for lipid trafficking in chloroplast biogenesis. *Progess in Lipid Research* 47:381-389.

Chen T., Liu H., Lu P., & Xue L. 2009. Construction of *Dunaliella salina* heterotrophic expression vectors and identification of heterotrophically transformed algal strains. *Chinese Journal of Biotechnology* 25:392-398.

Chisti Y. 2007. Biodiesel from microalgae. *Biotechnology Advances*. 25:294-306.

Chitlaru E. & Pick U. 1991. Regulation of glycerol synthesis in response to osmotic changes in *Dunaliella Plant Physiology* 96:50-60.

Danquah M.K., Gladman B., Moheimani N., & Forde G.M. 2009. Microalgal growth characteristics and subsequent influence on dewatering efficiency. *Chemical Engineering Journal* 151:73-78.

European Parliament. 2009. Directive 2009/28/EC, L 140. pp. 16-59

Feng S., Xue L., Liu H. and Lu P. 2009. Improvement of efficiency of genetic transformation for *Dunaliella salina* by glass beads method. *Molecular Biology Reports* 36:1433-1439.

Fortman J.L., Chhabra S., Mukhopadhyay A., Chou W., *et al.* 2008. Biofuel alternatives to ethanol: pumping the microbial well. *Trends in Biotechnology* 26:375-381.

Geng D., Wang Y., & Wang P. 2003. Stable expression of hepatitis B surface antigen gene in *Dunaliella salina. J Appl Phycol* 15:451-456.

Gordon J.M. & Polle J.E.W. 2007. Ultrahigh bioproductivity from algae. *Appl Microbiol Biotechnol* 76:969-975.

Gouveia L. and Oliveira A.C. 2009 Microalgae as raw material for biofuels production. *J Ind Microbial Biotechnol* 36:269-274.

Goyal A. 2007. Osmoregulation in *Dunaliella,* Part II: Photosynthesis and starch contribute carbon for glycerol synthesis during a salt stress in Dunaliella tertiolecta. *Plant Physiology and Biochemistry* 45:705-710.

Hallmann A. 2007. Algal Transgenics and Biotechnology. *Transgenic Plant Journal* 1:81-98.

Hallmann A. & Rappel A. 1999. Genetic engineering of the multicellular green alga *Volvox*: a modified and multiplied bacterial antibiotic resistance gene as a dominant selectable marker. *The Plant Journal* 17:99-109.

Harris E.H. 2001. Chlamydomonas as a Model Organism. *Annu Rev Plant Physiol Plant Mol Biol* 52:363-406.

Hejazi M.A., Holwerda E. & Wijffels R.H. 2004a. Milking microalga *Dunaliella salina* for β-carotene production in two phase bioreactors. *Biotechnol Bioeng* 85:475-481.

Hejazi M.A., Kleinegris D. & Wijffels R.H. 2004b. Mechanisms of extraction of β-carotene from *Dunaliella salina* in two-phase bioreactors. *Biotechnol Bioeng* 88:593-600.

Hoerlein G. 1994. Glufosinate (phosphinothricin), a natural amino acid with unexpected herbicidal properties. *Rev Environ Contam Toxicol* 138:73-145.

Hoffert M.I., Caldeira K., Jain A.K., Haites E.F., *et al.* 1998. Energy Implications of future stabilization of atmospheric CO2 content. *Nature* 395:881-884.

Hu Q., Guterman H. & Richmond A. 1996. A flat inclined modular photobioreactor for outdoor mass cultivation of photoautotrophs. *Biotechnol Bioeng* 51:51-60.

Huntely M.E. & Redalje D.G. 2007. CO_2 mitigation and renewable oil from photosynthetic microbes: a new appraisal. *Mitigation and Adaptation Strategies for Global Change* 12:573-608.

Jakobiak T., Mages W., Scarf B., Babinger P., *et al.* 2004. The bacterial paromomycin resistance gene, aphH, as a dominant selectable marker in *Volvox carteri. Protist* 155:381-393.

Jamers A., Blust R., & Coen W.D. 2009. Omics in algae: Paving the way for a systems biological understanding of algal phenomena? *Aquatic Toxicology* 92:114.

Janssen M., Tramper J., Mur L.R., & Wijffels R.H. 2003. Enclosed outdoor photobioreactors: light regime, photosynthetic efficiency, scale-up, and future prospects. *Biotechnol Bioeng* 81:193-210.

Jia Y., Xue L., & H. Liu J.L. 2009a. Characterization of the glyceraldehyde-3-phosphate dehydrogenase (GAPDH) dene from the halotolerant alga Dunaliella salina and inhibition of its expression by RNAi. *Current Microbiology* 58:426-431.

Jia Y., Xue L., Li J., & Liu H. 2009b. Isolation and proteomic analysis of the halotolerant alga *Dunaliella salina* flagella using shotgun strategy. Journal 2009.

Jiang G.-Z., Lu Y.-M., Niu X.-L., & Xue L.-X. 2005. The actin gene promoter-driven bar as a dominant selectable marker for nuclear transformation of *Dunaliella salina. Acta Cenetica Sinica* 32:424-433.

Jiang G.Z., Nu X.L., Lu Y.M., Xie H., *et al.* 2003. Cloning and characterization of Hsp70a cDNA fragment of *Dunaliella salina. Yi Chuan* 25: 573-576.

Jin E.S., Polle E.W., & Melis A. 2001. Involvement of zeaxanthin and of the Cbr protein in the repair of photosystem II from photoinhibition in the green alga *Dunaliella salina. Biochim Biophys Acta* 1506:244-259.

Lamers P.P., Janssen M., Vos R.C.H.D., Binoand R.J., *et al.* 2004. Exploring and exploiting carotenoid accumulation in *Dunaliella salina* for cell-factory applications. *Trends Biotechnol* 26:631-638.

Leon-Banares R., Gonzalez-Ballester D., Galvan A., & Fernandez E. 2004. Transgenic microalgae as green cell factories. *Trends in Biotechnology* 22:45-52.

Li J., Xue L., Yan H., Wang L., *et al.* 2007. The nitrate reductase gene-switch: a system for regulated expression in transformed cells of *Dunaliella salina. Gene* 403:132-142.

Liska A.J., Shevchenko A., Pick U., & Katz A. 2004. Enhanced photosynthesis and redox energy production contribute to salinity tolerance in *Dunaliella* as revealed by homology-based proteomics. *Plant Physiology* 136:2806-2817.

Liu H.T., Zang W.D., Lu Z.M., Wang N., *et al.* 2005. Cloning and analysis of psaB cDNA of *Dunaliella salina. Chinese Journal of Biotechnology* 21:643-645.

Long S.P., Humphries S., & Falkowski P.G. 1994. Photoinhibition of photosynthesis in nature. *Annu Rev Plant Physiol Plant Mol Biol* 45:633-662.

Lu P., Yan H., Li J., Liu H., *et al.* 2009. An optimal electroporation system for *Dunaliella salina. Chinese Journal of Biotechnology* 25:520 525.

Lumbreras V., Stevens D.R., & Purton S. 1998. Efficient foreign gene expression in *Chlamydomonas reinhardtii* mediated by an endogenous intron. *Plant J* 14:441-447.

Miller S.M. 2002. Taming the fierce roller: an "enhanced" understanding of cellular differentiation in Volvox. *Bioessays* 24:3-7.

Miron A.S., Gomez A.C., Camacho F.G., Grima E.M., *et al.* 1999. Comparative evaluation of compact photobioreactors for large-scale monoculture of microalgae. *Journal of Biotechnology* 70:249-270.

Muller-Feuga A. 2004. Microalgae for Aquaculture: The current global situation and future trends, Blackwell Science.

Mussgnug J.H., Thomas-Hall S., Rupprech J., Foo A., *et al.* 2007. Engineering photosynthetic light capture: impacts on improved solar energy to biomass conversion. *Plant Biotechnol J* 5:802-814.

Nitschke W.R. and Wilson C.M. 1965. Rudolph Diesel, pioneer of the age of power. The University of Oklahoma Press, Norman, OK.

Olmos J., Paniagua J., & Contreras R. 2000. Molecular identification of *Dunaliella* sp. utilizing the 18S rDNA gene. *Letters in Applied Microbiology* 30:80-84.

Polle J.E.W. & Qin S. 2009. Development of genetics and molecular tool kits for species of the unicellular green alga *Dunaliella* (Chlorophyta) Science Publishers (USA), Enfield, NH.

Polle J.E.W., Struwe L., & Jin E. 2008. Identification and characterization of a new strain of the unicellular green alga *Dunaliella salina* (Teod.) from Korea. *J Microbiol Biotechnol* 18:821-827.

Qiu B., Liu J., Liu Z., & Liu S. 2002. Distribution and ecology of the edible cyanobacterium Ge-Xian-Mi (*Nostoc*) in rice fields of Hefeng County in China. *Journal of Applied Phycology* 14:423-429.

Raja R., Hemaiswarya S., Kumar N.A., Sridhar S., *et al.* 2008. A perspective on the biotechnological potential of microalgae. *Critical Review in Microbiology* 34:77-88.

Shaish A., Ben-Amotz A., & Avron M. 1992. Biosynthesis of beta-carotene in *Dunaliella. Methods Enzymol* 213:439-444.

Sheehan J., Dunahay T., Benemann J., & Roessler P. 1998. A Look Back at the U.S. Department of Energy's Aquatic Species Program: Biodiesel from Microalgae. TP-580-24190. National Renewable Energy Lab, Golden, Colorado.

Schiedlmeier B., Schmitt R., Muller W., Kirk M.M., Gruber H., Mages W., & Kirk D.L. 1994. Nuclear transformation of *Volvox carteri. Proc Natl Acad Sci USA*, 91:5080-5084.

Shelef G., Sukenik A., & Green M. 1984. Microalgae harvesting and processing: a literature review. Solar Energy Research Institute, pp. 1-65.

Schenk P.M., Thomas-Hall S.R., Stephens E., Marx U.C., Mussgnug J.H., Posten C., Kruse O., Hankamer B. 2008. Seconds generation biofuels: high-efficiency microalgae for biodiesel production. *Bioenergy Research.* 1:20-43.

Sillers R., Show A., Tracy B., & Papoutsakis E.T. 2008. Metabolic engineering of the non-sporulating, non-solventogenic *Clostridium acetobutylicum* strain M5 to produce butanol without acetone demonstrate the robustness of the acid-formation pathways and the importance of the electron balance. *Metabolic Engineering* 10:321.

Sizova I., Fuhrmann M., & Hegemann P. 2001. A *Streptomyces rimosus* aphVIII gene coding for a new type phosphotransferase provides stable antibiotic resistance to *Chlamydomonas reinhardtii. Gene* 277:221-229.

Smith D.R., Lee R.W., Cushman J.C., Magnuson J.K., Tran D., & Polle J.E.W. 2010. The *Dunaliella salina* organelle genomes: large sequences, inflated with intronic and intergenic DNA. *BMC Plant Biology*, 10:83.

Smolinska J. 2008. EU Economic Report. ACEA - European Automobile Manufacturers' Association.

Spolaore P., Joannis-Cassan C., Duran E., & Isambert A. 2006. Commercial applications of microalgae. *Journal of Bioscience and Bioengineering* 101:87-96.

Sun G., Zhang X., Sui Z., & Mao Y. 2008. Inhibition of pds gene expression via the RNA interference approach in *Dunaliella salina* (Chlorophyta). *Mar Biotechnol* 10:219-226.

Sun Y., Yang Z., Gao X., Li Q., *et al.* 2005. Expression of foreign genes in *Dunaliella* by electroporation. *Mol Biotechnol* 30:185-192.

Tan C., Quin S., Zhang Q., Jiang P., *et al.* 2005. Establishment of a micro-particle bombardment transformation system for *Dunaliella salina*. *J Microbiol* 43:361-365.

Thompson C.J., Movva N.R., Tizard R., Crameri R., *et al.* 1987. Characterization of the herbicide-resistance gene *bar* from *Streptomyces hygroscopicus*. *EMBO J* 6:2519-2523.

Thompson G.A. 2005. Mechanisms of osmoregulation in the green alga *Dunaliella salina*. *Journal of Experimental Zoology* 268:127-132.

Tredici M.R. and Zittelli G.C. 1998. Efficiency of sunlight utilization: tubular versus flat photobioreactors. *Biotechnol Bioeng* 57:187-197.

Ugwu C.U., Aoyagi H., & Uchiyama H. 2008. Photobioreactors for mass cultivation of algae. *Bioresource Technology* 99:4021-4028.

U.S. Energy Information Administration. Annual Energy Outlook 2009 with Projections to 2030. DOE/EIA-0383(2009).

Vemuri G.N. & Aristidou A.A. 2005. Metabolic engineering in -omics era: elucidating and modulating regulatory networks. *Microbiology and Molecular Biology Reviews* 69:197-216.

Walker T.L., Becker D.K., & Collet C. 2004. Characterisation of the *Dunaliella tertiolecta* RbcS genes and their promoter activity in *Chlamydomonas reinhardtii*. *Plant Cell Rep* 23:727-735.

Walker T.L., Collet C., & Purton S. 2005. Algal transgenics in the genomic era. *Journal of Phycology* 41:1077-1093.

Wallace R., Ringer M., Pienkos P. 2008. National Renewable Energy Laoratory (NREL) Report to Congress: Algal Biofuels.

Wang T., Xue L., Hou W., Yang B., *et al.* 2007. Increased expression of transgene in stably transformed cells of *Dunaliella salina* by matrix attachment regions. *Applied Microbiology and Biotechnology* 76:651-657.

Wang T., Xue L., Xiang J., Li J., *et al.* 2009. Cloning and characterization of the 14-3-3- protein gene from the halotolerant alga *Dunaliella salina*. *Molecular Biology Reports* 36:207-214.

Weldy C.S. & Huesemann M. 2007. Lipid production by *Dunaliella salina* in batch culture: effects of nitrogen limitation and light intensity. *U.S. Department of Energy Journal of Undergraduate Research* 7:115-122.

Weyer K., Bush D., Darzins A., & Wilson D. October 24, 2008. Theoretical Maxium Algal Oil Production. *Algae Biomass Summit* Solix.

Wu Q. & Miao X. 2006. Biodiesel production from heterotrophic microalgal oil. *Bioresource Technology* 97:841-846.

Xie H., Xu P., Jia Y., Li J., *et al.* 2007. Cloning and heterologous expression of nitrate reductase genes from *Dunaliella salina*. *J Appl Phycol* 19:497-504.

Ye Z.-W., Jiang J.-G., & Wu G.-H. 2008. Biosynthesis and regulation of carotenoids in *Dunaliella*: Progresses and prospects. *Biotechnol Advances* 26:352-360.

13

Rheological Characterization of Bio-Oils from Pilot Scale Microwave Assisted Pyrolysis

Chinnadurai Karunanithy and Kasiviswanathan Muthukumarappan
Agricultural and Biosystems Engineering, South Dakota State South Dakota USA

1. Introduction

Renewable energy is gaining importance in satisfying environmental concerns and addressing economical concerns over fossil fuel usage. Lignocellulosic materials are the most abundant renewable resources on earth (Lynd et al., 2005). Energy can be obtained from biomass either biochemically or thermochemically. In the biochemical process, pretreatment of biomass is a necessary and the first step in opening up structure of the biomass cell wall to permit the access of enzymes to cellulose and hemicellulose. Pyrolysis, gasification, and combustion are the three main thermochemical processes to get energy from biomass. Combustion has a maximum efficiency of more than 30% (Yu et al., 2007). Because gasification offers higher efficiency compared to combustion, it has attracted a high level of interest (Bridgwater, 2004). According to Wornat et al (1994), the burning of bio-oils produced through the pyrolysis of biomass is more efficient. Bio-oil also offers advantages in storage and transport and in its versatility as an energy carrier and as a source of chemicals (Bridgwater, 2004).

The thermochemical process can convert a low-carbohydrate or non-fermentable biomass to alcohol fuels, thus adding technological robustness to efforts to achieve the 30 x 30 goal. Pyrolysis is an endothermic reaction wherein thermal decomposition occurs in the absence of oxygen. It is always the first step in gasification and combustion, wherein partial or total oxidation of the substrate occurs. Gas is the main product (85%) in gasification, whereas bio-oil (70-80%) is the main product in most types of pyrolysis. The yield of pyrolysis products such as syngas/ producer gas (mixture of CO and H_2), bio-oil, and bio-char (charcoal) would vary depending upon the pyrolysis methods (conventional, fast, vacuum, flash, and ultra), biomass characteristics (feedstock type, moisture content, particle size), and reaction parameters (rate of heating, temperature, and residence time). Bridgwater (2003) listed four essential features to get bio-oil from fast pyrolysis: very high heating rates (1000°C/s), high heat transfer rates (600-1000 W/cm^2), short vapor residence times (typically <2 s), and rapid cooling of pyrolysis vapors and aerosols. Because the heart of a fast pyrolysis process is the reactor, during the last two decades several different reactor designs to meet the rapid heat-transfer requirements have been explored. Achieving very high heating and heat transfer rates during pyrolysis usually require a finely ground biomass feed.

Pyrolysis using microwave irradiation is one of the many ways of converting biomass into high value products and chemicals. Not only does microwave assisted pyrolysis (MAP) not require a high degree of grinding (e.g., large chunk of wood logs can be used) as required in other fast pyrolysis, it also can handle mixed feedstocks (e.g., municipal solid wastes). Fast internal heating by microwave irradiation has an advantage over conventional heating. Moreover, microwave energy deposition in the dielectric loss mode of heating can cause spatially uniform heating (Miura et al., 2000) and is easy to control. The other technical advantages of MAP over conventional pyrolysis include the lack of necessity of size reduction (powder form) and the absence of need for agitation and fluidization; thus the resulting pyrolytic gas and bio-oils that are cleaner than those from gasification and conventional pyrolysis. The absence of a carrier gas for fluidization results in a higher heating value of the syngas produced. Because microwave heating is a mature technology, developing a microwave heating system for biomass pyrolysis would have a low cost.

Both Bioenergy (International Energy Agency) studies and work performed in Finland have estimated bio-oil to be the lowest-cost liquid biomass-based fuel. The targeted final application would dictate the desirable quality of bio-oils. For example, the calorific value, viscosity, density, surface tension, and distillation characteristics are of critical importance for fuel applications (Garcia-Perez et al., 2006a). The above characteristics can be achieved if bio-oil has (i) a low solid content, (ii) good homogeneity and stability, and (iii) a reasonably high flash point. Solantausta et al (1994) concluded that the use of bio-oil in gas turbines can be increased by optimizing their physical and chemical properties (ash content, alkali content, heating value, and viscosity). The viscosity of bio-oil affects the spray pattern and droplet size. A high viscosity of bio-oil results in high line pressure drops, thereby requiring the fuel pump to work harder in order to maintain a constant flow rate. Doll et al (2008) derived an equation describing automation characteristics of the fuel consisting of terms like characteristic number (K), density (ρ), Weber number (We), Reynolds number (Re), surface tension (γ), and kinematic viscosity (v). In order to get the desired low characteristic number (K), two physical parameters of the fuel that must be governed are the kinematic viscosity (v) and surface tension (γ). Moreover, viscosity is considered as the more important of these two factors. The viscosity of bio-oils can vary over a wide range (35-1000 cP at 40 °C) depending on the feedstock and process conditions (Bridgwater, 2004; Czernik & Bridgwater, 2004). According to Diebold (2002), an efficient collection of volatile components during bio-oil production results in a bio-oil with more low-molecular weight components with lower viscosity, better solvency properties, and possibly better storage properties.

In general, the production of bio-oil through pyrolysis is a thermodynamically non-equilibrium process. This process requires only a short residence time in a high temperature zone followed by rapid thermal quenching to produce a bio-oil that is also not at equilibrium (Ringer et al., 2006). The presences of many reactive organic compounds in the bio-oil interact to achieve equilibrium during storage. The reactions result in the formation of larger molecules and consequently increase the viscosity of the bio-oil (Diebold & Czernik, 1997; Ringer et al., 2006). Because of the high oxygen (40-50 wt %) and water content (15-30 wt %) and the low H/C ratios, bio-oils cannot be used as transportation fuels directly without prior upgrading. As mentioned earlier additional obstacles are the limited stability of the bio-oils under storage conditions due to the presence of unsaturated compounds and their minor miscibility with conventional liquid fuels (Samolada et al., 2000). Catalytic biomass pyrolysis is a promising approach due to the elimination of costly condensation and re-evaporation procedures prior to bio-oil upgrading (Samolada et al., 2000; Lu et al., 2009a).

Several studies have indicated that the viscosity of bio-oil depends on the type of feedstocks, type of pyrolyzer, and pyrolysis conditions. The type of feedstock is the main variable that affects the quality of the bio-oil apart from the postproduction processing techniques. In order to gain a better understanding of the effect of feedstock on product quality, a comparison of feedstocks is needed (Oasmaa et al., 2005a). Accordingly, the current study was undertaken to compare the viscosity of bio-oils produced from different feedstocks through microwave pyrolysis and to characterize them using storage and loss moduli.

1.1 Significance of bio-oil viscosity

Viscosity of a bio-oil is the measure of its internal friction which resists the flow of it. Viscosity is an important fuel property that should be considered when attempting to design and select handling, processing and transportation equipment. Viscosity of bio-oil affects the operation of fuel injection equipment, particularly when the increase in the viscosity affects the fluidity of fuel at low temperatures. Again, the quality and practical application of bio-oil as fuel is closely dependent on its viscosity and the elemental compositions i.e., the lower viscosity and oxygen content is desirable (Ertas & Alma, 2010). In general, bio-oil has high viscosity as compared to crude oil and diesel fuel (Onay & Kockar, 2006; Parihar et al., 2007; Pootakham & Kumar, 2010a,b). Pootakham and Kumar (2010a) reported that the loading equipment of the petroleum product such as gasoline and diesel fuels operates between 0.9 and 1.3 m^3/min, whereas they can be operated for bio-oil at a volume flow rate of 0.6 m^3/min and an operating pressure of 205 kPa (or 30 psi) for safety (Jones & Pujadó, 2006). Bio-oil is more viscous than crude oil at room temperature; however its viscosity is very similar to that of crude oil in a temperature range of 35-45°C, (Bridgewater, 1999; Thamburaj, 2000; Pootakham & Kumar, 2010a,b). In order to transport the bio-oil in pipeline, the temperature of the pipeline should be maintained in the range of 35-45°C to keep the viscosity similar to that of crude oil (Pootakham & Kumar, 2010a, b).
According to Thangalazhy-Gopakumar et al (2010), viscosity of bio-oil is relatively higher than that of diesel (0.011 Pa.s) and gasoline (0.006 Pa.s). In general, high viscosity fuel results in poor atomization and incomplete combustion, formation of excessive carbon deposits on the injection nozzles and the combustion chamber, and contamination of the lubricating oil with unburnt residues. The viscosity of the fuel directly influences atomization and mixing in the combustion chamber. In fuel application, the lower the viscosity, the easier it is to pump and to atomize and achieve finer droplets (Ji-Lu, 2008). Hence bio-oils in their original form are not suitable for use in modern diesel engines (Özaktas et al., 1997). Because of their high acidity, low thermal stability, low calorific value, high viscosity, and poor lubrication characteristics limit their use as transportation fuel (Garcia-Perez et al., 2006b). Oasmaa et al (2005) stated that for engine application, the viscosity should be in the range of 10-20 cSt with a solids content of less than 0.1 wt%. As is known, bio-oils are entirely different from petroleum fuels. There is a necessity to establish fuel specifications for commercial application of bio-oils as liquid fuels. The specifications should include the most critical properties such as viscosity, lubricity, homogeneity, stability, heating value, pH, water, flash point, solids, and ash (Qiang et al., 2008).
The viscosity of bio-oil varies depending on the temperature, feedstock, water content of the oil, amount of light ends that have been collected and the extent to which the pyrolysis oil has aged (Ji-Lu, 2008). For example, bio-oil produced from *P. indicus* and *F. mandshurica* had a kinetic viscosity of 70-350 mPa s and 10-70 mPa s separately, and bio-oil produced from rice straw had a minimum kinetic viscosity about 5-10 mPa s, which is mainly due to high water content in bio-oil from rice straw (Luo et al., 2004). The presence of water has both

negative and positive effects on the storage and utilization of bio-oils. The negative effects are, it lowers heating values, causes phase separation, increases ignition delay, and reduces combustion rates and adiabatic flame temperatures during the combustion process. Further, it leads to premature evaporation and subsequent injection difficulties during the preheating process. The positive effects are, it reduces viscosity, facilitates atomization, and reduces pollutant emissions during combustion (Calabria et al., 2007). Moreover, OH radicals from water can inhibit the formation of soot and can also accelerate its oxidation.

According to Senso¨z and Kaynar (2006), viscosity of the bio-oils is related to fatty acid chain length and number of saturated bonds. In general, the density of bio-oil is higher than that of water confirms that it contains heavy fractions (Sensöz et al., 2006). The lignin content of original feedstock has a positive influence on molecular weight and viscosity of bio-oil (Fahmi et al., 2008). Recently, Ertas and Alma (2010) compared the average molecular weight and molecular weight distribution of laurel extraction residues bio-oil (664 g mol/l and 1.52) and found they were very close to those of switchgrass bio-oil of 658 g mol/l and 1.49, respectively (He et al., 2009a). Viscosity of bio- oil increases during storage, due to slow polymerization and condensation reactions, the increase in viscosity is enhanced by higher temperature. The presence of inhibitors (hydroquinone) can dramatically reduce the rate of increase in bio-oil viscosity, due to the suppression of thermal polymerization reactions by the inhibitors (Ji-Lu, 2008). Garcia-Perez et al (2010) observed that the increase in viscosity of bio-oils is due to the solubilization of lignin derived oligomers. The condensation reactions occur ageing increases the water content in bio-oil with time (Garcia-Perez et al., 2002). The instability may be attributed to the presence of alkali metals in the ash, which are being carried over/entrained by the char particles with the vapors. These alkali metals catalyse the polymerization reactions and thereby increase the viscosity (Diebold, 2002).

Simple methods such as addition of polar solvents, diesel or other fuels can address some of the undesired bio-oil characteristics. Polar solvents, such as methanol or ethanol, can improve the volatility and heating value and decrease the viscosity and acidity. The addition of ethanol improves the volatility, stability and heating value and decreases the viscosity, acidity and corrosivity (Ji-Lu & Yong-Ping, 2010). The blending of diesel or other fuels can rearrange the viscosity of bio-oil (Onay & Kockar, 2006). In order to improve fuel properties of bio-oils, many methods are under investigation such as emulsification, hydrotreating, and catalytic cracking, which is beyond the scope of this chapter.

2. Materials and methods

2.1 Biomass selection

The low density, strength, and stiffness of aspen make it unsuitable for many structural applications (Mackes & Lynch, 2001). However, aspen has an alternative use as bedding material or feedstock for biofuel or bio-oil.The amount of biomass produced per unit area by canola depends on irrigation and varies from 5 to 10 tons/ha (Enayati et al., 2009). According to the US Canola Association, canola was cultivated on 3.2 million ha during 2009, resulting in 16-32 million tons of canola straw. Early in the US history corn cobs were an important feedstock for heating houses, farm buildings, and small businesses. Now, corn cobs are reemerging as a potential feedstock for direct combustion, gasification, and cellulosic ethanol due to numerous advantages (dense and relatively uniform size, high heat value, and low N and S contents) over many competing feedstocks. The average cob yield is about 14 % of the grain yield and represents about 16 % of the corn stover biomass in a field on a dry matter basis (Roth & Gustafson, 2010). According to Blaschek and Ezeji (2010), 15% of corn stover is corn cob. Various sources have independently estimated that anywhere from 200 to 250 million dry tons of corn stover are produced per year (Sokhansanj et al., 2002; Kadam & McMillan, 2003). Based on the 15-16% of stover as cobs, the US annual production of corn cob is estimated at 30-40 million metric tons.

2.2 Bio-oil production

All the bio-oils were produced either in batch or continuous microwave assisted pyrolysis (MAP) processes at the University of Minnesota (Figure 1). About 250 g samples were put in a 1000 mL quartz flask, which in turn was placed inside the microwave cavity (Panasonic NN-SD797S). The power level was set at 8 (the output power is about 1000 W). After the sample was microwaved for around 12 minutes, the volatile pyrolyzates were condensed with cool water at temperature of around 4-5 °C. The fraction collected from bottles connected to the bottom of the condensers was termed as the bio-oil. The condensates adhering to the interior wall of the tubes were then washed with ethanol and concentrated at 80°C using a vacuum rotovap (Buchi R-141, Flawil, Switzerland) to a near constant weight, and the concentrate was added to the bio-oil.

Fig. 1. Batch and continuous microwave assisted pyrolysis (Courtesy: Dr. R. Ruan, University of Minnesota, MN, USA)

As mentioned earlier aspen, canola, and corn cob were the feedstocks selected for the bio-oil production. Aspen pellets were used for the production of bio-oil in a batch MAP process. Canola compost pellets were also used in a batch MAP process. All corn cobs were ground to less than 1 cm in size before MAP. Corn cob 1 and corn cob 2 bio-oils were produced using a continuous MAP, where heavy fractions of the bio-oil were not collected. Corn cob 3 bio-oil was produced using a batch MAP, where heavy fractions of the bio-oil were collected. Corn cob 4 was the bio-oil produced in a batch MAP of corn cob pretreated with 4% sulfuric acid; the bio-oil contained more water and furfural than other bio-oil samples. Fig. 2 shows bio-oils from different feedstocks produced through MAP. Heavy fractions of

Fig. 2. Samples of bio-oil produced through MAP from different feedstocks

the bio-oils were collected for all batch processes. The bio-oil separated into two phases: about top one-third of the bio-oil was in oily phase and the bottom two-thirds were in aqueous phase. The oily phase is a relatively stable, light emulsion containing water soluble chemicals and light oily components; whereas the aqueous phase is a large molecule oily mixture characterized by high viscosity and water insolubles (Yang et al., 2010). The aqueous phase yield varied between batch (aspen and canola >60%) and continuous (corn cobs 10-20%) production. The aqueous phase yield of corn cobs were in agreement with corn stover (Yang et al., 2010).

2.3 Bio-oil characterization

The viscosity measurement and dynamic rheology were performed on Rheometer (ATS Rheosystems, Rheologica Instruments Inc, NJ) using cup and bob assembly. Approximately 15 mL of sample was filled in cup and shear rate was applied through the bob. The viscosity of bio-oils was tested in rheometer at 20°C with shear rate up to 1 to 200/s. The stress and viscosity was calculated for the shear rate applied using the software. In order to identify the linear viscoelastic region, stress sweep was performed at a constant frequency. For frequency sweep (1 to 100 Hz), stress was selected from the linear region identified in the stress sweep test. Temperature sweep (20-100°C) was then performed by employing constant stress and frequency identified in the previous steps. The viscoelastic modulii (G′ – storage modulus and G″ – loss modulus) were measured as a function of frequency for all the samples. A graph was plotted on logarithmic scales to identify the linear viscoelastic region.

3. Results and discussion

3.1 Effect oftemperature on viscosity of bio-oils from different feedstocks

The viscosities of bio-oils produced from different feedstocks were measured at 20, 40, 50, 60, 80, and 100°C and are shown in Fig. 3. In general, the viscosity of bio-oil at 20°C was higher than that of 100°C, irrespective of the feedstocks and with or without catalyst. This observation was in agreement with viscosity of bio-oil from pine wood chips (Thangalazhy-Gopakumar et al., 2010). Bio-oils used in this study had a higher viscosity at lower shear rate and the viscosity decreased exponentially at higher shear rate (>10/s) and similar result was reported for bio-oil from pine wood chips (Thangalazhy-Gopakumar et al., 2010). All the bio-oils used in this study showed non-Newtonian behavior as evident from Fig. 1. As mentioned in the introduction section, viscosity plays an important role in atomization through influencing inertial and aerodynamic instabilities. The Sauter mean diameter (SMD) of spray increases with viscosity for Newtonian fluid, whereas elasticity or shear thinning behavior of non-Newtonian fluid would affect the SMD. Thus, it is important to examine the non-Newtonian or viscoelastic nature of bio-oil since it may exhibit these effects during the application. According to Lu et al (2009b), most of the bio-oils behave as Newtonian fluids at temperatures lower than 80°C, whereas all the bio-oils used in this study even at high temperature (100°C) behaved as non-Newtonian fluid. A prevalent shear thinning behavior was observed at 50 and 80°C by Tzanetakis et al (2008) and similar behaviors were observed for all the bio-oils irrespective of the type of process (batch or continuous),with or without catalyst and kinds of feedstocks.

The viscosity of bio-oil between 20, 40, and 80°C were statistically different for all the feedstocks. Although the viscosities of bio-oil from corn cob were different at 20°C, the differences vanished at 80°C. In general, the viscosity of bio-oils produced from different feedstocks decreased with an increase in temperature. Similar trends were reported for bio-oils produced from different feedstocks such as softwood bark (Boucher et al., 2000a, b), sugarcane bagasse (Garcia-Perez et al., 2002), rice husk (Zhang et al., 2006), switchgrass (Boateng et al., 2007), corn stover (Yu et al., 2007), hardwood (Tzanetakis et al., 2008), pine and oak wood and bark (Ingram et al., 2008), pine wood chips (Thangalazhy-Gopakumar et al., 2010), and rice husk (Ji-Lu & Yong-Pong, 2010). When temperature was increased from 20 to 40°C, viscosity of bio-oil from canola showed a minimum decrease of 9% and bio-oil from corn cob 1 showed a maximum decrease of 25%. A further increase in temperature to 80°C resulted in viscosity decrease of 26 and 52%, respectively, for the bio-oil produced from canola and corn cob 1. Bio-oil derived from hardwood showed a similar behavior; however, the decrease was seven fold (Tzanetakis et al., 2008). The bio-oil viscosity measured at 40°C in this study was ten-fold lower than the viscosity (0.02 Pa.s) of the bio-oil produced from (heterotrophic) microalgae (Miao & Wu, 2004). The viscosity of bio-oils produced from different feedstocks though MAP was lower than the light fuel viscosity of 4 cSt (Mohan et al., 2006), the heavy fuel oil viscosity of 50 cSt (Czernik and Bridgwater, 2004), the US #4 fuel oil viscosity of 5.5-24 cSt (Oasmaa et al., 2009), commercial automotive #2 diesel viscosity of 2-4.5 cSt (Islam et al., 2010), diesel viscosity of 0.011 Pa s (Thangalazhy-Gopakumar et al.,

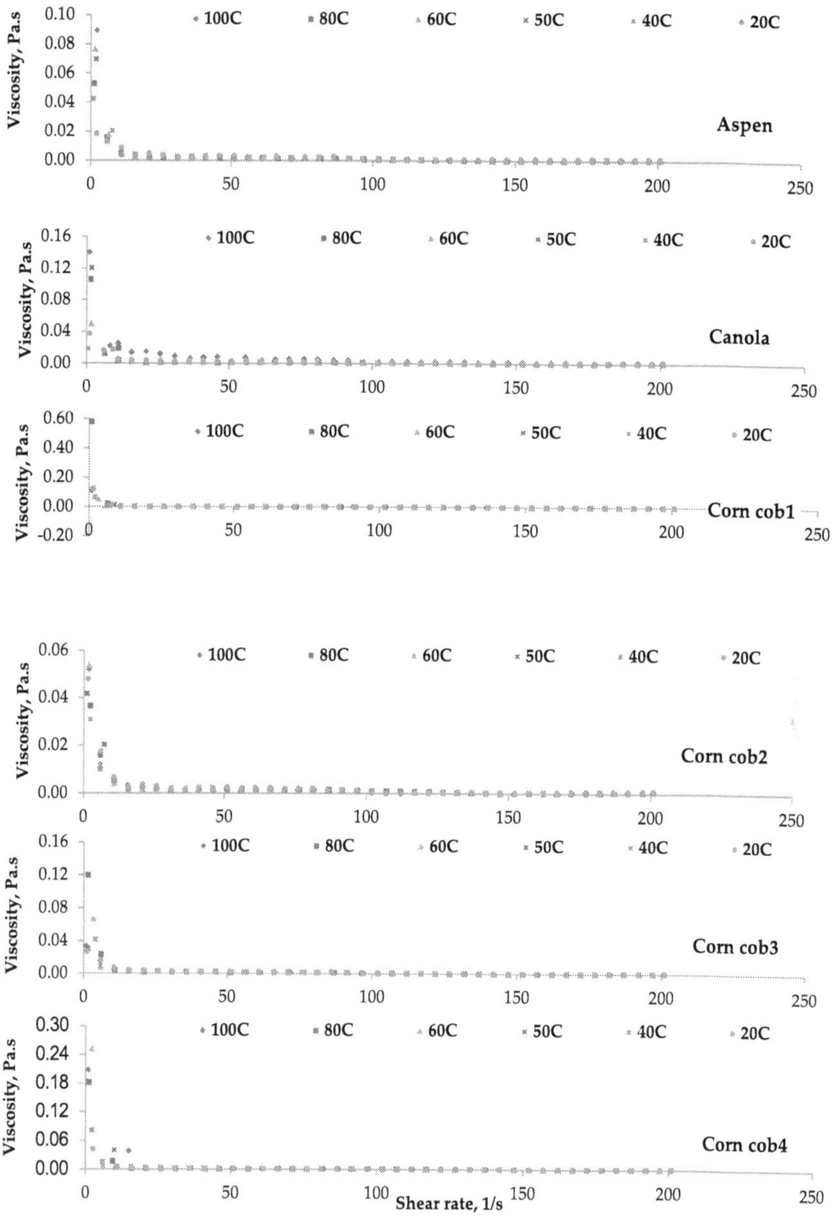

Fig. 3. Viscosity of bio-oils produced from different feedstocks at indicated temperatures

2010), and was higher than JP4 viscosity of 0.88 cSt (Chiaramonti et al., 2007) and gasoline viscosity of 0.006 Pa s (Thangalazhy-Gopakumar et al., 2010) at a temperature of 40°C. Considering the viscosity criteria (15 cSt at 35-45°C and 21.5 cSt at 30°C) presented by researchers (Pootakham & Kumar 2010a; Islam et al., 2010) for loading/handling and pipe transportation, the bio-oils from different feedstocks produced through MAP can be easy to load using existing petroleum loading equipments and easy to transport through pipe also. According to ASTM burner fuel standard, the bio-oil can have a maximum viscosity of 125 cSt at 40°C without filtering (Oasmaa et al., 2009). Considering this limit, the viscosity of the bio-oils used in this study had a much low viscosity and these bio-oils can be used as burner fuel.

Czernik and Bridgwater (2004) reported that the viscosity of bio-oil produced from wood would vary between 40 and 100 cP at 50°C, whereas the viscosity of the heavy fuel oil is 180 cP. As evident from Fig. 3, the viscosity of bio-oils produced from different feedstocks through batch or continuous MAP with or without catalyst had a significantly lower viscosity (1.5-2.2 cP) than the viscosity values reported by Czernik and Bridgwater (2004) and the viscosity of bio-oil (33 cP) from hardwood at 50°C (Tzanetakis et al., 2008). The viscosity of bio-oils used in this study was lower than that of bio-oil from sugarcane bagasse (12.1-28 cSt) measured at 60 and 80°C (Das et al., 2004). This result indicates that the bio-oils produced through MAP can be easily atomized. One possible reason for low viscosity of the bio-oils in this study was the absence of agitation and fluidization in MAP resulted in a clearer bio-oil (free from fine char/particles) than that of conventional pyrolysis.

3.2 Comparison of viscosity from different feedstocks

As noted in Fig. 3, bio-oil from canola had lower viscosity than that of aspen. Among the bio-oils produced from corn cob, the viscosity of bio-oil with catalyst (corn cob 4) was lower than other bio-oils due to more water content. The viscosity of bio-oil produced from corn cob without catalyst was similar to that of liquid phase of canola and it was lower than that of liquid phase of aspen.

All the bio-oils from different feedstocks were behaved as a non-Newtonian fluid. Similar behaviors have been reported for bio-oils (upper layer) obtained from forest residues (Garcia-Perez et al., 2006a) and pine and oak bark (Ingram et al., 2008). However, Ingram et al (2008) reported Newtonian behavior at 25°C for the bio-oils produced from pine and oak wood through auger reactor but at higher temperatures (50 and 80°C) they showed mild shear thinning behaviors. Rheological data of shear rate and shear stress of the bio-oils were fitted according to the Power-Law model

$$\mu = k\gamma^{n-1} \tag{1}$$

where, μ is the viscosity (Pa-s), k is the consistency coefficient (Pa.s^n), γ is the shear rate, 1/s and n is the flow behavior index of the fluid (dimensionless). The power law conststants for different bio-oils are presented in Table 1. The flow behavior indexes n less than 1 suggests that presence of the pseudoplastic behavior (shear thinning). A possible reason might be breakdown of (waxy) structure would result in low viscosity at high shear rate. In general, the values of flow behavior index are more reliable than that of consistency coefficient (Johnson, 1999).The deviation of flow behavior index from 'unity' indicates the degree of deviation from Newtonian behavior. For shear thinning, the index value can be anywhere between 0 and 1. The smaller the value of n, the greater is the degree of shear thinning (Chhabra & Richardson, 1999). Considering the above points, canola aqueous phase exhibited strong shear thinning behavior than that of the rest. In general, the bio-oils from corn cob approaches Newtonian behavior as the n values were close to unity.

		Temperature, °C					
Bio-oil	Power law constants	20	40	50	60	80	100
Aspen	k	0.019	0.030	0.040	0.071	0.063	0.153
	n	0.456	0.229	0.126	0.003	0.031	0.298
Canola	k	0.065	0.105	0.094	0.093	0.128	0.223
	n	0.132	0.077	0.085	0.056	0.179	0.058
Corn cob 1	k	0.030	0.047	0.040	0.058	0.123	0.103
	n	0.289	0.107	0.144	0.034	0.165	0.175
Corn cob 2	k	0.061	0.097	0.039	0.046	0.067	0.103
	n	0.084	0.074	0.143	0.056	0.083	0.173
Corn cob 3	k	0.040	0.070	0.048	0.05	0.107	0.070
	n	0.250	0.010	0.095	0.08	0.131	0.093
Corn cob 4	k	0.059	0.066	0.089	0.091	0.046	0.048
	n	0.043	0.044	0.063	0.146	0.008	0.001

Table 1. Power law constants of the bio-oils at different temperatures

3.3 Comparison of viscosity viscosity measurements and bio-oil viscosities

Accurate measurement of the viscosity of bio-oil/fuel is essential for the proper operation of fuel supply systems and atomisers. The viscosity of bio-oil can be measured according to the ASTM D 445 using the following equation

$$\eta = \pi \Pr^4 t / 8lv = \pi h\rho g r^4 t / 8lv \tag{2}$$

where η is the viscosity (dynes/cm^2 or poise), v is the volume of liquid (c.c.), t is the liquid flowing time (s), r is the radius of narrow tube (cm), l is the length of narrow tube (cm). This is most widely followed method, as evident from table 2. The viscosity of bio-oil can be measured using capillary or rotational viscometers and they are reported as kinematic (cSt) or dynamic viscosity (mPa.s). The kinematic viscosity of the bio-oil can be converted into dynamic viscosity if the density (kg/dm^3) of bio-oil is known at a given temperature using the following formula

$$\eta(\text{mPa.s}) = \eta\ (\text{cSt}) * \rho\ (\text{kg}/\text{dm}^3) \tag{3}$$

Pyrolysis type and conditions	Feedstock	Viscosity method	Temp, °C	Viscosity, cSt	Density method	Density at temp kg/ m³	Ref
Fluidized bed fast pyrolysis, 480°C	Switchgrass	ASTM D445	40	22.14			Boateng et al., 2007
		-do-	50	13.11			
		-do-	100	2.54			
Fluidized fast pyrolysis, 420-540°C	Rice husk	ASTM D445-88	20	128		1190	Ji-lu, 2007
Fluidized bed reactor, 300-600°C	Jute stick		30	2.34 cP	Density measurement bottle	1.11 g/ mL at 30°C	Asadullah et al., 2008
Fluidized bed reactor, 350-575C	Mallee wood	Fluids spectrometer RFS II rheometer	22	40-98 cP			Garcia-Perez et al (2008)
Fluidized bed fast pyrolysis, 450-550°C	Switchgrass	ASTM D445 Schott Ubbelohde capillary voscometer	40	17-28.2	ASTM D1217	1.11-1.17	He et al., 2009a
Fluidized bed fast pyrolysis	Switchgrass	ASTM D445	40	37.6	ASTM D4052	1190 at 15°C	He et al.,2009b
Bubbling fluidized bed pyrolyzer, 450-550°C	Guayule		60	2472.7		1.1382 g/ mL	Boateng et al., 2009
Auger reactor, 450°C	Pine wood	Stony Brook Model PDVa-100	50	55.2	Std	1.18 g/ mL	Hassan et al., 2009a
Bubbling fluidized bed fast pyrolysis , 500°C	Barley straw	Graber Mini VisII automatic microviscometer	40	23.5			Mullen et al., 2010
	Barley hull		40	10.2			
Fast pyrolysis VTT, 520°C, 1-2s	Pine		20	100-200		1100-1300	Oasmaa et al., 2004
Fast pyrolysis VTT, 520°C, 1-2s	Forest residue		40	15-50			Oasmaa et al., 2005b
Fast pyrolysis VTT, 520°C, 1-2s	Forest residue	ASTM D445	20	100			Oasmaa & Kuoppala, 2003
		-do-	40	30			
Fluidized bed flash pyrolysis, 530°C	Wheat straw	ASTM D445	50	11	ASTM D 4052	1.186 kg/ dm³ at 15°C	Siplia et al., 1998

	Pine wood	-do-		46		1.266 kg/ dm^3	
	Hard wood	-do-		50		1.233 kg/ dm^3	
Fast pyrolysis, 420-540°C	Rice husk	Glass capillary viscometer	40	13.2	Densitometer	1.14 g/ mL at 30°C	Lu et al., 2008
Fast pyrolysis, 300-600°C	Sugarcane bagasse	Capillary viscometer		7.1 cP	Wt/ Volume	1167	Anto & Thomas, 2009
Attrition free pyrolysis, 340-540°C, 300-900s	Beach wood chips	Dynamic stress rheometer (SR 200), shear rate 5/ s	40	3.6-6.9 cP	Wt/ Volume	1200-1230 at room temp	Guillain et al., 2009
Ablative pyrolysis, 625°C, 130 m/ s	Oak wood	Brookfield digital viscometer model LVTD	40	159 cP		1.29 g/ cm^3	Czernik et al., 1994
Slow pyrolysis, 650°C, 15°C/ min, 30 min	Rape seed cake	ASTM D88	50	42.6	ASTM 1744	900	Çulcuoglu et al., 2005
Slow pyrolysis, 600°C, 5°C/ min, 2 h	Oil palm empty fruit bunch	ASTM	50	13.52	ASTM	1.031 g/ cm^3	Khor et al., 2009
Vacuum pyrolysis, 500°C, 14 kPa	Softwood bark	ASTM D445-88 Cannon Fenske upflow viscometer (modified) Bohlin rheometer (0.5 Pa)	25	128	ASTM D369	1.066 g/ mL at 20°C	Boucher et al., 2000a
		-do-	40	38		1.051 g/ mL	
		-do-	50	24		1.043 g/ mL	
		-do-	60	15		1.034 g/ mL	
		-do-	80	7.5			
Vacuum pyrolysis, 500°C, 14 kPa	Softwood bark	ASTM D445 Cannon Fenske upflow viscometer	30	85			Boucher et al., 2000b
		-do-	50	24			
		-do-	80	8			
Vacuum pyrolysis, 530°C, 40 kPa	Softwood bark	ASTM D445-88 Brookfield viscometer model LVDV III+	50	62	ASTM D369	1.188	Ba et al., 2004
		-do-	80	15			
Vortex reactor-NREL	Hybrod poplar	Brookifield viscometer #18 spindle, shear rate 79/ s	40	22 cP		1.2 g/ cc	Diebold& Czernik, 1997
	Bagasse		60	510.3		1.0904 g/ mL	

Auger reactor, 500°C, 5.9 min	Pine chips	Brookfield synchrometric LVT viscometer	25	125.6 cP	Densitometer	1.183	Garcia-Perez et al., 2007
	Pine pellets	-do-	25	44.8 cP		1.236	
Auger reactor, 450°C, 30 s	Pine wood	Brookifield viscometer LV-DVI+, Rheometer, TA instruments Model 1000N	50	60.9 cP	Pycnometer ASTM D4052	1.19 g/ cc	Ingram et al., 2008
	Oak wood	-do-	50	41.6 cP		1.20 g/ cc	
	Pine bark	-do-	80	70 cP		1.17 g/ cc	
	Oak bark	-do-	80	131 cP		1.20 g/ cc	
Auger reactor, 450°C	Pine wood	Stony Brook 200 5B viscometer		80.7	ASTM D4052	1.19 g/ mL	Bhattacharya et al., 2009
Auger reactor, 450°C	Pine wood	Stony Brook viscometer model PDVa-100	40	30			Hassan et al., 2009b
	Pine bark	-do-		137.9			
	Cottonwood	-do-		77			
	Cottonwood bark	-do-		472.7			
Microwave, 600W, 2450Hz, 40 min	Corn stover	ASTM D445 Rotational viscometer Brookfield DV-E	20	1270 cP		1.25 g/ mL	Yu et al., 2007
		-do-	40	185 cP			
		-do-	50	60 cP			
		-do-	80	34 cP			
Fast pyrolysis 530 °C, 12 kPa	Sugarcane bagasse	ASTM D445-88 Cannon Fenske upflow viscometer	20	116.5	ASTM 369 Piconometer	1211	GarciaPerez et al., 2002
			40	26.7		1195	
			60	11.2		1180	
			80	5.4		1160	
Slow pyrolysis 500°C, 10°C/ min	Olive waste	ASTM D445-88	50	51	ASTM D1298	1070 at 15°C	SensÖz et al., 2006
Fixed bed reactor 500°C, 40°C/ min, 30 min	Rapeseed	ASTM D445-88	50	43	ASTM D1298	918 at 30°C	SensÖz et al., 2000
500°C, 50°C/ min , N_2 flow rate of 100 cm^3/ min	Safflower seed press cake	ASTM D445-88	50	225	ASTM D4052	1079 at 15°C	SensÖz & Angin, 2008

Fixed bed fire tube heating pyrolysis, 475°C	Bagasse	ASTM D445		21.5	ASTM D189	1150	Islam et al., 2010
Fixed bed Pyrolysis 500°C, N_2 flow rate of 200 mL/ min	Bagasse	ASTM D445	20	2.25-3.90 cP	Density measurement bottle	1050-1130 at 20°C	Asadullah et al., 2007
Continuous autothermal fast pyrolysis 120 kg/ h, 475°C	Rice husk	Glass capillary viscometer	40	13.2	Densitometer	1.14 g/ ml at 30°C	Quang et al., 2008
Fixed bed reactor 500°C, [illegible]0°C/ min, N_2 flow rate of 100 mL/ min	Laurel	ASTM D445-88	40	61	ASTM D4052	1133 at 15°C	Ertas &Alma, 2010
Fixed bed reactor 550°C, N_2 flow rate of 173 cm^3/ min	Bagasse	ISO 3904 Cannon 5R-33 viscometer	40 100	232.3 15.63	ASTM 5002	1254.2 at 20°C	Parihar et al., 2007
Flash pyrolysis under N_2 500°C	Rapeseed	ASTM D445-88	50	36	ASTM D1298	984 at 30°C	Onay & Kockar, 2006
Fixed bed reactor 400°C, 50°C/ min	Soybean oil cake	ASTM D445-88	50	72.38	ASTM D4052	1107 at 15°C	Sensoz & Kaynar, 2006

Kinematic viscosity- cSt; Dynamic viscosity-cP

Table 2. Comparison of viscosity measurement and bio-oil viscosity produced from various feedstocks through different pyrolysis

According to ASTM D445, the viscosity of standard fuels, which are Newtonian fluids, is typically measured as kinematic viscosity. Leroy et al (1988) conducted extensive studies on rheological characterization of several bio-oils from wood and concluded that those bio-oils exhibited an essentially Newtonian behavior at the shear rate range of 1 to 1000/s. In contrary, bio-oils used in this study showed shear thinning behavior. For Newtonian fluid, the viscosity remains constant with increasing shear rate. Radovanovic et al (2000) reported a procedure to measure bio-oil viscosity using falling ball viscometer. Recently, Osamaa et al (2009) recommended using Cannon-Fenske viscometer tubes because the flow direction in these tubes compared to Ubbelohde tubes ensured more accurate results with dark coloured liquids. No prefiltration of the sample is required if the bio-oil is visually homogenous. Elimination of air bubbles before sampling and an equilibration time of 15 min are essential for viscosity measurement at a given temperature. Comparison of viscosity measurement and viscosity of bio-oils from produced different feedstocks through different pyrolysis reactors are presented in Table 2. Bio-oil density measurement and density values are also included for converting the viscosity from kinematic to dynamic units. According to ASTM D445-88, viscosity should be measured at 20 and 40°C, as seen from the table the viscosity was reported at different temperatures ranging from 20 to 100°C. The viscosity of bio-oils used in this study had lower viscosity than the viscosities listed in the table irrespective of temperatures.

3.4 Stress and frequency sweep

In general, the linear limits of viscoelastic behavior are determined by identifying the range of stress values over which G′ and G″ are constant and thus independent of stress. Storage and loss modulii of bio-oil from different feedstocks as a function of shear stress are depicted in Fig. 4. Shear stress from 10-100 Pa exhibited the linear regions for all the bio-oils as noted in Fig. 4. A shear stress of 55 Pa was selected for frequency and temperature sweep. As evident from the figure, the storage modulus (G′) was predominant than that of loss modulus irrespective of the feedstocks, whereas Ba et al (2004) observed the loss-modulus as dominant behavior of the bio-oil produced from softwood bark through vacuum pyrolysis. In another study, Garcia-Perez et al (2008) reported that the storage modulus was lower than that of loss modulus for the bio-oil produced from mallee woody biomass. Ba et al (2004) identified four different regions including two plateaus for G′, which was not observed in this study. The bio-oils from corn cob (3 and 4) produced in a batch MAP had a high storage modulus than that of bio-oils from corn cob produced in continuous MAP as evident from Fig. 5. The bio-oil produced from batch MAP (corn cob 3 and 4) had lower loss modulus than the bio-oil from continuous MAP of corn cobs and batch MAP of other feedstocks. The storage and loss moduli were similar for the bio-oils produced from corn cob in a batch MAP with or without catalyst (Fig. 5).

The frequency sweep was conducted in the range of 0.1-100 Hz and found the linear region between 1 and 100 Hz. Accordingly, the frequency sweep experiments were repeated and identified the linear region between 10 and 100 Hz as shown in Fig. 3. As G′ approaches a slope of more than 2, which confirms the existence of linear viscoelastic region (Tzanetakis et al., 2008). Similarly, G″ approaches a slope of more than 2 (less than slope of G′), which is also consistent with linear viscoelastic behavior. A frequency of 50 Hz was selected from the linear range depicted in Fig. 4 for temperature sweep. The frequency sweep was also confirmed that the storage modulus was predominant than that of respective loss modulus of each bio-oil. Tzanetakis et al (2008) reported that the loss modulus was ten times higher than that of storage modulus for the bio-oil produced from hardwood, and it was an opposite observation. Similar to stress sweep, the bio-oil produced from corn cob 3 and 4 in a batch MAP had lower loss modulus than the bio-oil from other feedstocks including corn cob. The storage and loss moduli were overlapping between the bio-oils at a low frequency (10Hz) and a clear difference was observed as frequency increases. The bio-oil from aspen and canola had a higher storage and loss moduli than that of bio-oils from corn cob.

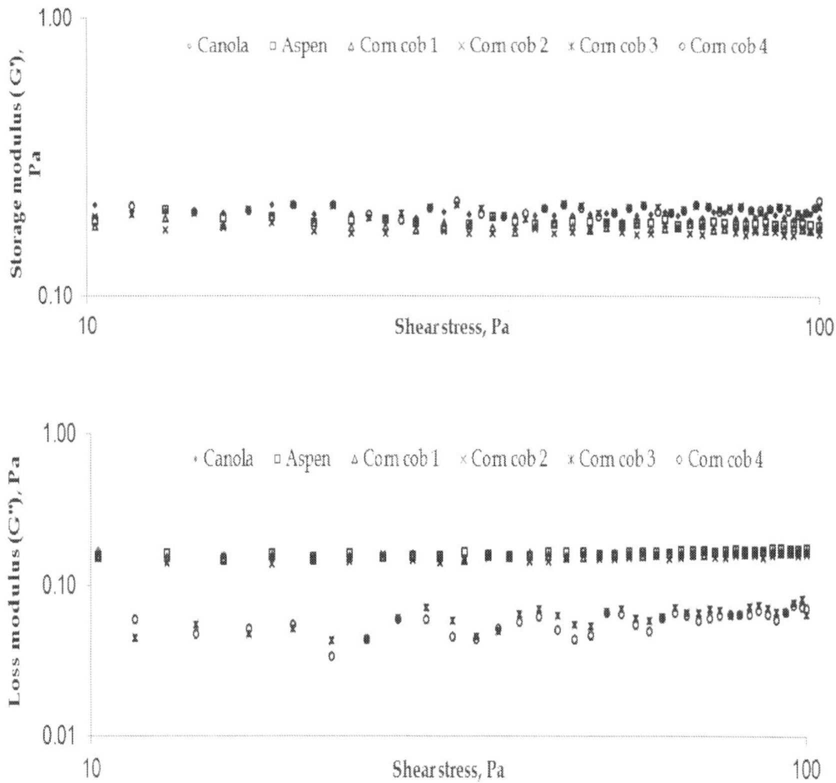

Fig. 4. Stress sweep of bio-oils from produced from different feedstocks

3.5 Temperature sweep

Temperature sweep of the bio-oils were conducted with a shear stress of 55 Pa and a frequency of 50 Hz. The rate of temperature increase was maintained at 10°C/min. The storage and loss modulii trend for all the bio-oils are shown in Fig. 6. The moduli of bio-oil from canola showed a decreasing trend once the temperature reached 80°C. Again, the temperature sweep also confirmed that the storage modulus was more predominant as compared to loss modulus of the bio-oils from different feedstocks. Typically bio-oils are Newtonian fluids (Oasmaa & Peacocke, 2001; Oasmaa et al., 2003); however, in Fig. 6 storage and loss moduli lines confirm that the behavior of bio-oils as non-Newtonian fluids irrespective of the feedstocks.

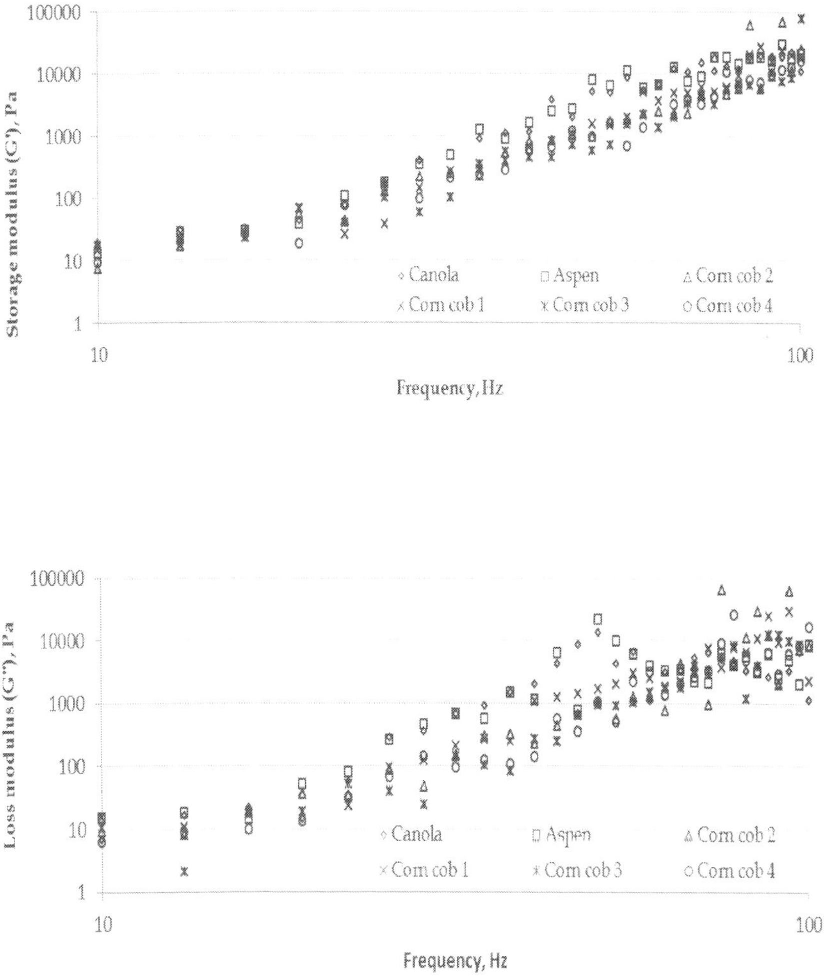

Fig. 5. Frequency sweep of bio-oils from produced from different feedstocks

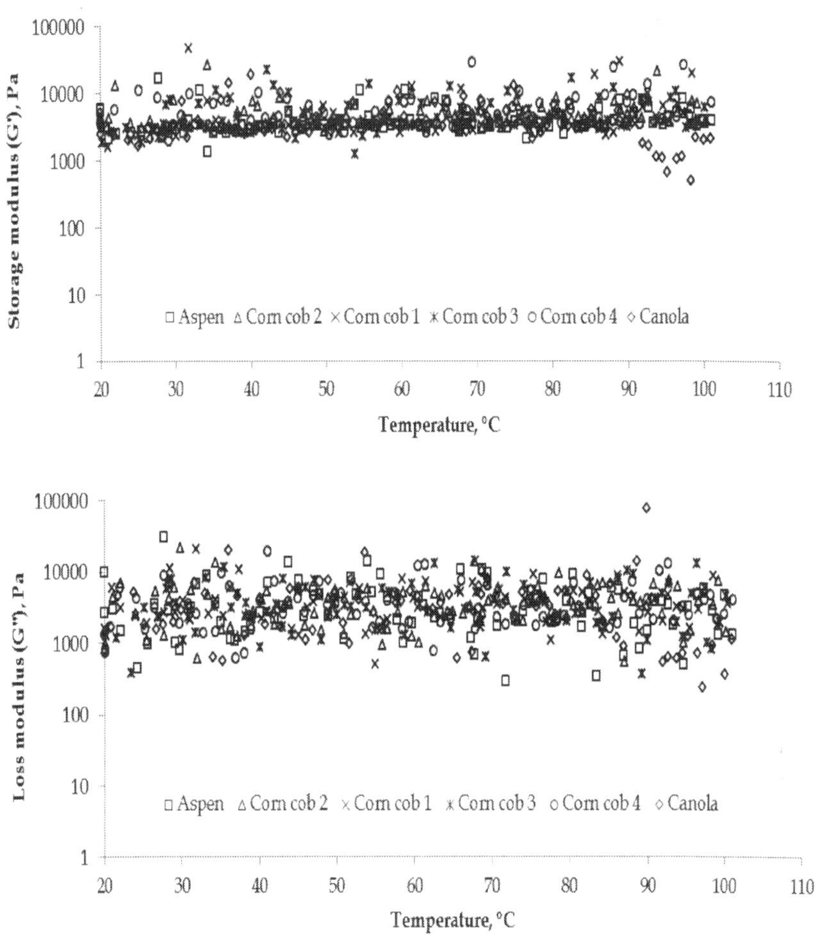

Fig. 6. Temperature sweep of bio-oils produced from different feedstocks

4. Conclusions

When handling bio-oils, viscosity is an important parameter. A thorough rheological characterization was performed on bio-oils produced from aspen, canola, and corn cobs through MAP. Results from simple viscosity and dynamic rheology indicated that the bio-oils tested in this study behaved as non-Newtonian fluids. The viscosities of the bio-oils were in the range of 0.00272 - 0.00196 Pa.s at 20°C and 0.00236 - 0.00163 Pa.s at 40°C. These viscosities are much less than the viscosities of bio-oils derived from different feedstocks reported in the previous literature. Moreover, the bio-oils used in this study had a lower viscosity than that of light and heavy fuel oil, the US #4 fuel oil, and ASTM burner fuel standard indicating varieties of potential application. The viscosity range of the bio-oils indicates that they are easy to handle and processing; however, viscosity is not the only factor deciding the application of bio-oil. Therefore, other factors should be investigated to assess the suitability of these bio-oils.

5. Acknowledgements

This research was supported by funding from the Agricultural Experiment Station and North Central Sun Grant Center at South Dakota State University through a grant provided by the US Department of Transportation, Office of the Secretary, Grant No.DTOS59-07-G-00054. Also, Bio-oils provided by Dr. Roger Ruan, University of Minnesota for conducting this study was greatly appreciated.

6. References

Anto, L. P., & Thomas, S. (2009). Production of bio-oil from pyrolysis of bagasse. Proceedings of International Conference on Energy and Environment ISSN: 2070-3740, pp 558-559, March 19-21, 2009.

Asadullah, M., Rahman, M. A., Ali, M. M., Rahman, M. S., Motin, M. A., Sultan, M. B., & Alam, M. R. (2007). Production of bio-oil from fixed bed pyrolysis of bagasse. *Fuel* 86, 2514-2520, 0016-2361

Asadullah, M., Rahman, M. A., Ali, M. M., Motin, M. A., Sultan, M. B., Alam, M. R., & Rahman, M. S. (2008). Jute stick pyrolysis for bio-oil production in fluidized bed reactor. *Bioresource Technology* 99, 44-50, 0960-8524

Ba, T., Chaala, A ., Garcia-Perez, M., Rodrigue, D., & Roy, C. (2004). Colloidal properties of bio-oils obtained by vacuum pyrolysis of softwood bark. Characterization of water-soluble and water-insoluble fractions. *Energy & Fuels* 18, 704-712, 0887-0624

Bhattacharya, P., Steele, P. H., Hassan, E. M., Mitchell, B., Ingram, L., & Pittman Jr, C. U. (2009). Wood/plastic copyrolysis in an auger reactor: Chemical and physical analysis of the products. *Fuel* 88, 1251-1260, 0016-2361

Blaschek, H. P., & Ezeji, T. C. (2010). Science of Alternative Feedstocks. http://www.farmdoc.illinois.edu/policy/research_reports/ethanol_report/ethanol%20report%20-%20ch%207.pdf. Accessed on April 15, 2010

Boateng, A. A., Daugaard, D. E., Goldberg, N. M., & Hicks, K. B. (2007). Bench-scale fluidized-bed pyrolysis of switchgrass for bio-oil production. *Industrial & Engineering Chemistry Research* 46, 1891-1897, 0888-5885

Boateng, A. A., Mullen, C. A., Goldberg, N. M., Hicks, K. B., McMahan, C. M., Whalen, M. C., & Cornish K. (2009). Energy-dense liquid fuel intermediates by pyrolysis of guayule (*Parthenium argentatum*) shrub and bagasse. *Fuel* 88, 2207-2215, 0016-2361

Boucher, M. E., Chaala, A., & Roy, C. (2000a). Bio-oils obtained by vacuum pyrolysis of softwood bark as a liquid fuel for gas turbines. Part I: Properties of bio-oil and its blends with methanol and a pyrolytic aqueous phase. *Biomass and Bioenergy* 19, 337-350, 0961-9534

Boucher, M. E., Chaala, A., Pakdel, H., & Roy, C. (2000b). Bio-oils obtained by vacuum pyrolysis of softwood bark as a liquid fuel for gas turbines. Part II: Stability and ageing of bio-oil and its blends with methanol and a pyrolytic aqueous phase. *Biomass and Bioenergy* 19, 351-361, 0961-9534

Bridgewater, A. V. (1999). Principle and practice of biomass pyrolysis process for liquid. *Journal of Analytical and Applied Pyrolysis* 51, 3–22, 0165-2370

Bridgwater, A. V. *(2003)*. Renewable fuels and chemicals by thermal processing of biomass. Chemical Engineering Journal 91, *87–102*, 1385-8947

Bridgwater, A. V. (2004). Biomass fast pyrolysis. *Thermal Science* 8(2), 21–49, 0354-9836

Calabria, R., Chiariello, F., & Massoli, P. (2007). Combustion fundamentals of Pyrolysis oil based fuels. *Experimental Thermal and Fluid Science* 31, 413–420, 0894-1777

Chhabra, R. P., & Richardson J. F. (1999). Non-Newtonian flow in the process industries: Fundamentals and Engineering Applications. Butterworth-Heinemann. p:10. ISBN: 0750637706.

Chiaramonti, D., Oasmaa, A., & Solantausta, Y. (2007). Power generation using fast pyrolysis liquids from biomass. *Renewable and Sustainable Energy Reviews* 11(6), 1056–1086, 1364-0321

Çulcuoglu, E., Ünay, E., Karaosmanoglu, F., Angin, D., & Şensöz, S. (2005). Characterization of the bio-oil of rapeseed cake. *Energy Sources* 27, 1217–1223, 0090-8312

Czernik, S., & Bridgwater, A. V. (2004). Overview of applications of biomass fast pyrolysis oil. *Energy & Fuels* 18, 590–598, 0887-0624

Czernik, S., Johnson, D. K., & Black, S. (1994). Stability of wood fast pyrolysis oil. *Biomass and Bioenergy* 7, 187–192, 0961-9534

Das, P., Ganesha, A., & Wangikar, P. (2004). Influence of pretreatment for deashing of sugarcane bagasse on pyrolysis products. *Biomass and Bioenergy* 27, 445–457, 0961-9534

Diebold, J. P., & Czernik, S. (1997). Additives to lower and stabilize the viscosity of pyrolysis oils during storage. *Energy & Fuels* 11, 1081–1091, 0887-0624

Diebold, J. P. (2002). A review of the chemical and physical mechanisms of the storage stability of fast pyrolysis biooils. In: Bridgwater AV, Editor. Fast pyrolysis of biomass: A handbook, vol. 2. UK: CPL Press. ISBN 1872691471

Doll, K. M., Sharma, B. K., Suarez, P. A. Z., & Erhan S. Z. (2008). Comparing biofuels obtained from pyrolysis, of soybean oil or soapstock, with traditional soybean biodiesel: Density, kinematic viscosity, and surface tensions. *Energy & Fuels* 22, 2061–2066, 0887-0624

Enayati, A. A., Hamzeh, Y., Mirshokraie, S. A., & Molalii, M. (2009). Paper from canola stalks. *BioResources* 4(1), 245–256, 1930-2126

Ertas, M., & Alma, H. (2010). Pyrolysis of laurel (*Laurus nobilis L.*) extraction residues in a fixed-bed reactor: Characterization of bio-oil and bio-char. *Journal of Analytical and Applied Pyrolysis* 88, 22–29, 0165-2370

Fahmi, R., Bridgwater, A. V., Donnison, I., Yates, N., & Jones, J. M. (2008). The effect of lignin and inorganic species in biomass on pyrolysis oil yields, quality and stability. *Fuel* 87, 1230– 1240, 0016-2361

Garcia-Perez, M., Chaala, A., & Roy, C. (2002). Vacuum pyrolysis of sugarcane bagasse. *Journal of Analytical and Applied Pyrolysis*. 65, 111-136, 0165-2370

Garcia-Perez, M., Chaala, A., Pakdel, H., Kretschmer, D., Rodrigue, D., & Roy, C. (2006a). Multiphase structure of bio-oils. *Energy & Fuels* 20, 364–375, 0887-0624

Garcia-Perez, M., Lappas, P., Hughes, P., Dell, L., Chaala, A., Kretschmer, D., & Roy, C. (2006b). Evaporation and combustion characteristics of bio-oils obtained by vacuum pyrolysis of wood industry residues. *IFRF combustion J.* Article No 200601,1562-479X

Garcia-Perez, M., Adams, T. T., Goodrum, J. W., Geller, D. P., & Das K. C. (2007). Production and fuel properties of pine chip bio-oil/biodiesel blends. *Energy & Fuels* 21, 2363–2372, 0887-0624

Garcia-Perez, M., Wang, X. S., Shen, J., Rhodes, M. J., Tian, F., Lee, W-J., Wu, H., & Li, C-Z. (2008). Fast pyrolysis of oil mallee woody biomass: effect of temperature on the yield and quality of pyrolysis products. *Industrial & Engineering Chemistry Research* 47, 1846–1854, 0888-5885

Garcia-Perez, M., Adams,T. T., Goodrum, J. W., Das, K. C., & Geller, D. P. (2010). DSC studies to evaluate the impact of bio-oil on cold flow properties and oxidation stability of bio-diesel. *Bioresource Technology* 101, 6219–6224, 0960-8524

Guillain, M., Fairouz, K., Mar, S. R., Monique, F., & Jacques, L. (2009). Attrition-free pyrolysis to produce bio-oil and char. *Bioresource Technology* 100, 6069–6075, 0960-8524

Hassan, E. M., Steele, P. H., & Ingram, L. (2009a). Characterization of fast pyrolysis bio-oils produced from pretreated pine wood. *Applied Biochemistry and Biotechnology* 154, 182–192, 0273-2289

Hassan, E. M., Yu, F., Ingram, L., & Steele, P. (2009b). The potential use of whole-tree biomass for bio-oil fuels. *Energy Sources, Part A* 31, 1829–1839, 1556-7036

He, R., Ye, X. P., English, B. C., & Satrio, J. A. (2009a). Influence of pyrolysis condition on switchgrass bio-oil yield and physicochemical properties. *Bioresource Technology* 100, 5305–5311, 0960-8524

He, R., Ye, X. P., Harte, F., & English, B. (2009b). Effects of high-pressure homogenization on physicochemical properties and storage stability of switchgrass bio-oil. *Fuel Processing Technology* 90, 415–421, 0378-3820

Ingram, L., Mohan, D., Bricka, M., Steele, P., Strobel, D., Crocker, D, Mitchell, B., Mohammad, J., Cantrell, K., & Pittman, Jr C. U. (2008). Pyrolysis of wood and bark in an auger reactor: Physical properties and chemical analysis of the produced bio-oils. *Energy & Fuels* 22, 614–625, 0887-0624

Islam, M. R., Parveen, M., & Haniu, H. (2010). Properties of sugarcane waste-derived bio-oils obtained by fixed-bed fire-tube heating Pyrolysis. *Bioresource Technology* 101, 4162–4168, 0960-8524

Ji-lu, Z. (2007). Bio-oil from fast pyrolysis of rice husk: Yields and related properties and improvement of the pyrolysis system. *Journal of Analytical and Applied Pyrolysis* 80, 30–35, 0165-2370

Ji-Lu, Z. (2008). Pyrolysis oil from fast pyrolysis of maize stalk. *Journal of Analytical and Applied Pyrolysis* 83, 205–212, 0165-2370

Ji-Lu, Z., & Yong-Ping, K. (2010). Spray combustion properties of fast pyrolysis bio-oil produced from rice husk. *Energy Conversion and Management* 51, 182–188, 0196- 8904

Johnson, A. T. (1999). Biological process engineering: an analogical approach to fluid flow, heat transfer, and mass transfer applied to biological systems. John Wiley & Sons, ISBN: 0471245447 p 208.

Jones, D. S. J., & Pujadó, P. P. (2006). Handbook of Petroleum Processing, first ed.Springer, Berlin. Chapter 13, p. 545

Kadam, K. L., & McMillan, J. D. (2003). Availability of corn stover as a sustainable feedstock for bioethanol production. *Bioresource Technology* 88, 17–25, 0960-8524

Khor, K. H., Lim, K. O., & Zainal, Z. A. (2009). Characterization of bio-oil: A by-product from slow pyrolysis of oil palm empty fruit bunches. *American Journal of Applied Sciences* 6 (9), 1647-1652, 1546-9239

Leroy, J., Choplin, L., & Kaliaguine, S. (1988). Rheological characterization of pyrolytic wood derived oils: Existence of a compensation effect. *Chemical Engineering Communications* 71(1), 157-176, 0098-6445

Lu, Q., Yang, X-L., & Zhu, X-F. (2008). Analysis on chemical and physical properties of bio-oil pyrolyzed from rice husk. *Journal of Analytical and Applied Pyrolysis* 82, 191–198, 0165-2370

Lu, Q., Zhu, X-F., Li, W. Z., Zhang, Y., & Chen, D. Y. (2009a). On-line catalytic upgrading of biomass fast pyrolysis products. *Chinese Science Bulletin* 54, 1941–1948, 1001-6538

Lu, Q., Li, W-Z., & Zhu, X-F. (2009b). Overview of fuel properties of biomass fast pyrolysis oils. *Energy Conversion and Management* 50, 1376–1383, 0196- 8904

Luo, Z., Wang, S., Liao, Y., Zhou, J., Gu, Y., & Cen, K. (2004). Research on biomass fast pyrolysis for liquid fuel. *Biomass and Bioenergy* 26, 455 - 462, 0961-9534

Lynd, L. R., van Zyl, W. H., McBride, J. E., & Laser, M. (2005). Consolidated bioprocessing of cellulosic biomass: An update. *Current Opinion Biotechnology* 16, 577–583, 0958-1669

Mackes, K. H., & Lynch, D. L. (2001). The effect of aspen wood characteristics and properties on utilization. USDA Forest Service Proceedings RMRS-P-18. 2001. Pp 429–440.

Miao, X., & Wu, Q. (2004). High yield bio-oil production from fast pyrolysis by metabolic controlling of *Chlorella protothecoides*. *Journal of Biotechnology* 110, 85–93, 0168-1656

Miura, M., Kaga, H., Tanaka, S., Takanashi, K., & Ando K. J. (2000). Rapid microwave pyrolysis of wood. Journal of *Chemical Engineering Japan* 33(2), 299–302, 0021-9592

Mohan, D., Charles, U. P., & Philip, H. S. (2006). Pyrolysis of wood/biomass for bio-oil: A critical review. *Energy & Fuels* 20, 848–889, 0887-0624

Mullen, C. A., Boateng, A. A., Hicks, K. B., Goldberg, N. M., & Moreau R. A. (2010). Analysis and comparison of bio-oil produced by fast pyrolysis from three barley biomass/byproduct streams. *Energy & Fuels* 24, 699–706, 0887-0624

Oasmaa, A. & Peacocke, C. (2001). A guide to physical property characterisation of biomass-derived fast pyrolysis liquids; VTT Publication 450; VTT: Espoo, Finland, 65 pp + appendices (34 pp).

Oasmaa, A., Kuoppala, E., Gust, S., & Solantausta, Y. (2003). Fast pyrolysis of forestry residue. 1. Effect of extractives on phase separation of pyrolysis liquids. *Energy & Fuels* 17(1), 1–12, 0887-0624

Oasmaa, A., & Kuoppala, E. (2003). Fast pyrolysis of forestry residue. 3. Storage stability of liquid fuel. *Energy & Fuels* 17, 1075–1084, 0887-0624

Oasmaa, A., Kuoppala, E., Selin, J-F, Gust, S., & Solantausta, Y. (2004). Fast pyrolysis of forestry residue and pine. 4. Improvement of the product quality by solvent addition. *Energy & Fuels* 18, 1578–1583, 0887-0624

Oasmaa, A., Peacocke, C., Gust, S., Meier, D., & McLellan, R. (2005). Norms and standards for pyrolysis liquids. End-user requirements and specifications. *Energy & Fuels A-I*, 0887-0624

Oasmaa, A., Peacocke, C., Gust, S., Meier, D., & McLellan, L. (2005a). Norms and standards for pyrolysis liquids: End-user requirements and specifications. *Energy & Fuels* 19, 2155–2163, 0887-0624

Oasmaa, A., Sipilae, K., Solantausta, Y., & Kuoppala, E. (2005b). Quality improvement of pyrolysis liquid: Effect of light volatiles on the stability of pyrolysis liquids. *Energy & Fuels* 19, 2556–2561, 0887-0624

Oasmaa, A., Elliott , D. C., & Muller, S. (2009). Quality control in fast pyrolysis bio-oil production and use. *Environmental Progress & Sustainable Energy* 28(3), 404–409, 1944-7442

Onay, O., & Kockar, O. M. (2006). Pyrolysis of rapeseed in a free fall reactor for production of bio-oil. *Fuel* 85, 1921–1928, 0016-2361

Özaktas, T., Cıg˘ızog˘lu, K. B., & Karaosmanog˘lu, F. (1997). Alternative diesel fuel study on four different types of vegetable oils of Turkish origin. *Energy Sources* 19, 173–181, 0090-8312

Parihar, M. F., Kamil, M., Goyal, H. B., Gupta, A. K., & Bhatnagar, A. K. (2007). An experimental study on pyrolysis of biomass. *Trans IChemE, Part B, Process Safety and Environmental Protection* 85(B5), 458–465, 0957-5820

Pootakham, T., & Kumar, A. (2010a). Bio-oil transport by pipeline: A techno-economic assessment. *Bioresource Technology* 101, 7137–7143, 0960-8524

Pootakham, T., & Kumar, A. (2010b). A comparison of pipeline versus truck transport of bio-oil. *Bioresource Technology* 101, 414–421, 0960-8524

Qiang, L., Xu-lai, Y., & Xi-Feng, Z. (2008). Analysis on chemical and physical properties of bio-oil pyrolyzed from rice husk. *Journal of Analytical and Applied Pyrolysis* 82, 191–198, 0165-2370

Radovanovic, M., Venderbosch, R. H., Prins, W., & van Swaaij, W. P. M. (2000). Some remarks on the viscosity measurement of pyrolysis liquids. *Biomass and Bioenergy* 18, 209–222, 0961-9534

Ringer, M., Putsche, V., & Scahill, J. (2006). Large-scale pyrolysis oil production: a technology assessment and economic analysis. NREL/TP-510-37779. National Renewable Energy Laboratory, Golden, Colorado.

Roth, G., & Gustafson, C. (2010). Corn cobs for biofuel production. http://www.extension.org/pages/Corn_Cobs_for_Biofuel_Production. Accessed on April 15, 2010.

Samolada, M. C., Papafotica, A., & Vasalos, I. A. (2000). Catalyst evaluation for catalytic biomass pyrolysis. *Energy & Fuels* 14, 1161–1167, 0887-0624

Sensöz, S., Angin, D., & Yorgun, S. (2000). Inuence of particle size on the pyrolysis of rapeseed (*Brassica napus L.*): Fuel properties of bio-oil.*Biomass and Bioenergy* 19, 271–279, 0961-9534

Sensöz, S., & Kaynar, I. (2006). Bio-oil production from soybean (*Glycine max L.*); fuel properties of bio-oil. *Industrial Crops and Products* 23, 99–105, 0926-6690

Sensöz, S., Demiral, I., & Ferdi-Gercel, H. (2006). Olive bagasse (*Olea europea L.*) pyrolysis. *Bioresource Technology* 97, 429–436, 0960-8524

Sensöz, S., & Angın, D. (2008). Pyrolysis of safflower (*Charthamus tinctorius L.*) seed press cake in a fixed-bed reactor: Part 2. Structural characterization of pyrolysis bio-oils. *Bioresource Technology* 99, 5498–5504, 0960-8524

Sipilaè, K., Kuoppala, E., Fagernaès, L., & Oasmaa, A. (1998). Characterization of biomass-based flash pyrolysis oils. *Biomass and Bioenergy* 14(2), 103–111, 0961-9534

Sokhansanj, S., Turhollow, A., Cushman, J., & Cundiff, J. (2002). Engineering aspects of collecting corn stover for bioenergy. *Biomass and Bioenergy* 23, 347–355, 0961-9534

Solantausta, Y., Nylund, N. O., & Gust, S. (1994). Use of pyrolysis oil in a test diesel engine to study the feasibility of a diesel power plant concept. *Biomass and Bioenergy* 7, 297–306, 0961-9534

Thamburaj, R. (2000). Dynamotive engineering. Fast pyrolysis of biomass for green power generation. <http://www.dynamotive.com> (accessed 20.06.06.).

Thangalazhy-Gopakumar, S., Adhikari, S., Ravindran, H., Gupta, R. B., Fasina, O., Tu, M., & Fernando, S. D. (2010). Physiochemical properties of bio-oil produced at various temperatures from pine wood using an auger reactor. *Bioresource Technology* 101(21), 8389-8395, 0960-8524

Tzanetakis, T., Ashgriz, N., James , D. F., & Thomson M. J. (2008). Liquid fuel properties of a hardwood-derived bio-oil fraction. *Energy & Fuels* 22, 2725–2733, 0887-0624

Wornat, M. J., Porter, B. J., & Yang, N. Y. (1994). Single droplet combustion of biomass pyrolysis oils. *Energy& Fuels* 8, 1131–1142, 0887-0624

Yang, C., Zhang, B., Moen, J., Hennessy, K., Liu, Y., Lin, X., Wan, Y., Lei, H., Chen, P., & Ruan, R. (2010). Fractionation and characterization of bio-oil from microwave-assisted pyrolysis of corn stover. *Internationational Journal of Agricultural & Biological Engineering* 3(3), 54-61, 1934-6344

Yu, F., Deng, S., Chen , P., Liu, Y., Wan, Y., Olson, A., Kittleson, D., & Ruan, R. (2007). Physical and chemical properties of bio-oils from microwave pyrolysis of corn stover. *Applied Biochemistry and Biotechnology* 136–140, 957–970, 0273-2289

Zhang, Q., Chang, J., Wang, T. J., & Xu, Y. (2006). Upgrading bio-oil over different solid catalysts. *Energy & Fuels* 20, 2717–2720, 0887-0624

14

Co-production of Bioethanol and Power

Atsushi Tsutsumi and Yasuki Kansha
Collaborative Research Centre for Energy Engineering, Institute of Industrial Science
The University of Tokyo, 4-6-1 Komaba, Meguro-ku, Tokyo
Japan

1. Introduction

Recently, biomass usage for fuel has attracted increased interest in many countries to suppress global warming caused mainly by the consumption of fossil fuels. (Mousdale, 2010). In particular, many researchers expect that bioethanol may be a substitute for petroleum. In fact, bioethanol loses less energy and exergy potential during chemical reactions, saccharification and fermentation for ethanol production, because it is produced merely through energy conversion by chemical reactions (Cardona et al. 2010). However, after fermentation, the product contains a large amount of water, which prevents maximizing the heat value of the product. Therefore, separation of the ethanol-water mixture is required to obtain pure ethanol for fuel (Zamboni et al. 2009a, 2009b, Huang et al. 2008). In practice, distillation is widely used for the separation of this mixture (Fair 2008). However, conventional distillation is well-known to be an energy-consuming process, and also pure ethanol fuel cannot be produced directly from a distillation column, because ethanol and water form an azeotropic mixture. To separate pure ethanol from ethanol-water mixtures by distillation, it is necessary to use an entrainer (azeotrope breaking agent), because the azeotropic mixture is one that vaporizes without any change in composition. Benzene, cyclohexane, or isopropyl alcohol can be used as entrainers for the ethanol-water mixture. Therefore, at least two separation units are required to produce pure ethanol, leading to further increases in energy consumption (Doherty& Knapp 2008). In fact, it is believed that about half of the heat value of bioethanol is required to distill the ethanol from the mixture. To reduce energy consumption during bioethanol production, many researchers have proposed membrane separations (Baker 2008, Wynn 2008) or pressure swing adsorption (PSA) (Mulla & Lang, 2008) as alternatives to azeotropic distillation, often successfully developing appropriate membranes or sorbents to achieve an efficient separation. However, in many cases, they have paid little attention to the overall process scheme or have developed heat integration processes based on conventional heat recovery technologies, such as the well known heat cascading utilization. As a result, the minimum energy requirement of the overall process has not been reduced, because changes to the condition of the process stream are constrained in conventional heat recovery technologies (Hallale 2008, Kemp 2007). Moreover, most cost minimization analyses for bioethanol plants have been conducted based on these conventional processes and technologies. Thus, the price of product bioethanol still remains high compared to fossil fuels.

Nowadays, by reconsidering the energy and production system from an improvement of energy conversion efficiency and energy saving point of view, the concept of co-production of energy and products has been developed. However, to realize co-production, it is necessary to analyze and optimize the heat and power required for production in each process. Therefore, the authors have developed self-heat recuperation technology based on exergy recuperation (Kansha et al. 2009) and applied it to several chemical processes for co-production (Fushimi et al. 2011, Kansha et al. 2010a, 2010b, 2010c, 2011, Matsuda et al.2010). In this chapter, self-heat recuperation technology is introduced and applied to the separation processes in bioethanol production for co-production. Moreover, the feasibility and energy balance for co-production of bioethanol and power using biomass gasification based on self-heat recuperation is discussed.

2. Energy balance for conventional bioethanol production

It assumed that the amount of energy in feed stock wet biomass is 100 and that 50% of this energy consists of that from reactant sugars, such as starch, cellulose and others. Thus, the amount of energy of the original component of sugar (50) transfers to ethanol (46) and heat (4) through chemical reactions (saccharification and fermentation) with water. This energy is estimated from the following calculation; the caloric value of sugar is 685 kcal/mol, the caloric value of ethanol is 316 kcal/mol and 2 mol ethanol is produced from 1 mol sugar through the above reaction. The pure ethanol product is then separated by distillation and additional heat energy (23) is required for this distillation work when azeotropic distillation is used for the separation. Non-reactants contain a large amount of water, for which the higher heat value is almost equal to the evaporation heat, leading to a net heat value of 0. The above energy relation is shown in Fig. 1. Beyond this, some additional energy is required to produce heat energy from the wet biomass for distillation (23). This additional energy (15) is used to dry the wet biomass in a heater to produce dry biomass that is used as fuel for distillation. Figure 2 shows the total energy balance including this additional energy. It is noted that 50-80% moisture content in wet biomass is assumed in this energy analysis, because many types of wet biomass exist in this range, such as those that originate from ligneous, garbage and sludge. It can be seen from Fig. 2 that 138 units of energy in the wet biomass feed stock is required to produce 46 energy units of ethanol and that about 1/3 of the energy of the wet biomass can be utilized as bioethanol for fuel. Thus, 2/3 of the wet biomass feed stock energy is wasted. Even though this wasted heat energy could potentially be heat sources for other processes, the exergy ratio and temperature of the waste heats are quite low. Thus, it is difficult to achieve energy saving from this by heat integration technologies such as cascading utilization. In fact, the highest required temperature during bioethanol production is normally at the distillation column reboiler and this temperature is lower than 150 °C. This heat is exhausted from the condenser at below 100 °C. To utilize the biomass energy more effectively, it is clear that the energy consumption during distillation for separating water and product ethanol and for drying of the wet biomass must be reduced. When an integrated system of distillation and membrane separation processes are utilized to substitute for azeotropic distillation, the energy required can be decreased from 23 to 12 units (8: distillation, 4: membrane separation). However, the pressure difference for membrane separation requires electric power. If we assume that the power generation efficiency from dry biomass is 25% and 75% of the energy for the membrane separation process is provided by electricity, 35 energy units from wet biomass are required for distillation and dehydration by membrane separation.

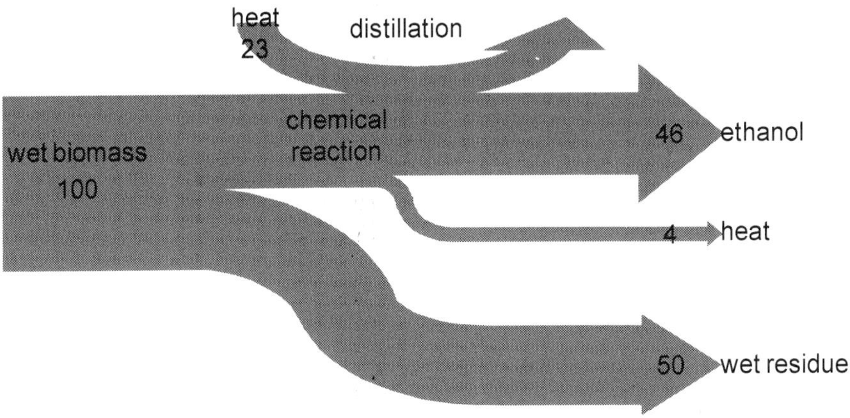

Fig. 1. Energy balance for bioethanol production

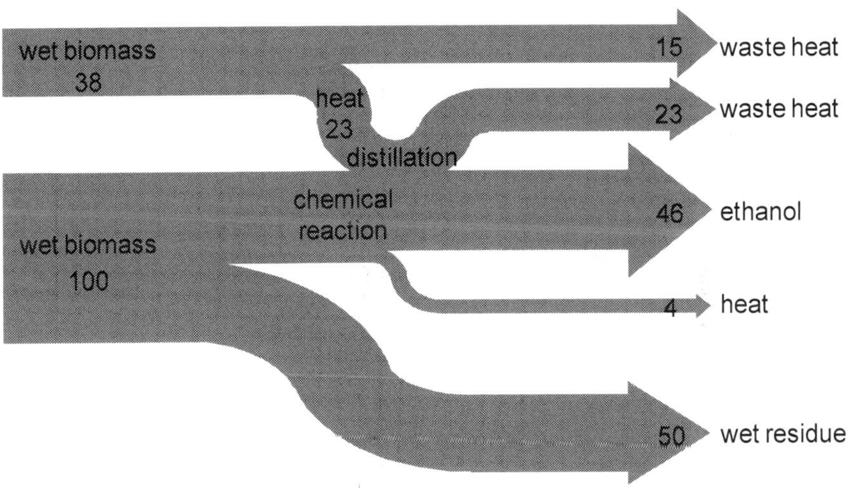

Fig. 2. Total energy balance for bioethanol production

3. Self-heat recuperation technology and self-heat recuperative processes

Self-heat recuperation technology (Kansha et al. 2009) facilitates recirculation of not only latent heat but also sensible heat in a process, and helps to reduce the energy consumption of the process by using compressors and self-heat exchangers based on exergy recuperation. In this technology, i) a process unit is divided on the basis of functions to balance the heating and cooling loads by performing enthalpy and exergy analysis, ii) the cooling load is recuperated by compressors and exchanged with the heating load. As a result, the heat of the process stream is perfectly circulated without heat addition, and thus, the energy consumption for the process can be greatly reduced. By applying this technology to each process (distillation and dehydration), the energy balance for the ethanol production can be changed significantly from that described above. In this section, the design methodology for self-heat recuperative processes is introduced by using a basic thermal process, and the self-heat recuperative processes applied to the separation processes are then introduced.

3.1 Self-heat recuperative thermal process

To reduce the energy consumption in a process through heat recovery, heating and cooling functions are generally integrated for heat exchange between feed and effluent to introduce heat circulation. A system in which such integration is adopted is called a self-heat exchange system. To maximize the self-heat exchange load, a heat circulation module for the heating and cooling functions of the process unit has been proposed, as shown in Figure 3 (Kansha et al. 2009).

Figure 3 (a) shows a thermal process for gas streams with heat circulation using self-heat recuperation technology. In this process, the feed stream is heated with a heat exchanger (1→2) from a standard temperature, T_0, to a set temperature, T_1. The effluent stream from the following process is pressurized with a compressor to recuperate the heat of the effluent stream (3→4) and the temperature of the stream exiting the compressor is raised to T_1' through adiabatic compression. Stream 4 is cooled with a heat exchanger for self-heat exchange (4→5). The effluent stream is then decompressed with an expander to recover part of the work of the compressor. This leads to perfect internal heat circulation through self-heat recuperation. The effluent stream is finally cooled to T_0 with a cooler (6→7). Note that the total heating duty is equal to the internal self-heat exchange load, Q_{HX}, without any external heating load, as shown in Fig. 3 (b).

In the case of ideal adiabatic compression and expansion, the input work provided to the compressor performs a heat pumping role in which the effluent temperature can achieve perfect internal heat circulation without any exergy dissipation. Therefore, self-heat recuperation can dramatically reduce energy consumption.

Figure 3 (c) shows a thermal process for vapor/liquid streams with heat circulation using the self-heat recuperation technology. In this process, the feed stream is heated with a heat exchanger (1→2) from a standard temperature, T_0, to a set temperature, T_1. The effluent stream from the subsequent process is pressurized with a compressor (3→4). The latent heat can then be exchanged between feed and effluent streams because the boiling temperature of the effluent stream is raised to T_b' by compression. Thus, the effluent stream is cooled through the heat exchanger for self-heat exchange (4→5) while recuperating its heat. The effluent stream is then depressurized by a valve (5→6) and finally cooled to T_0 with a cooler (6→7). This leads to perfect internal heat circulation by self-heat recuperation, similar to the above gas stream case. Note that the total heating duty is equal to the internal self-heat exchange load, Q_{HX}, without any external heating load, as shown in Fig. 3 (d). It can be understood that the vapor and liquid sensible heat of the feed stream can be exchanged with the sensible heat of the corresponding effluent stream and the vaporization heat of the feed stream is exchanged with the condensation heat of the effluent stream. As a result, the energy required by the heat circulation module is reduced to 1/22–1/2 of the original by the self-heat exchange system in gas streams and/or vapor/liquid streams.

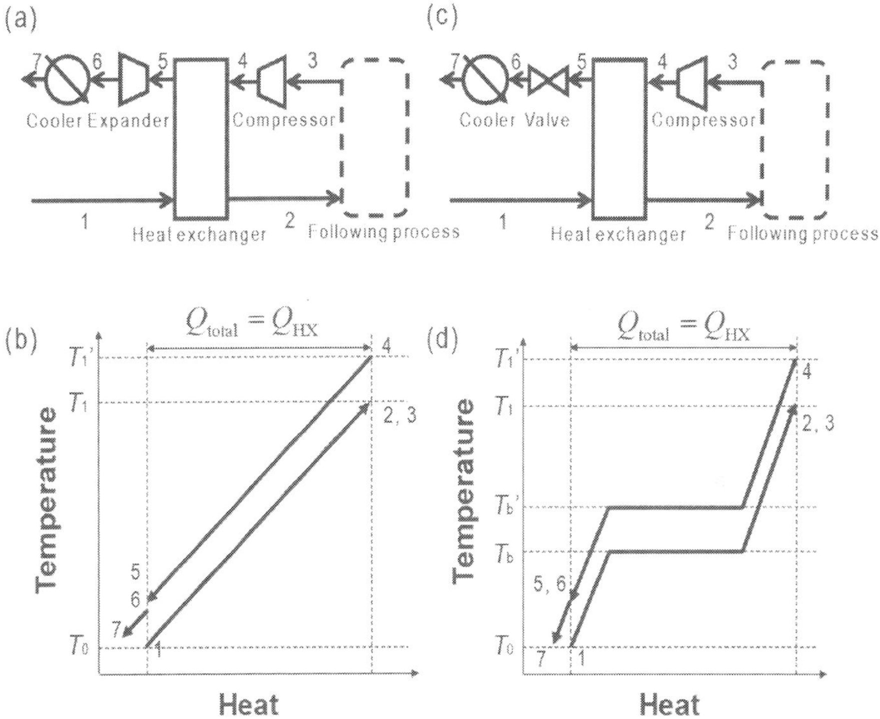

Fig. 3. Self-heat recuperative thermal process a) process flow of gas streams, b) temperature-heat diagram of gas streams, c) process flow of vapor/liquid streams, d) temperature-heat diagram of vapor/liquid streams

3.2 Self-heat recuperative distillation

Expanding the self-heat recuperative thermal process to distillation processes in particular (Kansha et al. 2010a, 2010b), a system including not only the distillation column but also the preheating section, is developed in order to minimize the required energy, as shown in Fig. 4. A distillation process can be divided into two sections, namely the preheating and distillation sections, on the basis of functions that balance the heating and cooling load by performing enthalpy and exergy analysis, and the self-heat recuperation technology is applied in these two sections. In the preheating section, one of the streams from the distillation section is a vapor stream and the stream to the distillation section has a vapor-liquid phase that balance the enthalpy of the feed streams and that of the effluent streams in the section. In balancing the enthalpy of the feed and effluent streams in the preheating section, the enthalpy of the streams in the distillation section is automatically balanced. Thus, the reboiler duty is equal to the condenser duty of the distillation column. Therefore, the vapor and liquid sensible heat of the feed streams can be exchanged with the sensible heat of the corresponding effluent streams and the vaporization heat can be exchanged with the condensation heat in each section.

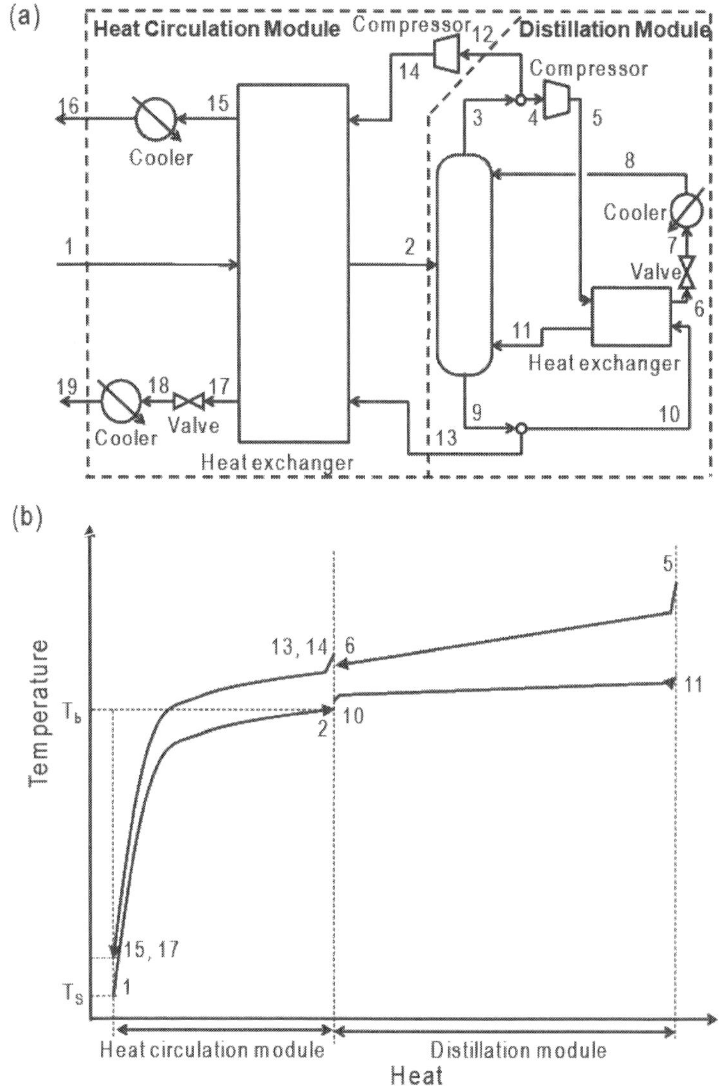

Fig. 4. Self-heat recuperative distillation process a) process flow diagram, b) temperature-heat diagram

Figure 4 (a) shows the structure of a self-heat recuperative distillation process consisting of two standardized modules, namely, the heat circulation module and the distillation module. Note that in each module, the summation of the enthalpy of the feed streams and that of the effluent streams are equal. The feed stream in this integrated process module is represented as stream 1. This stream is heated to its boiling point by the two streams independently recuperating heat of the distillate (12) and bottoms (13) by the heat exchanger (1→2). A distillation column separates the distillate (3) and bottoms (9) from stream 2. The distillate (3) is divided into two streams (4, 12). Stream 4 is compressed adiabatically by a compressor and cooled down by the heat exchanger (2). The pressure and temperature of stream 6 are adjusted by a valve and a cooler (6→7→8), and stream 8 is then fed into the distillation column as a reflux stream. Simultaneously, the bottoms (9) is divided into two streams (10, 13). Stream 10 is heated by the heat exchanger and fed to the distillation column (10→11). Streams 12 and 13 are the effluent streams from the distillation module and return to the heat circulation module. In addition, the cooling duty of the cooler in the distillation module is equal to the compression work of the compressor in the distillation module because of the enthalpy balance in the distillation module.

The effluent stream (12) from the distillation module is compressed adiabatically by a compressor (12→14). Streams 13 and 14 are successively cooled by a heat exchanger. The pressure of stream 17 is adjusted to standard pressure by a valve (17→18), and the effluents are finally cooled to standard temperature by coolers (15→16, 18→19). The sum of the cooling duties of the coolers is equal to the compression work of the compressor in the heat circulation module. Streams 16 and 19 are the products.

Figure 4 (b) shows the temperature and heat diagram for the self-heat recuperative distillation process. In this figure, each number corresponds to the stream numbers in Figure 4 (a), and T_s and T_b are the standard temperature and the boiling temperature of the feed stream, respectively. Both the sensible heat and the latent heat of the feed stream are subsequently exchanged with the sensible and latent heat of effluents in heat exchanger 1. The vaporization heat of the bottoms from the distillation column is exchanged with the condensation heat of the distillate from the distillation column in the distillation module. The heat of streams 4 and 12 are recuperated by the compressors and exchanged with the heat in the module. It can be seen that all the self-heat is exchanged. As a result, the exergy loss of the heat exchangers can be minimized and the energy required by the distillation process is reduced to 1/6–1/8 of that required by the conventional heat exchanged distillation process.

3.1.2 Self-heat recuperative azeotropic distillation for dehydration

Conventional azeotropic distillation processes, which have one distillation column for dehydration to separate ethanol and another to separate water from their mixture, are divided into three modules. The sum of the feed enthalpy is made equal to that of the effluent stream enthalpy in each module to analyze the heating and cooling loads of all process streams by following self-heat recuperation technology. According to this analysis, the recovery streams are selected and the internal heat of the process stream in each module can be recovered and recirculated using a compressor and heat exchanger through self-heat recuperation technology.

Figure 5 a) shows the structure of the self-heat recuperative azeotropic distillation module (Kansha et al. 2010c), consisting of three modules, namely, the first distillation module, the heat circulation module, and the second distillation module. In this self-heat recuperative distillation module, stream 1 represents a feed stream of the ethanol-water azeotropic mixture and stream 2 represents an entrainer (benzene and cyclohexane) feed stream. These streams are fed into the distillation column of the first distillation module. The vapor stream from the first distillation process is compressed adiabatically by a compressor (4→5). Subsequently, stream 5 is cooled in a heat exchanger (5→6), and the pressure and

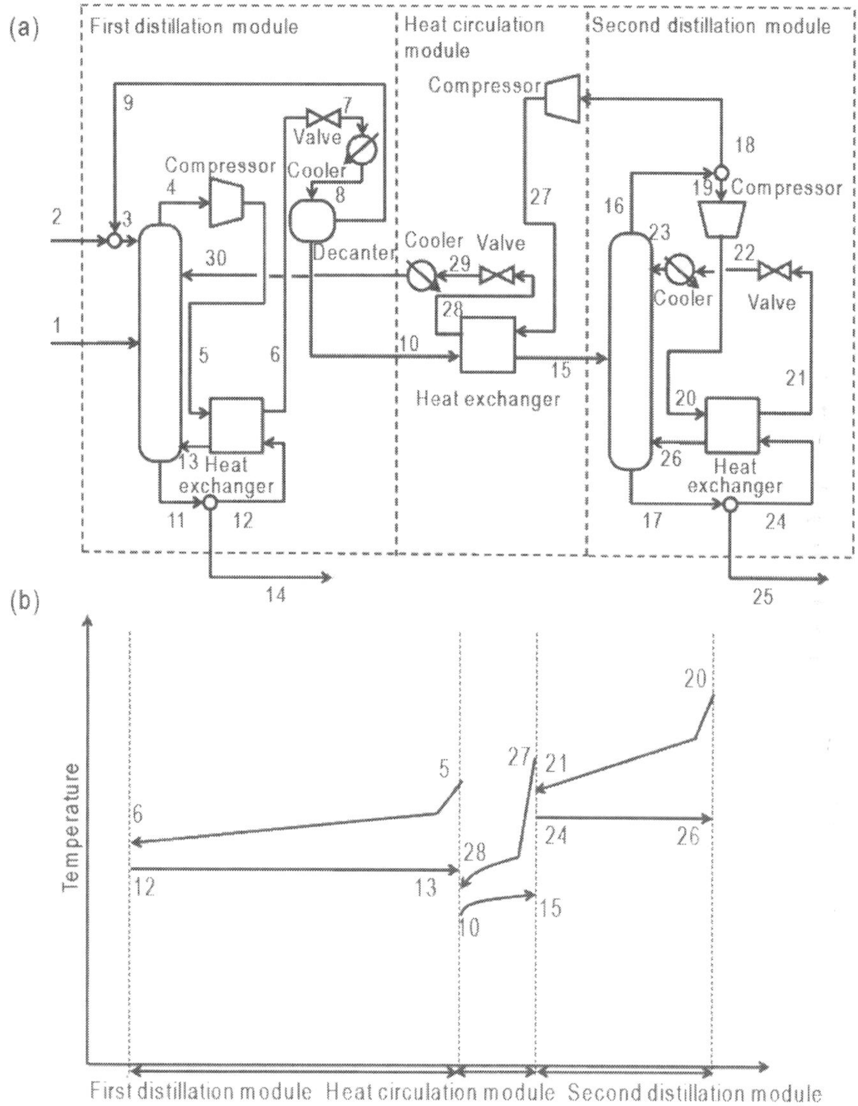

Fig. 5. Self-heat recuperative azeotropic distillation process for dehydration a) process flow diagram, b) temperature-heat diagram

temperature of stream 6 are adjusted by a valve and a cooler (6→7→8). The liquid stream (8) is divided into two streams (9 and 10) in a decanter. Stream 9 consists mainly of the entrainer, which is recycled to the feed benzene (3). The bottom (11) of the distillation column is divided into two streams (12 and 14). Stream 14 becomes a product stream (pure ethanol). Stream 12 is heated in the heat exchanger and fed into the distillation column. In the heat circulation module, the effluent stream (10) from the first distillation module is heated in a heat exchanger and fed to the distillation column in the second distillation module. At the same time, the recycled stream, which is the distillate stream from the second distillation module, is adiabatically compressed by a compressor (18→27) and cooled by exchanging heat in the heat exchanger (27→28). The pressure and temperature of stream 28 are adjusted by a valve and cooler (28→29→30) and stream 30 is fed into the distillation column of the first distillation module as the recycled stream. Next, in the second distillation module, the feed stream (15) is separated into the distillate (16) and the bottoms (17) by the distillation column. The vapor distillate (16) is divided into two streams (18 and 19) by a separator. Stream 18 is recycled to the heat circulation module, while stream 19 is adiabatically compressed (19→20) and exchanged with the heat in a heat exchanger (20→21). The temperature and pressure of stream 21 are adjusted by a valve and a cooler (21→22→23), and then the effluent stream is fed into the distillation column. Subsequently, the bottom stream (17) from the distillation column is divided into two streams (24 and 25). Stream 25 is the product water. The other stream (24) is vaporized in the heat exchanger and fed into the distillation column (26).

Figure 5 b) shows a temperature–heat diagram for the self-heat recuperative distillation module for azeotropic distillation. Note that numbers beside the composite curve correspond to the stream numbers in Figure 5 a). It can be seen that the latent heats of the effluent streams are exchanged with those of the feed streams, as well as the sensible heats in each module, leading to minimization of the exergy loss in the heat exchangers. From this figure, it can be understood that all of process heat is recirculated without any heat addition and the total heating duty was covered by internal heat recovery. All of the compression work in each module was discarded into coolers in each module, because the sum of enthalpy in the feed streams was equal to that of the effluent streams in each module when using internal heat recovery. As this relationship indicates, the compression work was used for inducing heat recovery and circulation in each module and exhausted as low exergy heat. As a consequence, the energy required of the self-heat recuperative distillation module for azeotropic distillation is 1/8 of that of the conventional azeotropic distillation process.

3.1.3 Self-heat recuperative drying

Biomass resources usually contain a large amount of moisture, leading to higher transportation costs, debasement during storage, and reduction of thermal efficiency during conversion. Drying is a key technology for utilizing the biomass (McCormick & Mujumdar 2008). In addition to the use of biomass for fuel, the energy required for drying occupies a large amount of energy in the production due to the large latent heat of water during evaporation. Moreover, this characteristic of the drying process is the same as for the thermal and distillation processes. Therefore, a drying process based on self-heat recuperation technology was recently proposed (Fushimi et al., 2011).

Figure 6 a) shows a schematic image of a self-heat recuperative drying process. The wet sample is heated in a heat exchanger (1→2). The heated wet sample and vapor are then fed into an evaporator (dryer) with dry gas to assist evaporation (16). The heat for evaporation is supplied by superheated steam and gas (7). The hot dry sample (3) is separated and cooled by the dry gas (15) (3→5). After eliminating the unseparated sample to prevent it from entering the compressor, the evaporated steam and gas (4) are compressed (7) by a

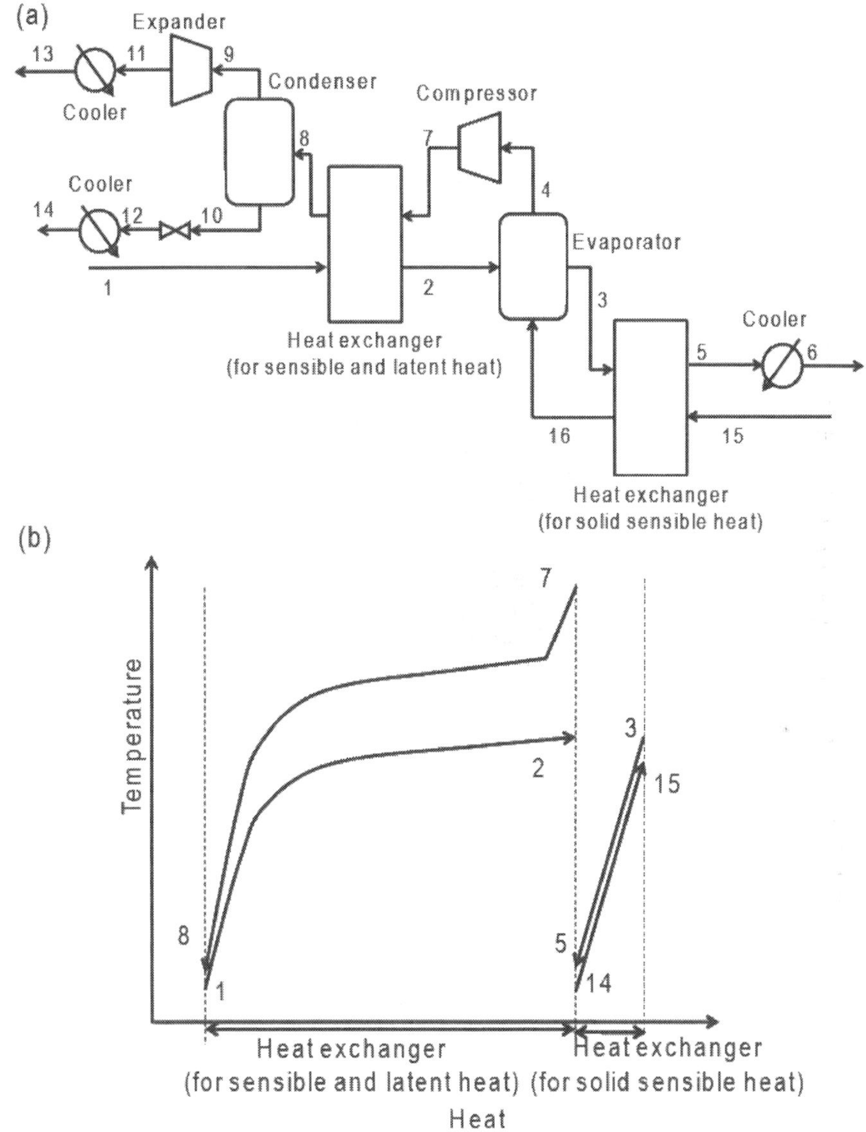

Fig. 6. Self-heat recuperative drying process for dehydration a) process flow diagram, b) temperature-heat diagram

compressor. The sensible and latent heats of the compressed steam and gas are exchanged in the heat exchanger (7→8) and fed into a condenser to separate the water and gas; the water is then drained (10). The pressure and temperature of drain water are adjusted by a valve and cooler (10→12→14). Simultaneously, the pressure energy of the gas (9) is partially recovered in an expander. The temperature of the gas is then cooled by a cooler (13). This exhausted gas can be recycled as the gas feed (15). To use this gas as the dry gas feed, makeup gas is necessary to compensate for the loss, because a small amount of gas dissolves in water in the condenser. Considering a real application for a drying process, the dried sample is separated immediately after the evaporation and reversed back to the heat exchanger for heat utilization. However, with the aim of reducing drying time (higher drying rate) and providing the driving force required in the drying process, gas that has been preheated by the sample enters the evaporator. It should be noted that an increase in gas flow rate causes an increase in the energy required for compression for the following reasons: (1) an excess amount of gas must be compressed and (2) a smaller partial pressure of steam requires larger compression pressure for condensation. Consequently, the gas flow rate should be optimized.

Figure 6 b) shows a temperature-heat diagram of the self-heat recuperative drying process. Note that the numbers beside the composite curve in this temperature-heat diagram correspond to the stream numbers in Figure 6 a). It can be seen that the condensation heat of the steam in the effluent stream (7→8) is exchanged with the evaporation heat of the feed stream (1→2), as well as the sensible heats in a heat exchanger. At the same time, the heat of solid sample after evaporation is exchanged with the heat of the gas stream in the other heat exchanger and this heat is supplied to the feed solid sample. These lead to minimization of the exergy loss in the heat exchangers. From this figure, it can be understood that all process heat is recirculated without heat addition, and that the total heating duty is covered by internal heat recovery. All of the compression work in each module was discarded into coolers, because the base conditions of the stream are fixed at standard conditions. As a consequence, to circulate the process stream heat in the process using heat exchangers and a compressor, the energy required for the self-heat recuperative drying process is 1/7 of that of the conventional heat recovered drying process.

4. Integration with biomass gasification

To adopt self-heat recuperative processes, it is necessary to generate power in substitution for heat energy. According to the energy balance shown in Figs. 1 and 2, much residue with insufficient heat value for utilization due to its high moisture content is produced during bioethanol production. By integrating the self-heat recuperative drying process with power generation, this wet biomass can be utilized for energy. In this section, an integrated system for self-heat recuperative bioethanol production with biomass gasification is introduced.

4.1 Biomass gasification and its impact on the system

One of the easiest ways to generate power from biomass is direct combustion of biomass in a boiler, wherein thermal energy is produced and power is generated from this thermal energy by using a steam turbine (boiler and turbine generator). However, energy conversion efficiency under this procedure is not good enough. To increase the conversion efficiency of energy from biomass to power, biomass gasification reaction is used. Gasification reactions can be divided into two mechanisms; pyrolysis and gasification by chemical reaction (partial oxidation, etc.) Biomass gasification normally passes through both of these. After passing through a series of gasification procedures, the gases are fed into a gas turbine, and then the power is generated. Gasification reactions are normally endothermic reactions, and must be provided with heat during reactions. However, the overall energy conversion efficiency will be increased compared with the boiler and turbine generator. In addition, a further increase in energy conversion efficiency, through a biomass-based integrated gasification combined cycle (IGCC) technology has been investigated (Bridgwater 1995).

It is currently assumed that the energy conversion efficiency of biomass through power generation and biomass gasification is 25%. The energy amount of the wet residue is 50 in Figs. 1 and 2. It assumed that half of the energy amount of this wet residue can be utilized for drying the biomass. According to the analysis of self-heat recuperative drying above, 1/8 of the amount of energy for water evaporation is required for power to dry this wet residue using self-heat recuperative drying. This means that power (8) can be generated and a part of this power (3) is used for drying, leading to 4% of the initial wet biomass being converted to power as net energy (5) from the wet residue as shown in Fig. 7.

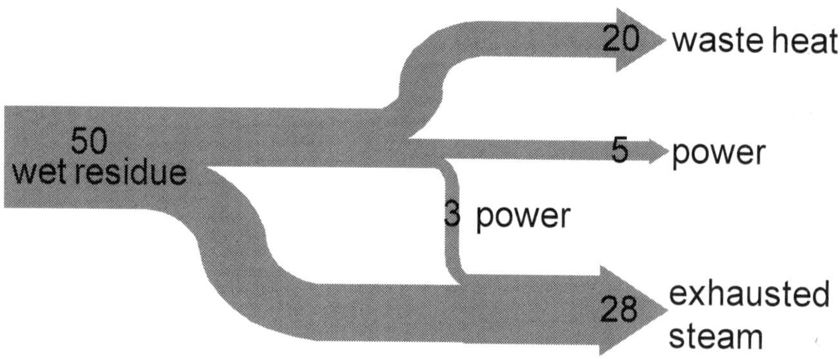

Fig. 7. Power generation from wet residue during bioethanol production

5. Energy balance for self-heat recuperative bioethanol production

The same assumption as for section 2 is assumed; the amount of energy in the wet biomass feed stock is 100, 50% of the energy value of the wet biomass consists of the energy value of reactant sugars such as starch, cellulose and others, and the amount of energy of the original sugar component (50) transfers to ethanol (46) and heat (4) through chemical reactions (saccharification and fermentation) with water.

By applying the self-heat recuperative distillation and azeotropic distillation process to the distillation and dehydration process, the additional heat energy for distillation is converted to power. At the same time, the energy (23) in Figure 1 is reduced to 4. This value was estimated from the energy reduction results from the self-heat recuperative processes in section 3.

By integrating the aforementioned biomass gasification in section 4 with the self-heat recuperative processes introduced in section 3, bioethanol (46) and power (1) can be produced as co-products from wet biomass (100) during bioethanol production, as shown in Fig. 8. Wet residue (non-reactants contain a large amount of water, for which the higher heat value is almost equal to the required evaporation heat, leading to net heat value of 0) in Figs.

1 and 2 can be utilized as the energy supply. Thus, it can be understood that 46% of the energy of the wet biomass is transferred to the bioethanol and 1% of the energy to power. Furthermore, the additional wet biomass (38) required to provide the distillation heat (23) is no longer necessary for this bioethanol production. Thus, power (4) can be generated from the additional wet biomass by using a self-heat recuperative drying process and biomass gasification, as shown in Fig. 9. As a result, 33% (= 46/138×100) of the energy of the wet biomass is transferred to bioethanol and 4% (= 5/138×100) is transferred to power for co-production. It can be said that this bioethanol production procedure achieves not only energy savings but also reduction of exergy dissipation for the whole process, leading to achievement of optimal co-production. In addition, substituting the azeotropic distillation process by dehydration uses a membrane separation. All of the self-heat recuperative processes and biomass gasification are applied to produce this energy. The energy required can be decreased to 4 as power, where the same assumptions as used for the results described above are used in the calculation, such that power generation efficiency from dry biomass is 25% and 75% of the energy required for the membrane separation process is provided by electricity. This value of power is the same as the energy required by applying self-heat recuperative processes to the distillation and dehydration processes. Although the energy required by membrane separation process is smaller than that of azeotropic distillation in the conventional processes, it becomes equal after applying the self-heat recuperative processes.

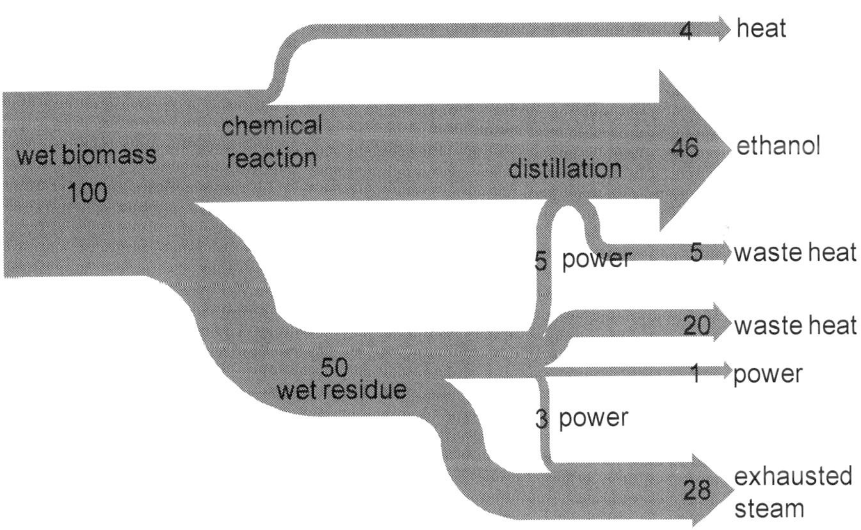

Fig. 8. Energy balance for bioethanol production with self-heat recuperation

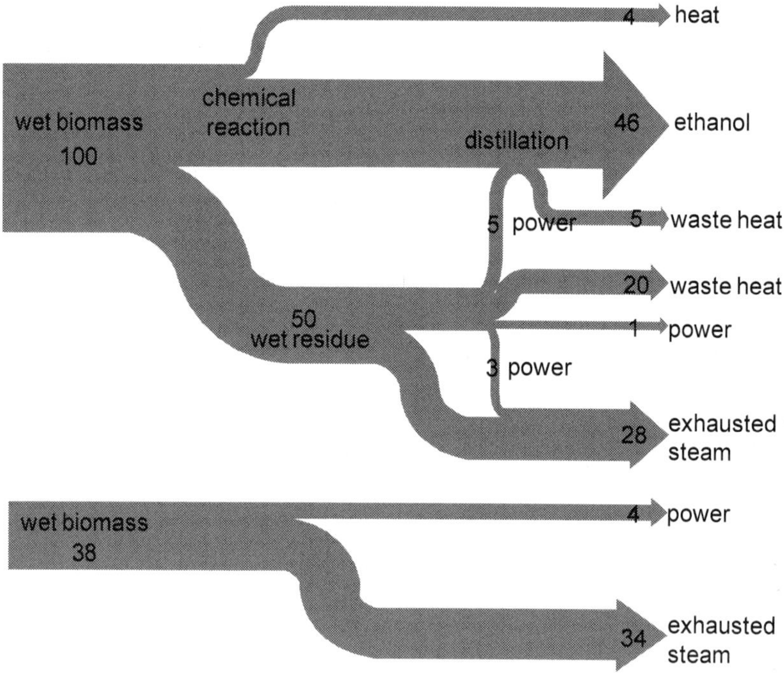

Fig. 9. Total energy balance for bioethanol production with self-heat recuperation

6. Conclusion

In this chapter, a newly developed self-heat recuperation technology is introduced and the feasibility of co-production of bioethanol and power by integration of self-heat recuperative processes and biomass gasification for power generation is examined based on energy balances. From analysis of the energy balance for the conventional bioethanol production processes, a large amount of energy is consumed for separation of water (distillation and drying) so that the operational costs for bioethanol production are high, limiting the potential contribution of bioethanol to society. However, by incorporating self-heat recuperative processes for distillation, azeotropic distillation and drying, not only are the energy requirements reduced dramatically due to heat circulation in the processes, but also wasted residue can be utilized as a power source through biomass gasification. Thus, it is shown that co-production of bioethanol and power is feasible, enabling the economic impact of the bioethanol product. Finally, this system is expected to help the uptake of bioethanol and decrease global CO_2 emissions.

7. References

Baker, R.W. (2008). Membrane Technology, Introduction, In: *Kirk-Othmer Separation Technology 2nd Ed. Vol. 2,* A. Seidel, (Ed.), 446-502, John Wiley & Sons, ISBN 978-0-470-12741-4, NJ, USA

Bridgwater, A.V. (1995). The technical and economic feasibility of biomass gasfication for generation, Fuel, Vol. 74, No. 5, pp. 631-653, ISSN 0016-2361

Cardona, C.A.; Sanchez, O.J. & Gutierrez, L.F. (2010). *Process Synthesis for Fuel Ethanol Production*, ISBN 978-1-4398-1597-7, FL, USA

Doherty, M.F. & Knapp, J.P. (2008). Distillation, Azeotropic and Extractive, In: *Kirk-Othmer Separation Technology 2nd Ed. Vol. 1,* A. Seidel, (Ed.), 918-984, John Wiley & Sons, ISBN 978-0-470-12741-4, NJ, USA

Fair J.R. (2008). Distillation, In: *Kirk-Othmer Separation Technology 2nd Ed. Vol. 1,* A. Seidel, (Ed.), 871-917, John Wiley & Sons, ISBN 978-0-470-12741-4, NJ, USA

Fushimi, C.; Kansha, Y.; Aziz, M.; Mochidzuki, K.; Kaneko, S.; Tsutsumi, A.; Matsumoto, K.; Yokohama, K.; Kosaka, K.; Kawamoto, N.; Oura, K.; Yamaguchi, Y. & Kinoshita, M. (2011). Novel drying process based on self-heat recuperation technology, *Drying Technology*, Vol. 29, No. 1, pp.105-110, ISSN 0737-3937

Hallale, N. (2008). Process Integration technology, In: *Kirk-Othmer Separation Technology 2nd Ed. Vol. 2,* A. Seidel, (Ed.), 837-871, John Wiley & Sons, ISBN 978-0-470-12741-4, NJ, USA

Huang, H.-J.; Ramaswamy, S.; Tschirner, U.W. & Ramarao, B.V. (2008). A review of separation technologies in current and future biorefineries, Separation and Purification Technology, Vol. 62, No. 1, pp.1-21, ISSN: 1383-5866

Kansha, Y.; Tsuru, N.; Sato, K.; Fushimi, C. & Tsutsumi, A. (2009). Self-heat recuperation technology for energy saving in chemical processes, *Industrial and Engineering Chemistry Research*, Vol. 48, No. 16, pp.7682-7686, ISSN 0888-5885

Kansha, Y.; Tsuru, N.; Fushimi, C.; Shimogawara, K. & Tsutsumi, A. (2010a). An innovative modularity of heat circulation for fractional distillation, *Chemical Engineering Science*, Vol. 65, No.1, pp.330-334, ISSN 0009-2509

Kansha, Y.; Tsuru, N.; Fushimi, C. & Tsutsumi, A. (2010b). Integrated process module for distillation processes based on self-heat recuperation technology, *Journal of Chemical Engineering of Japan*, Vol. 43, No. 6, pp. 502-507, ISSN 0021-9592

Kansha, Y.; Tsuru, N.; Fushimi, C. & Tsutsumi, A. (2010c). New design methodology based on self-heat recuperation for production by azeotropic distillation, *Energy & Fuels*, Vol. 24, No. 11, pp. 6099-6102, ISSN 0887-0624

Kansha, Y.; Kishimoto, A.; Nakagawa, T. & Tsutsumi, A (2011). A novel cryogenic air separation process based on self-heat recuperation, *Separation and Purification Technology*, Vol. 77, No. 3, pp. 389-396, ISSN 1383-5866

Kemp, I.C. (2007). *Pinch Analysis and Process Integration A User Guide on Process Integration for the Efficient Use of Energy 2nd Ed.*, Elsevier, ISBN 13 978-0-75068-260-2, Oxford, UK

Matsuda, K.; Kawazuishi, K.; Hirochi, Y.; Sato, R.; Kansha, Y.; Fushimi, C.; Shikatani, Y.; Kunikiyo, H. & Tsutsumi, A. (2010). Advanced energy saving in the reaction section of the hydro-desulfurization process with self-heat recuperation technology, Applied Thermal Engineering, Vol. 30, No. 16, pp. 2300-2305, ISSN 1359-4311

McCormick, P.Y. & Mujumdar, A.S. (2008). Drying, In: *Kirk-Othmer Separation Technology 2nd Ed. Vol. 1,* A. Seidel, (Ed.), 984-1032, John Wiley & Sons, ISBN 978-0-470-12741-4, NJ, USA

Modla, G. & Lang P. (2008). Feasibility of new pressure swing batch distillation methods, *Chemical Engineering Science*, Vol. 63, No. 11, pp. 2856-2874, ISSN 0009-2509

Mousdale, D.M. (2010). *Introduction to Biofules*, CRC Press, ISBN 978-1-4398-1207-5, FL, USA

Wynn, N.P. (2008). Pervaporation, In: *Kirk-Othmer Separation Technology 2nd Ed. Vol. 2,* A. Seidel, (Ed.), 533-550, John Wiley & Sons, ISBN 978-0-470-12741-4, NJ, USA

Zamboni, A.; Shah, N. & Bezzo, F. (2009a). Spatially Explicit Static Model for the Strategic Design of Future Bioethanol Production Systems. 1. Cost Minimization, *Energy & Fuels*, Vol. 23, No. 10, pp. 5121-5133, ISSN 0887-0624

Zamboni, A.; Shah, N. & Bezzo, F. (2009b). Spatially Explicit Static Model for the Strategic Design of Future Bioethanol Production Systems. 2. Multi-Objective Environmental Optimization, *Energy & Fuels*, Vol. 23, No. 10, pp. 5134-5143, ISSN 0887-0624

15

Conversion of Non-Homogeneous Biomass to Ultraclean Syngas and Catalytic Conversion to Ethanol

Stéphane C. Marie-Rose, Alexis Lemieux Perinet and Jean-Michel Lavoie
Industrial Research Chair on Cellulosic Ethanol
Department of Chemical and Biotechnological Engineering
Université de Sherbrooke, Sherbrooke, Québec
Canada

1. Introduction

Reducing greenhouse gas emissions, rising energy prices and security of supply are reasons that justify the development of biofuels. However, food prices recorded in 2007 and 2008 affected more than 100 million of people that became undernourished worldwide (Rastoin, 2008) The food crisis has been caused by several factors: underinvestment in agriculture, heavy speculation on agricultural commodities and competition of biofuels vs. food. It is estimated that by 2050, it will be essential to increase by 50% the food production to support the 9 billion people living on the planet (Rastoin, 2008).

Recycling the carbon from residual waste to produce biofuels is one of the challenges of this new century. Several companies have been developing technologies that are able to transform residual streams into syngas, which is subsequently converted into alcohols. "Green" ethanol plays an important role in reducing dependency toward petroleum and providing environmental benefit, through its role in the fuel additive market. Ethanol is an oxygenate and also serves as an octane enhancer. The waste-to-syngas approach is an alternative to avoid the controversy food vs. fuel whilst reducing landfills and increasing carbon recuperation. Using this approach, yields of ethanol produced are above 350 liters/dry tonne of feedstock entering the gasifier (Enerkem's technology is taken as example). Residual heat, also a product of the process, is used in the process itself and, as well, it can be used for outside heating or cooling. Enerkem Inc. is moving the technology from bench scale, to pilot, to demo to commercial implementation (a 12,500 kg/h of sorted and biotreated urban waste, is being constructed in Edmonton, Alberta). Economics of the process are favorable at the above commercial capacity, given the modular construction of the plant, reasonable operational costs and a tipping fee for the residue going into the gasifier.

The first part of this chapter will present feedstock preparation, gasification and gas conditioning. The characteristics of the heterogeneous feedstock will determine its performance during gasification for syngas production whose composition has the appropriate H_2/CO ratio for downstream synthesis. The second part of the chapter will be directed at the methanol synthesis in a three-phase reactor using syngas. The third and last part of the chapter will focus on the catalytic steps to convert methanol into bio-ethanol.

2. Synthesis gas (syngas) production by gasification

2.1 Characteristics and composition of heterogeneous wastes as feedstock

Biomass is defined as an organic material derived from plants or animals that contain potential chemical energy; for example wood, which was the first fire source used by mankind, and which is still used today by population for cooking and heating.

At world scale, biomass is now the fourth largest energy source, but it has the capacity to become the first. Photosynthesis can store up to 5-8 times more energy in biomass annually than the actual world energy consumption (Prins et al., 2005). The basic reaction of photosynthesis is as follow: carbon dioxide and water are converted to glucose and oxygen, an endothermic process for which the energy is supplied by photons. Examples of biomass are residues from agriculture or from the forest industry such as branches, straw, stalks, saw dust, etc. An important example of residual agricultural biomass is related to the ethanol production from sugar cane in Brazil which produces 280 kg of residual bagasse at 50% of dry solids. Lignocellulosic materials can be collected and recovered because this material has some energetic content (Ballerini and Alazard-Toux, 2006), however, leaving a part of this material on place is imperative since it keeps the soil fertile.

It is important for governments and citizens to realize that hydrocarbon-based waste material is another source of energy that should be taken advantage on. Waste can be solid or liquid form. It can be land filled, incinerated or converted. Municipal solid waste used electrical transmission poles and railroad ties treated with creosote, sludge from wastewater treatment and pulp and paper industries, wood from construction and demolition operations which contains paints and resins, etc., are all materials that contain carbon that can be valorized in bio-refineries such as the one Enerkem is constructing (2011) in Edmonton, AB.

Assessment of residual biomass or Municipal Solid Waste (MSW) as feedstock to produce bio-ethanol requires a basic understanding of feedstock composition and of the specific properties that dictates its performance as feed in the gasifier. The most important are: moisture content, ash content, volatile matter content, elemental composition and heating value.

The moisture content of biomass is the quantity of water in the material, expressed as percentage of material weight. This weight can be referred to on a wet basis or on a dry basis. If the moisture content is determined on a "wet" basis, the water's weight is expressed as a percentage of the sum of the weight of the water, ash, and dry- ash free matter. It is sometimes necessary to dry the feedstock to a certain level in order to maximize the gasification reaction. Indeed, more moisture is transferred by a higher consumption of oxygen in order to keep the ideal temperature in the gasifier. Temperature of gasification is crucial on the process efficiency. An optimum exists with just the right amount of oxygen needed to perform completely the gasification reaction. This represents a temperature of about 660°C for biomass with 20 % of moisture and about 695°C for biomass with 10 % of moisture (Prins et al., 2005). If more oxygen is added, formation of carbon dioxide will increase and gasification efficiency will drop; the heating value of the synthesis gas will thus decrease (van der Drift et al., 2001). However, not enough oxygen will promote reduction of carbon leading to an increase of methane formation.

The inorganic component (ash content) can be expressed the same way as the moisture content. In general, the ash content is expressed on a dry basis. Both total ash content and chemical composition are both important in regards of the gasification process. The composition of the ash affects its behaviour under high temperatures of combustion and gasification. For example, melted ash may cause problems in both combustion and gasification reactors. These problems may vary from clogged ash-removal caused by slagging ash to severe operating issue in fluidized bed systems. Measurement of ash melting point is thus crucial.

Volatile matter refers to the part of biomass that is released when the biomass is heated beyond its dehydration temperatures. During this heating process the biomass decomposes into volatile gases and solid char. Biomass typically has a high volatile matter content (up to 80%), whereas coal has a low volatile matter content (<20%).

Elemental composition of the ash-free organic component of residual biomass is relatively uniform. The major components are carbon oxygen and hydrogen. Most biomass also contains a small proportion of nitrogen and sulfur. Table 1 presents the elementary composition of biomass as derived from ultimate analyses.

Element	Wt% (dry basis)
Carbon	44 - 51
Hydrogen	5.5 - 6.7
Oxygen	41 - 50
Nitrogen	0.12 - 0.6
Sulfur	0 - 0.2

Table 1. Elementary composition of residual biomass

The heating value of a fuel is an indication of the energy chemically bound in the fuel with reference to a standardized environment. The standardization involves the temperature, state of water, and the combustion products. The calorific value is presented as the higher heating value and the lower heating value. The higher heating value represents the heat release per unit of mass when the material (at 25°C) is completely oxidized to carbon dioxide and water and then returned to 25°C. A calorimetric bomb is the standard instrument used to measure this value. The lower heating value is not taking into account the energy supplied by the condensation of water (latent heat of vaporization of water at 25°C which is 2440 kJ/kg). The water includes moisture from the feedstock and the product from the reaction between oxygen and hydrogen comprised in the raw material (Borman and Ragland, 1998). Basic and complementary information about biomass intended for gasification can be obtained via the proximate and ultimate analyses.

Proximate analysis measures moisture content, volatile matter, fixed carbon, ash content and calorific value.

Ultimate analysis provides information about elementary composition of the biomass in weight percentage of carbon, hydrogen, oxygen, sulphur and nitrogen. The carbon to hydrogen ratio in the feedstock has a direct impact on the syngas, more particularly on the ratio of H_2/CO (Higman and van der Burgt, 2008).

Table 2 (depicted below) presents a comparison of different feedstock properties. Moisture content was provided after drying of feedstock and compositions are approximated (Ciferno and Marono, 2002).

Analysis		Sawdust	Bagasse	Switchgrass	Straw	Bituminous Coal	Dry Sewage (MSW)
Utlimate Analysis (% w/w) dry basis	C	50	48	43	43,5	61.5	20.5
	H	6,3	6.0	5.6	4.2	4.2	3.2
	O	43	42	46	40.3	6	17.5
	N	0,8	-	0.5	0.6	1.2	2.3
	S	0,03	-	0.1	0.2	5.1	0.6
Proximate Analysis (% w/w) dry basis	Moisture	7.8	1	8.4	7.6	8.7	4.7
	Volatiles	74	80	73	68.8	36.1	41.6
	Fixed Carbon	25.5	15	13.5	13.5	42	2.3
	Ash	0.03	4	4.5	10.1	21.9	56
	HHV (MJ/kg)	19.3	17	15.4	17	27	8

Table 2. Properties of biomass feedstocks

Note that higher calorific value formulas, refering to equation 1, were developed from the composition of the feedstock. Dulong formula is one of them (Higman and van der Burgt, 2008). However, numerous correlations were developed and reported in literature. For instance, experiments on more than 200 species were performed at the Indian Institute of Technology (Bombay) and the following equation was derived from empirical data.

$$HHV\ (MJ/kg) = 0.3491C + 1.1783H - 0.1034\ O - 0.0211\ A + 0.1005S - 0.0151N \quad (1)$$

Where C is the weight fraction of carbon, H of hydrogen, O of oxygen, A of ash, S of sulfur and N of nitrogen. This correlation offers an average absolute error of 1.45 % (Channiwala and Parikh, 2002).

There are other feedstock properties that are critical to determine in order to treat the material accordingly: size, distribution and shapes of the material, porous structure and bulk density.

Gasification and incineration are both design for transforming hydrocarbon-based hazardous material to more simple by-products. Nonetheless, these two technologies are completely different in their operation and finalities. The goal of the combustion is to proceed to a complete oxidation of carbon in order to produce carbon dioxide and water. Thereby an excess of oxidizing agent is used. Sulfur and nitrogen comprised in the feedstock are oxidized to SO_x and NO_x. Whilst only gasification can lead to synthesis, both processes can lead to production of electricity. However, with the goal of generating added-value product, gasification has the possibility to convert the carbon of waste to biofuels instead of releasing the carbon as CO2. Gasification is a partial oxidation process with just enough oxygen to produce the heat to yield reduced carbon and hydrogen (Wetherold et al., 2000).

2.2 Syngas production and purification

Gasification produces a synthetic gas or syngas mainly composed of hydrogen, methane and carbon monoxide (Marie-Rose et al., 2011, Villano et al., 2010). Gasification started in the late 1800's and first applications were for town gas for street lights illumination and cooking purposes. Other applications were heating, production of raw material for chemical industry and also power generation. When the Second World War was raging and oil was limited, wood gasification was used for transportation and heat and power generation. The energy crisis of the 70's (1973 and 1979 oil shocks) also provided motivation for improving gasification technologies (Higman and van der Burgt, 2008).

The basics behind gasification are to react the feedstock containing carbon with steam and oxygen (could be air, enriched air or pure oxygen). The following chemical reactions are predominant during gasification (Higman and van der Burgt, 2008)

Thermal decomposition (i.e. pyrolysis), which covers dehydration as well as cracking reactions leading to gases, intermediate vapours and carbon structures known as "char".
Partial oxidation of the "char"which forms CO and CO2 generating heat for the otherwise endothermic reactions.

The partial oxidation

$$C + \tfrac{1}{2} O_2 = CO \qquad -111\ MJ/kmol \tag{2}$$

$$CO + \tfrac{1}{2} O_2 = CO_2 \qquad -283\ MJ/kmol \tag{3}$$

$$H_2 + \tfrac{1}{2} O_2 = H_2O \qquad -242\ MJ/kmol \tag{4}$$

Steam-carbon, i.e. the water-gas reaction, that converts carbon structures into H_2 and CO.

The water gas reaction

$$C + H_2O = CO + H_2 \qquad 131\ MJ/kmol \tag{5}$$

Steam reforming of intermediates formed by thermal decomposition.

The Steam methane reforming reaction

$$CH_4 + H_2O = CO + 3H_2 \qquad 206\ MJ/kmol \tag{6}$$

The steam reforming reaction

$$C_xH_y + xH_2O = x\ CO + (\tfrac{1}{2} y + x)\ H_2 \tag{7}$$

The dry reforming reaction

$$C_xH_y + xCO_2 = 2x\ CO + (\tfrac{1}{2} y + x)\ H_2 \tag{8}$$

Reactions involving CO_2 and H_2 with carbon and with intermediates are kinetically slower than the steam induced reactions at the conditions used in gasifiers.

The Boudouard reaction

$$C + CO_2 = 2CO \qquad 172\ MJ/kmol \tag{9}$$

The methanation reaction

$$C + 2H_2 = CH_4 \qquad -75\ MJ/kmol \tag{10}$$

Water gas shift reactions that lead to a desired H_2/CO ratio.

The Water gas shift reaction

$$CO + H_2O = CO_2 + H_2 \qquad -41\ MJ/kmol \tag{11}$$

Steam is hence chemically involved in the gasification, but it also served as moderating agent to limit the temperature. Carbon dioxide could also be used to this purpose. Other elements in feedstocks such as sulphur, nitrogen and chlorine will lead to different molecular species contaminating the syngas such as H_2S, COS, NH_3, HCN and HCl.
Different types and configurations of reactors can be used to produce syngas: fixed bed, entrained flow gasifier and bubbling fluidized bed. Fixed bed gasifier (essentially slowly moving-beds) can be updraft or downdraft.

The **Updraft fixed bed gasifier** is probably the oldest technology. Biomass or coal are fed from the top and move slowly by gravity. A grate supports the material while oxidizing the gas in its way upwards. The synthesis gas is withdrawn at the top of the reactor and ash from the bottom. There is a combustion zone in the bottom where char combusts in contact of oxygen to form CO_2. Char reacts with carbon dioxide to form two moles of carbon monoxide (equation 9). A zone of pyrolysis take place a little higher in the fixed bed because all oxygen has been consumed at this level. Finally, hot synthesis gas dries the feedstock at its entrance (Ciferno and Marono, 2002, Paes, 2005). This simple proven process has as drawback the production of a tar-rich syngas. Downdraft fixed bed gasifier employs co-current flow of oxidant and feedstock. This configuration has low tar formation as an outcome. Nevertheless, a part of the carbon remains unconverted (Ciferno and Marono, 2002).

Entrained flow gasifier utilizes a blast of oxidant in a co-current arrangement with the feed grounded to ensure carry-over. A high temperature is necessary because of the short residency time, counted in seconds. More oxygen is thus required to reach higher temperature (Higman and van der Burgt, 2008). With this type of gasifier, the particles entrained are passed to a cyclone and a riser to bring this material again through the bottom of the gasifier.

Bubbling fluidized bed gasifer is designed to keep the particles in suspension. This technology is based on a configuration where feedstock is added continuously to a bed of alumina, olivine (iron and magnesium orthosilicate) and/or other material serving as a heat carrier. This type of gasifier has high heat/mass transfer and conversion and is able to treat a large range of particles sizes and heterogeneous feedstock. Fluidized bed gasifier has the following advantages: flexibility and easiness for control and maintenance. Figure 1 illustrates an overview of the process developed by Enerkem Inc based on a bubbling fluidized reactor.

Feeding is assured by conveyor, lock hoppers, rotary valves and cooled feeding screw. Oxygen and steam are added through the bed in order to fluidize the medium. Low severity conditions are used in the bed, i.e. pressure lower than 4 atm and temperature lower than 750°C (Chornet et al., 2010). The reactor itself is made from carbon steel with insulation and refractory.

Ash is collected at the cyclones and larger inert material exits at the bottom of the gasifier. Ash can be composed of silica, alumina, potassium, phosphorus, sodium, magnesium, ferric oxide, and smaller amounts of titanium oxide and sulfur compounds (Higman and van der Burgt, 2008).

CaO/MgO is also introduced in the gasifier for neutralising a part of HCl, H_2S and COS formed. Literature reports that the use of calcined dolomite and NiO-loaded calcined dolomite helps in the reduction of tar and char formation (Corujo et al., 2010). This dolomite contained 24,6 % of Ca, 19,7 % of Mg and 0,035 % of Fe on a weight basis. Calcination of dolomite produces magnesium and calcium oxides. Corujo et al. have stated that the use of NiO-loaded calcined dolomite catalysts increased the total product gas volume by 30% and decreased the rate of char and tar formation leading to a higher gas energy yield. In fact, bed materials are selected for two criteria: attrition resistance and possible catalytic activity in hydrocarbon and tar reforming. Olivine was compared to silica sand, dolomite, ash, magnetite and iron ore. With the use of Olivine, higher conversion of toluene was observed during steam gasification at 900°C (Rauch et al., 2004).

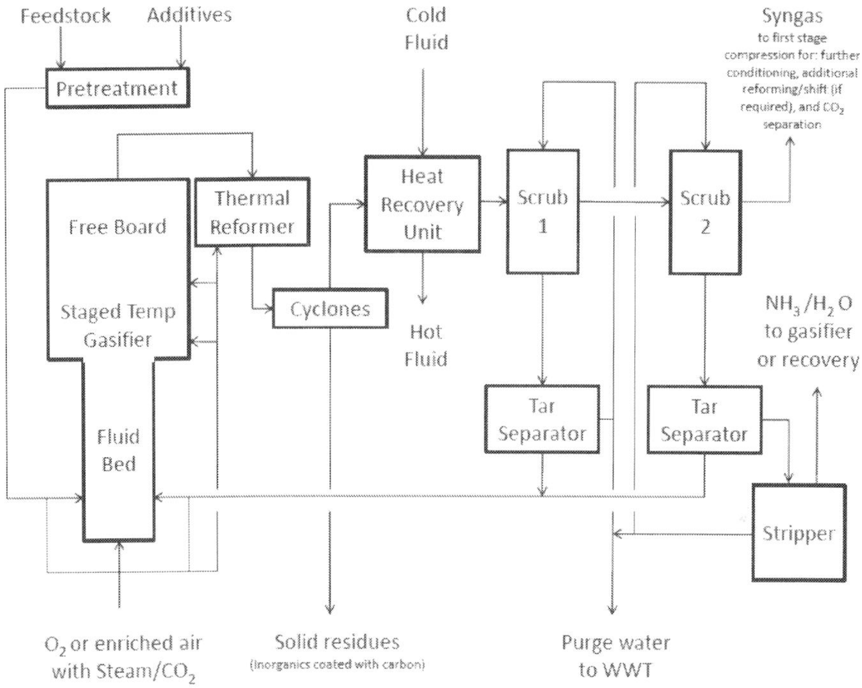

Fig. 1. Schematic diagram of Enerkem's process

Comparison of Ni/olivine catalyst with olivine alone demonstrated a higher activity, a higher selectivity to hydrogen and carbon monoxide and a lower carbon deposition. These observations were explained by an enhanced steam adsorption on magnesium oxide facilitating surface carbon gasification rates and by nickel dilution effect, associated with Fe-Ni alloys, ensuring stable catalyst activity (Swierczynski et al., 2007)

Leaving the freeboard of the gasifier, the syngas is composed of: H_2, CO, CO_2, H_2O N_2, hydrocarbons, tar, fines, char and contaminants. Tar is a complex blend of condensable hydrocarbons, which includes single to five-rings aromatic compounds along with oxygen-containing hydrocarbons and complex PAH (Yan et al., 2006). The freeboard is sized appropriately for disengagement of alumina and/or other solid fluidized bed materials.

A reforming zone is used for converting hydrocarbons, tar and char to obtain more syngas (see equations 8, 9, 10 and 11). Since thermal decomposition of hydrocarbons requires activation and that reforming reactions are endothermic, heating the gas, by direct or indirect manner, is essential. For example, if oxygen is injected directly, it must be added carefully to avoid lowering the product gas heating value and cold gas efficiency. However, with just the right amount of the oxidizing gas, selective oxidation of methane, ethylene and propylene will occur (Villano et al., 2010). Target temperatures are in the 750°C to 1200°C range. Table 3 depicts general compositions of syngas produced out of some previously mentioned biomass.

Biomass	MSW	MSW	Sludges*	Res. Forest/Agriculture	Typical biomass
Conditions	700 °C, Fixed bed gasifier, Dolomite (Catalyst),	900 °C, Fixed bed gasifier, Dolomite (Catalyst), Steam (oxidizing agent)	829 °C Fluidized bed	900 °C, Fixed bed gasifier, Dolomite (Catalyst), Steam (oxidizing agent)	O_2 / Steam / CO_2 low severity gasification + Thermal cracking and reforming at higher severity
Char (%w/w)	19.15	12.65	n.d	n.d	n.d
Tar (%w/w)	12.94	2.62	n.d	n.d	n.d
Gas (%w/w)	94.52	145.23	n.d.	n.d.	n.d.
Gas composition	% mol	% mol	% mol	% mol	% mol
H_2	16.92	36.98	17.23	46.2	23-27
CO	20.33	27.37	26.17	33.2	21-23
CO_2	35.28	20.78	45.64	16.1	38-44
CH_4	21.44	9.94	7.96	4.4	6-8
C_2	6.03	4.93	3.00	0.1	-
C2H4	n.d	n.d	n.d	n.d	Traces
C2-C5	n.d	n.d	n.d	n.d	0.2-0.5
C5-C10	n.d.	n.d.	n.d.	n.d.	Traces
Reference	(He et al., 2009)	(He et al., 2009)	(Van der Drift et al., 2001)	(Corujo et al., 2010)	(Marie-Rose et al. 2010)

* Demolition wood + paper residue sludge
n.d. not determined

Table 3. Solid, liquid and gas ratio as well as gas composition following gasification of common residual feedstocks.

Increasing temperature cause the decomposition of char and tar by thermal cracking and steam reforming reactions. Moreover, this has a direct impact on the production and composition of the gas as seen from table 3 (He et al., 2009)

Cyclones are used for entrained ash removal. The latter having typically a diameter greater than 10 microns. Efficiency of this cyclone system is from 90 to 95% and mostly recuperates char. Following the cyclone and recuperation of ash, the next unit operation is a heat exchanger. The latter cools the crude synthesis gas and allocates heat recovery for other application on the process using thermal oil or liquid water which turns to steam.

Note that the reforming step mentioned above could be by-passed directly to the cyclones if the feedstock contains inorganic materials that form eutectics with low melting point Indeed, high temperature in this situation would have as a consequence scaling and fouling of the walls of the system.

Quenching is required for further cooling and removal of condensable materials, tars and fines. In the following step, a wet venturi scrubber with alkaline water is used for neutralisation of acid gases as H_2S and HCl. A coalescer/demister at the exit removes fines particles and mist.

A second scrubbing step involving neutral or slightly acidic water is used in order to capture ammonia, trace tars, residual fines and impurities such as chlorine and metals. After this cooling step, the cooled purified syngas is around 30 °C.

Water is recirculating after proper handling. The water loop includes knockout drums and separation of tars and fines particles with air flotation, decanters and/or centrifuges. The skimmed phase recovered (overflow), enclosing tars and fines particles, is emulsified and recycled to the gasifier. The underflow, containing heavier organics and particles, is also re-injected to the gasifier. A part of the recirculated water is withdrawn from the system to maintain the balance of water and contaminants. This purge is subjected to water treatment in the objective of meeting below the required environmental standards. Note that ammonia is separated from the water by steam stripping above 100°C and at about 1 to 3 atm. The ammonia is sent back to the gasifier producing hydrogen and nitrogen.

An additional reforming step is then possible for converting light hydrocarbons to more CO and H_2 and final adjustment of the molar ratio of H_2/CO.
The syngas is then passed through filters and adsorbers for dehumidification, elimination of traces of chlorine, sulphur and other contaminants such as metal carbonyls. An utlraclean syngas is obtained and compression follows to remove carbon dioxide assisted either by amine-based processes or chilled methanol.

3. Syngas conversion into methanol

3.1 Thermodynamic consideration

The two major components of synthesis gas, hydrogen and carbon monoxide are the building blocks of what is often known as C1 chemistry. Conversion of *syngas* to liquid fuels as well as conversion rates is directly related to the composition of the catalyst. Syngas can be efficiently converted to different products as alcohols and aldehyde (Figure 2).

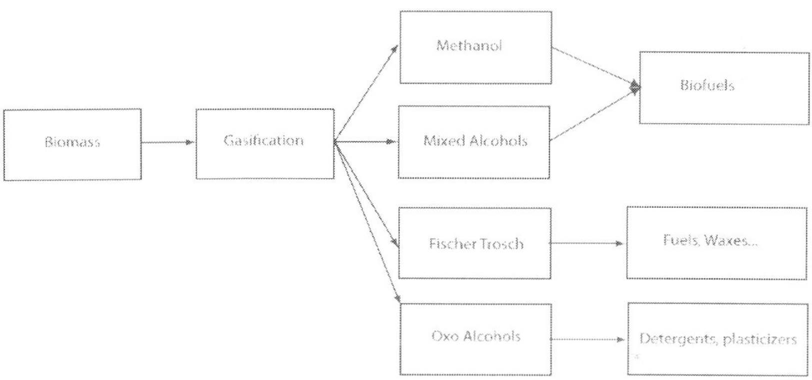

Fig. 2. Examples of application for syngas produced from biomass (Higman and van der Burgt, 2008)

Although many routes are available, the most promising route at industrial level is the production of methanol since the synthesis yields are the highest. The conversion to synthetic liquids as methanol is strongly influenced by thermodynamic factors (Rostrup-Nielsen, 2000).

$$CO + 2H_2 = CH_3OH \qquad -90.64\ KJ/mol \tag{12}$$

$$CO_2 + 3H_2 = CH_3OH + H_2O \qquad -49.67\ KJ/mol \tag{13}$$

$$CO + H_2O = CO_2 + H_2 \qquad -41.47\ KJ/mol \tag{14}$$

For methanol synthesis, a stoichiometric ratio defined as $(H_2-CO_2)/(CO+CO_2)$, of about 2 is preferred, which implies that there should be just the stoichiometric amount of hydrogen needed for methanol synthesis. For kinetic reasons and in order to control by-products formation, a value slightly above 2 is normally preferred (Dybkjr and Christensen, 2001).

Moreover, methanol synthesis is subjected to a thermodynamic equilibrium that limits the process to low conversion per pass and therefore implies a large recycle of unconverted gas. The reaction is strongly exothermic and consequently requires significant cooling duty.

Different phenomena exist at high and low pressure conditions. As example, when the pressure is relatively low, increasing temperature, CO conversion is not monotonic, and the trend is that of an increase followed by a decrease with the maximum conversion appearing near 250°C. This phenomenon is in agreement with many works in the literature (Li and Inui, 1996, Liaw and Chen, 2001, Wang et al., 2002). As the reaction temperature increases, the reaction rate gets higher and leads to the increase of CO conversion. However, methanol synthesis is an exothermic reaction and low temperature is more beneficial considering equilibrium. The conversion does not continue to increase due to the thermodynamic limitation and a decrease trend will even appear. When the system pressure is relative high, because of the relatively low CO conversion, the system is far from the thermodynamic equilibrium and under the control of reaction kinetics. In this case, CO conversion increases monotonically with an increase in temperature. From a mechanistic point of view for the methanol synthesis, two main reactions need to be in line with real world situation: CO hydrogenation and CO_2 hydrogenation.

3.1.1 CO hydrogenation as principal reaction for methanol synthesis

In this case, the main reaction leading to methanol is the hydrogenation of CO.

$$CO + 2H_2 = CH_3OH \quad -90.64\ KJ/mol \tag{15}$$

$$CO_2 + H_2 = CO + H_2O \quad 41.14\ KJ/mol \tag{16}$$

According to equation (15) methanol is predominantly synthesised via the direct hydrogenation of CO. The second reaction is the reverse of Water Gas Shift (RWGS). Experimental data involving typical syngas mixtures that contain 3 to 9% CO_2 show a decrease of its concentration in the reactor effluent stream (Sunggyu, 2007). It should be noted that the first reaction (methanol synthesis) is exothermic, whereas the second (RWGS) is endothermic. According to this depletion of carbon dioxide in the RWGS reaction, produce more reactant (CO) which overall boost the synthesis of methanol. Until 90s the role of CO_2 in the methanol synthesis was not clear. Deficiency of CO_2 in the feed composition can be extremely detrimental to the overall synthesis, very rapidly deactivating the catalysts and immediately lowering methanol productivity in the process. Typically, 2 to 4% of CO_2 is present in the syngas mixture for the vapor-phase synthesis of methanol, whereas this value is somewhat higher, with 4 to 9%, for liquid-phase synthesis (Cybulski, 1994).

3.1.2 CO_2 hydrogenation as principal reaction for methanol synthesis

$$CO_2 + 3H_2 = CH_3OH + H_2O \quad -49.67\ KJ/mol \tag{17}$$

$$CO + H_2O = CO_2 + H_2 \quad -41.14\ KJ/mol \tag{18}$$

In this case, the main reaction leading to methanol is hydrogenation of CO_2

It should be noted that according to this reaction, the synthesis of methanol proceeds predominantly by direct hydrogenation of CO_2 and not CO. It should also be noted that the

WGS reaction proceeds in the forward direction, consuming CO to produce the main reactants: CO_2 and H_2, thus boosting the eventual methanol productivity. A number of authors (Cybulski, 1994, Lee et al., 1989) have tried a variety of reaction to elucidate the true reaction pathways or mechanistic pathways, including isotope labelling studies and kinetic studies involving complete absence of one of the syngas components.

3.2 Catalysts used

The first high-temperature, high-pressure ever used methanol synthesis catalysts were ZnO/Cr_2O_3 and were operated at 350°C and 250-350 bar. Catalyst compositions contained 20-75 atom% Zn and these catalysts demonstrated high activity and selectivity for methanol synthesis and proved robust enough resist sulphur poisoning which is inherent when converting syngas from coal gasification. Over the years, as gas purification technologies improved, interest in the easily poisoned Cu catalysts for methanol synthesis was renewed. In 1966, ICI introduced a new, more active $Cu/ZnO/Al_2O_3$ catalyst was the first of a new generation of methanol production using lower temperature (220-275°C) and lower pressure (50-100 bar) than the established ZnO/ Cr_2O_3 catalysts. The last high temperature methanol synthesis plant was closed in the mid-1980s (Fiedler et al., 2003) and at the present, low temperature, low pressure processes based on Cu catalyst are used for all commercial production of methanol from syngas. The synthesis process has been optimised to the point that the modern methanol plants yield 1 kg of methanol/liter of catalyst/hr with >99.5% selectivity for methanol. Commercial methanol synthesis catalyst has lifetimes on the order of 3-5 years under normal operating conditions.

The Cu crystallites in methanol synthesis catalysts have been identified as the active catalytic sites although the actual state (oxide, metallic...) of the active Cu site is still being debated. The most active catalysts all have high Cu content, optimum about 60 wt% Cu on the catalyst that is limited by the need to have enough refractory oxide to prevent sintering of the Cu crystallites. Hindering agglomeration is why ZnO creates a high Cu metal surface area. ZnO also interacts with Al_2O_3 to form a spinel that provides a robust catalyst support. Acidic materials like alumina, are known to catalyse methanol dehydration reactions to produce DME. By interacting with the Al_2O_3 support material, the ZnO effectively improves methanol selectivity by reducing the potential for DME formation. Catalysts are typically prepared by the co-precipitation of metal salts with a variety of precipitation agents. It is important to avoid contaminating methanol catalysts with metals that have FT (Fischer Tropsch) activity (Fe or Ni) during the synthesis. Incorporation of alkali metal in the catalyst formulation should also be avoided because they catalyse the increase of higher alcohols production. Table 4 shows catalyst formulation from several commercial manufacturers.

Additional catalyst formulations have been presented in the literature with the purpose of improving per-pass methanol yields (Klier, 1982). The addition of Cs to Cu/ZnO mixtures has shown improved methanol synthesis yields. This only holds true for the heavier alkali metals, as the addition of K to methanol synthesis catalysts tends to enhance higher alcohols yields. The Cu/ThO_2 intermetallic catalysts have also been investigated for methanol synthesis (Klier, 1982). These catalysts have demonstrated high activity for forming methanol from CO_2-free syngas. Cu/Zr catalysts have proven active for methanol synthesis in CO-free syngas at 5 atm and 160-300°C (Herman, 1991). Supported Pd catalysts have also demonstrated methanol synthesis activity in CO_2-free syngas at 5-110 atm and 260-350°C (Spath and Dayton, 2003).

Manufacter	Cu (wt%)	Zn (wt%)	Al (wt%)	Other
IFP	45 - 70	15 - 35	4 - 20	Zr - 2-18
ICI	20 - 35	15 - 50	4 - 20	Mg
BASF	38.5	18.6	12.9	Rare Earth oxide - 5
Shell	71	24		
Sud Chemie	65	22	12	
Dupont	50	19	31	
Haldor Topsoe	>55	21 - 25	8 - 10	

Table 4. Commercial Methanol synthesis catalyst formulation (Spath and Dayton, 2003)

3.3 Methanol synthesis reactors

Today, the majority of the methanol is synthesised from syngas produced by steam reforming of natural gas (SMR). The synthesis can be done either with heat provided by a furnace where the tubular reactor is located, or by auto-thermal reforming (ATR) combined with steam reforming. Once the natural gas is reformed the resulting synthesis gas can be shift adjusted for its H2/CO ratio and the CO2 decreased to a few percentages as previously specified. The syngas is then fed to a reactor vessel in the presence of a catalyst to produce methanol and water vapor. This crude methanol, which usually contains up to 18% water, plus ethanol and higher alcohols is fed to a distillation plant that consists of a unit that removes the volatiles and a unit that removes the water and higher alcohols. The unreacted syngas is recirculated back to the methanol converter resulting in an overall conversion efficiency of 99%. A generic methanol synthesis process flow diagram (PFD) is shown in the Figure 3.

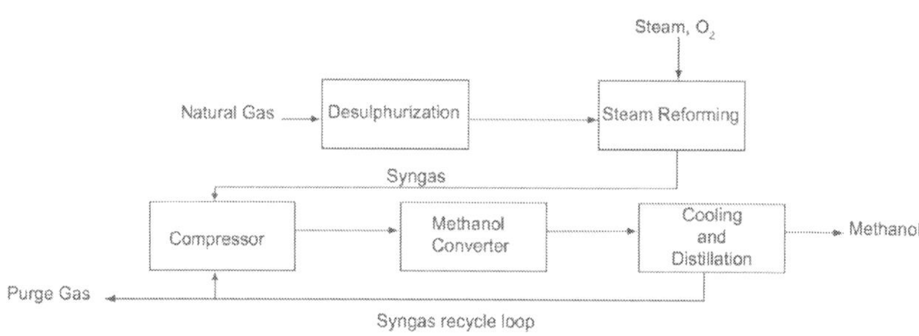

Fig. 3. Simplified Methanol Synthesis PFD (Spath and Dayton, 2003).

As is the case with Fisher Tropsch synthesis, one of the challenges associated with commercial methanol synthesis is removing the large excess heat of reaction. Methanol synthesis increases at higher temperatures so does the chance for competing side reactions. Controlling and dissipating the heat of reaction and overcoming the equilibrium constraint to maximise the per-pass conversion efficiency are the two main process features that are considered when designing the methanol synthesis reactor, commonly referred to as a methanol converter. Numerous methanol converter designs have been commercialised over the years and these can be roughly separated into two categories: adiabatic or isothermal reactors.

- Adiabatic reactors often imply multiple catalysts beds separated by gas cooling devices, either direct heat exchange or injection of cooled, fresh or recycled syngas.
- The isothermal reactors are designed to continuously remove the heat of the reaction so they operate essentially like a heat exchanger.

Haldor Topsoe low-pressure methanol synthesis process. This process is designed to produce methanol from natural gas or associated gas feedstocks, utilizing a two-step reforming process to generate a syngas mixture feed for the methanol synthesis (Sunggyu, 2007). Associated gas is natural gas produced with crude oil from the same reservoir. It is claimed that the total investment for this process is lower than with the conventional flow scheme based on straight steam reforming of natural gas by approximately 10%, even after considering an oxygen plant. As shown in figure 4, the two stage reforming is conducted by primary reforming in which a preheated mixture of natural gas and steam is reacted, followed by a secondary reforming which further converts the exit gas from the primary reformer with the aid of oxygen that is feed separately. The amount of oxygen required as well as the balance of conversion between the primary and the secondary reformers need to be properly adjusted so that a balanced syngas (in a stoichiometric ratio (2:1) of H_2/CO) is obtained with a low inert content.

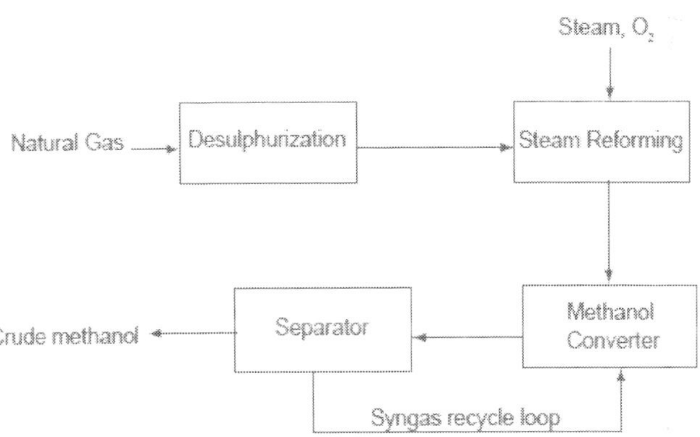

Fig. 4. Haldor Topsoe methanol synthesis process (Sunggyu, 2007).

Liquid-Phase methanol process. The liquid phase methanol process was originally developed by Chem. Systems Inc in 1975 (Cybulski, 1994). The R&D of this process was sponsored by the U.S. Department of Energy and Electric Power Research Institute. Commercialised by Air Products and Chemicals Inc and Eastman Chemical Co. in the 1990s, the process is based on the low-pressure methanol synthesis using coal as the source of syngas. Recently, in Québec (Canada) Enerkem Inc. has developed a liquid phase methanol process using syngas produced from biomass. The chemical reaction is carried out in a slurry reactor using a $Cu/ZnO/Al_2O_3$ catalyst at temperature ranging from 230 to 260 °C and 50 to 100 atm. The commercial reactor used a liquid entrained reactor in which fine grains of catalyst are slurried in an inert high-boiling oil typically white mineral oil. Pressurized gaseous reactants are dissolved in the oil and the dissolved molecular species are reacted on the catalytic surfaces of the grains present in a slurry. The figure 5 shows the schematic of liquid phase methanol process of Enerkem Inc.

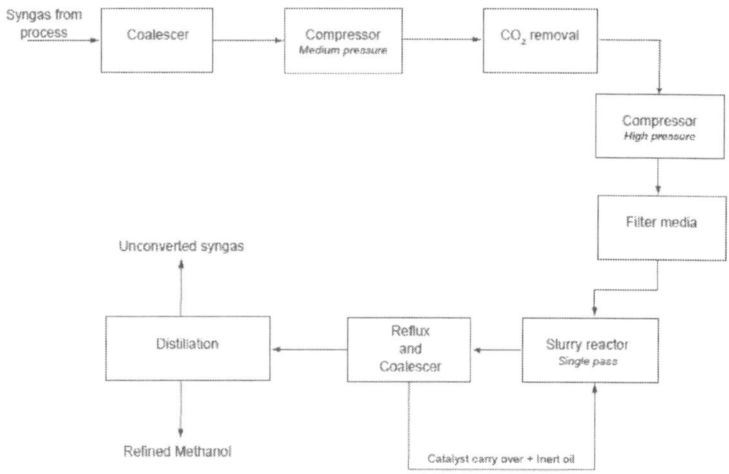

Fig. 5. Enerkem Inc. liquid phase methanol process

4. Ethanol synthesis

Ethanol can be readily produced by fermentation of simple sugars that are hydrolyzed form starch crop. Feedstocks for such fermentation include corn, barley, potato, rice and wheat (Cybulski, 1994). Sugar ethanol can be called *grain ethanol*, whereas ethanol produced from cellulose biomass is called *cellulosic or second generation ethanol*. Both grain ethanol and cellulosic are produced via biochemical processes, whereas *chemical ethanol* is synthesised by chemical synthesis routes that do not involve fermentation. In fact, thermo-catalysis process route can be a good alternative to the bioprocess route. In a recently reported process (Chornet et al., 2009) the ethanol is produced by a two steps process: first methanol carbonylation to produce methyl acetate and second hydrogenolysis of the latter as illustrated in the figure 6.

4.1 Methanol carbonylation: methyl acetate synthesis

Methyl acetate is produced industrially by methanol carbonylation in the acetic acid production processes. Methanol carbonylation reaction is a well known reaction since Monsanto in 1970 performed it in liquid phase using a rhodium-base catalytic system. which was later acquired and optimized by Celanese. Such system had been preceded by a cobalt-based system developed by BASF, which suffered from significantly lower selectivity and the necessity for much harsher conditions of temperature and pressure. Although the rhodium catalysed system has much better activity and selectivity, the R&D has continued for new catalysts which improve efficiency even further. The strategies employed have involved either modifications to the rhodium-based system or insertion of another metal, eventually Iridium (Haynes, 2006). In the homogenous system, acetic acid was used as solvent, so esterification leads to substantial conversion into methyl acetate. Methyl acetate formation is facilitated by the methyl iodide used in carbonylation as a mediator. Heterogenous systems for methanol carbonylation have been suggested for several years. Research has been primarily focus on two possible catalysts for this reaction: 1) rhodium supported polymers (Acetica process as an example) or 2) zeolites and a variety of metals supported on activated carbon (Merenov and Abraham, 1998). The choice of the support seems to play an important role in the activity of the catalyst (Yashima et al., 1979). Another way to produce methyl acetate is by an esterification reaction. Eastman Chemical Company, produced methyl acetate by the esterification of methanol with acetic acid in a reactive distillation column.

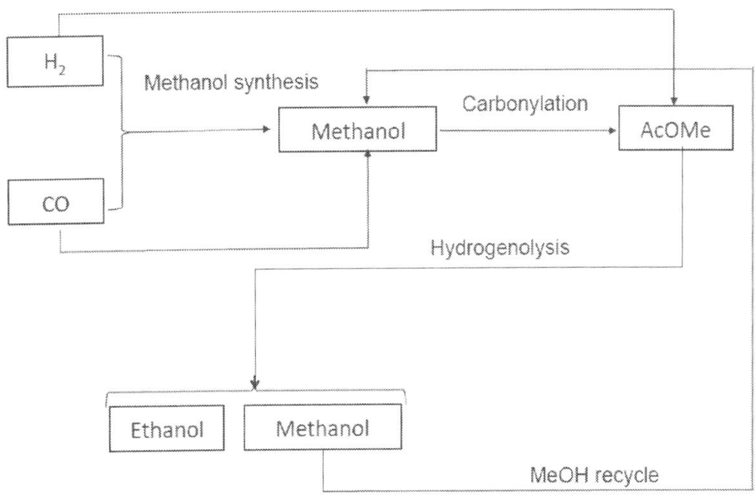

Fig. 6. Enerkem Inc. ethanol pathway production.

This company claimed the first commercial plant of reactive distillation (Zoeller, 2009). As illustrated by figure 7, in the reactive distillation there are three sections in the column, water/methanol stripping zone in the bottom, a reaction zone in the center, and methyl acetate enriching at the top. Descriptively, methanol (b.p. 65°C) is added at the bottom of the reaction zone and acetic acid (b.p. 118°C) and acid catalyst are added at the top of the reaction zone.

As methanol ascends up the column it encounters a consistently richer acetic acid concentration, which drives the equilibrium toward the products and prevents methanol from leaving column. Likewise, as acetic acid descends down the column, it also encounters a continuously enriched stream of methanol, which also pushes the equilibrium toward products and prevents acetic acid reaching the bottoms of the column. However, the acetic acid also serves as extractant for water, breaking the various azeotropes. A short section is left at the base of the distillation column to strip all the methanol and a short section is added at the top of the column to enrich methyl acetate and prevent acetic acid from being entrained as the product is distilled overhead in the column. By properly balancing the addition rates, the reaction provides methyl acetate in 100% yield based on both acetic acid and methanol.

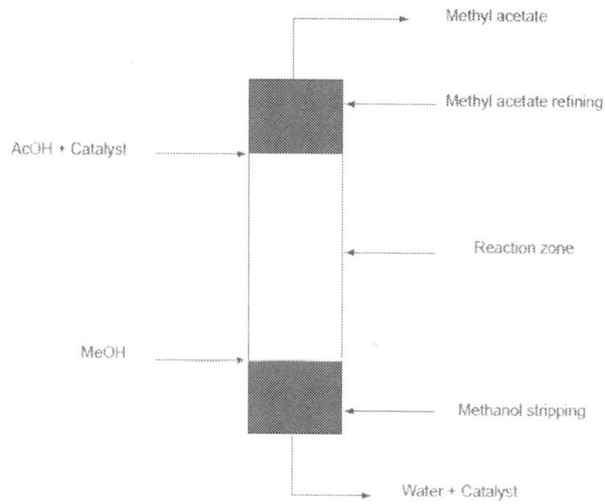

Fig. 7. Reactive distillation of methanol (MeOH) and acetic acid (AcOH) to form methyl acetate.

Enerkem Inc. has focused on the new gas phase approach to the carbonylation reaction. As acetic acid is not the desired end product of the carbonylation, the reaction conditions have been tailored to favour the methyl acetate directly according to the equation:

$$2CH_3OH + CO = CH_3COOCH_3 + H_2O \quad -138.08\ KJ/mol \tag{19}$$

This is advantageous for Enerkem Inc. since it reduces by one step their process leading to ethanol. Forcing the reaction towards methyl acetate also facilitates a strategy based on gas phase reactor. In the process, the methanol carbonylation is carried out maintaining methanol in the vapor phase using a fixed bed packed a rhodium-based catalyst. The methanol is vaporized under pressure and mixed with the CO-rich gas prior to flowing through the reactor. The methanol to CO molar ratio is comprised between 1 and 4. Methyl iodide (co-catalyst and mediator) is added to the system at a suitable mol ratio relative to methanol. The operating conditions are such that the GHSV (based on CO) varies up to 2,000 h^{-1}. At temperature comprised between 170 to 300°C and total pressure from 10 to 50 atm, it is found that the CO is converted at rate near 100% when the methanol:CO ratio is up to 2 (Chornet et al., 2009).

4.2 Methyl acetate hydrogenation

Esters (i.e. methyl acetate) are ubiquitous in nature and have vast industrial and commercial applications based on different substituent groups. An important class of industrial catalytic process for esters reduction is metal-catalyzed hydrogenolysis for the production of alcohols (e.i. ethanol) (Adkins and Folkers, 1931, Adkins et al., 1933, Turek et al., 1994a, Xu and Xu, 2010) New applications for ester hydrogenolysis are being found in catalytic upgrading and conversion of renewable resources such as lignocellulosic biomass into fuels or fine chemicals (Corma et al., 2007, Huber et al., 2006). Copper-based catalysts are widely used for ester hydrogenolysis (Adkins and Folkers, 1931, Cybulski et al., 2001, Thomas et al., 1992,

Turek et al., 1994b, Turek et al., 1994a). They show very good activity and selectivity for alcohols under high temperature and hydrogen pressure (500-700K, 200-300 bar). Precious metal-based alloys (Rh-Sn for example) show promise of lessening the severity of the reaction conditions (lowering hydrogen pressure to below 100 bar), but precious metal are costly and tend to favor hydrocarbon production.

In our approach, the recovered methyl acetate is maintained in liquid phase at 20°C. It is pumped against pressure ranging from 10 to 50 atm, through a heat exchanger that vaporizes it completely at temperature ranging 150 to 425°C. Hydrogen, preheated at the same temperature range is mixed with the methyl acetate vapor at the exit of the heat exchanger. The molar ratio hydrogen to methyl acetate is from 4 to 11. The hot mixture is flown through a catalytic bed where the catalyst CuO/Cr_2O_3 or $CuO/ZnO/Al_2O_3$ catalyst is placed. The CuO is reduced with H_2/N_2 mixtures prior to adding any methyl acetate (Claus et al., 1991). Methyl acetate is converted to ethanol and methanol according to the equation:

$$CH_3COOCH_3 + 2H_2 = CH_3OH + CH_3CH_2OH \quad -27\ KJ/mol \tag{20}$$

The GHSV (based on H_2) of the reaction is comprised between 1,000 and 2,000 h^{-1} and the methyl acetate conversion reached 95%, with a slightly higher selectivity towards ethanol.

5. Conclusion

Biomass is a renewable energy source whose conversion to biofuels is an option to reduce oil dependency and reduce the carbon dioxide footprint characteristic of fossil fuels.

This chapter has shown the chemical steps needed to convert non-homogeneous biomass into bioethanol via gasification. The syngas produced is catalytically turned into biofuels (methanol and ethanol). Such approach is practices by Enerkem's which uses non-homogeneous residual biomass including urban biomass as feedstock for the gasification. The syngas produced is cleaned prior used for catalytic synthesis of methanol. The reactor used for the methanol synthesis is a three phase reactor based on the Liquid-Phase methanol process. The yield of methanol is about 1 $kg_{MeOH}/kg_{cat}/h$ and the selectivity of the reaction into methanol is about 99%. The refined methanol produced is carbonylated to produce methyl acetate. The synthesis of methyl acetate is carried in a fixed bed reactor and uses a halide component as co-catalyst. The methyl acetate is hydrogenolized to form stoichiometrically one mole of ethanol and one mole of methanol. The methanol produced at the end of the process is recycled into the carbonylation unit.

6. Acknowledgement

The authors would like to acknowledge financial support of the sponsors of the Industrial Chair in Cellulosic Ethanol: the Quebec's government (MRNF) and its industrial partners: Greenfield Ethanol, CRB innovations and Enerkem. Additional support from NSERC's Synergy program is also acknowledged.

7. References

J.-L. Rastoin, (2008). Agrocarburant et sécurité alimentaire: l'alliance impossible, Vol.

M. J. Prins, K. J. Ptasinski and F. J. J. G. Janssen, (2005). Exergetic optimisation of a production process of Fischer-Tropsch fuels from biomass, Vol. 86,No. 4,pp. 375-389, 0378-3820

D. Ballerini and N. Alazard-Toux. (2006). Les biocarburants: état des lieux, perspectives et enjeux du développement, Technip, 9782710808695,

A. van der Drift, J. van Doorn and J. W. Vermeulen, (2001). Ten residual biomass fuels for circulating fluidized-bed gasification, Vol. 20,No. 1,pp. 45-56, 0961-9534

G. L. Borman and K. W. Ragland. (1998). Combustion engineering, McGraw-Hill, 9780070065673,

C. Higman and M. van der Burgt. (2008). Gasification (2nd Edition), Elsevier, 978-0-7506-8528-3,

J. P. Ciferno and J.J. Marano (2002). Benchmarking Biomass Gasification Technologies for Fuels, Chemicals and Hydrogen Production, report for US DOE.

S. A. Channiwala and P. P. Parikh, (2002). A unified correlation for estimating HHV of solid, liquid and gaseous fuels, Vol. 81,No. 8,pp. 1051-1063, 0016-2361

B. Wetherold, D. Orr, D. Maxwell (2000). A comparison of gasification and incineration Hazardous wastes. US DOE, DCN 99.803931.02

S. C. Marie-Rose, J.-M. Lavoie, D. Lynch and E. Chornet. (2011). From biomass-rich residues into fuels and green chemicals via gasification and catalytic synthesis, WIT Transaction on Ecoloy and the Environment, Vol. 143, pp. 123-132, 1743-3541

S. M. Villano, J. Hoffman, H.-H. Carstensen, A.M. Dean (2010). Selective Oxidation of Ethylene in a Model "Dirty Synthesis Gas Stream: An Experimental and Kinetic Modeling Study, ACS National meeting, San Francisco

T. Paes (2005). Modeling for control of biomass gasifier, Thesis of Endhoven University of Technology.

E. Chornet, B. Valsecchi, G. Drolet, M. Gagnon, B. Nguyen (2010). Production and conditioning of synthesis gas obtained from biomass. Patent Application number 20100051875.

A. Corujo, L. Yermán, B. Arizaga, M. Brusoni and J. Castiglioni, (2010). Improved yield parameters in catalytic steam gasification of forestry residue; optimizing biomass feed rate and catalyst type, Vol. 34,No. 12,pp. 1695-1702, 0961-9534

R. Rauch, H. Hofbauer, K. Bosch, I. Siefert, C. Aichernig, H. Tremmel, K. Voigtlaender, R. Koch, R. Lehner (2004). Steam Gasification of Biomass at CHP Plant in Guessing-Status of the Demonstration Plant. 2nd World Conference and Technology Exhibition on Bioamass for Energy, Italy.

D. Swierczynski, S. Libs, C. Courson and A. Kiennemann, (2007). Steam reforming of tar from a biomass gasification process over Ni/olivine catalyst using toluene as a model compound, Vol. 74,No. 3-4,pp. 211-222, 0926-3373

C. Yan, W. Yang, J. T. Riley and W.-P. Pan. (2006). A novel biomass air gasification process for producing tar-free higher heating value fuel gas, Elsevier, Amsterdam, PAYS-BAS

M. He, B. Xiao, Z. Hu, S. Liu, X. Guo and S. Luo, (2009). Syngas production from catalytic gasification of waste polyethylene: Influence of temperature on gas yield and composition, Vol. 34,No. 3,pp. 1342-1348, 0360-3199

J. R. Rostrup-Nielsen, (2000). New aspects of syngas production and use, Vol. 63,No. 2-4,pp. 159-164, 0920-5861

I. Dybkjr and T. S. Christensen. (2001). Syngas for large scale conversion of natural gas to liquid fuels, Studies in Surface Science and Catalysis, J. J. S. E. Iglesia and T. H. Fleisch,pp. 435-440, Elsevier, 0167-2991,

J. L. Li and T. Inui, (1996). Enhancement in methanol synthesis activity of a copper/zinc/aluminum oxide catalyst by ultrasonic treatment during the course of the preparation procedure, Vol. 139,No. 1-2,pp. 87-96, 0926-860X

B. J. Liaw and Y. Z. Chen, (2001). Liquid-phase synthesis of methanol from CO2/H2 over ultrafine CuB catalysts, Vol. 206,No. 2,pp. 245-256, 0926-860X

Y. W. Wang, Z. L. Wang, J. F. Wang and Y. Jin, (2002). Study on the methanol synthesis in slurry reactors, Vol. 31,No. 8,pp. 597-601, 10008144 (ISSN)

L. Sunggyu. (2007). Methanol Synthesis from Syngas, Handbook of Alternative Fuel Technology, Taylor & Francis Group,

A. Cybulski, (1994). Liquid-phase methanol synthesis: Catalysts, mechanism, kinetics, chemical equilibria, vapor-liquid equilibria, and modeling - A review, Vol. 36,No. 4,pp. 557-615, 01614940 (ISSN)

S. Lee, V. R. Parameswaran, I. Wender and C. J. Kulik, (1989). Roles of carbon dioxide in methanol synthesis, Vol. 7,No. 8,pp. 1021-1057,

E. Fiedler, G. Grossmann, D. B. Kersebohm, G. Weiss and C. Witte. (2003). Ullmann's Encyclopedia of Industrial Chemistry Release 2003, Methanol, W.-V. V. G. C. KGaA,pp.

K. Klier. (1982). Methanol Synthesis, Advances in Catalysis, H. P. D.D. Eley and B. W. Paul,pp. 243-313, Academic Press, 0360-0564,

R. G. Herman. (1991). Chapter 7 Classical and Non-Classical Routes for Alcohol Synthesis, Studies in Surface Science and Catalysis, L. Guczi,pp. 265-349, Elsevier, 0167-2991,

P. L. Spath, D.C. Dayton (2003). Preliminary Screening and Economy Assessment of Synthesis fuels and Cjemicals with Emphasis on the Potential for Biomass-Derived Syngas

A. Haynes. (2006). Acetic Acid Synthesis by Catalytic Carbonylation of Methanol, Catalytic Carbonylation Reactions, M. Beller,pp. 179-205, Springer Berlin / Heidelberg,

A. S. Merenov and M. A. Abraham, (1998). Catalyzing the carbonylation of methanol using a heterogeneous vapor phase catalyst, Vol. 40,No. 4,pp. 397-404, 0920-5861

T. Yashima, Y. Orikasa, N. Takahashi and N. Hara, (1979). Vapor phase carbonylation of methanol over RhY zeolite, Vol. 59,No. 1,pp. 53-60, 0021-9517

J. R. Zoeller, (2009). Eastman Chemical Company's "Chemicals from Coal" program: The first quarter century, Vol. 140,No. 3-4,pp. 118-126, 0920-5861

H. Adkins and K. Folkers, (1931). The catalytic hydrogenation of Esters to alcohols, Vol. 53,No. 3,pp. 1095-1097, 0002-7863

E. Chornet, B. Valsecchi, Y. Avilla, B. Nguyen, J-M. Lavoie, (2009). Production of ethanol to methanol, patent Application, WO2009032608O

H. Adkins, B. Wojcik and L. W. Covert, (1933). The Catalytic Hydrogenation of Esters to Alcohols. III, Vol. 55,No. 4,pp. 1669-1676, 0002-7863

T. Turek, D. L. Trimm and N. W. Cant, (1994a). The Catalytic Hydrogenolysis of Esters to Alcohols, Vol. 36,No. 4,pp. 645-683, 0161-4940

L. Xu and Y. Xu, (2010). Activation of methyl acetate on Pd(111), Vol. 604,No. 11-12,pp. 887-892, 0039-6028

A. Corma, S. Iborra and A. Velty, (2007). Chemical Routes for the Transformation of Biomass into Chemicals, Vol. 107,No. 6,pp. 2411-2502, 0009-2665

G. W. Huber, S. Iborra and A. Corma, (2006). Synthesis of Transportation Fuels from Biomass:,Äâ Chemistry, Catalysts, and Engineering, Vol. 106,No. 9,pp. 4044-4098, 0009-2665

A. Cybulski, J. Chrzszcz and M. V. Twigg, (2001). Hydrogenation of dimethyl succinate over monolithic catalysts, Vol. 69,No. 1-4,pp. 241-245, 0920-5861

D. J. Thomas, J. T. Wehrli, M. S. Wainwright, D. L. Trimm and N. W. Cant, (1992). Hydrogenolysis of diethyl oxalate over copper-based catalysts, Vol. 86,No. 2,pp. 101-114, 0926-860X

T. Turek, D. L. Trimm, D. S. Black and N. W. Cant, (1994b). Hydrogenolysis of dimethyl succinate on copperbased catalysts, Vol. 116,No. 1-2,pp. 137-150, 0926-860X

P. Claus, M. Lucas, B. Lücke, T. Berndt and P. Birke, (1991). Selective hydrogenolysis of methyl and ethyl acetate in the gas phase on copper and supported Group VIII metal catalysts, Vol. 79,No. 1,pp. 1-18, 0926-860X

16

Novel Methods in Biodiesel Production

Didem Özçimen and Sevil Yücel
Yıldız Technical University, Bioengineering Department, Istanbul Turkey

1. Introduction

The depletion of fossil fuels and their effects on environmental pollution necessitate the usage of alternative renewable energy sources in recent years. In this context, biodiesel is an important one of the alternative renewable energy sources which has been mostly used nowadays. Biodiesel is a renewable and energy-efficient fuel that is non-toxic, biodegradable in water and has lesser exhaust emissions. It can also reduce greenhouse gas effect and does not contribute to global warming due to lesser emissions. Because it does not contain carcinogens and its sulphur content is also lower than the mineral diesel (Sharma & Singh, 2009; Suppalakpanya et al., 2010). Biodiesel can be used, storaged safely and easily as a fuel besides its environmental benefits. Also it is cheaper than the fossil fuels which affect the environment in a negative way. It requires no engine conversion or fuel system modification to run biodiesel on conventional diesel engines.

Today, biodiesel is commonly produced in many countries of the world such as Malaysia, Germany, USA, France, Italy and also in Australia, Brazil, and Argentina. Biodiesel production of EU in 2009 was presented in Table 1 (European Biodiesel Board, July 2010). As can be seen from Table 1, 9 million tons biodiesel were produced in European Union countries in 2009. Germany and France are the leaders in biodiesel production. EU represents about 65% of worldwide biodiesel output. Biodiesel is also main biofuel produced and marketed in Europe. In 2009, biodiesel represented is about 75% of biofuels produced in Europe.

The world production of biodiesel between 1991 and 2009 was presented in Figure 1. From Figure 1, biodiesel production increased sharply after 2000s in the world.

Firstly in 1900, Rudolph Diesel showed that diesel engines could work with peanut oil. And then, the different kinds of methods such as pyrolysis, catalytic cracking, blending and microemulsification were used to produce biodiesel from vegetable oil for diesel engines (Sharma & Singh, 2009; Varma & Madras, 2007). Finally, transesterification process was developed as the most suitable method to overcome problems due to direct use of oil in diesel engines (Varma & Madras, 2007).

Biodiesel is generally produced from different sources such as plant oils: soybean oil (Kaieda et al., 1999; Samukawa et al., 2000; Silva et al., 2010; Cao et al., 2005; Lee et al., 2009; Yu et al., 2010), cottonseed oil (Köse et al., 2002; He et al., 2007; Royon et al., 2007; Hoda, 2010; Azcan & Danisman, 2007; Rashid et al., 2009), canola oil (Dube et al., 2007; Issariyakul et al., 2008), sunflower oil (Madras et al., 2004), linseed oil (Kaieda et al., 1999), olive oil (Lee et al., 2009), peanut seed oil (Kaya et al., 2009), tobacco oil (Veljkovic et al., 2006), palm oil (Melero et al., 2009), recycled cooking oils (Issariyakul et al., 2008; Rahmanlar, 2010; Zhang et al. 2003; Demirbaş, 2009) and animal fats (Da Cunha et al., 2009; Öner & Altun, 2009; Gürü et al., 2009; Gürü et al., 2010; Tashtoush et al., 2004; Teixeira et al., 2009; Chung et al., 2009).

The major economic factor to consider for input costs of biodiesel production is the feedstock. 90 % of the total cost of the biodiesel production is the resource of the feedstock. Studies to solve this economic problem especially focused on biodiesel production from cheaper raw material. Using agricultural wastes, high acid oils, soapstock, waste frying oil and alg oil as raw materials for biodiesel production are being reported in literature (Haas & Scott, 1996;Özgül & Türkay, 1993; Özgül & Türkay, 2002; Leung & Guo, 2006; Yücel et al., 2010; Özçimen & Yücel, 2010).

Country	Production (1000 Tons)	Country	Production (1000 Tons)
Austria	310	Italy	737
Belgium	416	Latvia	44
Bulgaria	25	Lithuania	98
Cyprus	9	Luxemburg	0
Czech Republic	164	Malta	1
Denmark/Sweden	233	Netherlands	323
Estonia	24	Poland	332
Finland*	220	Portugal	250
France	1959	Romania	29
Germany	2539	Slovakia	101
Greece	77	Slovenia	9
Hungary	133	Spain	859
Ireland*	17	UK	137
TOTAL:		9.046	

*Data include hydrodiesel production

Table 1. Biodiesel production of EU in 2009 (EBB 2010)

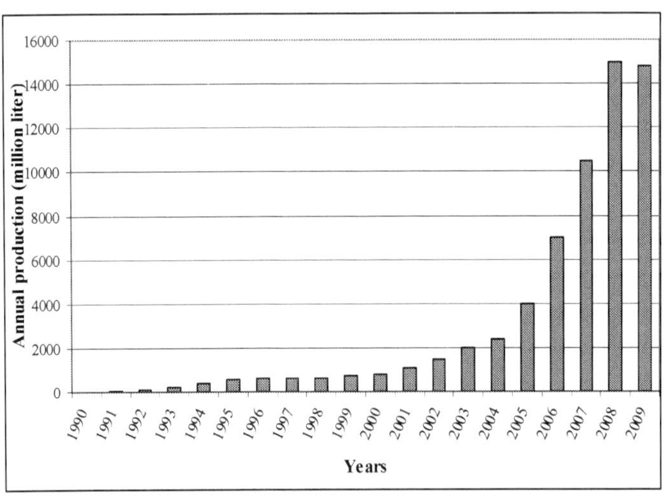

Fig. 1. The world production of biodiesel between 1991 and 2009 (Licht, 2009)

Transesterification process, as showed in Figure 2 (Barnard et al., 2007) is a conventional and the most common method for biodiesel production. In transesterification reaction homogeneous catalysts (alkali or acid) or heterogeneous catalysts can be used. The catalysts split the oil into glycerin and biodiesel and they could make production easier and faster.

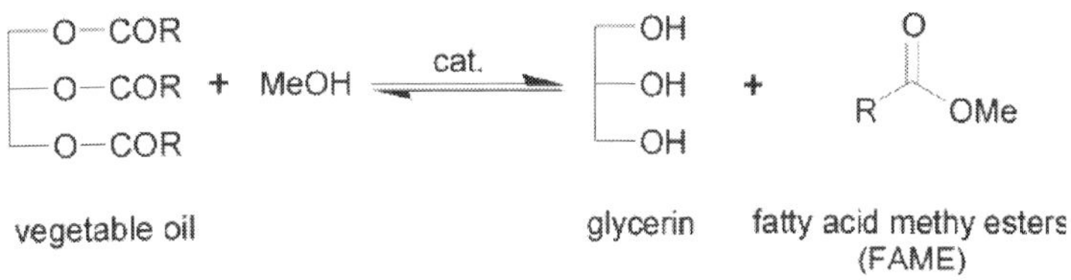

Fig. 2. Biodiesel production via transesterification reaction (Barnard et al., 2007)

In this method, fatty acid alkyl esters are produced by the reaction of triglycerides with an alcohol, especially ethanol or methanol, in the presence of alkali, acid or enzyme catalyst etc. The sodium hydroxide or potassium hydroxide, which is dissolved in alcohol, is generally used as catalyst in transesterification reaction (Dube et al., 2007). The products of the reaction are fatty acid methyl esters (FAMEs), which is the biodiesel, and glycerin (Vicente et al., 2004). Ethanol can be also used as alcohol instead of methanol. If ethanol is used, fatty acid ethyl ester (FAEE) is produced as product (Hanh et al., 2009b). Methyl ester rather than ethyl ester production was preferred, because methyl esters are the predominant product of commerce, and methanol is considerably cheaper than ethanol (Zhou & Boocock, 2003). However, methanol usage has an important disadvantage, it is petroleum based produced. Whereas ethanol can be produced from agricultural renewable resources, thereby attaining total independence from petroleum-based alcohols (Saifuddin & Chua, 2004; Encinar et al. 2007). Ethanol is also preferred mostly in ethanol producing countries. Propanol and butanol have been also used as alcohols in biodiesel production.

Alkali-catalyzed transesterification proceeds much time faster than that catalyzed by an acid and it is the one most used commercially (Dube et al., 2007; Freedman et al., 1984). The most commonly used alkali catalysts are NaOH, CH_3ONa, and KOH (Vicente et al., 2004). Potassium hydroxide (KOH) and sodium hydroxide (NaOH) flakes are inexpensive, easy to handle in transportation and storage, and are preferred by small producers. Alkyl oxide solutions of sodium methoxide or potassium methoxide in methanol, which are now commercially available, are the preferred catalysts for large continuous-flow production processes (Singh et al., 2006).

For acid-catalyzed systems, sulfuric acid has been the most investigated catalyst, but other acids, such as HCl, BF_3, H_3PO_4, and organic sulfonic acids, have also been used by different researchers (Lotero et al, 2005). But in alkali catalyzed method, glycerides and alcohol must be substantially anhydrous, otherwise it leads to saponification (Helwani et al., 2009). Due to saponification the catalytic efficiency decreases, the separation of glycerol becomes difficult and it also causes gel formation (Helwani et al., 2009). In homogeneous catalyzed reactions, separation of catalyst from the reaction mixture is hard and expensive. With this purpose, large amount of water is used to separate catalyst and product (Vyas et al., 2010). On the other hand, undesired by-product formation such as glycerin can be seen, the reaction lasts very long and energy consumption may be very high. Thus, researchers have focused on development of new biodiesel production methods and the optimization of the processes (Sharma et al., 2008). So, various processes such as supercritical process, microwave assisted method and ultrasound assisted method have recently developed. Alternative energy stimulants or non-classical energies have been used for many years to increase the reaction rate and to enhance the yield of particular reaction products. Novel methods or combining innovative methods and techniques are a challenge that can lead to unexpected advances in biodiesel production techniques (Nuechter et al., 2000). In this study, biodiesel production in supercritical conditions, in microwave and ultrasound techniques as novel methods through the years (2000-2011) was reviewed and presented in detail.

2. Supercritical process

Supercritical method is one of the novel methods in biodiesel production. Biodiesel production can be easily achieved by supercritical process without catalysts. A supercritical fluid is any substance at a temperature and pressure above its critical point. It can diffuse through solids like a gas, and dissolve materials like a liquid. These fluids are environment-friendly and economic. Generally, water, carbon dioxide and alcohol are used as supercritical fluids. Supercritical fluids have different application areas. One of these applications is the biodiesel production that is firstly achieved by Saka and Kusdiana in 2001. And many studies on biodiesel production in supercritical conditions were made since 2001. All studies in the literature since 2001 were reviewed and presented in Table 2. The biodiesel production have been studied by using supercritical process from different oils such as rapeseed oil (Kusdiana & Saka, 2001; Saka et al., 2010; Saka & Kusdiana, 2002; Minami & Saka, 2006; Yoo et al., 2010), algae oil (Patil et al., 2010b), chicken fat (Marulanda et al., 2010), jatropha oil (Hawash et al., 2009; Rathore & Madras, 2007; Chen et al., 2010), soybean oil (Cao et al., 2005; He et al., 2007 ; Cheng et al., 2010; Yin et al., 2008), waste cooking oil (Patil et al., 2010a; Demirbaş, 2009), sunflower oil (Demirbaş, 2007), cottonseed oil (Demirbaş, 2008), linseed oil (Demirbaş, 2009), hazelnut kernel oil (Demirbaş, 2002), coconut oil (Bunyakiat et al, 2006), palm oil (Gui et al., 2009 ; Tan et al., 2010c; Tan et al., 2009 ; Song et al., 2008).

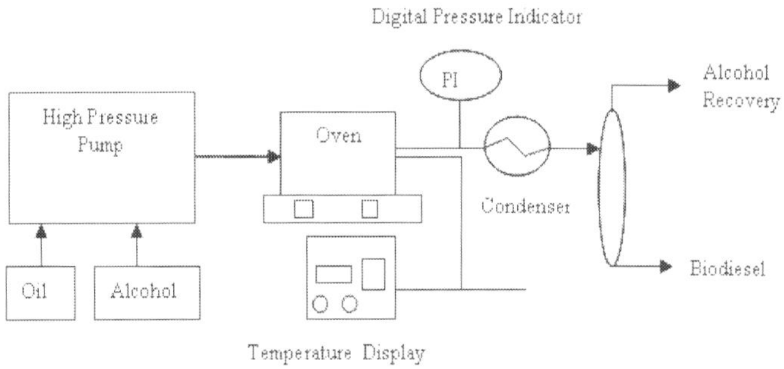

Fig. 3. Biodiesel production by continuous supercritical alcohol process

In Saka's study, rapeseed oil was converted to methyl esters with supercritical methanol (molar ratio of methanol to rapeseed oil: 42 to 1) at temperature of 350°C in 240 s. The methyl ester yield of the supercritical methanol method was higher than those obtained in the conventional method with a basic catalyst. Liquid methanol is a polar solvent and has hydrogen bonding between OH oxygen and OH hydrogen to form methanol clusters, but supercritical methanol has a hydrophobic nature with a lower dielectric constant, so non-polar triglycerides can be well solvated with supercritical methanol to form a single phase oil/methanol mixture. For this reason, the oil to methyl ester conversion rate was found to increase dramatically in the supercritical state (Saka & Kusdiana, 2001; Fukuda et al., 2001).

Main factors affecting transesterification via supercritical process are the effect of temperature, pressure and effect of molar ratio between alcohol and oil sample.

Temperature is the most important factor in all parameters that affects the transesterification under supercritical condition. In the study of Kusdiana & Saka, the conversion of triglyceride to methyl esters is relatively low due to the subcritical state of methanol at temperatures of 200 and 230^{0}C. In these conditions, methyl esters formed are most about 70 wt% for 1 h treatment. However, a high conversion of rapeseed oil to methyl esters with the yield of 95 wt% at 350^{0}C for 4 min reaction time (Kusdiana & Saka, 2001).

Pressure is also very important parameter, but, reaction pressure increases with the increase of temperature. Thus the effect of pressure on the transesterification is always correlated with temperature. High pressure increases the solubility of triglyceride, thus, a contact at the molecular level between alcohol and triglyceride become closer at high pressure (Lee & Saka, 2010).

The effect of molar ratio between alcohol and oil sample is the other important parameter in supercritical condition as mentioned before. Higher molar ratio between methanol and triglyceride is favored for transesterification reaction under supercritical condition. The reason can be that contact area between methanol and triglycerides are increased at the higher molar ratios of methanol. In Kusdiana's study, the effect of the molar ratio of methanol to rapeseed oil was studied in the range between 3.5 and 42 on the yield of methyl esters formed for supercritical methanol treatments. For a molar ratio of 42 in methanol, almost complete conversion was achieved in a yield of 95% of methyl esters, whereas for the lower molar ratio of 6 or less, incomplete conversion was apparent with the lower yield of methyl esters (Kusdiana & Saka, 2001).

Advantages of supercritical process are the shorter reaction time, easier purification of products and more efficient reaction.Although higher temperature, pressure and molar ratio between methanol and triglyceride are favored for transesterification reaction under supercritical condition, energy consumption, and excess amount alcohol usage are the disadvantages for the biodiesel production in supercritical conditions (Lee & Saka, 2010).

For biodiesel production, generally supercritical methanol and supercritical ethanol is used. However, supercritical carbon dioxide can be also used for this purpose since it is cheap, non-flammable and non-toxic (Varma & Madras, 2007). In recent years, two step transesterification processes such as both subcritical and supercritical, both enzyme and supercritical fluid conditions etc. were also developed (Saka & Isayama, 2009).

Kusdiana and Saka developed a two-step biodiesel production method "Saka-Dadan process (Kusdiana & Saka, 2004). Besides the same advantages as one-step supercritical methanol process, the two-step method is found to use milder reaction condition and shorter reaction time, which may further allow the use of common stainless steel for the reactor manufacturing and lower the energy consumption (Lee & Saka, 2010). Minami & Saka (2006), Saka et al. (2010) and Cao et al. (2005) used two-step supercritical method in their studies. Therefore, two- step method has advantages that are milder reaction conditions, high reaction rate, applicable to various feedstocks, easier separation, no catalyst needed there is no high equipment cost and high alcohol oil ratio.

Raw Material	Alcohol	Alcohol/oil molar ratio	Reaction temperature and pressure	Reaction time	Reactor type	Performance (%)	Ref.
Rapeseed oil	Supercritical methanol	42:1	350 °C,14 MPa	240 s	Batch-type vessel	35 (methyl ester yield)	Kusdiana & Saka, 2001
Wet algae	Supercritical methanol	9:1	255 °C, 1200 psi	25 min	Micro-reactor	90 (FAME yield)	Patil et al., 2010b
Rice bran oil Dewaxed-degummed rice bran oil	Supercritical methanol	27:1	300 °C, 30 MPa	5 min	Stainless steel reactor	51.28 94.84 (FAME yield)	Kasim et al., 2009
Chicken fat	Supercritical methanol	6:1	400 °C, 41.1 MPa	6 min	Batch reactor	88 (FAME yield)	Marulanda et al., 2010
Jatropha oil	Supercritical methanol + propane	43:1	593 K, 8.4 MPa	4 min	Bench-scale reactor	100 (FAME yield)	Hawash et al., 2009
Soybean oil	Supercritical methanol	24:1	280 °C, 12.8 MPa	10 min	Batch-type vessel	98 (methyl ester yield)	Cao et al., 2005
Refined palm oil	Supercritical ethanol	33:1	349 °C, P>6.38 MPa	30 min	batch-type tubular	79.2 (biodiesel yield)	Gui et al., 2009
Rapeseed oil	Supercritical methanol	42:1	350 °C, 19 MPa	4 min	Batch-type vessel	95 (methyl ester yield)	Kusdiana & Saka, 2001
Rapeseed oil	Supercritical methanol	42:1	350 °C, 30 MPa	240 s	Batch-type vessel	95 (conversion)	Saka & Kusdiana, 2001
Rapeseed oil	Supercritical methanol	42:1	350 °C, 35 MPa	240 s	Batch-type vessel	98.5 (conversion)	Saka & Kusdiana, 2002
Rapeseed oil	Subcritical acetic acid Supercritical methanol	54:1 14:1	300 °C, 20 MPa 270 °C, 17 MPa	30 min 15 min	Batch-type vessel	92 97 (FAME yield)	Saka et al., 2010
Waste cooking oil	Supercritical methanol	10:1-50:1	300 °C, 1450 psi	10-30 min	Micro-reactor	80 (biodiesel yield)	Patil et al., 2010a
Waste cooking oil	Supercritical methanol	41:1	560 K	1800 s	Cylindrical autoclave	100 (biodiesel yield)	Demirbaş, 2009
Sunflower oil	Supercritical methanol + calcium oxide (%3 wt)	41:1	525 K, 24 Mpa	6 min	Cylindrical autoclave	100 (methyl ester yield)	Demirbaş, 2007
Cottonseed oil	Supercritical methanol Supercritical ethanol	41:1 41:1	523 K 503 K	8 min 8 min	Cylindrical autoclave	98 70 (methyl ester yield)	Demirbaş, 2008
Linseed oil	Supercritical methanol Supercritical ethanol Supercritical methanol Supercritical ethanol	41:1 41:1 41:1 41:1	523 K 523 K 503 K 503 K	8 min 8 min 8 min 8 min	Cylindrical autoclave	98 89 70 65 (methyl ester yield)	Demirbaş, 2009
Hazelnut kernel oil	Supercritical methanol	41:1	350 °C	300 s	Cylindrical autoclave	95 (conversion)	Demirbaş, 2002
Jatropha oil	Supercritical methanol	40:1	350 °C, 200 bar	40 min	Small scale batch reactor	>90 (conversion)	Rathore & Madras, 2007
Soybean oil	Supercritical methanol	40:1	310 °C, 35 MPa	25 min	Tube reactor	96 (methyl ester yield)	He et al., 2007
Coconut oil and palm kernel oil	Supercritical methanol	42.1	350 °C, 19 MPa	400 s	Tubular reactor	95-96 (conversion)	Bunyakiat et al, 2006
Jatropha oil	Supercritical methanol	5:1	563 K, 11 MPa	15 min	Tubular reactor	100 (conversion)	Chen et al., 2010

Raw Material	Alcohol	Alcohol/oil molar ratio	Reaction temperature and pressure	Reaction time	Reactor type	Performance (%)	Ref.
R. sativus L. oil	Supercritical ethanol Supercritical methanol	42:1 39:1	590.5 K, 12.5 MPa 590 K, 14.1 MPa	29 min 27 min	Batch reactor	95.5 99.8 (ester yield)	Valle et al., 2010
Purified palm oil	Supercritical methanol Supercritical ethanol	40:1 33:1	372 °C, 29.7 MPa 349 °C, 26.2 MPa	16 min 29 min	Batch-type tube reactor	81.5 79.2 (biodiesel yield)	Tan et al., 2010c
Palm oil	Supercritical methanol	30:1	360 °C, 22 MPa	20 min	Batch-type tube reactor	72 (biodiesel yield)	Tan et al., 2009
Refined, bleached and deodorized palm oil	Supercritical methanol	45:1	350 °C, 40 MPa	5 min	Batch-type reactor	90 (FAME yield)	Song et al., 2008
Rapeseed oil	Subcritical water+Two-step supercritical methanol Supercritical methanol	1:1 (v/v) 1.8:1 (v/v) 1.8:1 (v/v)	270 °C, 20 MPa 320 °C, 20 MPa 380 °C, 20 MPa	60 min 10 min 15 min	Tubular reactor	90 (methyl ester yield) 80 (methyl ester yield)	Minami & Saka, 2006
Refined soybean oil	Supercritical methanol Supercritical methanol+hexane (co-solvent) Supercritical methanol+CO_2 (co-solvent) Supercritical methanol+ KOH	42:1	350 °C, 20 MPa 300 °C 300 °C 160 °C, 10 MPa	10 min 30 min 30 min 30 min	Cylindirical autoclave	95 85.5 90.6 98 (methyl ester yield)	Yin et al., 2008
Waste palm cooking oil Refined palm oil	Supercritical methanol	40:1	300 °C	20 min	Batch-type tube reactor	79 80 (biodiesel yield)	Tan et al., 2010a
Free fatty acids	Supercritical methanol	1.6:1	270 °C, 10 MPa	30 min	Batch reactor	97 (FAME yield)	Alenezi et al., 2010
Rapeseed oil	Supercritical methanol +metal oxide catalysts (ZnO)	40:1 % 1 (wt) ZnO	250 °C, 105 bar	10 min	Batch- type reactor system	95.2 (FAME yield)	Yoo et al., 2010
Soybean oil	Supercritical methanol	40:1	375 °C, 15 MPa	1000 s	Vertical tubular reactor	92 (methyl ester yield)	Cheng et al., 2010

Table 2. Biodiesel production studies in supercritical conditions

Both enzyme and supercritical fluid conditions were used in recent years (Table 3). No soap formation, no pollution, easier purification, catalyst reusable, no waste water are advantages for this mixed method. Enzymes represent an environmentally friendly alternative to chemical catalysts. Biodiesel production can further conform to environmental concerns if volatile, toxic, and flammable organic solvents are avoided and replaced enzyme with supercritical carbon dioxide (Wen et al., 2009). In recent years, it has been discovered that especially lipases can be used as catalyst for transesterification and esterification reactions. Enzyme catalyzed transesterification, using lipase as catalyst does not produce side products and involves less energy consumption (Fjerbaek et al., 2009). However, enzyme applications have also disadvantages that they are expensive and have stricted reaction conditions and some initial activity can be lost due to volume of the oil molecule (Marchetti et al., 2007).

Raw Material	Alcohol+enzyme	Alcohol/oil molar ratio	Reaction temperature and pressure	Reaction time	Reactor type	Performance (%)	Ref.
Sesame oil	Supercritical methanol Supercritical ethanol	40:1 40:1	350 °C, 200 bar 350 °C, 200 bar	40 min 40 min	Batch reactors	90 100	Varma et al., 2010
Mustard oil	Supercritical methanol Supercritical ethanol +Novozym 435 Candida antarctica	40:1 40:1	350 °C, 200 bar 350 °C, 200 bar	70 min 25 min		80 100 (conversion) 70 (conversion)	
Sunflower oil	Supercritical methanol + Novozyme 435 enzyme in supercritical CO_2 Supercritical ethanol +Novozyme 435 enzyme in supercritical CO_2	40:1 40:1	400 °C, 200 bar 400 °C, 200 bar	40 min 40 min	Batch reactor	96 99 (conversion)	Giridhar et al., 2004
Soybean oil Olive oil Sunflower oil Rapeseed oil Palm oil	Supercritical methanol + Candida antartica lipase enzyme in supercritical CO_2	40:1 40:1	45 °C, 130 bar	6 h	Batch reactor	58 65.8 50 60 59 (conversion)	Lee et al., 2009

Table 3. Enzyme usage in supercritical fluid conditions for biodiesel production

Raw Material	Solvent	Solvent/oil molar ratio	Reaction temperature and pressure	Reaction time	Reactor type	Performance (%)	Ref.
Rapeseed oil Oleic acid	Supercritical methyl acetate	42:1	350 °C, 20 MPa	45 min	Batch-type vessel	91	Saka & Isayama, 2009
Soybean oil Waste soybean oil Sunflower oil Jatropha curcas oil	Supercritical methyl acetate	42:1 42:1 42:1 42:1	345 °C, 20 MPa 345 °C, 20 MPa 345 °C, 20 MPa 345 °C, 20 MPa	50 min 50 min 50 min 50 min	Batch reactor	100 100 100 100	Campanelli et al., 2010
Purified palm oil	Supercritical methyl acetate	30:1	399 °C	59 min	Batch-type tube reactor	97.6 (biodiesel yield)	Tan et al., 2010b
Jatropha curcas oil	Sub-critical water+ Sub-critical dimethyl carbonate	217:1 14:1	270 °C, 27 MPa 300 °C, 9 MPa	25 min 15 min	Batch-type vessel	> 97 (methyl ester yield)	Ilham & Saka, 2010

Table 4. Different solvents instead of methanol in supercritical processes

In supercritical processes, as solvent not only methanol but also methyl acetate and dimethyl carbonate are now good candidates. However, further researches are needed for their practical applications. Saka & Isayama (2009), Tan et al. (2010b) and Campanelli et al. (2010) studied with supercritical methyl acetate for biodiesel production (Table 4). High products recovery and no glycerol produced are advantages, however, lower reactivity than methanol is the main disadvantage for these applications of supercritical biodiesel production processes (Lee&Saka2010).

3. Microwave assisted process

Generally, heating coils are used to heat the raw material in biodiesel production process. This treatment can be also done by microwave method. An alternative heating system "microwave irradiation" has been used in transesterification reactions in recent years. Microwaves are electromagnetic radiations which represent a nonionizing radiation that influences molecular motions such as ion migration or dipole rotations, but not altering the molecular structure (Fini & Breccia, 1999; Varma, 2001; Refaat et al., 2008). The frequencies of microwave range from 300 MHz to 30 GHz, generally frequency of 2.45 GHz is preferred in laboratory applications (Taylor et al., 2005). Microwave irradiation activates the smallest degree of variance of polar molecules and ions with the continuously changing magnetic field (Azcan& Danisman, 2007). The changing electrical field, which interacts with the molecular dipoles and charged ion, causes these molecules or ions to have a rapid rotation and heat is generated due to molecular friction (Azcan& Danisman, 2007; Saifuddin & Chua, 2004). The absorption of microwaves causes a very rapid increase of the temperature of reagents, solvents and products (Fini & Breccia, 1999).

Microwave process can be explained for the biodiesel production with transesterification reaction: the oil, methanol, and base catalyst contain both polar and ionic components. Microwaves activate the smallest degree of variance of polar molecules and ions, leading to molecular friction, and therefore the initiation of chemical reactions is possible (Nuechter et al., 2000). Because the energy interacts with the sample on a molecular level, very efficient and rapid heating can be obtained in microwave heating. Since the energy is interacting with the molecules at a very fast rate, the molecules do not have time to relax and the heat generated can be for short times and much greater than the overall recorded temperature of the bulk reaction mixture. There is instantaneous localized superheating in microwave heating and the bulk temperature may not be an accurate measure of the temperature at which the actual reaction is taking place (Barnard et al., 2007; Refaat et al., 2008).

When the reaction is carried out under microwaves, transesterification is efficiently accelerated in a short reaction time. As a result, a drastic reduction in the quantity of by-products and a short separation time are obtained (Saifuddin & Chua, 2004; Hernando et al., 2007) and high yields of highly pure products are reached within a short time (Nuechter et al., 2000). So, the cost of production also decreases and less by-products occurs by this method (Öner & Altun, 2009). Therefore, microwave heating compares very favorably over conventional methods, where heating can be relatively slow and inefficient because transferring energy into a sample depends upon convection currents and the thermal conductivity of the reaction mixture (Koopmans et al., 2006; Refaat et al., 2008). Microwave assisted transesterification process schematic diagram was presented in Figure 4.

There can be also a few drawbacks of microwave assisted biodiesel production, beside the great advantages. Microwave synthesis may not be easily scalable from laboratory small-scale synthesis to industrial production. The most significant limitation of the scale up of this technology is the penetration depth of microwave radiation into the absorbing materials, which is only a few centimeters, depending on their dielectric properties. The safety aspect is another drawback of microwave reactors in industry (Yoni & Aharon, 2008; Vyas et al., 2010). This survey of microwave assisted transformations is abstracted from the literature published from 2000 to 2011. And studies on microwave assisted method of transesterification reaction in the literature were summarized in Table 5. The biodiesel production have been studied by using microwave assisted method from different oils such as cottonseed oil (Azcan& Danisman, 2007), safflower seed oil (Düz et al., 2011), rapeseed oil (Hernando et al., 2007; Geuens et al., 2008), soybean oil (Hernando et al., 2007; Hsiao et al., 2011; Terigar et al., 2010), corn oil (Majewski et al., 2009), macauba oil (Nogueira et al., 2010), waste frying palm oil (Lertsathapornsuk et al., 2008), micro algae oil (Patil et al., 2011), karanja oil (Venkatesh et al., 2011), jatropha oil (Shakinaz et al., 2010), yellow horn oil (Zhang et al., 2010), canola oil (Jin et al., 2011), camelina sativa oil (Patil et al., 2009), castor oil (Yuan et al., 2009), waste vegetable oils (Refaat et al., 2008), maize oil (Öztürk et al., 2010) and sunflower oil (Han et al., 2008; Kong et al., 2009).

Fig. 4. Microwave assisted transesterification process shematic diagram

Raw material	Catalyst	Catalyst amount (wt%)	Type of alcohol	Alcohol/ oil molar ratio	Microwawe conditions	Reaction time	Reaction tempe-rature	Performance (%)	Ref.
Cotton seed oil	KOH	1.5	Methanol	6:1	21% of 1200 W	7 min	333 K	92.4 (yield)	Azcan& Danisman, 2007
Safflower seed oil	NaOH	1	Methanol	10:1	300 W	6 min	333 K	98.4 (conversion)	Düz et al., 2011
Rapeseed oil Soybean oil	NaOH	%1.3	Methanol	18:1 1.27 ml	300 W	1 min	60 °C	97 95 (yield)	Hernando et al., 2007
Corn oil Soybean oil	*Diphenylammonium salts:* DPAMs (Mesylate) DPABs(Benzenesulfonate) DPATs (Tosylate) DPAMs DPABs	20 (molar) 10 (molar) 10 (molar) 10 9	Methanol	5 g methenol / 2 g oil	-	20 min	150°C	100 96 100 92 97 (methyl ester yield)	Majewski et al., 2009

Raw material	Catalyst	Catalyst amount (wt%)	Type of alcohol	Alcohol/ oil molar ratio	Microwawe conditions	Reaction time	Reaction tempe-rature	Performance (%)	Ref.
Waste frying oil	NaOH	1	Methanol	6:1	600 W	5 min	64°C	93.36 (methyl ester content)	Yücel et al., 2010
Macauba oil	Novozyme 435 Lipozyme IM	2.5 5	Ethanol Ethanol	9:1 9:1	-	15 min 5 min	30°C 40°C	45.2 35.8 (conversion)	Nogueira et al., 2010
Waste frying palm oil	NaOH	3	Ethanol	12:1	800 W	30 s	-	97 (conversion)	Lertsathapornsuk et al., 2008
Rapeseed oil	KOH NaOH	1 1	Methanol Methanol	6:1 6:1	67 % of 1200 W	5min 3min	323 K 313 K	93.7 92.7 (yield)	Azcan & Danisman, 2008
Soybean oil	nano CaO (heterogeneous catalyst)	3	Methanol	7:1	-	60 min	338 K	96.6 (conversion)	Hsiao et al., 2011
Soybean oil Oleic acid	sulfated zirconia	5	Methanol	20:1	-	20 min	60 °C	90 (conversion)	Kim et al., 2011
Dry micro algae	KOH	2	Methanol	9:1	800 W	6 min	-	80.13 (conversion)	Patil et al., 2011
Crude karanja oil	KOH	1.33	Methanol	%33.4 (w/w)	180 W	150 s	-	89.9 (conversion)	Venkatesh et al., 2011
Jatropha oil	KOH	1.50	Methanol	7.5:1	-	2 min	65°C	97.4 (conversion)	Shakinaz et al., 2010
Crude palm oil	KOH	1.50	Ethanol	8.5:1	70 W	5 min	70°C	85 (yield) 98.1 (conversion)	Suppalakpanya et al., 2010
Yellow horn oil	Heteropolyacid (HPA)	1	Methanol	12:1	500 W	10 min	60°C	96.22 (FAMEs)	Zhang et al., 2010
Soybean oil	NaOH	1	Methanol	6:1	900 W	1 min	303 K	97.7 (conversion)	Hsiao et al., 2011
Canola oil	$ZnO/La_2O_2CO_3$ (heterogeneous catalyst)	< 1	Methanol	1:1 (w/w)		< 5 min	<100°C	> 95 (yield)	Jin et al., 2011
Camelina sativa oil	Heterogeneous metal oxide catalysts (BaO, SiO)	1.5 2	Methanol	9:1	800 W	-	-	94 80 (FAME yield)	Patil et al., 2009
Castor bean oil	Al_2O_3 / 50% KOH SiO_2 / 50% H_2SO_4 SiO_2 / 30% H_2SO_4	1 1 1	Methanol Methanol Ethanol	1:6 1:6 1:6	40 W 40 W 220 W	5 min 30 min 25 min	-	95 95 95 (conversion)	Perin et al., 2008
Castor oil	H_2SO_4 / C	5	Methanol	1:12	200 W	60 min	338 K	94 (yield)	Yuan et al., 2009
Triolein	KOH NaOH	5	Methanol	1:6	25 W	1 min	323 K	98 (conversion)	Leadbeater & Stencel, 2006

Raw material	Catalyst	Catalyst amount (wt%)	Type of alcohol	Alcohol/ oil molar ratio	Microwawe conditions	Reaction time	Reaction tempe-rature	Performance (%)	Ref.
Frying oil	NaOH	0.5	Ethanol	1:6	50% of 750 W	4 min	60°C	87 (conversion)	Saifuddin & Chua, 2004
Rapeseed oil	-	-	Supercritical 1-butanol	2.5:1	-	4 hour 80 bar	310°C	91 (fatty acid buthyl ester conversion)	Geuens et al., 2008
Domestic waste vegetable oil Restaurant waste vegetable oil Neat vegetable virgin sunflower oil	KOH	1	Methanol	6:1	500 W	1 h	65°C	95.79 94.51 96.15 (biodiesel yield)	Refaat et al., 2008
Safflower seed oil	NaOH	1	Methanol	10:1	300 W	16 min	60°C	98.4 (methyl ester content)	Düz et al., 2011
Soybean oil	NaOH	1	Methanol	6:1	600 W (Ultrasonic) 900 W (Microwave)	1 min 2 min	333 K	97.7 (conversion)	Hsiao et al., 2010
Maize oil	NaOH	1.5	Methanol	10:1	-	-	-	98 (conversion)	Öztürk et al., 2010
Soybean oil Rice bran oil	NaOH	0.6	Ethanol	5:1	-	10 min 73°C	73°C	99.25 99.34 (FAME yield)	Terigar et al., 2010
Jatropha curcas	NaOH	4	Methanol	30:1	-	7 min	328 K	86.3 (conversion)	Yaakob et al., 2008
Sunflower oil	H_2SO_4	0.05	Methanol	10:1	400W	45 min-		96.2 (conversion)	Han et al., 2008
Sunflower oil	TiO_2/SO_4	0.02	Methanol	12:1	300W	-25 min		94.3 (biodiesel yield)	Kong et al., 2009

Table 5. Microwave assisted method studies of transesterification reaction in the literature

4. Ultrasound assisted process

Ultrasonic waves are energy application of sound waves which is vibrated more than 20,000 per second. In another words, it can be defined as the sound waves beyond human hearing limit. Human hear can not hear sound waves with more high-pitched sound waves of an average of 10-12 kHz. Ultrasonic or ultrasound signals are in the order of 20 kHz- 100 kHz and above the limit of human hearing. Ultrasonic waves were used as the first for medical research and detectors in the 1930s and 1940s (Newman& Rozycki, 1998). Idea of the use of ultrasound, especially in the industry since the 1980s began to develop rapidly, and today a wide range of applications using ultrasonic waves appeared. At present, ultrasonic waves are used in areas such as *Atomization:* Water sprays for dust suppression and humidifiers, low velocity spray coating, spray drying nozzles. Cleaning and cleaning of engineering items, small electronic items and jeweler using aqueous based solvents. Cleaning and disinfection of medical instruments and food processing equipment. *Processing:* Dispersion of pigments and powders in liquid media and emulsification. *Extraction:* Essential oil, flavonoid, resin, *Crystallization* and *Filtration* (Cintas et al., 2010; Mason et al., 1996; Mason, 2000).

Ultrasonic irradiation has three effects according to the investigators. First one is rapid movement of fluids caused by a variation of sonic pressure. It causes solvent compression and rarefaction cycles (Mason, 1999). The second and the most important one is cavitation. If a large negative pressure gradient is applied to the liquid, the liquid will break down and cavities (cavitation bubbles) will be created. At high ultrasonic intensities, a small cavity may grow rapidly through inertial effects. So, bubbles grow and collapse violently. The formation and collapse of micro bubbles are responsible for most of the significant chemical effects (Kumar et al., 2010a). Cavitation is considered as a major factor which influences on reaction speed. Cavity collapse increases mass transfer by disrupting the interfacial boundary layers known as the liquid jet effect. The last effect of ultrasound is acoustic streaming mixing.

Ultrasound has been used to accelerate the rates of numerous chemical reactions, and the rate enhancements, mediated by cavitations, are believed to be originated from the build-up of high local pressures (up to 1000 atm) and temperatures (up to 5000 K), as well as increased catalytic surface areas and improve mass transfer (Yu et al., 2010). Low frequency ultrasonic irradiation is widely used for biodiesel production in recent years. In transesterification reaction, mixing is important factor for increasing biodiesel yield. Oil and methanol are not miscible completely in biodiesel processing. Ultrasonic mixing is an effective mixing method to achieve a better mixing and enchancing liquid–liquid mass transfer (Ji et al., 2006). Vigorous mixing increases the contact area between oil and alcohol phases with producing smaller droplets than conventional stirring (Mikkola & Salmi, 2001; Stavarache et al., 2006). Cavitation effects increase mass and heat transfer in the medium and hence increase the reaction rate and yields (Adewuyi, 2001). Ultrasonic cavitation also provides the necessary activation energy for initiating transesterification reaction.

Ultrasonic waves are produced with the power converter (transducer) which is piezoelectric material. Sound waves are converted to ultrasonic waves vibrating at high frequency with quartz crystal oscillator. If ultrasound waves are used in chemical reactions and processes it is called as sonochemistry. Industrial sonochemial reactors were designed more than 40 years ago by Sarocco and Arzono (Cintas et al., 2010). They showed that reactor geometry affected enormously the reaction kinetics. Later many rectors have been developed by researchers for different chemical reactions. For conventional biodiesel production, batch and continuous reactors have been developed in industry. Ultrasonic cleaning bath, ultrasonic probe which are usually operated at a fixed frequency are mainly used as ultrasonic apparatus. Frequency is dependent on particular type of transducer which is 20 kHz for probes and 40 kHz for bath. Figure 5 shows schematic diagram of biodiesel production via ultrasound assisted method.

Ultrasonic processing of biodiesel involves the following steps: 1. Mixing vegetable oil is with the alcohol (methanol or ethanol) and catalyst, 2. Heating the mixture, 3. The heated mixture is being sonicated inline, 4. Glycerin separation by using centrifuge. Alternative reactors have also been developed to lower energy consumption. Cintas et al., (2010) designed a flow reactor constituted by three transducers and showed that considerable energy saving could be achieved by large-scale multiple transducer sonochemical reactors operating in a continuous mode.

The factors affecting ultrasound assisted biodiesel production are: -Effect of catalyst type on ultrasound assisted biodiesel production, -Effect of alcohol type on ultrasound assisted biodiesel production, -Effect of ultrasonic power on biodiesel processing, -Frequency effect on ultrasonic assisted biodiesel production.

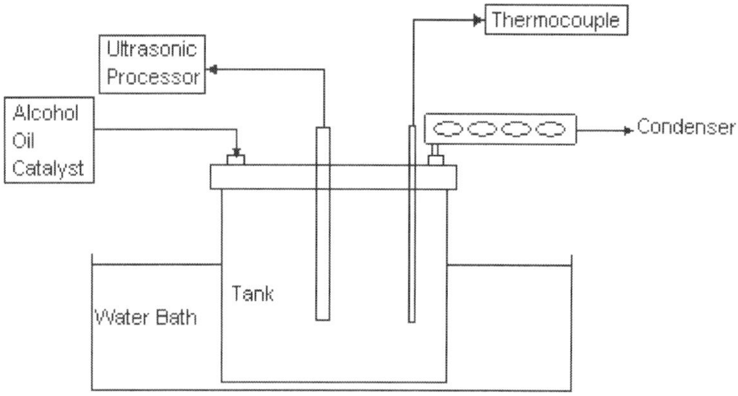

Fig. 5. Scheme of biodiesel production process via ultrasound assisted method

Effect of catalyst type on ultrasound assisted biodiesel production: In ultrasonic assisted biodiesel studies homogen (alkaline, acid), heterogen and enzyme catalyst were studied with many edible and nonedible oils under ultrasonic irradiation. Transesterification reactions have been studied with KOH catalyst for corn oil (Stavarache et al., 2007a; Lee et al., 2011), grape (Stavarache et al., 2007a), canola (Stavarache et al., 2007a; Thanh et al., 2010a; Lee et al., 2011), palm (Stavarache et al., 2007a), tung (Hanh et al., 2011), beef tallow (Teixeira et al.,2009), coconut (Kumar et al., 2010), soybean (Ji et al., 2006; Mahamuni & Adewuyi, 2009;Thanh et al., 2010a; Lee et al., 2011), triolein (Hanh et al., 2008; Hanh et al., 2009b), fish oil (Armenta et al.,2007),neat vegetable oil (Stavarache et al., 2005), waste cooking oil (Thanh et al., 2010b; Hingu et al.,2010).These studies were presented in Table 6 (one step transesterification), and Table 7 (two-step esterification). Generally KOH was preferred for transesterification reactions instead of NaOH. Soybean (Ji et al., 2006), neat vegetable oil (Stavarache et al., 2005), jatropha curcas L. (Deng et al., 2010) (in the second transesterification step) and triolein (Hanh et al., 2009b) were transesterified with NaOH. KOH and NaOH were used for ultrasound assisted transesterification of neat vegetable oil. They used 0.5%, 1% and 1.5 % alkali catalyst amount, 6:1 molar ratio methanol to oil and room temperature. The researchers reported that there were no great differences in the time to complete conversion between two types of catalyst (Stavarache et al., 2005).. 98% and 96% yields were achieved with 0.5 % NaOH and KOH catalyst, respectively. They also reported that when KOH was used, high yields were obtained even for 1.5% catalyst concentration. Potassium soap is softer, more soluble in water and does not make as much foam as sodium soap. The washing of esters when using potassium hydroxide is easier and the yields of isolated product are higher. In alkali catalyzed ultrasonic transesterification for biodiesel production (Tables 6 and 7), 0.3-1.5 % alkali catalyzed amounts were used. Apart from that, Cintas et al., (2010) developed a new ultrasonic flow reactor to scale up biodiesel from soybean oil in presence of (Na or K methoxide). Na and K methoxide, are alkaline metal alkoxides (as CH_3ONa for the methanolysis) are the most active catalysts because of stronger hydroxide group. In their reacton mixture of oil (1.6 L), methanol and sodium methoxide 30% in methanol (wt/wt ratio 80:19.5:0.5, respectively) was fully transesterified at about 45°C in 1 h (21.5 kHz, 600 W, flow rate 55 mL/min).

Heterogen catalysts were tried by researchers in a few studies (Ye et al., 2007; Salamatinia, 2010; Mootabadi et al., 2010;Kumar et al., 2010b). As it is known, ultrasound increase mixing of oil and alcohol with catalyst phases, as well as increase catalytic surface area. Catalyst can be broken into smaller particles by ultrasonic irradiation to create new sites of the subsequent reaction. Thus, solid catalyst is expected to last longer in the ultrasonic-assisted process (Mootabadi et al., 2010). Single component alkaline earth metal oxides (BaO, SrO, CaO) having lower solubility in alcohol catalyzed palm transesterification processes with methanol (Mootabadi et al., 2010). The catalytic activities of the three catalysts were correlated well with their basic strengths and found as the sequence of CaO < SrO < BaO. BaO catalyst achieved 95.2% of biodiesel yield within 60 min in the ultrasonic-assisted process while SrO catalyst generally demonstrated slightly lower result. CaO showed the lowest yield with 77.3%yield under optimum conditions. Although high activity of BaO as catalyst, this activity dropped severely in the BaO reusability test, especially under ultrasonic condition (compared to mechanical stirring). In another study, aluminum isopropoxide or titanium isopropoxide as heterogeneous transesterification catalysis are employed to produce nanoemulsions with large interfacial area for easy catalyst separation and enhanced reaction rate (Ye et al., 2007). These catalysts are produced by partial polymerization and metal alkoxides are connected by metal-oxygen bonds. Alkoxide parts in the polymer matrix catalyst gives the catalyst amphiphilic properties that help form and stabilize alcohol/ triglycerides nanoemulsion (Ye et al., 2007). The study showed that titanium isopropoxide also showed good catalytic activity and considerable amphiphilic properties in forming nanoemulsions. With aluminum isopropoxide or titanium isopropoxide, transparent alcohol/oil emulsions can be formed in less than four minutes and can significantly enhance the transesterification reaction rate. The micelle size was observed to be as low as 5.1 nm.

High acidity oils (Jatropha curcas L, waste frying oil) can be transesterified by two-step processes. In the first step, free fatty acids are converted to esters by direct esterification with acid catalyst. Eq. 1 shows esterification of fatty acids. In the second step, basic catalyst was used to esterify triglycerides as it was shown in Figure 2.

$$RCOOH + CH_3OH \longrightarrow RCOOCH_3 + H_2O \quad (1)$$

In production of biodiesel from Jatropha curcas L. oil (non edible oil) Deng et al., (2011) used a two-step process. The first step pretreatment (acid-esterification) of Jatropha oil was performed at 318 K an ultrasonic reactor for 1.5 h in their first study (Deng , et al., 2010). After reaction, the acid value of Jatropha oil was reduced to 0.7 mg KOH/g and 93.3% esterification rate was achieved. The second step, a base-catalyzed transesterification was performed with nano sized Mg/Al oxides under different conditions. At the optimized condition, (Table 6) 95.2% biodiesel yield was achieved, and the Jatropha oil biodiesel properties were found to be close to those of the German standard. It was reported that the catalyst could be reused for 8 times.

Although it is known that ultrasonic mixing has a significant effect on enzymatic transesterification there are a little study about using of lipases as enzyme catalyst. It has been reported that enzyme activity of Novozym 435 enhanced by ultrasound irradiation (Sinisterra, 1992; Lin & Liu, 1995). Novozym 435 (Candida antarctica lipase B immobilized on polyacrylic resin) was used in biodiesel production from soybean oil and methanol with a low frequency ultrasonic (40 kHz) waves to see enzyme activity and compare their overall

effects under two different conditions—ultrasonic irradiation and vibration (Yu et al., 2010). They investigated effects of reaction conditions, such as ultrasonic power, water content, organic solvents, ratio of solvent/oil, and ratio of methanol/oil, enzyme dosage and temperature on the activity of Novozym 435. Novozym 435 activity significantly increased by ultrasonic irradiation compared with vibration and reaction rate was further increased under the condition of ultrasonic irradiation with vibration (UIV). Yu et al (2010) indicated that 96% yield of fatty acid methyl ester (FAME) could be achieved in 4 h under the optimum conditions: 50% of ultrasonic power, 50 rpm vibration, water content of 0.5%, tert-amyl alcohol/oil volume ratio of 1:1, methanol/oil molar ratio of 6:1, 6% Novozym 435 and 40 °C. Since the lipase enzyme is expensive catalyst it is important to reuse the catalyst in biodiesel industrial productions. The researchers also pointed out that Novozym 435 was not deactivated under UIV, only 4 % enzyme activity slightly decreased after five cycles.
Effect of alcohol type on ultrasound assisted biodiesel production: Methanol was mostly used in transesterification reaction under ultrasonic irradiation with oils shown in Tables 6 and 7. High conversion and yields were obtained with methanol and ethanol using. Stavarache et al., (2007a) used methanol in transesterification of commercial edible oil, corn, grapeseed, canola and palm oil. Excellent yields (99%) were obtained for all type oils in 20 minutes with 6:1 methanol to oil molar ratio at 36 °C. As it is shown in Figure 6, triglycerides are converted to di and monoglycerides to produce biodiesel to produce biodiesel and glycerin. They also examined the transesterification reaction mechanism under low frequency (40 kHz) ultrasonically driven esterification.

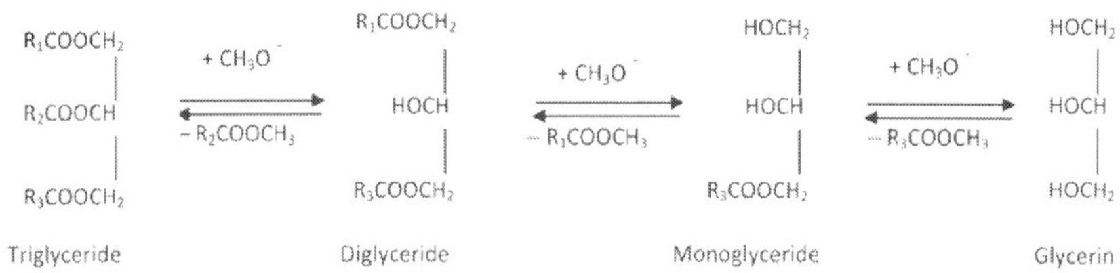

Fig. 6. Alkali catalyzed transesterication steps of triglyceride with methanol

They have reported that the major part of the transesterification took place in the first 3-10 minutes of reaction if not faster and the rate- determining reaction switches from diglyceride (DG) → monoglyceride (MG) (classical mechanic agitation) to MG + ROH→Gly + ME (ultrasonically driven transesterification). In another study, the conversion of FAME greater than 99.4 % was achieved after about 15 minutes at 40 °C with ultrasonic agitation for 6:1 methanol: oil molar ratio (Calucci et al., 2005). They have also concluded that hydrolysis rate constants of DG and TG are three to five times higher than those of mechanical agitation. Ji et al., (2006) used ultrasonic transesterification process for soybean oil transesterification with methanol and reported 99% yield at 10 min reaction time with 6:1 methanol to oil molar ratio at 45°C. Oleic acid, triolein, coconut were esterified with ethanol and 90% conversion, about 99% yield and >92% yields were achieved respectively (Hanh et al., 2009a; Hanh et al., 2009b; Kumar et al., 2010a). Table 8 shows the some biodiesel yield and conversion with various monoalcohols and comparing of the alcohols.
Stravarache et al., (2005) studied effects of alcohol type on transesterification of neat vegetable oil under ultrasonic and mechanical stirring. The results of transesterfication with primary, secondary and tertiary alcohols after 60 min of reaction were presented in Table 8.

Raw material	Catalyst	Catalyst amount (wt %)	Alcohol type	Alcohol /oil molar ratio	Reaction temp. (°C)	Reaction time	Reactor conditions	Performance (%)	Ref.
Oleic acid	H_2SO_4	5	Ethanol	3:1	60	2 hour	Ultrasonic cleaner 40 kHz, 1200 W	~90 (conversion)	Hanh et al., 2009a
Commercial edible oil Corn Grape seed Canola Palm	KOH	0.5	Methanol	6:1	36 ± 2	20 min	Ultrasonic cleaner 40 kHz,1200 W	~ 99 (conversion)	Stavarache et al., 2007a
Refined soybean oil	KOH	1.5	Methanol	6:1	40	15 min	20 kHz, 14.49 W	>99.4 (conversion)	Colucci et al., 2005
Soybean	NaOH	1	Methanol	6:1	45	10 min	197 kHz, 150W	99 (yield)	Ji et al., 2006
Soybean	KOH	0.5	Methanol	6:1	26 - 45	30 min<	611 kHz, 139 W	>90 (conversion)	Mahamuni & Adewuyi, 2009
Soybean	Na or K methoxide	0.15	Methanol	6:1	45	1 h	21.5 kHz, 600 W	Fully transesterified	Cintas et al., 2010
Canola Soybean Corn	KOH	1	Methanol	6:1	55	30 min	450 W	95 (yield) 95 (yield)	Lee et al., 2011
Tung and Blended oil (20% Tung, 30% canola, 50% palm)	KOH	1	Methanol	6:1	20-30	30 min	25 kHz, 270 W	91.15 (yield) 94.03 (yield)	Hanh et al., 2011
Beef Tallow	KOH	0.5	Methanol	6:1	60	70 s	40 kHz, 1200 W	>92 (conversion)	Teixeira et al., 2009
Triolein	KOH	1	Methanol	6:1	25	30 min	Ultrasonic cleaner 40 kHz, 1200 W	~99 (yield)	Hanh et al., 2008
Triolein	NaOH KOH	1	Methanol Ethanol	6:1	25	25 min	Ultrasonic cleaner 40 kHz, 1200 W	>95 (conversion)	Hanh et al., 2009b
Neat vegetable oil	NaOH KOH	0.5	Methanol	6:1	25	20 min	Ultrasonic cleaner 20 kHz 40 kHz 1200 W	98 (yield) 96 (yield	Stavarache et al., 2005
Coconut	KOH	0.75	Ethanol	6:1	-	7 min	24kHz, 200 W	>92 (yield)	Kumar et al., 2010a
Waste cooking oil	KOH	1	Methanol	6:1	45	40 min	20 kHz, 200 W	89 (conversion)	Hingu et al., 2010
Palm	KOH	-	Methanol	6:1	38–40	20 min	45 kHz,600 W	95 (yield)	Stavarache et al., 2007b
Palm	CaO SrO BaO	3	Methanol	9:1	65	60 min	30 kHz	77.3 (yield) 95 (yield) 95 (yield)	Mootabadi et al., 2010
Palm	BaO SrO	2.8	Methanol	9:1	65	50 min <	20 kHz, 200 W	>92 (yield)	Salamatinia et al., 2010
Canola	KOH	0.7	Methanol	5:1	25	50 min	20 kHz, 1000 W	>95 (conversion)	Thanh et al., 2010a
Soybean	$Ti(Pr)_4$ Al(Pr)3	3	Methanol	6:1	60	2 h	-	64 (yield)	Ye et al., 2007
Soybean	Novozym 435	6	Methanol	6:1	40	4h	40 kHz, 500 W	96 (yield)	Yu et al., 2010
Jatropha oil	Na/SiO_2	3	Methanol	9:1	50-70	15 min	24 kHz, 200W	98.5 (yield)	Kumar et al., 2010b
Fish oil	KOH C_2H_5ONa	1 0.8	Ethanol	6:1 6:1	20-60 20-60	>30 >30	25-35 kHz 25-35 kHz	>95 (conversion) >98 (conversion)	Armenta et al., 2007

Table 6. The studies for biodiesel production from various feedstocks at different conditions under ultrasound irradiation

Oil	Catalyst type	Catalyst amount (wt%)	Alcohol type	Alcohol: oil ratio	Reaction temperature (ºC)	Reaction time	Ultrasound conditions	Performance (%)	Ref.
Waste cooking	KOH	0.7 0.3	Methanol	2.5:1 (mol) 1.5:1	20-25	10 min 20 min	20 kHz, 1000W (For each step)	81 (yield) 99 (yield)	Thanh et al., 2010b
Jatropha curcas L.	H_2SO_4 (1. step) Mg/Al oxides (2. step)	1 (For each step)	Methanol	4:1(mol) (For each step)	40 (For each step)	1.5 h (For each step)	210W (For each step)	95.2 (total yield)	Deng et al., 2011
Jatropha curcas L.	NaOH H_2SO_4	1 (For each step)	Methanol	0.4 (v/v) 6:1 (mol)	60 (For each step)	1h 30 min	210W (For each step)	96.4 (total yield)	Deng et al., 2010

Table 7. Biodiesel production with two step transterification under ultrasound irradiation

Alcohol type	Neat vegetable oil a (Stavarache et al., 2005)	Triolein b (Hanh et al., 2009b)	Soybean oil c (Colucci et al., 2005)
	Performance (%) Stirring Ultrasonic	Conversion (%)	Conversion (%)
Methanol	80 (Yield) 98 (Yield) (60 min) (20 min)	98	99.3
Ethanol	79 (Yield) 88 (Yield) (20 min) (20 min)	~98	99.1
n- propanol	78 (Yield) 88 (Yield) (10 min) (10 min)	~93	-
Iso-propanol	No conversion Some conversion	3	29.2
n-butanol	83 (Yield) 92 (Yield) (>60 min) (>60 min)	~93	92.0
Iso- butanol	No conversion Some conversion	3	-
Tertiary- butanol	No conversion No conversion	-	

a Reaction conditions for neat vegetable oil: 0.5% (wt/wt) NaOH, 6:1 alcohol to oil molar ratio, 40 KHz,
b Reaction conditions for triolein: 25 min, 25 °C, 0.1% (wt/wt) KOH, 6:1 alcohol to triolein molar ratio, 40 KHz,
c Reaction conditions for soybean oil: 2h, 1.5% (wt/wt) KOH, 6:1 alcohol to oil molar ratio, 40 KHz

Table 8. The influence of alcohol on the ultrasound assisted transesterification of different oils for biodiesel production

N- chain alcohols (methanol, ethanol, n- propanol, and n-butanol) showed the high yields between 88-98% in 10-20 min reaction time. The yields of biodiesel in ultrasound activation were higher than mechanical stirring since ultrasound produce less soap. By using ultrasound the reaction time was found much shorter than mechanical stirring. The secondary alcohols showed some conversion while transesterification reaction took place under stirring. Tertiary-butanol had no conversion with both type of procedure. Hanh et al., (2009b) produced biodiesel with triolein and various alcohols (methanol, ethanol, propanol, butanol, hexanol, octanol and decanol). The productions were performed at molar ratio 6:1 (alcohol: triolein) and 25°C in the presence of base catalysts (NaOH and KOH) under ultrasonic irradiation (40 kHz) and mechanical stirring (1800 rot/min) conditions. The rate of ester formation depended on alcohol types; as the alcohol carbon number increased, reaction rate decreased. The secondary alcohols such as 2-propanol, 2-butanol, 2-hexanol, and 2-octanol showed 3% conversion, suggesting that the steric hindrance strongly affected the transesterification of triolein. N-propanol showed approximately 93% conversion under ultrasonic irradiation, while 75% conversion was obtained under mechanical stirring. Soybean oil was transesterifed with methanol, ethanol, n-butanol, and iso-propanol over 2 h reaction period with 1.5 % KOH as the catalyst and a 6:1 molar ratio of alcohol/oil at 60°C (Colucci et al., 2005). The similar results obtained with methanol, ethanol and n-butanol compared to other studies.

Effect of ultrasonic power on biodiesel processing: The effect of ultrasonic power on the biodiesel formation has been reported (Mahamuni& Adewuyi, 2009; Hingu et al., 2010; Lee et al.,2011). Biodiesel yield increased with increasing ultrasonic power in all the studies. Nahamuni& Adewuyi (2009) studied this effect for three different frequencies and various powers (181, 90, 181 W at 1300 kHz , 104, 139 ,68 W at 611 kHz, 181, 117, 81, 49 W at 581 kHz). The reactions were carried out for 60-180 minutes. The reaction rate increased with increasing ultrasound power at any given frequency and biodiesel yield was obtained above 90%. At start of the reaction, reaction rate is very low because of low interfacial area available for the reaction. As time increased the reaction rate increased. This increase is due to the amount and size of the emulsion formation varies because of ultrasonic cavitation. Ultrasonic cavitation produces finer and stable emulsion and following this higher mass transfer and hence, higher biodiesel formation. When the ultrasonic power increases acoustic amplitude increases. So, cavitation bubble will collapse each other violently resulting in high velocity and micromixing at the phase boundary between two immiscible phases. Ultrasonication can result in mean droplet sizes much lower than those generated by conventional agitation, and can be a more powerful tool in breaking methanol into small droplets (Wu et al., 2007). The emulsion droplet size of methanol/soybean oil dispersions for ultrasonic and mechanical stirring was investigated and was shown that emulsion droplet size in ultrasonic mixing 2.4 times lower than that of conventional agitation. The mean droplet sizes were 148 and 146 nm with ultrasonic energy at 50 and 70 W, respectively. However, the droplet size was about 340 nm with impeller at 1000 rpm.

Higher power levels usually gives lower conversions because of cushioning effect and hence lower cavitational activity (Ji et al., 2006; Hingu et al., 2010; Lee et al., 2011). Hingu et al. (2010) observed that while the biodiesel conversion was obtained around 66% at 150 W power 89% of conversion was obtained when the power dissipation was increased to 200 W. But further increase in power from 200 W to 250 W resulted in lower FAME conversion FAME conversion rate also depends on the emulsification degree of reaction system (Ji et al., 2006). These authors also noted that the order of affecting factors on FAME yield was substrate molar ratio > temperature > pulse frequency > ultrasonic power.

Ultrasound pulse (few seconds on followed by second off) effects the biodiesel conversion (Hingu et al., 2010; Ji et al., 2006). Higher conversion can be obtained when higher pulse is applied to system. For example, while biodiesel conversion was obtained for the pulse 2 s ON and 2 s OFF, the conversion were 65.5% for 5 s ON and 1 s OF (Hingu et al., 2010). For a pulse duration as 1 min ON and 5 s OFF, conversion of 89.5% was obtained because of better emulsification of the methanol and oil layers. The effect of horn position on biodiesel production was investigated by same researchers. They kept the reaction parameters constant such as 6:1, methanol to waste cooking oil molar ratio, 1% catalyst concentration, 45°C temperature, 200W power ad 1 min ON and 5 s OFF pulse. Cavitation intensity depends on some parameters physicochemical properties namely viscosity, surface tension and density. Cavitation is generated due to the presence of horn in oil or methanol. According to the horn position various results can be observed. Hingu et al. (2010) applied there different positions: in the oil phase, at the interface and in methanol. While maximum conversion was achieved as 89.5% when the horn was dipped in methanol rich layer, the lowest conversion was obtained as 8.5% when the horn is dipped in the oil phase. 58.5% conversion was observed when the horn is located at the interface of two phases. Maximum ester conversion was obtained since methanol contributed cavitating conditions significantly.

Frequency effect on ultrasonic assisted biodiesel production: The effect of ultrasonic frequency was studied on the yield of transesterification reaction of vegetable oils and shortchain alcohols (Stavarache et al., 2005). NaOH or KOH were used as base catalysts. It was observed that the reaction time gets shorter (the reaction fastens) as the ultrasonicirradiation increases but the yield slightly decreases. At 40 KHz, the reaction time was shorter than 28 KHz, but the yield was obtained higher when studied at 28 kHz.. This is because of the higher formation of soap at 40 KHz and higher quantity of soap makes the purification process harder. The more soap is formed, more esters gets trapped in the soap micelles and the yield of the reaction decreases at 40 KHz as a result.

General comparison of ultrasound irradiation with conventional stirring: Ultrasonic assisted transesterification of oil presents some advantages compared to conventional stirring methods such as; reducing reaction time, increase the chemical reaction speed and decrease molar ratio and methanol, increase yield and conversion. Ultrasound irradiation reduce the reaction time compared to conventional stirring operation (Stavarache et al., 2005; Ji et al., 2006; Hanh, et al., 2008; Mootabadi, et al., 2010; Hingu et al., 2010; Lee et al., 2011). Stavarache et al. (2005) studied transesterification of vegetable oil with short-chain alcohols, in the presence of NaOH, by means of low frequency ultrasound (28 and 40 kHz). By using ultrasounds the reaction time was found much shorter (10–40 min) than for mechanical stirring. The optimal conditions for triolein methanolysis was methanol/triolein molar ratio of 6/1, KOH concentration of 1 wt% and irradiation time of 30 min. But the optimal conditions for the conventional stirring method were found to be as were methanol/triolein molar ratio of 6/1, KOH concentration of 1 wt% and 4 h (Hanh et al.., 2008). In transesterification of waste cooking oil with methanol 89.5% conversion was obtained in 40 minutes whereas conventional stirring resulted in 57.5% conversion (Hingu et al., 2010). Palm oil was esterified with 95% yield in 60 minutes compared to 2-4 h with conventional magnetic stirring under optimal conditions. Ultrasonic irradiation method enabled to reduce the reaction time by 30 min or more comparing to conventional heating method in production of biodiesel from various vegetable oils. Also this method improved conversion rate (Hanh et al., 2007; Lee et al., 2011). In transesterification reaction, mixing is important factor for increasing biodiesel yield. Ultrasonic effect induces an effective emulsification and mass transfer compared to conventional stirring thus reaction rate increase (Hanh et al., 2009; Hingu et al., 2010). Comparison of yield and conversion of vegetable oilwith various alcohols was presented in Table 8 and also was explained in the effect of alcohol type on ultrasound assisted biodiesel production section.

Ultrasound assisted method has a similar effect as microwave assisted method that both of them reduce the separation time from 5 to 10 hours to less than 60 minutes compared to conventional transesterification method (Kumar et al., 2010). Also, during production of biodiesel via acid or base catalyst, ultrasound irradiation provides a fast and easy route (Yu et al., 2010) and the purity of glycerin increases.

The production of biodiesel from non-edible vegetable oil and waste cooking oil using ultrasonication allows under ambient operating conditions (Kumar et al., 2010a; Hingu et al., 2010). Also, biodiesel production works from vegetable oils given in Table 6 illustrates the applicability of ultrasonic irradiation under atmospheric and ambient conditions. The transesterification reaction with methanol is usually performed at 60°C with classical stirring. Roomtemperature is hardly competitive in terms of energy consumption. Room temperature is hardly competitive in terms of energy consumption. The production of biodiesel with ultrasound is effective and time and energy saving and economically functional method (Ji et al., 2006; Kumar et al., 2010a; Hanh et al., 2011). Power ultrasonic method required approximately a half of the energy that was consumed by the mechanical stirring method (Ji et al., 2006). Special mixing devices can be used to increase mass transfer. It was reported that sonochemical reactors consume only about one third the energy required for a specialty mixer for same conversion (Lifka & Ondruschka, 2004). All these results clearly indicate that ultrasonic method inexpensive, simple and efficient and would be promising to the conventional stirring method.

Type of alcohol		28 kHz	40 kHz	Mechanical stirring
Methanol	Reaction time (min)	10	10	10
	Yield (%)	75	68	35
Ethanol	Reaction time (min)	20	10	10
	Yield (%)	75	30	47
n-propanol	Reaction time (min)	20	10	10
	Yield (%)	75	78	79
n-butanol	Reaction time (min)	40	20	20
	Yield (%)	87	90	89

Table 9. The yields and reaction times of FAMEs as a result of different frequencies of ultrasonic irradiation and mechanical stirring in the presence of NaOH catalyst (1.5% wt))(Stavarache et al., 2005)

As seen from the Table 9, the length of the alcohol chain affects the yield of the reaction, as the frequency of the ultrasonic irradiation affects the reaction time. In longer alcohol chains, the yield of the reaction is higher. The longer alcohol chains increases the solubility (miscibility) of alcohol into the oil. 40 kHz of ultrasonic irradiation is preferable if faster reaction is needed but it has to be taken into account that the yield decreases as the reaction fastens because of the higher formation of soap in faster reactions. In conclusion, miscibility of oil and alcohol is better under the control of ultrasonic waves. This effect increases the surface area and higher yields of isolated methyl esters can be achieved. The mass transfer is better so that the soap formation is lower resulting as better and easier isolation of methyl esters. Power of the ultrasonic irradiation makes the reaction faster, as the yield slightly decreases under higher frequencies (40 kHz).

5. Conclusion

Due to the growing energy necessity and environmental problems the studies focused on renewable alternative energy sources. Biodiesel is one of the important renewable energy sources used in many countries in the world as an alternative diesel fuel. Biodiesel is generally produced transesterfication reaction of vegetable and animal oils with catalyst under conventional stirring with batch and continuous processes. Because of the economical causes, choosing efficient transesterification method for biodiesel production has become important in recent years. In this context, the researchers have been investigating different new processes such as supercritical, microwave assisted and ultrasound assisted process to avoid inefficient processes. It is found that these methods have several distinctions compared to conventional methods. Homogenous catalyst (sulfuric acid, sodium hydroxide, potassium hydroxide, sodium and potassium metoxide etc.), heterogeneous catalyst (ZnO, SiO, MgO, BaO, SrO etc.) and enzymatic catalyst (lipase) are also easily being used in microwave and ultrasonic assisted processes. However, supercrital transesterfication reaction of vegetable oils is a noncatalytic reaction and higher yields can be obtained with compared to conventional methods. New methods for biodiesel production offer more advantages but these methods have also some negative effects. For example, energy consumption, excess amount alcohol usage are the disadvantages of supercritical process. Microwave synthesis is still in lab-scale synthesis and it is not viable in large scale for industrial production due to penetration depth of microwave radiation into the absorbing materials. The safety aspect is another drawback of microwave reactors for industry. Ultrasonic biodiesel production could be advantageous for small producers, but in large scale processing maybe challenging because of necessity of many ultrasound probes. Although there are some disadvantages of novel methods in biodiesel production, these methods give several important advantages for the transesterification of oils such as: reducing reaction time and reaction temperature, unwanted by-products; and increasing ester yields, conversion easier compared to conventional method. In conclusion, these methods with their important advantages can be more preferred than conventional method anymore.

6. References

Adewuyi, Y.G. (2001). Sonochemistry: Environmental Science and Engineering Applications. *Industrial & Engineering Chemistry Research*, Vol. 40, pp. 4681-4715.

Alenezi, R., Leeke, G.A., Winterbottom, J.M., Santos, R.C.D.,& Khan, A.R. (2010). Esterification kinetics of free fatty acids with supercritical methanol for biodiesel production. *Energy Conversion and Management*, Vol. 51, pp. 1055-1059.

Armenta, R.E., Vinatoru, M., Burja, A.M., Kralovec, J.A. & Barrow, C.J. (2007). Transesterification of fish oil to produce fatty acid ethyl esters using ultrasonic energy. *Journal of American Oil Chemists' Society*, Vol. 84, pp. 1045-1052.

Azcan , N. & Danisman, A. (2007). Alkali catalyzed transesterification of cottonseed oil by microwave irradiation. *Fuel*, Vol. 86, pp. 2639–2644.

Azcan, N. & Danisman, A. (2008). Microwave assisted transesterification of rapeseed oil. *Fuel*, Vol. 87, pp. 1781-1788.

Baig, S., Faroog, A. & Rehman, F. (2010). Sonochemistry and its industrial applications. *World Applied Sciences Journal*, Vol. 10, No. 8, pp. 936-944, ISSN 1818-4952.

Barnard, T.M., Leadbeater, N.E., Boucher, M.B., Stencel, L.M. & Wilhite, B.A. (2007). Continuous-flow preparation of biodiesel using microwave heating. *Energy & Fuels*, Vol.21, No. 3, pp. 1777-1781.

Bunyakiat, K., Makmee, S., Sawangkeaw, R. & Ngamprasertsith, S. (2006). Continuous production of biodiesel via transesterification from vegetable oils in supercritical methanol. *Energy & Fuels*, Vol. 20, pp. 812-817.

Campanelli, P., Banchero, M. & Manna, L. (2010). Synthesis of biodiesel from edible, non-edible and waste cooking oils via supercritical methyl acetate transesterification. *Fuel*, Vol. 89, pp. 3675-3682.

Cao, W., Han, H. & Zhang, J. (2005). Preparation of biodiesel from soybean oil using supercritical methanol and co-solvent. *Fuel*,Vol. 84, pp. 347–351.

Chen C-H., Chen W-H., Chang C-M J., Lai S-M. & Tu C-H. (2010). Biodiesel production from supercritical carbon dioxide extracted Jatropha oil using subcritical hydrolysis and supercritical methylation. *Journal of Supercritical Fluids*, Vol. 52, pp. 228-234.

Cheng, Z., Cunwen, W., Weiguo, W., Yuanxin, W, Faquan, Y, Ru'an, C. & Junfeng, Z. (2010). Continuous production of biodiesel from soybean oil using supercritical methanol in a vertical tubular reactor: I. phase holdup and distribution of intermediate product along the axial direction. *Chinese Journal of Chemical Engineering*, Vol. 18, pp. 626-629.

Chung, K.H., Kim, J. & Lee, K.Y. (2009). Biodiesel production by transesterification of duck tallow with methanolonalkalicatalysts. *Biomass and Bioenergy*, Vol.33, pp. 155–158.

Cintas, P., Mantegna, S., Gaudino, E.C. & Cravotto, G. (2010). A new pilot flow reactor for high-intensity ultrasound irradiation. Application to the synthesis of biodiesel. *Ultrasonics Sonochemistry*, Vol. 17, pp. 985-989.

Colucci, J.A., Borrero, E.E. & Alape, F. (2005). Biodiesel from an alkaline transesterification reaction of soybean oil using ultrasonic mixing. *Journal of American Oil Chemists' Society*, Vol. 82, No. 7, pp. 525-530.

Da Cunha, M.E., Krause, L.C., Moraes, M.S.A., Faccini, C.S., Jacques , R.A., Almeida, S.R., Rodrigues, M.R. A. & Caramão, E.B. (2009). Beef tallow biodiesel produced in a pilot scale. *Fuel Processing Technology*, Vol. 90, pp. 570–575.

Demirbaş, A. (2002). Biodiesel from vegetable oils via transesterification in supercritical methanol. *Energy Conversion Management*, Vol. 43, pp. 2349-4356.

Demirbaş, A. (2007). Biodiesel from sunflower oil in supercritical methanol with calcium oxide. *Energy Conversion Management*, Vol. 48, pp. 937-941.

Demirbaş, A. (2008). Studies on cottonseed oil biodiesel prepared in non-catalytic SCF conditions. *Bioresource Technology*, Vol. 99, pp. 1125–1130.

Demirbaş, A. (2009). Biodiesel from waste cooking oil via base-catalytic and supercritical methanol transesterification. *Energy Conversion and Management*, Vol. 50, pp. 923–927.

Deng, X., Fang, Z. & Liu, Y. (2010). Ultrasonic transesterification of Jatropha curcas L. oil to biodiesel by a two-step process. *Energy Conversion and Management*, Vol. 51, pp. 2802-2807.

Deng, X., Fang, Z., Liu, Y. & Yu, C.L. (2011). Production of biodiesel from Jatropha oil catalyzed by nanosized solid basic catalyst. *Energy*, Vol. 36, pp. 777-784.

Dube, M.A., Tremblay, A.Y. & Liu, J. (2007). Biodiesel production using a membrane reactor. *Bioresource Technology*, Vol. 98, pp. 639–647.

Düz, M.Z., Saydut, A. & Öztürk, G. (2011). Alkali catalyzed transesterification of safflower seed oil assisted by microwave irradiation. *Fuel Processing Technology*, Vol. 92, pp. 308-313.

Encinar, J.M., Juan, F., Gonzalez, J.F. & Rodriguez-Reinares, A. (2007). Ethanolysis of used frying oils: Biodiesel preparation and characterization. *Fuel Processing Technology,* Vol. 88, pp. 513-522.

Fini, A. & Breccia, A. (1999). Chemistry by microwaves. *Pure and Applied Chemistry,* Vol. 71, No. 4, pp. 573-579.

Fjerbaek, L., Christensen, K.V. & Norddahl, B. (2009). A review of the current state of biodiesel production using enzymatic transesterification. *Biotechnology and Bioengineering,* Vol. 102, No. 5, pp. 1298–1315

Freedman, B., Pryde, E.H. & Mounts, T.L. (1984). Variables affecting the yields of fatty esters from transesterified vegetable oils. *Journal of American Oil Chemists' Society,* Vol. 61, pp. 1638–1643.

Fukuda, H., Kondo, A. & Noda, H. (2001). Biodiesel Fuel Production by Transesterification of Oils. *Journal of Bioscience and Bioengineering,* Vol. 92, No. 5, pp. 405-416.

Geuens, J., Kremsner, J.M., Nebel, B.A., Schober, S., Dommisse, R.A., Mittelbach, M., Tavernier, S., Kappe, C.O. & Maes, B.U.W. (2008). Microwave-assisted catalyst-free transesterification of triglycerides with 1-butanol under supercritical conditions. *Energy & Fuels,* Vol. 22, No. 1, pp. 643-645.

Giridhar, M., Kolluru, C. & Kumar, R. (2004). Synthesis of biodiesel in supercritical fluids. *Fuel,* Vol. 83, pp. 2029-2033.

Gui, M. M., Lee, K. T. & Bhatia, S. (2009). Supercritical ethanol technology for the production of biodiesel: Process optimization studies. *Journal of Supercritical Fluids,* Vol. 49, pp. 286–292.

Gürü, M., Artukoğlu, B.D., Keskin, A. & Koca, A. (2009). Biodiesel production from waste animal fat and improvement of its characteristics by synthesized nickel and magnesium additive. *Energy Conversion and Management,* Vol.50, pp. 498–502.

Gürü, M., Koca, A., Can, Ö., Cınar, C. & Şahin, F. (2010). Biodiesel production from waste chicken fat based sources and evaluation with Mg based additive in a diesel engine. *Renewable Energy,* Vol. 35, pp. 637–643.

Haas, M.J. & Scoot, K.M. (1996). Combined nonenzymatic-enzymatic method for the synthesis of simple alkyl fatty acid esters from soapstock. , *Journal of the American Oil Chemists' Society,* Vol. 73, pp. 1393-1401.

Han, X., Chen, L., Peng, Q. (2008). Preparation of biodiesel from sunflower oil under microwave irradiation by ionic liquids H_2SO_4, *Journal of Zhengzhou University (Engineering Science),* Vol.4.

Hanh, H.D., Dong, N.T., Okitsu, K., Maeda, Y. & Nishimura, R. (2007). Effects of molar ratio, catalyst concentration and temperature on transesterification of triolein with ethanol under ultrasonic irradiation. *Journal of the Japan Petroleum Institute,* Vol. 50, pp. 195–199.

Hanh, H.D., Dong, N.T., Starvarache, C., Okitsu, K., Maeda, Y. & Nishimura, R. (2008). Methanolysis of triolein by low frequency ultrasonic irradiation. *Energy Conversion and Management,* Vol. 49, pp. 276-280.

Hanh, H.D., Dong, N.T., Okitsu, K., Nishimura, R. & Maeda, Y. (2009a). Biodiesel production by esterification of oleic acid with short-chain alcohols under ultrasonic irradiation condition. *Renewable Energy,* Vol. 34, pp. 780-783.

Hanh, H.D., Dong, N.T., Okitsu, K., Nishimura, R. & Maeda, Y. (2009b). Biodiesel production through transesterification of triolein with various alcohols in an ultrasonic field. *Renewable Energy,*Vol. 34, pp. 766-768.

Hawash, S., Kamal, N., Zaher, F., Kenawi, O. & Diwani, G. (2009). Biodiesel fuel from Jatropha oil via non-catalytic supercritical methanol transesterification. *Fuel,* Vol. 88, pp. 579-582.

He, C., Baoxiang, P., Dezheng, W. & Jinfu, W. (2007). Biodiesel production by the transesterification of cottonseed oil by solid acid catalysts. *Frontiers of Chemical Engineering in China,* Vol. 1, No. 1, pp. 11-15.

Helwani, Z., Othman, M.R., Aziz, N., Kim, J. & Fernando, W.J.N. (2009). Solid heterogeneous catalysts for transesterification of triglycerides with methanol: A review. *Applied Catalysis A: General,* Vol. 363, pp. 1–10.

Hernando, J., Leton, P., Matia, M.P., Novella, J.L. & Alvarez-Builla, J. (2007). Biodiesel and FAME synthesis assisted by microwaves. Homogeneous batch and flow processes. *Fuel,* Vol. 86, No. 10–11, pp.1641–1644.

Hingu, S.M., Gogate, P.R. & Rathod, V.K. (2010). Synthesis of biodiesel from waste cooking oil using sonochemical reactors. *Ultrasonics Sonochemistry,* Vol. 17, pp. 827-832.

Hoda, N. (2010). Optimization of biodiesel production from cottonseed oil by transesterification using NaOH and methanol. *Energy Sources, Part A: Recovery, Utilization, and Environmental Effects,*Vol.32, pp. 434- 441.

Hsiao, M.C., Lin, C.C., Chang, Y.H. & Chen, L.C. (2010). Ultrasonic mixing and closed microwave irradiation-assisted transesterification of soybean oil. *Fuel,* Vol. 89, No: 12, pp. 3618-3622.

Hsiao, M.C., Lin, C.C., & Chang, Y.H. (2011). Microwave irradiation-assisted transesterification of soybean oil to biodiesel catalyzed by nanopowder calcium oxide. *Fuel, Article in press.*

Ilham, Z. & Saka, S. (2010). Two-step supercritical dimethyl carbonate method for biodiesel production from Jatropha curcas oil. *Bioresource Technology,* Vol. 101 pp. 2735–2740.

Issariyakul, T., Kulkarni, M.G., Meher, L.C., Dalai, A.K. & Bakhshi, N.N. (2008). Biodiesel production from mixtures of canola oil and used cooking oil. *Chemical Engineering Journal,* Vol. 140, pp. 77–85.

İşcan, E. (2006). Biodiesel production in supercritical conditions. MSc Thesis, Yıldız Technical University, İstanbul.

Saka, S. & Kusdiana, D. (2001). Biodiesel fuel from rapeseed oil as prepared in supercritical methanol. *Fuel,* Vol. 80, pp. 225-231.

Ji, J., Wang, J., Li, Y., Yu, Y. & Xu, Z. (2006). Preparation of biodiesel with the help of ultrasonic and hydrodynamic cavitation. *Ultrasonics,* Vol. 44, pp. 411-414.

Jin, L., Zhang, Y., Dombrowski, J.P., Chen, C., Pravatas, A., Xu, L., Perkins, C. & Suib, S.L. (2011). $ZnO/La_2O_2CO_3$ layered composite: A new heterogeneous catalyst for the efficient ultra-fast microwave biofuel production. *Applied Catalysis B: Environmental, Article in press.*

Kaieda, M., Samukawa, T., Matsumoto, T., Ban, K., Kondo, A., Shimada, Y., Noda, H., Nomoto, F., Ohtsuka, K., Izumoto, E. & Fukuda H. (1999). Biodiesel fuel production from plant oil catalyzed by Rhizopus oryzae lipase in a water-containing system without an organic solvent. *Journal of Bioscience and Bioengineering*, Vol. 88, No. 6, pp. 627-631.

Kasim, N.S., Tsai T-H., Gunawan S. & Ju Y-H. (2009). Biodiesel production from rice bran oil and supercritical methanol. *Bioresource Technology*, Vol. 100, pp. 2399–2403.

Kaya, C., Hamamci, C., Baysal, A., Akba, O., Erdogan, S. & Saydut, A. (2009). Methyl ester of peanut (Arachis hypogea L.) seed oil as a potential feedstock for biodiesel production. *Renewable Energy*, Vol. 34, pp. 1257–1260.

Kim, D., Choi, J., Kim, G., Seol, S.K., Ha, Y.C., Vijayan, M., Jung, S., Kim, B.H., Lee, G.D. & Park, S.S. (2011). Microwave-accelerated energy-efficient esterification of free fatty acid with a heterogeneous catalyst. *Bioresource Technology*, Vol. 102, pp. 3639-3641.

Kong, J., Han, X., Chen, L. & Huo, J. (2009). Preparation of Biodiesel under Microwave Irradiation from Sunflower Oil by Solid Super Acid TiO_2/SO_4. *Guangzhou Chemical Industry* , Vol.2.

Koopmans, C., Iannelli, M., Kerep, P., Klink, M., Schmitz, S., Sinnwell, S. & Ritter, H. (2006). Microwave-assisted polymer chemistry: Heckreaction, transesterification, Baeyer–Villiger oxidation, oxazoline polymerization, acrylamides, and porous materials. *Tetrahedron*, Vol. 62, No. 19, pp. 4709–4714.

Köse, Ö., Tüter M. & Aksoy, H.A. (2002). Immobilized Candida antarctica lipase-catalyzed alcoholysis of cotton seed oil in a solvent-free medium. *Bioresource Technology*, Vol. 83, No. 2, pp. 125-129.

Kumar, D., Kumar, G. & Singh, P.C.P. (2010a). Fast, easy ethanolysis of coconut oil for biodiesel production assisted by ultrasonication. *Ultrasonics Sonochemistry*, Vol. 17, pp. 555-559.

Kumar, D., Kumar, G. & Singh, P.C.P. (2010b). Ultrasonic-assisted transesterification of Jatropha curcus oil using solid catalyst, Na/SiO_2. *Ultrasonics Sonochemistry*, Vol. 17, pp. 839-844.

Kusdiana, D. & Saka, S. (2001). Kinetics of transesterification in rapeseed oil to biodiesel fuel as treated in supercritical methanol. *Fuel*, Vol. 80, pp. 693-698.

Kusdiana, D. & Saka, S. (2004). Effects of water on biodiesel fuel production by supercritical methanol treatment. *Bioresource Technology*, Vol. 91, pp. 289–295.

Leadbeater, N.E. & Stencel, L.M. (2006). Fast, easy preparation of biodiesel using microwave heating. *Energy & Fuels*, Vol. 20, No. 5, pp. 2281-2283.

Lee, J.H., Kwon, C.H., Kang, J. W., Park, C., Tae, B. & Kim, S.W. (2009). Biodiesel production from various oils under supercritical fluid conditions by Candida antarctica lipase B using a stepwise reaction method. *Applied Biochemistry and Biotechnology*, Vol. 156, pp. 454–464.

Lee, J.S. & Saka, S. (2010). Biodiesel production by heterogeneous catalysts and supercritical technologies. *Bioresource Technology*, Vol. 101, pp. 7191–7200.

Lee, S.B., Lee, J.D. & Hong, I.K. (2011). Ultrasonic energy effect on vegetable oil based biodiesel synthetic process. *Journal of Industrial and Engineering Chemistry*, Vol. 510, pp. 1-6.

Lertsathapornsuk, V., Pairintra, R., Aryusuk, K. & Krisnangkur, K. (2008). Microwave assisted in continuous biodiesel production from waste frying palm oil and its performance in a 100 kW diesel generator. *Fuel Processing Technology*, Vol. 89, pp. 1330-1336.

Leung, D.Y.C. & Guo, Y. (2006). Transesterification of neat and used frying oil: optimization for biodiesel production. *Fuel Process Technology,* Vol. 87, pp. 883–90.

Licht, F.O. (2009). World-Biodiesel Production, World Ethanol and Biofuels Report, (4), 16:365.

Lifka, J. & Ondruschka, B. (2004). Influence of mass transfer on the production of biodiesel, *Chemical Engineering & Technology,* Vol. 27, No. 11, pp. 1156–1159.

Lin, G. & Liu, H. (1995). Ultrasound promoted lipase catalyzed reactions. *Tetrahedron Letters,* Vol. 36, No. 34, pp. 6067-6068.

Lotero, E., Liu, Y., Lopez, D.E., Suwannakarn, K., Bruce, D.A. & Goodwin J.G. (2005). Synthesis of Biodiesel via Acid Catalysis. *Industrial & Engineering Chemistry Research,* Vol. 44, pp. 5353-5363.

Madras, G., Kolluru, C. & Kumar R. (2004). Synthesis of biodiesel in supercritical fluids. *Fuel,* Vol. 83, pp. 2029–2033.

Mahamuni, N.N. & Adewuyi, Y.G. (2009). Optimization of the synthesis of biodiesel via ultrasound-enhanced base-catalyzed transesterification of soybean oil using a multifrequency ultrasonic reactor. *Energy & Fuels.* Vol. 23, pp. 2757-2766.

Majewski, M.W., Pollack, S.A. & Curtis-Palmer, V.A. (2009). Diphenylammonium salt catalysts for microwave assisted triglyceride transesterification of corn and soybean oil for biodiesel production. *Tetrahedron Letters,* Vol. 50, pp. 5175-5177.

Marchetti, J.M., Miguel, V.U. & Errazu, A.F. (2007). Possible methods for biodiesel production. *Renewable and Sustainable Energy Reviews,* Vol. 11, pp. 1300–1311.

Marulanda, V.F., Anitescu, G. & Tavlarides, L.L. (2010). Biodiesel fuels through a continuous flow process of chicken fat supercritical transesterification, *Energy & Fuels,* Vol. 24, No. 1, pp. 253-260.

Mason, T.J., Paniwnyk, L. & Lorimer, J.P. (1996). The uses of ultrasound in food Technology. *Ultrasonics Sonochemistry,* Vol. 3, pp. 253-260.

Mason, T.J. (1999). *Sonochemistry,* Oxford University Press, NewYork, pp. 2–30.

Mason, T.J. (2000). Large scale sonochemical processing: aspiration and actuality. *Ultrasonics Sonochemistry,* Vol. 7, No:4, pp. 145-149.

Manh, D.V., Chen, Y.H., Chang, C.C., Chang, M.C. & Chang, C.Y. (2011). Biodiesel production from Tung oil and blended oil via ultrasonic transesterification process. *Journal of the Taiwan Institute of Chemical Engineers,* Vol. 297, pp. 1876-1070.

Melero, J.A., Bautista, L., F., Morales, G., Iglesias, J. & Sánchez-Vázquez, R. (2009). Biodiesel production from crude palm oil using sulfonic acid-modified mesostructured catalysts. *Chemical Engineering Journal, Article in press.*

Mikkola, J.P. & Salmi, T. (2001). Three-phase catalytic hydrogenation of xylose to xylitolprolonging the catalyst activity by means of on-line ultrasonic treatment. *Catalysis Today,* Vol. 64, pp. 271–277.

Minami, E. & Saka S. (2006). Kinetics of hydrolysis and methyl esterification for biodiesel production in two-step supercritical methanol process. *Fuel,* Vol. 85, pp. 2479-2483.

Mootabadi, H., Salamatinia, B., Bhatia, S. & Abdullah, A.Z. (2010). Ultrasonic-assisted biodiesel production process from palm oil using alkaline earth metal oxides as the heterogeneous catalysts. *Fuel,* Vol.89, pp. 1818-1825.

Newman, P.G. & Rozycki, G.S. (1998). The history of ultrasound. *Surgical Clinics of North America,* Vol. 78, No: 2, pp. 179-195.

Nogueira, B.M., Carretoni, C., Cruz, R., Freitas, S., Melo, P.A., Costa-Felix, R., Pinto, J.C. & Nele, M. (2010). Microwave activation of enzymatic catalysts for biodiesel production. *Journal of Molecular Catalysis B: Enzymatic,* Vol. 67, pp. 117-121.

Nuechter, M., Ondruschka, B., Jungnickel, A. & Mueller, U. (2000). Organic processes initiated by non-classical energy sources. *Journal of Physical Organic Chemistry,* Vol. 13, pp. 579–586.

Öner, C. & Altun, Ş. (2009). Biodiesel production from inedible animal tallow and an experimental investigation of its use as alternative fuel in a direct injection diesel engine. *Applied Energy,* Vol. 86, pp. 2114–2120.

Özçimen, D. & Yücel, S. (2010). Biodiesel production from algae. *Proceedings of 101st AOCS Annual Meeting & Expo.,* Phoneix-USA, May, 2010.

Özgül, S. & Türkay, S. (1993). In-situ esterification of rice bran oil with methanol and ethanol. , *Journal of the American Oil Chemists' Society,* Vol. 70, pp. 145-147.

Özgül-Yücel, S. & Türkay, S. (2002). Variables affecting the yields of methyl esters derived from in- situ esterification of rice bran oil. , *Journal of the American Oil Chemists' Society,* Vol. 79, pp. 611-614.

Öztürk, G., Kafadar, A.B., Duz, M.Z., Saydut, A. & Hamamcı, C. (2010). Microwave assisted transesterification of maize (*Zea mays* L.) oil as a biodiesel fuel. *Energy, Exploration & Exploitation,* Vol. 28, No. 1, pp. 47-58, ISSN 0144-5987.

Patil, P.D., Gude, V.G., Mannarswamy, A. & Deng, S. (2009). Biodiesel Production from Jatropha Curcas, Waste Cooking, and Camelina Sativa Oils, *Industrial&Engineering Chemistry Research,* Vol. *48,* No:24, pp. 10850–10856.

Patil, P., Deng, S., Rhodes, J. I. & Lammers, P.J. (2010a). Conversion of waste cooking oil to biodiesel using ferric sulfate and supercritical methanol processes. *Fuel,* Vol. 89, pp. 360–364.

Patil, P.D., Gude, V.G., Mannarswamy, A., Deng, S., Cooke, P., Munson-McGee S., Rhodes I., Lammers, P. & Nirmalakhandan, N. (2010b). Optimization of direct conversion of wet algae to biodiesel under supercritical methanol conditions. *Bioresource Technology,* Vol. 102, pp. 188-122.

Patil, P.D., Gude, V.G., Mannarswamy, A., Cooke, P., Munson-McGee, S., Nirmalakhandan, N., Lammers, P. & Deng, S. (2011). Optimization of microwave-assisted transesterification of dry algal biomass using response surface methodology. *Bioresource Technology,* Vol. 102, pp. 1399-1405.

Perin, G., Alvaro, G., Westphal, E., Viana, L.H., Jacob, R.G., Lenardao, E.J. & D'Oca, M.G.M. (2008). Transesterification of castor oil assisted by microwave irradiation. *Fuel,* Vol. 87, pp. 2838-2841.

Rahmanlar, İ. (2010). Alcoholysis of frying oils with microwave method. MSc. Thesis, Yildiz Technical University, Istanbul.

Rashid, U., Anwar, F. & Knothe, G. (2009). Evaluation of biodiesel obtained from cottonseed oil. *Fuel Processing Technology,* Vol. 90, pp. 1157–1163.

Rathore, V. & Madras, G. (2007). Synthesis of biodiesel from edible and non-edible oils in supercritical alcohols and enzymatic synthesis in supercritical carbon dioxide. *Fuel,* Vol. 86, pp. 2650-2659.

Refaat, A.A., El Sheltawy, S.T. & Sadek, K.U. (2008). Optimum reaction time, performance and exhaust emissions of biodiesel produced by microwave irradiation.

17

Pyrolysis Oil Stabilisation by Catalytic Hydrotreatment

Venderbosch R.H.[1] and Heeres H.J.[2]
[1]*BTG Biomass Technology Group B.V.*
[2]*RijksUniversiteit Groningen*
The Netherlands

1. Introduction

Being the only sustainable product containing carbon, biomass is the only alternative for fossil derived crude oil derivatives. Research on the use of biomass for first generation biofuels is rapidly expanding (e.g. bio-ethanol from sugar sources and starches and bio-diesel from pure plant oils). Biomass, and in a particular ligno-cellulosic material, is difficult to convert easily into transportation fuels. Research is focused on indirect routes like: i) the fractionation of biomass and fermentation of the cellulosic and hemi-cellulosic fraction to ethanol, and ii) the destructive gasification of the complete biomass to produce syngas for further upgrading to e.g. methanol or Fischer-Tropsch diesel. Conventional refinery scales (up to 100 t/hr crude oil equivalence) are preferred for economic reason, but problematic for biomass resources, as biomass is scattered and collection is costly. In addition, various types of biomass are very different in structure and composition, have a low energy density compared to many fossil resources, and often contain significant amounts of water and ash. Such disadvantages can be overcome if the biomass is first de-centrally restructured, densified at a scale of 2 to 10 t/hr, and (preferably simultaneously) 'decontaminated'. The intermediate product is then transported to a large central processing unit where it is transformed to the final product (at a scale of say 50 to 200 t/hr). A potentially attractive technology for this purpose is fast pyrolysis (see recent review of Venderbosch&Prins, 2010). Pyrolysis liquids contain negligible amounts of ash, and have a volumetric energetic density 5 to 20 times higher than the original biomass. However, the oil is acidic in nature, polar and not miscible with conventional crude oil. In addition, it is unstable, as some (re)polymerisation of organic matter in the oil causes an increase in viscosity in time. Pyrolysis oil should thus not be regarded as an oil in a historic perspective, but merely as a carbohydrate rich syrup. Proof for this statement is given by a solvent fractionation method, where the largest part of the oil, up to 70%, can be easily extracted by water. A typical example is presented in Figure 1 (and analysis in Table 1), where pyrolysis oil is divided into several groups of compounds, differing in oxygen functionality and molecular size (Oasmaa et al. 2003; Oasmaa, 2003). Interestingly, the large ether insoluble fraction contains significant amounts of sugar-like components, and has the appearance of a syrup-like liquid. The high levels of oxygen in the pyrolysis oil are reported to be the main cause for the unstable character. However, it may be apparent that not the oxygen itself renders the pyrolysis oil unstable, but the nature/reactivity of the oxygen containing functional chemical groups. Especially the carbonyl compounds (aldehydes, ketones) seem to be responsible for the thermal instability and transformation into less reactive organic groups (f.i. to the corresponding alcohols) seems a viable option.

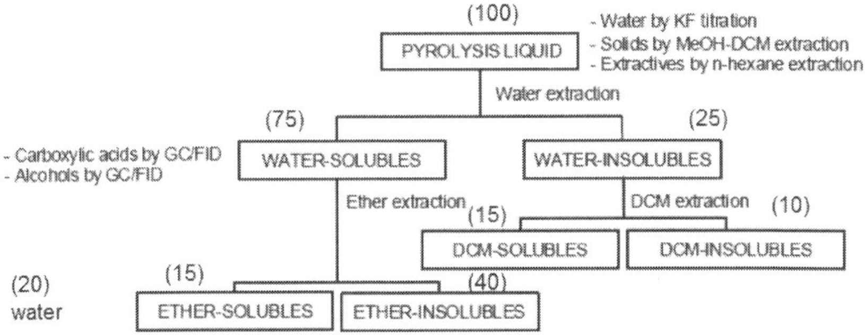

Fig. 1. Fractionation scheme (Oasmaa, 2003)

	wt %	COMPOUND TYPES	C	H	N	O
	(wet basis)		wt % (dry basis)			
WATER-SOLUBLES	75-85					
Acids, alcohols	5-10	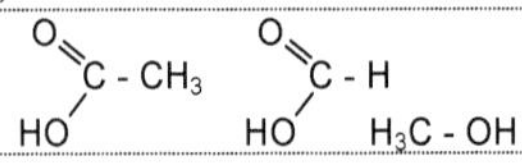	36.0	6.0	0	58.0
Ether-solubles	5-15	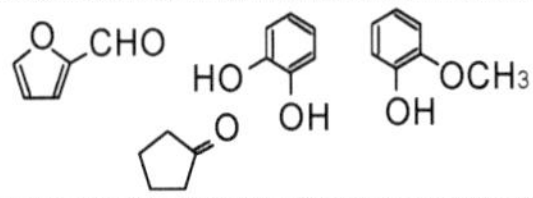	60.0	6.0	0.1	33.9
Ether-insolubles	30-40	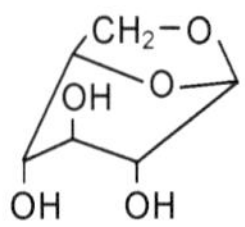	46.0	6.3	0.3	47.4
Water	20-30		0	11.1	0	88.9
WATER-INSOLUBLES	15-25		66.2	6.6	0.3	26.9
n-Hexane-solubles	2-6	$CH_3 - (CH_2)_n - CO_2H$ n = 10-30 $HOCH_2 - (CH_2)_n - CO_2H$ n = 10-28 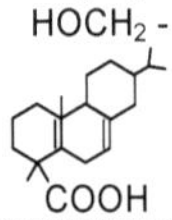	77.4	10.4	0	12.2
DCM-solubles	5-10	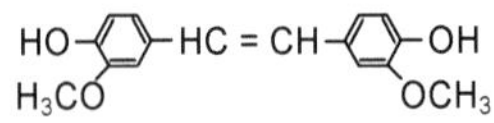	68.1	6.7	0.4	24.7
DCM-insolubles	2-10	degraded lignin	64.1	5.9	1.5	28.4

Table 1. Chemical composition of reference pine oil and its fractions (Oasmaa, 2003)

A number of catalytic approaches have been proposed to upgrade and improve the product properties of fast pyrolysis oil. A well known example is catalytic cracking of pure biomass and/or pyrolysis oil to oxygen-free products. However, this approach is accompanied by a significant amount of coke production (up to 40 wt.% of the biomass) (Horne &Williams, 1996; Vispute et al., 2010) and this issue needs to be resolved.

A number of studies have been performed with the objective to remove the bound oxygen in the form of CO and/or CO_2 by decarbonylation and decarboxylation reactions, either thermally or catalytically. The thermal process is known as the HPTT process, a high pressure temperature treatment (De Miguel Mercader et al., 2010). However, oxygen removal beyond a level of 10% appears very difficult with this approach, even when using catalysts, and the product is highly viscous, limiting its application potential.

Catalytic hydroprocessing or hydrotreating of fast pyrolysis oil is a more promising option (Conti, 1997; Elliott, 2007; Elliott et al.; 2009; Kaiser 1997; Rep et al., 2006). In this process, the fast-pyrolysis oil is treated with hydrogen in the presence of a heterogeneous catalyst with the aim to hydro(deoxy)genate the pyrolysis oil to a product with improved product properties.

A potentially attractive outlet for the upgraded oils is its use as co-feed in existing oil refinery units. This enables partial substitution of the fossil carbon in liquid transportation fuels by renewable carbon from biomass in existing infrastructure as proposed in the European Biocoup project. As such, it is expected to lower the investment costs and to facilitate other barriers for introduction of the fast pyrolysis technology. Crude pyrolysis oil, however, cannot be used for co-feeding purposes. It is immiscible with typical petroleum feeds (such as vacuum gas oil) due to the presence of polar components, but moreover, it is highly acidic due to the presence of organic acids (up to 10 wt%) (Oasmaa, 2003) leading to corrosion issues and possibly detrimental effects on typical catalysts in the refinery units (e.g. zeolites in the FCC process) (Dimitrijevic et al., 2006). Furthermore, it has a strong tendency for coking at elevated temperatures.

An alternative product outlet for upgraded pyrolysis oils is the direct use as a green transportation fuel in internal combustion engines. This will require deep hydrotreatment of the pyrolysis oil to an upgraded oil with very low oxygen levels (preferably below 1%) in order to mimic the product properties of conventional hydrocarbon fuels. Hydrogen consumption is expected to be high for this case, and will have a negative impact on the process economics.

This chapter on catalytic hydrotreatment of fast pyrolysis oil is divided in a number of subtopics. A short overview of typical catalysts for the hydrotreatment of pyrolysis oils will be given in part 2, followed by an overview of typical process requirements. Subsequently, in depth process studies using Ru/C as catalyst for the upgrading of fast pyrolysis oil will be provided (part 3). The effect of process conditions on yields, hydrogen uptake and relevant product properties will be discussed and rationalized by a reaction network. On the basis of these findings, improved catalyst formulations have been designed and their performance will be discussed and evaluated (part 5).

2. Catalyst studies on the hydrotreatment of fast pyrolysis oil

Well known catalysts for the hydrotreatment of pyrolysis oil are conventional hydrodesulfurisation catalysts such as sulfided NiMo/Al_2O_3, CoMo/Al_2O_3, and NiMo/Al_2O_3-SiO_2 (Ferrari et al., 2002; Maity et al., 2000) to cope with the harsh reaction conditions (T > 200 C, aqueous environment with organic acids). Carbon-supported CoMo, or Mo supported on TiO_2, ZrO_2 and TiO_2-ZrO_2 mixed oxides (Satterfield&Yang, 1983; Lee&Ollis, 1984) have been tested as well. The latter catalysts all require sulphur in the feed to maintain activity. Pyrolysis oils contain only limited amounts of sulphur (Furimsky, 2000) and sulphur addition would be required during processing to maintain catalytic activity. This will lead to sulfur emissions e.g. in the flue gas, which is not preferred for an environmental point of view.

Non-sulfided, non noble metal catalysts have also been tested for the catalytic hydrotreatment of pyrolysis oil, but to a far lesser extent. The hydrodeoxygenation of phenol over a Ni/SiO_2 catalyst has been reported and showed low HDO activity. Horne&Williams (1996) tested ZSM-5 zeolites as the catalyst for the deoxygenation of model compounds such as anisole. Their results, though, showed that anisole is mainly converted into phenol and methyl substituted phenols, and not to the oxygen-depleted compounds. Xu, et al. (2010) tested $MoNi/\gamma\text{-}Al_2O_3$ (reduced prior to reaction) for the mild hydrotreatment of pyrolysis oil (3 MPa and 200 °C).

Noble metal based catalysts have also been evaluated as substitutes for sulfided catalysts. Examples are Pd on zeolite carriers (Horne&Williams, 1996), Pd on mesoporous CeO_2 and ZrO_2 (Senol et al., 2005), and Rh on zirconia (Gutierrez et al., 2008). Carbon supports have also been explored (Wildschut et al., 2009b; Wildschut et al., 2010a;, Wildschut et al., 2010b; Venderbosch et al., 2010). We have recently reported exploratory catalyst studies on the upgrading of fast pyrolysis oil by catalytic hydrotreatment using a variety of heterogeneous noble metal catalysts (Ru/C, Ru/TiO_2, Ru/Al_2O_3, Pt/C and Pd/C) and the results were compared to typical hydrotreatment catalysts (sulfided $NiMo/Al_2O_3$ and $CoMoAl_2O_3$). The reactions were carried out at temperatures in the range of 250 and 350 °C and pressures between 100 and 200 bar. The Ru/C catalyst appears superior to the classical hydrotreating catalysts with respect to oil yield (up to 60 %-wt.) and deoxygenation level (up to 90 %-wt.) (Figure 2) (Wildschut et al, 2009).

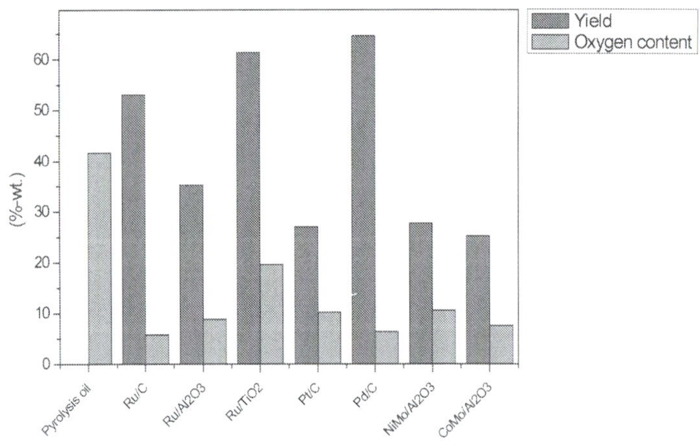

Fig. 2. Catalyst screening studies for noble metal catalysts on various supports

Unfortunately, noble metal catalysts are very expensive and this limits their application potential. The catalytic hydrotreatment of fast pyrolysis oil using much cheaper bimetallic $NiCu/\delta\text{-}Al_2O_3$ catalysts with various Ni/Cu ratios (0.32 - 8.1 w/w) at a fixed total metal intake of about 20 wt% was reported recently (Ardiyanti et al., 2009). Hydrotreatment reactions of fast pyrolysis oil (batch autoclave, 1 h at 150 °C followed by 3 h at 350 °C, at 200 bar total pressure) were performed and highest catalyst activity (in terms of H_2 uptake per g of Ni) was observed for the catalyst with the lowest Ni loading (*5.92*Ni*18.2*Cu). Product oils with oxygen contents between 10 and 17 wt% were obtained and shown to have improved product properties when compared to the feed.

Recently, we have also shown the potential of homogeneous Ru catalysts for the catalytic upgrading of fast pyrolysis oil fractions (Mahfud et al., 2007a). Fractions were obtained by treatment of the pyrolysis oil with dichloromethane or water. The former approach leads to a dichloromethane layer enriched with the lignin fraction of pyrolysis oil. The dichloromethane fraction was hydrogenated in a biphasic system (dichloromethane/water) using a homogeneous water soluble Ru/tri-phenylphosphine-tris-sulphonate (Ru-TPPTS) catalyst. Analysis revealed that the oxygen content of was lowered considerably and that particularly the amounts of reactive aldehydes were reduced substantially.

The water-soluble fraction of pyrolysis oil, obtained by treatment of fast pyrolysis oil with an excess of water was hydrogenated in a biphasic system with a homogeneous Ru-catalyst (Ru/tri-phenylphosphine, Ru-TPP) dissolved in an apolar solvent (Mahfud et al., 2007b). Initial experiments were conducted with representative, water soluble model compounds like hydroxyacetaldehyde and acetol. The effects of process parameters (e.g. temperature, initial H_2 pressure, and initial substrate concentration) were investigated and quantified for acetol using a kinetic model. The hydrogenation of the pyrolysis oil water-soluble fraction using the biphasic system at optimized conditions was performed and significant reductions in the amounts of reactive aldehydes such as hydroxyacetaldehyde and acetol were observed, demonstrating the potential of homogeneous Ru-catalysts to upgrade pyrolysis oils.

3. Process studies on the catalytic hydrotreatment of fast pyrolysis oils

3.1 Introduction

In the past, it was assumed that the catalytic hydrotreatment of pyrolysis oils shows strong resemblances with conventional hydrotreatment processes of fossil feeds (hydrodesulphurisation, hydrodenitrogenation) and typical catalysts, process conditions and reactor configuration were taken from these conventional processes (Elliott, 2007). For all, the objective is a reduction of certain elements (oxygen, sulphur, nitrogen) in the feed by the action of hydrogen and a catalyst. However, common reaction conditions for the hydroprocessing of crude oil derivatives cannot be adopted directly for pyrolysis oils. For instance, pyrolysis oil cannot be treated as such at temperatures exceeding 300 °C because of its high charring tendency.

Numerous papers on hydrotreating pyrolysis oil in packed beds and autoclaves were published in the eighties and nineties, and now very recently as well (Elliott, 2007). Specifically the older literature seems rather phenomenological in nature, and merely focuses on 'fact-finding'. Large differences in operating conditions like in pressure, temperature, residence times and hydrogen consumption are reported. In any case, without active catalyst or (high pressure) hydrogen, significant charring of pyrolysis oil occurs. Reduction of charring is possible by applying a (relatively) low temperature catalytic hydrotreatment at 175 to 250 °C, in which reactive components in the oil are 'stabilized'. Subsequently, the stable product is further processed at higher temperatures (> 300 °C) and pressures (> 150 bar). High pressures are believed to be essential to keep the water in the pyrolysis oil feed in a liquid state (and as such probably reduce the charring reactions), to promote the solubility of hydrogen in the initially polar bio-oil, and to increase the rate of the actual hydrogenation reactions.

3.2 Process studies using Ru/C

3.2.1 Batch studies

We have recently performed an in depth study in a batch set-up to determine the effects of process conditions on the catalytic hydrotreatment of fast pyrolysis oil using a Ru/C catalyst 350 °C and 200 bar pressure (Wildschut et al., 2010a). The liquid product after reaction consisted of three different phases, a slightly yellow aqueous phase and two brown oil phases, one with a density higher than water and one with a density lower than water. Furthermore, substantial amounts of solids (coke/char, about 5 wt% on pyrolysis oil intake) and gas phase organics (CO, CO_2, CH_4) were formed as well. The oil yield and elemental compositions of the product phases were shown to be a strong function of the reaction time. Highest oil yields (65 %-wt.) were obtained after 4 h using a 5 %-wt. intake of catalyst on fast pyrolysis oil (Figure 3). Longer reaction times lead to a reduction of the oil yield due to the formation of gas phase components (methane, ethane, propane, CO/CO_2). Hydrogen uptake after 4 h reaction time was about 400 Nm^3/t of pyrolysis oil (dry basis).

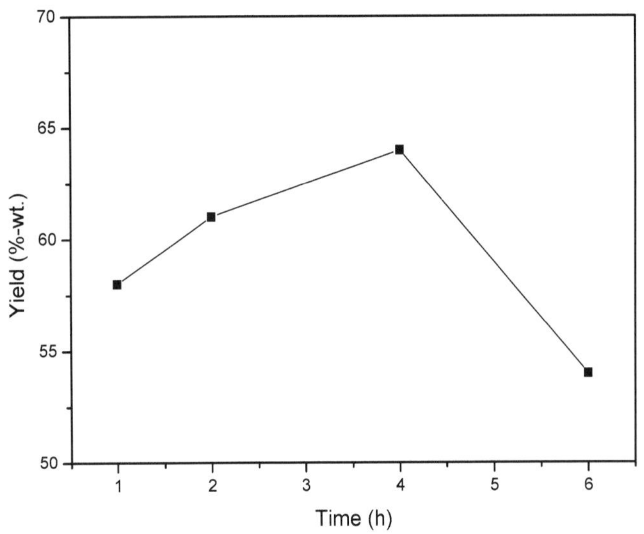

Fig. 3. Total oil yield (dry basis) versus reaction time for the hydrotreatment of pyrolysis oil (350 °C and 200 bar, Ru/C).

A solvent-solvent extraction scheme based on the work of Oasmaa et al. (2003) was used to gain insight in the reactivity of various component classes (fractions) in the fast pyrolysis oil during the catalytic hydrotreatment process. The fractionation scheme (Figure 1) was applied to the original fast pyrolysis oil and product oils obtained at different reaction times (350 °C and 200 bar). The results are given in Figure 4.

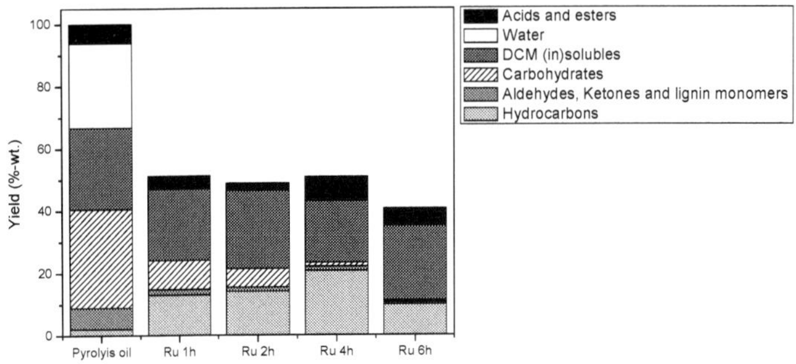

Fig. 4. Composition of fast pyrolysis oil and hydrotreated product oils (Ru/C, 350 °C, 200 bar) at various reaction times using solvent-solvent extraction

It shows the amounts of the various fractions (carbohydrates, aldehydes/ketones/lignin monomers, hydrocarbons, acids and esters) as a function of the reaction time. A fast decline in the carbohydrate fraction versus time is visible. Almost complete conversion to other components within 6 h reaction time is observed, an indication of the high reactivity of this fraction.

3.2.2 Experimental studies with Ru/C in continuous set-ups

Recently, in depth catalytic hydrotreatment experiments with the Ru/C catalyst in a continuous packed bed set-up were reported (Venderbosch et al., 2010). The results of this study will be provided in detail in the following as it provides detailed insights in the effect of process conditions on product yields, product properties and the various reactions taking place on a molecular level. Some experiments were carried out in the absence of catalysts to probe thermal reactions.

The catalytic hydrotreatment reactions were carried out in a set-up consisting of 4 packed bed reactors in series. The temperature in each reactor may be varied independently, allowing experiments at different temperature profiles over the length of the reactor. Typical pressures were between 150 and 300 bar, temperatures between 150 and 400 °C and WHSV's between 2-10 kg/kg.cat.h. In the following, the thermal reactions will be discussed, followed by catalytic hydrotreatment reactions at different temperature levels.

3.2.2.1 Thermal reactions

To study the thermal, non-catalytic reaction in detail, pyrolysis oil was pumped through the reactor (without catalyst) at pressures of up to 300 bar and temperatures of maximum 350 °C for residence times in the order of tenths of second - minutes. Typically under these conditions, a single-phase pyrolysis oil is converted into a viscous organic liquid, an aqueous phase and a gas phase. The carbon content of the viscous phase is about 60 wt.% (starting with 40 wt.% in the original oil), and the oxygen content about 32 wt.%. Additional water is produced, up to 30 % compared to the water initially present in the pyrolysis oil. The water is distributed over the two layers, but most of it ends up in the aqueous phase. Energetically, 80% of the thermal energy in the pyrolysis oil is transferred to the viscous product, less than 20% and 1 % is retained by the water phase and gas phase, respectively. The gas phase in such experiments consists of CO and CO_2 in a ratios varying from 1:10 to

1:3 (depending on temperature, pressure, residence time), and in yields of almost 4 wt.% of the pyrolysis feed.

Although it is unknown at a molecular level which reactions actually take place, at least two parallel pathways can be distinguished, viz. a reaction causing the formation of gas (here referred to as decarboxylation / decarbonylation, yielding CO and/or CO_2), and the other causing dehydration (likely by condensation (polymerisation) reactions). Possible sources of these gases are the organic acids in the oil. For all aqueous (and organic) samples produced the pH, however, is almost similar to the pyrolysis oil feed. This indicates that either the acids are not converted or the acids are converted and simultaneously produced as well. A detailed acid analysis of the products is not available, and the precise events taking place and mechanism however remain unclear. It seems that dilution of the pyrolysis oils with 'inert' solvents suppresses the re-polymerisation. Additionally, the gas yield becomes independent of the temperature and the residence time after a certain threshold in the residence time, while the amount of water produced is increasing. This indicates that the reaction mechanism for the formation of gas is different than the polymerisation reactions.

Phase separation of the oil at these conditions may have a number of causes, e.g. an overall increase in the water content due to the formation of water by condensation reactions. It is known (but not fully explained yet) that above a certain water content pyrolysis oils phase separate into an aqueous phase and a rather nonpolar phase. Repolymerisation of some molecules / fractions in the oil is also a plausible reason, as it renders the products less soluble in water, for example caused by transformation of the polar sugar constituents behaving as bridging agents in the dissolution of hydrophilic lignin material (Diebold 2002).

3.2.2.2 Catalytic hydrotreatment reactions

The catalytic hydrotreatment reactions were carried out at three process severity levels, a mild hydrogenation at either 175 or 225 °C, a mild hydrodeoxygenation (HDO) at 225 - 275 °C and a deep hydrodeoxygenation. For the latter, samples from the mild HDO were first allowed to phase separate completely, after which the organic fraction (containing about 3 wt.% water) was treated at temperatures ranging from 350 °C in the first two reactor segments, to 400 °C in the last two.

3.2.2.3 Visual appearances of liquid phase after reaction

The catalytic hydrotreatment reaction at 175 °C resulted in a single phase oil with a visual appearance close to that of the original feed. Thus, at this temperature, phase separation does not occur. This may be related to the limited production of water at this temperature. The product has a considerable sweeter smell/odor than the original pyrolysis oil. The mild hydrogenation at 225 °C gives two liquid phases, an organic and a water rich phase. The water phase has a higher density than the aqueous phase. A similar situation was observed for experiments at higher process severities (mild HDO), see Figure 5 for details. The second stage HDO product oil has even a lower density than the aqueous phase.

The organic product yields for the various process severities are given in Figure 6. Here, the severity is expressed in terms of hydrogen consumption, and high severity is associated with high hydrogen consumption. The yield is a clear function of the temperature. A drop in the yield to about 40% is observed at about 200 °C due to the occurrence of phase separation and transfer of part of the carbon and oxygen to the aqueous phase. A further slight reduction in yield is observed at higher severities, presumably due to gasification reactions and further net transfer of components from the organic to aqueous phase.

Fig. 5. Pictures of pyrolysis oil (left), mild HDO (middle) and 2nd stage HDO (right) products

Oxygen contents of the product oils are a function of the process severity, see Figure 6 for details. Phase separation between 175 and 225 °C results in a dramatic drop in the oxygen content. This is due to the loss of water and the transfer of very polar highly oxygenated components to the aqueous phase. At the highest severity, the oxygen content is about 15%, compared to about 40% for the original pyrolysis oil.

The hydrogen consumption ranges between 65 and 250 Nm^3/t pyrolysis oil. Higher process severities lead to higher hydrogen uptakes (Figure 6).

A useful representation to assess the changes in the elemental composition of the product oils at various process severities is a van Krevelen diagram. Here, the ratio between O/C and H/C of the products are plotted together in a single diagram. In Figure 7, a typical plot is provided for selected literature data on pyrolysis oil hydroprocessing (Elliott, 2007; Venderbosch et al., 2010) and our results with Ru/C at different severities. Presented here are data points from e.g.:

- wood and pyrolysis oil, and for the four cases referred to in this paper (HPTT, hydroprocessing at 175 and 225 °C, Mild HDO and 2nd stage HDO);
- A selection of data points derived from literature studies (Baldauf et al. 2007; Churin et al., 1988; Conti, 1997; Diebold, 2002; Kaiser 1997; Samolada et al., 1998). Some of these data are derived from various oils from a variety of resources and processed in different reactors, different catalysts and at different conditions.

The plot also contains curves to represent the changes taking place in elemental composition during hydroprocessing, a theoretical curve for the dehydration of pyrolysis oil, and trend lines for the thermal (HPTT) route and hydroprocessing routes based upon the experimental data points.

Based on our work on the Ru/C catalysts and supported by the literature points in Figure 7, several reaction pathways can be distinguished:

a. Essentially repolymerisation of the pyrolysis oil (no catalyst, no hydrogen, 'HPTT');
b. Merely hydrogenation of the pyrolysis oil at mild conditions (up to 250°C, with catalyst and hydrogen, referred to as mild hydrogenation),
c. Dehydration of the oil at temperatures near 250-275 °C, and
d. Hydroprocessing of pyrolysis oil at temperatures up to 400 °C

Upon thermal treatment, the principal reactions are rejection of oxygen as water. Some CO_2 and CO is released as well, which shifts the trend line to slightly higher H/C ratios (but decarboxylation / decarbonylation is limited to approx. 10 wt.% of the feed). A high conversion (i.e. at high temperatures and residence times) eventually leads to a hydrogen-depleted solid material (and probably similar to conventional carbonisation processes, charcoal).

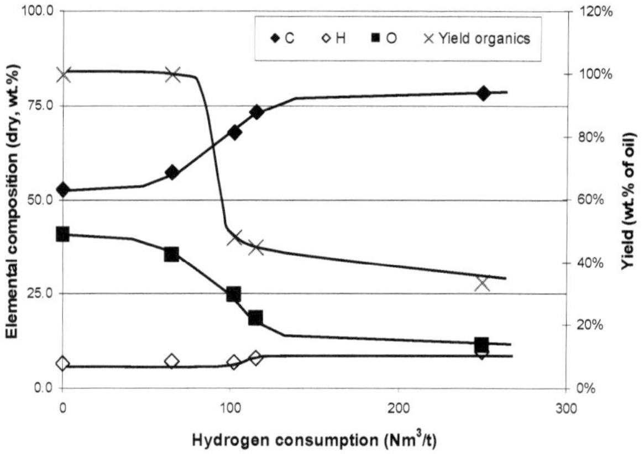

Fig. 6. The elemental composition of the organic oil product (dry basis) versus the hydrogen consumption for pyrolysis oil, mild hydrogenation, mild HDO and 2nd stages HDO

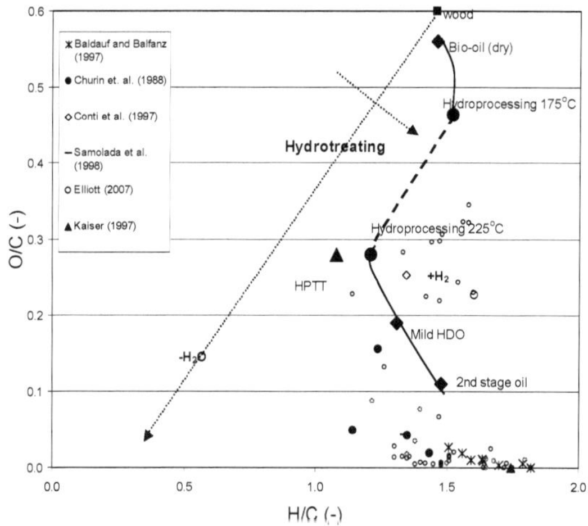

Fig. 7. The Van Krevelen plot for oils derived from the thermal pathway (HPTT), mild hydrogenation, mild HDO and 2nd stages HDO, including relevant literature data points

To obtain a liquid product with a higher H/C ratio, additional hydrogen is thus required. This path is shown in Figure 7 and includes the mild hydroprocessing step, at around 175 °C (no phase separation) and 225 °C (phase separation), followed by further hydrodeoxygenation (and hydrocracking).

3.2.3 Product oil fractionation; insights in molecular changes

The various organic products were subjected to a standardized liquid-liquid fractionation protocol (Oasmaa, 2003, Figure 1) to gain insights on the severity of the hydrotreatment process on product composition. The results are compiled in Figure 8 and show major changes in composition upon reaction. The pyrolysis oil feed mainly consist of ether solubles, ether insolubles and water. The components in these fractions originate from the cellulose and hemi-cellulose fraction in the biomass feed and particularly the ether insoluble fraction is rich in carbohydrates. The amounts of DCM solubles and insolubles, from the lignin fraction of the biomass feed, are by far lower and are about 20% in total.

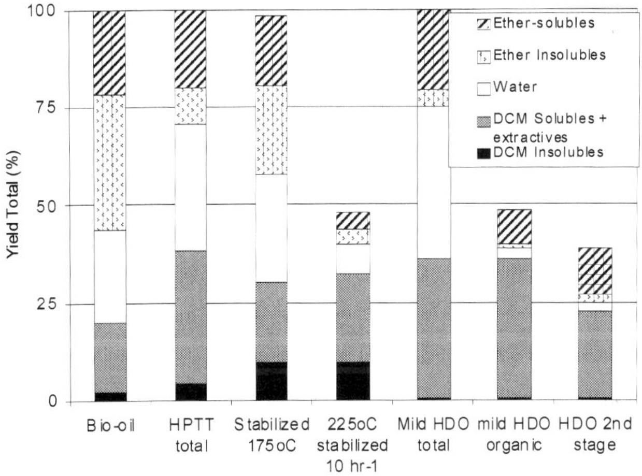

Fig. 8. Comparison of the fractionation results for various process severities

3.2.3.1 Thermal reactions

When comparing the composition of the pyrolysis oil feed with the product from the thermal route, it is clear that the ether insolubles are converted to DCM-solubles and -insolubles, and additional water. A similar change occurs in wood oils, stored for several months or years, where water insoluble products are produced at the expense of the sugar fraction (Oasmaa&Kuoppala, 2003). At higher temperatures and residence times, especially this sugar fraction is responsible for charring, likely through the formation of first DCM solubles and subsequently DCM insolubles ('char'). Solids production upon heating aqueous solution of C-6 sugars (e.g. D-glucose, D-mannose) to temperatures up to 400 °C is well known. Thermal decomposition, either catalytic (mostly by acids) or non-catalytic, leads to solid products referred to as humins (Girisuta et al., 2006; Watanabe et al., 2005a; Watanabe et al., 2005b). The proposed reaction pathway consists of C-6 sugar conversion to

5-hydroxymethyl furfural (HMF) and subsequently levulinic acid (LA) and formic acid (FA). Both reactions also accompanied by solids (humin) formation (Scheme 1).
Solids formation is highly undesirable and limits the yields of the two promising biobased chemicals LA and HMF. Despite large research efforts, it has so far not been possible to avoid solids/humin formation when performing the reactions in aqueous media.

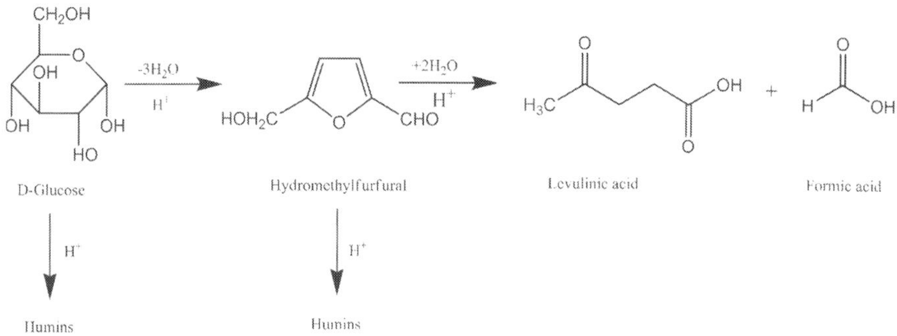

Scheme 1. Decomposition reactions of D-glucose at elevated temperatures.

Higher temperatures and the presence of acid catalysts (homogeneous and heterogeneous) increase the rate of D-glucose decomposition (Girisuta et al., 2006). Such reactions may also occur in the fast pyrolysis oil matrix. The oil is acidic in nature due to the presence of organic acids and these will catalyse the depolymerisation of oligmeric sugars to D-glucose and other C-6 sugars followed by the reaction to solids and hydroxymethylfurfural and levulinic acid/formic acid.
Knezevic et al. (2009) studied the thermal decomposition of D-glucose in hot compressed water under conditions of relevance for the catalytic hydrotreatment of pyrolysis oil (240-374 °C). It was shown that D-glucose decomposes mainly to char and some gaseous components (primarily CO_2), while only a limited number of components remained in the water phase (for example formaldehyde). At these conditions, the reactions are very fast and decomposition to char takes place on the time scale of seconds to minutes.

3.2.3.2 Catalytic hydrotreatment reactions

The composition of the product from a mild hydrogenation at 175 °C (see Figure 8) differs considerably from that of the original pyrolysis oil. The amount of water increased slightly (from 25 up to about 30 wt.%), which appears insufficient for phase separation. In addition, the ether solubles (aldehydes, ketones, acids, etc) are converted, but in smaller amounts compared to HPTT. The ether insoluble (sugar fraction) is reduced considerably from 35 down to 24 wt.%, while the water insoluble fraction is increased accordingly. Simultaneously, the increase of the DCM insoluble fraction is about 8%, while the DCM soluble fraction increases with only 3 wt.%.
Similar to HPTT, we assume that the sugar fraction in the oils is (partially) converted to more water insolubles and some additional water. However, the actual components formed during mild hydrotreatment are different in nature than the HPTT oils and particularly the amount of DCM insolubles is higher.

The results of the fractionation of the product oil derived from an experiment 225 °C (mild hydrogenation) are provided in Figure 8. Phase separation occurs and as such the amount of product oil is reduced considerably. As a result, the amounts of water, ether solubles and ether insolubles in the organic phases are lowered and imply that components have been transferred to the water phase. Figure 8 also shows the result for the mild HDO reaction. Compared with the oil samples obtained at lower temperatures, the DCM insoluble fraction is now almost completely converted to DCM soluble components, evidence that some hydrocracking reaction have taken place here as well. In the 2nd stage hydroprocessing the amount of ether solubles increases, at the expense of DCM solubles and the extractives.

3.3 Product characteristics

In all hydrogenation experiments except those at temperatures below 200°C, the product obtained consisted of two liquid phases, viz. an aqueous phase and brown-red organic phase. For all of them, relevant (basic) characteristics were determined, viz. elemental composition (vide supra, Figure 7), water content and average molecular weight. Additionally, to get some insights in the coking tendency, the samples were analyzed using thermogravimetric analysis (TGA). Here, the residual weight of the sample, heated under N_2 up to about 900°C, was taken as a measure of coking. A high residue indicates a high tendency for coking and thus a low thermal stability at elevated temperature. The residue after a TGA measurement is a strong function of the process severity, see Figure 9 for details.

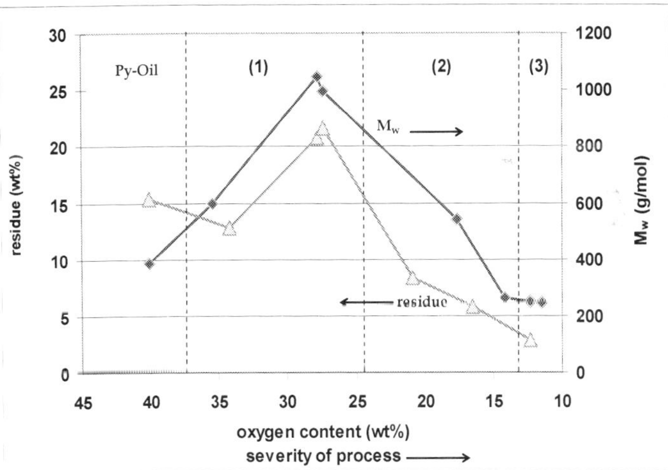

Fig. 9. Mass average molecular weight and TGA residue of products from (1) stabilisation, (2) mild hydrotreating, and (3) 2-stage hydrotreating.

At low process severities, the TGA residue increases and the highest value (22%) is observed at intermediate severities. A further increase in severity leads to a strong reduction in the TGA residue. Thus, it may be concluded that intermediate severities lead to product oils with a high TGA residue and consequently have a higher tendency for coking and may be less suitable as a refinery feedstock.

The organic products were analyzed using gel permeation chromatography (GPC) to determine the average molecular weights and the results are given in Figure 9. The molecular weight of the product oils increases compared to the pyrolysis oil feed at low severity hydrotreatment reactions. Apparently, polymerisation occurs and this has also been observed when heating up pyrolysis oil to 275°C in the absence of catalysts (HPTT process) (Rep et al., 2006). A further increase in the severity (higher temperatures, shorter WHSV's) leads to a reduction of the molecular weight and a value of less than 300 is observed at the highest severities.

Of particular interest is the relation between the molecular weight of the products and the TGA residue. Products with a higher Mw also lead to higher TGA residues and this may be rationalized by assuming that the higher molecular weight fragments in the products are precursors for coke formation.

4. Proposed reaction pathways and implications

4.1 Reaction pathways

A schematic and simplified representation of relevant reactions assumed on basis of this work is presented in Figure 10. In the initial phase of the hydrotreatment process, catalytic hydrogenation and thermal, non-catalytic repolymerisation occur in a parallel mode.

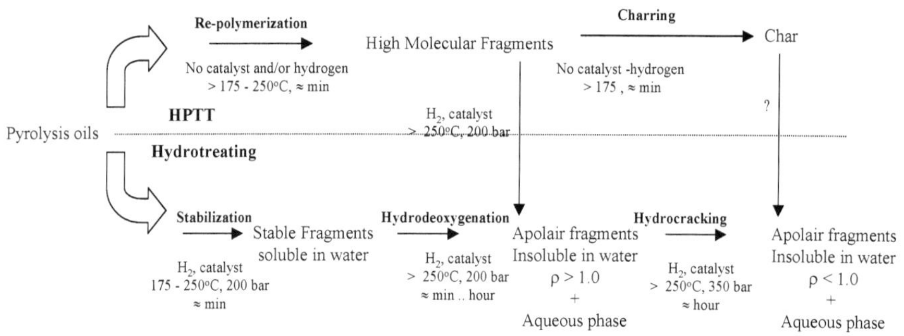

Fig. 10. Proposed pathways for the catalytic hydrotreatment of pyrolysis oils

Repolymerisation leads to the formation of soluble higher molecular weight fragments which upon further condensation reactions give char. This route is as such not preferred and the rate of the polymerisation reactions should be reduced as much as possible. The preferred pathway involves hydrogenation of the thermally labile components in the pyrolysis oil feed to stable molecules that are not prone to polymerisation. Subsequent reactions (hydrogenations and hydrocracking) on a time scale of hours lead to products with reduced oxygen contents and ultimately to higher H/C ratio's (Figure 7). The observed molecular weight of the organic phase as a function of the process severity (Figure 9) implies that upon the use of Ru/C as the catalyst, the repolymerisation step cannot be avoided, and a slight increase in molecular weight is observed at low process severities. However, higher severities lead to a reduction in the average molecular weight, an indication that soluble higher molecular weight fragments may also be (partly) depolymerised by the action of hydrogen and a catalyst.

As stated earlier, pyrolysis oil contains large amounts of oligo- and monomeric sugars, arising from the cellulose and hemi-cellulose fraction of the lignocellulosic biomass feed. As such, it is of interest to compare the reaction pathways provided in Figure 10 for pyrolysis oil with that of typical hydrogenation and thermal reactions occurring for carbohydrates at various process severities.

Thermal decomposition of various monomeric sugars in aqueous media has been studies in detail and is known to lead to oligomerisation to soluble and subsequently to insoluble humins (Girisuta et al., 2006). As an example, Knezevic et al. (2009) studied the thermal decomposition of D-glucose in hot compressed water at elevated temperatures (240-374 °C), giving solids (char, humins) and some gaseous components (primarily CO_2). At these conditions, the reactions are very fast and decomposition to char takes place on the time scale of seconds to minutes.

Catalytic hydrotreatment of carbohydrates using heterogeneous catalysts has been reported extensively in the literature. The main focus is on the hydrogenation of D-glucose to D-sorbitol, a well-known chemical with use in the pharmaceutical and the food industry (Kusserow et al., 2003). Catalytic hydrotreatment of D-glucose over Ni, Ru based and Pd based heterogeneous catalysts at 80 °C, 80 bar yields D-sorbitol in high yields (Crezee et al., 2003; Makkee et al., 1985) (Scheme 2). The hydrogenation reactions at these low temperature levels may be considered as the stabilisation step in fast pyrolysis oil upgrading.

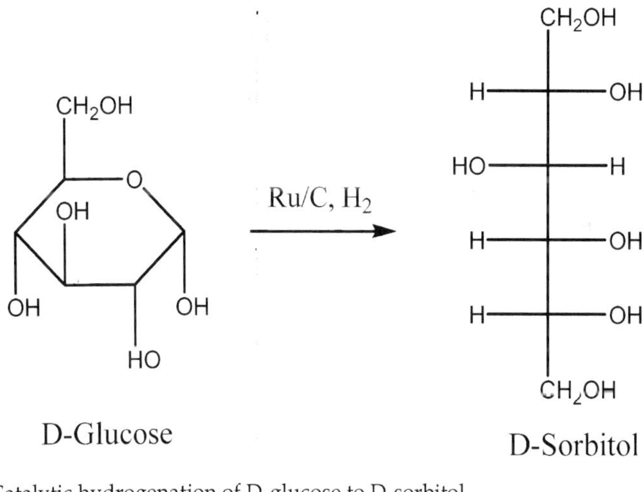

Scheme 2. Catalytic hydrogenation of D-glucose to D-sorbitol

In the presence of hydrogen and a catalyst, D-sorbitol is not inert at elevated temperatures (above 180 °C) and may be converted to a variety of products. For instance, Huber et al. (2004) showed that D-sorbitol can be converted to n-hexane in high yield using Pd and Pt catalyst on SiO_2 or Al_2O_3 (225-265 °C and 26-58 bar). Over Ru/SiO_2, hydrogenolysis of D-sorbitol at 180-240 °C and 80-125 bar hydrogen pressure yields mainly glycerol and 1,2-propanediol (Sohounloue et al., 1982) (Scheme 3). This implies that C-C bond cleavage occurs readily, leading to the formation of lower molecular weight products.

These reactions are likely also occurring upon the catalytic hydrotreatment of fast pyrolysis and may explain the formation of more apolar lower molecular weight products at higher process severities.
Thus, it may be concluded that the typical reaction pathways for pyrolysis oils at typical low severity hydrotreatment conditions mimic those of low molecular weight sugars viz. repolymerisation reactions to solids (humins) and hydrogenation/C-C bond cleavage reactions to for instance polyols and finally to hydrocarbons. This strengthens our initial hypothesis that pyrolysis oil should be regarded as a carbohydrate rich "syrup" and not a conventional fossil derived hydrocarbon liquid.

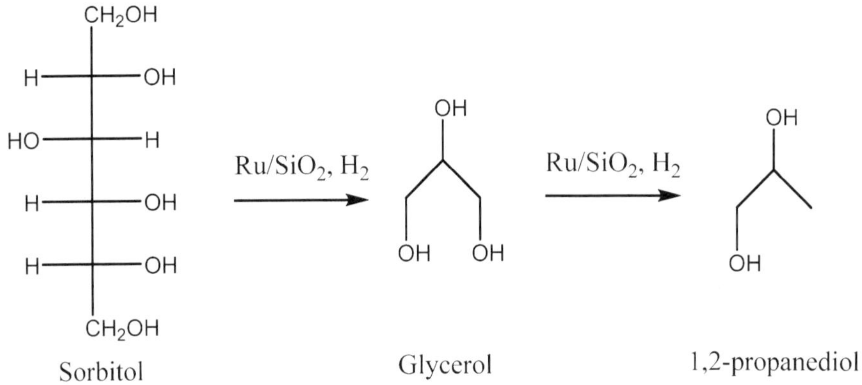

Scheme 3. Hydrogenolysis of D-sorbitol to glycerol and 1,2-propanediol.

4.2 Process implications

The proposed reaction pathway for catalytic hydrotreatment of pyrolysis oil (Figure 10) implies that the rate of the hydrogenation route should be much higher than the rate of the repolymerisation route to obtain good quality upgraded pyrolysis oil (low molecular weight, low viscosity, low coking tendency). An obvious solution is the development of highly active hydrogenation catalysts. These studies will be reported in the next paragraph of this paper. However, a smart selection of process conditions and reactor configurations may also be considered, particularly to enhance the rate of the hydrogenation/hydrodeoxygenation pathway compared to the repolymerisation pathway. In this respect, it is highly relevant to gain some qualitative insights in the factors that determine the rate of the individual pathways (hydrogenation versus repolymerisation).
A schematic plot is presented in Figure 11, where an envisaged reaction rate (arbitrary values, in mole reactant/min) is presented versus the actual reaction temperature. The lines drawn are taken in case (i) gas-to-liquid mass transfer determines the overall reaction rate (ii) the catalytic hydrotreatment reactions dominate, and (iii) polymerisation reactions prevail. Figure 11 is derived on basis of simplified kinetics for the glucose hydrogenation - polymerisation reactions, but a detailed outline of all assumptions made is beyond the scope of the presentation here. For this reason the exact values on the x- and y-axes are omitted. The following relations are taken into account to derive Figure 11:

- The conversion rate due to the hydroprocessing reactions R_H (mol/m^3_r.s) can be simplified as a product of the intrinsic kinetic rate expression k_R and the surface area available per reactor volume. Being a catalytic reaction, the influence of temperature can be rather high.

- The overall gas-to-solid mass transfer rate of hydrogen depends on reactor geometry and operating variables. In case of stirred tank reactors (including batch-wise operated autoclaves), the actual stirring rate is important, while in packed bed the catalyst particle wetness is relevant. In both, the concentration of hydrogen (thus hydrogen pressure) is important, together with catalyst particle size, and, to a limited extent, temperature.
- The rate of polymerisation, R_P, will depend largely on the temperature, and, being a reaction with order in reactant(s) > 1 (and probably up to 2 or 3), on the concentration of the reactant.

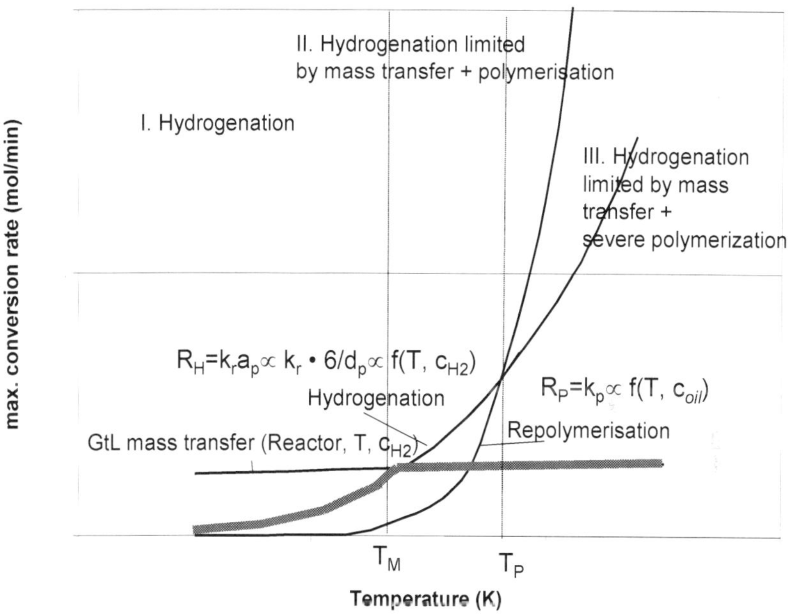

Fig. 11. Schematic indication of the conversion rates in case of mass transfer, hydrogenation and polymerisation reactions as a function of temperature. The blue solid line would be the net effective conversion rate.

A number of options may be envisaged to promote the hydrogenation pathway:

- Increase the hydroprocessing reaction rate, for instance by a higher catalyst intake or by an increase in the effective hydrogen concentration in the liquid (pressure, application of a solvent with a high hydrogen solubility).
- Reducing the polymerisation reaction, a.o. by performing the initial stabilisation step at a low temperature (< 100°C) and reduction of the concentration of the reactants (a.o. by dilution).

- Increase the overall gas-to-liquid mass transfer rate in case the reaction is performed in the gas-liquid mass transfer limited regime. This may be possible by increasing the mass transfer surface area in the reactor, higher mass transfer coefficient and / or increasing the concentration difference between the gas and the liquid.

5. Improved catalyst formulations for the catalytic hydrotreatment of fast pyrolysis oil

The development of highly active metal catalysts is of prime importance to reduce the tendency for repolymerisation during catalytic hydrotreatment. All data presented in this chapter so far are based on a Ru/C catalyst. Ru, however, is an expensive noble metal and there is an incentive to identify not only more active but also cheaper catalysts for the hydrotreatment reaction. A possibility is the use of cheaper bimetallic metal catalysts based on Ni. Ni is known to have high hydrogenation activity for a variety of organic functional groups and particularly for reactive ketones and aldehydes, and as such is a potential active metal for hydrotreatment reactions. However, monometallic Ni catalysts (on silicon oxide, γ- or δ-alumina, or other supports) at the typical temperature and pressures applied here are not suitable to be used as a hydrogenation catalyst. There are basically two reasons: 1) Ni requires high reduction temperature (typically 700 °C) for complete reduction, and 2) Ni catalysts are known to deactivate rapidly at elevated conditions by char deposition ("coking"). The carbon deposition can block the nickel surface, or the pore mouths, and, eventually leading to a strong reduction in the reaction rates. These two drawbacks regarding the use of Ni were solved a.o. by using another element (metal or non-metal), also designated as a promoter. One of these proprietary catalysts was studied in detail and will be referred to in the following as catalyst D.

Figure 12 shows the liquid phase after a hydrotreatment over Cat D versus the severity of the process, showing the original oil (left) and an oil derived at the most severe conditions tested here on the right. Interestingly, the product oils obtained over cat D are much more transparent than those derived from the Ru/C catalyst.

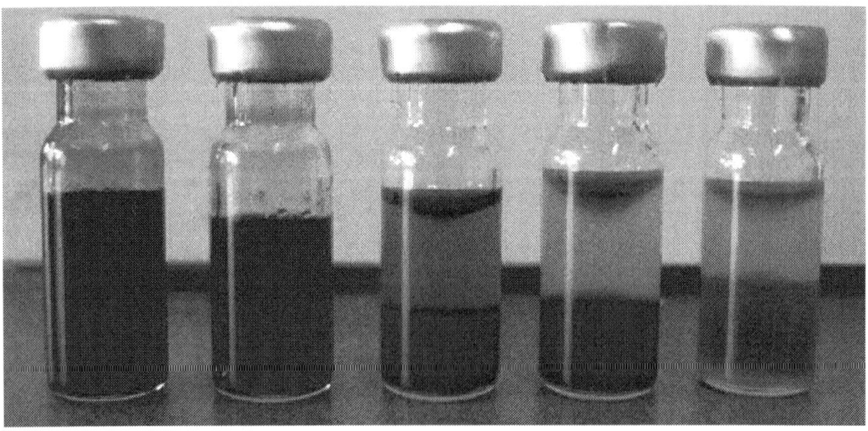

Fig. 12. Visual appearance of the liquid phase after hydrotreatment over catalyst D

A van Krevelen plot gives valuable insights in the difference in performance between catalyst D and Ru/C (Figure 13). A similar pattern for both is observed as a function of severity but the curve for Cat D is shifted to higher H/C values. Thus, at a similar oxygen content, the H/C ratio is higher for catalyst D. This is indicative for a higher hydrogenation rate for cat D and is known to be favorable regarding product properties.
Repolymerisation reactions appear to occur to a limited extent when using cat D instead of Ru/C. This is evident when comparing the average molecular weight of the final products (Figure 14a), as determined by GPC, for both Cat D and Ru/C. For Ru/C the average molecular weight shows a significant increase from 400 up to 1000 Da at low severities, but a constant value over the oxygen content interval of 400-450 Da for catalyst D is observed.
TGA residues of the product oils using cat D (Figure 14b) show carbon residues of around 5%. Surprisingly, and not expected on basis of test carried out using other catalysts, already at less severe operating conditions, a significant reduction in the TGA residues is achieved. Thus, products with a higher H/C ratio and a lower carbon residue were obtained with cat D indicating that the rate of the hydrogenation/hydrodeoxygenation reactions over catalysts D are higher than for Ru/C.

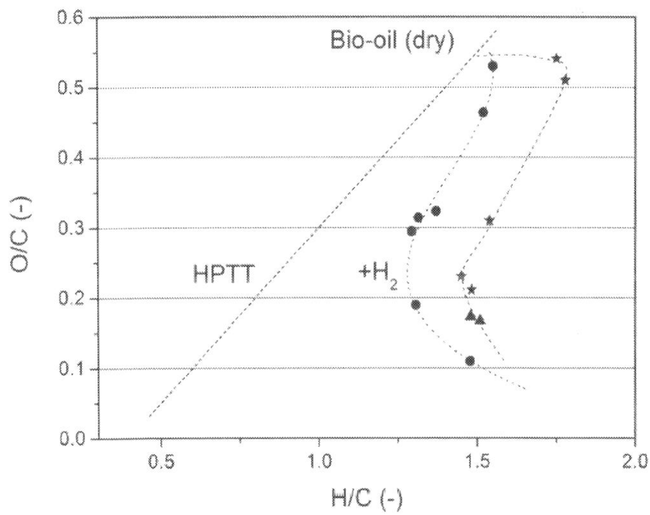

Fig. 13. Van Krevelen plot for oils derived over catalyst Ru/C (circles) and over catalyst D (stripes). Lines are trendlines.

An important product property for the upgraded oils is the viscosity. In Figure 15, the viscosity profile of the product oils versus the oxygen content is compared for conventional catalysts (Ru/C and NiMo, CoMo) and catalyst D. Clearly, the viscosity in the mid range of oxygen contents is much lower for cat D. Further testing at the extreme of low oxygen content will be required to grab the full picture but it is clear that cat D gives upgraded products with a lower viscosity than for conventional catalysts. The lower viscosity is likely the result of a lower average molecular weight of the products, as shown earlier and the result of higher hydrogenation/hydrodeoxygenation rates for Cat D compared to Ru/C.

On the basis of the product properties of the upgraded oils obtained with cat D, we can conclude that repolymerisation is not occurring to a considerable extent. As a result, product oils with a lower molecular weight and a concomitant lower viscosity, lower TGA residue is obtained. Thus, the reaction pathway for catalyst D may be simplified considerably, see Figure 16 for details.

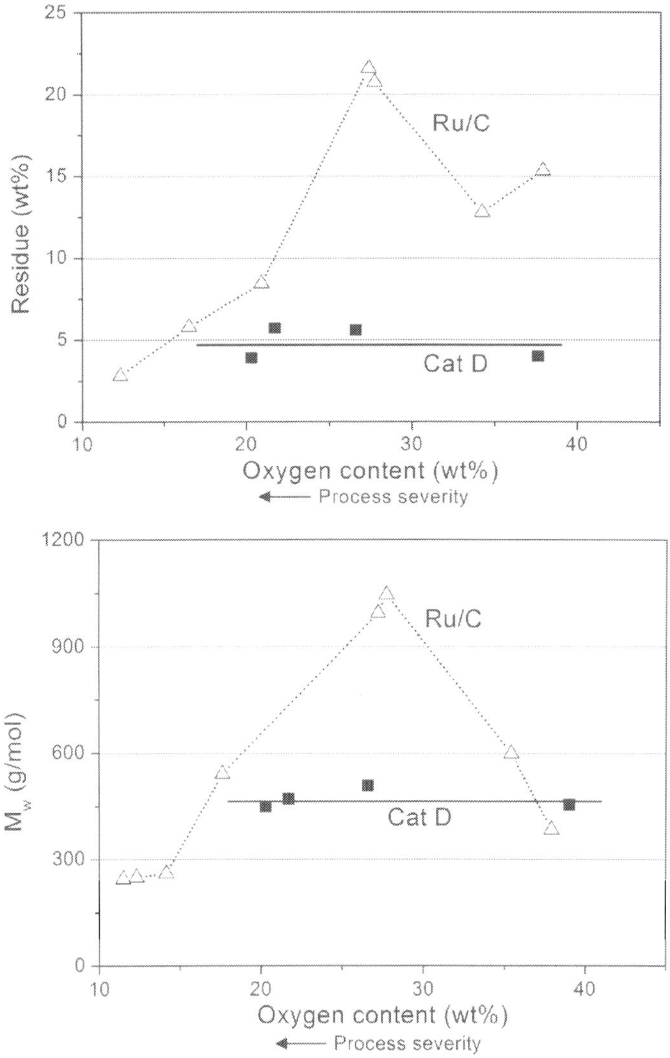

Fig. 14. TGA residue (wt%) for catalyst Ru/C and the catalyst D (top); average molecular weight of final product (bottom)

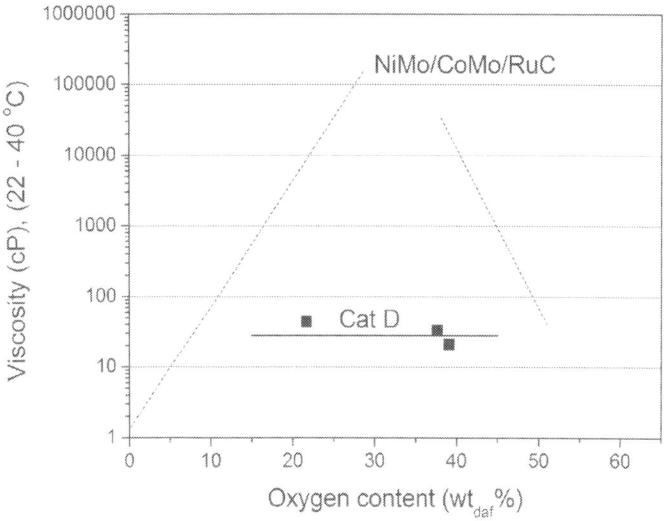

Fig. 15. Viscosity of the oil versus the oxygen content for conventional catalysts and for Cat D.

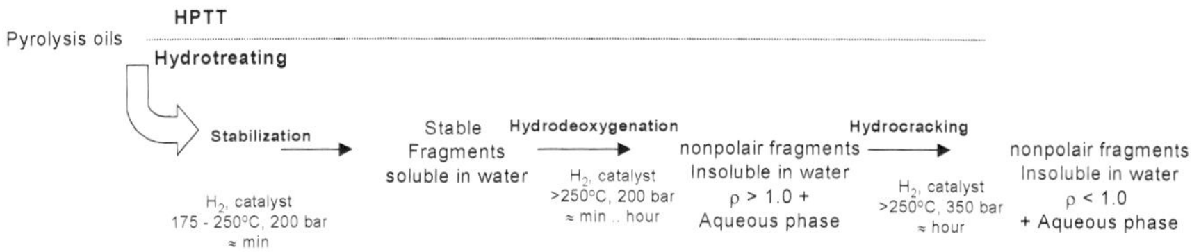

Fig. 16. Proposed pathways for the mild hydrotreating of pyrolysis oils over catalyst D.

6. Application potential of upgraded oils and critical product properties

The research described here is principally meant to produce pyrolysis oil derived products with the potential to be co-fed in crude oil refineries. It is of interest to discuss some of the relevant product properties that likely are crucial for this application. An important product property of the product oil is its tendency to produce coke upon heating, for example determined by the 'Conradson Carbon Residue' (CCR) or residue upon an thermographimetric analysis ('TGA residue', *vide supra*) (Furimsky, 2000). In general, pyrolysis-oils show CCR values around 20 to 30% (Samolada, 1998) and this limits its direct use as a co-feed. The TGA residues for upgraded oils obtained with Ru/C and cat D as presented in this paper, show values between 3 and 22 wt% (see Figure 14) and are tunable by process severity (temperature, residence/batch time). It should be realized that low TG residue values are accessible for hydrotreated products which still contain considerable amounts of bound oxygen (> 10 wt%). Thus, stable oils may be prepared despite relatively high bound oxygen contents. From a processing point of view this is also advantageous as it limits the hydrogen usage for the catalytic hydrotreatment process, a major variable cost contributor.

Recently investigations have been reported on the co-processing of hydrotreated pyrolysis oils obtained with a Ru/C catalyst (oxygen contents > 5 wt%) in a lab scale simulated FCC unit (MAT) (de Miguel Mercader et al., 2010). The hydrotreated products were successfully dissolved and processed in Long Residue (20 wt% upgraded oil). The yields of FCC gasoline (44-46 wt.%) and Light Cycle Oil (23-25 wt.%) were close to the base feed and an excessive increase of undesired coke and dry gas was not observed. Experiments with undiluted upgraded oils were less successful and dry gas and coke yield were significantly higher than in case of co-feeding. This clearly demonstrates that co-processing is necessary to obtain good product yields. This study also shows that, in contrast to initial thoughts, it is likely not necessary to aim for an upgraded oil with an oxygen content lower than 1 wt%.

Further MAT testing with product oils derived from catalyst D is in progress and will allow the establishment of detailed process-product relations for co-feeding purposes. The upgraded oils prepared with cat D have a much higher thermal stability than the original pyrolysis oil, as evident from the TGA residues (Figure 14). Preliminary investigations have shown that this allows distillation of the oil in various fractions without the formation of excessive amounts of char. Further detailed studies are in progress and will be reported in due course.

7. Conclusions

The upgrading of pyrolysis oil by catalytic hydrotreatment reactions using heterogeneous catalysts was studied in detail using a Ru/C catalyst. The investigations provided valuable insights in the chemical transformations occurring during catalytic hydrotreatment and include both thermal and hydrogenation/hydrodeoxygenation pathways. The repolymerisation pathway in which the oils are further condensed to soluble oligomers and eventually to char components competes with a catalytic hydrogenation reaction. In case H_2 is present with a proper catalyst, these soluble oligomers may be depolymerised to stabilized components that can be further upgraded. In this respect, the pyrolysis oils show reactivity typically observed for carbohydrates, which is rationalised by considering the high amounts of oligo- and monomeric sugars in the oil. New, highly active Ni-based catalysts have been developed, which show much better performance than conventional ones and provide products with improved properties.

Experimental work is foreseen to elucidate the reaction pathways occurring during catalytic hydrotreatment in more detail and to develop efficient processes to obtain a stabilized oil with the desired product properties at the lowest manufacturing costs. These include:

- Determination of the effects of reactor configuration on the reaction rates (including mass transfer issues) and subsequent reactor selection;
- Determination of relevant physical properties (e.g hydrogen solubility);
- Optimisation of the hydroprocessing conditions and particularly the required hydrogen levels
- Determination of product-process relations. The effective hydroprocessing severity required for further co-refining must be defined;
- The source and availability for hydrogen: perhaps even syngas is applicable;
- Effects of reaction exothermicity need to be determined.

8. Acknowledgement

The authors would like to acknowledge the EU for partial funding of the work through the 6th Framework Program (Contract Number: 518312). We also would like to thank the partners of BIOCOUP project, and especially D. Assink (BTG), A. Ardiyanti and J. Wildschut (RuG, The Netherlands) for performing (part of) the experiments and mutual interpretation of results, VTT (Finland) and particularly A. Oasmaa for the fractionation analysis and V. Yakovlev and S.A. Khromova of the Boreskov Institute for Catalysis (Russia) for catalyst preparation and stimulating discussions. Furthermore, financial support from SenterNovem (in CORAF, project no. EOSLT04018, and NEO-project, no. 0268-02-03-03-0001) is gratefully acknowledged.

9. References

Ardiyanti, A.R., Venderbosch, R.H., Heeres, H.J., (2009) Process-product studies on pyrolysis oil upgrading by hydrotreatment with Ru/C catalysts, Proceedings of the 2009 AIChE Spring National Meeting, Tampa.

Baldauf W., Balfanz U., Upgrading of fast pyrolysis liquids at Veba Oel AG, in: *Biomass gasification & pyrolysis: state of the art and future prospects,* Kaltschmidt M. and Bridgwater A.V, pp. 392-397, ISBN 1-872691-71-4.

Churin E., Maggi R., Grange P., Delmon B., (1988) Characterization and upgrading of a bio-oil produced by pyrolysis of biomass, in: *Research in Thermochemical biomass conversion,* Bridgwater A.V., Kuester J.L., pp. 896-909, Elsevier Science Publishers.

Conti L., (1997) Improvement of the flash pyrolysis process and pilot plant for bio-oils upgrading, Final Report Contract AIR 1-CT92-0216.

Crezee, E., Hoffer, B.W., Berger, R.J., Makkee, M., Kapteijn, F., Moulijn, J.A. (2003). Three-phase hydrogenation of D-glucose over a carbon supported Ruthenium catalyst-mass transfer and kinetics. *Appl. Catal. A-Gen.*, 251(1), pp. 1-17.

Diebold, J.P., (2002) A review of the chemical and physical mechanisms of the storage stability of fast pyrolysis biooils. In: *Fast Pyrolysis of Biomass: A Handbook, Vol. 2.,* Bridgwater, A.V., CPL Press, Newbury, UK.

Dimitrijevic R., Lutz, W., Ritzmann, A. (2006) Hydrothermal stability of zeolites: Determination of extra-framework species of H-Y faujasite-type steamed zeolite, *Journal of Physics and Chemistry of Solids,* 67, pp. 1741-1748.

Elliott D.C., (2007), Historical developments in hydroprocessing bio-oils, *Energy and Fuels,* 21, pp. 1792-1815.

Elliott D.C., Hart T.R., Neuenschwander G.G., Rotness L.J., Zacher A.H., (2009) Catalytic hydroprocessing of Biomass fast Pyrolysis Bio-Oil to produce hydrocarbon Products, *Environmental Progress & Sustainable Energy,* 28(3), pp. 441-449.

Ferrari, M., Delmon, B., Grange, P., (2002) Influence of the impregnation order of molybdenum and cobalt in carbon-supported catalysts for hydrodeoxygenation reactions, *Carbon,* 40, pp. 497-511.

Furimsky, E., (2000) Catalytic hydrodeoxygenation, *Applied Catalysis A: General* 199, pp. 147-190.

Girisuta, B., Janssen, L.P.B.M., Heeres, H.J. (2006). A kinetic study on the decomposition of 5-hydroxymethylfurfural into levulinic acid. *Green. Chem.*, 8(8), pp. 701-709.

Gutierrez, A., Kaila, R.K., Honkela, M.L., Slioor, R., Krause, A.O.I., (2009) Hydrodeoxygenation of guaiacol on noble metal catalysts, Catal. Today, 147, pp. 239-246.

Horne P., Williams P.T., (1996) Reaction of oxygenated biomass pyrolysis compounds over a ZSM-5 catalyst, *Renewable Energy*, 2, pp. 131-144.

Huber, G.W., Cortright, R.D., Dumesic, J.A. (2004). Renewable alkanes by aqueous-phase reforming of biomass-derived oxygenates. *Angew. Chem.*, 43, pp. 1549-1551.

Kaiser M., (1997), Upgrading of fast pyrolysis liquids by DMT, in: *Biomass gasification & pyrolysis: state of the art and future prospects*, Kaltschmidt M., and Bridgwater A.V., pp. 399-406, ISBN 1-872691-71-4.

Knezevic, D., van Swaaij, W. P. M., and Kersten, S. R. A. (2009). Hydrothermal conversion of biomass: I, Glucose conversion in hot compressed water. *Ind. Eng. Chem. Res.*, 48(10], pp. 4731-4743.

Kusserow, B., Schimpf, S., Claus, P. (2003). Hydrogenation of glucose to sorbitol over nickel and ruthenium catalysts. *Adv. Synth. Catal.*, 345, pp. 289-299.

Lee, C.L., Ollis, D.F. (1984) Interactions between Catalytic Hydrodeoxygenation of Benzofuran and Hydrodesulfurization of Dibenzothiophene, *Journal of Catalysis* 87, pp. 332-338.

Mahfud, F.H., Bussemaker, S., Kooi, B., ten Brink G., Heeres, H.J., (2007a), The application of water soluble Ruthenium catalysts for the hydrogenation of the dichloromethane soluble fraction of fast pyrolysis oil and related model compounds in a two phase aqueous-organic system. *J. Mol. Cat. A.* 277, pp. 127-136.

Mahfud, F.H., Ghijsen, F., Heeres, H.J., (2007b) Hydrogenation of fast pyrolysis oil and model compounds in L-L biphasic systems using homogeneous ruthenium catalysts, *J. Mol. Cat. A.*, 264, pp. 227-236.

Maity, S.K., Rana, M.S., Srinivas, B.N., Bej, S.K., Murali Dhar, G., Prasada Rao, T.S.R., (2000), Characterisation and evaluation of ZrO2 supported hydrotreating catalysts, *Journal of Molecular Catalysis A: Chemical*, 153, pp. 121-127.

Makkee, M., Kieboom, A.P.G., Van Bekkum, H. (1985). Hydrogenation of D-fructose and D-glucose mixtures. *Carbohydr. Res.*, 138(2), pp. 225-236.

Mercader, F.M., Groeneveld, M.J., Kersten, S.R.A., Way, N.W.J., Schaverien, J.C, Hogendoorn,J.A., (2010) Production of advanced biofuels: Co-processing of upgraded pyrolysis oil in standard refinery units, *Applied Catalysis B: Environmental*, 96, pp. 57-66.

Miguel Mercader F., Hogendoorn J.A., Venderbosch R.H., Kersten S.R.A., Groeneveld M.J., (2010) Pyrolysis oil upgrading by high pressure thermal treatment, *Fuel*, 89, pp. 2829-2837.

Oasmaa A., Kuoppala E., Solantausta Y., (2003), Fast pyrolysis of forestry residue: 2. Physicochemical composition of product liquid. *Energy & Fuels*, 17(2), pp. 433.

Oasmaa A., Kuoppala E. (2003) Fast pyrolysis of forestry residue. 3. Storage stability of liquid fuel, *Energy & Fuels*, 17(4), pp. 1075-1084.

Oasmaa A., (2003) Fuel Oil Quality Properties of Wood-Based Pyrolysis Liquids, thesis University of Jyväskylä, Finland.

Rep M., Venderbosch R.H., Assink D., Tromp W., Kersten S.R.A., Prins W., Van Swaaij W.P.M., (2006) De-oxygenation of bio-oils, in *Science in thermal and chemical biomass conversion*, Bridgwater A.V and Boocock D.G.B., pp. 1526-1535, CLP Press.

Samolada M.C., Baldauf W., Vasalos I.A., (1998) Production of a bio-gasoline by upgrading biomass flash pyrolysis liquids via hydrogen processing and catalytic cracking. *Fuel*, 77(14), pp., 1667-75.

Satterfield, C.N., Yang, S.H., (1983), Some Effects of Sulfiding a NiMo Catalyst on its Activity for Hydrodenitrogenation of Quinoline, *Journal of Catalysis*, 81, pp. 335-346.

Senol, O.I., Viljava, T.R., Krause, A.O.I, (2005) Hydrodeoxygenation of methyl esters on sulfided NiMo/Al_2O_3 and CoMo/Al_2O_3 catalysts, *Catalysis Today*, 100, pp. 331-335.

Sohounloue, D.K., Montassier, C., Barbier, (1982). Hydrogenolysis of sorbitol. *React. Kinet. Catal. Lett.*, 22, pp. 391-397.

Venderbosch R.H., Prins W.,(2010), Fast Pyrolysis Technologies, *Biofuels Bioproducts and Biorefining*, 4 (2), pp. 178 - 208.

Venderbosch, R.H., Ardiyanti, A.R., Wildschut, J., Oasmaa, A., Heeres, H.J. (2010) Stabilization of biomass derived pyrolysis oils, *Journal of Chemical Technology & Biotechnology*, 85, pp. 674-686.

Vispute, T.P.; Zhang, H.Y.; Sanna, A.; Xiao, R.; Huber, G.W.; (2010) Renewable Chemical Commodity Feedstocks from Integrated Catalytic Processing of Pyrolysis Oils, *Science* 330 (6008), pp. 1222-1227.

Watanabe, M., Aizawa, Y., Iida, T., Aida, T.M., Levy, C., Sue, K, Inomata, H. (2005a). Glucose reactions with acid and base catalysts in hot compressed water at 473 K. *Carbohydr. Res.*, 340(12), pp. 1925-1930.

Watanabe, M., Aizawa, Y., Iida, T., Levy, C., Aida, T.M., Inomata, H. (2005b). Glucose reactions within the heating period and the effect of heating rate on the reactions in hot compressed water. *Carbohydr. Res.*, 340(12), pp. 1931-1939.

Wildschut J., Arentz J., Rasrendra C.B., Venderbosch R.H., Heeres H.J., (2009a) Catalytic Hydrotreatment of Fast Pyrolysis oil: Model studies on Reaction Pathways for the Carbohydrate Fraction, *Environmental Progress & Sustainable Energy*, 28(3), pp. 450-460.

Wildschut, J., Mahfud, F.H., Venderbosch, R.H., Heeres, H.J., (2009b) Hydrotreatment of Fast Pyrolysis Oil Using Heterogeneous Noble-Metal Catalysts, *Industrial & Engineering Chemistry Research* 48, pp. 10324-10334.

Wildschut, J., Melián-Cabrera, I., Heeres, H.J., (2010a) Catalyst studies on the hydrotreatment of fast pyrolysis oil, *Applied Catalysis B: Environmental*, 99, pp. 298-306.

Wildschut, J., Iqbal, M., Mahfud, F.H., Melian Cabrera, I., Venderbosch, R.H., Heeres, H.J., (2010b) Insights in the hydrotreatment of fast pyrolysis oil using a Ru on C catalyst., *Energy and Environmental Science*, 3, pp. 962-970

Xu, Y., Wang, T., Ma, L., Zhang, Q., Liang W., (2010), Upgrading of the liquid fuel from fast pyrolysis of biomass over MoNi/γ-Al2O3 catalysts, *Applied Energy*, 87, pp. 2886-2891.

18

Biomass Feedstock Pre-Processing – Part 1: Pre-Treatment

Lope Tabil[1], Phani Adapa[1] and Mahdi Kashaninejad[2]
[1]Department of Chemical and Biological Engineering, University of Saskatchewan
[2]Department of Food Science & Technology, Gorgan University of Agricultural Sciences and Natural Resources Gorgan
[1]Canada
[2]Iran

1. Introduction

1.1 Sources of biomass

The two main sources of biomass for energy generation are purpose-grown energy crops and waste materials (Larkin et al., 2004). Energy crops, such as Miscanthus and short rotation woody crops (coppice), are cultivated mainly for energy purposes and are associated with the food vs. fuels debate, which is concerned with whether land should be used for fuel rather than food production. The use of residues from agriculture, such as barley, canola, oat and wheat straw, for energy generation circumvents the food vs. fuel dilemma and adds value to existing crops (Chico-Santamarta et al., 2009). In fact, these residues represent an abundant, inexpensive and readily available source of renewable lignocellulosic biomass (Liu et al., 2005).

1.2 Current issues related to biomass utilization

The main problem with agricultural straw is its relatively low density in its original or baled forms. The bulk density of loose and standard baled straw is approximately 40 kg/m^3 and 100 kg/m^3, respectively, compared with the bulk density of unprocessed wood residue, which is approximately 250 kg/m^3 (Demirbaş, 2001; Tripathi et al., 1998). The relative low density of straw makes it more expensive to transport compared to wood and coal because a lower mass of straw can be transported per unit volume. Additionally, a larger storage area/volume is required for baled straw compared to wood chip. Densification into pellets increases the bulk density of biomass (McMullen et al., 2005; Obernberger and Thek ,2004) and as a result, the net calorific content per unit volume is increased (Bhattacharya et al., 1989) and the storage, transport and handling of the material is easier and cheaper (Balatinecz, 1983; Bhattacharya et al., 1989; Kaliyan and Morey, 2006).

The quality of fuel pellet is usually assessed based on its density and durability. High bulk density increases storage and transport capacity of pellets (Adapa et al., 2007; Mani et al., 2003). Since feeding of boilers and gasifiers generally is volume-dependent, variations in bulk density should be avoided (Larsson et al., 2008). A bulk density of 650 kg/m^3 is stated as design value for wood pellet producers (Obernberger and Thek 2004). Low durability of pellets results in problems like disturbance within pellet feeding systems, dust emissions, and increased risk of fire and explosions during pellet handling and storage (Temmerman et al. 2006).

Densification of straw and determining the optimal parameters involved is an art in itself. The entire process involves securing of baled straw from agricultural fields, size reduction (chopping and grinding), application of pre-treatment (chemical, physico-chemical, and biological), determining the physical and frictional properties of straw grinds, lignocellulosic characterization of straw, lab-scale and pilot-scale densification of grinds into pellets to determine the effect of various independent parameters on quality (density and durability), and energy analysis/ balance (Fig. 1). This chapter will only address the effect and need of lignocellulose characterization, pre-treatment and size reduction, and physical properties on densification of agricultural straw.

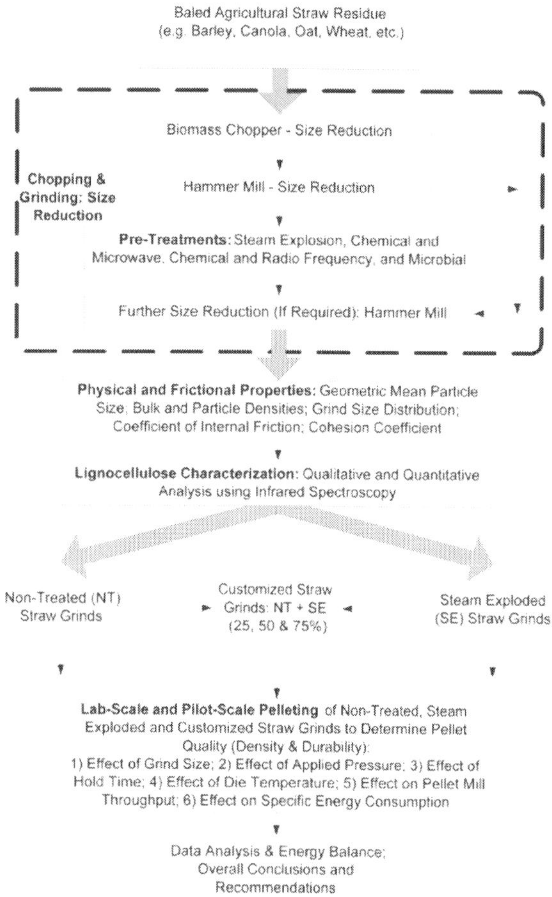

Fig. 1. Processing steps involved in converting straw from field to pelletized product.

2. Lignocellulosic biomass characterization

2.1 Structure of lignocellulosic material

Lignocellulosic material refers to plant biomass that is composed of cellulose, hemicellulose, and lignin (Fig. 2) (Lin and Tanaka, 2006). The major combustible component of non-food energy crops is cellulose, followed by lignin.

Cellulose: Cellulose is an organic polysaccharide consisting of a linear chain of several hundreds to over nine thousand β(1→4) linked D-glucose ($C_6H_{10}O_5$)n units (Crawford, 1981; Updegraff, 1969). Cellulose, a fibrous, tough, water-insoluble substance, is found in the cell walls of plants, particularly in the stalks, stems, trunks and all the woody portions of the plant body (Nelson and Cox, 2005). Cellulose comprises 40-60% of the dry weight of plant material (the cellulose content of cotton is 90% and that of wood is 50%) (Encyclopædia Britannica, 2008; USDE, 2006).

Zandersons et al. (2004) and Shaw (2008) reported that binding of wood material during hot pressing / densification is mainly dependent on the transition of cellulose into the amorphous state. According to Hon (1989), due to the semi-crystalline structure, hydrogen bonded cellulose cannot be dissolved easily in conventional solvents, and it cannot be melted before it burns; hence, cellulose itself is not a suitable adhesive. This can be overcome by breaking the hydrogen bonds, thus making the cellulose molecule more flexible (Hon 1989). Cellulose requires a temperature of 320°C and pressure of 25 MPa to become amorphous in water (Deguchi et al., 2006).

Hemicellulose: Hemicellulose is made of several heteropolymers (matrix polysaccharides) present in almost all plant cell walls along with cellulose (Fig. 2). While cellulose is crystalline, strong, and resistant to hydrolysis; hemicellulose has a random, amorphous structure with less strength. Hemicellulose is a polysaccharide related to cellulose and comprises 20-40% of the biomass of most plants. In contrast to cellulose, hemicellulose is derived from several sugars in addition to glucose, including especially xylose but also mannose, galactose, rhamnose and arabinose (Shambe and Kennedy, 1985). Branching in hemicellulose produces an amorphous structure that is more easily hydrolyzed than cellulose (Shaw, 2008). Also, hemicellulose can be dissolved in strong alkali solutions. Hemicellulose provides structural integrity to the cell. Some researchers believe that natural bonding may occur due to the adhesive properties of degraded hemicellulose (Bhattacharya et al., 1989).

Lignin: Lignin is a complex chemical compound most commonly derived from wood and is an integral part of the cell walls of plants (Lebo et al., 2001; Zandersons et al., 2004). The compound has several unusual properties as a biopolymer, not the least its heterogeneity in lacking a defined primary structure. Lignin fills the spaces in the cell wall between cellulose and hemicellulose (Fig. 2). It is covalently linked to hemicellulose and thereby crosslinks different plant polysaccharides, conferring mechanical strength to the cell wall and consequently to the whole plant structure (Chabannes et al., 2001).

Lignin acts as a binder for the cellulose fibres (Fig. 2). van Dam et al. (2004) have reported that lignin can be used as an intrinsic resin in binderless board production due to the fact that when lignin melts (temperatures above 140°C), it exhibits thermosetting properties. Lignin is the component that permits adhesion in the wood structure, and is a rigidifying and bulking agent (Anglès et al., 2001). Lehtikangas (2001) reported that water (8-15%) in pellets will reduce the softening temperature of lignin to 100-135°C by plasticizing the molecular chains. The adhesive properties of thermally softened lignin are thought to contribute considerably to the strength characteristics of briquettes made of lignocellulosic materials (Granada et al., 2002; Shaw, 2008).

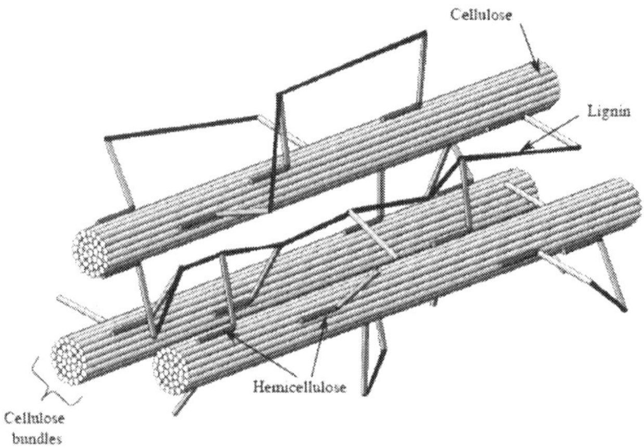

Fig. 2. Location and arrangement of cellulose microfibrils in plant cell walls (Murphy and McCarthy, 2005; Shaw, 2008).

2.2 Rapid characterization of lignocellulosic materials

The effect of various pre-processing and pre-treatment methods (Fig. 1) on the lignocellulosic matrix at the molecular level is not well understood. Applications of pre-processing methods such as size reduction or increasing porosity, and pre-treatment techniques such as steam explosion on agricultural biomass have demonstrated an improvement in pellet (compact) quality that can be attributed to the changes in the lignocellulosic components and distribution (Bagby, 1982; Focher et al., 1998). Therefore, it is critical to rapidly quantify the change in cellulose, hemicelluloses and lignin components of biomass due to application of pre-treatment methods.

Infrared spectroscopy has the potential to produce qualitative and quantitative analytical data for samples with minimum or no sample preparation, and at high speed and throughput (Adapa et al., 2011b and 2009; Budevska, 2002; Luypaert et al., 2003; Smola and Urleb, 2000; Tucker et al., 2000). Traditionally, chemical analyses of the individual components (e.g., lignin) of lignocellulosics have been performed by acid hydrolysis followed by gravimetric determination of lignin (Kelley et al., 2004). These methods can provide highly precise data. However, these methods are laborious, time-consuming, and, consequently, expensive to perform and sample throughput is limited.

2.3 Fourier transform infrared spectroscopy

Fourier Transform Infrared Spectroscopy (FTIR) can be used to rapidly characterize and quantify cellulose-hemicellulose-lignin composition prior to and after application of various methods of pre-processing and pre-treatment of biomass (Adapa et al., 2009). The quantitative analysis of FTIR absorption spectrometry is based on the Bouguer-Beer-Lambert law (Sherman Hsu, 1997). According to this law, the intensities of absorption bands are linearly proportional to the concentration of each component in a homogenous mixture or solution.

Regression equations to predict the lignocellulosic content of agricultural biomass can be developed using pure cellulose, hemicelluloses and lignin as reference samples, and subsequently mixing them in different proportions to determine the change in absorption intensity at characteristic peak height (Adapa et al., 2011b). An overview of the experimental procedure to characterize the lignocellulosic composition is provided in Figure 3.

Pure cellulose has five distinct characteristic/ prominent peaks at wavenumbers of 1431, 1373, 1338, 1319 and 1203 cm^{-1}. Similarly, hemicellulose (xylan) has prominent peaks at wavenumbers of 1606, 1461, 1251, 1213, 1166 and 1050 cm^{-1}. The lignin spectrum has characteristic peaks at wavenumber of 1599, 1511, 1467, 1429, 1157 and 1054 cm^{-1}. The intensity of absorption at characteristic peak heights of cellulose, hemicellulose and lignin were used to develop regression equations to predict lignocellulosic composition of any agricultural biomass (Table 1) (Adapa et al., 2011b).

Fig. 3. Experimental procedure followed to characterize lignocellulosic composition of agricultural straw (Adapa et al., 2011b).

Equation	% Mean Absolute Deviation
$\%Cellulose = -135.10 + 781.35(PH\_1319) - 795.57(PH\_1431) - 135.26(PH\_1203) + 436.11(PH\_1338) - 94.24(PH\_1373)$	7.5
$\%Hemicellulose = 1638.72 - 2581.71(PH\_1251 \times PH\_1461) - 1260.90(PH\_1213) - 2518.05(PH\_1166) + 1573.69(PH\_1213 \times PH\_1251) + 118.74(PH\_1050) + 3128.51(PH\_1166 \times 1251) + 2179.65(PH\_1461) + 92.36(PH\_1606) - 2294.15(PH\_1251) - 59.29(PH\_1461 \times PH\_1606)$	2.5
$\%Lignin = 7110.87 + 388.32(PH\_1511 \times PH\_1599) - 16440.93(PH\_1467) + 447.36(PH\_1599)^2 + 19572.82(PH\_1157 \times PH\_1467) + 18374.36(PH\_1157) + 15659.98(PH\_1054 \times PH\_1429) - 4952.80(PH\_1157 \times PH\_1599) + 800.20(PH\_1511) - 3032.75(PH\_1429)^2 - 11269.16(PH\_1429) - 948.04(PH\_1511)^2 + 3444.69(PH\_1599) - 12344.90(PH\_1054) - 16689.44(PH\_1157)^2$	3.8
Note: PH - Characteristic Peak Height (Photoacoustic Units)	

Table 1. Regression equations to predict the lignocellulosic composition of agricultural biomass (Adapa et al., 2011b).

3. Pre-treatment of lignocellulosic biomass

3.1 Need for pre-treatment

Upon densification, many agricultural biomass materials, especially those from straw and stover, result in a poorly formed pellets or compacts that are more often dusty, difficult to handle and costly to manufacture. This is caused by lack of complete understanding on the natural binding characteristics of the components that make up biomass (Sokhansanj et al., 2005). The natural binding characteristics of lignocellulosic biomass can be enhanced by modifying the structure of cellulose-hemicellulose-lignin matrix by application of pre-processing and pre-treatment methods (Sokhansanj et al. 2005). It is postulated that by disrupting the lignocellulosic matrix of biomass materials via application of various chemical, physico-chemical (steam explosion, microwave, and radio frequency heating), and biological pre-treatment, the compression and compaction characteristics can be improved (Shaw 2008; Kashaninejad and Tabil, 2011). When high molecular amorphous polysaccharides are reduced to low molecular components, the polymer becomes more cohesive in the presence of moisture (Chen et al., 2004). The cellulose-hemicellulose-lignin matrix can be broken down to smaller amorphous molecules through acid or alkaline hydrolysis as well as through steam explosion (Ladisch, 1989; Vlasenko, 1997). Alkaline or acid solutions are often used for pre-treatment of biomass and the effect of pre-treatment depends on the lignin content of biomass. When biomass is treated with dilute alkaline solution, the internal surface area of the material is increased by swelling. Swelling causes a decrease in the degree of polymerization, separation of structural linkages between lignin and carbohydrates and disruption of the lignin structure (Fan et al., 1987). Increased moisture content resulting from chemical and enzymatic treatments is a problem, as the treated biomass has to be dried prior to densification. Steam explosion results in the hemicelluloses being hydrolyzed and water soluble, the cellulose is slightly depolymerized, the lignin melts and is depolymerized, which aid in binding particles together during densification. Zandersons et al. (2004) stated that activation of lignin and changes in the cellulosic structure during the steam explosion process facilitate the formation of new chemical bonds. Lam et al. (2008) reported that the quality (durability) of compacts produced from steam exploded sawdust was 20% higher than non-treated sawdust.

3.2 Physico-chemical pre-treatments

3.2.1 Steam explosion

Steam explosion is one of the most applied pre-treatment processes owing to its low use of chemicals and limited energy consumption (Harmsen et al., 2010). During steam explosion pre-treatment process, the lignocellulosic biomass is heated with high pressure saturated steam having temperatures typically in the range of 180-230°C for 2-10 minutes. Subsequently, the substrate is quickly flashed to atmospheric pressure; as a result, the water inside the substrate vaporizes and expands rapidly, disintegrating the biomass (Grous et al., 1985; Kokta and Ahmed, 1998; Zimbardi et al., 1999). This causes great reduction in the particle size of the substrate (Fig. 4). The heart of the explosion pulping process is the reactor, which allows the use of high pressure during heating and cooking. The reactor can be of either the batch (Fig. 5) (Jin and Chen, 2006) or continuous type (Fig. 6) (Kokta and Ahmed, 1998; Adapa et al., 2010a).

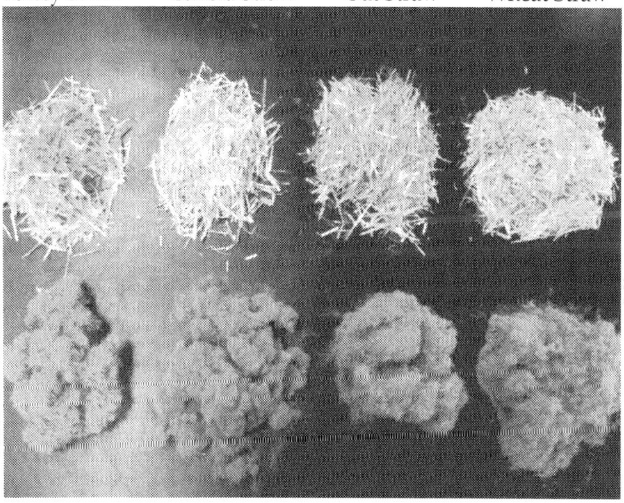

Fig. 4. Photographs showing the non-treated (30 mm hammer mill screen size) and steam exploded barley, canola, oat and wheat straw grinds.

The extent of chemical and structural modifications from steam-explosion pre-treatment depends on residence time, temperature, particle size and moisture content (Sun and Cheng, 2002). However, the severity (Ro) of steam explosion is quantified as a function of retention time and reaction temperature (Equation 1) (Overend and Chornet, 1987; Viola et al. 2008).

$$\text{Ro} = t \times \exp\left(\frac{T-100}{14.75}\right) \tag{1}$$

Where T is the temperature in °C and t is the time in minutes.
According to Zimbardi et al. (1999), the simplest way to carry out steam explosion is by batch procedure, hence widely reported in literature. However, the continuous reactors are of major interest for industrial applications. They have indicated that although the products obtained at the same treatment, severity in batch and continuous reactors are macroscopically different at first sight, there is still a lack of understanding to explain these differences. Consequently, they have developed experimental relationships between the two systems useful in making the data transfer straightforward (Equation 2).

$$log(Ro)_{Batch} = 1.50 \times (log(Ro)_{Continuous} - 1) \quad (2)$$

In addition, studies have been carried out to try to improve the results of steam explosion by addition of chemicals such as acid or alkali (Cara et al., 2008; Harmsen et al., 2010; Stenberg et al., 1998; Zimbardi et al., 2007). Limitations of steam explosion include the formation of degradation products that may inhibit downstream processes (Garcia-Aparicio et al., 2006).

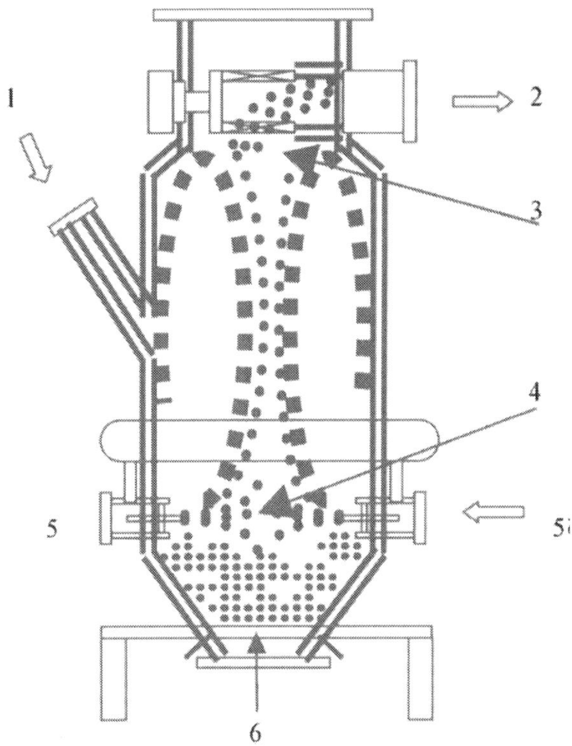

Fig. 5. Schematic diagram of the FJM-200 fluidized bed opposed Jet mill. 1, Infeed; 2, collection; 3, classification section; 4, grinding section; 5, compressed air; 6, discharge opening (Jin and Chen, 2006).

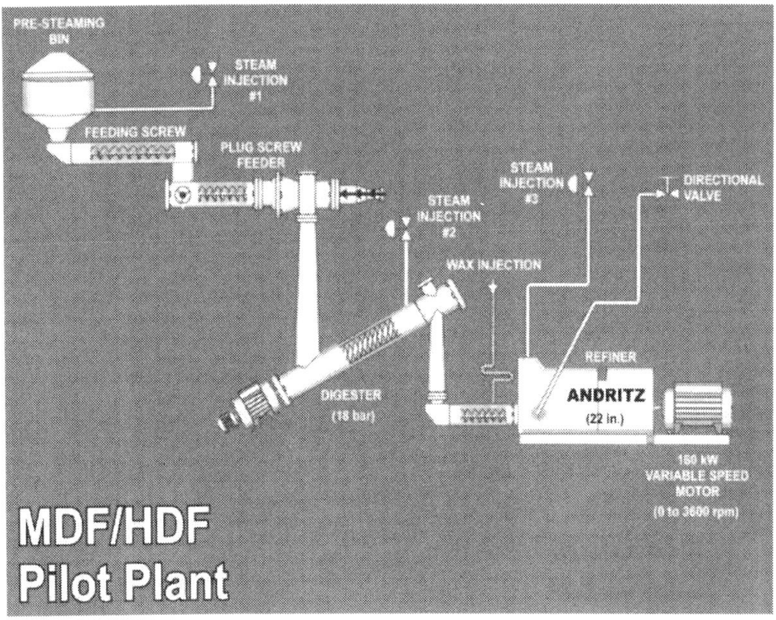

Fig. 6. The Andritz (ANDRITZ AG, Graz, Austria) continuous biomass steam explosion facility for manufacturing of Medium Density and High Density Fiberboards (MDF/HDF), Forintek pilot plant at the FPInnovations, Quebec City, Quebec (Adapa et al., 2010a).

3.2.2 Microwave and radio frequency (RF) heating

Dielectric heating is an alternative method to conventional heating. Unlike conduction/convection heating, which is based on superficial heat transfer, dielectric heating is based on volumetric and rapid heat transfer (de la Hoz et al., 2005). When lignocellulosic materials are placed in an electric field for dielectric heating pre-treatment, dipole molecules such as water or other dielectric materials, rotate vigorously to orient in the field. More polar components will absorb more energy, and thus, "hot spots" will be created in non-homogeneous materials. It is hypothesized that this unique heating feature results in an "explosion" effect in the particles and improves the disruption of the lignocellulosic structures. In addition, the non thermal effects of electromagnetic field accelerate the disintegration of the crystal structures (de la Hoz et al., 2005). Dielectric heating can be categorized as microwave or radio frequency depending on the wavelength used in the heating devices (Oberndorfer et al., 2000).

Microwave is electromagnetic waves between 300 MHz (wavelength 1 m) and 300 GHz (wavelength 1 mm). This range of spectrum lies between infrared and radio frequency radiation. Microwave irradiation has been extensively used in many processes because of its high heating efficiency and easy operation. Microwave energy can penetrate into materials and heat them quickly and uniformly. Microwave irradiation is considered to create thermal and non thermal effects. It has been applied as an efficient pre-treatment technique to enhance the hydrolysis of biomass materials. Some studies have demonstrated that microwave irradiation can change the structure of lignocellulosic materials and degrade or reduce lignin content, reduce cellulose crystallinity, and increase porosity and surface area of the materials (Azuma, 1984; Zhu et al., 2006b; Kashaninejad and Tabil, 2011).

This pre-treatment involves microwave irradiation of immersed biomass in an aqueous environment. Different types of lignocellulosic materials have been pre-treated using microwave irradiation including wheat straw (Ooshima et al., 1984; Zhu et al., 2006b; 2006c; Kashaninejad and Tabil, 2011), barley straw (Kashaninejad and Tabil, 2011), rice straw (Ooshima et al., 1984; Zhu et al., 2005; 2006a), rice hulls (Magara and Azuma, 1989), sugarcane bagasse (Ooshima et al., 1984; Magara and Azuma, 1989; Kitchaiya et al., 2003), switchgrass (Hu and Wen, 2008; Keshwani et al., 2007) and woody materials (Azuma et al., 1984). These materials were subjected to microwave pre-treatment of 2450 MHz in the range of 250 to 1000 W. The temperature of operation ranged from 70 to 230°C, while heating time varied from 5 to 120 minutes. However, higher microwave power with short pre-treatment time and the lower microwave power with long pre-treatment time had almost the same effect on chemical composition of lignocellulosic materials (Zhu et al., 2005; 2006b; Kashaninejad and Tabil, 2011).

In order to increase the efficiency of microwave heating pre-treatment, some researchers have combined microwave treatment with alkaline treatment such as NaOH or $Ca(OH)_2$. Some used alkaline solution during microwave heating treatment (Zhu et al., 2005; 2006a; 2006b; 2006c; Keshwani et al., 2007; Kashaninejad and Tabil, 2011) and some applied the alkaline solution before the lignocellulosic materials were subjected to microwave irradiation (Zhu et al., 2006a; Hu and Wen, 2008). Combination of microwave irradiation and alkali treatment improves the degradation of biomass by accelerating the reactions during the pre-treatment process compared with the conventional heating chemical pre-treatment process. Remarkable changes (Table 2) have been reported in the chemical composition of biomass samples after microwave-alkali pre-treatment, particularly in hemicellulose, lignin, and cellulose contents (Kashaninejad and Tabil, 2011). It has been reported that alkali treatment dissolves lignin and hemicellulose, and microwave irradiation facilitates dissolving these components in alkali solutions (Jackson, 1977; Kumar et al., 2009; Lesoing et al., 1980; Zhu et al., 2005). Biomass samples pretreated by microwave-alkali technique have lower lignin, hemicellulose, and cellulose than samples pretreated by microwave-distilled water or untreated samples. Moreover, degradation and depolymerisation of lignin to smaller phenolic components is another influence of microwave-alkali pre-treatment that could be considered as binder in densification process. The pellets made from microwave-chemical pre-treated biomass grinds have significantly higher density and tensile strength (Table 3) than the untreated or samples pre-treated by microwave alone (Kashaninejad and Tabil, 2011).

Radio frequency (RF) can penetrate more deeply into the materials compared with microwave heating because the radio frequency wavelength is up to 360 times greater than microwave (Marra et al., 2007). This unique characteristic is an advantage to treat large amount of material and it is easier to scale up the process. While radio frequency as a heating method has been widely applied in food-processing industries, there is not much report on application of radio frequency heating for lignocellulosic materials pre-treatment. Hu et al. (2008) used radio frequency heating in the NaOH pre-treatment of switchgrass to enhance its enzymatic digestibility. Because of the unique features of radio frequency heating (i.e., volumetric heat transfer, deep heat penetration of the samples, etc.), switchgrass could be treated on a large scale, at high solids content, and with a uniform temperature profile. At 20% solids content, radio frequency-assisted alkali pre-treatment (at 0.1 g of NaOH/g of biomass loading and 90°C) resulted in a higher xylose yield than the conventional heating pre-treatment. The optimal particle size and alkali loading in the radio frequency pre-treatment were determined to be 0.25-0.50 mm and 0.25 g of NaOH/g of biomass, respectively.

Treatment	Wheat straw						Barley straw					
	Protein	Lignin	Ash	Starch	Cellulose	Hemicellulose	Protein	Lignin	Ash	Starch	Cellulose	Hemicellulose
Untreated	1.99b	8.33a	6.33f	1.11d	44.99b	27.96a	1.61d	11.95a	6.03d	0.79c	46.93a	27.40a
Microwave-distilled water	2.24a	8.01c	8.87e	1.48b	39.69d	22.62b	2.01a	8.85b	6.28d	1.08b	45.25b	27.21a
Microwave-NaOH (1%)	1.41e	7.82d	17.32b	1.89a	35.82e	12.32d	1.80b	6.65e	16.96b	0.60e	40.81c	8.74c
Microwave-NaOH (2%)	1.36f	7.09f	34.77a	0.27f	34.77f	4.06f	1.62d	4.52f	41.43a	0.54f	35.22d	5.46d
Microwave-$Ca(OH)_2$ (1%)	1.85c	8.11b	12.24d	0.69e	45.66a	14.94c	1.81b	7.27d	13.21c	0.72d	41.01c	15.00b
Microwave-$Ca(OH)_2$ (2%)	1.52d	7.55e	15.89c	1.31c	42.56c	11.10e	1.68c	8.16c	16.73b	1.19a	41.24c	8.97c

Means with the same letters designation (a, b, c, d, and e) in a column are not significantly different at P = 0.05.

Table 2. Chemical composition (% dry basis) of untreated and microwave pretreated of wheat and barley straw at power 713 W.

Treatment	Wheat straw		Barley straw	
	Fracture load (N)	Tensile strength (MPa)	Fracture load (N)	Tensile strength (MPa)
Untreated	19.10±5.61	0.81±0.24	16.25±5.30	0.68±0.22
Microwave- distilled water	35.00±11.93	1.48±0.46	14.25±5.31	0.61±0.21
Microwave- NaOH (1%)	85.46±22.94	3.99±0.82	57.13±12.12	2.42±0.47
Microwave- NaOH (2%)	88.00±15.86	3.69±0.66	90.75±22.42	3.59±0.98
Microwave- $Ca(OH)_2$ (1%)	67.05±19.82	3.03±0.79	42.38±10.30	1.83±0.49
Microwave- $Ca(OH)_2$ (2%)	78.25±25.07	3.31±1.03	67.25±19.94	2.88±0.91

Table 3. Effect of microwave-chemical pre-treatments on fracture load and tensile strength of wheat and barley straw pellets at power 713 W.

3.2.3 Chemical pre-treatment

Different chemicals such as acids, alkalis, oxidizing agents and ozone have been used for chemical pre-treatment of lignocellulosic materials. Depending on the type of chemical used, pre-treatment could have different effects on structural components. Alkaline pre-treatment, ozonolysis, peroxide and wet oxidation pre-treatments were reportedly more effective in lignin removal, whereas dilute acid pre-treatment was more efficient in hemicellulose solubilization (Galbe and Zacchi, 2002; Sánchez and Cardona, 2008; Tomas-Pejo et al., 2008).

Acid Hydrolysis: Inorganic acids such as H_2SO_4 and HCl have been used for pre-treatment of lignocellulosic materials and have been used on a wide range of feedstocks ranging from hardwoods to grasses and agricultural residues. Acid hydrolysis can be classified as concentrated or dilute-acid hydrolysis based on the dose of acid used in the process. In the first case, the biomass is treated with high concentration of acids at ambient temperatures, which results in high conversion of lignocellulosic materials. Although concentrated acids are powerful agents for cellulose hydrolysis, they are toxic, corrosive, hazardous, and thus require reactors that are resistant to corrosion, making the pre-treatment process very expensive. In addition, the concentrated acid must be recovered after hydrolysis to make the process economically feasible (Galbe and Zacchi, 2002; Sun and Cheng, 2002).

Dilute-acid hydrolysis has been successfully developed for pre-treatment of lignocellulosic materials. Sulfuric acid at concentrations usually below 4% (wt) has been of the most interest in such studies as it is inexpensive and effective. Dilute H_2SO_4 pre-treatment can achieve high reaction rates and significantly improve cellulose hydrolysis (Esteghlalian et al., 1997). High temperature is favorable to attain acceptable rates of cellulose conversion. Despite low acid concentration and short reaction time, the use of high temperatures in dilute-acid hydrolysis accelerates the rate of hemicellulose sugar decomposition and increases equipment corrosion (Galbe and Zacchi, 2002; Taherzadeh and Karimi, 2007).

Alkali hydrolysis: Dilute alkali such as sodium, potassium, calcium, and ammonium hydroxides have been used for pre-treatment of lignocellulosic materials in alkali hydrolysis. The effectiveness of these agents depends on the lignin content of the materials. Temperature and pressure are lower in alkali pre-treatment compared with other pre-treatment methods (Mosier et al., 2005). Alkali pre-treatment can be conducted at ambient conditions, but process time is longer (hours or days instead of minutes or seconds). Compared with acid process, alkaline process causes less sugar degradation, and many of the caustic salts can be recovered and/or regenerated.

Sodium hydroxide has been studied more than other agents (Soto et al., 1994; Fox et al., 1989; MacDonald et al., 1983). Treatment of lignocellulosic materials using dilute NaOH results in swelling, leading to an increase in internal surface area, a decrease in the degree of polymerization, a decrease in crystallinity, separation of structural linkages between lignin and carbohydrates, and disruption of the lignin structure. However, calcium hydroxide (lime) is the least expensive hydroxide and has been shown to be an effective pre-treatment agent. The process of lime pre-treatment involves slurrying the lime with water, spraying it onto the biomass material, and storing the material in a pile for a period of hours to weeks. The particle size of the biomass is typically 10 mm or less. Elevated temperatures reduce contact time.

Oxidizing agents: In this pre-treatment, an oxidizing compound such as hydrogen peroxide (H_2O_2) or peracetic acid (CH_3CO_3H) is used to treat lignocellulosic materials and sometimes is applied in combination of an alkaline solution (e.g. NaOH) to improve effectiveness. This pre-treatment is usually carried out under mild temperature. This pre-treatment is more effective to increase crop residue digestibility compared with NaOH pre-treatment alone. Gould (1984) delignified agricultural residues using 1% H_2O_2 at 25°C for 18–24 h. Under this condition, more than half of the lignin and most of hemicellulose were solubilized. The pre-treatment of cane bagasse with H_2O_2 greatly enhanced its susceptibility to further hydrolysis. About 50% of the lignin and most of the hemicellulose were solubilized by 2% H_2O_2 at 30°C within 8 h, and a 95% efficiency of glucose production from cellulose was achieved in the subsequent saccharification by cellulase at 45°C for 24 h (Azzam, 1989).

Ozonolysis: In this process, ozone is used to change the structure of lignocellulosic materials and has been used for different materials such as wheat straw (Ben-Ghedalia and Miron, 1981), bagasse, green hay, peanut, pine (Neely, 1984), cotton straw (Ben-Ghedalia and Shefet, 1983) and poplar sawdust (Vidal and Molinier, 1988). Ozonolysis is carried out at room temperature and normal pressure. It can effectively remove the lignin without producing any toxic residues. In this process, hemicellulose is slightly affected, but no change in cellulose has been reported. The main restriction of this process is the large amount of ozone utilization that makes the process expensive (Sun and Cheng, 2002). Binder et al. (1980) reported 60% removal of the lignin from wheat straw using ozone pre-treatment. Enzymatic hydrolysis yield increased from 0% to 57% as the percentage of lignin decreased from 29% to 8% after ozonolysis pre-treatment of poplar sawdust (Vidal and Molinier, 1988). Garcia-Cubero et al. (2009) studied the ozonolysis pre-treatment of wheat straw in a fixed bed reactor at room conditions and concluded that enzymatic hydrolysis yield of up to 88.6% compared to 29% in non-ozonated sample.

3.3 Biological pre-treatment

Most pre-treatments require expensive instruments or equipment that require high energy requirements, depending on the process. In particular, physical and thermo-chemical processes require ample amount of energy to change the lignocellulosic structure of biomass. Biological pre-treatment using various types of rot fungi is a process that does not require high energy for lignin removal from a lignocellulosic biomass, despite extensive lignin degradation. Biological pre-treatments are safe, environmentally friendly and less energy intensive compared to other pre-treatment methods. However, the rate of hydrolysis reaction is very slow and needs a great improvement to be commercially applicable.

Biological pre-treatment comprises of using microorganisms such as brown-, white-, and soft-rot fungi for selective degradation of lignin and hemicellulose among which white-rot fungi seems to be the most effective microorganism (Fan et al., 1987). Brown rots mainly attack cellulose, while white and soft rots attack both cellulose and lignin. Lignin degradation occurs through the action of lignin-degrading enzymes such as peroxidases and laccase (Okano et al., 2005). These enzymes are regulated by carbon and nitrogen sources. The suitable fungi for biological pre-treatment should have higher affinity for lignin and degrade it faster than carbohydrate components.

Hatakka et al. (1983) studied the pre-treatment of wheat straw by 19 white-rot fungi and found that 35% of the straw was converted to reducing sugars by *Pleurotus ostreatus* in 5 weeks. Similar conversion was obtained in the pre-treatment by *Phanerochaete sordid* (Ballesteros et al., 2006) and *Pycnoporus cinnabarinus* (Okano et al., 2005) in 4 weeks. Akin et al. (1995) also reported the delignification of bermudagrass by white-rot fungi. The biodegradation of bermudagrass stems was improved by 29-32%, after 6 weeks, using *Ceriporiopsis subvermispora* and by 63-77% using *Cyathus stercoreus*.

4. Particle size reduction and physical properties

The application of pre-processing operations such as particle size reduction/ grinding is critical in order to increase the surface area of lignocellulosic biomass prior to densification (Mani et al. 2004). Particle size reduction increases the total surface area, pore size of the material and the number of contact points for inter-particle bonding in the compaction process (Drzymala, 1993). Size reduction is an important energy intensive unit operation essential for bioenergy conversion process and densification to reduce transportation costs (Bitra et al., 2009; Soucek et al., 2003). Energy consumption of grinding biomass depends on initial particle size, moisture content, material properties, feed rate of the material and machine variables (Lopo, 2002). A comprehensive summary of literature review on size reduction of lignocellulosic biomass is provided in Table 4.

Type of Choppers/ Grinders	Biomass	Parameters Measured and Correlations	Observations	Reference
Hammer Mill	Non-Treated and Steam Exploded Barley, Canola, Oat and Wheat Straw	• Hammer mill screen size (from 1.6 to 30.0 mm) on Specific Energy • Effect of geometric mean particle size on bulk density • Effect of geometric mean particle size on particle density • Analysis on ground particle size distribution	• Negative exponential and power correlation between geometric mean particle size with both bulk and particle density • Specific energy requirement is material dependent • Negative power correlation between hammer mill screen size and specific energy • Shapiro-Wilk Test for normality was performed	Adapa et al., 2011a
Tub Grinder	Rectangular Wheat Straw Bales, Round Rice Straw and Corn Stover Bales	• Effect of screen sizes (from 12.7 to 50.8 mm) on Specific Energy • Effect of screen sizes on grinding rate • Analysis on ground particle size distribution	• Positive correlation between screen size and grinding rate • Positive correlation between tub rotational speed and grinding rate • Specific Energy is material dependent • Negative power/ exponential correlation between screen size and specific energy	Arthur et al., 1982
Hammer Mill	Coastal Bermudagrass	• Effect of moisture content on Specific Energy • Effect of feed rate on Specific Energy	• Positive correlation between moisture content and specific energy • Positive correlation between feed rate and specific energy	Balk, 1964
Hammer Mill	Wheat Straw and Corn Stover	• Effect of operating speeds (from 2000 to 3600 rpm) on Specific Energy • Effect of hammer angles (90° and 30° hammers) on Specific Energy • Analysis on ground particle size distribution	• Positive correlation between operating speed and specific energy • Geometric mean particle diameter decreased with an increase in hammer mill speed • Specific energy increased with a decrease in hammer angle	Bitra et al., 2009
Knife and Hammer Mills	Hardwood Chips, Wheat Straw and Corn Stover	• Effect of screen size (from 1.6 to 12.7 mm) on specific energy	• Negative correlation between screen size and specific energy • Specific Energy is material dependent	Cadoche and López, 1989
Hammer Mill	Red Winter Wheat Straw	• Effect of screen size (from 1.59 to 4.76 mm) on specific energy • Effect of feed rate (from 1.5 to 2.5 kg/min) on specific energy	• Negative correlation was observed between screen size and specific energy • Feed rate did not have significant effect on specific energy	Fang et al., 1997

Type of Choppers/ Grinders	Biomass	Parameters Measured and Correlations	Observations	Reference
Hammer Mill	Wheat Straw	• Effect of screen size (3.2 and 1.6 mm) on Specific Energy • Analysis on ground particle size distribution	• Negative correlation between specific energy and screen size	Himmel et al., 1985
Tub Grinders	Round Bales of Corn Stover and Perennial Grasses	• Effect of screen size (from 19.1 to 127.0 mm) on Specific Energy • Effect of screen size on throughput • Effect of screen size on bulk and particle densities	• Negative correlation between screen size and bulk density • Positive correlation between screen size and particle density • Negative correlation between screen size and specific energy • Positive correlation between screen size and throughput	Kaliyan et al., 2010
Hammer Mill	Barley Straw, Corn Stover and Switchgrass	• Effect of Screen Sizes (3.2, 1.6 and 0.8 mm) on Specific Energy • Effect of Moisture Content (8% and 12% wb) on Specific Energy • Correlation between bulk and particle densities and geometric mean diameter • Analysis on ground particle size distribution	• Negative linear correlation between specific energy and hammer mill screen size at 8% mc • Quadratic correlation between specific energy and hammer mill screen size at 12% mc • Polynomial relations for bulk and particle densities with geometric mean particle diameters	Mani et al., 2004
Hammer Mill	Oat Straw, Rattle Grass and Miscanthus	• Effect of Screen size (from 1.0 to 10.0 mm) on specific energy • Effect of moisture content on specific energy	• Negative power correlation between screen size and specific energy • Positive correlation between moisture content and specific energy; significantly higher specific energy is required at smaller screen sizes	Soucek et al., 2003
Hammer Mill	Corn Stover	• Effect of hammer thickness (6.4 and 3.2 mm) on Specific Energy • Effect of hammer thickness (6.4 and 3.2 mm) on Grinding Rate • Effect of hammer tip speed (54 to 86 m/s) on Specific Energy	• Negative correlation between hammer thickness and Specific Energy • Negative correlation between hammer thickness and Grinding Rate • Positive correlation between hammer tip speed and Specific Energy	Vigneault et al., 1992
Knife Mill	Switchgrass, Corn Stover, and Wheat Straw	• Effect of rotational speed (from 250 to 500 rpm) on specific energy • Analysis on ground particle size distribution	• Positive correlation between rotational speed and specific energy • Screen size has significant effect on particle size distribution	Womac et al., 2007

Table 4. A comprehensive summary of literature review on size reduction of lignocellulosic biomass

4.1 Chopping

Baled agricultural biomass from the field does not have good flowing characteristics and may not flow easily into grinders such as hammer mills and disc refiners. Therefore, biomass needs to be chopped with a chopper (rotary shear shredder)/ knife mill/ tub grinder to accommodate bulk flow and uniformity of feed rate. A chopper, knife cutter, or knife mill is often used for coarse size reduction (>50 mm) of stalk, straw, and grass feed stocks (Bitra et al., 2009). Knife mills reportedly worked successfully for shredding forages under various crop and machine conditions (Cadoche and López, 1989).

Bitra et al. (2009) reported that the total specific energy (including energy to operate the knife mill) for agricultural biomass chopping increases with knife mill speed. The total specific energy for knife mill and tub grinder has been observed to have negative correlation with screen size and mass feed rate (Arthur et al., 1982; Bitra et al., 2009; Himmel et al., 1985). However, grinding rate (throughput) increases with an increase in screen size (Arthur et al., 1982).

For tub grinders, an increase in screen size results in an increase in geometric mean length of particles and throughput, but a decrease in bulk density of the particles and specific energy consumption (Kaliyan et al., 2010).

4.2 Hammer mill grinding

Typically, hammer mills are used in forage processing industry as they are relatively inexpensive, easy to operate and produces wide range of particles (Lopo, 2002). Hammer mills have achieved merit because of their ability to finely grind a greater variety of materials than any other machines (Scholten et al., 1985). The performance of a hammer mill is measured in terms of energy consumption and geometric mean diameter and particle size distribution of the ground product (Adapa et al., 2011a; Mani et al., 2004).

Screen Size: Hammer mill screen opening size was the most significant factor affecting mill performance (Fang et al., 1997) and also has significant effect on mean particle size (Pfost and Headley, 1971). The specific energy required to grind agricultural biomass significantly increases with a decrease in hammer mill screen size and shows a negative power correlation (Arthur et al., 1982; Soucek et al., 2003). Similarly, Adapa et al. (2011a) reported negative correlation between specific energy and particle size of biomass as affected by hammer mill screen sizes. However, two other studies reported a second-order polynomial relationship between the specific energy requirements for grinding biomass (Mani et al. 2004; Sitkei, 1986). Usually, the mean geometric particle size for any particular biomass decreases with a decrease in hammer mill screen size (Adapa et al., 2011a). It has been reported that wider particle size distribution is suitable for compaction (pelleting/ briquetting) process (Adapa et al., 2011a; Mani et al., 2004). During compaction, smaller (fine) particles rearrange and fill in the void space of larger (coarse) particles producing denser and durable compacts (Tabil, 1996).

Operating Speed (Peripheral Velocity): The speed has a significant effect on mean particle size (Pfost and Headley, 1971). The total specific energy of hammer mill grinding has direct correlation to an increase in hammer tip speed (Bitra et al., 2009; Vigneault et al., 1992). High speed hammer mills with smaller diameter rotors are good for fine or hard-to-grind material. However, at high tip speeds, the material moves around the mill parallel to the screen surface making the openings only partially effective. At slower speeds, the material impinges on the screen at a greater angle causing greater amounts of coarser feed to pass through (Balk, 1964).

Hammer Angles and Thickness: The direct energy input for grinding also depends on hammer angles. In general, the specific energy for grinding decreases with an increase in hammer degrees (Bitra et al., 2009). In addition, the specific energy for grinding increases with an increase in hammer thickness (Vigneault et al., 1992).

Material Moisture Content and Feed Rate: A positive correlation has been reported between moisture content and specific energy consumption for grinding of agricultural biomass (Balk, 1964; Mani et al., 2004; Soucek et al., 2003). Feeding rate also has significant effect on specific energy consumption during hammer mill grinding and has positive correlation (O'Dogherty, 1982).

Bulk and Particle Densities, and Geometric Mean Particle Size: Usually, the bulk and particle density of agricultural straw significantly increases with a decrease in hammer mill screen size (Adapa et al., 2011a). The geometric mean particle size of pre-treated straw is usually smaller than that of the non-treated straw. This could be due to the fact that application of pre-treatment disrupts/ disintegrates the lignocellulosic structure of the biomass (Sokhansanj et al., 2005) leading to lower shear strength (easier to grind the straw).

4.3 Physical and frictional properties of biomass

Bulk Density: The goal of densification is to increase the bulk density of agricultural straw to facilitate economic storage, transportation and handling of the material. In addition, densification results in an increase in the net calorific content per unit volume. The bulk density of agricultural biomass depends on the type of biomass, moisture content, grind size, and pre-treatment (Mani et al., 2006). Lower bulk densities, and concerns with uneven and low flowability of straw grinds are critical issues to sustainable production of pellets using pellet mills (Adapa et al., 2010c; Larsson et al., 2008).

Typically, the bulk density of ground straw increases with a decrease in hammer mill screen size. Also, pre-treatments usually results in a decrease in bulk density since the organized lignocellulosic structure of biomass is disturbed/ disintegrated. In addition, the bulk density and geometric mean particle size of material is correlated by either power or exponential relations (Adapa et al., 2010b; Mani et al., 2004). Table 5 shows a summary of average bulk density of various agricultural biomasses ground using a hammer mill.

Particle Density: Particle size of the grinds will have direct effect on the final pellet density. Theoretically, the density of pellet can be as high as the particle density of the ground biomass. Similar to bulk density, particle density also depends on the type of biomass, moisture content, grind size, and pre-treatment (Adapa et al., 2010b). The particle density is observed to have negative correlation with hammer mill screen size. Application of pre-treatment increases the particle density since disturbance/ disintegration of lignocellulosic structure results in finer components (Adapa et al., 2010b). Table 5 shows a summary of average particle density of various agricultural biomasses ground using a hammer mill.

Geometric Mean Particle Size and Distribution: It has been reported that wider particle size distribution is suitable for compaction (pelleting/briquetting) process (Mani et al., 2004a). During compaction, smaller (fine) particles rearrange and fill in the void space of larger (coarse) particles producing denser and durable compacts (Tabil, 1996). Therefore, ideally the grinds should be normally distributed, should have near zero skewness and lower peak than expected for the normal and wider distribution of data (negative Kurtosis values). In addition, a decrease in the biomass grind size has been observed to have a positive effect on pellet mill throughput (Adapa et al., 2004).

Frictional Properties: Prior to densification, biomass grinds need to be efficiently stored, handled and transported. Physical and frictional properties of biomass have significant effect on design of new and modification of existing bins, hoppers and feeders (Fasina et al., 2006). The frictional behavior of biomass grinds in all engineering applications is described by two independent parameters: the coefficient of internal friction, and the coefficient of wall friction. The former determines the stress distribution within particles undergoing strain, and the latter describes the magnitude of the stresses between the particle and the walls of its container (Seville et al., 1997). The classic law of friction states that frictional force is directly proportional to the total force that acts normal to the shear surfaces (Chancellor, 1994; Chung and Verma, 1989; Larsson, 2010). Frictional force depends on the nature of the materials in contact but is independent of the area of contact or sliding velocity (Mohsenin, 1970). Material properties such as moisture content and particle size affect the frictional properties and densification performance of an individual feedstock (Larsson 2010; Shaw and Tabil, 2006). In addition, the determination of coefficient of friction is essential for the design of production and handling equipment and in storage structures (Adapa et al., 2010a; Puchalski and Brusewitz, 1996). A comprehensive summary of literature review on coefficient of internal friction and cohesion of agricultural biomass is provided in Table 6.

Predominantly, a linear correlation exists between normal and shear stress for agricultural straw grinds (Adapa et al., 2010a; Chevanan et al., 2008; Richter, 1954) at any specific hammer mill screen size. An increase in hammer mill screen size significantly decreases the shear stress for ground straw at any specific normal stress (Adapa et al., 2010a).

The correlation for coefficient of internal friction and cohesion with average geometric mean particle sizes for agricultural straw grinds is provided in Adapa et al. (2010a). These correlations can be used to predict the coefficient of internal friction (slope of the linear plot) and the cohesion (intercept of the linear plot) for various geometric mean particle sizes. In general, the coefficient of internal friction for ground agricultural straw decreases with an increase in average geometric mean particle diameter. The coefficient of cohesion for straw grinds increases with an increase in average geometric mean particle size (Adapa et al., 2010a).

Biomass	Hammer Mill Screen Size (mm)	Moisture Content (%, wb)	Geometric Mean Particle Diameter (mm)	Bulk Density (kg/m3)	Particle Density (kg/m3)	Reference
Barley	6.4	8.9	0.88	96	1046	Adapa et al., 2011a
	3.2	5.3	0.46	149	1089	
	1.6	7.8	0.46	155	1149	
Canola	6.4	12.6	0.89	144	1019	Adapa et al., 2011a
	3.2	9.2	0.52	190	1192	
	1.6	8.3	0.37	203	1309	
Corn Stover	3.2	6.22	0.41	131	1170	Mani et al., 2004
	1.6	6.22	0.26	156	1330	
	0.8	6.22	0.19	158	1340	
Oat	6.4	10.9	0.94	111	873	Adapa et al., 2011a
	3.2	9.4	0.57	156	1093	
	1.6	7.7	0.40	196	1240	
Wheat	6.4	9.5	0.99	107	1078	Adapa et al., 2011a
	3.2	9.5	0.72	141	1225	
	1.6	8.6	0.45	154	1269	

Table 5. Average bulk and particle densities of various agricultural biomasses ground using a hammer mill

Biomass	Experimental Conditions	Observations	Reference
Corn Stover	• Normal stress from 10 to 200 kPa • Hammer mill screen sizes of 6.35 and 3.18 mm • Moisture contents of 7, 11 and 15% (wb) • Galvanized steel	• Coefficient of wall friction increased from 0.18 to 0.26 with an increase in moisture from 7 to 15% • No clear trend observed for the adhesion coefficient	Mani et al., 2004
Peat hull, Switchgrass and Poultry Litter	• Consolidating stress from 1.5 to 12.0 kPa • Screen sizes from 0.79 to 3.20 mm • Ring shear test	• No effect of screen size on internal friction and cohesive properties	Fasina et al., 2006
Peat Moss, Wheat Straw, Oat Hulls and Flax Shives	• Normal stress from 10 to 400 kPa • Geometric mean particles sizes of 0.74 (peat moss), 0.65 (wheat straw), 0.47 (oat hulls) and 0.64 (flax shives) mm • Moisture content 9-10% (wb) • Mild steel surface	• Coefficient of wall friction were 0.68 (peat moss), 0.45 (wheat straw), 0.39 (oat hulls), and 0.41 (flax shives) • Adhesion coefficients were 0.2635 kPa (peat moss), 10.687 kPa (wheat straw), 4.719 kPa (oat hulls), and 16.203 kPa (flax shives)	Shaw and Tabil, 2006
Alfalfa, Barley Straw, Wheat Straw and Whole Green Barley	• Normal stress from 200 to 735 kPa • Moisture content for alfalfa, barley straw from 12.0 to 45.7%, and for wheat straw at 10.0% and whole green barley at 51.0% (wb) • Chop size from 10 to 90 mm • Polished steel surface	• Coefficient if friction on steel surface for alfalfa and barley straw increased with moisture content and was from 0.15 to 0.26, and 0.14 and 0.27, respectively • Coefficient of friction for wheat straw and whole green barley were 0.13 and 0.21, respectively • No effect of chop size on coefficient of internal friction on barley straw	Afzalinia and Roberge, 2007
Switchgrass, Wheat straw, and Corn Stover	• Normal stresses from 1.23 to 4.92 kPa • Chop size of 7.81 and 13.50 mm for switchgrass, 7.09 and 10.39 mm for wheat straw, 7.80 and 14.89 mm for corn stover	• No effect of chop size on friction coefficients • Coefficient of internal friction varied from 0.765 to 1.586	Chevanan et al., 2008
Reed Canary Grass	• Normal stresses of 0.52-7.52 kPa (low) and 23-275 MPa (high) • Moisture contents from 6.7 to 17.1% (wb) for low normal stress, and 8.9 to 27.2% for high normal stress • Screen sizes of 4.0 mm • Ring shear test •	• High friction value of 0.6 was obtained at normal stress of 50 MPa and lower • At low normal stresses, the coefficient of kinematic wall friction (ratio of shear stress and normal stress) was positively correlated with moisture content and negatively correlated to normal stress • At high normal stresses, the coefficient of kinematic wall friction was negatively correlated to both moisture content and normal stress	Larsson, 2010
Non Treated and Steam Exploded Barley, Canola, Oat and Wheat Straw	• Normal stress from 9.8 to 39.2 kPa • Moisture content of 10% (wb) • Screen size from 1.6 to 6.4 mm	• No effect of screen size on shear stress values • Steam exploded straw had higher coefficient of internal friction than non-treated straw grinds • Coefficient of friction for non-treated barley, canola, oat and wheat straw were in the range of 0.505 to 0.584, 0.661 to 0.665, 0.498 to 0.590, and 0.532 to 0.591, respectively. • Coefficient of friction for steam exploded barley, canola, oat and wheat straw were in the range of 0.562 to 0.738, 0.708 to 0.841, 0.660 to 0.860, and 0.616 to 1.036, respectively	Adapa et al., 2010a

Table 6. A comprehensive summary of literature review on coefficient of internal friction and cohesion of agricultural biomass

5. Summary

The current chapter has explored the effects of pre-treatment (chemical, physico-chemical, and biological) and pre-processing (size reduction) techniques on densification of agricultural straw resulting in high quality (density and durability) pellets. It has been determined that an increase in bulk density of biomass also increases the net calorific content per unit volume of pellets, and facilitates easy and economical storage, transport and handling of the biomass. Pre-treatment and pre-processing methods disintegrate the basic lignocellulosic structure of biomass, and change the relative composition of lignin, cellulose and hemicelluloses in the material. In addition, physical and frictional properties of agricultural straw are altered. The application of pre-treatments breaks the long-chain hydrogen bond in cellulose, making hemicelluloses amorphous, and loosening the lignin out of the lignocellulosic matrix, resulting in better quality (physically) pellets. During this process, the high molecular amorphous polysaccharides are reduced to low molecular components to become more cohesive in the presence of moisture during densification process. Particle size reduction increases the total surface area, pore size of the material and the number of contact points for inter-particle bonding in the compaction process.

It has been shown that the Fourier Transform Infrared Spectroscopy (FTIR) can be used to rapidly characterize and quantify cellulose-hemicellulose-lignin composition prior to and after application of various pre-processing and pre-treatment methods. Regression equations were developed to predict the lignocellulosic content of agricultural biomass using pure cellulose, hemicelluloses and lignin as reference samples.

6. References

Adapa, P.K., Tabil L.G. & Schoenau, G.J. (2011a). Grinding Performance and Physical Properties of Non-Treated and Steam Exploded Barley, Canola, Oat and Wheat Straw. *Journal of Biomass and Bioenergy*, doi:10.1016/j.biombioe.2010.10.004, 35(2011), pp. 549-561.

Adapa, P.K., Tabil L.G., Schoenau, G.J., Canam, T., & Dumonceaux, T. (2011b). Quantitative Analysis of Lignocellulosic Components of Non-Treated and Steam Exploded Barley, Canola, Oat and Wheat Straw using Fourier Transform Infrared Spectroscopy. *The Journal of Agricultural Science and Technology*, 5(7), (In-Press).

Adapa, P.K., Tabil L.G. & Schoenau, G.J. (2010a). Physical and Frictional Properties of Ground Non-Treated and Steam Exploded Barley, Canola, Oat and Wheat Straw Grinds. *Journal of Powder Technology*, doi:10.1016/j.powtec.2010.03.038, 201(2010), pp. 230-241.

Adapa, P.K., Tabil L.G. & Schoenau, G.J. (2010b). Pelleting Characteristics of Selected Biomass with and without Steam Explosion Pre-treatment. *International Journal of Agricultural and Biological Engineering*, 3(3), pp. 62-79.

Adapa, P.K., Karunakaran, C. Tabil, L.G., & Schoenau, G.J. (2009). Potential Applications of Infrared and Raman Spectromicroscopy for Agricultural Biomass. *Agricultural Engineering International: the CIGR Ejournal*, Manuscript 1081, XI(February), pp. 1-25.

Adapa, P.K., Schoenau, G.J., Tabil, L.G., Arinze, E.A., Singh, A.. & Dalai, A.K. (2007). Customized and value-added high quality alfalfa products - a new concept.

Agricultural Engineering International: the CIGR Ejournal, Manuscript FP 07 003, IX(June), pp. 1-28.

Adapa, P.K., Tabil, L.G. Schoenau G.J., & Sokhansanj, S. (2004). Pelleting characteristics of fractionated sun-cured and dehydrated alfalfa grinds. *Applied Engineering in Agriculture, ASAE Transactions*, 20(6), pp. 813-820.

Afzalinia, S., & Roberge, M. (2007). Physical and mechanical properties of selected forage materials. *Canadian Biosystems Engineering*, 49, pp. 2.23-2.27.

Akin, D.E., Rigsby, L.L., Sethuraman, A., Morrison, W.H., Gamble, G.R., & Eriksson, K.E.L. (1995). Alterations in structure, chemistry, and biodegradability of grass lignocellulose treated with the white rot fungi Ceriporiopsis subvermispora and Cyathus stercoreus. *Applied Environmental Microbiology*, 61, pp. 1591–1598.

Anglès, M.N., Ferrando, F., Farriol X., & Salvadó, J. (2001). Suitability of steam exploded residual softwood for the production of binderless panels. Effect of the pre-treatment severity and lignin addition. *Biomass and Bioenergy*, 21, pp. 211-224.

Arthur, J.F., Kepner, R.A., Dobie, J.B., Miller, G.E., & Parsons, P.S. (1982). Tub grinder performance with crop and forest residues. *Transactions of the ASAE*, 25, pp. 1488–1494.

Azuma, J. I., Tanaka, F., & Koshijima, T. (1984). Enhancement of enzymatic susceptibility of lignocellulosic wastes by microwave radiation. *Journal of Fermentation Technology*, 62(4), pp. 377-384.

Azzam, M. (1989). Pre-treatment of cane bagasse with alkaline hydrogen peroxide for enzymatic hydrolysis of cellulose and ethanol fermentation. *Journal of Environmental Science and Health*, B., 24 (4), pp. 421–433.

Bagby, M. O. (1982). In International Symposium on Ethanol from Biomass. H. E. Duckworth and E. A. Thompson, eds. 561 -569. Winnipeg, MB: The Royal Society of Canada.

Balatinecz, J.J. (1983). The Potential of Densification in Biomass Utilization. In: W.A. Cote. ed. Biomass utilization. London: Plenum Press: 181–189 p.

Balk, W.A. (1964). Energy requirements for dehydrating and pelleting coastal Bermuda grass. *Transactions of the ASAE*, 4, pp. 349-1, 355.

Ben-Ghedalia, D., & Miron. J. (1981). The effect of combined chemical and enzyme treatment on the saccharification and in vitro digestion rate of wheat straw. *Biotechnology and Bioengineering*, 23, pp. 823–831.

Ben-Ghedalia, D., & Shefet, G. (1983). Chemical treatments for increasing the digestibility of cotton straw. *Journal of Agricultural Sciences*, 100, pp. 393–400.

Bhattacharya S., Sett, S., & Shrestha, R M. (1989). State of the Art for Biomass Densification. *Energy Sources*, 11, pp. 161–182.

Binder, A., Pelloni, L., & Fiechter, A. (1980). Delignification of straw with ozone to enhance biodegradability. *Applied Microbiology and Biotechnology*, 11, pp. 1–5.

Bitra, V.S.P., Womac, A.R., Chevanan, N., Miu, P.I., Igathinathane, C., Sokhansanj, S., & Smith, D.R. (2009). Direct mechanical energy measured of hammer mill comminution of switchgarss, wheat straw, and corn stover and analysis of their particle size distributions. *Powder Technology*, 193(2009), pp. 32-45.

Budevska, B.O. 2002. Handbook of Vibrational Spectroscopy, Vol. 5, edited by J. M. Chalmers and P. R. Griffiths, New York: John Wiley and Sons: 3720–3732 p.

Cadoche, L. & López, G.D. (1989). Assessment of size reduction as a preliminary step in the production of ethanol for lignocellulosic wastes. *Biological Wastes*, 30, pp. 153–157.

Cara, C., Ruiz, E., Ballesteros, M., Manzanares, P., Jose Negro, M., & Castro, E. (2008). Production of fuel ethanol from steam-explosion pretreated olive tree pruning. *Fuel*, 87(6), pp. 692-700.

Chabannes, M., Ruel, K., Yoshinaga, A,. Chabbert, B., Jauneau, A., Joseleau, J.P., & Boudet, A.M. (2001). In situ analysis of lignins in transgenic tobacco reveals a differential impact of individual transformations on the spatial patterns of lignin deposition at the cellular and subcellular levels. *Plant Journal*, 28, pp. 271-282, doi:10.1046/j.1365-313X.2001.01159.x

Chancellor, W.L. (1994). Friction between soil and equipment materials. ASAE Paper No. 941034. St. Joseph, MI.: American Society of Agricultural and Biological Engineers.

Chen, W., Lickfield, G.C., & Yang, C.Q. (2004). Molecular modeling of cellulose in amorphous state. Part I: model building and plastic deformation study. *Polymer*, 45, pp. 1063-1071.

Chevanan, N., Womac, A.R. Bitra, V.S. & Yoder, D.C. (2008). Flowability parameters for chopped switchgrass, wheat straw and corn stover. ASABE Annual International Meeting, Rhode Island, Paper No. 084162. St. Joseph, MI: American Society of Agricultural and Biological Engineers, June 29 - July 2, 12 pages.

Chico-Santamarta, L., Humphroes, A., White, D., Chaney, K., & Godwin, R.J. (2009). Effect of Pre- and Post-Pelletization Storage of Canola (Oilseed Rape) Straw on the Quality and Properties of Pellets. ASABE Annaul International Meeting, Grand Sierra Resort and Casino, Reno, Nevada, Paper No. 096105, June 21 - 24, 17 pages.

Chung, J.H., & Verma, L.R. (1989). Determination of friction coefficients of beans and peanuts. *Transactions of the ASAE* 32(2), pp. 745–750.

Crawford, R. L. (1981). Lignin biodegradation and transformation. New York: John Wiley and Sons. ISBN 0-471-05743-6.

de la Hoz, A., Diaz-Ortiz, A. & Moreno, A. (2005). Microwaves in organic synthesis: Thermal and non-thermal microwave effects. *Chemical Society Reviews*, 34, pp. 164–178.

Deguchi, S., Tsujii, K., & Horikoshi, K. (2006). Cooking Cellulose in Hot and Compressed Water. *Chemical Communications*, pp. 3293 - 3295, doi:10.1039/b605812d

Demirbaş, A. (2001). Biomass Resource Facilities and Biomass Conversion Processing for Fuels and Chemicals. *Energy Conversion and Management*, 41, pp. 1357-1378.

Drzymala, Z. (1993). Industrial briquetting - fundamentals and methods. Studies in mechanical engineering, Vol. 13 Warszawa: PWN-Polish Scientific Publishers.

Encyclopædia Britannica. (2008). Cellulose. Retrieved May 23, 2008, from Encyclopædia Britannica Online.

Esteghlalian, A., Hashimoto, A.G., Fenske, J.J., & Penner, M.H. (1997). Modeling and optimization of the dilute-sulfuric-acid pre-treatment of corn stover, poplar and switchgrass. *Bioresource Technology*, 59, pp. 129–136.

Fan, L. T., Gharpuray, M. M., & Lee, Y. H. (1987). In Cellulose Hydrolysis. *Biotechnology Monographs*, 57, pp. 149-187. New York, NY: Springer-Verlag.

Fang, Q., Bolloni, I., Haque, E., & Spillman, C.K. (1997). Comparison of energy efficiency between a roller mill and a hammer mill. *Applied Engineering in Agriculture*, 13(35), pp. 631-635.

Fasina, O.O, Bransby D., & Wood, C.W. (2006). Flow properties of biomass feedstock. ASABE Annual International Meeting, Portland, Oregon, Paper No. 066149, July 9-12, 12 pages.

Focher, B., Marzetti, A., Marsano, E., Conio, G., Tealdi, A., Cosani, A., & Terbojevich, M. (1998). Regenerated and Graft Copolymer Fibres from Steam-Exploded Wheat Straw: Characterization and Properties. *Journal of Applied Polymer* Science, 67, pp. 961-974.

Fox, D.J., Gray, P.P., Dunn, N.W., & Warwick, L.M. (1989). Comparison of alkali and steam (acid) pre-treatments of lignocellulosic materials to increase enzymic susceptibility: Evaluation under optimized pre-treatment conditions. *Journal of Chemical Technology & Biotechnology*, 44, pp. 135–146.

Galbe, M., & Zacchi, G. (2002). A review of the production of ethanol from softwood. *Applied Microbiology and Biotechnology* 59, pp. 618–628.

Garcia-Cubero, M.T., Gonzalez-Benito, G. Indacoechea, I. Coca, M. & Bolado, S. (2009). Effect of ozonolysis pre-treatment on enzymatic digestibility of wheat and rye straw. *Bioresource Technology*, 100, pp. 1608–1613.

Garcia-Aparicio, M. P., Ballesteros, I., Gonzalez, A., Oliva, J.M., Ballesteros, M., & Negro, M.J. (2006). Effect of inhibitors released during steam-explosion pre-treatment of barley straw on enzymatic hydrolysis. *Applied Biochemistry and Biotechnology*, 129(1-3), pp. 278-288.

Gelbrich, J., Mai, C. & Militz, H. (2009). Evaluation of Bacterial Wood Degradation by Fourier-Transform-Infrared (FTIR) Measurements, The study has been done within the EU-project BACPOLES (EVK4-CT-2001-00043) at the University of Gottingen (2009), 6 pages.

Gould, J.M. (1984). Alkaline peroxide delignification of agricultural residues to enhance enzymatic saccharification. *Biotechnology and Bioengineering*, 26, pp. 46–52.

Granada, E., López González, L.M., Míguez J.L., & Moran, J. (2002). Fuel lignocellulosic briquettes, die design and products study. *Renewable Energy*, 27, pp. 561-573.

Grous, W.R., Converse, A.O., & Grethlein, H.E. (1985). Effect of Stem Explosion Pre-treatment on Pore Size and Enzymatic Hydrolysis or Poplar. *Enzyme Microbial Technology*, 1986(8), pp. 274-280.

Harmsen, P.F.H., Huijgen, W.J.J., Bermudez Lopez, L.M., & Bakker, R.R.C. (2010). Literature Review of Physical and Chemical Pre-treatment Processes for Lignocellulosic Biomass. Energy Research Centre of the Netherlands. Report accessed on January 2010. Web Address: http://www.ecn.nl/docs/library/report/2010/e10013.pdf

Hatakka, A.I. (1983). Pre treatment of wheat straw by white-rot fungi for enzymic saccharification of cellulose. *Applied Microbiology and Biotechnology*, 18, pp. 350– 357.

Himmel, M., Tucker, M., Baker, J., Rivard, C., Oh, K., & Grohmann, K. (1985). Comminution of biomass: Hammer and knife mills. *Biotechnology and Bioengineering Symposium*, 15, pp. 39-58.

Hon, D.N.-S. (1989). Cellulosic adhesives. In Adhesives from Renewable Resources, Eds. R.W. Hemmingway and A.H. Conner, 289-304. Washington, DC: American Chemical Society.

Hu Z., & Wen, Z. (2008). Enhancing enzymatic digestibility of switchgrass by microwave assisted alkali pre-treatment. *Biochemical Engineering Journal*, 38, pp. 369-378

Hu, Z., Wang, Y. & Wen, Z. (2008). Alkali (NaOH) Pre-treatment of switchgrass by radio frequency-based dielectric heating. *Applied Biochemistry and Biotechnology*, 148, pp. 71–81.

Igathinathane, C., Womac, A.R., Sokhansanj, S., & Narayan, S. (2008). Knife grid size reduction to pre-process packed beds of high- and low-moisture switchgrass. *Bioresource Technology*, 99, pp. 2254–2264.

Jackson, M. G. (1977). The alkali treatment of straws. *Animal Feed Science and Technology*, 2, pp. 105-130.

Jin, S. & Chen, H. (2006). Superfine Grinding of Steam-Exploded Rice Straw and its Enzymatic Hydrolysis. *Biochemical Engineering Journal*, 30(2006), pp. 225-230.

Kaliyan, N., Schmidt, D.R., & Morey, R.V. (2010). Commercial Scale Grinding of Corn Stover and Perennial Grasses. ASABE Annual International Meeting, David L. Lawrence Convention Centre, Pittsburgh, Pennsylvania, June 20-23, Paper No. 1009062, 2950 Niles Road, St. Joseph, MI 49085-9659 USA.

Kaliyan, N., & Morey, R.V. (2006). Factors affecting strength and durability of densified products. ASABE Annual International Meeting, American Society of Agricultural and Biological Engineers, Portland, Oregon July 9-12, Paper Number 066077, 2950 Niles Road, St. Joseph, MI 49085-9659 USA.

Kashaninejad, M., & Tabil. L.G. (2011). Effect of microwave-chemical pre-treatment on compression characteristics of biomass grinds. *Biosystems Engineering*, 108, pp. 36-45.

Kelley, S.S., Rowell, R.M., Davis, M., Jurich, C.K., & Ibach, R. (2004). Rapid Analysis of the Chemical Composition of Agricultural Fibres Using Near Infrared Spectroscopy and Pyrolysis Molecular Beam Mass Spectrometry. *Biomass and Bioenergy*, 27(2004), pp. 77-88.

Keshwani, D.R., Cheng, J.J., Burns, J.C., Li, L., & Chiang, V. (2007). Microwave pre-treatment of switchgrass to enhance enzymatic hydrolysis. ASABE Annual Meeting Paper No. 077127, Minneapolis, MN, USA.

Kitchaiya, P., Intanakul, P., & Krairiksh, M. (2003). Enhancement of enzymatic hydrolysis of lignocellulosic wastes by microwave pre-treatment under atmospheric pressure. *Journal of Wood Chemistry and Technology*, 23(2), pp. 217- 225.

Kokta, B.V., & Ahmed, A. (1998). Chapter 6: Steam Explosion Pulping. Environmentally Friendly Technologies for the Pulp and Paper Industry, edited by Raymond A. Young and Masood Akhtar, ISBN 0-471-15770-8, John Wiley and Sons, Inc.

Kumar, P., Barrett, D.M., Delwiche, M.J., & Stroeve, P. (2009). Methods for pre-treatment of lignocellulosic biomass for efficient hydrolysis and biofuel production. *Industrial and Engineering Chemistry Research*, 48, pp. 3713–3729.

Ladisch, M. R. (1989). Hydrolysis. In: Biomass Handbook. O. Kitani and C. W. Hall, eds. 434-451. New York: Gordon and Breach.

Lam, P.S., Sokhansanj, S., Bi, X., Lim, C.J., Mani, S., & Melin, S. (2008). Energetic feasibility of biomass densification with steam explosion pretreatment. CSBE Annual General Meeting; July 14-16, Poster Number CSBE08305, Vancouver, BC.

Larkin, S., Ramage, J., & Scurlock, J. (2004). Bioenergy. In: G. Boyle. ed. Renewable energy. Power for a sustainable future: 106-146 p.

Larsson, S.H. (2010). Kinematic Wall Friction Properties of Reed Canary Grass Powder at High and Low Normal Stresses. *Powder Technology*, 198, pp. 108-113.

Larsson, S.H., Thyrel, M., Geladi, P., & Lestander, T.A. (2008). High Quality Biofuel Pellet Production from Pre-Compacted Low Density Raw Materials. *Bioresource Technology*, 99(2008), pp. 7176-7182.

Lebo, S. E. Jr., Gargulak, J.D., & McNally, T.J. (2001). "Lignin". Kirk-Othmer Encyclopedia of Chemical Technology. John Wiley & Sons, Inc. DOI:10.1002/0471238961.12090714120914.a01.pub2. Retrieved on 2007-10-14.

Lehtikangas, P. (2001). Quality properties of palletized sawdust, logging residues and bark. *Biomass and Bioenergy*, 20, pp. 351-360.

Lesoing, G., Klopfenstein, T., Rush, I., & Ward, J. (1980). Chemical treatment of wheat straw. *Journal of Animal Science*, 51, pp. 263-269.

Lin, Y., & Tanaka, S. (2006). Ethanol fermentation from biomass resources: current state and prospects. *Applied Microbiology and Biotechnology*, 69.6, pp. 627-642.

Liu, R., Yu, H., and Huang, Y. (2005). Structure and Morphology of Cellulose in Wheat Straw. *Cellulose*, 12, pp. 25-34.

Lopo, P. (2002). The right grinding solution for you: roll, horizontal or vertical. *Feed Management*, 53(3), pp. 203-6.

Luypaert, J., Zhang, M.H., & Massart, D.L. (2003). Feasibility Study for the Use of Near Infrared Spectroscopy in the Qualitative and Quantitative Analysis of Green Tea. *Analytca Chimica Acta*, 478(2003), pp. 303-312.

MacDonald, D.G., Bakhshi, N.N., Mathews, J.F., Roychowdhury, A., Bajpai, P., & Moo-Young, M. (1983). Alkali treatment of corn stover to improve sugar production by enzymatic hydrolysis. *Biotechnology and Bioengineering*, 25, pp. 2067–2076.

Magara K., & Azuma, J. (1989). Microwave-irradiation of lignocellulosic materials X: Conversion of microwave-irradiated agricultural wastes into ethanol. *Wood Research*, 76, pp. 1–9

Mani, S., Tabil, L.G., & Sokhansanj, S. (2006). Effects of Compressive Force, Particle Size, and Moisture Content on Mechanical Properties of Biomass Pellets from Grasses. *Biomass and Bioenergy*, 30(2006), pp. 648-654.

Mani, S., Tabil, L.G., & Sokhansanj, S. (2004). Grinding performance and physical properties of wheat and barley straws, corn stover and switchgrass. *Biomass and Bioenergy*, 27(2004), pp. 339-52.

Mani, S., Tabil, L.G., & Sokhansanj, S. (2003). An overview of compaction of biomass grinds. Powder Handling & Process, 15(3), pp. 160-168.

Marra, F., Lyng, J., Romano, V., & McKenna, B. (2007). Radio-frequency heating of foodstuff: Solution and validation of a mathematical model. *Journal of Food Engineering*, 79, pp. 998–1006.

McMullen, J., Fasina, O.O., Wood, C.W., & Feng, Y. (2005). Storage and handling characteristics of pellets from poultry litter. *Applied Engineering in Agriculture*, 21(4), pp. 645-651.

Mohsenin, N.N. (1970). Physical Properties of Plant and Animal Materials. New York, N.Y.: Gordon and Breach Science Publishers.

Mosier, N., Wyman, C., Dale, B., Elander, R., Lee, Y.Y., Holtzapple, M., & Ladisch, M. (2005). Features of promising technologies for pre-treatment of lignocellulosic biomass. *Bioresource Technology*, 96, pp. 673–686.

Murphy, J.D and K. McCarthy. (2005). Ethanol production from energy crops and wastes for use as a transport fuel in Ireland. *Applied Energy*, 82, pp. 148-166.

Neely, W. C. (1984). Factors affecting the pre-treatment of biomass with gaseous ozone. *Biotechnology and Bioengineering*, 26, pp. 59–65.

Nelson, D.L., & Cox, M.M. (2005). Lehninger Principles of Biochemistry. New York, NY: W.H. Freeman and Company.

Obernberger, I., & Thek, G. (2004). Physical characterization and chemical composition of densified biomass fuels with regard to their combustion behavior. *Biomass and Bioenergy*, 27, pp. 653-669.

Oberndorfer, C., Pawelzik, E., & Lucke, W. (2000). Prospects for the application of dielectric heating processes in the pre-treatment of oilseeds. *European Journal of Lipid Science and Technology*, 102, pp. 487–493.

O'Dogherty, M.J. (1982). A review of research on forage chopping. *Journal of Agricultural Engineering Research*, 27(4), pp. 267-289.

Okano, K., Kitagawa, M., Sasaki, Y., & Watanabe, T. (2005). Conversion of Japanese red cedar (Cryptomeria japonica) into a feed for ruminants by white-rot basidiomycetes. *Animal Feed Science and Technology*, 120, pp. 235–243

Ooshima, H., Aso, K. Harano, Y., & Yamamoto, T. (1984). Microwave treatment of cellulosic materials for their enzymatic hydrolysis. *Biotechnology Letters*, 6(5), pp. 289-294.

Overend, R.P., & Chornet, E. (1987). Fractionation of lignocellulosics by steam-aqueous pre-treatment. *Philosophical Transactions of the Royal Society*, London A 321, pp. 523–536.

Pfost, H.B., & Headley, V.E. (1971). Use of logarithmic normal distribution to describe hammer mill performance. *Transactions of the ASABE*, 14(3), pp. 531-5.

Puchalski, C., & Brusewitz, G.H. (1996). Coefficient of friction of watermelon. *Transactions of the ASABE*, 39(2), pp. 589-594.

Sánchez, Ó.J., & Cardona, C.A. (2008). Trends in biotechnological production of fuel ethanol from different feedstocks. *Bioresource Technology*, 99, pp. 5270–5295.

Schell, D.J., & Harwood, C. (1994). Milling of Lignocellulosic Biomass: Results of Pilot-Scale Testing. *Applied Biochemistry and Biotechnology*, 45(46), pp. 159-168.

Scholten, R.L., & McEllhiney, R.R. (1985). The effects of prebreaking in hammer mill particle size reduction. American Society of Agricultural Engineers; Paper no. 85-3542, St. Joseph MI.

Seville, J.P.K., Tuzun, U., & Clift, R. (1997). Processing of Particulate Solids. London, U.K.: Chapman and Hall.

Shambe, T., & Kennedy, J.F. (1985). Acid and enzymatic hydrolysis of chaotropically pretreated millet stalk, acha and rice straws and conversion of the products to ethanol. *Enzyme and Microbial Technology*, 7, pp. 115-120.

Shaw, M. (2008). Feedstock and Process Variables Influencing Biomass Densification. M.Sc. Thesis, Department of Agricultural and Bioresource Engineering, University of Saskatchewan. Website: http://library2.usask.ca/theses/available/etd-03092008-115636/unrestricted/Thesis.MarkShaw.FinalR3.pdf

Shaw, M.D., & Tabil, L.G. (2006). Mechanical properties of selected biomass grinds. ASABE Annual International Meeting, Oregon Convention Center, Portland, Oregon, Paper Number 066175, July 9-12: 12 pages.

Sherman Hsu, C.P. (1997). Chapter 15: Infrared Spectroscopy, in: F. Settle, Editor, Handbook of Instrumental Techniques for Analytical Chemistry, Prentice-Hall, NJ 270.

Sitkei, G. (1986). Mechanics of agricultural materials. Amsterdam: Elsevier.

Smola, N., & Urleb, U. (2000). Qualitative and Quantitative Analysis of Oxytetracycline by Near-Infrared Spectroscopy. *Analytica Chimica Acta*, 410(2000), pp. 203-210.

Sokhansanj, S., Mani, S., Bi, X., Zaini, P., & Tabil, L.G. (2005). Binderless pelletization of biomass. ASAE Annual International Meeting, Tampa Convention Centre, Tampa, Florida; July 17-20, Paper Number 056061, 2950 Niles Road, St. Joseph, MI 49085-9659 USA.

Soto, M.L., Dominguez, H., Nunez, M.J., & Lema, J.M. (1994). Enzymatic saccharification of alkali-treated sunflower hulls. *Bioresource Technology*, 49, pp. 53–59.

Soucek, J., Hanzlikova, I., & Hulta, P. (2003). A fine disintegration of plants suitable for composite biofuels production. *Research in Agricultural Engineering*, 49(1), pp. 7-11.

Stenberg, K., Tengborg, C., Galbe, M., & Zacchi, G. (1998). Optimisation of steam pre-treatment of SO2-impregnated mixed softwoods for ethanol production. *Journal of Chemical Technology and Biotechnology*, 71(4), pp. 299-308.

Sun, F., & Chen, H. (2007). Evaluation of enzymatic hydrolysis of wheat straw pretreated by atmos-pheric glycerol autocatalysis. *Journal of chemical technology and biotechnology*, 82, pp. 1039-1044.

Sun, Y., & Cheng, J. (2002). Hydrolysis of lignocellulosic materials for ethanol production: a review. *Bioresource Technology* 83, pp. 1-11.

Tabil, L.G. (1996). Pelleting and binding characteristics of alfalfa. Unpublished Ph.D. thesis, Department of Agricultural and Bioresource Engineering, Saskatoon, SK Canada: University of Saskatchewan.

Taherzadeh, M.J., & Karimi, K. (2007). Acid-based hydrolysis processes for ethanol from lignocellulosic materials: a review. *Bioresources*, 2, pp. 472–499.

Temmerman, M., Rabier, F., Jensen, P.D., Hartmann, H., & Bohm, T. (2006). Comparative Study of Durability Test Methods for Pellets and Briquettes. *Biomass and Bioenergy*, 30(11), pp. 964.

Tomas-Pejo, E., Oliva, J.M., & Ballesteros, M. (2008). Realistic approach for full-scale bioethanol production from lignocellulose: a review. *Journal of Scientific & Industrial Research*, 67, pp. 874–884.

Tripathi, K., Iyer, P.V.R., and Kandpal, T.C. (1988). A Techno-Economic Evaluation of Biomass Briquetting in India. *Biomass and Bioenergy*, 14, pp. 479–488.

Tucker, M.P., Mitri, R.K., Eddy, F.P., Nguyen, Q.A., Gedvilas, L.M., & Webb, J.D. (2000). Fourier Transform Infrared Quantification of Sugars in Pretreated Biomass Liquors. *Applied Biochemistry and Biotechnology*, 84-86(2000), pp. 39-50.

United States Department of Energy (USDE). (2006). Energy Efficiency and Renewable Energy - Biomass Program. Website Accessed on January 21, 2011. Web Address: http://www1.eere.energy.gov/biomass

Updegraff, D.M. (1969). Semimicro Determination of Cellulose in Biological Materials. *Analytical Biochemistry*, 32, pp. 420 - 424.

van Dam, J.E.G., van den Oever, M.J.A., Teunissen, W., Keijsers, E.R.P., & Peralta, A.G. (2004). Process for production of high density/high performance binderless boards from whole coconut husk. Part 1: lignin as intrinsic thermosetting binder resin. *Industrial Crops and Products*, 19, pp. 207-216.

Vazquez, G., Gonzalez, J., Freire, S., & Antorrena, G. (2002). Characterization of Eucalyptus Globules and Pinus Pinaster Acetosolv Pulps Prebleached with O2 by FTIR and DRIFT Spectroscopy. *Holz als Roh- und Werkstoff*, 60(2002), pp. 25-30.

Vlasenko, E.Y., Ding, H., Labavitch J.M., & Shoemaker, S.P. (1997). Enzymatic Hydrolysis of Pretreated Rice Straw. *Bioresource Technology*, 59, pp. 109-119.

Vidal, P.F., & Molinier, J. (1988). Ozonolysis of lignin-improvement of in vitro digestibility of poplar sawdust. *Biomass*, 16, pp. 1-17.

Vigneault, C., Rothwell, T.M., & Bourgeois, G. (1992). Hammer mill grinding rate and energy requirements for thin and conventional hammers. *Canadian Biosystems Engineering*, 34, pp. 203-6.

Viola, E., Zimbardi, F., Cardinale, M., Cardinale, G., Braccio, G., & Gambacorta, E. (2008). Processing Cereal Straws by Steam Explosion in a Pilot Plant to Enhance Digestibility in Ruminants. *Bioresource Technology*, 99(2008), pp. 681-689.

Womac, A.R., Igathinathane, C., Bitra, P., Miu, P., Yang, T., Sokhansanj, S., & Narayan, S. (2007). Biomass Pre-Processing Size Reduction with Instrumented Mills. ASABE Annual International Meeting, Minneapolis Convention Centre, Minneapolis, June 17-20, Paper No. 076046, 2950 Niles Road, St. Joseph, MI 49085-9659 USA.

Zandersons, J., Gravitis, J., Zhurinsh, A., Kokorevics, A., Kallavus, U., & Suzuki, C.K. (2004). Carbon materials obtained from self-binding sugar cane bagasse and deciduous wood residues plastics. *Biomass and Bioenergy*, 26, pp. 345-360.

Zimbardi, F., Viola, E., Nanna, F., Larocca, E., Cardinale, M., & Barisano, D. (2007). Acid impregnation and steam explosion of corn stover in batch processes. *Industrial Crops and Products*, 26(2), pp. 195-206.

Zimbardi, F., Viggiano, D., Nanna, F., Demichele, M., Cuna, D., & Cardinale, G. (1999). Steam Explosion of Straw in Batch and Continuous Systems. *Applied Biochemistry and Biotechnology*, 77-79, pp. 117-125.

Zhu, S., Wu, Y., Yu, Z., Zhang, X., Li, H., & Gao, M. (2006a). The effect of microwave irradiation on enzymatic hydrolysis of rice straw. *Bioresource Technology*, 97, pp. 1964-1968.

Zhu, S., Wu, Y., Yu, Z., Chen, Q., Wu, G., Yu, F., Wang, C., & Jin, S. (2006b). Microwave-assisted alkali pre-treatment of wheat straw and its enzymatic hydrolysis. Biosystems Engineering, 94, pp. 437-442.

Zhu, S., Wu, Y., Yu, Z., Zhang, X., Wang, C., Yu, F., & Jin, S. (2006c). Production of ethanol from microwave-assisted alkali pretreated wheat straw. *Process Biochemistry*, 41, pp. 869 - 873.

Zhu, S., Wu, Y., Yu, Z., Liao, J., & Zhang, Y. (2005). Pre-treatment by microwave/alkali of rice straw and its enzymic hydrolysis. *Process Biochemistry*, 40, pp. 3082-3086.

19

Biomass Feedstock Pre-Processing – Part 2: Densification

Lope Tabil[1], Phani Adapa[1] and Mahdi Kashaninejad[2]
[1]*Department of Chemical and Biological Engineering, University of Saskatchewan*
[2]*Department of Food Science & Technology, Gorgan University of Agricultural Sciences and Natural Resources Gorgan,*
[1]*Canada*
[2]*Iran*

1. Introduction

1.1 The need for densification

Agricultural biomass residues have the potential for the sustainable production of bio-fuels and to offset greenhouse gas emissions (Campbell et al., 2002; Sokhansanj et al., 2006). Straw from crop production and agricultural residues existing in the waste streams from commercial crop processing plants have little inherent value and have traditionally constituted a disposal problem. In fact, these residues represent an abundant, inexpensive and readily available source of renewable lignocellulosic biomass (Liu et al., 2005). New methodologies need to be developed to process the biomass making it suitable feedstock for bio-fuel production. In addition, some of the barriers in the economic use of agricultural crop residue are the variable quality of the residue, the cost of collection, and problems in transportation and storage (Bowyer and Stockmann, 2001; Sokhansanj et al., 2006).

In order to reduce industry's operational cost as well as to meet the requirement of raw material for biofuel production, biomass must be processed and handled in an efficient manner. Due to its high moisture content, irregular shape and size, and low bulk density, biomass is very difficult to handle, transport, store, and utilize in its original form (Sokhansanj et al., 2005). Densification of biomass into durable compacts is an effective solution to these problems and it can reduce material waste. Densification can increase the bulk density of biomass from an initial bulk density of 40-200 kg/m^3 to a final compact density of 600-1200 kg/m^3 (Adapa et al., 2007, Holley, 1983; Mani et al., 2003; McMullen et al., 2005; Obernberger and Thek, 2004). Biomass can be compressed and stabilized to 7–10 times densities of the standard bales by the application of pressures between 400–800 MPa during the densification process (Demirbas and Sahin, 1998). Because of their uniform shape and size, densified products may be easily handled using standard handling and storage equipment, and they can be easily adopted in direct-combustion or co-firing with coal, gasification, pyrolysis, and utilized in other biomass-based conversions (Kaliyan and Morey, 2006a) such as biochemical processes. Upon densification, many agricultural biomass materials, especially those from straw and stover, result in a poorly formed pellets or compacts that are more often dusty, difficult to handle and costly to manufacture. This is caused by lack of complete understanding on the natural binding characteristics of the components that make up biomass (Sokhansanj et al., 2005).

1.2 Fuel pellet quality parameters

The quality of fuel pellet is usually assessed based on its density and durability. High density of pellet represents higher energy per unit volume of material, while durability is the resistance of pellets to withstand various shear and impact forces applied during handling and transportation. High bulk density increases storage and transport capacity of pellets. Since feeding of boilers and gasifiers generally is volume-dependent, variations in bulk density should be avoided (Larsson et al., 2008). A bulk density of 650 kg/m^3 is stated as design value for wood pellet producers (Obernberger and Thek, 2004). Low durability of pellets results in problems like disturbance within pellet feeding systems, dust emissions, and an increased risk of fire and explosions during pellet handling and storage (Temmerman et al., 2006). Other quality factors of biomass for thermo-chemical conversion include (FAO, 2011; Rajvanshi, 1986):

- *Energy content:* The choice of a biomass for energy conversion will in part be decided by its heating value. The method of measurement of the biomass energy content will influence the estimate of efficiency of a given gasifier. The only realistic way of presenting fuel heating values for gasification purposes is to give lower heating values (excluding the heat of condensation of the water produced) on an ash inclusive basis and with specific reference to the actual moisture content of the fuel.
- *Moisture content:* High moisture contents reduce the thermal efficiency since heat is used to drive off the water and consequently this energy is not available for the reduction reactions and for converting thermal energy into chemical bound energy in the gas. Therefore, high moisture contents result in low gas heating values during thermo-chemical processes.
- *Volatile matter:* The amount of volatiles in the feedstock determines the necessity of special measures (either in design of the gasifier or in the layout of the gas cleanup train) in order to remove tars from the product gas in engine applications.
- *Ash content and slagging characteristics:* The mineral content in the biomass that remains in oxidation form after complete combustion is usually called ash. The ash content of a fuel and the ash composition have a major impact on trouble free operation of a gasifier or a burner. Slagging or clinker formation in the reactor, caused by melting and agglomeration of ashes, at the best will greatly add to the amount of labour required to operate the gasifier. If no special measures are taken, slagging can lead to excessive tar formation and/or complete blocking of the reactor.
- *Reactivity:* The reactivity is an important factor determining the rate of reduction of carbon dioxide to carbon monoxide in a gasifier. Reactivity depends in the first instance on the type of fuel. For example, it has been observed that fuels such as wood, charcoal and peat are far more reactive than coal.
- *Size and size distribution:* Low bulk density feedstock may cause flow problems in the gasifier or burner as well as an inadmissible pressure drop over the reduction zone and a high proportion of dust in the gas. Large pressure drops will lead to reduction of the gas load, resulting in low temperatures and tar production. Excessively large sizes of particles or pieces give rise to reduction in reactivity of the fuel, resulting in start-up problems and poor gas quality, and to transport problems through the equipment. A large range in size distribution of the feedstock will generally aggravate the above phenomena. Too large particle sizes can cause gas channelling problems. Fluidized bed gasifiers are normally able to handle fuels with particle diameters varying between 0.1 and 20 mm (FAO, 2007).

- *Bulk density:* Fuels with high bulk density are advantageous because they represent a high energy-for-volume value. Consequently, these fuels need less bunker space for a given refuelling time. Low bulk density fuels sometimes give rise to insufficient flow under gravity, resulting in low gas heating values and ultimately in burning of the char in the reduction zone. Inadequate bulk densities can be improved by briquetting or pelletizing.

All of the abovementioned biomass properties could be altered by subjecting raw biomass to various processing methods and forming composites. Before choosing a gasifier, it is important to ensure that the individual biomass meets the requirements of the gasifier or that it can be treated to meet these requirements.

1.3 Effect of moisture content on pellet quality

The moisture in biomass both acts as a facilitator of natural binding agents and a lubricant (Kaliyan and Morey, 2006a). Many studies have indicated that the production of high quality pellets is possible only if the moisture content of the feed is between 8 and 12% (wb). Moisture contents above or below this range would lead to lower quality pellets (Hill and Pulkinen, 1988; Kashaninejad et al., 2011; Li and Liu, 2000; Obernberger and Thek, 2004; Shaw and Tabil, 2007). In general, an increase in moisture content from 10 to 44% could result in up to 30-40% decrease in pellet densities of biomass (Chancellor, 1962; Grover and Mishra, 1996; Gustafson and Kjelgaard, 1963; Kaliyan and Morey, 2006a; Mani et al., 2002 and 2006b; Smith et al., 1977). However, the percentage decrease in density depends on the type of biomass. Therefore, a moisture content of 10% (w.b.) is considered as optimal moisture content to obtain high density and durability pellets.

1.4 Effect of grind size on pellet quality

In general, finer grinds produces higher quality pellets since they can readily absorb moisture than large particles, and therefore, undergo a higher degree of conditioning. In addition, finer grinds have higher surface area of contact to form bonds/solid bridges during the compaction processes. Also, large particles are fissure points that cause cracks and fractures in compacts (MacBain, 1966). A reduction in hammer mill screen size from 3.2 to 0.6 mm can result in an increase in pellet densities from 5 to 16% (Kaliyan and Morey, 2006b; Kashaninejad et al., 2011; Mani et al., 2002 and 2004a). However, no significant trend in change in density were observed at geometric mean particles size of 0.6 mm and lower (Kaliyan and Morey 2006b; Mani et al., 2002). The change in pellet density depends on the type of biomass.

This chapter will address various factors that directly or indirectly effect densification of agricultural biomass residue into high quality pellets. The compression and compaction characteristics of ground biomass will be dealt in detail that will provide a comprehensive understanding of the behaviour of biomass as influenced by various factors. The compression studies will explore the affect of independent variables such as biomass, treatment, grind size, and moisture content on pellet density and durability, while compaction studies will study the effect of various machine variables on the pellet quality. In addition, overall specific energy requirements will be established and techno-economic models will be explained.

2. Lignocellulosic composition and higher heating values

The experimental lignocellulosic composition of agricultural straw can be determined using the modified NREL LAP method for "Determination of Structural Carbohydrates and Lignin in Biomass" (Table 1) (Adapa et al., 2011; Sluiter et al., 2008). This procedure uses a

two-step acid hydrolysis to fractionate the biomass into forms that are more easily quantified. During this process, the lignin fractionates into acid insoluble material and acid soluble material, while the polymeric carbohydrates are hydrolyzed into the monomeric forms, which are soluble in the hydrolysis liquid and subsequently can be measured using HPLC. The Percentage cellulose in the samples can be measured using the percentage glucan content, while the percentage hemicelluloses can be measured by adding the percentage mannose, galactose, xylose and arabinose content in the biomass samples.
Table 1 shows the lignocellulosic composition and higher heating values of non-treated and steam exploded barley, canola, oat and wheat straw samples. In general, the cellulose, hemicelluloses and lignin content of steam exploded straw was higher than non-treated straw. This may be due to other components (soluble lignin, loosely-bound sugars) being washed away during steam explosion, thereby leaving the proportion of insoluble lignin, cellulose and hemicellulose in the resulting dried sample higher than for the non-treated samples (i.e. higher percent of dry mass).

Properties of Biomass	Barley Straw		Canola Straw		Oat Straw		Wheat Straw	
	NT	SE	NT	SE	NT	SE	NT	SE
Composition (% of dry matter)								
Cellulose[b]	22.7 ± 0.9[a]	25.3 ± 1.8	22.4 ± 0.8	27.5 ± 1.1	25.4 ± 1.0	27.4 ± 2.4	27.1 ± 1.0	29.9 ± 1.4
Hemicellulose[c]	21.2 ± 0.5	21.0 ± 1.4	16.9 ± 0.5	20.2 ± 0.7	21.7 ± 0.9	18.8 ± 1.2	21.1 ± 0.5	19.7 ± 0.9
Galactose	0.9 ± 0.0	0.7 ± 0.0	1.0 ± 0.0	0.9 ± 0.1	0.8 ± 0.0	0.7 ± 0.0	0.8 ± 0.0	0.9 ± 0.1
Mannose	1.6 ± 0.2	1.5 ± 0.0	2.3 ± 0.1	1.9 ± 0.4	1.4 ± 0.0	1.7 ± 0.1	1.6 ± 0.1	2.8 ± 0.2
Xylose	14.4 ± 0.3	15.3 ± 1.0	11.5 ± 0.5	14.3 ± 0.2	15.1 ± 0.8	13.3 ± 1.0	14.9 ± 0.4	13.5 ± 0.4
Arabinose	4.4 ± 0.2	3.5 ± 0.5	2.0 ± 0.1	3.2 ± 0.0	4.4 ± 0.2	3.1 ± 0.2	3.9 ± 0.1	2.6 ± 0.2
Total Lignin[d]	21.0 ± 0.6	21.6 ± 0.6	19.6 ± 0.6	22.3 ± 0.2	19.5 ± 0.6	23.7 ± 0.2	22.5 ± 0.7	24.2 ± 0.3
Soluble Lignin	1.6 ± 0.1	1.4 ± 0.1	1.6 ± 0.1	1.2 ± 0.1	1.5 ± 0.1	1.3 ± 0.1	1.4 ± 0.0	1.0 ± 0.1
Insoluble Lignin	19.4 ± 0.6	20.2 ± 0.6	18.0 ± 0.6	21.1 ± 0.1	17.9 ± 0.7	22.4 ± 0.1	21.0 ± 0.7	23.3 ± 0.4
Higher Heating Values (MJ/kg of dry matter)								
HHV (MJ/kg)	16.4±0.3‡†	17.4±0.1	16.7±0.3	18.3±0.0	16.4±0.1	17.8±0.0	17.0±0.2	17.8±0.0

DM - Dry Matter; NT - Non-Treated; SE - Steam Exploded; [a] Average and standard deviation of 3 replicates at 95% confidence interval; [b]%Cellulose = %glucan; [c]%Hemicellulose = %(mannose + galactose + xylose + arabinose);
[d]%Total Lignin = %(soluble lignin + insoluble lignin); HHV - Higher Heating Values (measured using Parr 1281 Bomb Calorimeter); ‡3 replicates; † 95% confidence interval

Table 1. Lignocellulosic composition and higher heating values of non-treated and steam exploded agricultural straw (Adapa et al., 2011)

The calorific (heating) value of biomass feedstocks are indicative of the energy they possess as potential fuels. The gross calorific value (higher heating value, HHV) and the net calorific value (lower heating value, LHV) at constant pressure measures the enthalpy change of combustion with and without water condensed, respectively (Demirbaş, 2007). A bomb calorimeter can be used to determine the HHV of the non-treated and steam exploded straw in MJ/kg. In addition, the ASTM Standard D5865-03 (ASTM, 2003) test method for gross calorific value of coal and coke, can be used as a guideline for heating value testing (Table 1).

Cellulose, hemicelluloses and lignin are major components of a plant biomass. Therefore, a change in composition could potentially lead to change in HHV of the biomass (Adapa et al., 2010a). The Net combined percentage change of cellulose, hemicelluloses and lignin in steam exploded barley, canola, oat and wheat straw is 5%, 19%, 5% and 4% higher than non-treated straw, respectively. As a result, the average HHV of steam exploded barley, canola, oat and wheat straw was 6%, 10%, 9% and 5% higher than non-treated straw, respectively (Table 1).

3. Lab-scale pelleting of agricultural biomass

3.1 Compression test

A compression apparatus having a close fit plunger die assembly can be used to make a single compact in one stroke of the plunger from ground straw samples (Adapa et al., 2006 and 2010a; Mani et al., 2004). The compression test should be performed to study the effect of independent variables such as biomass, treatment, grind size, and moisture content on pellet density and durability. In order to simulate frictional heating during commercial pelleting operation, the compression die should be maintained at pre-heat temperatures of 75 to 100°C (Adapa et al., 2006; Kaliyan and Morey, 2009; Mani et al., 2006). Different levels of pre-set compressive forces can be applied using the Instron testing machine. Typical pre-set loads in the range of 31.0 to 150.0 MPa are applied to make pellets. Figure 1 represents the photographs of pellets made from barley, canola, oat and wheat straw grinds from hammer mill screen sizes of 3.2, 1.6 and 0.8 mm (Adapa et al., 2010a).

3.2 Single-pellet density

The density of pellet is calculated from the mass and volume (measuring the length and diameter) of compacts. In general, the density of pellets from agricultural straw significantly increases with an increase in applied pressure at any specific hammer mill screen size. An increase in pressure results in plastic deformation of ground particles and consequently leads to pellets that have densities closer to their respective particle densities (Adapa et al., 2010a; Kaliyan and Morey, 2009; Mani et al., 2004). The Application of pre-treatment has been observed to significantly increase the pellet density since pre-treated straw has lower geometric particle diameters and significantly higher particle densities (Adapa et al., 2010a; Kashaninejad and Tabil, 2011). Usually, it has been reported that an increase in moisture content from 10% and up results in a significant decrease in pellet quality (Hill and Pulkinen, 1988; Li and Liu, 2000; Obernberger and Thek, 2004; Shaw and Tabil, 2007). In general, a decrease in hammer mill screen size results in an increase in pellet density (Adapa et al., 2010a; Kaliyan and Morey, 2009; Kashaninejad et al., 2011; Mani et al., 2004). A comprehensive literature on various single-pellet compression test data is provided in Table 2.

Adapa et al. (2010a) reported that the type of agricultural biomass did not have any significant effect on pellet density, while steam explosion pre-treatment, applied pressure and screen size had significant effects. In addition, correlation for pellet density with applied pressure and hammer mill screen size having highest R^2 values were developed (Table 3). Similarly, Kaliyan and Morey (2009) indicated that the pellet density of corn stover or switchgrass briquettes was significantly affected by pressure, particle size, moisture content and preheating temperature. Kashaninejad and Tabil (2011) also indicated that the pellets made from microwave-chemical pretreated biomass grinds had a significantly higher density and tensile strength than the untreated or samples pretreated by microwave-distilled water.

The densities of pellets should also be measured after a storage period of one week to one month to ascertain its dimensional stability, and associated handling and storage costs (Adapa et al., 2010b; Kaliyan and Morey, 2009). Adapa et al. (2010b) reported that a reduction in pellet density is usually expected due to relaxation of grinds in the pellet after release of pressure. They have observed that the relaxation was higher for larger hammer mill screen sizes and lower applied pressures. In some cases, the average reduction in density was negative giving the impression that pellet density actually increased during storage period. However, these negative values are primarily due to higher standard deviations in pellet density measurements. Therefore, from a practical manufacturing point of view, these values should be considered as a zero percent change in pellet density (Adapa et al., 2010b).

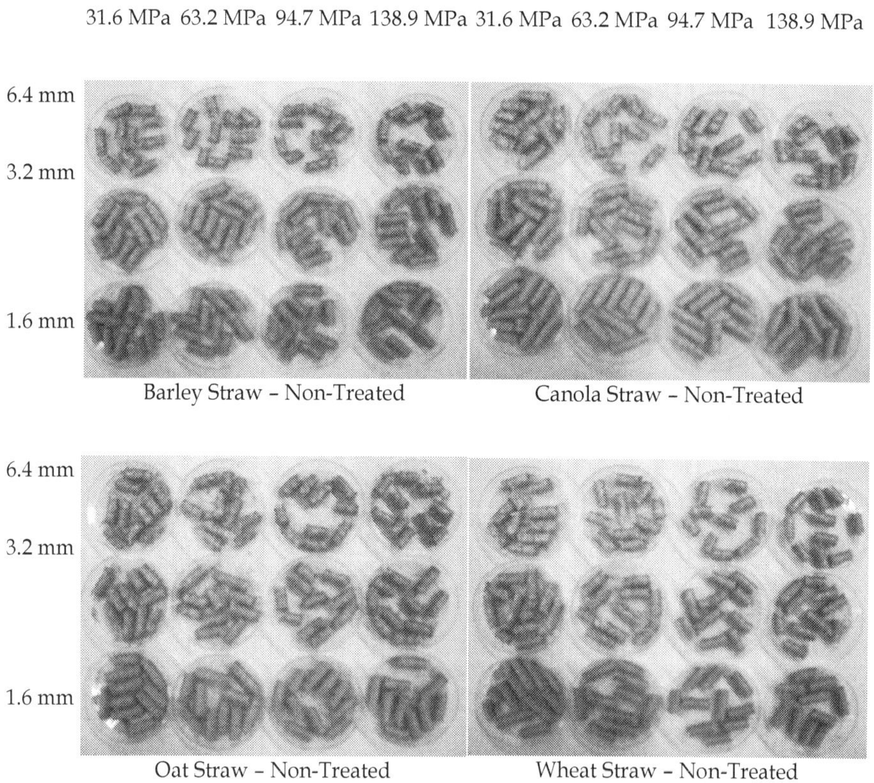

Fig. 1. Photograph of pellets made from barley, canola, oat and wheat straw grind from hammer mill screen sizes of 3.2, 1.6, and 0.8 mm.

Biomass	Independent Variables	Maximum Pellet Density	Reference
Barley, Canola, Oat and Wheat Straw	Hammer Mill Screen Size: 3.2, 1.6 and 0.8 mm Applied Pressure: 31.6, 63.2, 94.7 and 138.9 MPa Moisture Content: 10% (w.b.) Treatment: Non-Treated (NT) and Steam Exploded (SE)	Barley-NT 1003 kg/m^3 Barley-SE 1169 kg/m^3 Canola-NT 1035 kg/m^3 Canola-SE 1165 kg/m^3 Oat-NT 1024 kg/m^3 Oat-SE 1165 kg/m^3 Wheat-NT 1009 kg/m^3 Wheat-SE 1180 kg/m^3	Adapa et al., 2010a
Corn Stover and Switchgrass	Hammer Mill Screen Size: 3.0 and 4.6 mm Applied Pressure: 100 and 150 MPa Pre-Heat Temperature: 25, 75 and 100°C Moisture Content: 10 and 15% (w.b.)	Corn Stover 1197 kg/m^3 Switchgrass 1098 kg/m^3	Kaliyan and Morey, 2009
Barley and wheat straw	Hammer Mill Screen Size: 1.6 mm Applied Pressure: 126.3 MPa Moisture Content: 12% (w.b.) Treatment: Non-Treated (NT), Microwave Pretreated (MT) and Microwave-Chemical Pretreated (MCT)	Barley-NT 995 kg/m^3 Barley-MT 984 kg/m^3 Barley-MCT 1440 kg/m^3 Wheat -NT 950 kg/m^3 Wheat -MT 1032 kg/m^3 Wheat -MCT 1431 kg/m^3	Kashaninejad and Tabil, 2011
Barley and Wheat Straw, Corn Stover, and Switchgrass	Hammer Mill Screen Size: 3.2, 1.6 and 0.8 mm Applied Pressure: 31.6, 63.2, 94.7, 126.3 and 138.9 MPa Moisture Content: 6.22 to 8.30% (w.b.)	Barley Straw 1245 kg/m^3 Wheat Straw 1344 kg/m^3 Corn Stover 1399 kg/m^3 Switchgrass 1173 kg/m^3	Mani et al., 2004
Poplar Wood and Wheat Straw	Hammer Mill Screen Size: 3.2 and 0.8 mm Applied Pressure: 31.6, 63.2, 94.7 and 126.3 MPa Moisture Content: 9 and 15% (w.b.) Pre-Heat Die Temperature: 70 and 100°C Treatment: Non-Treated (NT) and Steam Exploded (SE)	Poplar-NT 1100 kg/m^3 Poplar-SE 1341 kg/m^3 Wheat-NT 1005 kg/m^3 Wheat-SE 1324 kg/m^3	Shaw, 2008

Table 2. Comprehensive literature review on single-pellet compression tests for agricultural biomass as feedstock for biofuel

Independent Variables and Interactions	Estimated Coefficients	R^2 value	Coefficient of Variation	Root Mean Square Error
$\rho_{BarleyNT} = 587.19 + 6.29(P) - 0.025(P)^2$				
Intercept	587.19	0.91	2.99	27.59
P	6.29			
P*P	-0.025			
$\rho_{BarleySE} = 727.38 + 7.41(P) - 0.03(P)^2 - 0.08(P \times S) + 2.84(S)^2$				
Intercept	727.38	0.89	3.30	34.89
P	7.41			
P*P	-0.03			
P*S	-0.08			
S*S	2.84			
$\rho_{CanolaNT} = 545.50 + 7.41(P) - 0.03(P)^2 - 0.11(P \times S)$				
Intercept	545.50	0.89	3.49	32.62
P	7.41			
P*P	-0.03			
P*S	-0.11			
$\rho_{CanolaSE} = 738.77 + 7.54(P) - 0.03(P)^2 + 0.11(P \times S) - 28.84(S)$				
Intercept	738.77	0.92	3.20	22.55
P	7.54			
P*P	-0.03			
P*S	0.11			
S	-28.84			
$\rho_{OatNT} = 666.24 + 5.79(P) - 0.02(P)^2 - 0.12(P \times S) + 3.54(S)^2$				
Intercept	666.24	0.87	3.13	29.00
P	5.79			
P*P	-0.02			
P*S	-0.12			
S*S	3.54			
$\rho_{OatSE} = 818.41 + 6.56(P) - 0.03(P)^2 + 0.12(P \times S) - 32.48(S)$				
Intercept	818.41	0.92	2.66	28.37
P	6.56			
P*P	-0.03			
P*S	0.12			
S	-32.48			
$\rho_{WheatNT} = 700.76 + 5.99(P) - 0.02(P)^2 - 36.06(S) + 3.48(S)^2$				
Intercept	700.76	0.90	3.01	27.91
P	5.99			
P*P	-0.02			
S	-36.06			
S*S	3.48			
$\rho_{WheatSE} = 717.60 + 6.77(P) - 0.03(P)^2$				
Intercept	717.60	0.91	2.95	31.55
P	6.77			
P*P	-0.03			

Note: ρ - Density, kg/m^3; NT - Non-Treated; SE - Steam Exploded; P - Pressure, MPa; S - Hammer Mill Screen Size, mm

Table 3. Correlation for pellet density (ρ, kg/m^3) with applied pressure (P, MPa) and hammer mill screen size (S, mm) for non-treated and steam exploded straw grinds.

3.3 Durability

Durability represents the measure of shear and impact forces that a pellet could withstand during handling, storing and transportation process. The durability of pellets is usually measured following the ASABE Standard S269 (ASABE, 2007), which require about 50-100 g of pellets/ compacts. However, due to the limited number of pellets obtained during single-pellet compression test, it is not feasible to use this method. Instead, the durability of pellets can be measured by following the drop test method (Al-Widyan and Al-Jalil, 2001; Khankari et al., 1989; Sah et al., 1980; Shrivastava et al., 1989), where a single pellet is dropped from a 1.85 m height on a metal plate. The larger intact portion of the mass retained is expressed as the percentage of the initial weight.

Adapa et al. (2010) reported that the type of agricultural biomass, steam explosion pretreatment, applied pressure and screen size all had significant effect on pellet durability. Statistically, no significant correlation (R^2 values) was obtained for change in durability with applied pressure and hammer mill screen sizes. In general, pellet durability increases with an increase in applied pressure and grind size, and application of pre-treatment. Similarly, Kaliyan and Morey (2009) indicated that the durability of corn stover or switchgrass briquettes was significantly affected by pressure, moisture content and preheating temperature, while particle size did not have any significant effect. Kashaninejad et al. (2011) also reported the mean durability of pellets made of giant wild rye and mixed forage increased from 63.08 to 89.26% and from 61.47 to 89.21%, respectively when the hammer mill screen size increased from 0.8 to 3.2 mm. This could be primarily due to mechanical interlocking of relatively long fibers at higher grind sizes. They also indicated that at any specific compressive load, the pellet durability of biomass grinds with 12% moisture content was significantly higher than samples with 9 and 15% and demonstrates the moisture contents above or below 12% would lead to lower quality pellets.

3.4 Specific energy for compaction and extrusion of pellet

During the compression and extrusion processes of individual biomass compacts, the force-displacement data is recorded and can be used to calculate the specific compression and extrusion energies following the methodology reported by Adapa et al. (2006) and Mani et al. (2006). The area under the force-displacement curve can be integrated using the trapezoid rule (Cheney and Kincaid, 1980); when combined with the pellet mass, the specific energy values in MJ/t can be calculated.

During single-pellet compression and extrusion, the pellets are prepared by densifying material against a base plate (representing the specific energy required to overcome friction within the straw grinds) as opposed to commercial operation where compacts are formed due to back pressure effect in the die. Therefore, the specific energy required to extrude the compact should be included, which will closely emulate the specific energy required to overcome the friction between the ground compressed biomass and the die. Mani et al. (2006) have indicated that the extrusion (frictional) energy required to overcome the skin friction was roughly half of the total energy (12-30 MJ/t) for corn stover. Mewes (1959) showed that roughly 40% of the total applied energy was used to compress the materials (straw and hay) and the remaining 60% was used to overcome friction. Faborode and O'Callaghan (1987) studied the energy requirement for compression of fibrous agricultural materials. They reported that chopped barley straw at 8.3% (wb) moisture content consumed 28-31 MJ/t of energy, while un-chopped material consumed 18-27 MJ/t. Shaw (2008) reported that between 95 and 99% of the total specific energy was required to compress the grinds, whereas between 1 and 5% of the total specific energy was required to extrude the compact in single pellet tests.

Shaw (2008) also reported that the mean values of specific compression energy ranged from 7.2 (pretreated wheat straw using steam explosion) to 39.1 MJ/t (wheat straw). Kashaninejad and Tabil (2011) indicated that microwave-distilled water and microwave-NaOH pre-treatments significantly increased the specific energy required for compression of wheat straw grinds so that it increased from 16.60 MJ/t to as high as 29.04 and 27.84 MJ/t after pre-treatment by microwave-distilled water and microwave-NaOH, respectively. They also reported less specific energy was required to compress wheat straw pre-treated by combination of microwave and $Ca(OH)_2$. More specific energy was required to eject the pre-treated wheat straw grinds than the untreated wheat straw grinds and it increased from 3.20 MJ/t to 23.08 after pre-treatment by microwave-NaOH. Data analysis showed that the total energy required for compression and ejection of wheat straw grinds pre-treated by microwave-distilled water or microwave-alkaline was higher than untreated samples.

Adapa et al. (2010b) reported that the type of agricultural biomass, steam explosion pretreatment, applied pressure and screen size all had significant effect on specific energy required to form a pellet. In addition, they have developed correlations for specific energy with applied pressure and hammer mill screen size having highest R^2 values for barley, canola, oat and wheat straw (Table 4). In general, the total and compression specific energy for compaction of non-treated and steam exploded barley, canola, oat and wheat straw at any particular hammer mill screen size significantly increased with an increase in applied pressure and significantly decreased with a decrease in hammer mill screen size.

Adapa et al. (2010b) also reported that the specific energy values obtained from the single-pellet compression tests should be used to compare the densification variables. However, these values may not have practical applications since the energy consumed by commercial densification machines / pilot-scale pellet mills may be higher.

4. Compression characteristics of biomass

4.1 Compression mechanism

The compression characteristics of ground agricultural biomass vary under various applied pressures. It is important to understand the fundamental mechanism of the biomass compression process, which is required in the design of energy efficient compaction equipment to mitigate the cost of production and enhance the quality of the product (Mani et al., 2004). To a great extent, the strength of manufactured pellets depends on the physical forces that bond the particles together (Tabil and Sokhansanj, 1996). These physical forces come in three different forms during pelleting operations: a) thermal; b) mechanical; and c) atomic forces (Adapa et al., 2002).

Pellets are formed by subjecting the biomass grinds to high pressures, wherein the particles are forced to agglomerate. It is generally accepted that the compression process is categorized in several distinct stages and difficult to let one simple monovariate equation to cover the entire densification region (Sonnergaard, 2001). Compression of grinds is usually achieved in three stages (Holman, 1991). In the first stage, particles rearrange themselves under low pressure to form close packing. The particles retain most of their original properties, although energy is dissipated due to inter-particle and particle-to-wall friction. During the second stage, elastic and plastic deformation of particles occurs, allowing them to flow into smaller void spaces, thus increasing inter-particle surface contact area and as a result, bonding forces like van der Waal forces become effective (Rumpf, 1962; Sastry and Fuerstenau, 1973; Pietsch, 1997). Brittle particles may fracture under stress, leading to mechanical interlocking (Gray, 1968). Finally, under high pressure the second stage of compression continues until the particle density of grinds has been reached. During this phase, the particles may reach their melting point and form very strong solid bridges upon cooling (Ghebre-Sellassie, 1989). Figure 2 shows the deformation mechanisms of ground particles under compression (Comoglu, 2007; Denny, 2002).

Independent Variables and Interactions	Estimated Coefficients	R^2 value	Coefficient of Variation	Root Mean Square Error
$\rho_{BarleyNT} = 587.19 + 6.29(P) - 0.025(P)^2$				
Intercept	587.19	0.91	2.99	27.59
P	6.29			
P*P	-0.025			
$\rho_{BarleySE} = 727.38 + 7.41(P) - 0.03(P)^2 - 0.08(P \times S) + 2.84(S)^2$				
Intercept	727.38	0.89	3.30	34.89
P	7.41			
P*P	-0.03			
P*S	-0.08			
S*S	2.84			
$\rho_{CanolaNT} = 545.50 + 7.41(P) - 0.03(P)^2 - 0.11(P \times S)$				
Intercept	545.50	0.89	3.49	32.62
P	7.41			
P*P	-0.03			
P*S	-0.11			
$\rho_{CanolaSE} = 738.77 + 7.54(P) - 0.03(P)^2 + 0.11(P \times S) - 28.84(S)$				
Intercept	738.77	0.92	3.20	22.55
P	7.54			
P*P	-0.03			
P*S	0.11			
S	-28.84			
$\rho_{OatNT} = 666.24 + 5.79(P) - 0.02(P)^2 - 0.12(P \times S) + 3.54(S)^2$				
Intercept	666.24	0.87	3.13	29.00
P	5.79			
P*P	-0.02			
P*S	-0.12			
S*S	3.54			
$\rho_{OatSE} = 818.41 + 6.56(P) - 0.03(P)^2 + 0.12(P \times S) - 32.48(S)$				
Intercept	818.41	0.92	2.66	28.37
P	6.56			
P*P	-0.03			
P*S	0.12			
S	-32.48			
$\rho_{WheatNT} = 700.76 + 5.99(P) - 0.02(P)^2 - 36.06(S) + 3.48(S)^2$				
Intercept	700.76	0.90	3.01	27.91
P	5.99			
P*P	-0.02			
S	-36.06			
S*S	3.48			
$\rho_{WheatSE} = 717.60 + 6.77(P) - 0.03(P)^2$				
Intercept	717.60	0.91	2.95	31.55
P	6.77			
P*P	-0.03			

Note: ρ - Density, kg/m³; NT - Non-Treated; SE - Steam Exploded; P - Pressure, MPa; S - Hammer Mill Screen Size, mm

Table 4. Correlation for pellet density (ρ, kg/m³) with applied pressure (P, MPa) and hammer mill screen size (S, mm) for non-treated and stem exploded straw grinds.

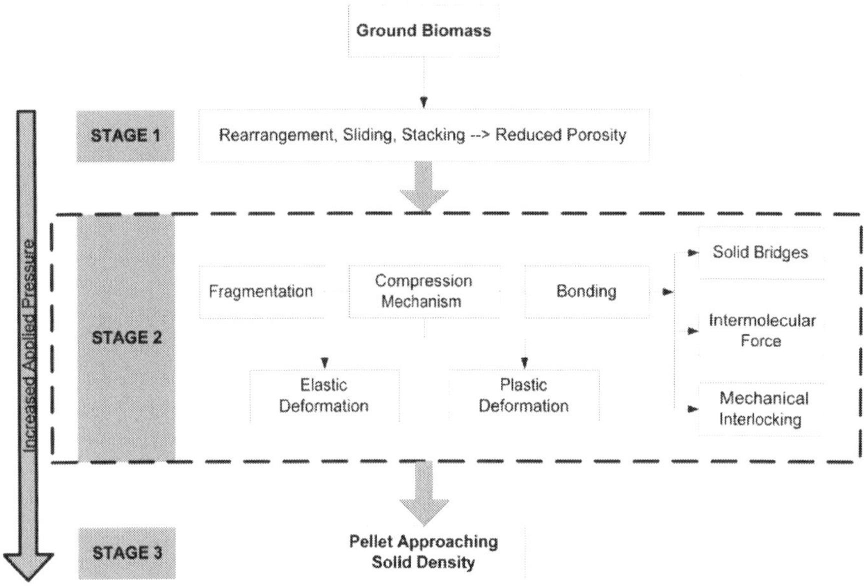

Fig. 2. The deformation mechanisms of ground particles under compression (Comoglu, 2007; Denny, 2002)

4.2 Compression characteristics models

Densification or compaction of agricultural biomass grinds into pellets is an essential process towards production of biofuels. Ground biomass particles behave differently under different applied pressures (Adapa et al., 2002 and 2009a). Therefore, it is important to investigate the change in compact density and volume with pressures. One of the main purposes of fitting experimental data to an equation is usually to develop linear plots in order to make comparisons easier between different sets of data (Comoglu, 2007). A majority of compression models applied to biomass materials have been discussed and reviewed in detail by Adapa et al. (2002 and 2009a), Denny (2002) and Mani et al. (2003). Adapa et al. (2009a) reported that Kawakita and Ludde (1971), Cooper and Eaton (1962) and Jones (1960) models provided the best compression and deformation characteristics of agricultural biomass.

4.2.1 Jones model

Jones (1960) expressed the density-pressure data of compacted powder in the form of equation 1.

$$ln(\rho) = mln(P) + b \tag{1}$$

where, ρ is bulk density of compact powder mixture, kg/m^3, P is applied compressive pressure, MPa; m and b are model constants

The constants b and m are determined from the intercept and slope, respectively, of the extrapolated linear region of the plot of $ln(\rho)$ vs $ln(P)$. The constant m has been shown to be equal to the reciprocal of the mean yield pressure required to induce plastic deformation (York and Pilpel, 1973). A large m value (low yield pressure) indicates the onset of plastic deformation at relatively low pressure, thus, the material is more compressible.

4.2.2 Cooper-Eaton model

Cooper and Eaton (1962) studied the compaction behavior of four ceramic powders. In each case it, was assumed that compression is attained by two nearly independent probabilistic processes, namely, the filling of voids having equal size as particles and filling of voids smaller than particles. Based on these assumptions, the following equation (2) was given:

$$\frac{V_O - V}{V_O - V_S} = a_1 e^{-\frac{k_1}{P}} + a_2 e^{-\frac{k_2}{P}} \tag{2}$$

where, V_O = volume of compact at zero pressure, m^3; V = volume of compact at pressure P, m^3; V_S = void free solid material volume, m^3; a_1, a_2, k_1, and k_2 = Cooper-Eaton model constants.

The difficulty in practical use of equation (2) is the assignment of some physical significance to the constant parameters. In addition, another drawback of this model is its applicability to only one-component system (Comoglu, 2007).

4.2.3 Kawakita-Ludde model

Kawakita and Ludde (1971) performed compression experiments and proposed an equation for compaction of powders based on observed relationship between pressure and volume (Equation 3).

$$\frac{P}{C} = \frac{1}{ab} + \frac{P}{a} \tag{3}$$

Where,

$$C = \frac{V_O - V}{V_O}$$

C = degree of volume reduction or engineering strain; a and b = Kawakita-Ludde model constants related to characteristic of the powder.

The linear relationship between P/C and P allows the constants to be evaluated graphically. This compression equation holds true for soft and low bulk density powders (Denny, 2002; Kawakita and Ludde, 1971), but particular attention must be paid on the measurement of the initial volume of the powder. Any deviations from this expression are sometimes due to fluctuations in the measured value of V_O. The constant a is equal to the values of $C = C_\infty$ at infinitely large pressure P.

$$C_\infty = \frac{V_O - V_\infty}{V_O}$$

Where, V_∞ = net volume of the powder, m^3.

It has been reported that the constant a is equal to the initial porosity of the sample, while constant $1/b$ is related to the failure stress in the case of piston compression (Mani et al., 2004).

4.3 Compressibility of different biomass

The constant m in the Jones (1960) model can provided valuable information about the onset of plastic deformation of the ground agricultural biomass. It has been observed that ground particles obtained from larger hammer mill screen sizes has higher compressibility. In addition, application of pre-treatment also improves the compressibility of the agricultural biomass (Adapa et al., 2010).

The dimensionless coefficients, a_1 and a_2 in Cooper and Eaton (1962) model represent the densification of powdered material by particle rearrangement and deformation, respectively. If the sum of coefficients ($a_1 + a_2$) is less than unity, it is an indication that other process must become operative before complete compaction is achieved. For agricultural biomass grinds, the a_1 values were higher than a_2 values, hence the material was primarily densified through the process of particle rearrangement. Occasionally, the sum of coefficients ($a_1 + a_2$) for agricultural biomass was observed to be above unity. The phenomenon of having sum of coefficient more than unity was also observed by Adapa et al. (2002 and 2009a), and Shivanand and Sprockel (1992), which implies that the densification could not be fully attributed to the two mechanisms of compression as assumed by the Cooper and Eaton (1962) model (Adapa et al., 2010a).

In the Kawakita and Ludde (1971) model, constant a represents the initial porosity of the sample. It has been reported that the porosity and hammer mill screen sizes (corresponding geometric mean particle diameter) are positively correlated. In addition, porosity increases with application of pre-treatment since organized lignocellulosic structure of biomass disintegrates during this process. The parameter $1/b$ in the Kawakita-Ludde model indicates the yield strength or failure stress of the compact. In general, the yield strength has negative correlation with hammer mill screen sizes. Also, application of pre-treatment lowers the yield strength of ground agricultural biomass. Statistically, the Kawakita and Ludde (1971) model has been observed to provide accurate representation of the compression and deformation characteristics of agricultural biomass (Adapa et al., 2010a).

5. Pilot-scale pelleting of agricultural biomass

Pilot-scale densification of biomass is required to demonstrate the feasibility of production of pellets by application of various variables studied during single-pellet experiments. A pilot-scale pellet mill such as CPM CL-5 pellet mill (Figure 3) (California Pellet Mill Co., Crawfordsville, IN) can be used for processing of agricultural straw grinds into pellets. The pellet mill usually consists of a corrugated roller and ring die assembly, which compacts and extrudes the biomass grinds from the inside of a ring-shaped die by pressure applied by rolls where either the die or the roll suspension is rotating. Rolls are mounted close to the die surface, but still leaving room for a compacted feed layer to enter the roll gap. Friction between feed layer and rolls makes the rolls rotate (Larsson et al., 2008). In addition to variables indicated in the single-pellet testing, the quality of pellets also depends on machine variables such as the ring die size (radius), length (thickness, l), ring hole diameter (d), l/d ratio, and the rotational speed of the pellet mill (Adapa et al., 2004; Hill and Pulkinen, 1988; Tabil and Sokhansanj, 1996). A monitoring study of commercial pellets was done by Hill and Pulkinen (1988), on variables such as die geometry, conditioning temperatures, natural moisture of the grind, forage quality, bulk density of the grinds, and the use of binding agents. Similarly, Larsson et al. (2008) studied the effect of raw material moisture content, steam addition, raw material bulk density, and die temperature on production of high quality pellets. Also, Serrano et al. (2011), determined the effect of grind size, moisture content and customization of barley straw by adding pine dust to the mixture (blended pellets).

The feed rate of ground biomass to the pellet mill can be controlled using a vibratory feeder (Figure 3). The feed rate should be optimized according to the pellet mill capacity, which will directly affect the throughput. The pilot-scale pelleting test should be performed for a predefined period and the manufactured pellets should be collected and weighed to determine the pellet mill throughput (kg/h). In addition, the pellet mill energy consumption (kWh) should be recorded in real time using a data logger connected to a computer and should be used to calculate the specific energy (MJ/t) required to manufacture pellets from ground agricultural biomass.

Raw materials causing uneven pellet production have low bulk density compared to other milled biofuel pellet raw materials. Low raw material bulk density will put higher demands on the die feeding system of the pelletizer with greater volume throughput for maintained production level. Larsson et al. (2008) investigated the pre-compaction of reed canary grass as an alternative to avoid low and intermittent production of biofuel pellets. They have observed that the process of pre-compaction can increase the bulk density of raw material from 150 kg/m^3 to 270 kg/m^3, which resulted in the continuous production of pellets at a moisture content of 13.8% (w.b.). Pressurized steam conditioners are used in the feed pellet industry to decrease raw material porosity and to improve pellet hardness/ durability (Thomas et al., 1997). Adapa et al. (2010b) were unable to produce any pellets due to the low bulk density of both non-treated and pre-treated agricultural straw grinds at 10% moisture content (w.b.). Therefore, they have added moisture and oil to increase bulk density of grinds to a level of 17.5% (w.b.) and 10% (by weight), respectively, which resulted in production of pellets. Similar observation was made by Serrano et al. (2011) where they have to increase the grind moisture content in the range of 19-23% (w.b.) to produce pellets in a pellet mill. However, addition of pine sawdust to barley straw resulted in high quality pellets at a lower moisture content of 12% (w.b.).

Testing and, if required improving the durability of pellets is important for the industry to evaluate pellet quality and minimize losses during handling and transportation. The concept is not to add any external binders to enhance pellet quality, but rather activate the natural binders in the agricultural biomass by application of various variables, pre-processing techniques and pre-treatments. Biomass pellets can be customized based on proximate analysis data to make them suitable for direct combustion and thermo-chemical conversion applications. Customization can be achieved by forming composites of different straws to control important variables such as energy and ash content of pellets. Similarly, addition of biomass having good binding characteristics to straw with less cohesive characteristics may enhance particle bonding resulting in durable pellets.

Adapa et al. (2010b) reported pellet mill tests on both non-treated and steam exploded agricultural biomass at different hammer mill screen sizes. They have successfully produced pellets from ground non-treated barley, canola, oat and wheat straw at hammer mill screen sizes of 0.8 and 1.6 mm having moisture content of 17.5% (wb) and flax seed oil of 10% by weight. The non-treated ground straw at 3.2 and 6.4 mm screen size did not produce pellets. Similar pelleting process was followed for ground steam exploded straw. Due to very low bulk density and poor flowability, the steam exploded grinds did not produce pellets at any of the hammer mill screen sizes used in the investigation. However, the customized barley, canola, oat and wheat straw having 25% steam exploded material by weight at 0.8 mm screen size successfully produced pellets. Addition of higher percentage of steam exploded straw and customization at screen sizes of 1.6, 3.2, and 6.4 mm did not produce pellets, which could be due to the fact that adding steam exploded (having very low bulk density) to non-treated straw (having relatively higher bulk density) decreased the overall bulk density and flowability of the grinds, thus hindering the production of pellets in the pilot scale mill. The pilot scale pellet mill in this test is constrained with a small motor (3.7 kW (5 hp)) running it, whereas in a commercial pellet mill, the motors are much bigger and more tolerant to changes in feed bulk density. Shaw et al. (2007) reported similar trends where the quality of wheat straw pellets increased with an increase in moisture content to 15.9% (wb). Figure 4 shows the photograph of pellets manufactured from barley, canola, oat and wheat straw from non-treated grinds at 0.8 and 1.6 mm screen sizes, and customized straw grinds at 0.8 mm having 25% steam exploded straw by weight (Adapa et al., 2010b).

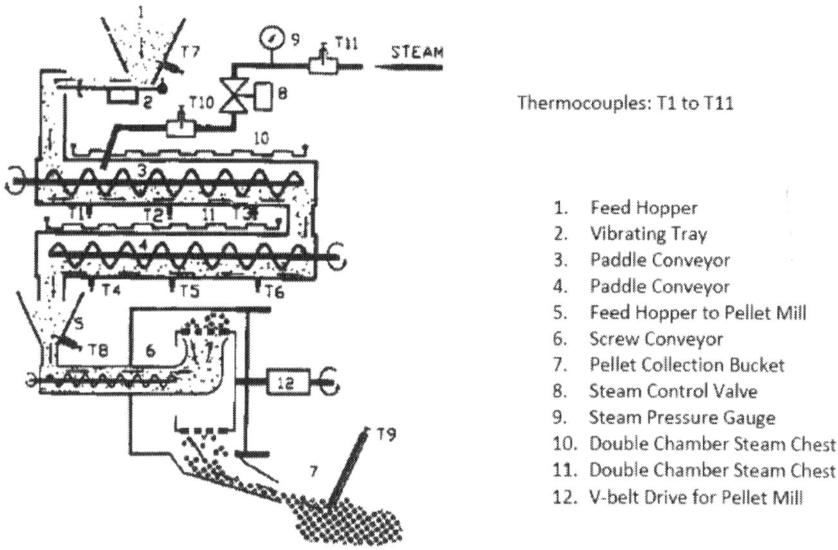

Fig. 3. Schematic diagram of CPM CL-5 pellet mill

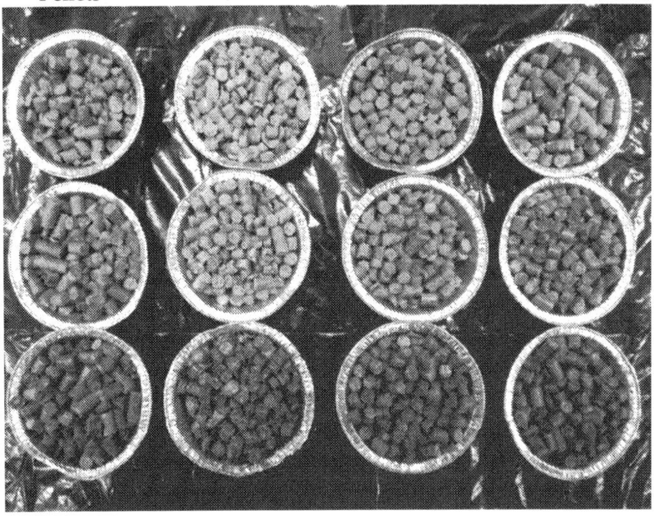

Fig. 4. Photograph of pellets manufactured using a pilot scale pellet mill for non-treated (NT) straw at 1.6 and 0.8 mm hammer mill screen size, and customized grinds at 0.8 mm screen size having 25% steam exploded (SE) straw.

5.1 Pellet bulk density

The mass, length and diameter of individual pellets should be used to determine individual pellet density in kg/m^3. The bulk density of manufactured pellets can be calculated by measuring the mass of pellets filled in a cylindrical container of known volume.

Larsson et al. (2008) reported that the most influential factor for the pellet bulk density was raw material moisture content and showing a negative correlation. Similarly, two other studies have observed that the bulk density of wheat straw, big bluestem grass, corn stover, sorghum stalk and switchgrass decreased with an increase in moisture content (Colley et al., 2006; Theerarattananoon, et al., 2011). Larsson et al. (2008) did not find any correlation between pellet bulk density and die temperature, which contradicts to the observations made by Hill and Pulkinen,1988, and Tabil and Sokhansanj (1996). Serrano et al. (2011) did not observe any significant effect of hammer mill screen size (4 mm and 7 mm) on pellet density. However, pellet density decreased with an increase in moisture content.

Adapa et al. (2010b) reported pellet density obtained from non-treated straw samples at 1.6 and 0.8 mm, and customized sample having 25% steam exploded straw at 0.8 mm screen size (Table 5). In general, pellet density increased with a decrease in screen size from 1.6 to 0.8 mm. However, no significant differences in density values were observed for non-treated samples at 0.8 mm and customized samples, except for canola and oat straw. This could be due to large fluctuation in individual pellet density values. All of the pellet density values reached near individual biomass particle densities at respective grind sizes (Adapa et al., 2010b).

Bulk density of pellets from barley, canola, oat and wheat straw showed significant difference with grind size and customization, except for wheat straw pellets at 0.8 mm for non-treated and customized samples (Table 5). In general, the average pellet bulk densities obtained for customized straw samples were higher (except for barley straw), which is consistent with increase in particle densities. The bulk densities of pellets manufactured were higher than the minimum design value of 650 kg/m^3 suggested by Obernberger and Thek (2004) for wood pellet producers.

5.2 Pellet durability

The durability of pellets can be measured following the ASABE Standard S269 (2007). The method states that 100 g of pellet sample should be weighed and placed in a dust-tight enclosure/ chamber, and tumbled for 10 min at 50 r/min. A 5.70 mm sieve should used to determine the fines produced by the pellets during the tumbling process. The mass of pellets left on the sieve, as percentage of the total mass of pellet sample used during the test, was considered as the durability of the pellets.

Hill and Pulkinen (1988) reported an increase in pellet durability by about 30 to 35% with an increase in pellet temperature from 60 to 104°C. A die length-to-diameter ratio (*l/d*) of 8 to 10 was also reported to be ideal for making high quality pellets. Similarly, Tabil and Sokhansanj (1996) conducted a study for improving the physical quality of alfalfa pellets by controlling and optimizing the manufacturing process. The process conditions investigated were steam conditioning temperatures, die geometry (length to diameter or *l/d* ratio), hammer mill screen sizes used in grinding dry chops, and die speed. They reported that a higher conditioning temperature (95°C) resulted in improved durability of processed pellets. The durability of samples was generally better using the smaller die (higher *l/d* ratio). The hammer mill screen size did not show any affect on pellet durability. Finally, they reported that high durable pellets are obtained at low die speed (250 rpm). Theerarattananoon et al. (2011) also observed that an increase in length / thickness of die resulted in significant increase in durability of biomass pellets.

Larsson et al. (2008) and Serrano et al. (2011) reported that the most influential factor affecting pellet durability was raw material moisture content and showing a positive correlation. The maximum durability was obtained at moisture content of 14.9% without steam addition, and at 13.7% (w.b.) with steam addition of 2.6% (Larsson et al., 2008). Serrano et al. (2011) found that the highest mechanical durability reached for barley straw pellets was 95.5% at moisture content of 19-23% (w.b.), while no pellets were formed below the 19% moisture content. In addition, they observed that the durability of barley straw pellets increased with addition of pine sawdust at 2, 7 and 12% bu weight. However, the percentage increase in pine sawdust did not have significant effect on durability. Colley et al. (2006) observed highest durability of 95.9% for pellets from switchgrass at a moisture content of 8.6% (w.b.). The moisture content in the range of 9-14%, 9-11% and 14-16% (d.b.) did not have any significant effect on maximum durability of 96.8%, 96.8% and 89.5% for pellets from wheat straw and corn stover, big bluestem grass, and sorghum stalk, respectively (Theerarattananoon, et al., 2011); however, further increasing the moisture content reduced pellet durability for respective agricultural biomass. Also, the durability was observed to be positively correlated to die temperature (Larsson et al., 2008).

Adapa et al. (2010b) reported that the durability of pellets obtained from non-treated straw samples at 1.6 and 0.8 mm, and customized sample having 25% steam exploded straw at 0.8 mm screen size were significantly different (Table 5). In general, higher durability values were observed for non-treated straw samples at 0.8 mm hammer mill screen size. The durability of pellets significantly increased with a decrease in grind size for non-treated samples from 1.6 to 0.8 mm. However, addition of steam exploded straw to non-treated straw at 0.8 mm screen size resulted in a decrease in durability, except for wheat straw. This could be due to the fact that steam exploded material has lower soluble lignin content and higher cellulose and hemicelluloses content compared to non-treated straw (Table 1). This observation is in contrast to Lam et al. (2008), who reported that the quality (durability) of pellets produced from steam exploded sawdust was 20% higher than non-treated sawdust. Though, it is important to note that high durability values (>80%) were obtained for all pilot scale pelleting tests.

5.3 Specific energy and energy balance during pelleting

The durability of pellets was negatively correlated to pellet mill throughput and was positively correlated to specific energy consumption (Table 5). The specific energy values obtained from pilot scale pellet mill are 10-25 times higher than reported by Mani et al. (2006b) and Adapa et al. (2010a and 2009b) for agricultural straw, using a single pellet Instron testing machine. The higher pellet mill specific energy numbers could be due to higher friction values and practical pelleting conditions, which are closer to industrial operations.

An overall specific energy analysis is desired in order to understand the net amount of energy available for the production of biofuels after postharvest processing and densification of agricultural straw. The specific energy analysis was performed for pilot-scale pelleting of non-treated and customized (75% non-treated + 25% steam exploded) barley, canola, oat and wheat straw at 1.6 and 0.8 mm hammer mill screen sizes (Table 6). The specific energy for grinding of straw at 0.8 mm was calculated using regression equations reported in Adapa et al. (2011b). The specific energy for chopping and grinding of biomass, production of pellets using pellet mill and higher heating values for straw were obtained from experimental data (Adapa et al., 2011b and 2010b). In addition, the specific energy required for operating the chopper, hammer mill and pellet mill were 337, 759 and 429 W, respectively. On average, the operation of biomass chopper required five times more energy than chopping of biomass. On the other hand, the grinding of biomass required on an average three times more energy than operation of hammer mill. Interestingly, almost the same amount of energy was required to operate the pellet mill and production of pellets. Total specific energy required to form pellets increased with a decrease in hammer mill screen size from 1.6 to 0.8 mm, however, the total specific energy for the process decreased for customized straw compared to non-treated straw at 0.8 mm screen size (Table 6). It has been determined that the net specific energy available for production of biofuel is a significant portion of original agricultural biomass energy (92-94%) for all agricultural biomass (Table 6).

5.4 Cost and life cycle assessment of biomass densification

Sultana et al. (2010) performed a techno-economic analysis and developed a model for a plant that can produce agricultural straw (barley, oat and wheat) pellets for 30 years. They have included the cost of obtaining the straw, transporting straw to the pellet plant, and producing pellets. Costs incurred by the plant for the production of pellets included capital cost, energy cost, labor cost, and consumable cost. The biomass procurement area was determined to estimate the transportation cost. The scale factors for all the equipment related to pellet production were determined based on the data of previous studies (Sultana et al., 2010). To develop the model, minimum, average and maximum yields of wheat, barley and oats straw in Western Canada were considered. They have determined that the cost of pellets does not change much for capacities over 70,000 tonnes per year (cost of production per tonne is $170.89). Therefore, the optimum size is the same for both average and maximum yield cases. In addition, it was observed that the total cost of pellet production is most sensitive to field cost followed by the transportation cost.

Life cycle assessment (LCA) study was performed on wheat straw production system and densification system in the Canadian Prairies using the LCA modelling software tool SimaPro 7.2 to determine the environmental burdens of manufacturing the wheat straw bale and wheat straw pellet (Li et al., 2011). The factors taken into consideration were greenhouse gas emission, acidification, eutrophication, ozone layer depletion, abiotic depletion, human toxicity, and photochemical oxidation. Li et al. (2011) reported that the production of biomass pellet has higher global warming effect than biomass bale, especially in CO_2 and CH_4 emissions from fossil fuel consumption, which is very high in densification system due to machinery usage. It was also reported that the production of wheat straw pellet has higher environmental impact on acidification, eutrophication, human toxicity and other categories than biomass bale. The dominant factors determining most environmental impacts in agricultural system are fertilizer use and production, while machinery use, manufacturing and energy consumption are main contributors to greenhouse gas emission and other environmental burdens in the densification system (Li et al., 2011).

Agricultural Biomass	Hammer Mill Screen Size (mm)	Pellet Density (kg/m^3)	Pellet Bulk Density (kg/m^3)	Durability (%)	Throughput (kg/h)	Specific Energy (MJ/t)
Barley Straw	1.6 (100% NT)	1158±109*†£ aD	665±01‡ aD	91±00‡ aD	4.88	293
	0.8 (100% NT)	1174±46 aD	700±07 bD	93±01 bD	4.21	353
	0.8 (75% NT + 25% SE)	1184±63 aD	714±02 cD	87±01 cD	3.46	301
Canola Straw	1.6 (100% NT)	1023±85 aE	629±01 aE	90±01 aD	3.86	385
	0.8 (100% NT)	1204±43 bDE	720±04 bE	95±00 bE	3.63	440
	0.8 (75% NT + 25% SE)	1144±50 cD	641±01 cE	82±00 cE	5.51	265
Oat Straw	1.6 (100% NT)	1140±63 abD	631±03 aE	89±01 aE	4.48	340
	0.8 (100% NT)	1188±78 aDE	649±02 bF	93±00 bD	3.81	344
	0.8 (75% NT + 25% SE)	1071±101 bE	676±06 cF	89±01 aF	4.03	335
Wheat Straw	1.6 (100% NT)	1163±57 aD	673±02 aF	94±01 aF	5.44	381
	0.8 (100% NT)	1278±136 bE	721±04 bE	95±01 bE	3.81	297
	0.8 (75% NT + 25% SE)	1213±88 abD	722±04 bG	95±00 cG	4.08	342

NT - Non-treated Straw Samples; SE - Steam Exploded Straw Samples; *10 replicates; ‡3 replicates; † 95% confidence interval; £ Student-Neuman-Keuls test at 5% level of significance for same sample biomass at various hammer mill screen sizes (a, b and c); at same hammer mill screen size for different sample biomass (D, E, F and G)

Table 5. Pellet density, durability, throughput and specific energy data for non-treated and steam exploded barley canola, oat and wheat straw at 17.5% moisture content (wb) and 10% flaxseed oil content

Treatment	Hammer Mill Screen Size (mm)	Specific Energy (MJ/t)				HHV (MJ/t)	Net Energy γ (MJ/t)
		Chopping Biomass	Grinding Biomass	Pilot-Scale Pelleting	Total£		
Barley							
NT*	1.6	11.3	90.4	293	924	16400	15476
NT	0.8	11.3	206.6	353	1100	16400	15300
75% NT + 25% SE*	0.8	11.3	189.3	301	1030	16650	15620
Canola							
NT	1.6	7.1	128.5	385	987	16700	15713
NT	0.8	7.1	363.3	440	1277	16700	15423
75% NT + 25% SE	0.8	7.1	341.6	265	1080	17100	16020
Oat							
NT	1.6	9.9	149.5	340	1029	16400	15371
NT	0.8	9.9	253.6	344	1137	16400	15263
75% NT + 25% SE	0.8	9.9	245.2	335	1120	16750	15630
Wheat							
NT	1.6	8.2	153.3	381	1048	17000	15952
NT	0.8	8.2	382.7	297	1194	17000	15806
75% NT + 25% SE	0.8	8.2	332.1	342	1188	17200	16012

*NT - Non-Treated; SE - Steam Exploded
£ Total Specific Energy = Specific Energy (Chopping Biomass + Operating Chopper + Grinding Biomass + Operating Hammer Mill + Pilot-Scale Pelleting)
γNet Energy = HHV - Total

Table 6. Overall specific energy analysis to show net energy available for production of biofuels after postharvest processing and densification of agricultural straw.

6. Summary

The densification of biomass into durable compacts is an effective solution to meet the requirement of raw material for biofuel production. The compression characteristics of ground agricultural biomass vary under various applied pressures. It is important to understand the fundamental mechanism of the biomass compression process, which is required to design an energy efficient compaction equipment to mitigate the cost of production and enhance the quality of the product. To a great extent, the strength of manufactured compacts depends on the physical forces that bond the particles together. These physical forces are generated in three different forms during compaction operations: a) thermal; b) mechanical; and c) atomic forces. To customize and manufacture high quality products that can withstand various forces during transportation and handling, it is essential to predict desirable and dependent quality parameters (density and durability) with respect to various independent variables (pre-treatment, grind size, applied pressure, hold time, die temperature, and moisture content). In addition, specific energy requirements of manufacturing biomass pellets should be established, which can assist in determining the economic viability of densification process.

The density of biomass pellet has been observed to significantly increase with an increase in applied pressure and a decrease in hammer mill screen size. In addition, application of pre-treatment has observed to significantly increase the pellet density since pre-treated straw has lower geometric particle diameters and significantly higher particle densities. Statistically, agricultural biomass did not have any significant effect on pellet density, while steam explosion pre-treatment, applied pressure, moisture content, pre-heat temperature and screen size had significant effect. A negative correlation has been observed between the pellet bulk density and moisture content, while a positive correlation exists between bulk density and pellet mill die temperature. In general, average pellet bulk densities obtained for customized straw samples is higher as a direct result of increase in particle densities.

Agricultural biomass, steam explosion pre-treatment, applied pressure, moisture content, pre-heat temperature and screen size all had significant effect on pellet durability. In general, durability of pellets increases with an increase in applied pressure and grind size, and application of pre-treatment. An increase in pellet mill die temperature, steam conditioning temperature and die thickness resulted in an increase in pellet durability. No specific trend in durability was observed with customization of straw by mixing non-treated and steam exploded straw grinds.

The specific energy required to form a pellet has been significantly affected by the type of agricultural biomass, steam explosion pre-treatment, applied pressure and screen size. The total and compression specific energy for compaction of non-treated and steam exploded barley, canola, oat and wheat straw at any particular hammer mill screen size significantly increased with an increase in applied pressure and significantly decreased with a decrease in hammer mill screen size. Durability of pellets was negatively correlated to pellet mill throughput and was positively correlated to specific energy consumption. An overall energy balance was performed, which showed that a significant portion of original agricultural biomass energy (92-94%) is available for the production of biofuels.

7. References

Adapa, P.K., Tabil L.G., Schoenau, G.J., Canam, T., & Dumonceaux, T. (2011a). Quantitative Analysis of Lignocellulosic Components of Non-Treated and Steam Exploded Barley, Canola, Oat and Wheat Straw using Fourier Transform Infrared Spectroscopy. *The Journal of Agricultural Science and Technology*, 5(7), (In-Press).

Adapa, P.K., Tabil, L.G., & Schoenau, G.J. (2011b). Grinding Performance and Physical Properties of Non-Treated and Steam Exploded Barley, Canola, Oat and Wheat Straw. *Journal of Biomass and Bioenergy, doi:10.1016/j.biombioe.2010.10.004*, 35(2011), pp. 549-561.

Adapa, P.K., Tabil, L.G., & Schoenau, G.J. (2010a). Compression Characteristics of Non-Treated and Steam Exploded Barley, Canola, Oat and Wheat Straw Grinds. *ASABE Journal of Applied Engineering in Agriculture*, 26(4), pp. 617-632.

Adapa, P.K., Tabil, L.G., & Schoenau, G.J. (2010b). Pelleting Characteristics of Selected Biomass with and without Steam Explosion Pretreatment. *International Journal of Agricultural and Biological Engineering*, 3(3), pp. 62-79.

Adapa, P.K., Tabil, L.G., & Schoenau, G.J. (2009a). Compression Characteristics of Selected Ground Agricultural Biomass. *Agricultural Engineering International: the CIGR Ejournal*, Manuscript 1347, XI(June), pp. 1-19.

Part 3

Process Optimization

20

Performances of Enzymatic Glucose/O_2 Biofuel Cells

Habrioux Aurélien[1], Servat Karine[1], Tingry Sophie[2] and Kokoh Boniface[1]
[1]LACCO "Equipe E-lyse", UMR 6503 CNRS-Université de Poitiers
[2]Institut Européen des Membranes, UMR 5635, ENSCM-UMII-CNRS
France

1. Introduction

Nowadays, the development of stable devices capable of converting chemical energy into electrical one both to supply implantable devices and microelectronic apparatus is related in numerous papers (Bullen et al., 2006; Davis & Higson, 2007; Minteer et al., 2007). It is actually of great interest since it could be at the origin of new insight concerning the treatment of illness such as diabetes, deafness or heart disease (Heller, 2004; Katz & Willner, 2003). Besides it could also be a cheap solution to provide energy for microelectromechanical systems (Calabrese Barton et al., 2004) or to treat wastewater (Fishilevich et al., 2009). In the case where at least one of the catalyst used comes from biological resources (enzyme, microorganisms), these devices are called "biofuel cells". As any fuel cell, a biofuel cell consists of two separated, or not, electrodes, an anode and a cathode. The research topic concerning biofuel cells is very vast and can be divided in three subsections. The first one includes microbial fuel cells which are bio-electrochemical systems that drive a current intensity by mimicking bacterial interactions found in nature. The second one deals with enzymatic biofuel cells which use enzymes as catalysts. In this kind of device the specificity of enzymes leads to the non-separation of each compartment of the cell, allowing to minimize the size of the system (Heller, 2004). The last kind of biofuel cells deals with hybrid biofuel cells that result in the combination of an enzymatic catalyst with an abiotic one. In these applications, it is possible to vary the operating conditions of the cell (pH value, concentration of reactants) so as to increase the power density. This chapter will only be focused on the second and third kind of biofuel cells. For the latter devices, the ideal fuel is obviously glucose which is present in all organic tissues and the oxidant is dioxygen. Nowadays there is a need for improvement in what concerns both the lifetime (in the range of a few days) (Calabrese Barton et al., 2004) and the power density (classically lower than 1 mW cm^{-2}) (Neto et al., 2010) of these apparatus. For this reason we describe herein both phenomena affecting stability and power density of biofuel cells and proposed solutions in terms of electrode assembly, catalysts used and design of the cells. Moreover, since the major way to increase enzymatic electrodes lifetime and efficiency is to improve the enzyme connection with the electrode surface, we will have a special look on the different immobilization techniques presently reported in literature. Besides, in the past few years it has noticeably been demonstrated that abiotic catalysts obviously increased the stability of the device (Kerzenmacher et al., 2008) and involved fast substrate conversion kinetic characteristics (Choi et al., 2009). Consequently, we will have a particular glance on new abiotic nanocatalysts and their use in hybrid biofuel cells.

2. Enzymatic electrodes for glucose/O_2 biofuel cells

Except the lack of stability of enzyme molecules due to their proteic nature, one of the major problems encountered with enzymatic electrodes concerns electron transfer between the enzyme and the electrode surface. In the next part we will describe the different electron transfer mechanisms occurring between an enzyme and the electrode as well as the immobilization techniques of the protein.

2.1 Electron transfer between enzyme and electrode

Enzymes are proteins which have high molecular weights. The active sites of these molecules are located in the organic matrix at a depth of several angstroms from the surface. It is thus easy to understand that kinetically fast electron transfer between enzymes and electrodes surface is difficult to obtain because of great insulation of the active centers (Armstrong et al., 1985). Different strategies have been used by the past to make efficient electrical connections between the enzyme and the electrode surface. Corresponding electron transfer mechanisms can be arranged in two different classes: mediated electron transfer (MET) and direct electron transfer (DET).

The major interest in directly transferring electrons between enzymes and electrodes is to reduce the electrode overpotential which is of particular importance for biofuel cells applications. DET is possible as soon as the distance between the active center of the enzyme and the electrode surface is in the order of a tunneling one (Degani & Heller, 1987). Different evidences for DET between enzymes classically used in glucose/O_2 biofuel cells and electrodes have already been given. Actually, laccase (Gupta et al., 2004), bilirubin oxidase (Shleev et al., 2005) and glucose oxidase (Wang et al., 2009) are capable of exhibiting non-negligible catalytic current densities without the presence of a redox mediator.

In the case of MET, a redox molecule acts as a substrate and is able to transfer electrons between the electrode surface and the active center of the enzymatic molecule. Let's notice that current densities obtained with MET are generally higher that what can be delivered in the case of DET. However, to get efficient MET, the redox mediator must possess some properties which can be deduced from Marcus theory as it was already mentioned by Rusling et al. (Rusling et al., 2008). This theory is used to describe outer sphere electron transfer between an electron donor (D) and an electron acceptor (A) as depicted in Fig. 1.

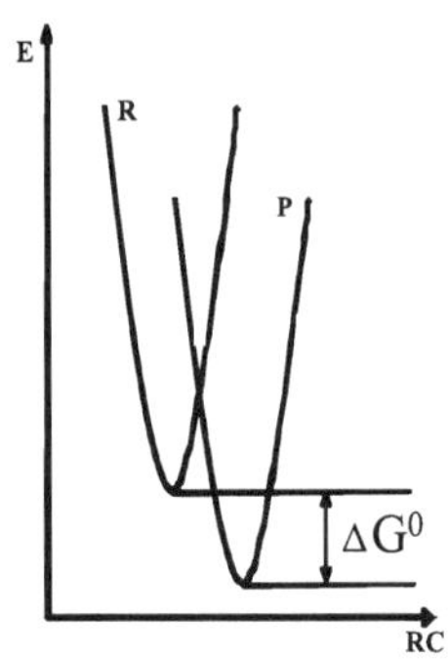

Fig. 1. Curves presenting potential energy of reactants (R) and products (P) ($A^{\delta-}$-$D^{\delta+}$) as a function of reaction coordinates (RC).

The rate (k) of electron transfer can be described as follows (Eq.1.) by an Arrhenius type law.

$$k = AKe^{\left(-\frac{(\Delta G^{0}-\lambda)^{2}}{4RT\lambda}\right)} \tag{1}$$

where A is the collision frequency, K is the electronic transmission factor, ΔG^0 is the Gibbs free energy, R is the gas constant and T the temperature. λ is the reorganization energy (energetic cost associated to the reorganization of both solvent and molecules and necessary to proceed in electronic transfer between the donor and the acceptor). From this relation it can be deduced that to have an efficient electron transfer between enzyme and mediator, it is essential that the redox mediator used presents a highly reversible redox system to minimize λ value. It is also fundamental to minimize the ΔG^0 value. Thus it is very important that formal potentials of mediator and enzyme are close. Moreover, since active centers of enzyme are greatly insulated in high molecular weight molecules it is necessary to use small mediator molecules to reduce the distance of electron transfer and to guarantee a high k value.

2.2 Immobilization of enzymes on electrode surfaces

It is of great interest to develop new non-damaging immobilization techniques of enzyme for the development of stable biofuel cells. In fact, it is very difficult to propose a technique that does not affect the stability of biomolecules. Enzymes are proteins which possess tridimensional structures in which active centers are insulated. To keep the stability of the molecule and to preserve its catalytic efficiency it is necessary not to modify this tridimensional structure and particularly not to affect the environment of the active center. Different combining techniques can be used to immobilize enzymes onto the surface of solid electrodes:

- immobilization into a polymer network
- adsorption on an electrode material
- covalent grafting to an electrode
- immobilization within a membrane.

The first technique consists in immobilizing enzyme in an electropolymerized thin film. It is a very simple technique since it only needs to dip the electrode into a solution containing both monomers and biomolecules. Then the growth of polymer film can be realized by different ways: chronoamperometry (Brunel et al., 2007), chronopotentiometry or cyclic voltammetry (Fei et al., 2007). Different monomers such as pyrrole (Habrioux et al., 2008), aniline (Timur et al., 2004) or phenol (Bartlett et al., 1992) can be electropolymerized. This kind of films can be either conductive or not. The main advantage in using tridimensional conductive films lies in their ability to transfer electrons. Moreover, to increase the number of enzymatic molecules immobilized close to the electrode surface, a first adsorption step of enzymatic molecules can be performed (Merle et al., 2009). Other non-electropolymerized films can be used for enzyme holding. Currently, both chitosan and Nafion® are commonly used (Habrioux et al., 2010; Klotzbach et al., 2008). These two polymers possess surfactant properties interesting to immobilize enzymes in micellar structures (Moore et al., 2004). Moreover, the hydrophobic/hydrophilic property of the polymers can be tuned by modifying the chemical structure of these molecules (Klotzbach et al., 2008; Thomas et al., 2003). It is also possible to simply use retention properties of the Nafion® film for buffering its sulfonic groups (Habrioux et al., 2010). The main problem associated with the use of these polymers lies in the non-control of the film thickness. One of the most promising immobilization techniques has been proposed by Heller's group. This approach consists in immobilizing enzymes in an osmium-based redox polymer (Mao et al., 2003) which is able to swell in contact with water. It acts both as an immobilizing network and an electrochemical mediator. The whole structure of the film leads to very fast electron transfer between the active centers of enzymes and the electrode surface. Another smart technique consists in covalent grafting of enzyme to the electrode surface. Thus Merle et al. (Merle et al., 2008) realized the grafting of amino groups on a carbon electrode before coupling these functions with amino-groups of enzymes using glutaraldehyde. This seems to confer a remarkable stability to the resulting electrode. Another well-known approach has been proposed by Willner et al. (Willner et al., 1996) that consisted in the reconstitution of the enzyme after the grafting of its active center on a gold electrode.

2.3 Enzymatic oxidation of glucose

The development of efficient enzymatic electrodes to oxidize glucose is of relevance for the development of implantable glucose/O_2 biofuel cells. Nowadays, enzymes classically used to perform efficient oxidation of glucose to gluconic acid are either glucose oxidase (GOD) or glucose dehydrogenase (GDH). In the next part the properties of these two enzymes will be explained in details.

2.3.1 Glucose oxidation catalyzed by glucose dehydrogenase

Contrary to GOD, GDH is non-sensitive towards oxygen (Zhang et al., 2007). This is an attractive property for its use in glucose oxygen biofuel cells. However, GDH is an NAD-dependant enzyme. It is well-known that the oxidation of glucose catalyzed by GDH is rather limited by oxidation kinetics of NADH into NAD^+. Even if the use of modified electrodes allows to reduce the overpotential associated to the oxidation of NADH into NAD^+ (Delecouls-Servat et al., 2001), the stability of the electrodes remains poor. Another solution based on the use of a pyrroloquinoline quinone (PQQ) cofactor proposes to suppress the NAD dependence. Nevertheless, the PQQ cofactor has a limited stability (Wang, 2007).

2.3.2 Glucose oxidation catalyzed by glucose oxidase

2.3.2.1 Properties of the enzyme

GOD is by far the most used anode catalyst in glucose/O_2 biofuel cells. Its molecular weight (155 kDa) and molecular size (60 Å × 52 Å × 77 Å) are high (Alvarez-Icaza et al., 1995). This constitutes a limitation for current densities obtained with a solid electrode since the footprint of the enzyme is great. This enzyme possesses two identical linked FAD (Flavine Adenine Dinucleotide) subunits which are responsible for β-D-glucose oxidation (Zhu et al., 2006) to gluconic acid (two electrons reaction product). The redox potential of FAD-$FADH_2$ cofactor is *ca.* -0.36 V *vs.* Ag/AgCl/KCl(sat.) at a pH value of 7.2 (Stankovich et al., 1978), which is of particular interest for biofuel cells applications since it allows low-potential glucose oxidation. In this reaction, dioxygen is the natural electrons acceptor. Therefore, during the oxidation process, dioxygen is reduced towards hydrogen peroxide. The formation of H_2O_2 leads to inhibition of the enzyme because it modifies the amino groups in the vicinity of the active center (Kleppe, 1966). The pH value and the temperature have also an effect on GOD performances. Temperatures higher than 40 °C lead to a drastic decrease of activity (Kenausis et al., 1997). The pH value which optimizes GOD activity greatly depends on the electron acceptor. This value is equal to 5.5 and 7.5 when oxygen (Kenausis et al., 1997) methylene blue (Wilson & Turner, 1992) are used, respectively.

2.3.2.2 Performances of GOD electrodes towards β-D-glucose oxidation

In the case of MET, the use of suitable electrochemical mediators is of importance to increase the rate of electron transfer between the enzyme and the electrode surface since it allows to raise current densities. The second interest lies in the possibility to inhibit the formation of peroxide. Actually, it is just necessary to use a mediator which is able to realize faster electron transfer with GOD than oxygen can do. One of the most efficient systems has been developed by Heller's group (Mano et al., 2005; Mao et al., 2003). It consists of a tridimensional matrix of an osmium based redox polymer containing GOD. The formal potential of the polymer is -195 mV *vs.* Ag/AgCl at pH 7.2. The covalent chain composed of thirteen atoms long allows the increase of the electron diffusion coefficient (Mao et al., 2003) by increasing the collision probability between reduced and oxidized forms of the osmium centers. The reticulation with PEGDGE (polyethyleneglycoldiglycydilether) allows the formation of a redox hydrogel capable of swelling in contact with water. It is probable that the matrix structure is responsible for a weak deformation of the protein structure. Such electrodes are able to deliver a catalytic current at potentials as low as -360 mV *vs.* Ag/AgCl in a physiologic medium containing 15 mM glucose (Mano et al., 2004).

2.4 Enzymatic reduction of oxygen to water

Generally, enzymes used to catalyze the reduction of oxygen into water are either laccase or bilirubin oxidase (BOD). The main property of these enzymes is their ability to directly reduce oxygen to water at potentials higher than what can be observed with platinum based electrodes (Soukharev et al., 2004). These two enzymes are classified in "multicopper oxidases" class and contain four Cu^{2+}/Cu^{+} active centers which are commonly categorized in three types: T_1, T_2 and T_3. T_1 site is responsible for the oxidation of the electron donor. The trinuclear center composed both of T_2 center and two equivalent T_3 centers is the place where oxygen reduction occurs (Palmer et al., 2001). The associated mechanism is proposed in Fig. 2.

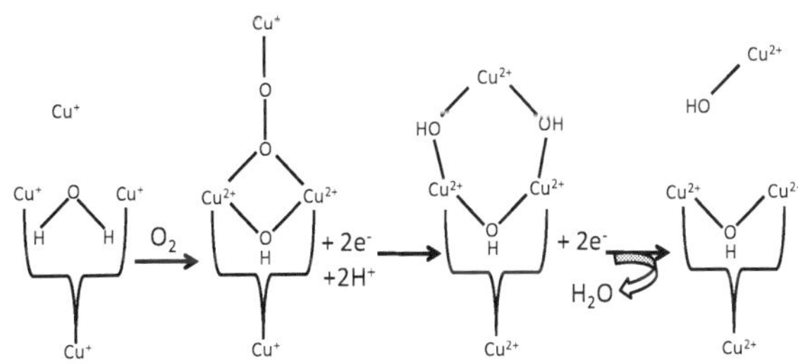

Fig. 2. Oxygen reduction catalyzed by "multicopper oxidases"

In the next part the different properties and performances of both laccase and BOD electrodes will be discussed.

2.4.1 Reduction of oxygen catalyzed by laccase

Laccase is able to oxidize phenolic compounds and to simultaneously reduce oxygen into water. The microorganism from which it is extracted greatly determines the redox potential of the T_1 site which can vary from 430 mV *vs.* NHE up to 780 mV *vs.* NHE (Palmore & Kim, 1999). Laccase from *Trametes versicolor* is the most attractive one since redox potential of its T_1 site is *ca.* 780 mV *vs.* NHE (Shleev et al., 2005). Nowadays, the best performances with laccase electrodes are obtained with osmium based polymers as redox mediators (Mano et al., 2006). Actually these electrodes are able to deliver a current density of 860 μA cm^{-2} at only -70 mV *vs.* O_2/H_2O at pH 5. In the same conditions, the identical current density is obtained at -400 mV *vs.* O_2/H_2O with a platinum wire as catalyst. Nevertheless, performances of laccase (from *Pleurotus Ostreatus*) electrodes drop drastically in the presence of chloride ions (Barton et al., 2002) what constitutes both a major problem and a great challenge for its use in implantable glucose/O_2 biofuel cells.

2.4.2 Reduction of oxygen catalyzed by bilirubin oxidase

BOD is naturally capable of catalyzing the oxidation of bilirubin into biliverdin and to simultaneously reduce dioxygen (Shimizu et al., 1999). BOD is very similar to laccase. Performances of BOD electrodes are greatly related to the amino-acids sequence around T_1 site of the enzyme (Li et al., 2004). It is clearly reported that the most efficient BOD enzyme comes from *Myrothecium verrucaria*. Redox potential of its T_1 site is included between 650 and 750 mV *vs.* NHE, and the enzyme is thermally stable up to 60 °C (Mano et al., 2002b). It is thus possible to use it at physiological temperature without denaturing the protein. To build efficient BOD electrodes intended in working at physiological pH value, it is judicious to use positively charged mediator molecules since the isoelectric point of BOD is close to pH = 4. Actually, during oxygen reduction reaction, the use of an osmium based redox polymer has lead to performances such as 880 μA cm^{-2} at 0.3 V *vs.* Ag/AgCl (physiological conditions) at a scan rate of 1 mV s^{-1} (Mano et al., 2002a). Additionally, the redox osmium based hydrogel conferred a very favorable environment to stabilize BOD since 95% of the initial activity of a BOD electrode can be preserved after three weeks storage (Mano et al., 2002a). This remarkable stability probably results in auspicious electrostatic interactions between the swelling matrix and the enzyme. Performances of BOD electrodes are furthermore unaffected in the presence of chloride ions. In fact BOD remains active for chloride concentrations lower than 1 M (Mano et al., 2002a). This property is of major interest for the development of implantable microscale glucose/O_2 biofuel cells using BOD as cathode catalysts. The major encountered problem with BOD electrodes is the relative lack of stability of the enzyme in physiological serum. Cupric centers of BOD are indeed capable of binding with one urea oxidation product, oxidation reaction catalyzed by the enzyme (Kang et al., 2004). This phenomenon can nevertheless be limited by spreading a Nafion® film on the catalyst (Kang et al., 2004). It is moreover reported that chemically modified Nafion® is capable of constituting a favorable environment to stabilize BOD (Topcagic & Minteer, 2006). Consequently, it seems of interest to immobilize BOD in Nafion® films. A promising technique for the development of efficient BOD electrodes has already been reported in literature (Habrioux et al., 2010). It consists in firstly adsorbing BOD/$ABTS^{2-}$ (2,2-azinobis-3-ethylbenzothiazoline-5-sulfonic acid) complex on a carbon powder, Vulcan XC 72 R in order to increase both enzyme loading, the stability of the protein and the quality of the percolating network in the whole thickness of the polymer film. Actually, to realize the electrochemical reaction, a triple contact point (between the catalytic system, the electrolyte and the electronic conductor) is required. Once the catalytic system is adsorbed, a buffered Nafion® solution is added. The whole system is then immobilized onto a solid carbon electrode (Fig. 3).

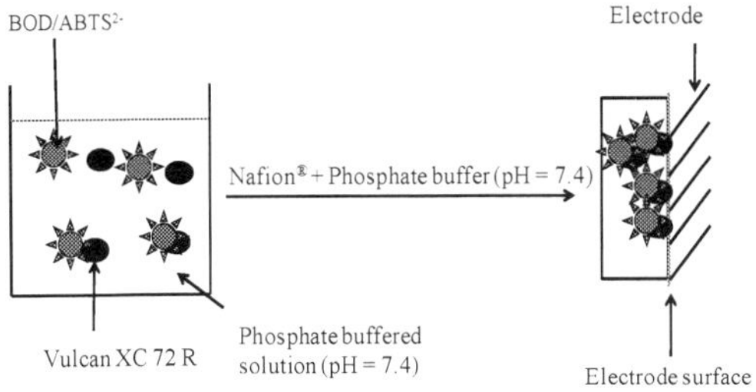

Fig. 3. Method used for the preparation of BOD cathodes according to the process described in Ref. (Habrioux et al., 2010)

Previous studies have shown the interest lying in the use of $ABTS^{2-}$ as redox mediator in combination with multicopper oxidases. One of them was carried out by Karnicka et al. who have shown that wiring laccase to glassy carbon through a $ABTS^{2-}$/carbon nanotube system was a very efficient pathway to reduce molecular oxygen into water (Karnicka et al., 2008). The combination of $ABTS^{2-}$ with BOD is also known to exhibit a high electrochemical activity towards oxygen reduction reaction (Tsujimura et al., 2001). These observations are confirmed by electrochemical studies performed on electrodes previously described (Fig. 3). Results are shown in Fig. 4.

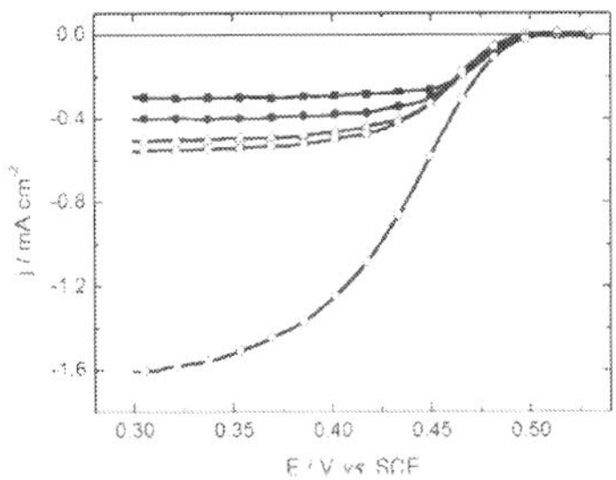

Fig. 4. Oxygen reduction reaction catalyzed by BOD/$ABTS^{2-}$/Nafion® electrode in a phosphate buffered solution (pH = 7.4, 0.2 M) at 25 °C. Curves registered at different rotation rates (Ω), in an air-saturated electrolyte at Ω = 100 rpm (■); Ω = 200 rpm (•); Ω = 400 rpm (Δ); Ω = 600 rpm (□) and in an oxygen saturated electrolyte at Ω = 600 rpm (○). Scan rate 3 mV s^{-1}.

Curves of Fig.4 clearly show the interest of such electrodes that exhibit a catalytic current from potentials as high as -50 mV *vs.* O_2/H_2O (0.536 V *vs.* SCE). Furthermore the half-wave potential is only 100 mV lower than the reversible redox potential of O_2/H_2O. This value is in good agreement with that reported by Tsujimura et al. (0.49 V *vs.* Ag/AgCl/KCl(sat.) at pH = 7.0) (Tsujimura et al., 2001). Let's notice that the half-wave potential value is very close to the redox potential of T_1 site of BOD (0.46 V *vs.* SCE). This has already been explained by the fact that the reaction between $ABTS^{2-}$ and BOD is an uphill one (Tsujimura et al., 2001).

Fig. 4 also shows that electrochemical performances of BOD/ABTS^{2-}/Nafion® clearly depend on the amount of oxygen dissolved in the electrolyte. The limiting current is a plateau and increases from 0.56 mA cm^{-2} in an air saturated electrolyte to 1.61 mA cm^{-2} in an oxygen saturated (at a rotation rate of 600 rpm). Dependence of limiting current with oxygen concentration in the electrolyte is presented in Fig.5. In this figure, current obtained at 0.2 V *vs.* SCE is plotted *versus* oxygen saturation.

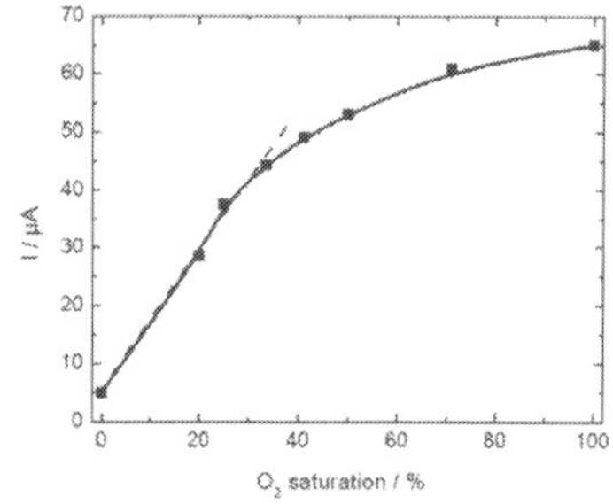

Fig. 5. Electrochemical activity of BOD/ABTS^{2-}/Nafion® electrode: dependence of the current value at 0.2 V *vs.* SCE with oxygen concentration

The current linearly increases with the oxygen concentration from low values to around 35%. This linearity suggests that the reaction is of a first order with oxygen concentration thereby, the Koutecky-Levich plots can be considered. Assuming that the rate determining step is an enzymatic intramolecular electron transfer step, it is possible to express the current density of a BOD/ABTS^{2-}/Nafion® electrode working in an air saturated solution as follows (Schmidt et al., 1999):

$$\frac{1}{|j|} = \frac{1}{|j_0|\frac{\theta}{\theta_c}e^{\left(\frac{\alpha nF}{RT}\right)|\eta|}} + \frac{1}{|j_L^{ads}|} + \frac{1}{|j_L^{film}|} + \frac{1}{|j_L^{diff}|} \tag{2}$$

In Eq.2, $j_L{}^{diff}$ represents the diffusion limiting current density expressed by Levich equation:

$$j_L^{diff} = 0.2nFD^{2/3}\nu^{-1/6}C_0\sqrt{\Omega} \tag{3}$$

In Eq.3, n is the number of electrons exchanged, D the diffusion coefficient, C_0 is the oxygen concentration, Ω is the rotation rate, F the Faraday constant and ν is the kinematic viscosity. Then, $j_L{}^{film}$ corresponds to the limitation due to oxygen diffusion in the catalytic film and $j_L{}^{ads}$ is the limiting current density due to oxygen adsorption on the catalytic site. Since these two last contributions to the total current density do not depend on Ω, it is impossible to separate them. They will be described according to Eq.4.

$$\frac{1}{j_L} = \frac{1}{j_L^{ads}} + \frac{1}{j_L^{film}} \tag{4}$$

In Eq.2, η is the overpotential (η = $E-E_{eq}$), j_0 the exchange current density, α the transfer coefficient, R = 8.31 J mol^{-1} K^{-1}, F=96500 C mol^{-1} and T the temperature. Θ and Θ_c are the covering rates of the active sites of the enzyme at E and E_{eq}, respectively. We will assume that $\theta \approx \theta_c$ for all potential values. From Eq.2, when $\Omega \rightarrow \infty$, the limit of $1/j$ can be expressed as follows:

$$\frac{1}{|j_k|} = \frac{1}{|j_L|} + \frac{1}{|j_0| e^{\left(\frac{\alpha nF}{RT}\right)|\eta|}} \tag{5}$$

In Eq.5, when $\eta \rightarrow \infty$, $1/j_K \rightarrow 1/j_L$. It is thus possible to determine j_L value by extrapolating and reporting the $1/j_K$ values as a function of the potential value E. Transforming Eq.5 (Grolleau et al., 2008), it becomes as follows:

$$\eta = E - E_{eq} = -b\left[\ln\frac{j_L}{j_0} + \ln\frac{j_k}{j_L - j_k}\right] \tag{6}$$

where $b = RT/\alpha nF$ is the Tafel slope. The plot of the η values *vs.* $ln(j_K/(j_L - j_K))$ (Fig. 6) permits the calculation of b and j_0 values.

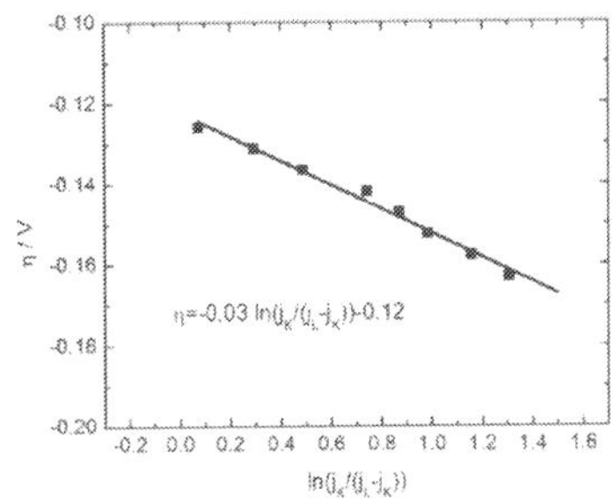

Fig. 6. Curve obtained from Koutecky-Levich treatment on oxygen reduction reaction catalyzed by BOD/ABTS^{2-}/Nafion® system.

Under these experimental conditions, calculated values for both Tafel slope and exchange current density are respectively of 69 mV/decade and 25 µA cm^{-2}. The high value obtained for j_0 confirms the ability of BOD/ABTS^{2-}/Nafion® system to activate molecular oxygen in a physiological type medium. Moreover, it also certifies that the oxygen reduction reaction starts at very high potentials. The reference catalyst classically used to reduce molecular oxygen is platinum. It can be noticed that under similar conditions, the exchange current density is only of 5 µA cm^{-2} when we used platinum nanoparticles as catalyst. This clearly shows the great interest lying in these electrodes to reduce oxygen in glucose/O_2 biofuel cells. Nowadays, the major problem encountered with these electrodes is the lack of stability of the redox mediator (ABTS^{2-}) (Tsujimura et al., 2001).

3. Abiotic catalysts for glucose/O_2 biofuel cells

In this part, a complete description of non-enzymatic catalysts which are used or potentially usable in glucose/O_2 biofuel cells systems is given. The major problem in employing abiotic catalyst in such applications lies in their lack of specificity. Consequently, their application in implantable microscale devices is difficult. Nevertheless, they often lead to fast substrate conversion kinetic characteristics and their stability is incomparably higher than enzymes one. Thus, they can be used as catalysts in biocompatible devices intended in supplying long-term high power densities.

3.1 Non-enzymatic oxidation of glucose

3.1.1 Different offered possibilities

A promising approach consists in using metallophtalocyanines to realize glucose oxidation. Particularly, cobalt phtalocyanines seem to exhibit interesting properties (Zagal et al., 2010). Furthermore, reactivity of these electrodes can be modulated by simple modification of the complex structure what is of interest for the development of electrodes. These catalysts could be used for glucose electrooxidation in glucose/O_2 biofuel cells but it is not still developed.

The other approach lies in the use of metallic nanomaterials as catalysts. Oxidation of glucose on metallic surfaces has extensively been studied. Among all these investigations, numerous ones have been devoted to the understanding of catalytic effect of platinum on glucose oxidation process (Kokoh et al., 1992a; Kokoh et al., 1992b; Sun et al., 2001). Experiments led to conclude that the major oxidation product is gluconic acid (Kokoh et al., 1992b; Rao & Drake, 1969). Actually, the oxidation process involves dehydrogenation of the anomeric carbon of glucose molecule (Ernst et al., 1979). The major interest in including platinum in the catalyst composition lies in its ability to oxidize glucose at very low potentials (lower than 0.3 V *vs.* RHE). However, it is also well-known that platinum surfaces are particularly sensitive to poisoning with chemisorbed intermediates (Bae et al., 1990; Bae et al., 1991). To solve this problem, different heavy atoms (Tl, Pb, Bi and W) have been used as adatoms to modify platinum surfaces to raise electrochemical activity of platinum (Park et al., 2006). Other studies relate glucose oxidation on platinum alloys in which the second metal can be Rh, Pd, Au, Pb (Sun et al., 2001), Bi, Ru and Sn (Becerik & Kadirgan, 2001). It appears that the most efficient catalysts are Pt-Pb or Pt-Bi (Becerik & Kadirgan, 2001). However, these catalysts are sensitive to dissolution of the second metal which prevents their use in fuel cells systems. Moreover most of the materials previously cited are toxic. The only one which could be environmentally friendly is gold even if the oxidative stress caused by nanoparticles on living cells is not well-known. Besides, synthesis of alloyed materials allows increasing significantly catalytic activity of pure metals by synergistic effect. This has noticeably been observed with platinum-gold nanoalloys (Möller & Pistorius, 2004).

3.1.2 Oxidation of glucose on gold-platinum nanoparticles

The oxidation of glucose on gold-platinum nanoparticles has been investigated in numerous studies (Habrioux et al., 2007; Sun et al., 2001). Jin and Chen (Jin & Chen, 2007) examined glucose oxidation catalyzed by Pt-Au prepared by a co-reduction of metallic salts. An oxidation peak of glucose was visible at much lower potentials than on gold electrode. Moreover, they showed that both metals favored the dehydrogenation of the glucose molecule. They concluded that the presence of gold prevents platinum from chemisorbed poisonous species. The efficiency of such catalysts towards glucose oxidation is thus not to be any more demonstrated, and greatly depends on the synthesis method used to elaborate the catalytic material.

3.1.2.1 Synthesis of gold-platinum nanoparticles

Various gold-platinum nanoparticles synthesis methods have been already studied: Polyol (Senthil Kumar & Phani, 2009), sol-gel (Devarajan et al., 2005), water-in-oil microemulsion

(Habrioux et al., 2007), electrodeposition (El Roustom et al., 2007) and Bönnemann (Atwan et al., 2006). Among all these methods, the water-in-oil microemulsion technique produces particles that exhibit high catalytic activity towards glucose electrooxidation (Habrioux et al., 2007). It consists in mixing two microemulsions, one containing the reducing agent in the aqueous phase and the other containing one or several metallic precursors in the aqueous phase. Collisions of water nanodroplets permit to obtain metallic nanoparticles which can be then cleaned and dispersed onto a carbon support. The choice of the different components of the microemulsions is not unique and influences the physical properties of the obtained nanoparticles. Actually, both surfactant molecules and oil-phase chemical nature have an effect on interfacial tension of the surfactant film that determines water solubility in micelles (Paul & Mitra, 2005). This greatly affects intermicellar exchanges. Moreover, the chemical nature of the reducing agent controls the rate of the nucleation step and subsequently the kinetic of particles formation. In the system described herein, n-heptan is used as oil phase, non-ionic polyethyleneglycol-dodecylether as emulsifier molecule and sodium borohydride as reducing agent. The synthesized particles have been dispersed onto Vulcan XC 72 R and then washed several times with acetone, ethanol and water, respectively to remove surfactant from their surface (Habrioux et al., 2009b). The removal of surfactant molecules from all the catalytic sites without modifying structural properties of the catalyst is currently a great challenge (Brimaud et al., 2007). Since electrocatalysis is a surface phenomenon depending on the chemical nature of the surface of the catalyst, on its crystalline structure and on the number of active sites, it is useful to precisely know the physico-chemical properties of the used nanoparticles to understand their electrochemical performances.

3.1.2.2 Electrochemical behaviour of gold-platinum nanoparticles towards glucose electrooxidation

This part aims at showing the importance to realize a correlation between the structural properties of the catalysts and their electrocatalytic activities towards glucose oxidation. The use of nanocatalysts indeed involves a deep structural characterization of the nanoparticles to fully understand the whole of the catalytic process. Therefore, in order to show the presence and the proportion of gold and platinum at the surface of the catalysts, electrochemical investigations have been carried out (Burke et al., 2003). It is indeed possible to quantify surface compositions of the catalysts by using cyclic voltammetry and by calculating the amount of charge associated with both reduction of platinum and gold oxides (Woods, 1971). The charge calculated for pure metals was 493 µC cm^{-2} and 543 µC cm^{-2} for Au and Pt, respectively, for an upper potential value of 250 mV *vs.* MSE (Habrioux et al., 2007) in a NaOH (0.1 M) solution. The atomic ratio between gold and platinum can be thus determined according to Eq. 7 and Eq. 8 assuming that for all bimetallic compositions, the oxidation takes place only on the first atomic monolayer.

$$\%Au = \frac{S_{Au}}{S_{Au} + S_{Pt}} \times 100 \qquad (7)$$

and

$$\%Pt = \frac{S_{Pt}}{S_{Au} + S_{Pt}} \times 100 \qquad (8)$$

Both voltammograms used and results of the quantification are shown in Fig. 7. Mean diameter of the different nanoparticles weighted to their volume (obtained from transmission electron microscopy measurements) as well as their mean coherent domain size weighted to the volume of the particles (obtained from X-ray diffraction measurements) are also presented in Fig. 7.

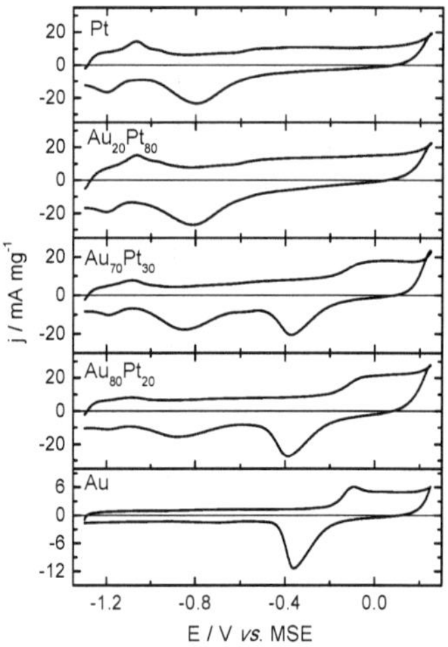

% At. Pt	%At. Au	D_V (nm)	L_v (nm)
100	0	4.7	3.5
100	0	4.6	4.0
44	56	5.2	4.3
29	71	5.3	5.0
0	100	9.4	14.1

Fig. 7. Voltammograms (after 19 cycles) of gold-platinum nanoparticles recorded at 25 °C in alkaline media (0.1 M NaOH). Scan rate = 20 mV s^{-1}. The surface composition of the used catalyst is given on the right of the corresponding voltammogram.

In Fig. 7 it is noticed that for all compositions, desorption of oxygen species occurs in two peaks. The reduction of the gold surface takes place at -0.38 V *vs*. MSE whereas the potential for which platinum surface is reduced depends on the amount of gold in the alloy. Indeed, for pure platinum nanoparticles this potential is *ca*. -0.8 V *vs*. MSE (reduction of platinum oxides). The potential at which oxygen species desorption occurs, shifts to lower potentials when the atomic ratio of gold increases in the composition of alloys. The deformation of this peak increases with the amount of gold probably because of the formation of more complex platinum oxides. The quantification realized on the different bimetallic compositions, clearly shows a platinum enrichment of nanoparticles surfaces. Desorption of gold oxides is indeed invisible for low gold containing samples (*i.e.* with gold content lower than 40%). These nanoparticles exhibit a typical core-shell structure composed of a gold core and a platinum shell (Habrioux et al., 2009b), while high gold content samples (*i.e.* with gold content higher than 80%) possess a surface composition that is close to the nominal one. This results in a purely kinetic effect. Actually, reduction of gold precursor is considerably faster than reduction of platinum cation. Consequently, there is firstly formation of a gold seed on which platinum reduction occurs. So, the natural tendency of these systems is to form core-shell particles. Furthermore, let's notice that both mean diameter of nanoparticles weighted to their volume and their mean coherent domain size weighted to their volume increase with gold content but ever stay in the nanometer range. That is only the result of differences in reduction kinetics of the particles since the ratio water to surfactant remains constant whatever the synthesized sample. To correlate surface composition with efficiency to oxidize glucose for all gold-platinum catalysts compositions, voltammograms were first recorded in alkaline medium. Results are shown in Fig. 8.

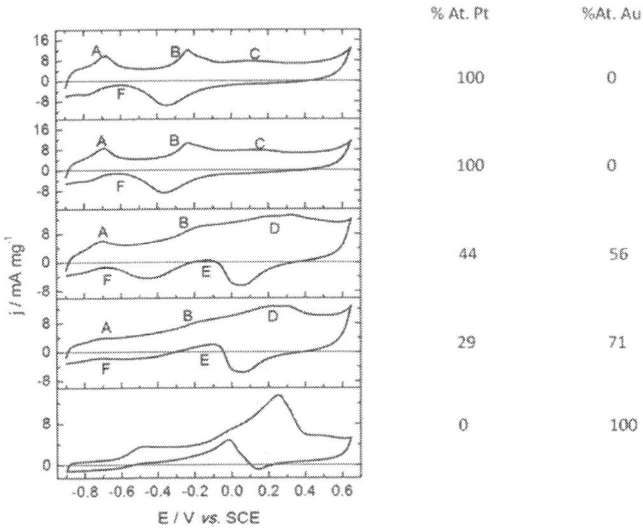

Fig. 8. Voltammograms (after 19 cycles) of gold-platinum nanoparticles recorded at 3 °C in alkaline medium (0.1 M NaOH) in the presence of 10 mM glucose. Scan rate = 20 mV s^{-1}. Surface composition of the used catalyst is given on the right of the corresponding voltammogram.

In Fig. 8, different oxidation peaks appear during the oxidation process on gold-platinum nanocatalysts. When platinum content decreases in the bimetallic surface composition, intensity of peak A, located at *ca.* -0.7 V *vs.* SCE, diminishes. For pure gold catalyst, this peak is furthermore invisible. It is thus related to the oxidation phenomenon on platinum. It has already been attributed to dehydrogenation of anomeric carbon of glucose molecule (Ernst et al., 1979). Peaks B and C correspond to the direct oxidation of glucose molecule (Habrioux et al., 2007) and are located both in gold and platinum oxides region. In the case of catalysts with nominal compositions such as $Au_{70}Pt_{30}$ or $Au_{80}Pt_{20}$, the different oxidation peaks located between -0.3 V *vs.* SCE and 0.4 V *vs.* SCE are not well-defined. For these catalysts, the presence of platinum at their surface allows a low potential oxidation of glucose molecule, which starts earlier than on pure gold. Moreover, on these catalysts, after the dehydrogenation step, current densities raise rapidly. Furthermore, in the potential region where formation of both gold hydroxides and platinum oxides occurs, current densities are very high (*i.e.* 12 mA mg^{-1} at 0.2 V *vs.* SCE). This is the result of a synergistic effect between the two oxidized metals at the bimetallic catalyst surface (Habrioux et al., 2007). Such effect between gold and platinum has already been observed for CO oxidation (Mott et al., 2007). On these catalysts, during the negative going scan, two oxidation peaks, E and F, are visible. During the reduction of both oxidized gold and platinum clusters, oxygenated species are desorbed from the surface and stay at its vicinity. Subsequently, there is desorption of adsorbed lactone from the electrode surface what implies the formation of both peak E and peak F (Beden et al., 1996). Fig. 9 shows the reactions involving in the oxidation of glucose on the catalyst surface.

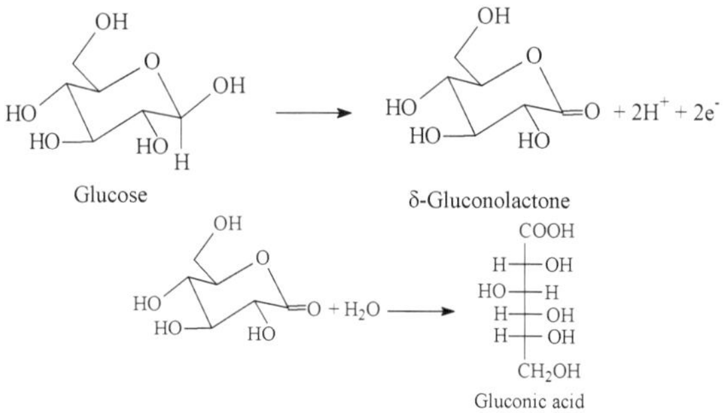

Fig. 9. Oxidation of glucose on gold-platinum catalysts

The remarkable electrocatalytic activity of both $Au_{80}Pt_{20}$ and $Au_{70}Pt_{30}$ nanocatalysts towards glucose electrooxidation is probably the result of a suitable surface composition combined with a convenient crystallographic structure. An X-ray diffraction study (Fig. 10) based on Warren's treatment of defective metals and previously described (Vogel et al., 1998; Vogel et al., 1983) combined with high resolution transmission electron microscopy (HRTEM) measurements allowed to exhibit the peculiar structure of high gold content catalysts (Habrioux et al., 2009b).

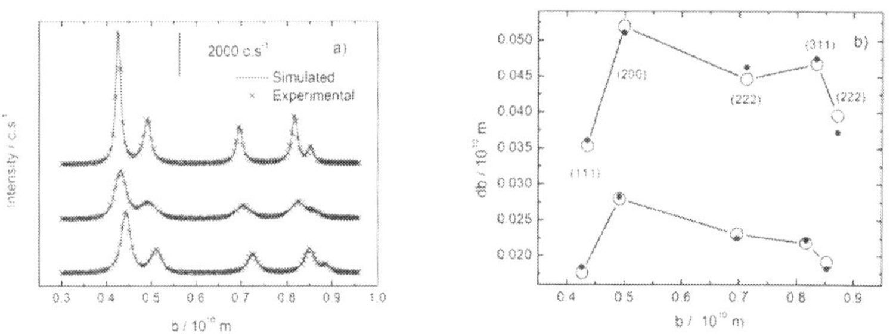

Fig. 10. a) Experimental and simulated diffractograms obtained with Au, $Au_{70}Pt_{30}$ and Pt nanoparticles (from top to bottom), b) Experimental (•) and simulated (○) Williamson-Hall diagrams obtained with $Au_{30}Pt_{70}$ and Au nanoparticles (from top to bottom).

Each experimental diffractogram has been fitted with five Pearson VII functions what gives two important parameters: the accurate peak position b ($b = 2sin\theta/\lambda$) and the integral line width db. The value of db is plotted versus b in Fig.10b. As a result of best fits, it can be assumed that line profiles of diffractograms are lorentzian. This implies that all contributions to the integral line width can be added linearly and can be expressed as follows:

$$db = db_{size} + db_{stacking\ fault} + db_{strain} \quad (9)$$

with
$$db_{size} = \frac{1}{L_v} \quad (10)$$

$$db_{stacking\ fault} = \frac{\alpha V_{hkl}}{a} \quad (11)$$

and
$$db_{strain} = \frac{2\sigma b}{E_{hkl}} \quad (12)$$

where L_v is the mean coherent domain size weighted to the volume of the particles, a the stacking fault probability, V_{hkl} a parameter depending on the miller indexes, σ the mean internal stress and E_{hkl} the young modulus. The fit of Williamson-Hall diagrams with the expression given by Eq.7 leads to the determination of L_v, a and σ for each catalyst. It has been concluded that for catalysts with nominal compositions of $Au_{70}Pt_{30}$ and $Au_{80}Pt_{20}$, both σ and a values were high (Habrioux et al., 2009b). For $Au_{80}Pt_{20}$, these values were indeed of 510 $N.mm^{-2}$ and 8.2%, respectively for σ and a. In the case of $Au_{70}Pt_{30}$, these values were of 490 $N.mm^{-2}$ and 7.4%. HRTEM observations have confirmed the results of the fit since the observed particles present numerous twins and stacking faults, as shown in Fig. 11.

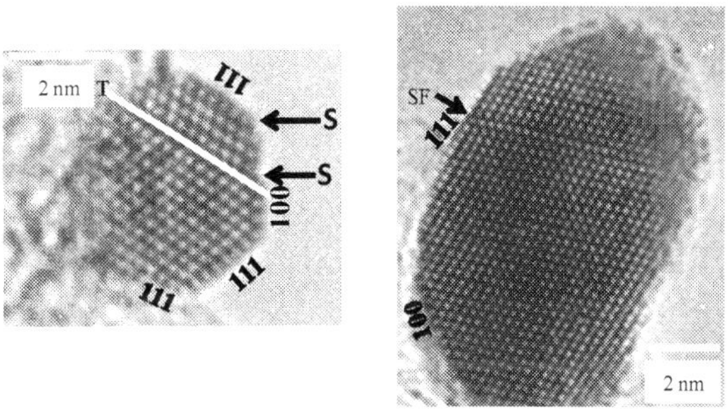

Fig. 11. HRTEM observations of $Au_{70}Pt_{30}$ nanoparticle (left image) and Au nanoparticle (right image).

As a result of the high internal mean strain existing in these particles, there is an important strain energy which leads to the formation of twins and stacking faults. Consequently the equilibrium shape of the particles is modified and the interaction between the different surface atoms is changed. Accordingly, the catalytic behaviour of these particles is greatly affected. This can also explain the remarkable activity of these particles towards glucose oxidation both in alkaline medium as shown in Fig. 8, and in physiological type medium, as shown in Fig. 12. Let's notice that at low potential values, current densities obtained with $Au_{70}Pt_{30}$ and Pt catalysts are similar. Competitive adsorption between phosphate species and glucose molecules can be involved to explain this phenomenon. Actually, de Mele et al. (de Mele et al., 1982) showed that phosphate species are capable of creating oxygen-metal bonds with platinum surfaces and thus inhibiting glucose oxidation. This engenders the low current density observed at low potentials on pure platinum. On $Au_{70}Pt_{30}$ catalyst, it is possible that modification of 5d band center of platinum due to the presence of gold allows discriminating the adsorption of phosphate species. Furthermore, the oxidation of glucose on high gold content catalysts starts at a very low potential value (*i.e.* -0.5 V *vs.* SCE), which can easily be compared with values observed for catalysts such as Pt-Bi, Pt-Sn (Becerik & Kadirgan, 2001) or Pt-Pd (Becerik et al., 1999).

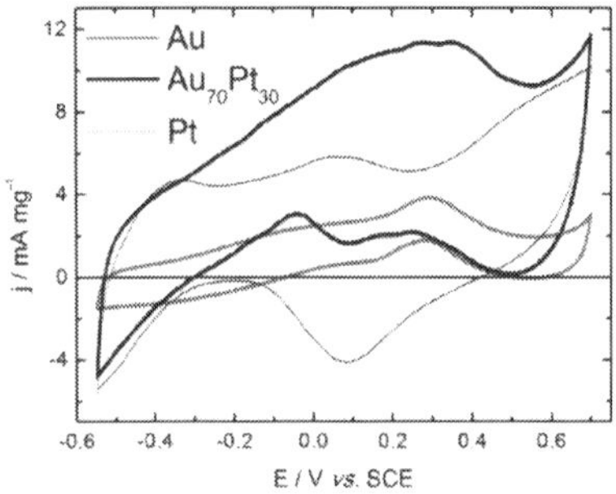

Fig. 12. Voltammograms (after 19 cycles) of gold-platinum nanoparticles recorded at 37 °C in a phosphate buffered solution (0.1 M pH 7.4) in the presence of 10 mM glucose. Scan rate = 20 mV s^{-1}.

3.2 Oxygen reduction reaction on abiotic catalysts

It is difficult to tailor non-enzymatic catalyst, capable of exhibiting electrochemical performances similar to those shown by laccase or BOD in physiological type media. The major problem with enzymes lies in the natural lack of stability of the proteins. One of the possibilities to tailor new efficient and stable cathode catalysts for glucose/O_2 biofuel cells is to artificially reproduce active centers of enzymes and to stabilize their environment by mimicking the structure of enzymatic proteins and by removing all organic parts responsible for instability of enzymes. The possibility of designing this kind of catalyst has already been discussed (Ma & Balbuena, 2007).

4. Design of glucose/O_2 biofuel cells

The global reaction associated to the glucose/O_2 biofuel cell can be described according to Eq. 13 :

$$C_6H_{12}O_6 + \frac{1}{2}O_2 \rightarrow C_6H_{10}O_6 + H_2O \tag{13}$$

Gibbs free energy associated to this reaction is Δ_rG^0 = -251 kJ mol^{-1}. This implies that the theoretical cell voltage is E^0 = 1.3 V (Kerzenmacher et al., 2008). Furthermore, when the cell delivers a current j, the cell voltage $E(j)$ can be expressed as follows:

$$E(j) = E_{eq} - \eta_a - \eta_c - Rj \tag{14}$$

where η_a is the anodic overvoltage, η_c the cathodic one, R the cell resistance and E_{eq} the equilibrium cell voltage. In Eq.14, it clearly appears that both values of η_a, η_c and R must be very low in order to increase the cell performances.

Since the development of the first biofuel cell realized by Yahiro et al. (Yahiro et al., 1964) that consisted in a two-compartment anionic membrane cell in which two platinum foils were used as conducting supports, numerous progress have been realized in designing devices. Nowadays, four main designs are developed. The first one has been developed by Heller's group. It simply consists in using two carbon fibers of 7 µm diameter as electrode materials. On these fibers, enzymes are immobilized in a redox osmium based hydrogel capable of immobilizing enzymes. These two electrodes are directly dipped into the electrolyte. In a physiological medium containing 15 mM glucose, the device was primarily able to deliver a power density of 431 µW cm^{-2} at a cell voltage of 0.52 V (Mano et al., 2002c). The device exhibited a high stability, since after one week of continuous working, it was still capable of delivering 227 µW cm^{-2}. Based on this study, and by replacing carbon fibers by newly engineered porous microwires comprised of assembled and oriented carbon nanotubes, Mano's group (Gao et al., 2010) recently made the most efficient glucose/O_2 biofuel cell ever designed. It indeed achieved a remarkably high power density of 740 µW cm^{-2} at a cell voltage of 0.57 V. The success of the experiment probably lies in the increase of the mass transfer of substrates. Other promising but presently less performing designs of glucose/O_2 biofuel cells have been developed in the recent past years. The first one consists in using a microfluidic channel to build a glucose/O_2 biofuel cell. The laminar flow obtained in the channel at low Reynold's number prevents the electrodes from depolarization phenomena and/or from degradation. The mixing of the reactants indeed occurs only on a very small distance in the middle of the channel. The development of such glucose/O_2 biofuel cells seems of great interest for various applications. It is very simple to use abiotic and non-specific materials as catalysts. Moreover, it offers the possibility of working with two different pH values for the catholyte and the anolyte what can be interesting to improve electrochemical performances of each electrode (Zebda et al., 2009a). Nowadays, these devices are capable of delivering 110 µW cm^{-2} for a cell voltage of 0.3 V (Zebda et al., 2009b) by using GOD and laccase as catalysts. Glucose/O_2 biofuel cells realized with classical fuel cell stacks have also been carried out (Habrioux et al., 2010). Both the used system and the obtained performances are described in Fig. 13.

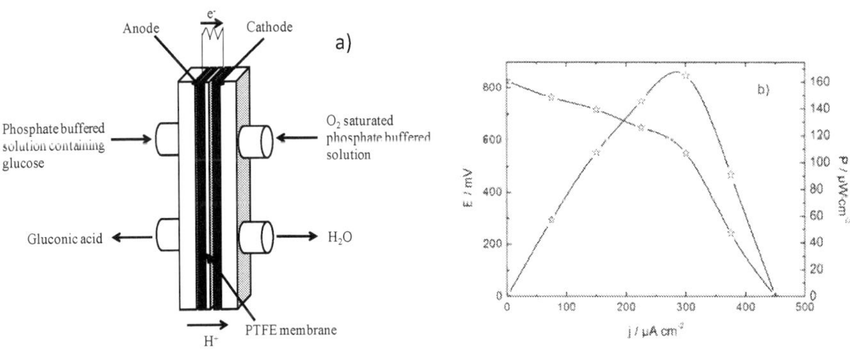

Fig. 13. a) Description of the glucose/O_2 biofuel cell design, b) Characteristic E vs. j of glucose/O_2 cell performed at 20 °C: anode ($Au_{70}Pt_{30}$/Vulcan XC 72R, metal loading 40%); cathode (BOD/ABTS/Vulcan XC 72 R system). Test realized in the presence of a phosphate buffered solution (0.2 M; pH 7.4) containing 0.3 M glucose. The cathodic compartment contains an oxygen saturated phosphate buffered solution (pH 7.4; 0.2 M).

Fig. 13 shows that the maximum power density obtained is 170 μW cm^{-2} for a cell voltage of 600 mV. However, let's notice that performances of the biofuel cell rapidly decrease for current densities higher than 300 μA cm^{-2}. This is clearly due to a very low ionic exchange rate between the two compartments of the cell since this value is too weak to correspond to mass transfer limitation of glucose molecule. The last design of glucose/O_2 biofuel cell developed in the last past years is the concentric device (Habrioux et al., 2008; Habrioux et al., 2009a). It is based on concentric carbon tubes as electrodes and operates at physiological pH. An oxygen saturated solution circulates inside the internal tube composed of porous carbon, which is capable of providing oxygen diffusion. The whole system is immersed in a phosphate buffered solution (pH 7.4, 0.1 M) containing various glucose concentrations. Oxygen consumption occurs at the cathode such that no oxygen diffuses towards the anode. This allows to use in this device both abiotic and enzymatic materials as anode and cathode catalysts, respectively. BOD/ABTS/Vulcan XC 72 R system is immobilized on the internal surface of the inner tube whereas Au-Pt nanocatalysts are immobilized on the internal surface of the outer tube. The surfaces of the cathode and anode were 3.14 and 4.4 cm^2, respectively. The system is fully described in Fig. 14.

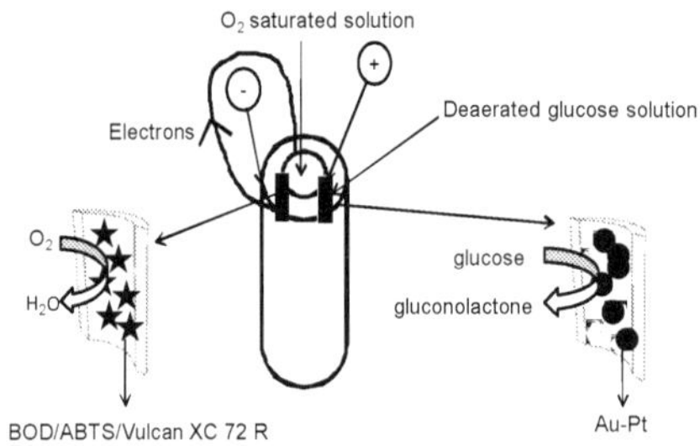

Fig. 14. Schematic view of the glucose/O_2 biofuel cell system

Different fuel cell tests realized by using various nominal compositions of Au-Pt nanomaterials have been realized. The best performances are obtained with $Au_{70}Pt_{30}$ as anodic catalyst. Actually, the maximum power density achieved is approximately of 90 μW cm^{-2} for a cell voltage of 0.45 V. Results are shown in Fig. 15.

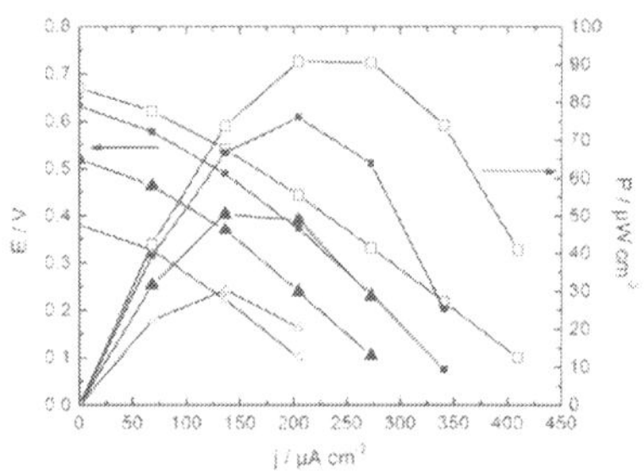

Fig. 15. Fuel cell performances obtained with Au (▲), $Au_{80}Pt_{20}$ (■), $Au_{70}Pt_{30}$ (□) and Pt (Δ) nanoparticles as anode catalysts. These performances were obtained in a phosphate buffered solution (0.2 M, pH 7.4) containing 10 mM glucose at 37 °C. A saturated oxygen solution circulated in the inner tube of the device.

When $Au_{80}Pt_{20}$ is used as anode catalyst, the open circuit voltage is lower (*i.e.* 0.64 V). This is clearly explained by the surface composition of the catalyst which only contains 29 at.% of platinum. In the case of pure platinum, the open circuit voltage is very low due to strong competitions between phosphate species and glucose for adsorption. Such competition also occurs on other Au-Pt catalysts but the presence of gold allows a weaker interaction between phosphate species and the metallic surface. Consequently, higher glucose concentrations were used so as to improve biofuel cell performances. The obtained results are given in Fig. 16.

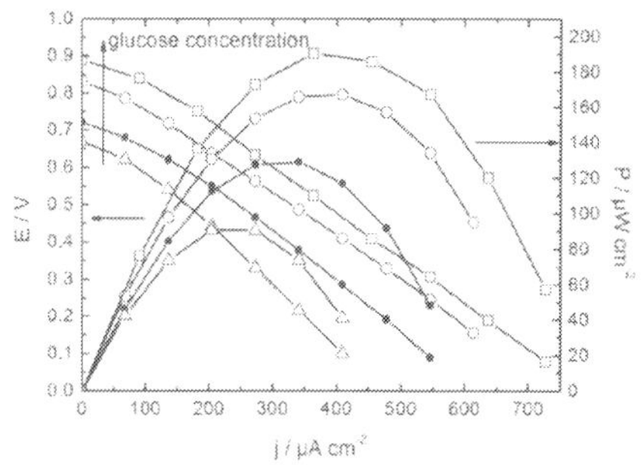

Fig. 16. Fuel cell performances obtained with 10 mM glucose (Δ), 100 mM glucose (•), 300 mM glucose (○) and 700 mM glucose (□), with $Au_{70}Pt_{30}$ nanoparticles as anode catalyst. Performances obtained in a phosphate buffered solution (0.2 M, pH 7.4) at 37 °C. A saturated O_2 solution circulated in the inner tube.

The data show a strong increase in cell voltage with glucose concentration. The raise observed in cell voltage between 0.1 M and 0.3 M can be attributed to the slow adsorption of phosphate species due to the presence of a higher glucose concentration. The maximum power density was also increased from 90 µW cm^{-2} (for a glucose concentration of 10 mM) up to 190 µW cm^{-2} (for a glucose concentration of 0.7 M). Nevertheless, in all cases, the fuel cell performances are greatly limited by resistance of the cell.

5. Conclusion

In this chapter we clearly show the importance of both electrodes assembly and global design of the cell on power output of the glucose/O_2 biofuel cell. Moreover, it seems that a suitable choice of well-characterized nanocatalysts materials can lead both to an increase of the cell performances and to an improvement of their lifetime resulting in the abiotic nature of these materials. The approach, which consists of the utilization of an abiotic anode catalyst and an enzyme for a four electrons reduction, can undoubtedly open new outlooks for biofuel cells applications. This hybrid biofuel cell combines the optimized fuel electrooxidation, as developed in classical fuel cells, with the complete reduction of dioxygen to H_2O without H_2O_2 production. Moreover, a concentric membrane-less design associated with an appropriate immobilization of the catalysts can avoid a costly separator of the cell events. Nevertheless, progresses to develop an efficient cell design are still necessary.

6. References

Alvarez-Icaza, M.; Kalisz, H.M.; Hecht, H.J.; Aumann, K.D.; Schomburg, D. & Schmid, R.D. (1995). The design of enzyme sensors based on the enzyme structure. *Biosensors and Bioelectronics,* Vol. 10, No. 8, pp. 735-742, ISSN 0956-5663.

Armstrong, F.A.; Hill, H.A.O. & Walton, N.J. (1985). Reactions of electron-transfer proteins at electrodes. *Quartely Reviews of Biophysics,* Vol. 18, No. 03, pp. 261-322, ISSN 0033-5835.

Atwan, M.H.; Macdonald, C.L.B.; Northwood, D.O. & Gyenge, E.L. (2006). Colloidal Au and Au-alloy catalysts for direct borohydride fuel cells: Electrocatalysis and fuel cell performance. *Journal of Power Sources,* Vol. 158, No. 1, pp. 36-44, ISSN 0378-7753.

Bae, I.T.; Xing, X.; Liu, C.C. & Yeager, E. (1990). In situ Fourier transform infrared reflection absorption spectroscopic studies of glucose oxidation on platinum in acid. *Journal of Electroanalytical Chemistry and Interfacial Electrochemistry,* Vol. 284, No. 2, pp. 335-349, ISSN 0022-0728.

Bae, I.T.; Yeager, E.; Xing, X. & Liu, C.C. (1991). In situ infrared studies of glucose oxidation on platinum in an alkaline medium. *Journal of Electroanalytical Chemistry and Interfacial Electrochemistry,* Vol. 309, No. 1-2, pp. 131-145, ISSN 0022-0728.

Bartlett, P.N.; Tebbutt, P. & Tyrrell, C.H. (1992). Electrochemical immobilization of enzymes. 3. Immobilization of glucose oxidase in thin films of electrochemically polymerized phenols. *Analytical Chemistry,* Vol. 64, No. 2, pp. 138-142, ISSN 0003-2700.

Barton, S.C.; Pickard, M.; Vazquez-Duhalt, R. & Heller, A. (2002). Electroreduction of O_2 to water at 0.6 V (SHE) at pH 7 on the "wired" Pleurotus ostreatus laccase cathode. *Biosensors and Bioelectronics,* Vol. 17, No. 11-12, pp. 1071-1074, ISSN 0956-5663.

Becerik, I. & Kadirgan, F. (2001). Glucose sensitivity of platinum-based alloys incorporated in polypyrrole films at neutral media. *Synthetic Metals,* Vol. 124, No. 2-3, pp. 379-384, ISSN 0379-6779.

Becerik, I.; Süzer, S. & Kadirgan, F. (1999). Platinum-palladium loaded polypyrrole film electrodes for the electrooxidation of d-glucose in neutral media. *Journal of Electroanalytical Chemistry,* Vol. 476, No. 2, pp. 171-176, ISSN 1572-6657.

Beden, B.; Largeaud, F.; Kokoh, K.B. & Lamy, C. (1996). Fourier transform infrared reflectance spectroscopic investigation of the electrocatalytic oxidation of -glucose: Identification of reactive intermediates and reaction products. *Electrochimica Acta,* Vol. 41, No. 5, pp. 701-709, ISSN 0013-4686.

Brimaud, S.; Coutanceau, C.; Garnier, E.; Léger, J.M.; Gérard, F.; Pronier, S. & Leoni, M. (2007). Influence of surfactant removal by chemical or thermal methods on structure and electroactivity of Pt/C catalysts prepared by water-in-oil microemulsion. *Journal of Electroanalytical Chemistry,* Vol. 602, No. 2, pp. 226-236, ISSN 1572-6657.

Brunel, L.; Denele, J.; Servat, K.; Kokoh, K.B.; Jolivalt, C.; Innocent, C.; Cretin, M.; Rolland, M. & Tingry, S. (2007). Oxygen transport through laccase biocathodes for a membrane-less glucose/O_2 biofuel cell. *Electrochemistry Communications,* Vol. 9, No. 2, pp. 331-336, ISSN 1388-2481.

Bullen, R.A.; Arnot, T.C.; Lakeman, J.B. & Walsh, F.C. (2006). Biofuel cells and their development. *Biosensors and Bioelectronics,* Vol. 21, No. 11, pp. 2015-2045, ISSN 0956-5663.

Burke, L.D.; Moran, J. & Nugent, P. (2003). Cyclic voltammetry responses of metastable gold electrodes in aqueous media. *Journal of Solid State Electrochemistry,* Vol. 7, No. 9, pp. 529-538, ISSN 1432-8488.

Calabrese Barton, S.; Gallaway, J. & Atanassov, P. (2004). Enzymatic Biofuel Cells for Implantable and Microscale Devices. *Chemical Reviews,* Vol. 104, No. 10, pp. 4867-4886, ISSN 0009-2665.

Choi, Y.; Wang, G.; Nayfeh, M.H. & Yau, S.-T. (2009). A hybrid biofuel cell based on electrooxidation of glucose using ultra-small silicon nanoparticles. *Biosensors and Bioelectronics,* Vol. 24, No. 10, pp. 3103-3107, ISSN 0956-5663.

Davis, F. & Higson, S.P.J. (2007). Biofuel cells--Recent advances and applications. *Biosensors and Bioelectronics,* Vol. 22, No. 7, pp. 1224-1235, ISSN 0956-5663.

de Mele, M.F.L.; Videla, H.A. & Arvia, A.J. (1982). Potentiodynamic Study of Glucose Electro-Oxidation at Bright Platinum Electrodes. *Journal of The Electrochemical Society,* Vol. 129, No. 10, pp. 2207-2213, ISSN 1945-7111.

Degani, Y. & Heller, A. (1987). Direct electrical communication between chemically modified enzymes and metal electrodes. I. Electron transfer from glucose oxidase to metal electrodes via electron relays, bound covalently to the enzyme. *The Journal of Physical Chemistry,* Vol. 91, No. 6, pp. 1285-1289, ISSN 0022-3654.

Delecouls-Servat, K.; Bergel, A. & Basseguy, R. (2001). Surface-modified electrodes for NADH oxidation in oxidoreductase-catalyzed synthesis. *Journal of applied electrochemistry,* Vol. 10, No. 31, pp. 1095-1101, ISSN 1572-8838.

Devarajan, S.; Bera, P. & Sampath, S. (2005). Bimetallic nanoparticles: A single step synthesis, stabilization, and characterization of Au-Ag, Au-Pd, and Au-Pt in sol-gel derived silicates. *Journal of Colloid and Interface Science,* Vol. 290, No. 1, pp. 117-129, ISSN 0021-9797.

El Roustom, B.; Siné, G.; Fóti, G. & Comninellis, C. (2007). A novel method for the preparation of bi-metallic (Pt–Au) nanoparticles on boron doped diamond (BDD) substrate: application to the oxygen reduction reaction. *Journal of applied electrochemistry,* Vol. 37, No. 11, pp. 1227-1236, ISSN 0021-891X.

Ernst, S.; Heitbaum, J. & Hamann, C.H. (1979). The electrooxidation of glucose in phosphate buffer solutions : Part I. Reactivity and kinetics below 350 mV/RHE. *Journal of Electroanalytical Chemistry,* Vol. 100, No. 1-2, pp. 173-183, ISSN 0368-1874.

Fei, J.; Song, H.-K. & Palmore, G.T.R. (2007). A biopolymer composite that catalyzes the reduction of oxygen to water. *Chemistry of Materials,* Vol. 19, No. 7, pp. 1565-1570, ISSN 0897-4756.

Fishilevich, S.; Amir, L.; Fridman, Y.; Aharoni, A. & Alfonta, L. (2009). Surface display of redox enzymes in microbial fuel cells. *Journal of the American Chemical Society,* Vol. 131, No. 34, pp. 12052-12053, ISSN 0002-7863.

Gao, F.; Viry, L.; Maugey, M.; Poulin, P. & Mano, N. (2010). Engineering hybrid nanotube wires for high-power biofuel cells. *Nature Communications,* Vol. 1, No. 1, pp. 1-7, ISSN 2041-1723.

Grolleau, C.; Coutanceau, C.; Pierre, F. & Léger, J.M. (2008). Effect of potential cycling on structure and activity of Pt nanoparticles dispersed on different carbon supports. *Electrochimica Acta,* Vol. 53, No. 24, pp. 7157-7165, ISSN 0013-4686.

Gupta, G.; Rajendran, V. & Atanassov, P. (2004). Bioelectrocatalysis of oxygen reduction reaction by laccase on gold electrodes. *Electroanalysis,* Vol. 16, No. 13-14, pp. 1182-1185, ISSN 1521-4109.

Habrioux, A.; Merle, G.; Servat, K.; Kokoh, K.B.; Innocent, C.; Cretin, M. & Tingry, S. (2008). Concentric glucose/O_2 biofuel cell. *Journal of Electroanalytical Chemistry,* Vol. 622, No. 1, pp. 97-102, ISSN 1572-6657.

Habrioux, A.; Napporn, T.; Servat, K.; Tingry, S. & Kokoh, K.B. (2010). Electrochemical characterization of adsorbed bilirubin oxidase on Vulcan XC 72R for the biocathode preparation in a glucose/O_2 biofuel cell. *Electrochimica Acta,* Vol. 55, No. 26, pp. 7701-7705, ISSN 0013-4686.

Habrioux, A.; Servat, K.; Tingry, S. & Kokoh, K.B. (2009a). Enhancement of the performances of a single concentric glucose/O_2 biofuel cell by combination of bilirubin oxidase/Nafion cathode and Au-Pt anode. *Electrochemistry Communications,* Vol. 11, No. 1, pp. 111-113, ISSN 1388-2481.

Habrioux, A.; Sibert, E.; Servat, K.; Vogel, W.; Kokoh, K.B. & Alonso-Vante, N. (2007). Activity of Platinum-Gold Alloys for Glucose Electrooxidation in Biofuel Cells. *The Journal of Physical Chemistry B,* Vol. 111, No. 34, pp. 10329-10333, ISSN 1520-6106.

Habrioux, A.; Vogel, W.; Guinel, M.; Guetaz, L.; Servat, K.; Kokoh, K.B. & Alonso-Vante, N. (2009b). Structural and electrochemical studies of Au-Pt nanoalloys. *Physical Chemistry Chemical Physics,* Vol. 11, No. 18, pp. 3573-3579, ISSN 1463-9076.

Heller, A. (2004). Miniature biofuel cells. *Physical Chemistry Chemical Physics,* Vol. 6, No. 2, pp. 209-216, ISSN 1463-9076.

Jin, C. & Chen, Z. (2007). Electrocatalytic oxidation of glucose on gold-platinum nanocomposite electrodes and platinum-modified gold electrodes. *Synthetic Metals,* Vol. 157, No. 13-15, pp. 592-596, ISSN 0379-6779.

Kang, C.; Shin, H.; Zhang, Y. & Heller, A. (2004). Deactivation of bilirubin oxidase by a product of the reaction of urate and O_2. *Bioelectrochemistry,* Vol. 65, No. 1, pp. 83-88, ISSN 1567-5394.

Karnicka, K.; Miecznikowski, K.; Kowalewska, B.; Skunik, M.; Opallo, M.; Rogalski, J.; Schuhmann, W. & Kulesza, P.J. (2008). ABTS-Modified Multiwalled Carbon Nanotubes as an Effective Mediating System for Bioelectrocatalytic Reduction of Oxygen. *Analytical Chemistry,* Vol. 80, No. 19, pp. 7643-7648, ISSN 0003-2700.

Katz, E. & Willner, I. (2003). A Biofuel Cell with Electrochemically Switchable and Tunable Power Output. *Journal of the American Chemical Society,* Vol. 125, No. 22, pp. 6803-6813, ISSN 0002-7863.

Kenausis, G.; Chen, Q. & Heller, A. (1997). Electrochemical glucose and lactate sensors based on "wired" thermostable soybean peroxidase operating continuously and stably at 37 °C. *Analytical Chemistry,* Vol. 69, No. 6, pp. 1054-1060, ISSN 0003-2700.

Kerzenmacher, S.; Ducrée, J.; Zengerle, R. & von Stetten, F. (2008). An abiotically catalyzed glucose fuel cell for powering medical implants: Reconstructed manufacturing protocol and analysis of performance. *Journal of Power Sources,* Vol. 182, No. 1, pp. 66-75, ISSN 0378-7753.

Kleppe, K. (1966). The effect of hydrogen peroxide on glucose oxidase from Aspergillus Niger. *Biochemistry,* Vol. 5, No. 1, pp. 139-143, ISSN 0006-2960.

Klotzbach, T.L.; Watt, M.; Ansari, Y. & Minteer, S.D. (2008). Improving the microenvironment for enzyme immobilization at electrodes by hydrophobically modifying chitosan and Nafion® polymers. *Journal of Membrane Science,* Vol. 311, No. 1-2, pp. 81-88, ISSN 0376-7388.

Kokoh, K.B.; Léger, J.M.; Beden, B.; Huser, H. & Lamy, C. (1992a). "On line" chromatographic analysis of the products resulting from the electrocatalytic oxidation of d-glucose on pure and adatoms modified Pt and Au electrodes-Part II. Alkaline medium. *Electrochimica Acta,* Vol. 37, No. 11, pp. 1909-1918, ISSN 0013-4686.

Kokoh, K.B.; Léger, J.M.; Beden, B. & Lamy, C. (1992b). "On line" chromatographic analysis of the products resulting from the electrocatalytic oxidation of d-glucose on Pt, Au and adatoms modified Pt electrodes-Part I. Acid and neutral media. *Electrochimica Acta,* Vol. 37, No. 8, pp. 1333-1342, ISSN 0013-4686.

Li, H.; Webb, S.P.; Ivanic, J. & Jensen, J.H. (2004). Determinants of the Relative Reduction Potentials of Type-1 Copper Sites in Proteins. *Journal of the American Chemical Society,* Vol. 126, No. 25, pp. 8010-8019, ISSN 0002-7863.

Ma, Y. & Balbuena, P.B. (2007). Designing oxygen reduction catalysts: Insights from metalloenzymes. *Chemical Physics Letters,* Vol. 440, No. 1-3, pp. 130-133, ISSN 0009-2614.

Mano, N.; Kim, H.-H. & Heller, A. (2002a). On the Relationship between the Characteristics of Bilirubin Oxidases and O_2 Cathodes Based on Their "Wiring". *The Journal of Physical Chemistry B,* Vol. 106, No. 34, pp. 8842-8848, ISSN 1520-6106.

Mano, N.; Kim, H.-H.; Zhang, Y. & Heller, A. (2002b). An Oxygen Cathode Operating in a Physiological Solution. *Journal of the American Chemical Society,* Vol. 124, No. 22, pp. 6480-6486, ISSN 0002-7863.

Mano, N.; Mao, F. & Heller, A. (2002c). A Miniature Biofuel Cell Operating in A Physiological Buffer. *Journal of the American Chemical Society,* Vol. 124, No. 44, pp. 12962-12963, ISSN 0002-7863.

Mano, N.; Mao, F. & Heller, A. (2004). Electro-oxidation of glucose at an increased current density at a reducing potential. *Chemical Communications,* Vol. No. 18, pp. 2116-2117, ISSN 1359-7345.

Mano, N.; Mao, F. & Heller, A. (2005). On the parameters affecting the characteristics of the "wired" glucose oxidase anode. *Journal of Electroanalytical Chemistry,* Vol. 574, No. 2, pp. 347-357, ISSN 1572-6657.

Mano, N.; Soukharev, V. & Heller, A. (2006). A laccase-wiring redox hydrogel for efficient catalysis of O_2 electroreduction. *The Journal of Physical Chemistry B,* Vol. 110, No. 23, pp. 11180-11187, ISSN 1520-6106.

Mao, F.; Mano, N. & Heller, A. (2003). Long Tethers Binding Redox Centers to Polymer Backbones Enhance Electron Transport in Enzyme "Wiring" Hydrogels. *Journal of the American Chemical Society,* Vol. 125, No. 16, pp. 4951-4957, ISSN 0002-7863.

Merle, G.; Brunel, L.; Tingry, S.; Cretin, M.; Rolland, M.; Servat, K.; Jolivalt, C.; Innocent, C. & Seta, P. (2008). Electrode biomaterials based on immobilized laccase. Application for enzymatic reduction of dioxygen. *Materials Science and Engineering: C,* Vol. 28, No. 5-6, pp. 932-938, ISSN 0928-4931.

Merle, G.; Habrioux, A.; Servat, K.; Rolland, M.; Innocent, C.; Kokoh, K.B. & Tingry, S. (2009). Long-term activity of covalent grafted biocatalysts during intermittent use of a glucose/O_2 biofuel cell. *Electrochimica Acta,* Vol. 54, No. 11, pp. 2998-3003, ISSN 0013-4686.

Minteer, S.D.; Liaw, B.Y. & Cooney, M.J. (2007). Enzyme-based biofuel cells. *Current Opinion in Biotechnology,* Vol. 18, No. 3, pp. 228-234, ISSN 0958-1669.

Möller, H. & Pistorius, P.C. (2004). The electrochemistry of gold-platinum alloys. *Journal of Electroanalytical Chemistry,* Vol. 570, No. 2, pp. 243-255, ISSN 1572-6657.

Moore, C.M.; Akers, N.L.; Hill, A.D.; Johnson, Z.C. & Minteer, S.D. (2004). Improving the environment for immobilized dehydrogenase enzymes by modifying Nafion with tetraalkylammonium bromides. *Biomacromolecules,* Vol. 5, No. 4, pp. 1241-1247, ISSN 1525-7797.

Mott, D.; Luo, J.; Njoki, P.N.; Lin, Y.; Wang, L. & Zhong, C.-J. (2007). Synergistic activity of gold-platinum alloy nanoparticle catalysts. *Catalysis Today,* Vol. 122, No. 3-4, pp. 378-385, ISSN 0920-5861.

Neto, S.; Forti, J. & De Andrade, A. (2010). An Overview of Enzymatic Biofuel Cells. *Electrocatalysis,* Vol. 1, No. 1, pp. 87-94, ISSN 1868-2529.

Palmer, A.E.; Lee, S.K. & Solomon, E.I. (2001). Decay of the Peroxide Intermediate in Laccase: Reductive Cleavage of the O-O Bond. *Journal of the American Chemical Society,* Vol. 123, No. 27, pp. 6591-6599, ISSN 0002-7863.

Palmore, G.T.R. & Kim, H.-H. (1999). Electro-enzymatic reduction of dioxygen to water in the cathode compartment of a biofuel cell. *Journal of Electroanalytical Chemistry,* Vol. 464, No. 1, pp. 110-117, ISSN 1572-6657.

Park, S.; Boo, H. & Chung, T.D. (2006). Electrochemical non-enzymatic glucose sensors. *Analytica Chimica Acta,* Vol. 556, No. 1, pp. 46-57, ISSN 0003-2670.

Paul, B.K. & Mitra, R.K. (2005). Water solubilization capacity of mixed reverse micelles: Effect of surfactant component, the nature of the oil, and electrolyte concentration. *Journal of Colloid and Interface Science,* Vol. 288, No. 1, pp. 261-279, ISSN 0021-9797.

Rao, M.L.B. & Drake, R.F. (1969). Studies of Electrooxidation of Dextrose in Neutral Media. *Journal of The Electrochemical Society,* Vol. 116, No. 3, pp. 334-337, ISSN 1945-7111.

Rusling, J.F.; Wang, B. & Yun, S.E., (2008). Electrochemistry of redox enzymes, In: *Bioelectrochemistry : Fundamentals, Experimental Techniques and Applications,* P.N. Bartlett, (Ed.), 39-85, Wiley, ISBN 978-0-470-84364-2, New-York, U. States.

Schmidt, T.J.; Gasteiger, H.A. & Behm, R.J. (1999). Rotating Disk Electrode Measurements on the CO Tolerance of a High-Surface Area Pt/Vulcan Carbon Fuel Cell Catalyst. *Journal of The Electrochemical Society,* Vol. 146, No. 4, pp. 1296-1304, ISSN 1945-7111.

Senthil Kumar, S. & Phani, K.L.N. (2009). Exploration of unalloyed bimetallic Au-Pt/C nanoparticles for oxygen reduction reaction. *Journal of Power Sources,* Vol. 187, No. 1, pp. 19-24, ISSN 0378-7753.

Shimizu, A.; Kwon, J.-H.; Sasaki, T.; Satoh, T.; Sakurai, N.; Sakurai, T.; Yamaguchi, S. & Samejima, T. (1999). Myrothecium verrucaria Bilirubin Oxidase and Its Mutants for Potential Copper Ligands. *Biochemistry,* Vol. 38, No. 10, pp. 3034-3042, ISSN 0006-2960.

Shleev, S.; Tkac, J.; Christenson, A.; Ruzgas, T.; Yaropolov, A.I.; Whittaker, J.W. & Gorton, L. (2005). Direct electron transfer between copper-containing proteins and electrodes. *Biosensors and Bioelectronics,* Vol. 20, No. 12, pp. 2517-2554, ISSN 0956-5663.

Soukharev, V.; Mano, N. & Heller, A. (2004). A four-electron O_2-electroreduction biocatalyst superior to platinum and a biofuel cell operating at 0.88 V. *Journal of the American Chemical Society,* Vol. 126, No. 27, pp. 8368-8369, ISSN 0002-7863.

Stankovich, M.T.; Schopfer, L.M. & Massey, V. (1978). Determination of glucose oxidase oxidation-reduction potentials and the oxygen reactivity of fully reduced and semiquinoid forms. *Journal of Biological Chemistry,* Vol. 253, No. 14, pp. 4971-4979, ISSN 0021-9258.

Sun, Y.; Buck, H. & Mallouk, T.E. (2001). Combinatorial Discovery of Alloy Electrocatalysts for Amperometric Glucose Sensors. *Analytical Chemistry,* Vol. 73, No. 7, pp. 1599-1604, ISSN 0003-2700.

Thomas, T.J.; Ponnusamy, K.E.; Chang, N.M.; Galmore, K. & Minteer, S.D. (2003). Effects of annealing on mixture-cast membranes of Nafion® and quaternary ammonium bromide salts. *Journal of Membrane Science,* Vol. 213, No. 1-2, pp. 55-66, ISSN 0376-7388.

Timur, S.; Pazarlioglu, N.; Pilloton, R. & Telefoncu, A. (2004). Thick film sensors based on laccases from different sources immobilized in polyaniline matrix. *Sensors and Actuators B: Chemical,* Vol. 97, No. 1, pp. 132-136, ISSN 0925-4005.

Topcagic, S. & Minteer, S.D. (2006). Development of a membraneless ethanol/oxygen biofuel cell. *Electrochimica Acta,* Vol. 51, No. 11, pp. 2168-2172, ISSN 0013-4686.

Tsujimura, S.; Tatsumi, H.; Ogawa, J.; Shimizu, S.; Kano, K. & Ikeda, T. (2001). Bioelectrocatalytic reduction of dioxygen to water at neutral pH using bilirubin oxidase as an enzyme and 2,2'-azinobis (3-ethylbenzothiazolin-6-sulfonate) as an electron transfer mediator. *Journal of Electroanalytical Chemistry,* Vol. 496, No. 1-2, pp. 69-75, ISSN 1572-6657.

Vogel, W.; Bradley, J.; Vollmer, O. & Abraham, I. (1998). Transition from five-fold symmetric to twinned FCC gold particles by thermally induced growth. *The Journal of Physical Chemistry B,* Vol. 102, No. 52, pp. 10853-10859, ISSN 1520-6106.

Vogel, W.; Tesche, B. & Schulze, W. (1983). X-ray structure and optical properties of silver films prepared by condensation of pre-formed Ag-clusters. *Chemical Physics,* Vol. 74, No. 1, pp. 137-142, ISSN 0301-0104.

Wang, J. (2007). Electrochemical Glucose Biosensors. *Chemical Reviews,* Vol. 108, No. 2, pp. 814-825, ISSN 0009-2665.

Wang, Z.; Liu, S.; Wu, P. & Cai, C. (2009). Detection of glucose based on direct electron transfer reaction of glucose oxidase immobilized on highly ordered polyaniline nanotubes. *Analytical Chemistry,* Vol. 81, No. 4, pp. 1638-1645, ISSN 0003-2700.

Willner, I.; Heleg-Shabtai, V.; Blonder, R.; Katz, E.; Tao, G.; Backmann, A.F. & Heller, A. (1996). Electrical wiring of glucose oxidase by reconstitution of FAD-modified monolayers assembled onto Au-electrodes. *Journal of the American Chemical Society,* Vol. 118, No. 42, pp. 10321-10322, ISSN 0002-7863.

Wilson, R. & Turner, A.P.F. (1992). Glucose oxidase: an ideal enzyme. *Biosensors and Bioelectronics,* Vol. 7, No. 3, pp. 165-185, ISSN 0956-5663.

Woods, R. (1971). The surface composition of platinum-gold alloys. *Electrochimica Acta,* Vol. 16, No. 5, pp. 655-658, ISSN 0013-4686.

Yahiro, A.T.; Lee, S.M. & Kimble, D.O. (1964). Bioelectrochemistry: I. Enzyme utilizing bio-fuel cell studies. *Biochimica et Biophysica Acta (BBA) - Specialized Section on Biophysical Subjects,* Vol. 88, No. 2, pp. 375-383, ISSN 0926-6577.

Zagal, J.H.; Griveau, S.; Silva, J.F.; Nyokong, T. & Bedioui, F. (2010). Metallophthalocyanine-based molecular materials as catalysts for electrochemical reactions. *Coordination Chemistry Reviews,* Vol. 254, No. 23-24, pp. 2755-2791, ISSN 0010-8545.

Zebda, A.; Renaud, L.; Cretin, M.; Innocent, C.; Pichot, F.; Ferrigno, R. & Tingry, S. (2009a). Electrochemical performance of a glucose/oxygen microfluidic biofuel cell. *Journal of Power Sources,* Vol. 193, No. 2, pp. 602-606, ISSN 0378-7753.

Zebda, A.; Renaud, L.; Cretin, M.; Pichot, F.; Innocent, C.; Ferrigno, R. & Tingry, S. (2009b). A microfluidic glucose biofuel cell to generate micropower from enzymes at ambient temperature. *Electrochemistry Communications,* Vol. 11, No. 3, pp. 592-595, ISSN 1388-2481.

Zhang, M.; Mullens, C. & Gorski, W. (2007). Coimmobilization of dehydrogenases and their cofactors in electrochemical biosensors. *Analytical Chemistry,* Vol. 79, No. 6, pp. 2446-2450, ISSN 0003-2700.

Zhu, Z.; Momeu, C.; Zakhartsev, M. & Schwaneberg, U. (2006). Making glucose oxidase fit for biofuel cell applications by directed protein evolution. *Biosensors and Bioelectronics,* Vol. 21, No. 11, pp. 2046-2051, ISSN 0956-5663.

21

Quantifying Bio-Engineering: The Importance of Biophysics in Biofuel Research

Patanjali Varanasi[1,2,3], Lan Sun[1,2,3], Bernhard Knierim[1,2], Elena Bosneaga[2,4], Purbasha Sarkar[2,4], Seema Singh[1,3] and Manfred Auer[1,2,4]

[1]*Joint BioEnergy Institute, Physical Biosciences Division Lawrence Berkeley National Laboratory, Emeryville, CA*

[2]*Life Sciences Division, Lawrence Berkeley National Laboratory, Berkeley, CA*

[3]*Sandia National Laboratories, Biomass Science and Conversion Technology Department, Livermore, CA*

[4]*Energy Biosciences Institute, UC Berkeley, CA*

United States

1. Introduction

The decreasing availability and the increasing demand for fossil energy sources as well as concerns of irreversible climate change have sparked a quest for alternative energy sources, including carbon-neutral transportation fuels. One such alternative to fossil fuels is biofuel produced from currently unused plant biomass. Since lignocellulosic biomass -unlike corn-starch- cannot be used as a food source for humans it constitutes an ideal source for the production of biofuels. Currently, the production of biofuels from this unutilized biomass is not economically viable and crippled due to high costs involved in the conversion of biomass to sugars, and the limited repertoire of current generation microbes to produce a host of necessary transportation fuels beyond the simple fermentation into ethanol. Recently extensive research efforts are underway to overcome the bottlenecks for an economically viable lignocellulose biofuel industry. Advances must be made in the area of feedstocks engineering, optimization of deconstruction processes, including chemical, enzymatic or microbial pretreatment and saccharification approaches, as well as in the area of fuels synthesis, and require a variety of biophysical approaches, some of which are discussed here in more detail.

2. ThermoGravimetric Analysis (TGA) and Differential Scanning Calorimetry (DSC)

Thermogravimetric analysis is a thermoanalytical technique that measures the change in mass of a substrate as a function of temperature. The temperature of the sample and a blank are increased/decreased at a constant rate and the change in weight is measured as a function of sample temperature. As these experiments are highly temperature sensitive they are conducted in highly insulated chambers. The sample holders used in TGA experiments have to be highly conducting to avoid any lag between temperature measured outside and inside the sample holder. There is practically no sample preparation required and the technique can be used for solids or liquids or a mixture of both. A small mass of the sample (<10 mg) is weighed and its mass change is measured against the mass change of an empty sample holder. The sample holders come with and without a hole in the lid thus allowing (or not) the flow of gases. The TGA experiments may be conducted under either inert (Nitrogen or Argon) or under oxidative conditions (Oxygen or Air). The decrease or increase in weight of the sample as the temperature is increased may indicate the loss of the material due to vaporization, decomposition or oxidation. The weight loss curve generated by the TGA instrument is representative of this mass loss (or increases) as a function of temperature changes. A derivative of the weight loss curves is usually used to easily read the TGA curves. The differential thermogravimetric (dTG) curves are useful for finding the temperature of vaporization or decomposition of both pure compounds and mixtures. TGA curves taken under different temperature ramp rates are used to find the order and activation energies of the decomposition reactions. TGA curves may also be used to find the total moisture content of the samples based on the weight loss in the region where water evaporates (80°C to 120°C). TGA/DSC techniques are particularly useful for mixtures with different constituents. Based on the temperature at which each of the constituents decompose or vaporize we may define weight loss region for each of the constituents. We may then use weight percent loss in each region to find the composition changes of the individual constituents in the original sample.

Differential scanning calorimetry (DSC) is also a thermoanalytical technique and is often used in conjunction with TGA. It measures the energy required to increase or decrease the sample temperature against a blank. Like TGA this technique also requires minimum sample preparation and can be used for solid, liquids or a mixture of both. Similar sample holders and similar sample weights are used in DSC as in TGA. Experimental conditions like, temperature ramp rate and gaseous atmosphere can be changed based on the kind of analysis. A DSC cure is highly data-rich as it gives us crucial information such as enthalpy of melting, boiling, decomposition and oxidation. By comparing both TGA and DSC curves one can determine the temperature at which the system boils, decomposes or oxidizes. An endothermic mass loss can be attributed to boiling or endothermic decomposition. An exothermic mass change is usually due to exothermic decomposition or oxidation of the substrate. Thus, DSC data is also used to find the caloric value of the decomposition and oxidation reactions. DSC data is used to find the endothermic energy required for dehydrating a sample in the moisture loss region. Such data may be used to find if the moisture is physically absorbed to the substrate or is chemically complexed with the substrate. In case of polymer substrates, DSC curves may also be used to find the glass transition temperatures and thus degree of polymerization of the samples (DP) (Couchman 1981). Amorphous to crystalline transitions can also be determined based on the temperature of weight loss and the enthalpy of decomposition. Amorphous substrates decompose at lower temperature when compared to crystalline substrates of the same material (Kim, Eom, and Wada 2010). Amorphous materials undergo crystallization before melting. Based on ratio of the energy required for crystallization and the energy required for melting we may also find the percent crystalline material in the sample.

TGA/DSC techniques are particularly useful for biomass characterization as biomass is a complex mixture of polymers (cellulose, hemicelluloses and lignin). Serapiglia et al. (2008); Kaloustian et al. (2003); Jaffe, Collins, and Mencze (2006) have shown that the dTG curves from biomass can be divided into three weight loss regions for hemicelluloses followed by cellulose and lignin. TGA is a very sensitive technique and can be used to differentiate between mutants of similar kinds of feedstocks. Serapiglia et al. (2008) have used high throughput TGA to find the differences between various mutants of shrub willow. TGA was also used to find the total lignin content of various feedstocks (Ghetti et al. 1996). TGA/DSC curves were used to differentiate and understand the effects of dilute acid steam explosion on Eucalyptus (Emmel et al. 2003). They reported that lignin fragmented and recondensated during the process. A decrease in softening temperature was reported for steam exploded bamboo lignin (Shao et al. 2009), resulting in a lower molecular weight polymer. DSC was also used to differentiate between hardwoods and softwoods, based on the glass transition temperature of the lignin extracted from the woods (Kubo and Kadla 2005). Lignin decomposition by white-rot fungi was studied using DSC (Reh et al. 1987). They show that as lignin-carbohydrate bonds are broken the peaks in each region separate and become sharper. The differences between lignin carbohydrate linkages in biomass can also be found using DSC (Reh et al. 1987; Tsujiyama and Miyamori 2000). Along with these kinds of qualitative measurements, TGA and DSC can be used to calculate the enthalpy, activation energy (Flynn and Wall 1966; Paul et al. 2010) and the order of the reaction as well as the percent cellulose crystallinity of the samples.

3. Cellulose crystallinity measurement

Measuring the percent crystallinity of the biomass samples is particularly important as it affects rate and yield of enzymatic saccharification (Hall et al. 2010). It is challenging to measure the percent crystallinity of the biomass using DSC, as cellulose decomposes even before it undergoes melting (Paul et al. 2010; Kaloustian et al. 2003; Soares et al. 1995). As enthalpy of crystallization and enthalpy of melting are required to find the percent crystallinity of the sample, it is not possible to measure percent crystallinity of the biomass based on this technique. However, the enthalpy of dehydration of cellulose and biomass samples has been correlated to percent cellulose crystallinity of the samples based on the hydrogen bonds formed by the crystalline and amorphous cellulose. A detailed method for measuring cellulose crystallinity (CC) by DSC has been reported by Bertran and Dale (1986). They have reported that DSC provides a better way of measuring CC than other traditional approaches like XRD, especially for substance with very low crystallinity index. The amount of moisture absorbed by cellulose containing substances is dependent on the amount of crystalline and amorphous cellulose present in the substance. Amorphous cellulose absorbs a high amount of moisture while pure crystalline cellulose has no absorption capacity. Hence the endothermic dehydration peak appearing in DSC thermograms can be used to estimate the percent crystallinity of the cellulose present in the substances. Using completely amorphous cellulose measured under the same experimental conditions, the percent CC of the sample can be estimated from the following equation where ΔH_0 is the heat of dehydration for a completely amorphous cellulose sample and ΔH_s is the heat required to dehydrate the sample.

$$\%CC = \frac{\Delta H_0 - \Delta H_s}{\Delta H_0} x100$$

In spite of the many advantages of using TGA and DSC analysis they cannot be used to find bond specific information about the substrate. Knowledge of the chemical composition of the substrate is required ahead of time. Analysis through this technique always requires one to correlate the original substrate to the pure components of the substrate. The changes to DP of a polymer can only be visualized but cannot be quantitatively correlated to changes in the molecular weight of the polymer. Also if the substrates decompose very close to each other, we cannot differentiate between the increase of one of the substrate or decrease in DP of the polymer degrading at a higher temperature. As these techniques affect the samples physically and chemically the sample used in this technique cannot be recovered or reused.

4. Pyro-GC/MS and GC/MS

Gas Chromatography/ Mass Spectrometry (GC/MS) is a very popular technique used to separate and identify each organic compound from mixtures. It provides both qualitative and quantitative information about the sample. Gas Chromatography (GC) separates the molecules based on their molecular weight and their volatility. Gas chromatography uses a conventional oven with a column of very small internal diameter. The temperature of the oven is changed and based on the volatility of the compounds they travel through the column at different rates. Hence they separate and reach the mass spectrometer at different times. The time a compound takes to reach the mass spectrometer from the time it is injected into the column is called retention time. The retention time of a compound depends on its molecular weight and structure (thus its volatility). The retention time of the compounds is dependent on the initial temperature and temperature ramp rate of the GC. The compounds injected into the GC reach the MS at different times based on their volatility and complex mixtures are thus separated, but this separation is not enough to understand their composition. Two compounds of different molecular weight and structure may reach the MS at the same time if they have similar volatility. A mass spectrometer is used to identify these compounds eluting from the GC column. The mass spectrometer charges the compounds, accelerates them under high magnetic fields and then breaks them into ions. Based on their mass and charge, the ions hit the detector at various positions and are thus detected. Unlike it is the case for TGA and DSC mass spectrometry analysis does not require a a thorough understanding of the mixture composition of the mixture is not necessary before analysis. However confirming the identity of the predicted compound, after the analysis of the unknown sample, with a known standard is necessary. The mass spectrometer is usually run in scan mode where it detects all the ions that hit the detector. However, scans of samples containing known compounds can be made very sensitive by forcing the mass spectrometer to detect the ions specific to the compounds of interest (selective ion mode). Based on the fragmented ions and the retention time, the chemical structure of the compound can be determined.

The mass spectrometer is very sensitive and detects even a small concentration of impurities. Hence the sample needs to be extremely pure and free of any oxygen or moisture. This implies the need for tedious sample preparation and purification step before the introduction of the sample into the instrument. It is also important to maintain the instrument itself and the column used in the GC. There is good chance of introduction of moisture and column degradation products into the sample without proper care. The quadruple of the MS itself needs to be cleaned and tested for its sensitivity frequently.

In liquid GC/MS the samples are usually extracted into volatile organic solvents and are then injected into the GC. The compounds thus detected are biased towards non-polar organic molecules of low molecular weight. However, it is also possible to detect polar organic molecules like organic acids after derivatization thus rendering them non polar. For quantitative analysis one injects a know quantity of a compound called internal standard (IS). An internal standard is a compound, which is similar in all respects to the compounds of interest but can be separated from the original compounds in the column. Deuterated counterparts of the original compounds are usually used as internal standards for the GC/MS analysis. Internal standards are injected into the original solid or liquid prior to extraction. The compounds of interest and the original solids are then extracted into a volatile organic solvent through soxhlet extraction or liquid-liquid extraction. Specific ASTM (American Society for Testing Materials) guidelines exist for the extraction procedures for quantitative analysis of various volatile organic pollutants. A standard mixture of known quantities of the internal standard and the compound of interest is made in similar solvent. This standard mixture is then analyzed in GC/MS using the same protocol used for the sample and the standard.

Differences in efficiency of the extraction of both the compounds and the internal standard can lead to erroneous results such as incorrect concentration predictions. Only samples which contain extractible substances of lower molecular weight can be analyzed using this technique.

These shortcomings can be overcome by using pyrolysis GC-MS. In this technique, the sample is pyrolyzed at high temperatures for very short amount of time and the products thus produced are directly injected into GC/MS. This technique is usually used to find the structures of heteropolymers (polystyrene and PVC blends, lignin, etc.). The primary advantage of this technique is that it requires a small sample volume and no sample preparation, but due to measurement errors of such very small volumes pyrolysis GC-MS is difficult to quantify and inaccurate. Hence the ratio of breakdown products is often a better measurement estimate to describe the original sample.

Both GC/MS and pyrolysis GC/MS are used extensively in biofuel research. Lignin as a heteropolymer present in the biomass is often analyzed for its chemical composition using pyrolysis GC/MS (Galletti et al. 1997; Mills et al. 2009). Lignin typically contains three major constituents called *p*-hydroxyphenyl lignin (H-lignin), guaiacyl lignin (G-lignin) and syringyl-lignin (S-lignin). The structure and strength of lignin is dependent on the ratio of these three kinds of lignin (Boudet et al. 1998). Lignin rich in G-lignin is supposed to be highly condensed when compared to lignin rich in S- lignin (Chiang and Funaoka 1990). Ralph and Hatfield (1991) were one of the first ones to analyze the composition of lignin using pyrolysis GC/MS. They show that the pyrolysis products of lignin can be a variety of products derived from H, G and S-lignin. They suggest using a ratio of the total amount of H to the total amount of G to the total amount of S products to find the H:G:S ratio. The S: G ratio of a feedstock has been shown to affect the efficiency of both the pretreatment technology and the enzymatic saccharification. Pyro-GC/MS has been used to find the changes in S:G ratio due to different pretreatments (Samuel et al. 2010; Huyen et al. 2010; Ibarra et al. 2007; Jung et al. 2010; Papatheofanous et al. 1995). The S:G ratio decreases as a result of kraft pulping (Ibarra et al. 2007) and thus hardwoods (S:G ratio >1) have always been a choice for the paper and pulp industry. Ammonia treated Miscanthus showed a decrease in the S:G ratio as result of pretreatment (Huyen et al. 2010). Samuel et al. (2010) have shown that the S:G ratio of switchgrass decreased due to dilute acid pretreatment of biomass, showing it is easier to break S-lignin when compared to G-lignin. Based on this hypothesis genes have been identified which increase the amount S-lignin present in the plant biomass (Marita et al. 1999; Ralph 2007). The mechanism of how lignin is formed and how each gene effects the formation of these individual components of lignin has been found by both NMR and pyro-GC/MS. Pyro-GC/MS is used to differentiate between various plant mutants with different S:G ratios. The changes to S:G ratio as function of harvesting time has also been identified using pyro-GC/MS (Huyen et al. 2010). Feedstocks with similar lignin content have been shown to have different saccharification yields based on their S:G ratio (Davison et al. 2006). Feedstocks with higher amount S-lignin were easier to pretreat but had lower xylose yields when compared to feedstocks with higher G-lignin (Davison et al. 2006). The differences in S:G ratio has also been correlated to in-*rumen* digestibility of plant biomass (Guo et al. 2001; Baucher et al. 1999; Vailhé et al. 1998). As the S:G ratio decreased the in-*rumen* digestibility increased, showing that feedstocks such as grass are ideal for animal feed. The lignin degradation products produced during the pretreatment process are also dependent on the S:G ratio of the original lignin polymer (Chiang and Funaoka 1990). S:G ratio, as measured from pyro-GC/MS, offers a platform to correlate its effects on different biofuel production steps and thus helps us design better feedstocks.

Lignin and some of its monomers have been shown to inhibit both cellulases during enzymatic saccharification (Sewalt et al. 1997; Ximenes et al. 2010; Ximenes et al. 2011) and microbes during fuel production (Mills et al. 2009). These lignin degradation products produced during the pretreatment may have an inhibitory role on the downstream processes. These products can be identified and quantified using GC/MS (Pecina et al. 1986). Identifying these lignin degradation products is also important for the biofuel industry as it gives us an idea about the byproducts produced during the pretreatment process itself (Ehara et al. 2005). As pyro-GC/Ms can give a ratio of the amount of G-lignin and S-lignin present in the lignin, derivatization followed by the reductive cleavage (DFRC) method has been developed by Ralph et al. to measure the quantitatively the various constituents of lignin in the biomass (Lu and Ralph 1997; Lu and Ralph 1998; Lu and Ralph 1999; Peng *et al.*1998; Ralph and Lu 1998). This method uses GC/MS to quantify various derivatized and reduced products. Genetic pathway engineering, also known as synthetic biology, is gaining popularity as it aims to produce microbes which may in turn act as biofuel factories to produce the desired kind of fuel from the glucose units. Qualitative analysis of the composition of the fuel allows the identification of the right modification of the microbes to produce fuel. Quantification of the products is also necessary to understand the most efficient mechanism and the toxicity limit of the fuel until which the microbes can sustain. GC/MS is also used to qualitatively and quantitatively measure the fatty acids and fuels produced from the microbes ((Akiyama et al. 2008; Atsumi et al. 2008; Dai et al. 2007; Lu et al. 2008; Tang et al. 2007).

5. Mechanical strength testing of plant stems

As a successful feedstock will have the ability to grow tall and hence provide high yield per area of land used, it is key to measure the mechanical stability for any plant derived by breeding, mutagenesis or targeted cell wall engineering. Mechanical strength is a very important property for materials and defines the use of materials as support systems. A tensile testing instrument measures the mechanical strength, the breaking force required and the strain that the material underwent during the tensile test. A simple tensile testing instrument consists of a motor to apply the tensile force and a force sensor to find the tensile force applied by the motor. The approximate stress and strain during the tensile testing is calculated based on the total length and cross sectional area of the sample before the tensile testing. Tensile stress is calculated as a ratio of force to the cross-sectional area. In ductile materials, as the cross sectional area changes due to deformation under tensile stress the measurement of tensile stress using the above method gives only an approximate value for stress. As brittle materials undergo little or no deformation during tensile testing, the above method provides a better approximation for the stress in case of brittle materials. The total strain that a material underwent during tensile testing can be calculated as the ratio of the total change in length of the material to the initial length of the material. The relative brittle or ductile nature of the samples can be inferred from the total strain they undergo during tensile testing. A brittle material undergoes little deformation and shows a lower amount of strain when compared to ductile materials, which undergo a higher deformation and hence show higher amount of strain. The breaking strength of material, defined as the mechanical stress at the breaking point, is also an important property. The maximum mechanical strength and the breaking strength of a material need not be the same. During tensile testing a material elongates and reaches a maximum stress value, this is the maximum force the material can endure. The material may not break at this point, but on further increasing the strain, the material becomes weaker and breaks at a force lower than the maximum strength. This difference in behavior is also related to brittle and ductile nature of the materials. In ductile materials, like metals, the crystals lattices expand and rearrange when stress is applied, but brittle materials like ceramics are very closely packed and have no room for rearrangement when stress is applied. For the same reason, in case of brittle materials the maximum tensile strength and the breaking strength are usually the same. Tensile testing of the materials thus provides in depth knowledge about maximum mechanical strength, total strain and the nature of the material. Standard ASTM procedures have been developed for measuring the mechanical strength and the above parameters for a variety of materials (Czichos et al. 2006).

Physical pretreatment or grinding is a major part of the pretreatment that would be applied to the biomass during the production of biofuel in order to provide higher surface area, which can increase the efficiency of enzymatic saccharification (Li et al. 2010) and improve the efficiency of the chemical pretreatment by providing smaller particles. Materials of lower mechanical strength are easier to handle and degrade compared to materials of higher mechanical strength. Thus the energy required during the grinding process can be co-related the mechanical strength of the biomass and this property plays an important role for biofuel production. Studies have been conducted which find the mechanical strength of the plant stems and total mechanical energy required for grinding the plant stems to be correlated. Decreasing the mechanical strength of the plant stems would in turn decrease the total energy required during the grinding process. Yu et al. (2006) studied the tensile stress and ultimate shear stress profile of switchgrass as a function of the plant maturity, the time after harvest and the moisture content. The tensile stress was reported to increase with an increase in the elapsed time after harvest and a decrease in the moisture content. Based on this they recommend tensile dominant size reduction to be carried out just after the harvest process of biomass rich in moisture. Igathinathane et al. (2008) showed the dependence of ultimate shear stress on the total moisture content of the switchgrass.

The measurement of the mechanical strength of plant mutants has been studied to find the effect of genetic manipulations on the overall mechanical strength of the plants. An in-house instrument was developed by Prat and Paresys (1989) for measuring different mechanical strength properties of plant stems. A decreased mechanical strength may be desirable for both reduced grinding energy and easier chemical pretreatment of the feedstocks (Wilkinson and Santillana 1978; Lin et al. 1985; Grethlein and Converse 1991; Ibrahim and Pearce 1983).

Lignin present in the plant cell walls provides the plant with support. It is reported to be the most resistant to pretreatment processes (Vinzant et al. 1997; Vailhé et al. 1998; Chen and Dixon 2007). Lignin also inhibits enzymatic saccharification (Chen and Dixon 2007; Jackson et al. 2008) and microbial growth during the fuel production process (Mills et al. 2009). Hence there is a major effort by plant biologists to understand and modify lignin in plant cell wall to ease the conversion of cellulose to fuel (Himmel et al. 2007; Chen and Dixon 2007; Lee et al. 2009). In this effort, mutants are developed with modified/ lowered lignin content (Davison et al. 2006). These modifications may have a direct effect on the plant cell wall and hence on the systems that provide mechanical strength to the plant. However, the decreased mechanical strength can have an adverse effect on the growth of the plant and its ability to sustain under natural conditions. A mutagenized plant may be too weak to sustain a strong gust of wind or even a heavy downpour. This initiates the need to reach an optimum level for mechanical strength to ease the pretreatment process and not lose the ability to sustain natural calamities. Mechanical strength can also be used to find the condition under which a plant would or would not sustain under natural environmental conditions (Coops and Van der Velde 1996). An Arabidopsis mutant grown in the presence of dichlorobenzonitrile, which inhibited the production of cellulose, was also reported to have lower strength than the wild type (Ryden et al. 2003). Ethyl methanesulfonate-mutagenized Arabidopsis populations showed the absence of interfascicular fibers in stems, decreased L-Fucose content and reduced mechanical strength only in the inflorescence part of the stem (Zhong et al. 1997; Reiter et al. 1993). Zhong et al. (2004) showed that the *fragile fiber3* (*fra3*) gene of Arabidopsis thaliana, which encodes type II inositol polyphosphate 5-phosphatases (*5PTases*), resulted in a decrease in secondary cell wall thickness and hence a reduction in stem strength. An expression of the *Heynh root hair defective3* (*RHD3*) gene in an Arabidopsis *fragile fiber4* (*fra4*) mutant also exhibited reduction in cell wall thickness and alteration in cell wall composition (Hu et al. 2003). This also resulted in lowered mechanical strength of the plant stems. Burk et al. (2001) isolated a group of weaker Arabidopsis plant mutants based on their mechanical strength and found a katanin-like protein, which regulated cell wall biosynthesis and cell elongation. Arabidopsis thaliana *fragile fiber8* (*fra8*), which was defective in xylan synthesis, was also found to have lower mechanical strength than the wild type. Arabidopsis thaliana stems down regulated in *secondary cell wall-associate NAC domain protein1* (*SND1*) and *NAC secondary wall thickening promoting factor1* (*NST1*) showed loss of secondary cell wall formation (Zhong et al. 2007), and as a result had lower amount of cellulose, hemicelluloses and lignin. They also showed a considerable decrease in mechanical strength. *Fragile fiber* (*fra1*) mutation of Arabidopsis exhibited a dramatic reduction in mechanical strength while no defects were found in cell wall thickness or composition or in the organization of fiber cells (Zhong et al. 2002). This lowered mechanical strength was attributed to differences in cellulose microfibril orientation in the walls of *fra1*. *Irregular xylem8* (*irx8*), *irx9*, and *irx14* mutations of Arabidopsis thaliana resulting in defects in glucuronoxylan biosynthesis (Keppler and Showalter 2010; Pena et al. 2007; Lee et al. 2010). These defects lead to decreased mechanical strength of the plant stems. An overexpression of these genes rescued the plants of only *irx8* and *irx9* (but not *irx14*) and the mechanical strength of the resulting plants was comparable to the wild type plants (Lee et al. 2010; Pena et al. 2007). A classic rice mutant *brittle clum1* was shown to have reduced mechanical strength and the mutants were characterized to display increased lignin content, a reduced cell wall thickness and decreased cellulose content (Li et al. 2003). The consequence of crowding on the structural behavior and response from marsh plants was studied by Harley and Bertness (1996). They reported that due to crowding the total amount of plant biomass (above the ground) decreased and resulted in thinner and weaker plants.

The effect of various gene expressions and the environmental conditions on the overall plant strength can be easily determined by mechanical strength analysis. It is also a very sensitive technique, which responds to the changes in the orientation of fibers, in plants with no changes in cell wall composition and thickness. The data from these tests would be very useful for the development of feedstock in biofuel research, which are environmentally sustainable, with desired cell wall composition and lowered grinding energies. However, this is a physical characterization technique and can be only be used to understand the effects of various parameters (like genetic modifications or natural effects) on the plant strength. It cannot be used to understand the reason for the observed increase or decrease in the mechanical strength. We are currently using this approach to swiftly characterize the mechanical properties of plants derived from a random-mutagenesis screen.

6. Cell wall analysis by vibrational spectroscopy and microscopy

Vibrational spectroscopy and microscopy techniques have been broadly used to analyze chemical and biological materials. The two main vibrational spectroscopic techniques, infrared spectroscopy and Raman spectroscopy, detect vibrations including both simple bond vibrations and group vibrations in molecules and thus identify these molecules by their spectral fingerprints originated from various vibrational modes. Infrared absorption typically involves photon absorption with the molecule excited to a higher vibrational energy level when the photon energy matches the energy difference between the two vibrational energy levels. This process depends on changes of dipole moments, and hence asymmetric vibrations cause the most intense infrared absorption. On the other hand, Raman scattering is the inelastic scattering of a photon that interacts with molecular vibrations, resulting in an energy shift of the exciting photon. This process depends on changes in polarizability of the electron cloud around the vibrating bonds or groups. Usually, symmetric vibrations cause the largest polarizability changes and thus render the greatest scattering. Therefore, these two techniques often provide complementary information about the molecules (Smith and Dent, 2005). All major cell wall biopolymers are both IR and Raman active. Nowadays both techniques are widely used in plant cell wall research (Dokken et al., 2005; Gierlinger and Schwanninger, 2007).

Unlike most of the current techniques for cell wall compositional analysis, such as wet chemistry assays, chromatography methods, mass spectrometry etc., which are destructive and involve breakdown of the cell wall components or extensive chemical treatment of the plant cell walls, IR and Raman spectroscopy can characterize cell wall components in their native form with minimal requirement for sample preparation. Moreover, a reliable measurement of a single sample can usually be completed from seconds to a few minutes. With the aid of a high throughput platform, such as 96-well plates, rapid screening of a large number of samples can be realized to meet the requirements of bioenergy feedstock development and biomass conversion process optimization for efficient biofuel production.

To acquire cellular level understanding of plant cell walls, various microscopic techniques have been employed, such as bright/dark field microscopy (D'Haeze et al., 2007), polarized light microscopy (Baskin et al., 2004), transmission electron microscopy (Fromm et al., 2003), scanning electron microscopy (Persson et al., 2007b), etc. However, to localize molecules of interest, histochemical and cytochemical staining and labeling methods have to be applied (Cavalier et al., 2008; Grünwald et al., 2002; Persson et al., 2007a). Although autofluorescence of lignin can be utilized to visualize distribution of lignin in the cell wall by fluorescence microscopy (Cavalier et al., 2008; De Micco and Aronne, 2007; Singh et al., 2009), chemical information of cellulose cannot be obtained without additional techniques and quantitative analysis is difficult. On the other hand, using IR and Raman microspectroscopy, chemical maps of specific cell wall components can be generated based on their spectroscopic fingerprints without any disruption of plant tissues and necessities of staining or any other extensive treatment of the cell walls. Localized and tissue/cell specific chemical information can be revealed, which enables acquisition of chemical information on the ultrastructure of plant cell walls. Thus, IR and Raman imaging can detect important compositional changes in cell walls by mutations not necessarily reflected in their average contents and spatial chemical changes during processing, which is very difficult to achieve by other chemical analysis and microscopic methods. Between these two vibrational spectroscopic techniques, IR spectroscopy has been established for decades as a useful tool for plant cell wall research. Integrated with attenuated total reflectance (ATR) technique, synchrotron sources, focal plane array (FPA) infrared detector, and chemometric analysis, IR technique has become increasingly powerful. However, this technique is limited by low sensitivity due to non-background-free detection, low spatial resolution associated with the long infrared wavelengths, and water absorption of the infrared light (Evans and Xie, 2008). The Raman technique, on the other hand, does not have these limitations. Recent advancement in laser technology, optics and detectors has led to a rapid growth in the applications of the Raman technique. Although auto-fluorescence from lignin may sometimes interfere with Raman measurement, the use of near infrared or UV excitation or a fluorescence quencher and effective baseline correction afterwards can alleviate this problem generally. In this section, the applications of IR and Raman techniques for cell wall analysis will be discussed.

Applications of IR microspectroscopy in plant cell wall research: The most prevalent type of IR spectrometer is a Fourier Transform Infrared Spectrometer (FTIR) by recording the raw data as an interferogram and then using Fourier transform to turn this raw data into a spectrum. Attenuated total reflection (ATR) objectives containing an internal reflection element usually made of ZnSe, Ge, or diamond are often used with IR spectroscopy to study plant materials due to faster sampling, improved reproducibility and less impedance by water absorption. IR spectra are most commonly obtained from 4000 to 400 cm^{-1}, the mid-infrared region, where the peaks in the spectra can be associated with fundamental vibrations and the peak intensities are proportional to concentrations. Chemical compositions of various lignocellulosic biomass have been studied by mid-IR spectroscopy, such as wood samples, grasses and herbaceous plants (Dokken et al., 2005). Peak assignments for major cell wall components of representative biomass were summarized by Adapa et al. (2009). Compositional changes by biochemical and chemical treatments of cell walls were also investigated by FTIR. For example, behaviors of cell wall components of oak wood and barley straw treated with cellulase, acidic sodium chlorite, acid and base were studied (Stewart et al., 1995). Dilute acid pretreatment and ionic liquid pretreatment were compared using switchgrass as a model bioenergy feedstock (Li et al., 2010). Moreover, with the aid of chemometric techniques, FTIR can be used as a rapid method for cell wall mutant screening (Chen et al., 1998; Mouille et al., 2003). In addition, using polarizers with FTIR, the orientation of particular functional groups or cell wall components can be determined (McCann et al., 1993; Wilson et al., 2000).

The chemical imaging capability has made IR microscopy a powerful tool to reveal spatial distribution of cell wall components by collecting spectra at each spatial position in the defined area for imaging. FTIR microscopy was also demonstrated by Gierlinger et al. (2008a) to monitor *in situ* the enzymatic degradation of poplar wood and fast and selective degradation of the gelatinous layer in tension wood was observed. This method could be used for enzyme screening and working condition optimization for enzymes. The source intensity from conventional IR thermal sources can only provide a spatial resolution of tens of micrometers, which is limited by both signal-to-noise (S/N) ratio and diffraction, thus restricting plant analysis to the tissue level. Coupling a synchrotron IR source with a small effective source size to IR microscopy can overcome this difficulty due to the high source brightness that allows smaller regions to be probed with acceptable S/N, and thus only diffraction controls spatial resolution in this case (Carr, 1999). Using the synchrotron radiation-based FTIR microspectroscopy (SR-FTIR), imaging structures of plant tissue at cellular level with high S/N at ultraspatial resolutions (3-10 μm) was achieved (Dokken and Davis, 2007; Yu et al., 2003). The main drawbacks of SR-FTIR are the expense and limited access to synchrotrons and slow imaging acquisition by the point-by-point process. A more recent technique known as focal plane array (FPA) based FTIR imaging becomes available and promises many advantages. The FPA-FTIR imaging technique is laboratory based and employs two-dimensional detector arrays to collect spectra at marked positions simultaneously. Thus, FPA-FTIR imaging can acquire the chemical map in a fraction of time required by SR-FTIR imaging. Heraud et al. (2007) compared results produced by FPA-FTIR imaging and SR-FTIR imaging using *Eucalyptus botryoides* leaves as a model sample. They found that the two methods produced similar infrared images allowing differentiation of all tissue types in the leaves. While SR-FTIR imaging provided superior S/N ratio and better spatial resolution, it only took approximate 2 min for FPA-FTIR to map a 350 μm^2 of tissue area, which took approximate 8 h for the SR-FTIR imaging to complete.

In addition to mid-IR spectroscopy, near infrared (NIR) spectroscopy has also been used as a useful tool for compositional analysis of lignocellulosic materials. Absorption spectra in the NIR region, i.e. 14000 - 4000 cm^{-1}, are derived from the overtone or harmonic vibrations. Krongtaew et al. (2010a) has summarized the NIR peak assignments for major cell wall components. Nevertheless, multivariate methods are often implemented for NIR data analysis and both qualitative and quantitative information can be derived. For example, a partial least-squares regression (PLSR) model was developed based on the NIR spectra to assess lignin composition (*p*-hydroxyphenyl to guaiacyl ratio) in maritime pine wood (Alves et al., 2006). Key properties influencing the enzymatic hydrolysis yield and rates, such as lignin content, hemicellulose content, and cellulose crystallinity, influenced by different pretreatment methods were resolved in FT-NIR spectra and successfully evaluated by principal component analysis (Krongtaew et al., 2010a). Total residual lignin content, enzymatically released reducing sugars, total solids, volatile solids, and biogas yield can be assessed quantitatively by FT-NIR spectroscopy combined with partial least-squares regression models (Krongtaew et al., 2010b).

Applications of Raman microspectroscopy in plant cell wall research: Classical dispersive Raman spectrometers are usually composed of laser with wavelength in the visible range for excitation, a dispersive spectrometer and a charge coupled device detector (CCD) for detection. This system is often coupled to a confocal microscope equipped with objectives with high numerical apertures to achieve high spatial resolution (Gierlinger and Schwanninger, 2007; Smith and Dent, 2005). For example, a Raman spectrometer with a 514.5 nm laser was used to study the concentration of lignocellulosics in the cell corner middle lamella of both birch and spruce (Tirumalai et al., 1996). However, for plant cell wall research, the strong autofluorescence from lignocellulosic materials may mask the Raman spectra. Therefore, near infrared Fourier Transform Raman spectrometers (NIR-FT Raman) with laser radiation at 1064 nm coupled with interferometers is often utilized for cell wall analysis, because fluorescence is much less in this region. FT-Raman was used to characterize cell wall components of milled black spruce wood (Agarwal and Ralph, 1997) and in various anatomical parts of flax (Himmelsbach and Akin, 1998). Alternatively, the excitation laser can be shifted to the UV region (below 300 nm) where fluorescence is nearly absence. The utilization of UV excitation also leads to the resonance enhancement of aromatic structures and thus is very sensitive for lignin analysis. Nuopponen et al. (2004) has used UV resonance Raman (UVRR) spectroscopy to analyze the extractable compounds and solid wood samples of Scots pine. Saariaho et al. (2003) has characterized Raman peaks for *p*-hydroxyphenyl, guaiacyl and syringyl structures of lignin using UVRR. Raman peak assignments for major cell wall components were summarized by Agarwal and Ralph (1997) and Adapa et al. (2009). In addition to compositional analysis, a Raman spectroscopy-based method was also developed to obtain mechanical properties of plant cell walls (Ryden et al., 2003), which may serve as an indicator for the ease of cell wall deconstruction during pretreatment or enzymatic saccharification.

Like IR microspectroscopy, one of the major advantages of Raman technique exists in its chemical mapping capabilities. With a much higher lateral spatial resolution than IR (~ 1 μm), ultrastructure of plant cell walls with the corresponding compositional information can be revealed. Using confocal Raman imaging, the distribution of lignin and cellulose in black spruce wood was investigated (Agarwal, 2006) and changes of molecular composition in secondary plant cell wall tissues of poplar wood were illustrated (Gierlinger and Schwanninger, 2006). Raman imaging of *Arabidopsis thaliana*, one of the most important model plants, was recently demonstrated (Schmidt et al., 2009). Principal component analysis (PCA) and partial least square (PLS) modeling can be incorporated in image analysis to provide a more detailed comparison of cell wall compositions at different mapped regions (Gierlinger et al., 2008b). Raman imaging technique was also used to compare lignification in wild type and lignin-reduced 4-coumarate-CoA ligase (4CL) transgenic *Populus trichocarpa* stem wood (Schmidt et al., 2009). Raman imaging was further implemented to provide a more complete picture of the effects of alkaline treatment on *Miscanthus x giganteus*, a potential energy crop and a model lignocellulosic material. Longitudinal and transversal-section images of the parenchyma cells were generated, which revealed that lignin is removed preferentially from the inner surface of the cell wall and that cellulose is largely undisturbed (Chu et al., 2010).

Very recently our laboratory has developed a Raman imaging method to provide complete tissue/cell type specific compositional information for the first time (Sun et al., 2011). The method was demonstrated on stem sections of corn stover ranging from the epidermis to the pith area by both one-dimensional and two-dimensional chemical mapping. Lignin and cellulose abundance was determined in various cell types in the following order: sclerenchyma cells and tracheids (~5 times) > epidermal cells (~3 times) > bundle sheath cells > parenchyma cells. Unlike other Raman imaging work only showing the total lignin content, a Raman characterization study was performed to assign peaks for lignin compositions in terms of *p*-hydroxyphenyl, guaiacyl and syringyl units. Our imaging results have shown that significant amount of *p*-hydroxyphenyl units are present in the tracheids of corn stover stem, but not in the tracheids of the *Eucalyptus globulus* stem, which was corroborated by literature data (Galletti et al., 1996; Pinto et al., 2005).

Polarized Raman is another very useful tool for plant cell wall research by including polarizers in the optical path. Cao et al. (2006) has demonstrated a Raman study on the net orientation of biomacromolecules in the outer epidermal walls of mature wheat stems by comparing spectra collected with Raman light polarized perpendicular or parallel to the longitudinal axis of the cell. By changing the laser polarization direction in 3° steps, Gierlinger et al. (2010) investigated the dependency between cellulose and laser orientation direction and determined cellulose microfibril angle in S1 and S2 layers of wood samples, which was validated by X-ray diffraction measurement.

A separate category of nonlinear Raman techniques represented by coherent anti-Stokes Raman scattering (CARS) microscopy and stimulated Raman scattering (SRS) microscopy have emerged in recent years for plant cell wall research. CARS is orders of magnitude more sensitive, much faster in image acquisition and less affected by fluorescence compared with spontaneous Raman microscopy, and has the intrinsic capability of three-dimensional sectioning due to the nonlinear nature. CARS imaging of lignin in cell walls was demonstrated using corn stover (Evans and Xie, 2008). However, a CARS spectrum is different from its corresponding spontaneous Raman spectrum due to a nonresonant background, which causes difficulties in image interpretation. The major advantages SRS offers include an identical response to spontaneous Raman scattering, a linear dependence on the analyte concentration and fast image acquisition. Saar et al. (2010) has realized real-time monitoring of delignification reaction in corn stover using the acid chlorite method by SRS with high spatiotemporal resolution. However, some major problems associated with nonlinear Raman techniques are the cost and limited access to the instruments that are not commercially available and system optimization requirement for daily usage, which allows only very experienced people to operate the instruments.

7. Cell wall analysis by 2D and 3D electron microscopy

Plant cell walls are highly complex networks made of carbohydrates, lignins and some proteins. Apart from the complexity of individual plant cell walls, the walls can be very diverse in composition and organization, between different groups of plants, between different species and even within same plant, organ, tissue and cell type. Cell walls are also dynamic in nature and their ultrastructure alters with growth and differentiation (Carpita and Gibeaut, 1993; Niklas, 2004; Popper, 2008; Sarkar et al., 2009). Molecular resolution imaging of plant cell walls is needed to obtain a detailed structural knowledge of cell wall organization, which in turn is needed for rational engineering of cell walls for improving biofuel production from biomass. Electron microscopy allows ultrastructural analysis at molecular resolution, whereas optical microscopy techniques are typically limited by the diffraction limit of the optical microscope and the signal-to-noise ratios encountered in autofluorescent specimens.

Transmission electron microscopy (TEM) allows an in-depth analysis of cellular ultrastructure and has been used to study plant cell walls since 1940s yielding the first high-resolution ultrastructural insights (Preston et al., 1948). For TEM analysis samples need to be thin for the electron beam to penetrate (ideally ~ 100-300 nm) and must be examined a vacuum, resulting in the necessity for resin embedding, followed by ultrathin sectioning. The necessary chemical sample preparation steps employed in a typical conventional protocol results in limited preservation: chemical fixation, heavy metal postfixation and staining, as well as organic solvent dehydration, can lead to fixation staining and dehydration artifacts, such as the denaturation, aggregation and extraction of biological material as well as uneven or preferential staining. Moreover, even modern sample preparation protocols including specific staining techniques are predominantly optimized for cell membrane lipids, nucleic acids, and proteins, but not for carbohydrates and lignin. Some histochemical staining methods are used to stain cell wall components such as, (1) negative staining with uranyl acetate for cellulose; (2) PATCO (Periodic acid - Thiocarbohydrazide - Silver Proteinate) method for hemicelluloses; (3) ruthenium red for pectins, (4) potassium permanganate for lignins (Krishnamurthy, 1999). However, these stains are often not highly specific and stain multiple cell wall components to various degrees. Distinguishing between cell wall components accurately at high resolutions is difficult by sole differential staining.

Most of the early high-resolution TEM imaging of plant cell walls was done on samples prepared by metal shadowing and surface replication after freeze-fracturing or freeze-etching (Preston et al., 1948, McCann et al., 1990). While in principle no chemical fixatives, no dehydrating agents and no stains are used in this method, thus potentially retaining the samples closer to their native state, the images although potentially of high resolution are restricted to topological structural information in two-dimensions. Moreover, no chemical information can be obtained from this method as the imaging is done on the metal replica and not the biological sample. In recent years, sophisticated cryo-methods have been developed to minimize or completely overcome the limitations of conventional TEM sample preparation methods. Sophisticated cryo-methods such as high pressure freezing, followed by freeze-substitution and resin embedding typically display a superior quality of sample preservation in a much closer to native state (McDonald, 1999; McDonald & Muller-Reichert, 2002, McDonald & Auer 2006). High-pressure freezing followed by vitreous sectioning and cryo-TEM imaging offers preservation of biological samples closest to their native state (Al-Amoudi et al., 2004a, 2004b). These samples not only provide high-resolution structural information, but can also provide chemical information by specific staining or by immunolabeling with target-specific gold-conjugated antibodies. Several monoclonal antibodies and carbohydrate-binding modules (CBM) are being developed against different cell wall components in many laboratories around the world (Knox, 2008; Pattathil et al., 2010), which can be used with TEM analysis to localize the various chemical components of plant cell walls with high precision. All TEM sample preparation methods mentioned above are usually time-consuming and labor intensive, although automatic microwave tissue processors are now commercially available for rapid chemical fixation, dehydration, resin embedding and polymerization.

Apart from sample preparation issues, conventional TEM imaging runs into a few other problems. Due to a small field of view in a TEM, only small representative areas of any sample can be imaged at a time. The sections used in TEM are also very thin (under a micron), which makes imaging larger cellular structures in their entirety an almost impossible task. Use of high-resolution wide-field imaging (montaging) and imaging serial sections can help in covering relatively larger sample areas. Aligning serial sections is specimen-dependant due to the characteristics of each section. Individual sections in a series may have differences in scaling and/or may have non-linear deformation because of sectioning, folding, drying, specimen tilt, and optical distortions of the microscope (Stevens and Trogadis, 1984). A more complicated limitation of conventional TEM is that the images obtained are 2D projections of a 3D volume, which means multiple molecular layers of the sample contribute to the same layer of the image. Such images can be difficult to interpret if the structures of interest are only a few nanometers in dimension and are very closely packed. Plant cell walls are a good example of this imaging problem. They are made up of a tightly packed network of cellulose microfibrils, each microfibril being ~3 nm in diameter (Ohad & Danon, 1964; Frey-Wyssling, 1968; Heyn, 1969; Somerville et al., 2004; Ding and Himmel, 2006). The cellulose microfibrils are tightly surrounded by hemicelluloses such as xyloglucans or arabinoxylans that form hydrogen bonds with the cellulose microfibrils and form cross-links between two neighboring cellulose microfibrils. The matrix space in between cellulose and hemicelluloses is crowded with complex nano-scale molecules of pectins and/or lignins (McCann et. al., 1990; Carpita and Gibeaut, 1993; Somerville et al., 2004). It is extremely challenging to resolve the ultrastructure of plant cell walls *in situ* by most imaging techniques available currently, including conventional TEM. Atomic force microscopy (AFM) has been successfully used to image plant cell wall ultrastructure at high resolutions (Ding and Himmel, 2006) but the information available is two-dimensional and only topological.

Electron tomography overcomes some of the limitations faced by conventional TEM and AFM. In this method, several hundred two-dimensional TEM images of the same sample are collected by rotation of the sample along the central axis in small increments, which leads to a 3D volume of data, which can be visualized, segmented out to develop realistic models and quantitatively analyzed in 3D. Individual layers within the 3D volume can be separately visualized and analyzed in different planes using sophisticated image analysis software. Different types of algorithms for automated segmentation are being developed, though the currently available algorithms are only reliable for relatively simple image sets. Since individual components of the plant cell walls are typically spaced close to the resolution limit of the data set, manual segmentation is still the most reliable method for analyzing plant cell wall tomograms. EM tomography data of plant cell walls can be used to measure dimensions, orientations and spacing of the different cell wall components. This method paired with biochemical analysis methods like Raman imaging and immunolabeling has the potential to develop precise, comprehensive 3D ultrastructural cell wall model(s) at molecular resolution.

EM tomography has been applied to study the 3D organization of the cellulose microfibrils in the S2 layer of the secondary cell wall in *Pinus* wood tissue (Xu et al., 2007, 2011), however the samples used in this study had been harshly chemically treated. Cryo electron tomography of vitrified sections of plant tissue can provide preservation closest to their native state. The processing of tomographic datasets (reconstruction, filtering and segmentation) of plant cell wall cryo-sections and faithful model development however, is highly challenging, as the images obtained by cryo-tomography are extremely low in contrast. Furthermore, since plant cell walls contain densely packed nano-scale components, cryo-tomography is not suitable for high-throughput imaging of plant cell walls. Instead high pressure freezing, followed by freeze-substitution and resin embedding is a way to go and will be the key to obtain high-throughput realistic cell wall models. EM tomography of high-pressure frozen, freeze-substituted, resin embedded samples has been used to study at unprecedented resolution the dramatic structural changes during cytokinesis and the assembly process of cell plates during the final stage of cell division, (Seguí-Simarro et al., 2008).

8. Changes to biomass by acanning and transmission electron microscopy

Apart from an in-depth look at the molecular ultrastructure of different plant cell walls, **Scanning Electron Microscopy (SEM)** can be utilized to swiftly characterize biomass samples from different pretreatment protocols and while 2D in nature provides a 3D impression of the effect such protocols had on plant cell walls. SEM uses backscattering properties of heavy atom-coated surfaces to determine surface topologies. SEM is strictly a surface technique, and while not yielding information about the inside of an object, it provides fast overviews of large areas at fairly high resolutions with relatively simple sample preparation, allowing rapid screening of many samples. **Focused Ion Beam (FIB)/SEM** is a promising new method that can provide continuous 3D information over a large depth range though it requires a much more sophisticated sample preparation as compared to conventional SEM and will be discussed further below.

Regarding sample preparation, usually, the best approach is to work with cross sections of plant stems (e.g. *Arabidopsis* or *Brachopodium*) that can be cut using a vibrating blade microtome (e.g. Leica VT1000S, Leica Microsystems), resulting in reproducible cross sections. In order to ensure reproducible results multiple sections of each stem are typically cut in water, picked up with a brush or a pair of very fine tweezers and placed on prepared SEM sample stubs. The sample can then be dehydrated and critical-point dried or directly mounted on sample stubs and sputter coated, which allows for better conductivity of the sample and therefore better SEM imaging results.

TEM with plant stem cross sections requires significantly more preparation than SEM (see section above), but yields more quantitative information regarding cell wall thickness and can discriminate between primary and secondary cell wall. We (Li *et al.* 2009, Çetinkol *et al.* 2010) have used the described SEM techniques to visualize the effects of different pretreatment techniques designed to break up plant parts before they can be subjected to either enzymes or microbes thus converting them into sugars and subsequently second or third generation biofuels. Both TEM and SEM were used (Yin et al. 2011) to visualize the isolation of plant organelles or changes of the plant structure due to mutations, affecting cell wall properties and susceptibility to deconstruction for second or third generation biofuels synthesis. Furthermore, SEM analysis can be used to study the fracture pattern in samples derived from the mechanical stress screening in order to determine whether the fracture pattern indicate ductile or brittle behavior.

9. Examining microbial communities by advanced electron microscopy

A variety of microbial communities are thought to have unique lignocellulolytic capabilities making them of great interest for the development of second and third generation biofuels. Among the natural degraders of lignocellulose are microbial communities that are found e.g. in the gut of wood-degrading insects such as termites, in ruminants, in compost or soil. Electron microscopy with its ability to conduct an analysis at a spatial resolution of about a nanometer allows the investigation of the interplay between the different bacterial cells that constitute a community, its association with the plant material as well as their macromolecular inventory and degradation strategies.

SEM can only be conducted on surfaces, e.g. plant part from the digestive system of a cellulose degrading animal containing the microbial community, or microbes filtered out of a microbial culture suspension. Since SEM imaging typically requires sample fixation, dehydration and critical point drying, followed by sputter coating, the fine details tend to get lost. The microscopy on those samples in a high-resolution SEM (typically, 2 - 10 kV accelerating voltage are used) can visualize the shape of microorganisms, thus sometimes enabling an identification of certain species, and show both interaction between microbes and between microbes and the surface (e.g. plant material). It can thus be a valuable first insight into the composition and functioning of a microbial community, facilitated by the fact that the technique is rather quick and does not require extensive preparations. However, the technique does not enable a more detailed analysis of the microbial community and is limited to the surface of the samples, thus possibly representing only a small fraction of the sample that might not be typical for the sample in general.

If a more in-depth analysis of the microbial community is the goal, one has to use TEM instead of additional SEM. The analysis of a microbial community with TEM can visualize a lot of interesting traits in this community, including frequent cell-cell interactions, either directly or via a variety of microstructures such as pili and flagella. Even more interesting in this case are the interactions between microbial cells and the plant biomass, which is present in the sample. In studies of such communities (Knierim et al., in preparation) we have found different strategies of cell wall attachment and biomas digestion. An unresolved challenge is the direct identification of species in TEM or SEM. There have been recent advances in this field, combining TEM with Catalized Reporter Deposition Fluorescence In-Situ Hybridization (CARD-FISH), which can identify either groups of bacteria or even certain species (Knierim *et al.*, submitted). However, this technique is extremely tedious and requires some compromises of the TEM imaging quality in order to enable the identification via CARD-FISH.

The described electron microscopy techniques are not only applicable to the understanding of microbial communities but can also be utilized for the analysis of bacteria that have been engineered for certain capabilities. Such capabilities that are developed by synthetic biologists can be the increased production of fuels, the production of different fuels (especially those with longer carbon chains which are more valuable than ethanol) or the assessment of toxicity of the produced biofuels. When combined with tag-based labeling TEM also allows a rational monitoring of protein expression levels, cell-to cell variation and subcellular localization. Thus the application of SEM and TEM to those samples can facilitate the re-engineering of microorganisms that are needed for the production of second and third generation biofuels.

A natural limitation of TEM usually is its coverage of very small sample volumes covered by thin sections. To overcome this problem, FIB/SEM can be employed, which provides a three dimensional view of large volumes of a microbial community at a resolution comparable to electron tomography. For FIB/SEM a similar sample preparation protocol as described above for TEM can be used, but one has to ensure a high internal contrast inside the plastic blocks. We have had very good experience with tannic acid for this purpose. During

FIB/SEM imaging one must pay attention to the area of interest careful, as this technique is very time intensive and hence currently can only be applied to a limited number of volumes. Typically, volumes of 10 x 10 x 5 μm can be covered at an estimated resolution around 10-15 nm. The datasets that are produced this way easily get very large (in the several Gigabyte range) and require careful 3D reconstruction and analysis using software packages with those capabilities such as UCSF Chimera (http://www.cgl.ucsf.edu/chimera) or Amira (http://www.amira.com). We are currently drafting a manuscript on the FIB/SEM 3D analysis of the termite hindgut microbial community, where we classify and quantify constituent microbial community member according to size, shape and internal density characteristics, and map out their distribution with respect to the biomass (Knierim *et al.*, in preparation).

An understanding of the functioning of microbial communities that are capable of lignocellulose degradation will ultimately lead to the development of better techniques for lignocellulose degradation in an industrial setting such as a biorefinery for second or third generation biofuels. We are still in the early steps of this understanding due to the high complexity and variability of these microbial communities, but if we can reduce the complexity by identifying a small set of microbes with valuable capabilities, we may be able to speed up this process. As the principles that are present in these microbial communities have been developed by evolution over billions of years, we can assume that they are very energy efficient, thus providing a maximal energy output while taking up a small amount of energy themselves - an important challenge for the design of industrial processes for the production of biofuels

10. Outlook

Improving feedstocks properties as well as optimizing each step of the deconstruction process and the fuels synthesis production step will decrease the production cost and is therefore key for replacing fossil fuels with biofuels. To accomplish such needed technological advances, one needs to resort to a variety of different biophysical techniques, typically carried out by specialists, that can quantify the effect of experimental intervention and lead to a detailed understanding of the physical, chemical and biological processes of lignocellulosic biofuel production.

11. Acknowledgements

This work was part of the DOE Joint BioEnergy Institute (http://www.jbei.org) supported by the U. S. Department of Energy, Office of Science, Office of Biological and Environmental Research, through contract DE-AC02-05CH11231 between Lawrence Berkeley National Laboratory and the U. S. Department of Energy. This work was in part supported in part by the Energy Biosciences Institute grant 007-G18.

12. References

Adapa P.K., Karunakaran C., Tabil L.G., and Schoenau G.J. 2009. Potential applications of infrared and raman spectromicroscopy for agricultural biomass. *Agr Eng Inte: CIGR J* 11:1-25

Agarwal U. 2006. Raman imaging to investigate ultrastructure and composition of plant cell walls: Distribution of lignin and cellulose in black spruce wood (*Picea mariana*) *Planta* 224:1141-1153

Agarwal U.P., and Ralph S.A. 1997. Ft-raman spectroscopy of wood: Identifying contributions of lignin and carbohydrate polymers in the spectrum of black spruce (*Picea mariana*). *Appl Spectrosc* 51:1648-1655

Akiyama K., Chikayama E., Yuasa H., Shimada Y., Tohge T., Shinozaki K., Hirai M.Y., Sakurai T., Kikuchi J., and Saito K. 2008. PRIMe: a Web site that assembles tools for metabolomics and transcriptomics. *In Silico Biol* 8: 339-345

Al-Amoudi A., Chang J., Leforestie A., McDowall A., Salamin L. M., Norlen L., Richter K., Blanc N. S., Studer D., and Dubochet, J. 2004. Cryo-electron microscopy of vitreous sections. EMBO J 23: 3583–3588

Al-Amoudi A., Norlen L., and Dubochet, J. 2004. Cryo-electron microscopy of vitreous sections of native biological cells and tissues. *J Struct Biol* 148: 131–135

Alves A., Schwanninger M., Pereira H., and Rodrigues J. 2006. Calibration of nir to assess lignin composition (h/g ratio) in maritime pine wood using analytical pyrolysis as the reference method. *Holzforschung* 60: 29-31

Atsumi S., Hanai T., and Liao, J.C. 2008. Non-fermentative pathways for synthesis of branched-chain higher alcohols as biofuels. *Nature* 451: 86-89

Baskin TI, Beemster GTS, Judy-March JE, Marga F. 2004. Disorganization of cortical microtubules stimulates tangential expansion and reduces the uniformity of cellulose microfibril alignment among cells in the root of arabidopsis. Plant Physiol 135(4):2279-2290.

Baucher M., Bernard-vailhé, M.A., Chabbert, B., Besle, J. M., Opsomer, C., Van Montagu M., and Botterman J. 1999. Down-regulation of cinnamyl alcohol dehydrogenase in transgenic alfalfa (*Medicago sativa* L.) and the effect on lignin composition and digestibility. *Plant Mol Biol* 39: 437-447

Bertran, M. S., and Dale B. E. 1986. Determination of cellulose accessibility by differential scanning calorimetry. *J Appl Polymer Sci* 32: 4241-4253

Boudet A. M., Goffner D., Marque C., Teulières C., and Grima-Pettenati J. 1998. Genetic manipulation of lignin profiles: a realistic challenge towards the qualitative improvement of plant biomass. *Ag Biotech News Info* 10: 295N-304N

Burk D. H., Liu B., Zhong R., Morrison W. H., and Ye Z. H. 2001. A katanin-like protein regulates normal cell wall biosynthesis and cell elongation. *Plant Cell* 13: 807-828

Cao Y., Shen D., Lu Y., and Huang Y. 2006. A raman-scattering study on the net orientation of biomacromolecules in the outer epidermal walls of mature wheat stems (*Triticum aestivum*). *Ann Bot* 97:1091-1094

Carpita N. C., and Gibeaut D. M. 1993. Structural models of primary cell walls in flowering plants: consistency of molecular structure with the physical properties of the walls during growth. *Plant J* 3: 1–30

Carr G. L. 1999. High-resolution microspectroscopy and sub-nanosecond time-resolved spectroscopy with the synchrotron infrared source. *Vib Spectrosc* 19: 53-60

Cavalier DM, Lerouxel O, Neumetzler L, Yamauchi K, Reinecke A, Freshour G, Zabotina OA, Hahn MG, Burgert I, Pauly M and others. 2008. Disrupting two arabidopsis thaliana xylosyltransferase genes results in plants deficient in xyloglucan, a major primary cell wall component. Plant Cell 20(6):1519-1537.

Çetinkol, Ö. P., Dibble D. C., Cheng G., Kent M. S., Knierim B., Auer M., Wemmer, D. E., Pelton J. G., Melnichenko Y.B., Ralph J., Simmons B. A. and Holmes, B. M. 2010. Understanding the impact of ionic liquid pretreatment on eucalyptus. *Biofuels* 1: 33-46

Chen L., Carpita N. C., Reiter W. D., Wilson R. H., Jeffries C., and McCann M. C. 1998. A rapid method to screen for cell-wall mutants using discriminant analysis of fourier transform infrared spectra. *Plant J* 16: 385-392

Chen F., and Dixon R. A. 2007. Lignin modification improves fermentable sugar yields for biofuel production. *Nature Biotech* 25: 759-761

Chiang V. L., and Funaoka M. 1990. The difference between guaiacyl and guaiacyl-syringyl lignins in their responses to kraft delignification. *Holzforschung* 44: 309-313

Chu L. Q., Masyuko R., Sweedler J. V., and Bohn P. W. 2010. Base-induced delignification of *Miscanthus x giganteus* studied by three-dimensional confocal raman imaging. *Bioresour Technol* 101:4919-4925

Coops H. and Van der Velde G. 1996. Effects of waves on helophyte stands: mechanical characteristics of stems of *Phragmites australis* and *Scirpus lacustris*. *Aquatic Botany* 53: 175-185

Couchman P. R. 1981. The effect of degree of polymerization on glass-transition temperatures. *Polym Eng Sci* 21: 377-380

Czichos H., Saito T., and Smith L. 2006. Measurement methods for mechanical properties. In Materials measurement methods, Springer, 1: 283-387

Dai C. C., Tao J., Xie F., Dai Y., and Zhao M. 2007. Biodiesel generation from oleaginous yeast *Rhodotorula glutinis* with xylose assimilating capacity. *African J Biotech* 6: 2130-2134

Davison B. H., Drescher S. R., Tuskan G. A., Davis M. F., and Nghiem N. P. 2006. Variation of S/G ratio and lignin content in a Populus family influences the release of xylose by dilute acid hydrolysis. *Appl Biochem and Biotechnol* 130: 427-435.

De Micco V, Aronne G. 2007. Combined histochemistry and autofluorescence for identifying lignin distribution in cell walls. Biotech Histochem 82(4):209-216.

D'Haeze W, Gao M, De Rycke R, Van Montagu M, Engler G, Holsters M. 2007. Roles for azorhizobial nod factors and surface polysaccharides in intercellular invasion and nodule penetration, respectively. Mol Plant Microbe Interact 11(10):999-1008.

Ding, S. and Himmel, M. E. 2006. The maize primary cell wall microfibril: A new model derived from direct visualization. *J Agric Food Chem*, 54: 597-606

Dokken K. M., Davis L. C. 2007. Infrared imaging of sunflower and maize root anatomy. *J Agric Food Chem* 55:10517-10530

Dokken K. M., Davis L. C., and Marinkovic N. S. 2005. Use of infrared microspectroscopy in plant growth and development. *Appl Spectrosc Rev* 40: 301 - 326

Ehara K., Takada D., and Saka S. 2005. GC-MS and IR spectroscopic analyses of the lignin-derived products from softwood and hardwood treated in supercritical. *J Wood Sci* 51: 256-261.

Emmel A., Mathias A. L., Wypych F., and Ramos L. P. 2003. Fractionation of Eucalyptus grandis chips by dilute acid-catalysed steam explosion. *Bioresour Technol* 86: 105-115.

Evans C. L., and Xie X. S. 2008. Coherent anti-stokes raman scattering microscopy: Chemical imaging for biology and medicine. *Annu Rev Anal Chem* 1:883-909

Flynn J. H., and Wall L. A. 1966. A quick, direct method for the determination of activation energy from thermogravimetric data. *J Polym Sci Part B: Polym phys* 4: 323-328

Frey-Wyssling (1968) The ultrastructure of wood. *Wood Sci Tech* 2: 73-83

Fromm J, Rockel B, Lautner S, Windeisen E, Wanner G. 2003. Lignin distribution in wood cell walls determined by tem and backscattered sem techniques. J Struct Biol 143(1):77-84.

Galletti G. C., Bocchini P., Smacchia A. M., and Reeves J. B. 1996. Monitoring phenolic composition of maturing maize stover by high performance liquid chromatography and pyrolysis/gas chromatography/mass spectrometry. *J Sci Food Agric* 71:1-9

Galletti G. C., Reeves J. B., and Bocchini P. 1997. Analytical pyrolysis as a tool to determine chemical changes in maize stovers during growth. *J anal and appl pyrolysis* 39: 105-114

Ghetti P., Ricca L., and Angelini L. 1996. Thermal analysis of biomass and corresponding pyrolysis products. *Fuel* 75: 565-573

Gierlinger N., Goswami L., Schmidt M., Burgert I., Coutand C., Rogge T., and Schwanninger M. 2008a. In situ ft-ir microscopic study on enzymatic treatment of poplar wood cross-sections. *Biomacromolecules* 9:2194-2201

Gierlinger N., Luss S., König C., Konnerth J., Eder M., and Fratzl P. 2010. Cellulose microfibril orientation of picea abies and its variability at the micron-level determined by raman imaging. *J Exp Bot* 61:587-595

Gierlinger N., Sapei L., and Paris O. 2008b. Insights into the chemical composition of *Equisetum hyemale* by high resolution raman imaging. *Planta* 227:969-980

Gierlinger N., and Schwanninger M. 2006. Chemical imaging of poplar wood cell walls by confocal raman microscopy. *Plant Physiol* 140:1246-1254

Gierlinger N., and Schwanninger M. 2007. The potential of raman microscopy and raman imaging in plant research. *Spectrosc-Int J* 21:69-89

Goddard T.D., Huang C.C., and Ferrin T.E. 2004 Visualizing density maps with UCSF Chimera, *J. Struct Biol.* 157: 281-287

Grethlein H. E., and Converse A. O. 1991. Common aspects of acid prehydrolysis and steam explosion for pretreating wood. *Bioresour Technol* 36: 77-82

Grünwald C, Ruel K, Kim YS, Schmitt U. 2002. On the cytochemistry of cell wall formation in poplar trees. Plant Biol 4(1):13-21.

Guo D., Chen F., Wheeler J., Winder J., Selman S., Peterson M., and Dixon R. A. 2001. Improvement of in-rumen digestibility of alfalfa forage by genetic manipulation of lignin O-methyltransferases. *Transgenic Research* 10: 457-464

Hall M., Bansal P., Lee J. H., Realff M. J., and Bommarius A. S. 2010. Cellulose crystallinity - a key predictor of the enzymatic hydrolysis rate. *FEBS Journal* 277: 1571-1582

Harley C. D. G., and Bertness M. D. 1996. Structural interdependence: An ecological consequence of morphological responses to crowding in marsh plants. *Functional Ecology* 10: 654-661

Heraud P., Caine S., Sanson G., Gleadow R., Wood B. R., and McNaughton D. 2007. Focal plane array infrared imaging: A new way to analyse leaf tissue. *New Phytologist* 173:216-225

Heyn, A. N. J.1969 The Elementary Fibril and Supermolecular Structure of Cellulose in Soft Wood Fiber. *J. Ultra. Res.* 26: 52-68

Himmel M. E., Ding S., Johnson D. K., Adney W. S., Nimlos M. R., Brady J. W., and Foust T. D. 2007. Biomass recalcitrance: Engineering plants and enzymes for biofuels production. *Science* 315: 804 -807

Himmelsbach D. S., and Akin D. E. 1998. Near-infrared fourier-transform raman spectroscopy of flax (*Linum usitatissimum* l.) stems. *J Agric Food Chem* 46:991-998

Hu Y., Zhong R., Morrison W. H. III, and Ye Z. 2003. The Arabidopsis RHD3 gene is required for cell wall biosynthesis and actin organization. *Planta* 217: 912-921

Huyen T. L., Ramond C., Dheilly R.M., and Chabbert B. 2010. Effect of harvesting date on the composition and saccharification of Miscanthus x giganteus. *Bioresour Technol* 101: 8224-8231

Ibarra D., Chávez M. I., Rencoret J., Río J. C. D., Gutiérrez A., Romero J., Camarero S., Martínez M. J., Jiménez-Barbero J., and Martínez A. T. 2007. Lignin modification during eucalyptus globulus kraft pulping followed by totally chlorine-free bleaching: A two-dimensional nuclear magnetic resonance, fourier transform infrared, and pyrolysis–gas chromatography/mass spectrometry study. *J Agric Food Chem* 55: 3477-3490

Ibrahim M. N. M., and Pearce G. R. 1983. Effects of chemical treatments combined with high-pressure steaming on the chemical composition and in vitro digestibility of crop by-products. *Agricultural Wastes* 7: 235-250

Igathinathane C., Womac A. R., Sokhansanj S., and Narayan S. 2008. Knife grid size reduction to pre-process packed beds of high- and low-moisture switchgrass. *Bioresour Technol* 99: 2254-2264

Jackson L. A., Shadle G. L., Zhou R., Nakashima J., Chen F., and Dixon R. A. 2008. Improving saccharification efficiency of Alfalfa stems through modification of the terminal stages of monolignol biosynthesis. *Bioenerg Res* 1: 180-192

Jaffe M., Collins G., and Mencze J. 2006. The thermal analysis of fibers in the twenty first century: From textile, industrial and composite to nano, bio and multi-functional. *Thermochim Acta* 442: 95-99

Jung S., Foston M., Sullards M. C., and Ragauskas A. J. 2010. Surface characterization of dilute acid pretreated Populus deltoides by ToF-SIMS. *Energy Fuels* 24: 1347-1357

Kaloustian J., El-Moselhy T. F., and Portugal H. 2003. Chemical and thermal analysis of the biopolymers in thyme (Thymus vulgaris). *Thermochim Acta* 401: 77-86

Keppler B. D., and Showalter A. M. 2010. IRX14 and IRX14-LIKE, Two Glycosyl transferases involved in Glucuronoxylan biosynthesis and drought tolerance in Arabidopsis. *Molecular Plant* 1: 1-8

Kim U., Eom S. H., and Wada M. 2010. Thermal decomposition of native cellulose: Influence on crystallite size. *Polym Degrad Stab* 95: 778-781

Knierim B, Lin M., Desai M, van Leer B., Goddard T.D., Hugenholtz P., McDonald K.L., Webb R.I., Auer M. (2011) Multiscale three-dimensional Organization of the Termite Hindgut Elucidated by FIB/SEM, in preparation

Knierim B, Luef B., Wilmes P., Webb R.I., Auer M., Comolli L.R., Banfield J.F. (2011) Correlative microscopy for phylogenetic and ultrastructural characterization of microbial communities, submitted to ISME J.

Knox J.P. 2008 Revealing the structural and functional diversity of plant cell walls. *Curr Opin Plant Biol* 11: 308-313

Krishnamurthy KV. 1999. Methods in cell wall cytochemistry. Boca Raton, FL: CRC Press. pp: 190-220

Krongtaew C., Messner K., Ters T., and Fackler K. 2010a. Characterization of key parameters for biotechnological lignocellulose conversion assessed by ft-nir spectroscopy. Part i: Qualitative analysis of pretreated straw. *BioResources* 5: 2063-2080

Krongtaew C., Messner K., Ters T., and Fackler K. 2010b. Characterization of key parameters for biotechnological lignocellulose conversion assessed by ft-nir spectroscopy. Part ii: Quantitative analysis by partial least squares regression. *BioResources* 5: 2081-2096.

Kubo S., and Kadla J. F. 2005. Hydrogen Bonding in Lignin: A fourier transform infrared model compound study. *Biomacromolecules* 6: 2815-2821

Lee C., Teng Q., Huang W., Zhong R., and Ye Z. 2009. Down-regulation of PoGT47C expression in Poplar results in a reduced Glucuronoxylan content and an increased wood digestibility by cellulase. *Plant Cell Physiol* 50: 1075 -1089

Lee C., Teng Q., Huang W., Zhong R., and Ye Z. 2010. The Arabidopsis family GT43 Glycosyltransferases form two functionally nonredundant groups essential for the elongation of Glucuronoxylan backbone. *Plant Physiol* 153: 526-541

Li C., Knierim B., Manisseri C., Arora R., Scheller H. V., Auer M., Vogel K. P., Simmons B. A., and Singh S. 2010. Comparison of dilute acid and ionic liquid pretreatment of switchgrass: Biomass recalcitrance, delignification and enzymatic saccharification. *Bioresour Technol* 101: 4900-4906

Li Y., Qian Q., Zhou Y., Yan M., Sun L., Zhang M., Fu Z., Wang Y., Han B., Pang X., Chen M., and Li J. 2003. BRITTLE CULM1, which encodes a COBRA-Like protein, affects the mechanical properties of rice plants. *Plant Cell* 15: 2020-2031

Lin K. W., Ladisch M. R., Voloch M., Patterson J. A., and Noller C. H. 1985. Effect of pretreatments and fermentation on pore size in cellulosic materials. *Biotechnol Bioeng* 27: 1427-1433

Lu F., and Ralph J. 1997. DFRC method for lignin analysis. 1. New method for β-aryl ether cleavage: lignin model studies. *J Agric Food Chem* 45: 4655-4660

Lu F., and Ralph J. 1998. The DFRC method for lignin analysis. 2. Monomers from isolated lignins. *J Agric Food Chem* 46: 547-552

Lu F., and Ralph J. 1999. The DFRC method for lignin analysis. 7. Behavior of cinnamyl end groups. *J Agric Food Chem* 47: 1981-1987

Lu F.C., and Ralph J. 1997. Derivatization followed by reductive cleavage (DFRC method), a new method for lignin analysis: Protocol for analysis of DFRC monomers. J *Agric Food Chem* 45: 2590-2592

Lu X., Vora H., and Khosla C. 2008. Overproduction of free fatty acids in E. coli: Implications for biodiesel production. *Metabolic Eng* 10: 333-339

Marita J. M., Ralph J., Hatfield R. D., and Chapple C. 1999. NMR characterization of lignins in Arabidopsis altered in the activity of ferulate 5-hydroxylase. *Proceedings of the National Academy of Sciences of the United States of America* 96: 12328 -12332

McCann M, Stacey N, Wilson R, and Roberts K. 1993. Orientation of macromolecules in the walls of elongating carrot cells. *J Cell Sci* 106:1347-1356

McCann, M.C., Wells, B. and Roberts, K. 1990. Direct visualization of cross-links in the primary plant cell wall. *J Cell Sci* 96: 323–334

McDonald, K. 1999. High-pressure freezing for preservation of high resolution fine structure and antigenicity for immunolabeling. Electron Microscopy Methods and Protocols *(ed. by* N.Hajibagheri*)*. Humana Press, Totowa, NJ. *pp.* 77–97.

McDonald Kent L, and Auer Manfred 2006. High-pressure freezing, cellular tomography, and structural cell biology. Biotechniques. 41:137-143

McDonald, K. and Müller-Reichert, T. *(*2002*)* Cryomethods for thin section electron microscopy. *Meth. Enzymol.* 351: 96–123

Mills T. Y., Sandoval N. R., and Gill R. T. 2009. Cellulosic hydrolysate toxicity and tolerance mechanisms in Escherichia coli. *Biotechnol for Biofuels* 2: 1-11

Mouille G., Robin S., Lecomte M., Pagant S., and Höfte H. 2003. Classification and identification of arabidopsis cell wall mutants using fourier-transform infrared (ft-ir) microspectroscopy. *Plant J* 35: 393-404

Niklas K. J. 2004. The cell walls that bind the tree of life. *BioScience* 54: 831–841

Nuopponen M., Willför S., Jääskeläinen A. S., Sundberg A., and Vuorinen T. 2004. A uv resonance raman (uvrr) spectroscopic study on the extractable compounds of scots pine (pinus sylvestris) wood: Part i: Lipophilic compounds. *Spectrochim Acta A* 60: 2953-2961

Ohad, I. and Danon, D. (1964) On the dimensions of cellulose microfibrils. *J Cell Biol* 22: 302-305

Papatheofanous M. G., Billa E., Koullas D. P., Monties B., and Koukios E. G. 1995. Two-stage acid-catalyzed fractionation of lignocellulosic biomass in aqueous ethanol systems at low temperatures. *Bioresour Technol* 54: 305-310

Pattathil S., Avci U., Baldwin D., Swennes A. G., McGill J. A., Popper Z., Bootten T., Albert A., Davis R. H., Chennareddy C., Dong R., O'Shea B., Rossi R. Leoff C., Freshour G., Narra R., O'Neil M., York W. S. and Hahn M. G. 2010. A comprehensive toolkit of plant cell wall glycan-directed monoclonal antibodies. *Plant Physiol* 153: 514–525

Paul S. A., Oommen C., Joseph K., Mathew G., and Thomas S. 2010. The role of interface modification on thermal degradation and crystallization behavior of composites from commingled polypropylene fiber and banana fiber. *Polym Compos* 31: 1113-1123

Pecina R., Burtscher P., Bonn G., and Bobleter O. 1986. GC-MS and HPLC analyses of lignin degradation products in biomass hydrolyzates. *Fresenius' J Anal Chem* 325: 461-465

Pena M. J., Zhong R., Zhou G., Richardson E. A., O'Neill M. A., Darvill A. G., York W. S., and Ye Z. 2007. Arabidopsis irregular xylem8 and irregular xylem9: Implications for the Complexity of Glucuronoxylan Biosynthesis. *Plant Cell* 19: 549-563

Peng J., Lu F., and Ralph J. 1998. The DFRC method for lignin analysis. 4. lignin dimers isolated from DFRC-degraded Loblolly Pine wood. *J Agric Food Chem* 46: 553-560

Persson S, Caffall KH, Freshour G, Hilley MT, Bauer S, Poindexter P, Hahn MG, Mohnen D, Somerville C. 2007a. The arabidopsis irregular xylem8 mutant is deficient in glucuronoxylan and homogalacturonan, which are essential for secondary cell wall integrity. Plant Cell 19(1):237-255.

Persson S, Paredez A, Carroll A, Palsdottir H, Doblin M, Poindexter P, Khitrov N, Auer M, Somerville CR. 2007b. Genetic evidence for three unique components in primary cell-wall cellulose synthase complexes in arabidopsis. Proc Natl Acad Sci USA 104(39):15566-15571.

Petterson E.F., Goddard T.D., Huang C.C., Couch G.S., Greenblatt D.M., Meng E.C. Ferrin T.E. 2004 UCSF Chimera - a visualization systems for exploratory research and analysis.

Pinto PC, Evtuguin DV, Neto CP. 2005. Effect of structural features of wood biopolymers on hardwood pulping and bleaching performance. *Ind Eng Chem Res* 44: 9777-9784

Popper Z. A. 2008. Evolution and diversity of green plant cell walls. *Curr Opin Plant Biol* 11: 286–292

Prat R., and Paresys G. 1989. Multiple use apparatus for cell wall extensibility and cell elongation studies. *Plant Physiol Biochem* 27: 955-962

Preston, R. D., Nicolai, E., Reed, R. and Millard, A. 1948. An electron microscope study of cellulose in the wall of *Valonia ventricosa*. *Nature* 162: 665–667

Ralph J. 2007. Perturbing ligninfication. In *The Compromised Wood Workshop 2007,* K. Entwistle, P.J. Harris and J. Walker, Eds., Wood Technology Research Centre, University of Canterbury, New Zealand, Canterbury, pp. 85-112

Ralph J., and Hatfield R. D. 1991. Pyrolysis-GC-MS charecterization of forage ,aterials. *J Agric Food Chem* 39: 1426-1437

Ralph J., and Lu F. 1998. The DFRC Method for Lignin Analysis. 6. A Simple Modification for identifying natural acetates on lignins. *J Agric Food Chem* 46: 4616-4619

Reh U., Kraepelin G., and Lamprecht I. 1987. Differential scanning calorimetry as a complementary tool in wood biodegradation studies. *Thermochim Acta* 119: 143-150

Reiter W., Chapple C., and Somerville C. R. 1993. Altered growth and cell walls in a fucose-deficient mutant of arabidosis. *Science* 261: 1032-1035

Ryden P., Sugimoto-Shirasu K., Smith A. C., Findlay K., Reiter W., and McCann M. C. 2003. Tensile properties of Arabidopsis cell walls depend on both a Xyloglucan cross-linked microfibrillar network and Rhamnogalacturonan II-Borate complexes. *Plant Physiol* 132: 1033-1040

Saar B. G., Zeng Y., Freudiger C. W., Liu Y. S., Himmel M. E., Xie X. S., Ding S. Y. 2010. Label-free, real-time monitoring of biomass processing with stimulated raman scattering microscopy. *Angew Chem* 122: 5608-5611.

Saariaho A-M, Jääskeläinen A-S, Nuopponen M, Vuorinen T. 2003. Ultra violet resonance raman spectroscopy in lignin analysis: Determination of characteristic vibrations of p-hydroxyphenyl, guaiacyl, and syringyl lignin structures. *Appl Spectrosc* 57: 58-66

Sarkar P., Bosneaga E. and Auer M. 2009. Plant cell walls throughout evolution: towards a molecular understanding of their design principles. *J Exp Bot* 60: 3615-3635

Samuel R., Pu Y., Raman B., and Ragauskas A. J. 2010. Structural characterization and comparison of Switchgrass ball-milled lignin before and after dilute acid pretreatment. *Appl Biochem Biotechnol* 162: 62-74

Schmidt M., Schwartzberg A., Perera P., Weber-Bargioni A., Carroll A., Sarkar P., Bosneaga E., Urban J., Song J., Balakshin M., Capanema E. A., Auer, M., Adams P. D., Chiang V. L. and Schuck P. J. 2009. Label-free in situ imaging of lignification in the cell wall of low lignin transgenic populus trichocarpa. *Planta* 230: 589-597.

Schmidt M., Schwartzberg A. M., Carroll A., Chaibang A., Adams P. D., and Schuck PJ. 2010. Raman imaging of cell wall polymers in *Arabidopsis thaliana*. *Biochem Biophys Res Commun* 395: 521-523.

Seguí-Simarro, J.M., Otegui, M. S., Austin, J. R. and Staehelin, A. L. 2008. Plant cytokinesis - Insights gained from electron tomography studies. In: Verma DPS, Hong Z (eds) Cell division control in plants. Springer, Berlin/Heidelberg, pp 251–287

Serapiglia M. J., Cameron K. D., Stipanovic A. J., and Smart L. B. 2008. High-resolution thermogravimetric analysis for rapid characterization of biomass composition and selection of shrub willow varieties. *Appl Biochem Biotechnol* 145: 3-11

Sewalt V. J. H., Glasser W. G., and Beauchemin K. A. 1997. Lignin impact on fiber degradation. 3. reversal of inhibition of enzymatic hydrolysis by chemical modification of lignin and by additives. *J Agric Food Chem* 45: 1823-1828

Shao S., Jin Z., Wen G., and Liyama K. 2009. Thermo characteristics of steam-exploded bamboo (Phyllostachys pubescens) lignin. *Wood Sci Technol* 43: 643-652

Singh S, Simmons BA, Vogel KP. 2009. Visualization of biomass solubilization and cellulose regeneration during ionic liquid pretreatment of switchgrass. Biotechnol Bioeng 104(1):68-75.

Smith E, Dent G. 2005. Modern raman spectroscopy: A practical approach. Chichester, England: John Wiley & Sons Ltd.

Soares, S., Cammino G., and Levchick S. 1995. Comparative study of the thermal decomposition of pure cellulose and pulp paper. *Polym Degrad Stab* 49: 275-283

Somerville, C., Bauer, S., Brininstool, G., Facette, M., Hamann, T., Milne, J., Osborne, E., Paredez, A., Persson, S., Raab, T., Vorwerk, S., and Youngs, H. (2004) Toward a systems approach to understanding plant-cell walls. *Science* 306: 2206-2211

Stevens, J.K. and Trogadis, J. 1984. Computer-assisted reconstruction from serial electron micrographs: a tool for the systematic study of neuronal form and function. *Advan Cell Neurobiol* 5: 341–369

Stewart D, Wilson H. M., Hendra P. J., and Morrison I. M. 1995. Fourier-transform infrared and raman spectroscopic study of biochemical and chemical treatments of oak wood (*Quercus rubra*) and barley (*Hordeum vulgare*) straw. *J Agric Food Chem* 43: 2219-2225

Sun L., and Simmons B. A, Singh S. 2011. Understanding tissue specific compositions of bioenergy feedstocks through hyperspectral raman imaging. *Biotechnol Bioeng* 108: 286-295

Tang, Y., Pingitore F., Mukhopadhyay A., Phan R., Hazen T. C., and Keasling J. D. 2007. Pathway confirmation and flux analysis of central metabolic pathways in desulfovibrio vulgaris hildenborough using gas chromatography-mass spectrometry and fourier transform-ion cyclotron resonance mass spectrometry. *J Bacteriol* 189: 940-949

Tirumalai V, Agarwal U, Obst J. 1996. Heterogeneity of lignin concentration in cell corner middle lamella of white birch and black spruce. Wood Sci Technol 30(2):99-104.

Tsujiyama S., and Miyamori A. 2000. Assignment of DSC thermograms of wood and its components. *Thermochim Acta* 351: 177-181

Vailhé M. A. B., Besle J. M., Maillot M. P., Cornu A., Halpin C., and Knight M. 1998. Effect of down-regulation of cinnamyl alcohol dehydrogenase on cell wall composition and on degradability of tobacco stems. *J Sci Food Agric* 76: 505-514

Vinzant T., Ehrman C., Adney W., Thomas S., and Himmel M. 1997. Simultaneous saccharification and fermentation of pretreated hardwoods. *Appl Biochem Biotechnol* 62: 99-104

Wilkinson J.M., and Santillana R. G. 1978. Ensiled alkali-treated straw. I. Effect of level and type of alkali on the composition and digestibility in vitro of ensiled barley straw. *Anim Feed Sci Technol* 3: 117-132

Wilson RH, Smith AC, Kačuráková M, Saunders PK, Wellner N, Waldron KW. 2000. The mechanical properties and molecular dynamics of plant cell wall polysaccharides studied by fourier-transform infrared spectroscopy. Plant Physiol 124: 397-406

Ximenes E., Kim Y., Mosier N., Dien B., and Ladisch M. 2010. Inhibition of cellulases by phenols. *Enzyme Microb Technol* 46: 170-176

Ximenes E., Kim Y., Mosier N., Dien B., and Ladisch M. 2011. Deactivation of cellulases by phenols. *Enzyme Microbl Technol* 48: 54-60

Xu, P, Donaldson, L. A., Gergely, Z. R, and Staehelin, L. A. 2007. Dual-axis electron tomography: a new approach for investigating the spatial organization of wood cellulose microfibrils. *Wood Sc and Tech* 41: 101–116

Xu, P., Liu, H., Donaldson, L. A., and Zhang, Y. 2011. Mechanical performance and cellulose microfibrils in wood with high S2 microfibril angles. *J Mater Sci* 46: 534-540

Yin, L., Verhertbruggen Y., Oikawa A, Manisseri C, Knierim B, Prak L., Kruger Jensen J., Knox J.P., Auer M., Willats W.G.T. Scheller H. V. (2011). "The Cooperative Activities of CSLD2, CSLD3 and CSLD5 are Required for Normal Arabidopsis Development." *Molecular Plant*, in press.

Yu M., Womac A. R., Igathinathane C., Ayers P.D., and Buschermohle M.J. 2006. Switchgrass ultimate stresses at typical biomass conditions available for processing. *Biomass Bioenergy* 30: 214-219

Yu P, McKinnon JJ, Christensen CR, Christensen DA, Marinkovic NS, Miller LM. 2003. Chemical imaging of microstructures of plant tissues within cellular dimension using synchrotron infrared microspectroscopy. *J Agric Food Chem* 51: 6062-6067

Zhong R., Taylor J. J., and Ye Z. H. 1997. Disruption of interfascicular fiber differentiation in an Arabidopsis mutant. *Plant Cell* 9: 2159-2170

Zhong R., Burk D. H., Morrison W. H., and Ye Z. 2002. A kinesin-like protein is essential for oriented deposition of cellulose microfibrils and cell wall strength. *Plant Cell* 14: 3101-3117

Zhong R., Burk D. H., Morrison W. H., and Ye Z. 2004. FRAGILE FIBER3, an Arabidopsis gene encoding a Type II Inositol Polyphosphate 5-Phosphatase, is required for secondary wall synthesis and Actin organization in fiber cells. *Plant Cell* 16: 3242-3259

Zhong R., Richardson E., and Ye Z. 2007. Two NAC domain transcription factors, SND1 and NST1, function redundantly in regulation of secondary wall synthesis in fibers of Arabidopsis. *Planta* 225: 1603-1611

Part 4

Process Synthesis and Design

22

Kinetic Study on Palm Oil Waste Decomposition

Zakir Khan[1], Suzana Yusup[1], Murni M. Ahmad[1], Yoshimitsu Uemura[1], Vuoi S. Chok[2], Umer Rashid[1] and Abrar Inayat[1]
[1]*Universiti Teknologi PETRONAS, Perak*
[2]*Platinum Energy Sdn. Bhd., Kuala Lumpur*
Malaysia

1. Introduction

Malaysia is the largest producer of palm oil and contributes 43% of worldwide production (Shuit et al., 2009). Beside palm oil, palm oil industry generated 169.72 million metric tons solid wastes which contribute 85.5% of total biomass waste produced in the country (Khan et al., 2010). This huge amount of wastes can be converted into valuable chemical feed stocks and fuels due to environmental problems associated with conventional fossil fuels.

It is well known that lignocellulosic biomass mainly consists of hemicellulose, cellulose and lignin. The usual proportions (wt%) vary as 40-50% cellulose, 20-60% hemicellulose and 10-25% lignin (Yang et al., 2007). The thermal decomposition of these individuals is important since they influence the basics of thermochemical conversion processes such as pyrolysis, combustion and gasification. Decomposition of these components is intensively studied in the literature. Demirbas et al. (2001) observed the ease of lignocellulosic biomass components decomposition as hemicellulose > cellulose >>> lignin. Based on different reasoning, Yang et al. (2007) proposed different decomposition regions of 220-300 °C, 300-340 °C and >340 °C for hemicellulose, cellulose and lignin, respectively. Lignin is the last to decompose due to its heavy cross linked structure (Guo & Lua, 2001).

Several techniques are available to study the kinetics of biomass decomposition. Among these, thermogravimetric analysis (TGA) is the most popular and simplest technique (Luangkiattikhun et al., 2008), based on the observation of sample mass loss against time or temperature at a specific heating rate. TGA provides high precision (Várhegyi et al., 2009), fast rate data collection and high repeatability (Yang et al., 2004) under well defined kinetic control region.

Very few attempts have been carried out to study the kinetics of empty fruit bunch (EFB) and palm shell (PS) using TGA. Guo & Lua (2001) presented the effect of sample particle size and heating rate on pyrolysis process and kinetic parameters for PS. They concluded a first order reaction mechanism for the decomposition of PS at different heating rates. They also suggested higher heating rates for faster and easy thermal decomposition of PS. Yang et al. (2004) studied activation energy for decompositions of hemicellulose and cellulose in EFB and PS by considering different temperature region for first order kinetic reaction. They evaluated average activation energy and pre-exponential factor from single-step decompositions of hemicellulose and cellulose. Luangkiattikhun et al. (2008) considered the effect of heating rate and sample particle size on the thermogram behaviour and kinetic parameters for palm oil shell, fibre and kernel. They observed that there is no significant effect of particle size on the thermogram behaviour at lower temperature i.e. <320 °C for palm oil shell. They further proposed nth order reaction mechanism to evaluate the kinetic parameters based on different models.

Previous works reported on EFB and PS kinetics were based on single heating rate in which activation energy is only a function of temperature. The present work evaluate the kinetic parameters based on a method, which requires at least three sets of experimental data generated at different heating rates. This method allows the dependence of activation energy on temperature and conversion at a desired heating rate (Vyazovkin & Wight, 1999) Secondly, lignin decomposition in EFB and PS is not intensively studied at relatively high heating rates. Present work considers lignin decomposition in EFB and PS to understand the effect of lignin content on kinetic parameters and decomposition rate. Furthermore, pure lignin decomposition is studied based on its thermogram analysis and kinetic parameters.

In this work, the kinetics of biomass decomposition which includes EFB, PS, pure cellulose and lignin were investigated using TGA under non-isothermal conditions. The detail thermogram analysis was presented to understand the decomposition of cellulose, hemicellulose and lignin as major components in lignocellulosic biomass. The decomposition kinetics of cellulose and lignin were studied under single-step first order kinetic model. Meanwhile, the decomposition of EFB and PS were reported based on single-step nth order kinetic model. Activation energy, pre-exponential factor and order of reaction were determined and discussed in comparison to the values reported in the literature.

2. Materials and methods

2.1 Materials preparation and experimental procedure

Cellulose in fibrous powder form and lignin in brown alkali powder form were purchased from Sigma Aldrich Sdn. Bhd., Malaysia. EFB and PS were collected from local palm oil industry in Perak, Malaysia. Biomass samples were dried at 105°C and the weighted was monitored at one hr interval, until the readings became constant. Samples were then grinded to particle size of 150-250μm. The method for drying, characterization and analysis were given in previous work (Abdullah et al., 2010). The biomass, pure cellulose and lignin properties are given in Tables 1 and 2.

The biomass decomposition experiments were carried out in EXSTAR TG/DTA 6300 (SII, Japan). N_2 was used as inert gas with a constant flow rate of 100 ml/min for the entire range of experiments. The sample initial weight used in all experiments was within the range of 3-6 mg. TG experiments were performed at heating rate of 10, 30 and 50 °C/min. All samples were first heated from 50 °C to 150 °C where it was kept constant for 10 min to remove moisture content, and then heated up to the final temperature of 800 °C. All experiments were carried twice for reproducibility. No significant variations were observed in the second experimental measurements.

2.2 Kinetic parameters determination

The biomass decomposition rate under non-isothermal condition is described (Cai & Bi, 2009).

$$\frac{d\alpha}{dT} = \frac{A}{\beta}\exp\left(-\frac{E}{RT}\right) f(\alpha) \tag{1}$$

$f\ (\alpha)$ depends on the reaction mechanism as listed in Table 3 and α is the mass fraction reacted.

$$\alpha = \frac{w_0 - w}{w_0 - w_f} \tag{2}$$

Where, 0 and f shows initial and final sample weight

$$g(\alpha) = \int_0^\alpha \frac{d\alpha}{f(\alpha)} = \frac{A}{\beta}\int_0^T \exp(-\frac{E}{RT})dT = \frac{AE}{\beta R}p(\frac{E}{RT}) \tag{3}$$

p(E/RT) function has no exact analytical solution, and therefore different approximations are reported to evaluate the function (Budrugeac et al., 2000). The method developed by Flynn-Wall (Flynn & Wall, 1996), Ozawa (1965) using Doyle's approximation (1961) is the most popular and commonly used by several researchers for biomass decomposition (Cai & Bi, 2009; Hu et al., 2007; Zhouling et al., 2009).

$$p(\frac{E}{RT}) = 0.0048e^{-1.0516(\frac{E}{RT})} \tag{4}$$

(3) is then rearranged for β

$$ln\beta_i = ln(\frac{A_\alpha E_\alpha}{R\,g(\alpha)}) - 5.331 - 1.0516\frac{E_\alpha}{RT_{\alpha,i}} \tag{5}$$

Where; w = sample weight (mg); β = heating rate (K/min); R = universal gas constant (8.314×10^{-3} kJ mol^{-1} K^{-1})

To determine the activation energy, $\ln\beta_i$ vs. $1/T_{\alpha,i}$ is plotted for different α values and heating rates (i) to give a straight line and the slope of which gives the activation energy (Doyle, 1961; Ozawa, 1965; Zsakó, & Zsakó, 1980; Flynn & Wall, 1996).

Analysis	EFB	PS
Volatiles	84.61	81.03
Ash	5.50	4.10
Fixed Carbon (by difference)	9.89	14.87
C	40.73	49.65
H	5.75	6.13
N	1.40	0.41
S	0.22	0.48
O (by difference)	51.90	43.33
Cellulose[a]	38.30	20.8
Hemicellulose [a]	35.30	22.7
Lignin[a]	22.10	50.7

[a] Kelly-Yong et al. (2007)

Table 1. Biomass analysis (wt% dry basis)

Analysis	Cellulose	Lignin
C	43.09	47.71
H	5.96	4.53
N	0.13	0.04
S	0.14	4.24
O (by difference)	50.67	43.40

Table 2. Pure cellulose and Lignin properties (wt % dry ash free)

Function	f(α)	g(α)
First order reaction	$1-\alpha$	$-\ln(1-\alpha)$
Second order reaction	$(1-\alpha)^2$	$(1-\alpha)^{-1}-1$
Third order reaction	$1/2(1-\alpha)^3$	$(1-\alpha)^{-2}-1$
nth order reaction	$(1-\alpha)$	$1-(1-\alpha)^{1-n}/1-n$

Table 3. Different *f(α)* and *g(α)* values based on kinetic control regime (Ahmad et al., 2009)

2.3 Model for kinetic parameter determination

The following assumptions are considered for the decomposition of EFB, PS, pure cellulose and lignin.

- Reaction is purely kinetic controlled.
- The decompositions follow single-step processes.
- First order reaction kinetics is considered for pure cellulose and lignin and PS and EFB kinetics are assumed to be nth order.
- No secondary reaction takes place among the gaseous products.

3. Results and discussions

3.1 Thermogram analysis

The TG and DTG curves for cellulose, lignin, EFB and PS at different heating rates are shown in Figures 1-4. The effect of different heating rate can be described by a lateral shift appeared at high heating rates. These lateral shifts are due to the thermal lag effect between surrounding and biomass particles (Yang et al., 2004; Luangkiattikhun et al., 2007). As a result, conversions are delayed at high heating rates. Thermal lag effect is due to the small heat conductive property of biomass particles (Zhang et al., 2006).

In the DTG curves (Fig. 1-4, b) for all samples, high decomposition rate was observed at 50 °C/min, which shows the increase of thermal decomposition rate of biomass at high heating rates.

The investigated EFB exhibited the decomposition rate corresponds to -41 wt%/min which is higher than -33 wt%/min of PS at 50 °C/min (see Fig. 5). The high decomposition rate for EFB and PS appeared at 342 and 382 °C, respectively. It is important to consider that 60 wt% of EFB and PS is decomposed at 400 and 429 °C for 50 °C/min. These results depict relatively easy and fast decomposition for EFB as compared to PS. This fast decomposition of EFB may be attributed to the comparatively high volatiles matter and low lignin content present in EFB as compared to PS. Conversely, pure cellulose and lignin decomposition rate is the highest and lowest among all species which is -124 and -19 wt%/min at 50 °C/min, respectively. Furthermore, the highest decomposition rate for cellulose and lignin is observed at 386 and 418 °C.

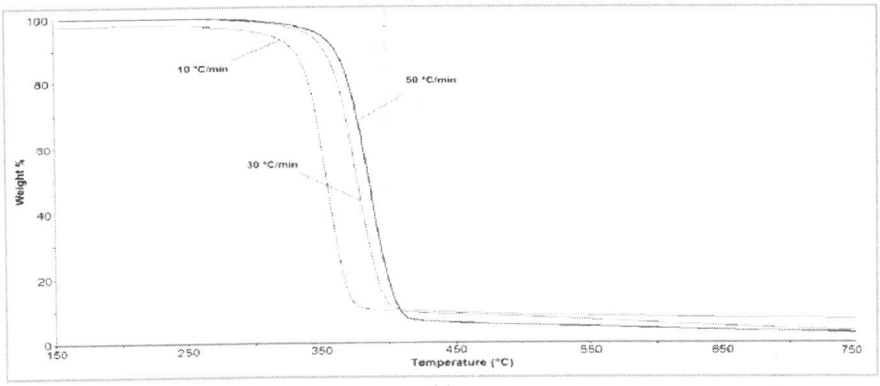

(a)

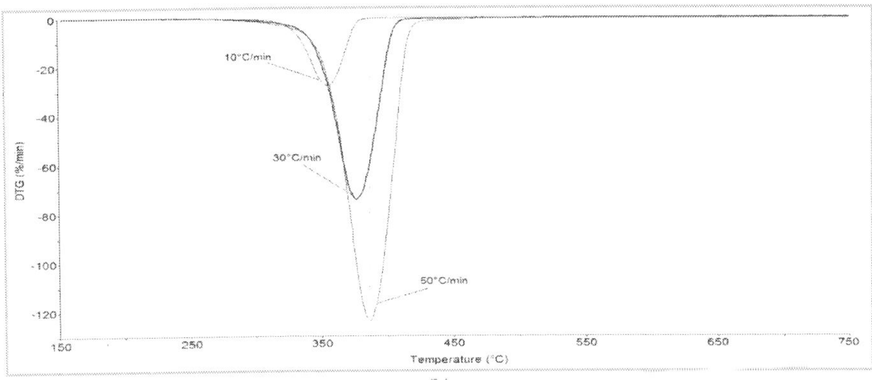

(b)

Fig. 1. Cellulose (a) TG and (b) DTG curves

The TG and DTG curves for EFB and PS are given in Figures 3-4. In these figures, the first peak represents the decomposition of hemicellulose. The second peak, which is sharper, gives the highest rate corresponds to the cellulose decomposition. The decomposition range of hemicellulose and cellulose of EFB is between 240-300 °C and 300-340 °C, respectively, at heating rate of 10 °C/min. Decomposition rate of hemicellulose in PS falls almost in the same temperature region as for EFB but higher decomposition range for cellulose (340-370 °C). It is important to consider that the cellulose decomposition rate in PS is in the same temperature region as pure cellulose (340-370 °C at 10 °C/min). The tail at high temperature shows lignin decomposition as found by Yang et al. (2004) and Luangkiattikhun et al. (2008). In the present study, at 10 °C/min, no lignin decomposition was observed for EFB and PS. Similar observation is reported by Yang et al. (2004) for heating rate of 10 °C/min at temperature >340 °C. At higher heating rates, there is some small lignin decomposition observed for EFB and PS which is in the range of 450-530 °C and 680-750 °C, respectively. Different region for lignin decomposition in EFB and PS may be due to different lignin structure and composition in both species.

Among all species, lignin decomposition produced highest residual fraction of ~40% followed by ~27% of EFB and PS and <7% for cellulose, respectively. High residual fraction for lignin shows its high resistance to thermal decomposition which can be seen by its lowest decomposition rate.

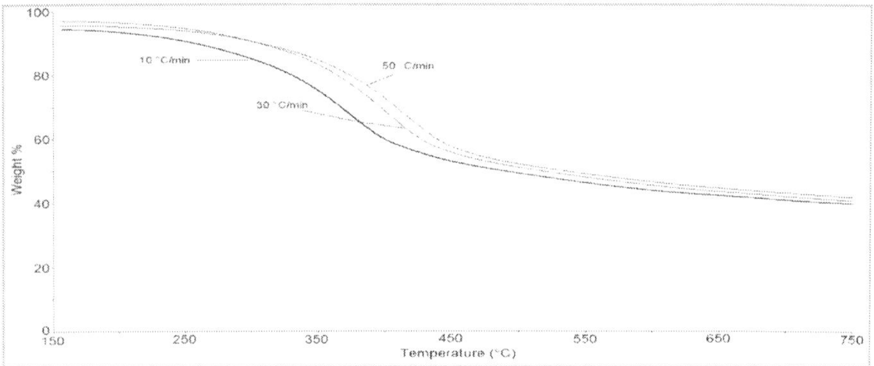

(a)

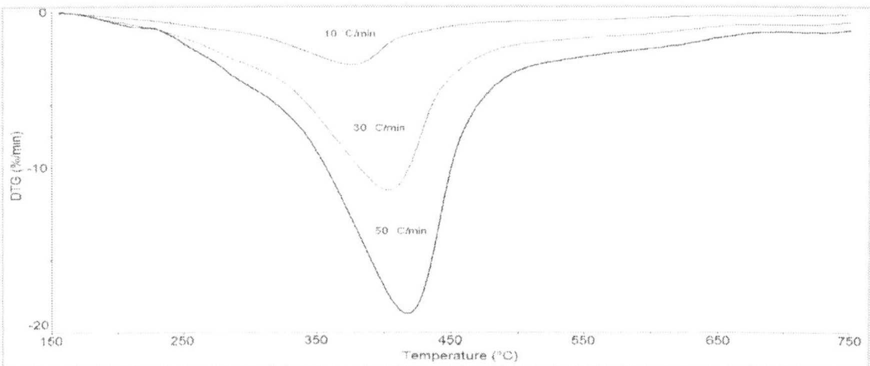

(b)

Fig. 2. Lignin (a) TG and (b) DTG curves

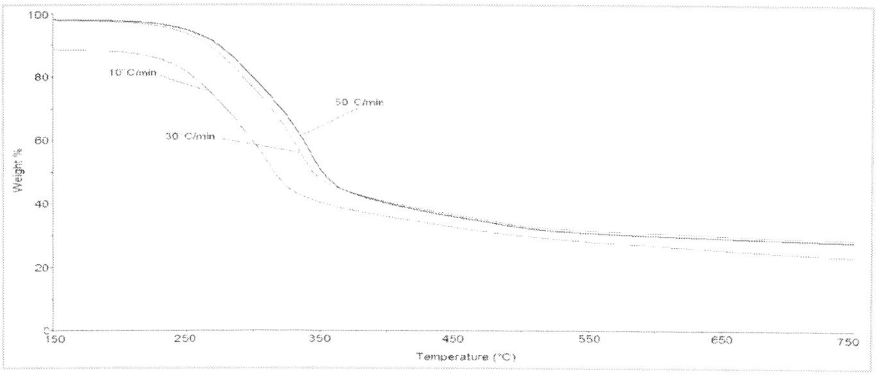

(a)

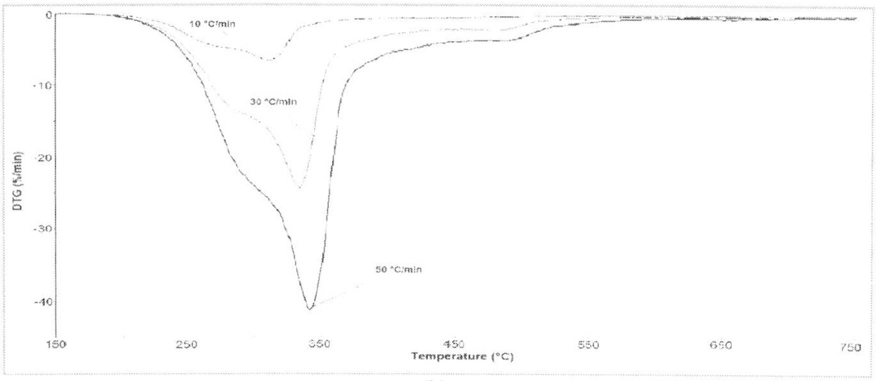

(b)

Fig. 3. EFB (a) TG and (b) DTG curves

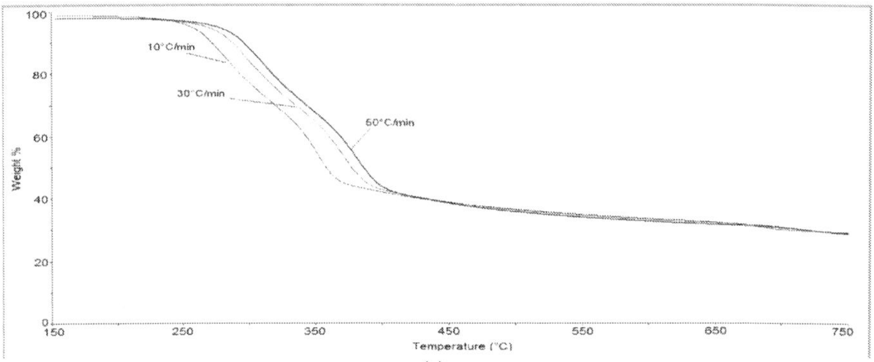

(a)

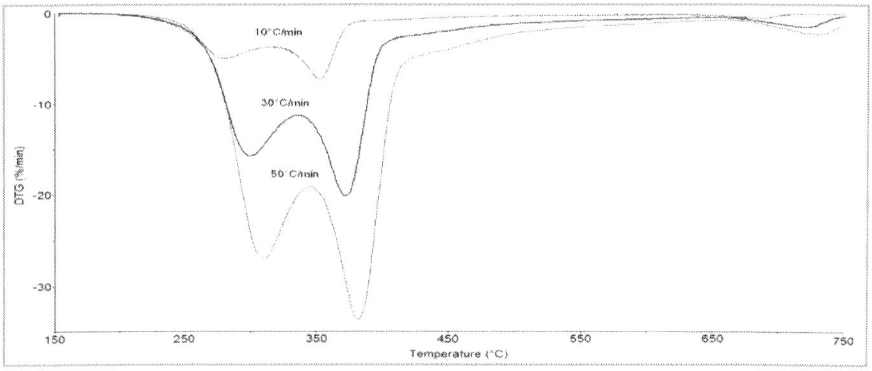

(b)

Fig. 4. PS (a) TG and (b) DTG curves

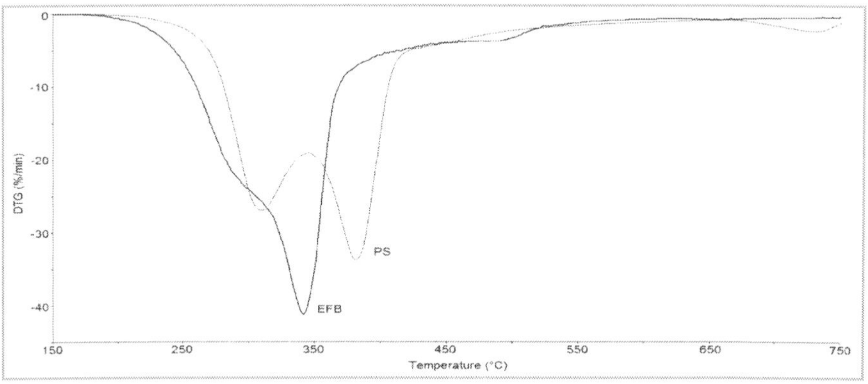

Fig. 5. EFB and PS DTG curves at 50 °C/min

3.2 Kinetic parameters

Kinetic parameters are evaluated using equation (5) at a heating rate of 10, 30 and 50 °C/min. Figures 6-9 shows the plot of $\ln\beta_i$ against $1/T$ at different mass fraction reacted (α) for cellulose, lignin, EFB and PS. The higher $\ln\beta_i$ values shows high heating rate (50 °C/min) followed by 30 and 10 °C/min. Kinetic parameters evaluated at each α are given in the Tables 4-7. For pure cellulose and lignin, the kinetic parameters are determined at α=0.1 to 0.8 and α=0.1 to 0.6, respectively. EFB and PS kinetic parameters are evaluated at α=0.1 to 0.7. Among all samples, pure lignin produced highest residual fraction and hence kinetic parameter determined up to α=0.6. The correlation coefficients (R^2) determined are higher than 0.991 for all cases.

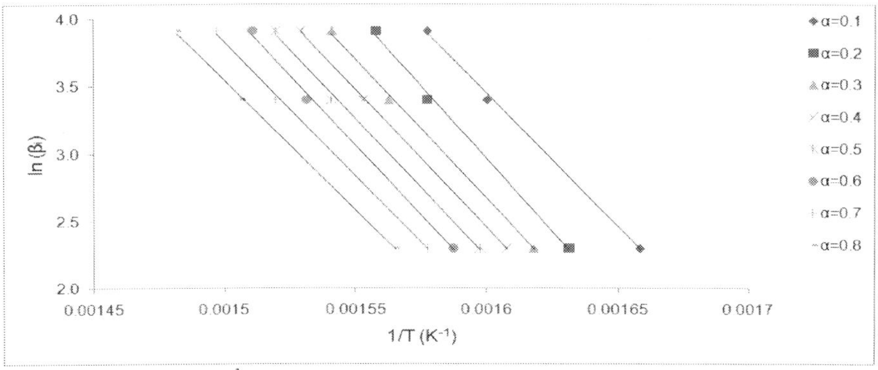

Fig. 6. The dependence of $\ln(\beta_i)$ on $1/T$ at a different α values of cellulose (solid lines shows linear fitting)

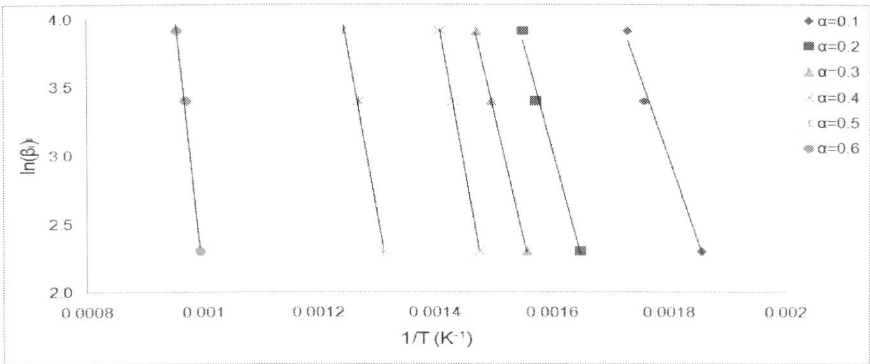

Fig. 7. The dependence of $\ln(\beta i)$ on $1/T$ at a different α values of lignin (solid lines shows linear fitting)

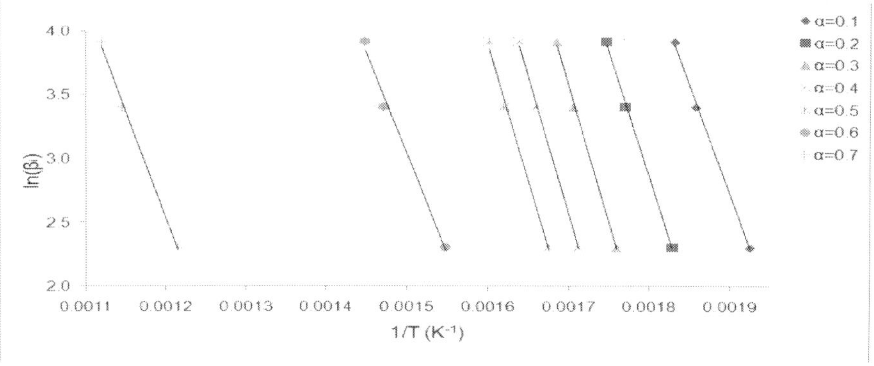

Fig. 8. The dependence of ln(βi) on 1/T at a different α values of EFB (solid lines shows linear fitting)

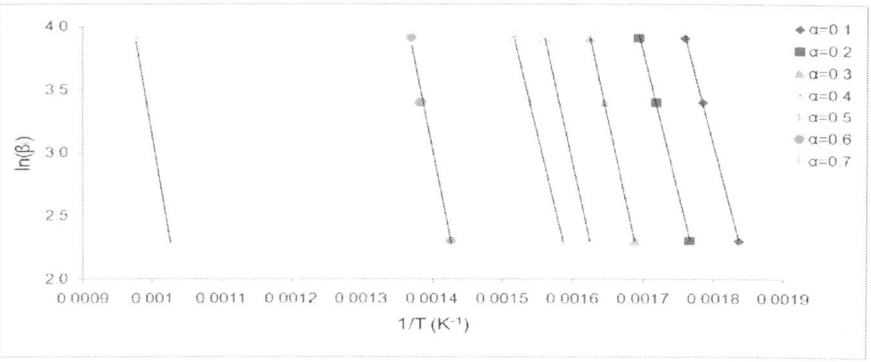

Fig. 9. The dependence of ln(βi) on 1/T at a different α values of EFB (solid lines shows linear fitting)

α	E (kJ/mol)	R^2	A (m^{-1})
0.1	155.31	0.996	3.9×10^{13}
0.2	170.14	0.996	3.9×10^{13}
0.3	162.95	0.998	1.1×10^{13}
0.4	161.12	0.999	9.2×10^{12}
0.5	161.64	0.997	1.1×10^{13}
0.6	163.66	0.997	1.8×10^{13}
0.7	157.03	0.998	5.3×10^{12}
0.8	151.53	0.991	1.9×10^{13}
Average	160	0.997	1.2×10^{13}

Table 4. Kinetic parameters of cellulose at different α values

α	E (kJ/mol)	R^2	A (m^{-1})
0.1	95.01	0.990	9.1E+10^7
0.2	123.28	0.990	4.4E+10^9
0.3	143.67	0.999	7.9E+10^{10}
0.4	183.89	0.999	3.8E+10^{13}
0.5	184.22	0.995	1.2E+10^{12}
0.6	318.48	0.992	1.2E+10^{16}
Average	175	0.994	2.06E+10^{15}

Table 5. Kinetic parameters of lignin at different α values

α	E (kJ/mol)	R^2	A (m^{-1})
0.1	135.88	0.998	3.9×10^{13}
0.2	154.59	0.996	6.0×10^{14}
0.3	169.01	0.998	4.5×10^{15}
0.4	171.96	0.998	4.7×10^{15}
0.5	169.92	0.993	9.1×10^{13}
0.6	125.60	0.997	9.4×10^{10}
0.7	131.69	0.997	4.0×10^{9}
Average	151	0.997	1.4×10^{15}

Table 6. Kinetic parameters of EFB at different α values

α	E (KJ/mol)	R^2	A (m^{-1})
0.1	165.20	0.999	6.5×10^{15}
0.2	176.23	0.999	1.9×10^{16}
0.3	200.13	0.999	6.2×10^{17}
0.4	194.15	0.998	5.4×10^{16}
0.5	186.27	0.999	7.6×10^{15}
0.6	214.97	0.992	5.9×10^{16}
0.7	255.68	0.998	4.0×10^{14}
Average	199	0.998	1.1×10^{17}

Table 7. Kinetic parameters of PS at different α values

The kinetic parameters for EFB, PS, pure cellulose and lignin are determined as listed in the Table 8 and compared with experimental works reported. The average activation energy of cellulose is 160 kJ/mol and pre-exponential factor is 1.2×10^{13} m^{-1} for first order kinetic model. Several researchers found that first order kinetics fit well for cellulose decomposition reaction (Várhegyi et al., 1997; Gronli et al., 1999; Hu et al., 2007).

Activation energy evaluated for cellulose in the present study is in good agreement with the work reported by Zhang et al. (2009) for a first order kinetic model. Nevertheless, the value is comparatively low corresponding to values reported by Varhegyi et al. (1997), Grønli et al. (1999) and Yang et al. (2004). The reason may be different source of cellulose and different method followed by the authors. Secondly, this may be due to relatively high heating rates used in the present study. Meanwhile, Hu et al. (2007) reported high activation energy (233 kJ/mol) using Flynn-Wall-Ozawa method at low heating rates of 2.5, 5, 10 °C/min. Grønli et al. (1999) observed the effect of different heating rates on activation energy for cellulose and found low activation energy at high heating rates. Similarly, much lower activation energies were found by Milosavljevic et al. (1995) at high heating rates. The same effect was also observed for pre-exponential factor. This is due to the heat transfer limitation between sample particles and the surroundings at high heating rates. Varhegyi et al. (1997) suggested utilization of low sample mass to minimize the heat transfer limitations at high heating rates.

Biomass	Kinetic Parameters			R^2	Reference
	E (kJ/mol)	A (m^{-1})	n (-)		
Cellulose	160	1.2×10^{13}	1	0.997	This study
	175	7.16×10^{12}	1	0.999	Zhang et al. (2009)
Lignin	175	2.06×10^{15}	1	0.994	This study
	171	2.73×10^{25}	1	-	Morugun et al. (2008)
EFB	151	1.4×10^{15}	5.3	0.997	This study
	61	3.14×10^{2}	1	0.991	Yang et al. (2004)
PS	199	1.1×10^{17}	5.0	0.998	This study
	111	5.27×10^{7}	2.54	-	Luangkiattikhun et al. (2008)

Table 8. Comparison of kinetic parameters

Pure lignin is decomposed with average activation energy of 175 kJ/mol and pre-exponential factor of 2.06×10^{15} m^{-1} for first order kinetics model. These kinetic parameters are in good agreement with work reported by Murugan et al. (2008).

EFB and PS are decomposed with the average activation energy of 151 and 199 kJ/mol and reaction order of 5 and 5.3, respectively. The corresponding average pre-exponential factors evaluated for EFB and PS are 1.4×10^{15} and 1.1×10^{17} m^{-1}, respectively. These values are somehow larger than those reported by Yang et al. (2004) for first order reaction kinetics. Luangkiattikhun et al. (2008) and Guo & Lua, (2001) observed lower activation energy and pre-exponential factor based on single-step nth order and first order kinetic model for PS. However, in these studies, comparatively high values were obtained using two step kinetic models.

Based on kinetic parameters, it is easy to decomposed EFB as compared to PS, pure cellulose and lignin. The order of decomposition from fast to slow is EFB > cellulose > lignin > PS.

4. Conclusion

A biomass decomposition study has been carried out to investigate different breakdown region and kinetic parameter evaluation for EFB and PS. As major components of biomass, pure cellulose and lignin decomposition kinetics were also studied. TG and DTG curves were studied in detail to understand the major decomposition region in EFB, PS, pure cellulose and lignin. The kinetic parameters of EFB and PS are found to be higher compared to reported values in the literatures. This difference may be due to the different methods for kinetic parameter determination and relatively high heating rates used in the present study. Based on the kinetic parameters, PS was difficult to decompose as compared to EFB. The possible reason is may be relatively high lignin content present in PS. This high lignin content was also responsible for low decomposition rate of PS as compared to EFB. Pure lignin had the lowest decomposition rate among all the species. Moreover, lignin content in PS was decomposed at high temperature as compared to EFB based on higher heating rates.

5. Acknowledgments

The author thanks the Petroleum Research Fund (PRF) of PETRONAS and Universiti Teknologi PETRONS for their financial support.

6. References

Abdullah, S.S. & Yusup, S. (2010). Method for Screening of Malaysian Biomass Based on Aggregated Matrix for Hydrogen Production through Gasification. *Journal of Applied Sciences,* Vol.10, No.24, pp. 3301-3306, ISSN 1812-5654

Ahmed, I.I. & Gupta, A.K. (2009). Hydrogen Production from Polystyrene Pyrolysis and Gasification: Characteristics and Kinetics. *International Journal of Hydrogen Energy,* Vol.34, No.15, pp. 6253-6464, ISSN 1520-5029

Budrugeac, P.; Homentcovschi, D. & Segal, E. (2000). Critical Analysis of the Isoconversional Methods for Evaluating the Activation Energy. I. Theoretical Background. *Journal of Thermal Analysis and Calorimetry,* Vol.63, No.2, pp. 457-463, ISSN 1388-6150

Cai, J. & Bi, L. (2009). Kinetic Analysis of Wheat Straw Pyrolysis using Isoconversional Methods. *Journal of Thermal Analysis and Calorimetry,* Vol.98, No.1, pp. 325-330, ISSN 1388-6150

Demirbas, A. (2001). Biomass Resource Facilities and Biomass Conversion Processing for Fuels and Chemicals. *Energy Conversion and Management,* Vol.42, No.11, pp. 1357-1378, ISSN 0196-8904

Doyle, C.D. (1961). Kinetic Analysis of Thermogravimetric Data. *Journal of Applied Polymer Science,* Vol.5, No.15, pp. 285-292, ISSN 0021-8995

Flynn, J.H. & Wall, L.A. (1966). A Quick, Direct Method for the Determination of Activation Energy from Thermogravimetric Data. *Journal of Polymer Science, Part B: Polymer Letter,* Vol.4, No.5, pp. 323-328, ISSN 1542 6254

Gronli, M.; Antal, M.J. & Varhegyi, G. (1999). A Round-Robin Study of Cellulose Pyrolysis Kinetics by Thermogravimetry. *Industrial Engineering and Chemistry Research,* Vol.38, No.6, pp. 2238-2244, ISSN 1520-5045

Guo, J. & Lua, A.C. (2001). Kinetic Study on Pyrolytic Process of Oil-Palm Solid Waste Using Two-Step Consecutive Reaction Model. *Biomass and Bioenergy,* Vol.20, No.3, pp. 223-233, ISSN 0961-9534

Hu, S.; Jess, A. & Xu, M. (2007). Kinetic Study of Chinese Biomass Slow Pyrolysis: Comparison of Different Kinetic Models. *Fuel,* Vol.86, No.17-18, pp. 2778-2788, ISSN 0016-2361

Kelly-Yong, T.L.; Lee, K.T.; Mohamed, A.R. & Bhatia, S. (2007). Potential of Hydrogen from Oil Palm Biomass as a Source of Renewable Energy Worldwide. *Energy Policy,* Vol.35, No.11, pp. 5692-5701, ISSN 0301-4215

Khan, Z.; Yusup, S.; Ahmad, M.M.; Chok, V.S.; Uemura, Y. & Sabil, K.M. (2010). Review on Hydrogen Production Technologies in Malaysia. *International Journal of Engineering & Technology*, Vol. 10, No.02, pp. 111-118, ISSN 2077-1185

Luangkiattikhun, P.; Tangsathitkulchai, C. & Tangsathitkulchai, M. (2008). Non-isothermal Thermogravimetric Analysis of Oil-Palm Solid Wastes. *Bioresource Technology*, Vol.99, No.5, pp. 986-997, ISSN 0960-8524

Milosavljevic, I. & Suuberg, E.M. (1995). Cellulose Thermal Decomposition Kinetics: Global Mass Loss Kinetics. *Industrial Engineering and Chemistry Research*, Vol.34, No.4, pp. 1081-1091, ISSN 1520-5045

Murugan, P.; Mahinpey, N.; Johnson, K.E. & Wilson, M. (2008). Kinetics of the Pyrolysis of Lignin Using Thermogravimetric and Differential Scanning Calorimetry Methods. *Energy & Fuels*, Vol.22, No. pp. 2720-2724, ISSN 0887-0624

Ozawa, T. (1965). A New Method of Analyzing Thermogravimetric Data. *Bulletin of the Chemical Society of Japan*, Vol.38, No.11, pp. 1881-1886, ISSN 0009-2673

Shuit, S.H.; Tan, K.T.; Lee, K.T. & Kamaruddin, A.H. (2009). Oil Palm Biomass as a Sustainable Energy Source: A Malaysian Case Study. *Energy*, Vol.34, No.4, pp. 1225-1235, ISSN 0360-5442

Várhegyi, G.B.; Chen, H. and Godoy, S. (2009). Thermal Decomposition of Wheat, Oat, Barley, and *Brassica carinata* Straws. A Kinetic Study. *Energy and Fuels*, Vol.23, No.2, pp. 646-652, ISSN 0887-0624

Várhegyi, G.; Antal, M. J.; Jakab, E. & Szabó, P. (1997). Kinetic Modeling of Biomass Pyrolysis. *Journal of Analytical and Applied Pyrolysis*, Vol.42, No.1, pp. 73-87, ISSN 0165-2370

Vyazovkin, S. & Wight, C.A. (1999). Model-Free and Model-Fitting Approaches to Kinetic Analysis of Isothermal and Nonisothermal Data. *Thermochimica Acta*, Vol. 340-341, pp. 53-68, ISSN 0040-6031

Yang, H.; Yan, R.; Chin, T.; Liang, D.T.; Chen, H. & Zheng, C. (2004). Thermogravimetric Analysis−Fourier Transform Infrared Analysis of Palm Oil Waste Pyrolysis. *Energy and Fuels*, Vol.18, No.6, pp. 1814-1821, ISSN 0887-0624

Yang, H.; Yan, R.; Chen, H.; Lee, D.H. & Zheng, C. (2007). Characteristics of Hemicellulose, Cellulose and Lignin Pyrolysis. *Fuel*, Vol.86, Vol.12-13, pp. 1781-1788, ISSN 0016-2361

Zhang, X.; de Jong, W. & Preto, F. (2009). Estimating Kinetic Parameters in TGA using B-spline Smoothing and the Friedman Method. *Biomass and Bioenergy*, Vol. 33, No.10, pp. 1435-1441, ISSN 0961-9534

Zhang, X.; Xu, M.; Sun, R. & Sun, L. (2006). Study on Biomass Pyrolysis Kinetics. *Journal of Engineering for Gas Turbines and Power*, Vol.128, No.3, pp. 493-496, ISSN 0742-4795

Zhouling.; Chang-qing, S.; Chao-yu, L.; En-cheng, J. & Xi-wei, X. (2009). Pyrolysis Characteristics and Kinetics of Cotton Stalk, *International Conference on Energy and Environment Technology*, pp. 768-772, ISBN 978-0-7695-3819-8, Guilin, China, October 16-18, 2009

Zsakó, J. & Zsakó, J. (1980). Kinetic Analysis of Thermogravimetric Data XIV. Three integral methods and their computer programs. *Journal of Thermal Analysis and Calorimetry*, Vol.19, No.2, pp. 333-345, ISSN 1388-6150

23

Biofuels and Energy Self-Sufficiency: Colombian Experience

Elkin Alonso Cortés-Marín and Héctor José Ciro-Velázquez
Universidad Nacional de Colombia- Sede Medellín
Colombia

1. Introduction

The non-renewable nature of fossil fuels combined with the high level of participation within transportation sector in the total consumption of primary energy and atmospheric pollution, have become the primary forces propelling research of alternative sources for vehicles, mainly those sources derived from biomass. This has resulted in an increased environmental consciousness that seeks to replace fossil fuels or to provide blends that reduce their overall consumption. Mainly searching for these sources in agribusiness, and taking into account that tropical countries play lead roles here in Colombia is where the greatest variety of plant species can be found and where the environmental conditions make production of these more advantageous.

The global energy problem leads to express the scope, opportunities and threats that the use and partial replacement of conventional fossil fuels by biofuels or agrofuels represent for the development of a country, focusing in Colombia as a case study. The growing importance of new energy sources, (which can be derived from a variety of crops) and raw materials, which demands high biomass amounts, must generate some level of concern about the possible harmful effects of deforestation, jungle loss and replacement of crop fields essential for human diet (food safety). Not to mention the challenges in the climatic, geographical and physical fields, i.e. on whole nations' economies (Cortés et al., 2009) .

Today, these new energy sources are the new financial, political and even environmental strategies. Their importance is such that currently there are more than 30 raw materials being tested worldwide. Despite this big boost they still do not provide a solution to the global energy crisis (Cortés & Álvarez, 1998).

The possibility of using biofuels in the development of cars and engines, has been considered from the very beginning, but only as a result of the current energy and environmental situation, do conditions exist for the shaping of a global biofuels industry. The development of alternate energy has allowed the concepts of biofuel and energy crops gain importance every day, with greater strength in agricultural and energy policies of both industrialized and developing countries. The motivating factors have been, among others, the evident depletion of fossil fuels, the periodic oil crisis and the so-called greenhouse effect caused by the accumulation of CO_2 in the atmosphere. Despite of this, it is important to recognize that biofuels will not end industrialized countries oil dependency, because there will not be enough land and water to meet the energy requirements of the automotive industry.

As a result, in order to prevent irreversible changes and reduce the impact of greenhouse gases on Earth's climate, many countries, including Colombia, have developed strategies to diversify energy production by using renewable sources. The first strategy has been replacing Oil-Derived Fuels with biofuels thus defining a reduction of CO2 emissions generated by mobile sources. It is therefore imperative to begin using alternative energies, that is to say those considered clean and renewable. For this reason biofuels could be a valid choice. Therefore, the use of renewable energy sources as an alternative to fossil fuels is a key strategy to reducing greenhouse gas emissions (Consejo Nacional de Política Económica y Social (CONPES, 2008)

It is at this stage that recently a dialogue, debate and confrontation regarding biofuels have been facilitated which allows for the development of new technologies and refineries to produce them. Such importance is not only the result of a sudden leap in scientific knowledge, although that has taken place, but rather it is a leap in governments funding, which seem concerned about oil prices rising and geostrategic dependence on them. Whatever the reason, if funding continues, in the short term a new generation of biofuels could be available.

Despite the enthusiasm, promotion and advocacy, there is a question: are biofuels a technical and economically viable energy and environmental option for replacing future fuel imports? But at the same time with the promotion and incentives (legal, regulatory, fiscal and financial framework), of alcohol fuels and bio-oils, employment rates will see a positive impact in farming regions. It is necessary not only to encourage biofuels production but also define programs that support the new refineries' biomass needs, so that the price of raw materials with dual purpose (food and biofuel) is not affected.

Certainly for Colombia it is necessary to diversify its raw materials portfolio for anhydrous alcohol and biodiesel production, incentivize the research and development of proprietary technology programs that produce biofuels at competitive levels for the domestic demand in the short and medium term and, in the long term to start exports.

In order to delay the depletion of reserves, to avoid the rising cost of imports, reduce gas emissions and the impact of particulate matter into the atmosphere, the policies of replacing energy sources, for the Colombian biofuel industry, provide an excellent opportunity due to the oil rising price, i.e. energy vulnerability risk decreasing.

In general, the text aims to illustrate the production and replacement of fossil fuels with bioenergy (ethanol, biodiesel), the progress, uncertainties and problems resulting from these processes for new energies generation and use, mainly with regards to the food safety and environmental consequences of poor countries. In particular, it presents the political, regulatory and legal framework, by which the Colombian government promotes the production and use of biofuels.

2. Energy replacement and consumption

For operating the machines, during this century and part from the previous: man has used greater amounts of energy from fuels. Fuels coming from forest resources like wood and natural energy coming from sun have not been enough to satisfy the energy appetite that this modern civilization has (Silvestrini, 2000).

This growing demand has created an energy crisis that has led to create programs, proposals, funding and research, to find processes for greater efficiency in energy generation, consumption reduction and uncover alternative sources that allow a diverse offering and increase life span of existing resources.

As long as societies develop, it is clear that energy dependence increases. As a result, the search for more efficient, cleaner, more economic new energy sources and fuels becomes a social need. Therefore, in many processes and business it is possible and feasible to replace oil with natural gas; electric power with solar power; firewood with biogas; tractors with animal traction, etc. At the moment, there is an increase and trend for the conversion of internal combustion engines (ICE) that use gasoline by replacing them with CNG (compressed natural gas). Several CNG dispensers have been installed in the main cities of Colombia. The introduction and claimed benefits have been linked with reductions in pollution and operation costs.

Mankind's welfare has always been associated with high-energy availability; unfortunately a society's progress is directly related to its level of consumption. Man needs energy to survive, manage his environment and produce goods. In distant times, at least, he needed energy to keep from starving. Over millions of years, the time it took to get from primitive forms to its present forms; mankind's evolution is closely linked to the different forms and quantities of energies that were available during each period. The energy consumption growth per capita and energy control has been a constant feature. Evolution changed Man from a gathering society, to a hunting society, an agriculture society, until today becoming a technological society, where per capita consumption in the U.S.A is close to 450 MBTU/inhab. Perhaps primitive man during his tenure in a gathering society consumed no more than 10 BTU/inhab. This growth in consumption is an indication of progress and risks as well as the complexities of the social organizations in which we live.

Although the safety and quality of our energy supply is of real concerns, we must avoid viewing the energy problem from a local perspective, surrounding countries, and in short term perspectives, now and in the near future. A realistic and deep approach regarding the energy issue must consider that a third of humanity lacks supply of electricity and any other form of advanced energy. It must take into account the safety of supplies for future generations and must be aware of the consequences of the environmental impact that energy production and consumption are having on the planet being passed on to our descendants. The amount of energy needed in the future mainly depends on its efficient production and use. With the purpose of evaluating energy efficiency, energy intensity may be applied as an indicator, i.e. energy consumption per GDP unit in each country. Along the way a trend in energy intensity reduction can be observed alongside the increase in economic development (Pérez, 2002).

Since the 70s it has been common to correlate the Gross Domestic Product (GDP) with countries' energy consumption. In the beginning it was an indicator of a developing country to see their energy consumption growing: more infrastructures and vehicles, appliances, heating and air conditioning systems, more power and better quality demand. It went from a traditional economy based on biomass combustion to an economy that makes a general use of electricity and fossil fuels, both for automotive and industrial use. On the other hand, increased consumption means more man-made fibers to fulfill clothing demands, larger workplaces and leisure trips, more housing space and in general more services. A country's welfare for better or worse is associated with increased income and this leads to an energy consumption increasing per capita (Valero, 2004).

In the last thirty years, global energy use has increased almost 70%, but this growth is uneven, because developing countries have almost tripled their energy consumption, while industrialized countries increased by 21%. The world's energy consumption grows faster than wealth or population because developing countries hold the 77% of the world's

population. On the other hand, since 1973 the countries belonging the Organization for Economic Co-operation and Development (OECD) have reduced their energy intensity or energy use by 24%per GDP unit. Energy demand has annually increased more than 2% since the 1973 crisis until today, and if current trends persist, this rate will continue over the next fifteen years (Brown et al., 2000).
Because the population is increasing and countries are getting industrialized, the global energy demand grows. That growth's projection classified by economic sectors can be seen in Figure 1. In addition, fossil fuels that are used are the main source of pollution in the atmosphere. The important role that energy plays in all human activities is widely recognized. Energy is not only transformed but it also transforms society. With its many developments and uses, energy has substantially changed modern life, by creating new services related to technological progress, and becoming the principal supply for the development and progress of any society.

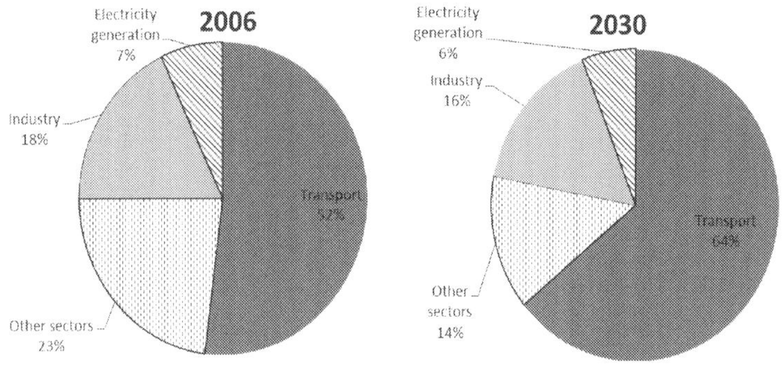

Source: Perspectives énergétiques mondiales à long terme - le cas spécifique des transports. Agence Internationale de l'Energie. July 2007

Fig. 1. World oil consumption participation by sectors 2006 and 2030

According to the International Energy Agency (2006), dynamic growth in the biofuels market is associated with the evolution of the world's demand for primary energy, where fossil fuels have increased participation. Energy demand depends on factor's performance such as: a) world population increasing b) economic growth c) technological developments which allow maximize efficiency production and use d) implementation of measures towards climate change such as the development of alternative energy sources. Thus, the IEA estimates that by 2020 energy demand will be 16,000 Mtoe (Million Tonnes of Oil Equivalent), it means an annual growth rate of 1.7%.
The World Energy Council considers that in the next twenty years, world energy consumption shall approximately increase by 50%, which means it could be possible to provide energy to 4,000 million people (2,000 million that currently do not have it, plus the other 2,000 expected during this period).
But at the moment, energy has six considerable issues that must not be ignored in the scope any global economic analysis (Casilda, 2000).
First problem: production and consumption in unequal distribution around the world: large areas of primary energy production are different from the major consumption areas.

Second problem: has to do with the limited energy sources currently in use. Still nowadays, about 80% of the world's primary energy production comes from fossil fuels (coal, oil and natural gas), i.e. non-renewable sources and limited reserves.

Third problem: is the dominant role that oil plays within energy supply, oil where there is huge gap between production and consumption.

Fourth problem: comes from the relationship between energy and its development. The current energy consumption per capita is varied, as are the geographic levels of development. If in the coming years part of the developing world were come close to the level of energy consumption of industrialized countries, the world would face a long energy crisis, and Latin America would have an uneven behavior, as countries like Mexico and Venezuela would be favored, the two largest oil producers in the region and others would not, well most of them, due to their dependence . Likewise, developed countries would also be seriously affected.

Fifth major problem: production and energy consumption pose serious environmental problems that affect other productive resources worldwide, which could lead to climate change of irreparable consequences.

Sixth problem: is the ability to increase energy supply, which directly depends on the capital to be allocated for this purpose. In many Latin American countries capital is restricted due to regulations and low energy prices.

The use of different types of (electricity, motion, light or heat) and forms of energy (fossil fuels like coal, oil and natural gas, hydropower, nuclear energy or alternative energy) necessary for technologic and economic advancement, has produced the energy crisis that, since 1970, questions the possibility of keeping the current development model, in addition to other harmful effects, both because of the uneven development and the environmental consequences (pollution, global warming, etc.)

The nature of non-renewable fossil fuels and the high level of participation within transport sector of the total primary energy consumption and air pollution, have become the force behind promoting research of alternative sources for vehicles, mainly those sources derived from biomass. This has resulted in an increasing environmental consciousness, and seeks to replace fossil fuels or provide blends that reduce their consumption, mainly searching for those sources in agribusiness. Tropical countries play lead roles because it is where the greatest variety of plant species can be found and where the environmental conditions make production of these more advantageous.

Add to this the geopolitical crisis existing with the U.S.A and Iran, Iraq and Venezuela, and the decrease in global reserves that have influenced oil prices (over 75 U.S.D $/barrel), threatening the stability of many economies that do not have energy self-sufficiency. Under these circumstances, biofuels have become a strategic matter for the U.S.A: in 10 years they are seeking to replace 20% of oil consumption, and so does the European continent.

A technological solution to this problem is the use of biofuels; however, the price competitiveness of these compared to liquid fuels is still disputable. One of biofuels' restrictions is the cost of raw material and its transformation; so low-cost substrates are required to reach competitive price levels.

In practice, different raw materials can be used for industrial alcohol and biodiesel production; however, it is fair to consider that production cost of each liter or gallon strongly depends on geopolitical crisis the characteristics of the raw materials used and the type of process or technology used for their production.

Mainly in industrialized countries, biofuels supply behavior has been due to: biomass availability for their production, production costs and subsidies, and incentives for their production and use. In such respect, demand has been associated with fossil fuels dynamics and the growing interest for the reduction of greenhouse gases (GHGs), from major fossil fuel consuming countries, the latter being an opinion that is under discussion worldwide and must be evaluated in the country.

The contribution of biomass, wood, agricultural, livestock and urban-world residues to energy consumption is currently limited to traditional use as a fuel, especially in Least Developed Countries. However, advanced technologies like gasification, fermentation and anaerobic digestion for biomass exploitation, are increasing its important role as a sustainable energy source like liquid fuel or electricity production. When biomass is grown for burning , there are no net CO2 emissions in the whole process.

Current challenge with biomass is its sustainability management. By 2050 the calculated potential of energy production from biomass is about 10 times the current production, which would be enough to meet the current global energy needs. However, there are several factors that restrict this potential, among them the principal being water availability (UNDP, UNDESA, WEC, "World Energy Assessment, United Nations Development Program, 2000).

According to the growth projections from the International Energy Agency, it is estimated that biofuels participation within the energy market will be 4% by 2030 in comparison with the current 1%. The United States, the European Union and Brazil will be the leading countries in biofuels demand. On the other hand, developing countries will have an important role in such energy supply, so that it is estimated Asian and African countries participation for ethanol production, and Malaysian and Indonesian for biodiesel. Considering this, the expected biofuels demand by 2020 is estimated in 50 Mtoe, which is an annual increasing of 6.3%. Thus, it is estimated that ethanol production will be 524 mmba and biodiesel will be mmba 397.

It becomes important to reconsider current energy consumptions, given the evidence that free energy flows will be renewed as long as earth is habitable, unlike stored energy (oil, coal, gas), which is finite, or in constant decrease with increasing cost. This reason justifies the systematic capture of free energy flows (solar, wind, hydraulic, geothermal) or the use of biomass for generating biofuels. Thus, as long as man and society dominate and know the physical phenomena and natural contributions, the more energy that can be exploited. Therefore, it is not whether there is or is not enough energy to sustain human development, but the physical limits of energy use are going to be or have already been overcome.

Recently, within the scenario of energy generation from diversified sources, rises the clash of the feasibility for replacing crops for human consumption or for biofuels production in the automotive industry. The burden on agriculture for energy production seems to be very hard and biofuels promotion measures in many countries may trigger a very serious food conflict, with still unknown impacts on the poorest countries. In short term they could bring higher food costs.

The world faces complex challenges, it can not be considered the solution to life survival on Earth based on the alternative of renewable energy from biofuels, as it would increase food crops replacement by monocultures, deforestation for energy crops, would boost diversity extinction, cause reduction of fertile lands and water; plus social consequences caused by population displacement.

The use of a specific type of energy depends on two factors: availability for potential energy and technological capacity to turn it into heat or work. Of these two conditions, Colombia has enough natural energy resources (coal, oil, gas, water falls, solar radiation, wind, biomass), being researched and developed in very early stages for its use. Therefore, replacing some conventional energy sources by renewable energies, with endless reserves, aims to reduce pollutant loads and is more environmentally friendly.

The kind or type of energy source used together with the application or purpose, varies over time within the countries, because the use of one or another depends on demand, pricing, availability in the same country or importation, technical reasons, international political situations, etc. In Colombia it can be said that most of current energy comes from burning fossil fuels, which are petroleum products processed in refineries.

Fossil fuels' finite and non renewable nature, plus the high air pollution from major population centers, automotive sector which has a high proportion in the total primary energy consumption; all this has promoted new energy or biofuels from organic resources (plants or animals). Colombia is a vulnerable country in regards to oil due to limited exploratory programs, lack of important new discoveries and decreasing reserves. As a result, the country is about to lose its oil self-sufficiency in the short term (2020), in the midst of high prices unfavorable situation. In addition there it is of poor quality due to its high sulfur content, especially in diesel which exceeds the international standards. Diversification of energy sources, preventing degraded climate change, and promoting social rejecting of polluting energy sources, constitute the challenges faced in order for science and technology to make a contribution for a development that is sustainable, recognized and supported by citizens

A condition for sustainable economic development is to ensure self-sufficiency in energy supply, which must be supported in a flexible and diversified energy structure; being at the same time, energy policy components. This policy should consider the agricultural sector's characteristics, pressed by other energy consumption patterns. Energy as a transformation, transference and accumulation process, should lead those communities to a more efficient use of the different available sources.

It must be a national goal to plan actions for replacing and optimizing the use of different energy forms. External energy inputs replacement by original sources in agricultural production units is one of the self-sufficiency energy objectives through the use of digesters, animal traction, solar, wind, mini stations and alternative fuels (alcohol, vegetable oils, natural gas, etc.)

The prospects resulting from energy optimization and replacement, will allow greater coverage of efficiency and economy of incorporated social benefits. Promoting this energy replacement, due mainly to higher oil prices in recent years, an increase that represents an onerous burden on the world's poorest countries' economies.

An energy program must be an integrated set of studies or research projects within biomass production, processing and the areas of energy consumption, linked to socio-economic studies and systems analysis, aimed to develop techniques and technologies on various energy sources (biogas digesters, gasifiers, solar cells, windmills, micro stations), bioenergy systems in rural areas, use of waste and design systems. Productivity and profitability paradigms that favor: high production values per unit of time, area and, invested money, must be accompanied by processes that also favor energy efficiency, in terms of consumption and savings.

Bioenergy is a rather broad term covering all energy products obtained from agricultural commodities or animal waste conversion processes, mainly required to meet automotive sector's demand in developed countries. Biofuels are a reality in several countries, including

Colombia. Biofuels have become an agro-industrial development model, which attempts to take part in the economic landscape and for this it has stimulus and a broad policy framework as well as multiple advocates and detractors.

Given the biofuels impressive growth (biodiesel and bioethanol) as a renewable energy source, it is necessary to consider in a more analytical and balanced way its options; benefits and limitations in a society that also requires increased food production. In any case, the use of these alternative fuels has negative effects both in natural and socio-economic levels. Despite some partial benefits, from the incorporation of new lands for the production of energy crops and agro-industrial development processes with the collateral effects on employment and possible exports as an added tax product. It is also necessary to face the nearly inevitable competition possibility between food and biofuels production, that within a high poverty levels society, it is worrying that production displacement in our food and agricultural system.

Biofuels present both opportunities and risks. The outcome would depend on the specific context of the country and the policies adopted. Current policies tend to favor producers in some developed countries over producers in most of developing countries. The challenge is to reduce or manage the risks while sharing the opportunities more widely. Biofuel production based on agricultural commodities increased more than threefold from 2000 to 2007, and now covers nearly two percent of the world's consumption of transport fuels. The growth is expected to continue, but the contribution of liquid biofuels (mostly ethanol and biodiesel) to transport energy, and even more so, for global energy use will remain limited. Despite the limited importance of liquid biofuels in terms of global energy supply, the demand for agricultural feedstocks (sugar, maize, oilseeds) for liquid biofuels will continue to grow over the next decade and perhaps beyond, putting upward pressure on food prices (FAO, 2008).

Today Colombia seeks different alternatives for solving the increasing difficulties set out by its development, mainly its population's diet in a natural sources panorama, especially when related to water and soil which may be eroded and contaminated. In order to prevent irreversible changes and reduce the impact of greenhouse gases on Earth's climate, many countries have decided to put their hopes on diversification of energy production strategies by using renewable sources. The first strategy has been replacing petroleum fuels with biofuels, achieving thereby a reduction in CO2 emissions generated by mobile sources (1). Therefore it is urgent to start using alternative energies, i.e. clean and renewable. For this reason biofuels can be a choice, not without question.

It is becoming more urgent to use other fuels in vehicles, mainly for replacing gasoline and diesel, diesel being the one that is gaining preference because of the costs (2). Therefore, it is advisable to research diesel replacements as opposed to gasoline.

The feasibility of using alcohol or biodiesel as automobile fuel in greater proportions than those currently stipulated is a proven fact. Even though there are great expectations, regarding hydrocarbon reserves existence underground in Colombian, the possibility that in a few decades the country has used all its oil must be. This consideration makes it necessary to pay attention to Brazil's experience, which aims to replace oil with other fuels. With optimism of our economy recovery, if new options are not found, the country will: increase its fossil fuels consumption of high priced products within the international market, that in the near future could consume our currency. No matter what, its price tends to increase as a result of the world demand and conflicts with governments in the Middle East and Venezuelan. For that reason, it is advisable to have a replacement from organic sources available, such as alcohol and biodiesel.

Unlike the oil industry, the new agro-energy industry involves a productive chain that impacts different economic sectors more directly, especially with regard to employment generation and agricultural and agribusiness development. The addition of Biofuels to blends helps to mitigate oil import requirements. This supports the national biofuel policy facing the balance of energy trade, and to some extent, defines security settings on the supply level.

Colombia has a great potential for establishing a big biofuel industry. Developing this industry gives the country an opportunity for exploiting its comparative advantages, as a tropical country with an agricultural tendency (biomass production) and suitable soils for feedstocks. Moreover, it becomes a technologically scientific challenge for research groups to focus their efforts on achieving proprietary technological developments by working together and interdisciplinary with public and private sectors. Unfortunately the substitution of conventional fuels with biofuels (ethanol and biodiesel) is also having ecological consequences.

Because of the current high demand, which is expected to increase in the coming years, many countries are cutting down large areas of tropical forests to cultivate biofuels. Although this is not the case in Colombia, as long as a favorable competition for energy feedstocks are shown (given the high government incentives and subsidies), against food without any government protection and also competing with the different free trade agreements. This is worrying given the large expanses of land necessary for feedstocks.

In conclusion, to really consolidate a coherent policy on new energy matter within the Colombian bucket, the following remarks must be considered:

- Environmental Ethics.
- The physical boundaries (finite resources).
- Climatic and geographical conditions.
- Crop Yields (kg/ha, l/ha, l/t).
- Energy intensity and energy return rate.
- Water requirements.
- Self Sufficient Process.
- Technology, return on investment (ROI) and profitability.

Not forget that compressed natural gas - natural gas vehicle (CNG-NGV) is another strong rival among the already diverse energy supply.

In that context, the National Government has promoted research and of new renewable energy sources that are sustainable with the increasing pace of life, while oil and its derivatives are partially replaced in different applications, mainly in transport sector. This promotion must also consider the implications when allocating millions of hectares for bioenergy production. This reality highlights the urgent need of fulfilling food needs or allocating those lands and feedstocks to meet the automotive industry energy requirements.

Energy is, undoubtedly, both a solution and a problem for sustainable development. As such, the hope is focused on the dilemma for fulfilling growing consumption needs while minimizing the social impact and ensuring resources. The world energy problem, fossil fuel reserves reduction, the air pollution problem and global warming, are matters of great importance for humanity without a global solution until now.

3. Precedents

Potential environmental and social benefits, among them; climate change mitigation and energy security contribution are mentioned as major support reasons from public sector for biofuel industries, where growth has been fast. Therefore, in recent years fuel production as an f alternative to fossil fuels, from renewable materials, has acquired in recent years, a global push because global production of biofules has doubled and in the medium term it is expected to have strong growth due to high demand. Motivations have been, among others, the inevitable decrease in fossil fuels, the regular oil crisis and the greenhouse effect caused by the CO2 accumulation in the atmosphere.

In particular, in our country, diesel is in higher demand when in comparison to the quantities supplied by the crude oil processed in the refinery. For that reason it must be partially imported already manufactured. In addition to high oil prices, the agriculture crisis and low international oil trade rates are some of the factors that have contributed to give an additional role to biodiesel.

Liquid biofuels also known as biofuels, which are derived from agricultural raw materials, are products that are being used as replacements for gasoline and diesel in vehicles, they can also be used as blends. At the moment many countries encourage the idea of planting their own fuel, to lessen dependency on imports or exhaustible reserves, while generating stable and top-quality jobs. Biofuels are alcohols, ethers, esters and other chemical products made from cellulosic biomass such as grasses and woods, agricultural and forest waste and most of the municipal and industrial waste; as wells as vegetable oils. The term Biofuel refers both to fuels for electricity or transportation. Unlike oil which is a nonrenewable resource, biofuels are renewable and represent an inexhaustible fuel source. Both commercial and noncommercial, these fuels should always be considered as valuable products which can meet the growing demand.

The most important biofuels are ethanol (from sugar cane, sugar beet and corn) and biodiesel (from oil palm, soybeans and other oilseeds). Biofuel promoters highlight the ecological character of these fuels: they are renewable and are apparently environmentally friendly and produce less greenhouse gas emissions (GHG) in comparison to petroleum fuels.

It has been extensively spoken about biofuel models for more than a decade, and the opportunities and challenges that these oil substituting fuels can offer. This potential is not only related to environmental improvement, but it also includes economical, cultural and social aspects (Cortés, 2007).

In many countries (e.g. Brazil, USA and some European), including Colombia, have implemented policies favoring biofuels. Now these actions are having an effect: more and more agricultural commodities are for biofuel production. For this reason demand grows and causes these agricultural commodities in world market have high prices.

Biofuels apparently help to reduce greenhouse gas emissions, but most of the time it is not considered that for biofuels production, fossil fuels are used (diesel for machinery and products transportation, inputs production, etc.) It is also left aside that the expansion of agricultural frontiers (caused by the increasing demand of agricultural commodities) does not reduce greenhouse gas emissions. In the contrary, the forest is a carbon reserve and turning it into cropland releases this carbon as CO2 (the most important GHG). However, with the production of biofuels there are side effects that disturb an apparent prosperity. There are several scientists who defend that for true CO2 reduction re-forestation is vital as opposed to de-forestation for farming.

We must also consider that in large scale certain that biofuel inputs are produced by an agro industrial agriculture. This type of agriculture is supported in large monocultures; abuse of agrochemicals and the soil fertility overexploitation provokes water pollution (with pesticides), soil erosion, air pollution and loss of biodiversity.

On a national level, within the context of energy replacement policies, a delay in exhaustion of reserves, prevention in the rise of import costs and reduction in the impact of gas emissions and particulate matter into the atmosphere, present a great opportunity for the biofuel industry due to in part to the rise in oil prices. This opportunity is supported by a regulatory and legal framework of agro-energy production, including Act 693, of 2001 that proposes an initial 5% replacement of gasoline with alcohol. Later increased to 10% by 2010 and 12% by 2012; with similar proportions for replacing diesel with biodiesel. The same for the use of suitable lands for energy crops such as: sugar cane, cassava, sugar beet, oil palm , castor oil plant and jatropha, all of them with studies and different productivity levels (ton/ha, l/ton). Without denying the possibility of cellulosic biofuels from different sources. In addition to the aforementioned, the National Government has promoted the development and search of new renewable energy sources, sustainable with the increasing pace of life, by partially replacing oil or its derivatives in different uses, especially within the transport sector. This promotion must also consider the implications when allocating millions of hectares for bioenergy production. This reality shows the urgent need for meeting food demands or allocating lands and feedstocks to meet the energy requirements of the automotive industry(Cortés, 2007). Within this context it is proposed that it will promote competition between the different biofuels, with criteria for financial sustainability and energy supply. For these purposes the feasibility and advisability of releasing biofuels prices and promoting removal of import duties on these products. Notwithstanding the aforementioned, the National Development Plan states that in any case the current pricing scheme based on opportunity costs of such energy, their replacements and raw materials used in its production must be considered. While simultaneously promoting strategies for prevention and control of air contamination, by promoting cleaner fuels, including those derived from crops with production potential for biodiesel and alcohol fuels.

The different types of biodiesel from the oil palm, castor oil plant, jatropha and sacha inchi, considered by the Ministerio de Agricultura y Desarrollo Rural (English: Ministry of Agriculture and Rural Development) as strategic for Colombia, have a wide range of chemical compositions and qualities. Oil palm biodiesel given its highly saturated chemical nature has excellent ignition and chemical stability qualities, but it has limitations as for its ease of flow at low temperatures. Oil palm biodiesel cloud point, i.e. the temperature at which crystals formation is visualized, is around 16°C. Crystals emerging and further agglomeration can clog fuel filters preventing the fuel to reach the engine. Castor oil biodiesel is characterized by high alkyl esters content from ricinoleic acid (about 90%), which are monounsaturated and have a hydroxyl group in its structure, giving the biodiesel a high viscosity. Jatropha biodiesel holds 80% of unsaturated alkyl esters from which 34.8% are di-unsaturated. Sacha inchi oil can reach an unsaturated level up to 94%, and it is the most unsaturated oil according to technical literature. Therefore, with jatropha biodiesel and, the oil obtained from sacha inchi, there could be problems in its chemical stability (Ministerio de Minas y Energía, 2007; Mesa, 2006).

Crops' biggest problem is that they can be expensive as raw materials, which makes the final product price high; so the State must allocate considerable tax resources to make these energies competitive. Aside from being expensive, these raw materials are crops' primary products, and only recently has the use of waste for biofuel production been taken advantage of. Faced with this adversity, many countries are researching and developing methods for producing ethanol from agricultural, forest and industrial waste, which are abundant and very cheap.

In this case, sugars shall be extracted from plant waste cellulose, such as banana or lumber industry. Today in Brazil there are technologies for the use of husks through fermentation processes. Are then biofuels a technically economic and environmental viable energy output for the nation, with sights to replace future fuel imports? Although after the laws regarding alcohol fuels and biological oils came to be, it can be surmised that employment rates in growing regions shall be positively influenced. It is necessary that the country not only encourage production of efficient biofuels and that from a cost perspective it can compete in the international market, but define programs that support need of new refineries for biomass production, so that the price of raw materials with dual purpose is not affected (for both, food and biofuel) (Cortés, 2007).

It is also necessary to define agricultural land management strategies in order to preserve forest areas and not turning them into biomass growing areas. According to previous observations, the country's agricultural industry can be articulated with the energy industry, without affecting food industry through price increases of raw materials, as it has been with sugar, wheat and corn.

3.1 Different energy issues

The correlation between development and energy consumption is well known. This is quite reasonable as we can consider the gross energy consumption from a society, as a way to amplify the human effort. Likewise, technological change allows for great development with a modest increased consumption of primary energy.

Energy access and its use strongly affect and are affected by population growth, urbanization, or development possibilities and poverty alleviation of a great part of the population. For example, energy consumption patterns of a third of humanity that use biomass as the sole source of energy, tend to reinforce their extreme poverty situation. Hundreds of millions of people, especially children and women, spend several hours a day looking for firewood or carrying water from considerable distances, this causes them to have fewer opportunities for education or more productive activities.

Current development and consumption model along with increased energy waste from rich people as well as consumption patterns from the most disadvantaged creates pollution and destruction that leads to poverty, and this poverty at the same time pollutes and destroys. This is the vicious circle: consumption - pollution - poverty. This is a complex relationship network, not always obvious, in which certain phenomena are cause and effect at the same time and no element can be considered separate or isolated. One of the most important challenges humanity has to face is to find how to produce and use energy in ways that in long term human development, in all its dimensions, is promoted: social, economic and environmental (Pérez, 2002).

It is expected that in the course of this century, the use of oil for producing electricity will be replaced by gas, clean burning coal , energy from renewable sources, and nuclear energy. However, the largest contributors to oil consumption and increased pollution is the transport sector, where in the medium term there is not yet a replacement, since current petroleum products are characterized by their high calorific value, they are also easy to store, carry and use.

The creation of liquid biofuels has been one of the ways that science has developed for replacing gasoline and diesel and preserving the environment. Modern bioenergy technologies are renewable energy sources that produce transportation fuels and are advancing very fast, mainly towards ethanol made from maize or sugar cane, which is blended with gasoline in order to reduce both oil consumption and pollution. To use ethanol as fuel by blending it with gasoline, it is necessary to remove water for purity close to 99.9%, which requires special distillation methods.

Bioenergy competitiveness is associated with oil price, if price keeps current trends, there will be options for those trends. It must be considered if benefits and efficiency of these new fuels could survive without stimuli (subsidies) that now favor them. In a realistic framework it is necessary to avoid ambiguous positions that require a choice between biofuels or food production. It is important to combine both processes and add technology that improves production. But food safety issues cannot be jeopardized.

Security of supply is synonymous with the availability of all the energy needed, at an affordable price and for a long time, in fact indefinitely so it can be sustainable. If supply security is considered from a national perspective; dependence on external resources and the uncertainty of this non native supply becomes an important aspect.

"Opportunities for developing countries to take advantage of biofuel demand would be greatly advanced by the removal of the agricultural and biofuel subsidies and trade barriers that create an artificial market and currently benefit producers in OECD countries at the expense of producers in developing countries "said the director general of FAO, Jacques Diouf.

In the coming decades several challenges must be faced because of the energy and environmental problems arisen from it:

Energy efficiency: It will be necessary to radically increase the energy efficiency of processes and systems.

New technologies: It is necessary to develop and incorporate new technologies that achieve the above goal.

Diversify energy sources: At the moment we strongly depend on hydrocarbons as a primary energy source. It is necessary to incorporate new energy sources: natural gas, biomass, wind power, micro-hydro, solar energy and others.

Cogeneration: cogeneration as a means for energy efficiency and savings is not new. But incorporating it into the system on a grand scale may have a very large positive impact.

Then it is important to recognize that biofuels will not end industrialized countries' dependency on oil -not even Colombia-, because there will not be enough land and water to meet the high energy demand. In spite of this, the development of a national biofuel industry is an opportunity for the country. There are a number of technological, regulatory, economic and environmental restrictions or challenges that may affect critical links in the chain of biofuel production, and if they are not superseded they could lead to failure.

3.2 Energy Impact on the environment

There is no doubt that development and energy use are closely linked. In the coming years, a key matter will be how to ensure that energy sources are economical and reliable enough to guarantee us an adequate level of development. Energy availability is an obstacle for development; but environmental impacts may also limit or put the development at risk. However, this is not the entire problem. It is clear that all activities will have an impact on the environment. The issue is when this impact becomes negative or even irreversible. Throughout history there are a lot of examples of societies that made their environment collapse and in turn they collapsed.

At the global level the impact of energy activities on climate change must be highly considered, the so-called greenhouse effect. Climate change is not indeed the only global threat to environmental sustainability, but many agree it is the most important. Its extent, complexity and direct connection with energy activities make climate change a paradigm. For example, the success or failure in the implementation of the Kyoto Protocol is still an excellent indicator of global community and each country's commitment with sustainable development (Pérez, 2002).

The impact of the production and use of energy has been observed for a long time. Deforestation of many areas and pollution associated with industrial processes are well known cases. But, although they were serious, it was about local impacts. In the last hundred years local effects have become global threats. The recognition of the association of energy with global environmental problems is a recent event that is affecting human health and quality of life, but especially that of future generations.

Undoubtedly, human activity has a big impact with respect to the environment. Today, Sustainable Development is mentioned a lot; in fact this is a contradiction of terms. Indeed, if we consider that development (which involves the permanent increase in the use of resources) must always increase. It is inevitable that within a long or short term we find crucial restrictions for development because of the inevitable shortages of resources. It is, therefore, essential to know the difference between growth and development. Indeed, a country can have strong growth, all the while achieving a high level of development simply at a slower pace. It is also possible to have large development increases with low growth rates. It is important to tell the difference between these two concepts as it allows taking a look at the evolution of a country from another perspective. In the last decade it is clear that Colombia has had enough growth; however our degree of development has been much lower. It is also clear that a finite growth is not viable because it involves having unlimited resources, which is not the case in our planet.

During the last century, impacts of human activity have been higher had taken place since the beginning of civilization. Footprints of human activity are changing the world at a rapid pace and energy has a lot to do with this impact.

Instead of detesting the technological development, it is necessary that some philosophical, ethical and social principles guide it. Not everything is good, cars have not given humans more freedom: because now travel time has increased, people need to work longer hours to pay it, cities have become uninhabitable, it is literally asphalting the living space, social conflicts are increasing because of the lack of communication, and due to noise and urban stress increasing, we move away from downtown and then our dependence on the car and our isolation increases. From that overview, it doesn't look like the car is synonymous with social development and welfare (Valero, 2004).

3.3 Biofuel, a sustainable energy?

In Colombia due to the act (693/2001), production of biofuel is it is expected to increase for use it in transportation as a measure to reduce greenhouse gas emissions. A target by 2010 is to replace 10% of fuels, used today for transportation, with biofuels that will increase up to 20%. With that, reduce CO2 emissions to the atmosphere that cause the greenhouse effect.

One of the main problems is that it causes higher prices for basic foods. Biofuels are produced from products like maize (corn), sugar cane, wheat or soybeans; and when the availability of them decreases in the food market, it drives up prices. At the same time, available area for food crops also decreases because most of these crops are for biofuels, which results in increasing the agricultural commodity's price in general.

Economic effects from this rising food prices are being felt in recent years, especially in some poor countries. For example the crisis that took place in Mexico because of the maize tortillas rising price, which has doubled in recent years (similar to what is happening in our country with the rising price of bread and milk).

Replacement of conventional fuels with biofuels is also generating adverse ecological consequences. Most of the feedstocks needed for processing take place in developing countries, mainly in Latin America and Asia; most of these countries are cutting down large areas of tropical forests for growing biofuels.

For the production of clean fuels it is necessary to use dirty fuels as energy source. For instance, intensive sugar cane crops (for ethanol) or other oil crops (for biodiesel) will need petroleum products: fertilizers, insecticides, fuel pumps, transport and industrial processing. Therefore it is possible that pollution levels increase by using dirty energy sources for producing and exporting clean energy sources.

So far it is clear that bio in biofuels must have a question mark. Then, it can not be neither justified nor adopted policies for biofuels promotion and support, based on ecological arguments, or in industrialized countries (where people want to use agro-energy) or in developing countries (where people want to produce them).

To classify biofuels as bio, it would necessary to grow in degraded and poor soils that are unsuitable for food production (the so-called second generation biofuels). This prevents the rising prices of food and deforestation. An international certification scheme could ensure the sustainability of agricultural practices for the production of raw materials for biofuels.

In order to reduce possible impacts caused by biofuel production, certification for procedures of its production have been developed around the world; this is how the Dutch government, among others, aims that imported biofuels are certified according to environmental and social criteria. Certification of the entire process shall be necessary to ensure the world sustainability production and the use of biofuels (Testing framework for sustainable biomass, 2007).

Likewise, one of the most important factors for defining biofuel production feasibility is energy balance (the comparison between the energy used for producing biofuels and energy production). From the energy perspective, it is not enough to take into account the energy generated by a fuel, but it also must be considered the global energy balance, considering energy expenditures for fuel production and energy derived from it. Undoubtedly, for the production of that fuel to be profitable, the balance must be positive, i.e. it must generate more energy than consume.

Again the usefulness of biofuels as potential replacement for fossil fuels in the reduction of greenhouse gas emissions has been questioned. Several specialists have shown that the cultivation and use of, is not as efficient as a measure to slow down climate change as their advocates say.

Specifically deforestation, caused because of these feedstocks expansion, can have devastating effects in terms of climate, as well as from the ecological perspective. According to studies, forests from a particular area can reduce CO2 emissions nine times more than a biofuel feedstock with the same area, as well as the subsequent use of those biofuels for transportation.

If that wasn't enough, along the acquisition process of these fuels (including cultivation, processing, transportation and distribution), more CO2 is released than those crops can absorb while growing. This is because large amounts of fossil fuels are needed; resulting in emission of greenhouse gases, that in the case of bioethanol, these plants cannot entirety absorb. This, linked with the high water consumption required for producing them, especially biodiesel (for one liter of biodiesel 12 liters of water are consumed), makes them a non-sustainable alternative compared to fossil fuels.

Given the multiple problems shown by first generation biofuels, once again a technological solution is offered: liquid biofuels production (BtL, Biomass to Liquid), which can be obtained from lignocellulosic biomass such as straw or wood chips. These include bio-ethanol produced by hydrolyzed biomass fermentation and biofuels obtained via thermo chemistry, such as bio-oil obtained from pyrolysis (carbonization), gasoline and diesel oils produced by Fischer-Tropsch Synthesis, among others.

3.4 Biofuel programs in Colombia: objectives

It is mainly to promote the diversification of the energy basket through the use of biofuels, with the following criteria (Mesa, 2006):

- Environmental sustainability.
- Favor lignocellulosic crops replacement.
- Agricultural employee maintenance and development.
- Energy self-sufficiency.
- Agro-industrial development.
- Improving the quality of country's fuels, as a result of a blending between biofuels and fossil fuels.

To achieve these goals, Colombia faces the challenge of moving into strategic areas, among them are: a) consolidation of an institutional framework for the formulation of actions related to the handling of biofuels; b) reduction in the production of biofuels in the most critical points of the production chain, c) increasing the productivity of biofuels throughout all the production chain, d) research and development looking towards increasing biomass crop yields, develop new varieties adapted to different growing conditions and resistant to plagues, and develop changing processes of first and second generation e) price regulation in order to encourage the efficient production of biofuels, and f) differentiation of the Colombian product in order make easier the access to international markets by adding strategic environmental and social variables, besides food safety protection measures (Consejo Nacional de Política Económica y Social (CONPES, 2008).

As stated by the Consejo Nacional de Política Económica y Social (CONPES) (in English: National Council of Economic and Social Policy) of the Colombian government: This will enable the ability to take advantage, in a competitive and sustainable way, of economic and social development opportunities offered by biofuels emerging markets. At the same time it will allow: increasing competitive sustainable biofuels production by contributing to employment generation, rural development and population welfare; promoting an alternative productive development to the reliable rural land occupation; contributing to the formal employment generation within the rural sector; diversifying the country's energy basket throughout biofuels efficient production, by using current and future technologies; ensuring an environmentally sustainable performance throughout the addition of environmental variables when making decisions in the chain of biofuel production.

4. The most common raw materials

Energy crops are those developed only for fuel. These crops include fast growing trees, shrubs and grasses. These can be grown in agricultural land not needed neither for food, nor pasture nor fibers. In addition, farmers can grow energy crops along the banks of rivers, around lakes or in farms areas including, natural forests or swamps, for creating habitat for wildlife, renewing and improving soil biodiversity. Trees can be grown for a decade and then being cut down for energy.

Thus, bioenergy covers all forms of energy derived from organic fuels (biofuels) form biological origin used for producing energy. It includes both crops intended to produce energy which are particularly grown and multipurpose crops and by-products (residues and wastes). The term By-products includes solid, liquid and gaseous byproducts derived from human activities. It can be considered biomass as a sort of converted solar energy.

It can be said that biodiesel production tends to come mainly from oils extracted from oilseeds plants, but any material containing triglycerides can be used for biodiesel production (sunflower, rapeseed, soybean, oil palm, castor oil, used cooking oils, animal fat). Here are the main raw materials for biodiesel production (Mesa, 2006).

Conventional vegetable oils: raw materials traditionally used for biodiesel production have been: oils from oilseeds such as sunflower and rapeseed (in Europe), soybeans (in The United States) and coconut (in The Philippines), and oils from oilseeds fruits such as oil palm (in Malaysia, Indonesia and Colombia).

Alternative vegetable oils: in addition to traditional vegetable oils, there are other species adapted to the conditions from the country where they are developed and better positioned within the field of energy crops: Jatropha curcas oil (Ministerio de Minas y Energia, 2007).

Biofuels have become very important because of the variety of crops from which they can be derived, but this energy supply demands a high production of them. This would have harmful effects because of the destruction of forests and jungles and replacement of crops that are essential to human diet; besides the drawbacks shown in the following fields: climatic, geographical and physical. The main supply sources of raw materials for biofuels production are shown in Table 1 and Figures 2, 3 and 4.

Crop	Efficiency (l/ha/year)	Efficiency (ton/ ha)	Estimated barrel price (US $)
Sugar Cane	9	100	45
Cassava	4,5	25	NA
Sugar Beet	5,000	NA	100
sweet sorghum	1,189	NA	NA
Cellulose	NA	NA	305
Maize	3,2	10	83
Oil palm	5,55	NA	NA
Coconut	4,2	NA	NA
Castor oil	2,6	NA	NA
Avocado	2,46	NA	NA
Jatropha	1,559	NA	43
Rapeseed	1,1	NA	NA
Peanut	990	NA	NA
Soybeans	840	NA	122
Rapeseed	NA	NA	125
Wheat	NA	NA	125
Sunflower	890	NA	NA
Oil	NA	NA	70-80

Table 1. Raw materials for biofuel production: Source: Ministerio de Agricultura y Desarrollo Rural, MADR (English: Ministry of Agriculture and Rural Development); Portafolio: Goldaman Sachs (2007)

But not all the questions are clear and therefore the UN declares: if growing fields for biofuels production increase disproportionately, food and the environment could be at risk. Increased logging. Also food prices could increase.

For major producing countries, costs of ethanol production range between 32 and 87 USD/barrel (International Energy Agency, 2006). According to the available information, about 47% and 58% of this cost is raw materials, about 13% and 24% for inputs, about 6% and 18% for operation and maintenance costs and, about 11% and 23% to capital costs. It can be said that production costs widely vary between countries due to agro-climatic factors, land availability and labor cost that affect the kind of biomass used as raw material; this factor affects transformation technologies selection.

Figure 2 shows sources of raw materials sources for alcohol and biodiesel production and the corresponding efficiency. Figure 3 and 4 show ethanol efficiency from biomass sources in countries outstanding in their production. There is higher ethanol efficiency from sugar beet, in comparison with sugar cane and corn.

For every ton of cassava, 200 liters of ethanol can be obtained, when making the cassava calculations as a yield base of 25 ton/ha it can be obtained a yield of 5000 liters/ha can be obtained which is lower in comparison to sugar beet but higher compared to corn and sugar cane. With fertilization programs and cassava crops mechanization, yields can be increased to values of 70 ton/ha, which will triple cassava yield in liters/ha (Altin et al., 2001).

Another important factor is that biofuels do not work as well as petroleum fuels. In order to increase their production most of the fertile lands would have to be assigned for farming them, which could be counterproductive in a world where hungry and desertification are two problems with difficult solution.

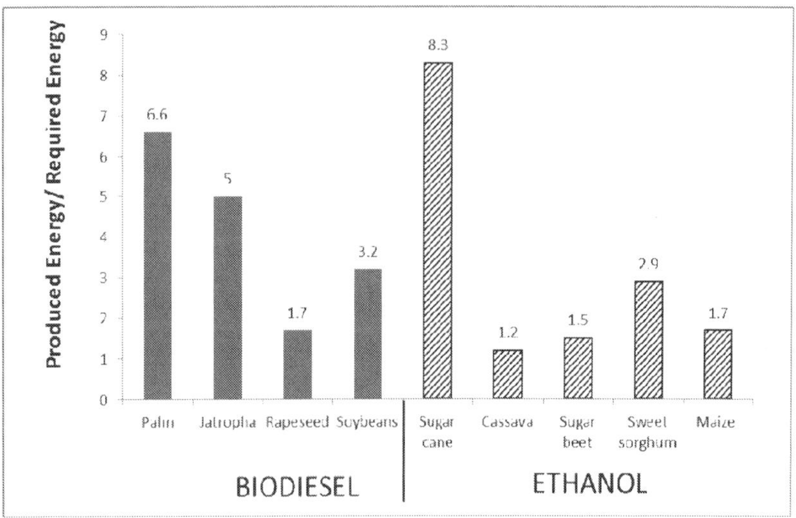

Source: Ministerio de Minas y Energía (English: Ministry of Mines and Energy), based on Goldman Sachs and LMC

Fig. 2. Energy efficiency in biofuel production

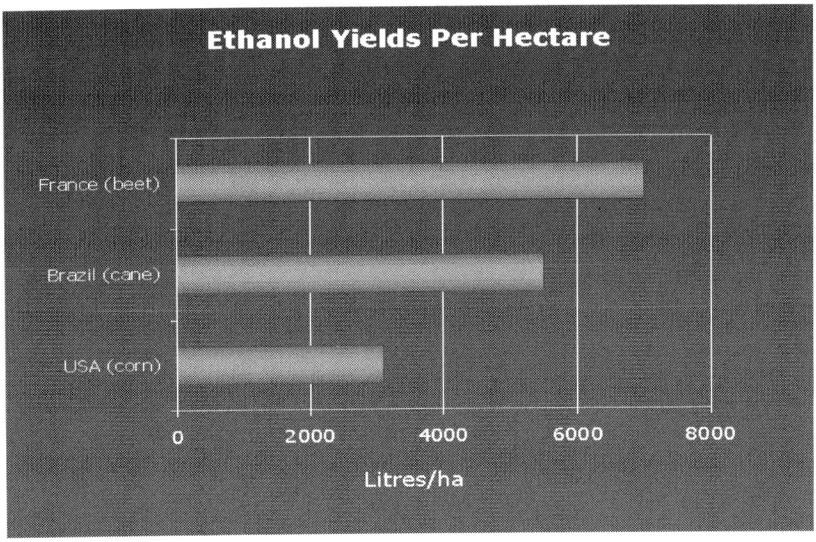

Fig. 3. Ethanol yields from biomass (Source: FAO, 2007)

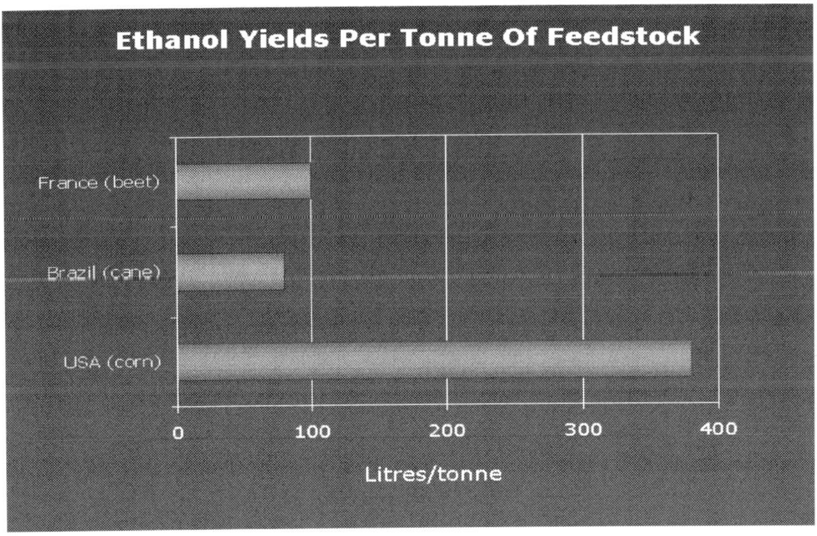

Fig. 4. Ethanol yields in liters per Tone of Feedstock. (Source: FAO, 2007)

5. Technical considerations

Biodiesel use in diesel engines is more limited. As well as ethanol, biodiesel is produced by fast pyrolysis of lignocellulosic biomass and mainly fermentation, because fast pyrolysis is a more expensive way (Bridgwater et al., 2002), it is a renewable oxygenated fuel with low cetane components (Ikura et al., 2003). Its heating value is about 60% of ethanol, but its high density makes up for its percentage. When using biodiesel in machines and engines there are some problems (Lopez & Salva, 2000) because of its higher viscosity and acidity, tar and fine particles resulting during working hours and solid residues during the combustion.

Following the direction of ethanol research, attempts have been made to overcome these problems by blending bio-oil with diesel to form an emulsion (Chiaramonti et al., 2003). In some success these efforts solve the operation with these fuels, however it is necessary to prove the feasibility and the additional cost of surfactant required to stabilize the blending which is a barrier for using it.

It must be considered that the blending of biodiesel and ethanol makes a stable blend and a fast pyrolysis, without using additives and surfactants. Current research on these blends is limited to gas turbines (López & Salva, 2000) and their use in these engines has shown positive results. Biodiesel blended with ethanol shall not exceed the problems of direct ethanol use in diesel engines without modification. However, using modified engines to use ethanol blends of ethanol/biodiesel could overcome the problems related to pure biodiesel combustion. As all new fuels, it is necessary to solve technical problems such as fuel storage, material compatibility, and procedures for turning engines on and off and long operation periods (Nguyen & Honnery, 2008).

5.1 Benefits

However, in Colombia, promotion of biofuels production may represent several benefits:

Energy sustainability: it will help to reduce the use of fossil fuels, thus protecting oil reserves. That is, a decreased risk of energy vulnerability. According to the Ministerio de Minas y Energía (English: Ministry of Mines and Energy) estimates show if new deposits are not found, known reserves will support the demand only for a few years. In this context, adding 10% of ethanol to gasoline helps to support fuel needs. Furthermore, Colombia has set the goal of increasing that percentage to 25% by 2020, which requires the new projects for ethanol production and the use of biomass sources other than sugar cane. In the short term the national program for Biofuels, seeks to improve fuel trade balance, and thus avoid wasting foreign reserves and spending at high prices by importing oil and petroleum products, that now tare close to 100 USD/barrel).

Environmental: biofuels are biodegradable, 85% is degraded in about 28 days.

Ethanol is a compound free of aroma, benzene and sulfur components, so the blending produces less smoke (particulates) and generate lower emissions (Stern, 2006). By using a 10% ethanol blending there is a reduction in CO emissions between 22% and 50% in carbureted vehicles, and a decrease of total hydrocarbons between 20% and 24% (Lopez & Salva, 2000).

With only a 10% blending of ethanol with gasoline, in new cars, 27% of carbon monoxide emissions decrease. In typical Colombian cars with 7-8 years of use it decreases 45%, and there is 20% reduction in hydrocarbons emissions. The effects of these reductions shall be reflected in the environmental emissions indices (Kumar, 2007), and in improve the citizens' living conditions, for example Bogotá where acute respiratory diseases are public health problems. Diesel blending decreases vehicle emissions such as particulate matter, polycyclic aromatic hydrocarbons, carbon dioxide and sulfur dioxide (U.S. Environmental Protection Agency, 2003).

Biodiesel is biodegradable, nontoxic and sulfur and aromatic components free, no matter the source of the oil used in its production. It reduces the soot emission in 40%-60%, and CO between 10% and 50%. Biodiesel can replace diesel (diesel fuel) without changes in ICE. Emissions with primary pollutants; with the exception of nitrogen oxides NOx. Despite these obvious benefits, there is not enough information about the solution to by-products and waste generated from biethanol-vinasses-and biodiesel-glycerin production processes, which are a source of future contamination if they are not properly disposed.

Agricultural development: biofuels production from agricultural raw materials, can guarantee both jobs growth and the possibility of crops diversification, including those for biofuel production. Export expectation, if there is pipe dream with Free Trades Agreement implementation, where Colombia supposedly is able to export bioenergy to poor energy countries, or that require large amounts of fuel for supporting economic growth.

Advantages of Colombia: As a reference, the abundance and variety of raw materials could be pointed out; several regions suitable for cultivation in all the country; guaranteed domestic market; government incentives and appropriate legal framework; high-yield crops, uninterrupted interest in research and development.

5.2 Regulations

Colombia, in order to reduce gasoline and diesel consumption, has implemented policies to encourage domestic production of biofuels. This purpose is economically boosted compared to fuels consumption reduction by the automotive industry and the best environmental indicators of mobile source emissions given the oxygenating effect of biofuels in combustion. For that reason in 2001 it is passed the Act N° 693 and in 2004 the Act N° 939, which states regulations on alcohol fuels and vegetable oils in the country, and creates incentives for their production, marketing and consumption.

In this regard, the Government has promoted development of biofuels through different measures to encourage their production and use. In this matter there is a broad regulatory and incentives for bioenergy production in Colombia, namely (Ministerio de Minas y Energía, 2007; Cala, 2003):

Act 693/2001: the regulations about the use of alcohol fuels are thereby stated; Incentives are created for their production, marketing and consumption. This act makes obligatory the use of oxygenated components in fuels for vehicles from cities with more than 500,000 inhabitants. A deadline of 5 years was established for gradual implementation of this regulation.

Act 788/2002: tax reform where exemptions were introduced to the Value Added Tax (VAT), the income tax and surcharge on alcohol fuel blended with gasoline engine.

Act 939/2004: defines the legal framework for the use of biofuels, by which the production and commercialization of biofuels of plant or animal origin, are thereby encouraged for use in diesel engines and other purposes. Exempts biodiesel from VAT and the income tax and establishes a net income exemption for 10 years to new oil palm plantation. This exemption applies to all plantations to be developed before 2015.

Act 1111/2006: establishes a 40% income tax deduction of investments in real productive fixed assets of industrial projects, including financial leasing.
Act 1083 2006: some regulations on sustainable urban planning and other provisions are thereby stated.
Resolution 1289/2005: establishes biofuels criteria quality for their use in diesel engines, states the date of January 1st 2008 as a blending start of 5% of biodiesel with diesel fuel.
Resolution No. 180127/2007: the heading "MD" in Act 4 from Resolution 82439 from December 23th, 1998 is thereby amended and amends Act 1st from Resolution 180822 from June 29th, 2005 and, states the provisions relating to Diesel Fuel pricing structure.
Decree 383/2007: Amends the Foreign-Trade Zones Decree 2685 of 1999, regulates the set up of Special Foreign-Trade Zones for high economic and social impact.
Decree 3492/2007: Act 939 of 2004 is thereby regulated.
Decree 2328/2008: The Intersectoral Commission for Biofuels Management is thereby created.
Decree 4051/2007: Permanent Foreign-Trade Zones area requirements is thereby stated; requirements for stating the existence of a Special and Permanent Foreign-Trade Zone and Industrial User recognition.
Resolution No. 180158/2007: clean fuels are stated thereby in accordance with the Paragraph in Article 1, Act 1083.
Resolution No. 180782/2007: biofuels quality criteria for use in diesel engines as a component of the blending with fossil diesel fuel in combustion processes are thereby amended.
Resolution No. 180212/2007: Resolution 181780 December 29th, 2005 is thereby partially amended, regarding the pricing structure of diesel fuels blended with biofuel for their use in diesel engines.
Decree 2629/2007: provisions for promoting the use of biofuels in the country are thereby stated, as well as applicable measures for vehicles and other motorized devices that use fuels. From January 1st, 2010 timetable is thereby set up for extending the mandatory blending of biofuels of 10% and, 20% from 2012 as well as the requirement that from January 1st 2012, new vehicle parc and other new motorized devices should be Flex-fuel at least 20%, for both E-20 blending (80% of gasoline from fossil fuel, with 20% of alcohol fuel) and B-20 (80% of diesel fuel with 20% of biofuels).
Decree 1135/2009: In connection with the use of alcohol fuels in the country and applicable measures to motor vehicles using gasoline, decree 2629, 2007 is thereby amended. And which states in its article 1: from January 1st, 2012 motor vehicles up to 2000 cm3 manufactured, assembled, imported, distributed and marketed in the country and requiring gasoline to operate, must be soup up so that their engines run Flex-fuel system (E85), i.e. they can work normally by using either basic gasoline or blends composed of basic fossil fuel with at least 85% alcohol fuel. To meet the above, each brand shall sell vehicles in the Colombian market according to the following schedule and provisions:
From January 1st, 2012: 60% of its annual supply must support E85.
From January 1st, 2014: 80% of its annual supply must support E85.
From January 1st, 2016: 100% of its annual supply must support E85.
From January 1st, 2013: vehicles with engine cubic capacity greater than 2000 cm3 from all brands and models shall bear E85.
It is worth mentioning CONPES-3510/2008 document (in English: National Council for Economic and Social Policy document 3510/2008), where a policy to promote the

production of sustainable biofuels in Colombia is thereby established, by taking advantage of economic and social development opportunities which are offered by biofuels emerging markets. Thus, it intends to expand the known biomass crops in the country and diversify the energy basket within a framework of production that is financially, socially and environmentally efficient and sustainable, that makes possible to compete in domestic and international markets.

Likewise the promotion of biofuels is also done through: the National Development Plan (NDP), the establishment of a regulatory framework and the development of financial and tax incentives. Also, the National Government has policy guidelines in areas such as: agriculture, research and development, infrastructure and environment that influence biofuels development.

There are also other complementary policy developments in the form of decrees and ministerial decisions that define the technical regulations, quality standards, as well as pricing, margins and rate parameters for fuel ethanol and biodiesel transport. There is an applicable regime in the Foreign-Trade Zone and several soft loan sources for agricultural development (González, 2008).

Among them, in the framework of Agro Ingreso Seguro Program (AIS), financial instruments that provide soft loan sources for growing crops that produce biomass for ethanol and biodiesel production have been implemented. In addition, through the Incentivo a la Capitalización Rural, ICR (in English: Rural Capitalization Incentive) it is promoted, among others, oil palm crops establishment and renewal, and the construction of infrastructure for biomass processing (Consejo Nacional de Política Económica y Social (CONPES, 2008)

Despite this broad regulatory framework, there is uncertainty about changes in: regulation, raw material prices and emerging new technologies. In particular, with gallon prices as defined by state intervention (subsidies), that generates the discussion about how much does it mean for the national treasury, and whether it is advisable or heavy subsidies is fair to benefit a minority that supply biofuels, for even small domestic market and one that is difficult to be exported.

As shown, the Colombian Government has a fairly strong policy and information that allows for investment in projects, sustainable energy and biofuels plans and programs through a set of tools, studies and institutional strengthening.

Therefore, the Colombian Government has promoted assessments that seek to: a) study the implications of the biofuel industry, from planting crops for biofuel production to the final consumers of ethanol or biodiesel (flex-fuel or normal vehicles); b) analyze the current infrastructure requirements for the expansion of the biofuel market; c) know the sector current status, as well as the economic instruments, regulatory elements, policies and tax incentives required or recommended for promoting renewable energy, energy efficiency and biofuels; d) analyze the renewable energy potential, energy efficiency and carbon credits through the Clean Development Mechanism. Likewise, institutional strengthening assessment required by the Ministerio de Minas y Energía (English: Ministry of Mines and Energy) (MME), in energy efficiency, renewable energy, bioenergy and carbon financing.

This set of measures that promote the enthusiasm for liquid biofuels such as the mandatory blending of biofuels with fossil fuels and tax incentives, have created a fast artificial growth in biofuel production. These incentives have broad social impacts, as they are resources that do not come into the State, and are taken for solving important issues such as health, education and basic sanitation.

These measures entail high economic, social and environmental costs and should be monitored promptly.

5.3 Current projects under construction

Ethanol: In compliance with the provisions of Act 693/01, the country began to implement initiatives for alcohol fuel from sugar cane. At the moment 5 ethanol plants are running: Incauca, Providencia, Manuelita, Mayaguez and Risaralda refineries that produce about 1,050,000 liters of alcohol fuel a day and this production is mainly to supply the domestic market. It is estimated a domestic demand close to 1,500,000 liters per day to cover the 10% of blending needs.

Likewise, in the country several alcohol production projects are being implemented in several departments: Antioquia, Boyacá, Santander and the coast, derived from different raw materials such as sugar cane, sugar beet, banana and cassava.

Unfortunately, due to the economic crisis there is absence of new plants. Projects are standstill, and Ecopetrol plant would only come into operation in 2011, starting with a production of 385,000 liters a day. At the moment there is another project being developed in Magdalena, where an international company sowed a very large sugar cane area for producing an average of 300,000 liters a day. With this, the 20% blending could be reached by in 2012 without any problem.

Biodiesel: At the moment there are five projects under construction for producing biodiesel from oil palm (Oleoflores - already in production, Odin Energy, Biocombustibles Sostenibles del Caribe) and two in the eastern region (Biocastilla, Bio D. SA). In addition, they are other projects under development, one in the central region (ECOPETROL), one in the eastern region (Manuelita), one in the west region and another in the north region. In 2008 it is expected they shall enter into production, with a total amount of 400,000 t/year (19).

How are investments for biodiesel production doing? Construction of the Ecopetrol plant in Barrancabermeja is almost over. With this in total there will be seven plants in the country. A total installed capacity of 526,000 biodiesel tonnes a year may be achieved.

6. Conclusions

It must be accepted that the so-called modern man now has the same challenge our ancestors solved centuries ago, that life is not over. Availability of natural resources and the way we use them, force us to shape a scenario of technological innovations and social coexistence, in which the ethics of life prevails over money; this becomes more valid in this global world that requires new economic, lifestyle, consumption and value models.

Society needs energy for its development, but development does not necessarily imply a waste of energy. In any productive process, materials and water may or may not be wasted, but it is certain that it will consume energy and that energy consumption will be associated with a real environmental impact. If energy production takes on all costs, it would be much more expensive.

New energy sources are the new economic, political and even environmental strategy. Their importance is such that currently over 30 raw materials are being tested worldwide. Despite this big boost, they do not yet provide a solution to the global energy problems.

Biofuels should not be taken as the solution to the energy and environmental problem, but as part of a complex human and energy project where leading countries still disagree on a solution. If Bioenergy is properly used, it provides a historic opportunity to contribute to the growth of many of the world's poorest countries.

A reality must be emphasized; alcohol fuel is more expensive than gasoline and biodiesel. It is not good business that a market economy develops isolated and organically; the market must be intervened so these alternatives are viable, because rival fuel is cheaper. Oil is in the reservoir, while cassava, sugar cane, oil palm or other crops used as raw material must be planted, and in expensive lands. Then, by definition, we talk about a project that is viable only if the State intervenes so it can be operated outside the framework of the market.

The world faces complex challenges and life's survival on the planet can not be supported on the solution to the renewable energy alternative based on biofuels, as it would grow the replacement of food crops with monocultures, deforestation for energy crops, while it would boost the diversity extinction, fertile lands and water reduction, and the social consequences population displacement causes.

In that sense the FAO has declared: Biofuel policies and subsidies should be urgently reviewed in order to preserve the goal of world food security, protect poor farmers, promote broad-based rural development and ensure environmental sustainability. But also states: Growing demand for biofuels and the resulting higher agricultural commodity prices offer important opportunities for some developing countries. Agriculture could become the growth engine for hunger reduction and poverty alleviation, production of biofuel feedstocks may create income and employment, if particularly poor small farmers receive support to expand their production and gain access to markets.

It also requires a certification system that ensures that biofuels will be marketed only if they have the necessary environmental requirements.

Colombia is not and cannot be indifferent to the global market trend for crude oil and its derivatives. This fact gives the opportunity for goods production, such as biofuels, that allow diversity in the energy basket available in the domestic market and that can be exported to international markets. However, a necessary condition for competing in the international market is efficient conditions for the production of these goods.

Colombia has enough available land for growing biofuels, from 14 million hectares for agriculture business and 20 for livestock, only 5 million are currently in use and the remaining is for extensive cattle ranching; a better use could be biofuels which would provide plant cover and rural income opportunities. It also holds high productivity in sugar production from sugar cane, but such activity has been focused on agribusiness models, where production is held in few companies from renowned economic groups.

Although in Colombia ethanol has been a biofuel pioneer, biodiesel projects are gaining strength and this fuel can have a greater impact and national coverage.

In the country there is a poor use of natural resources and a high dependence on them; there is not full agreement between vocation or fair and the use of resources. Productivity paradigm boosts to predatory models and the economic efficiency and profitability fallacy as sole indicator, productive projects that do not consider social and environmental benefits are presented.

Then, in the previous horizon, it is required to develop a long-term sustainable agriculture that is compatible with the environment. The aim of this is a critical reassessment of the current modernizing model, taking into account that different technological offers, articulated to a diverse set of socio-economic and environmental factors, require different technological solutions. Consequently, decisions about biofuels should consider the food safety situation but also land and water availability.

Energy has deep and broad relations with the three sustainability dimensions (economic, social and environmental); i.e., it must go into the integration, harmonization and optimization. The services energy provides help to meet several basic needs such as: water supply, lighting, health, ability for producing, transporting and processing food, mobility and information access so that access to a certain amount of advanced forms of energy such as electricity or liquid fuels and gaseous fuels, should be included among the inalienable human rights in the XXIst century. Energy supply safety and energy prices are crucial for economic development. On the other hand, it is clear that many ways of producing and consuming can reduce environmental sustainability. We must ask: is the current energy production and consumption sustainable? One of the most important challenges humanity faces is to find the way to produce and use energy so that in the long term human development is promoted, in all its dimensions: social, economic and environmental.

Finally, to balance the enthusiasm with objectivity: it is necessary to carefully study the economic, social and environmental bioenergy impact before deciding how fast it is desired to be developed, and what technologies, policies, investment and research strategies to follow.

7. References

Altin, Recep, Çetinkaya, Selim & Serdar Yücesu, Hüseyin. (2001). The potential of using vegetable oil fuels as fuel for diesel engines, Energy Conversion and Management, Vol. 42, No. 5, pp. 529-538, ISSN 01968904

Bridgwater, A.V., Toft A.J. & Brammer, J.G. (2002). A techno-economic comparison of power production by biomass fast pyrolysis with gasification and combustion, Renewable Sustainable Energy Reviews, Vol. 6, No. 3, pp. 181-246, ISSN 13640321

Brown, Lester R., Renner, Michael & Halweil, Brian. (2000). Signos vitales 2000: Las tendencias que guiarán nuestro futuro, GAIA Proyecto 2050, ISBN 9788493023225, Spain.

Cala, David F. (2003). Proyecto para producción de biodiesel a partir de palma africana en Colombia, Corporación para el Desarrollo Industrial de la Biotecnología y Producción limpia (CORPODIB), Bogotá, Colombia

Casilda, Béjar Ramón. (2002). Energía y desarrollo económico en América Latina. Boletín ICE Económico, No.2750, (December 2002), pp.31-43, ISSN 02102633

Chiaramonti, D., Bonini, M., Fratini, E., Tondi, G., Gart, K., Bridgwater, A.V., Grimm, H.P., Soldaini, I., Webster, A. & Baglioni, P. (2003). Development of emulsions from biomass pyrolysis liquid and diesel and their use in engines, part 1: Emulsion production, Biomass and Bioenergy, Vol. 25, pp. 85-99, ISSN 09619534

Consejo Nacional de Política Económica y Social (CONPES 3510). (2008). Lineamientos de política para promover la producción sostenible de biocombustibles en Colombia, Bogotá, Colombia

Cortés, Elkin Marin, Suarez, Héctor Mahecha & Pardo, Sandra Carrasco. (2009). Biocombustibles y autosuficiencia energética, Dyna, Vol.76, Nro. 158, pp. 101-110, ISSN 00127353

Cortés, M. E. (2007). Biocombustibles: ¿alternativa para la agricultura colombiana?, Memorias Agroexpo, Bogotá, Colombia

Cortés, Elkin & Álvarez, Fernando. (1998). Consumo energético y desarrollo agrícola sostenible, Ponencias de la III semana técnica Nacional de Ingeniería Agrícola, Medellín, Colombia, October 1998.

Food and Agriculture Organization of the United Nations (FAO). (2008). The state of food and agriculture, FAO, Retrieved from <ftp://ftp.fao.org/docrep/fao/011/i0100e/i0100e.pdf>

Gonzáles, M. César. (2008). Peor el remedio, In: Portafolio, 10.12.2010, Available from: http://www.portafolio.com.co/archivo/documento/MAM-2833067

Ikura M., Stanciulescu, M. & Hogan, E. (2003). Emulsification of pyrolysis derived bio-oil in diesel fuel, Biomass and Bioenergy, Vol. 24, No. 3, pp. 221 - 232, ISSN 09619534

International Energy Agency. (March 2006). Word Energy outlook 2006 In: International Energy Agency, 08.09.2010, Available from: http://www.iea.org/textbase/nppdf/free/2006/weo2006.pdf

Kumar, A.A. (2007). Biofuels (alcohols and biodiesel) applications as fuels for internal combustion engines, Progress in Energy and Combustion Science, Vol. 33, No.3, pp. 233–271, ISSN 0360-1285

López, Juste G. & Salva Monfort, J. J. (2000). Preliminary test on combustion of wood derived fast pyrolysis oils in a gas turbine combustor, Biomass Bioenergy, Vol. 19, No.2, pp. 119–128, ISSN 09619534

Mesa, Jens. (2006). Biocombustibles y Agricultura, Primer Congreso Grupo Empresarial del Campo, Bogotá, Colombia, November 15-17, 2006.

Ministerio de Minas y Energía-Unidad de Planeación Minero Energética. (2007). Desarrollo y consolidación del mercado de biocombustibles en Colombia, Bogotá, Colombia.

Nguyen, D. & Honnery, D. (2008). Combustion of bio-oil ethanol blends at elevated pressure, Fuel, Vol. 87, No. 2, pp. 232–243, ISSN 00162361

Pérez A., José Ignacio. (2002). In: Energía y desarrollo sostenible, 10.12.2010, Available from: http://www.oei.es/decada/portadas/MedAmbPDF4.pdf

Silvestrini, Vittorio. (2000). Qué es la entropía?, Colección Milenio/Norma, ISBN 9580438757, Colombia

Stern, D.I. (2006). Reversal of the trend in global anthropogenic sulfur emissions, Global Environmental Change, Vol. 16, No. 2, pp. 207–220, ISSN 09593780

Testing framework for sustainable biomass. (February 2007). In: Final report from the project group"Sustainable production of biomass", 16.11.2010, Available from: http://www.lowcvp.org.uk/assets/reports/070427-cramer-finalreport_en.pdf

U.S. Environmental Protection Agency. (2003). National Air Quality and Emissions Trends Report: 2003 Special Studies Edition, North Carolina, USA

Valero, Antonio. (March 2004). In: Energía y Desarrollo Social, 9.10.2010, Available from: http://habitat.aq.upm.es/boletin/n32/aaval.html.

24

Enzyme-Based Microfluidic Biofuel Cell to Generate Micropower

A.Zebda[1], C. Innocent[1], L. Renaud[2], M. Cretin[1], F. Pichot[3], R. Ferrigno[2] and S. Tingry[1]
[1]*Institut Européen des Membranes*
[2]*Institut des Nanotechnologies de Lyon*
[3]*Centrale de Technologies en Micro et Nanoélectronique, Université de Montpellier 2*
France

1. Introduction

Enzymatic biofuel cells (BFCs) employ enzymes to catalyse chemical reactions, thereby replacing traditional electrocatalysts present in conventional fuel cells. These systems generate electricity under mild conditions through the oxidation of renewable energy sources (Calabrese Barton et al., 2004). At the anode side the fuel is oxidized and the electrons, which are released by the oxidation reaction, are used to reduce the oxidant at the cathode side (Fig. 1).

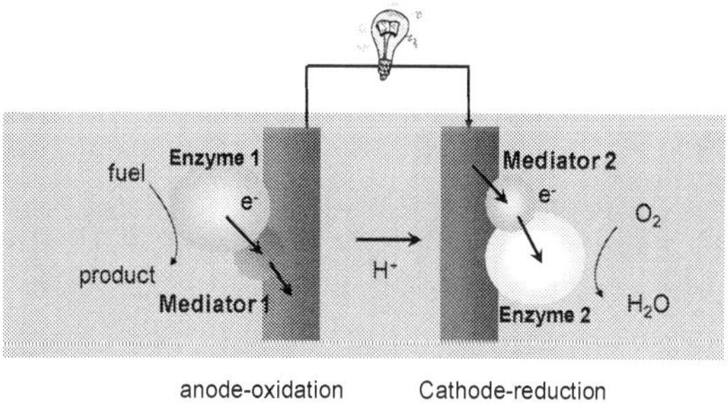

Fig. 1. Enzymatic biofuel cell principle.

Efficient connection is achieved by the use of appropriate redox mediators that are typically dyes or organometallic complexes, responsible for transferring the electrons from the enzymes to the electrode surface. The advantages of biocatalysts are reactant selectivity, activity in physiological conditions at room temperature, and manufacturability, compared to precious metal catalysts. Abundant organic raw materials such as sugars, low aliphatic alcohols, and organic acids can be used as substrates for the oxidation process, and mainly molecular dissolved oxygen acts as the substrate being reduced. The concept of biochemical fuel cell appeared in 1964 with the works of Yahiro and co-workers (Yahiro et al., 1964), which described the construction of a methanol/O_2 cell. In the nineties, BFCs have come in to prominence with the recent advancements in novel electrode chemistries developed by Katz and Willner (Wilson, 2002), and Heller (Degani et al., 1987). The most studied biofuel cell operates with glucose as fuel and oxygen as oxidant (Service, 2002). At the anode, glucose is oxidized to gluconolactone by the enzymes glucose oxidase (GOx) or glucose dehydrogenase, and at the cathode, dioxygen is reduced to water by the enzymes laccase or Bilirubin oxidase (BOD), which are multicopper oxidases.

Typical enzymatic fuel cells demonstrate powers in the range of microwatt to milliwatt. However, the tests are often performed under quite different conditions (concentration, temperature, pH, mass transport conditions, *etc.*), which complicates the comparison between different configurations in literature. Compared to conventional fuel cells, BFCs show relatively low power densities and short lifetime related to enzyme stability and electron transfer rate (Bullen et al., 2006). The improvement of the performance requires optimization of the components and solutions are described in literature in terms of catalysts, enzymatic electrode assemblies and design (Davis et al., 2007; Ivanov et al., 2010).

Nowadays, the explosive growth of portable, wireless consumer electronics and biomedical devices has boosted the development of new micro power sources able to supply power over long periods of time. Miniature BFCs are considered as promising alternative (Gellett et al., 2010) to power supply in wireless sensor networks (WSN). However, the miniaturization of these devices imposes significant technical challenges based on fabrication techniques, cost, design of the device, and nature of the materials. These devices must provide similar performances to larger biofuel cells in terms of efficiency and power density while using less reagents, space and time consumption.

The number of miniaturized biofuel cells mentioned in literature is mainly restricted to a few devices. These devices include conventional devices that have been miniaturized, as well as micro-devices that use completely novel methods of energy conversion. Therefore, the present chapter presents the recent advances in the miniaturization of BFCs. Miniature conventional BFCs will be first presented but, here, we will focus the discussion on the development of microfluidic enzymatic BFCs, where microfluidics plays a direct and essential role. These micro-devices operate within the framework of a microfluidic chip. They exploit the laminar flow of fluids that limits the convective mixing of fuel and oxidant within a microchannel, eliminating the need for a membrane. As a result, the reaction kinetics can be optimized for both the cathode and anode independently by adjusting the composition of the fuel and oxidant stream. A discussion on the parameters affecting the performances of the microfluidic BFCs is proposed and directed towards interesting theoretical and experimental works. Finally, issues that need to be considered are presented to improve microfluidic device performances for desirable solution in the energy conversion process.

2. Miniature BFCs

In this section we briefly describe conventional BFCs that have been miniaturized. The number of miniaturized BFCs mentioned in literature is mainly restricted to a few devices working from glucose and O_2 that have mostly been designed by reducing the electrode size and cell volume. Different strategies have been used to miniature BFCs design.

Heller and co-workers have successfully demonstrated the efficiency of original miniature membraneless BFCs functioning under physiological conditions. They have developed the first handmade miniature device containing only two components, an anode and a cathode of 7-μm diameter and 2-cm long carbon fibers, placed in a polycarbonate support. The anode was modified by GOx and the cathode was modified by either laccase or BOD, within and mediated by redox osmium-based hydrogels (Mao et al., 2003). In 2001, they developed the first miniature membraneless BFC that delivered 140 μW cm^{-2} at 0.4V (Chen et al., 2001). This simple device suggested that the goal of a miniature autonomous sensor–transmitter system could be realistic (Bullen et al., 2006; Heller, 2004). After further developments based on the improved redox polymer connecting the reaction centers of enzymes to the electrodes, the devices delivered higher power densities of 431 μm cm^{-2} at 0.52 V (Mano et al., 2002) and 440 μm cm^{-2} at 0.52 V (Mano et al., 2003), in pH 7.2, 37 °C and 15 mM glucose. The high power density delivered by these devices comes from the cylindrical mass transport at the carbon fibers and the use of efficient redox polymers to transport electron. They also showed that this system produced a power density of 240 μm cm^{-2} at 0.52V when implanted in a living organism, near the skin of a grape. Later, by replacing carbon fibers by engineered porous microwires made of oriented carbon nanotubes, the most efficient glucose/O_2 BFC ever designed was developed (Gao et al., 2010) and delivered high power density of 740 μW cm^{-2} at a cell voltage of 0.57 V. The success of the experiment probably results in the increase of the mass transfer of substrates.

Another miniature devices presently lower performance described stacked biofuel cell designs. One work described a stacking structure composed of six cells connected in series on a chip (Nishizawa et al., 2005), composed of GOx anode and polydimethylsiloxane-coated Pt cathode. The performance of the arrayed cells on the chip was 40 μW cm^{-2} in air-saturated buffer solution containing 5 mM glucose. Another work reported the development of a miniature BFC with a footprint of 1.4 cm^2, by adopting the design of stackable proton exchange membrane (PEM) fuel cells (Fischback et al., 2006). This device consisted of an air-breathing cathode and an enzymatic anode composed of crosslinked GOx clusters on the surface of carbon nanotubes. This study demonstrated the important role of buffer solution in determining the performance and stability of miniature BFCs. In buffered fuel solution the initial performance was very high (371 μW cm^{-2}), but quickly dropped due to a deactivation of the proton exchange membrane. However, in unbuffered solution, the initial performance was lower (117 μW cm^{-2}) due to low pH condition, but its performance was very stable for 10 hours. This work suggested that the use of miniature system and unbuffered fuel solution will be a benefit to practical applications.

Currently, in such miniature devices, current density and delivered power output are mainly limited by the diffusion of fuel to the electrode surface. One interesting innovation to maximize the transport efficiency is to use hydrodynamic flow and to pump the fuel to the electrode.

3. Microfabricated devices

3.1 Advantages of microfluidics

An alternative approach towards the miniaturization of energy conversion devices is the use of microfabrication techniques. Microchemical systems have inherent advantages over macrosystems, including increased rates of mass transfer, low amount of reagents, increased safety as a result of smaller volumes, and coupling of multiple microreactors. Microfluidic techniques are ideal for miniaturization of devices featured with typical scale of channels of submillimeter in height and with laminar flow. Application of microfluidics to fuel cells has been developed rapidly since the years 2000 (Ferrigno et al., 2002; Choban et al., 2004). In such devices, all functions and components related to fluid delivery and removal, reactions sites and electrodes structures are confined to a microfluidic channel. In the channel, as illustrated in Fig. 2, the flow of streams of fuel (colored pink) and oxidant (colored blue) is kept near-parallel, which ensures minimal diffusional mixing between the streams. The only way that molecules in opposite streams can mix is by molecular diffusion across the interface of the two fluid streams. The lack of convective mixing promotes laminar flow of fluids.

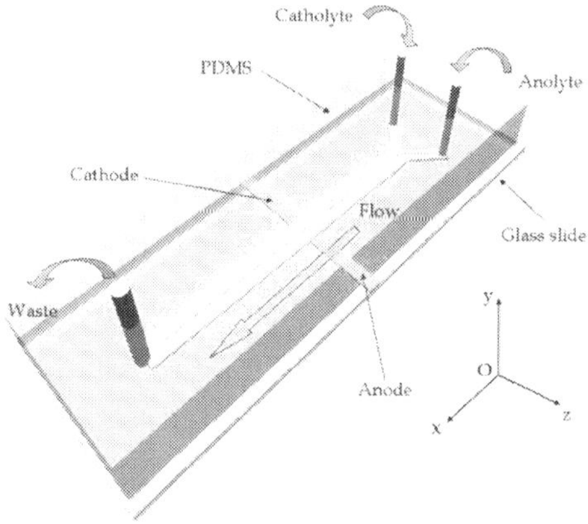

Fig. 2. Laminar flow of streams in a microfluidic channel.

The electrochemical reactions take place at the anode and cathode located within the respective streams, without needing a membrane to minimize the ohmic drop, what maximises the current density. Protons diffuse through the liquid-liquid interface created by the contacting streams of fuel and oxidant. The cathode and the anode are connected to an external circuit. The technique to force the fluid through microchannels is the pressure driven flow, in which the fluid is pumped through the device *via* positive displacement pumps, such as syringe pumps.

As summarized by authors (Luo et al., 2005; Gervais et al., 2006; Sun et al., 2007), the limiting factors in laminar flow-based microfluidic fuel cells that influence the performances are (i) cross-diffusional mixing of fuel and oxidant at the interface between the two streams, and (ii) the formation of depletion boundary layers at the surface of the electrodes as the result of the reaction of fuel and oxidant. Interesting papers have presented theoretical and experimental works to describe how to prevent or reduce these phenomena by concentring research efforts on designs, electronic and ionic conductivity, and electron-transfer kinetics in microfluidic fuel cells (Lee et al., 2007). The role of flow rate, microchannel geometry, and location of electrodes within microfluidic systems was also studied (Choban et al., 2005; Sun et al., 2007; Amatore et al., 2007; Chen et al. 2007).

Similarly to microfluidic fuel cells, advanced microfabrication techniques can be applied to build components of microfluidic enzymatic BFCs. The number of devices presented to date is limited. The devices have been developed based on both laminar flow within a microchannel and biological enzyme strategies. Indeed, the advantage of the co-laminar flow is to choose the composition of the two oxidant and fuel streams independently for optimum enzymatic activity and stability to improve reaction rates and current density (Zebda et al., 2009a).

3.2 Microfluidic biofuel cell fundamentals

In a microfluidic channel, the relationship between the fluid velocity and the absolute pressure for an incompressible viscous liquid is given by the classical fluid dynamics theory and the well-known Navier-Stokes equation:

$$\frac{\partial \vec{v}}{\partial t} + \left(\vec{v} \cdot \vec{\nabla}\right) \vec{v} = -\vec{\nabla}\left(\frac{P}{\rho}\right) + \mu \, \Delta \vec{v} \tag{1}$$

Where $\vec{v}$ stands for the fluid velocity vector with components (u, v, w), each expressed for a set of Euler components (x, y, z, t), P is the absolute pressure, ρ is the relative density, and μ is the kinematic viscosity.

In the case of a microfluidic horizontal straight channel (x-direction), the flow is always laminar under low pressure drop (typically a few bar), leading thus to a unidirectional flow and a uniform absolute pressure in the cross-section. For a fixed pressure drop ΔP between the inlet and the outlet of the channel, Eq. 1 simplifies to:

$$\frac{\partial u}{\partial t} = -\frac{1}{\rho}\frac{\Delta P}{L} + \mu\left(\frac{\partial^2 u}{\partial y^2} + \frac{\partial^2 u}{\partial z^2}\right) \tag{2}$$

Where L is the length of the microchannel. When the permanent flow is reached, the time derivative term becomes zero and Eq. 2 simplifies to:

$$\frac{-\Delta P}{\eta L} + \frac{\partial^2 u}{\partial y^2} + \frac{\partial^2 u}{\partial z^2} = 0 \tag{3}$$

Where η is the dynamic viscosity (10^{-3} Pa.s for water at 20°C), defined as the product of the dynamic viscosity and the relative density ρ. Due to the very large aspect ratio of the rectangular cross-section of the microchannel, a 2D approach is usually considered that leads to a pseudo infinite-plate flow (except in the borders). The directions along the length and height of the microchannel are indicated as x and y coordinates, respectively (see fig. 2). A typical parabolic rate profile is obtained for pressure driven flow:

$$u(y) = \frac{\Delta P}{2.\eta.L}\left(\frac{h^2}{4} - y^2\right) \tag{4}$$

Where h is the height of the microchannel and y is defined as $y=0$ at the middle of the microchannel and $y= \pm$ h/2 at the upper and under walls. By considering a rectangular microchannel (with l the width of the microchannel) in Eq. 4, the flow rate, Q, in laminar regime, is deduced and is proportional to the applied pressure (Eq. 5):

$$Q = \frac{\Delta P.l.h^3}{12.\eta.L} \tag{5}$$

In electrochemical laminar flow systems, the mass transport is achieved by both diffusion and convection transport. In the case of Y-shaped microchannel, the mixing between the two laminar streams occurs by transverse diffusion. Microscale devices are generally characterized by high Péclet number, Pe, ($Pe = U_{av}h/D$, with U_{av} the average velocity of the flow, h the height of the microchannel and D the diffusion coefficient of the molecule). In this condition, the transverse diffusion is much lower than the convection, and the diffusive mixing of the co-laminar streams is restricted to a thin interfacial width, δ_{mix}, in the center of the channel (Fig. 3) that grows as a function of the downstream position (x) and the flow rate, determined from Eq. 6 (Ismagilov et al., 2000):

$$\delta_{mix} = \left(\frac{(Dhx)}{U_{av}} \right)^{\frac{1}{3}} \tag{6}$$

Where D is the diffusion coefficient for ions of type i and U_{av} is the average flow velocity defined as:

$$U_{av} = \frac{\Delta P.h^2}{12.\eta.L}$$

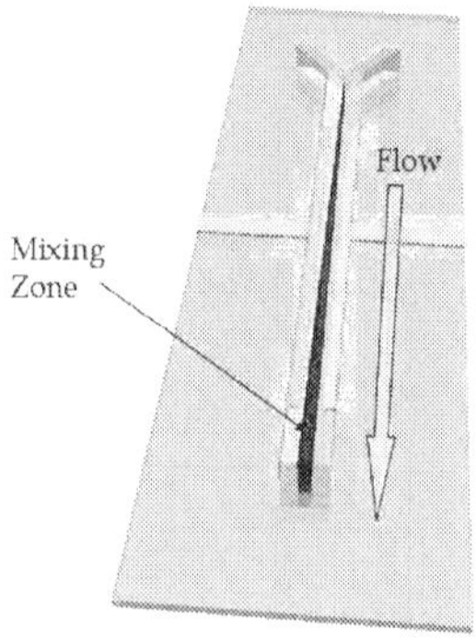

Fig. 3. Schematic of a laminar flow in a microchannel with the formation of the diffusion region during operation of a microfluidic BFC.

For fast electron transfer and in excess of supporting electrolyte, the kinetics of a simple electrochemical redox reaction is controlled by diffusion and convection. The concentration profiles of the chemical species involved in the reaction are determined by solving the convective diffusion equation (Eq. 8):

$$\frac{\partial c_i}{\partial t} + \vec{\nabla}\left(-D_i \vec{\nabla} c_i\right) + \vec{v}\vec{\nabla} c_i + R_i = 0 \tag{8}$$

Where c_i is the concentration of species i, D_i its diffusion coefficient, t the time, $\vec{v}$ the fluid velocity vector (given by Eq. 1) and R_i a term describing the rate of net generation or consumption of species i formed by homogeneous chemical reaction.
In the case of a microfluidic biofuel cell as described in this work, Eq. 8 can be simplified into a 2-dimensionnal cartesian steady state (Eq. 9):

$$-D_i \left(\frac{\partial^2 c_i}{\partial x^2} + \frac{\partial^2 c_i}{\partial y^2} \right) + u(y)\frac{\partial c_i}{\partial x} = 0 \tag{9}$$

The boundary conditions associated are usually: (i) $c_i = c_i^\circ$ (bulk concentration) at the inlet of the microchannel, (ii) $c_i = 0$ at the electrode surface and (iii) no flux at the other walls (no electrochemical reaction).
Those simulations were exploited in order to calculate the diffusive flux at the electrode, defined as:

$$J_{diff}(x) = D\left.\frac{\partial C}{\partial y}\right|_{electrode} \tag{10}$$

And, therefore, the total current is expressed as:

$$I_{tot} = \int_{electrode} n.F.J_{diff}(x)dx = n.F.D. \int_{electrode} (\left.\frac{\partial C}{\partial y}\right|_{y=0})dx \tag{11}$$

Where n is the number of electron exchanged, and F is the Faraday constant.
The pumping power $W_{pumping}$, required to sustain steady laminar flow in the microchannel by the syringe pump, is estimated on laminar flow theory (Bazylak et al., 2005) as the pressure drop multiplied by the flow rate:

$$W_{pumping} = \frac{8.\eta.L.Q^2}{l.h^3} \tag{12}$$

One can note that the contributions from inlet and outlet feed tubes are not included, because they are negligible.
The fuel utilization (FU) is estimated by the following Eq. 13, defined as the current output divided by the flux of reactant entering the channel (Bazylak et al., 2005):

$$FU = \frac{I}{n.F.C.Q} \tag{13}$$

The fuel utilization is maximized for the lowest flow rate and decreases with flow rate. Typical fuel utilization for microfluidic fuel cell is ~ 1% (Hayes et al., 2008).

3.3 Manufacturing technology

3.3.1 Fabrication of the microchannel

Microchannel architecture typically represents a T- or Y-channel configuration. There are mainly two fabrication techniques. The first one utilizes conventional chip-manufacturing techniques of semiconductor industries. Silicon wafers are patterned by lithography step followed by etched step in order to get the desired form of the channel (Moore et al., 2005; Lee et al., 2007). The second technique allows the fabrication of microchannels by rapid prototyping using standard soft lithography procedure to build the channel in poly(dimethylsiloxane) (PDMS) (Duffy et al., 1998). PDMS is relatively inert and compatible with most solvents and electrolytes (Kjeang et al., 2008). Besides it is permeable to gases, which is essential for biofuel cells working from enzymes with oxygen as the cofactor. Typically, a pretreated microscope glass slide or a silicon wafer is coated with a thin layer of photoresist by spin-coating and exposed directly to UV light through a photomask that defines the desired channel structure. Several thick photoresist layers are sequentially laminated on the first layer to get the desired channel depth, and then exposed to UV light. The structure is then developed by spraying an aqueous solution of sodium carbonate (1% wt) and hardened by a final irradiation. The result is a master with a positive pattern defined by the master. The channel structure is thus obtained by pouring PDMS monomer over the master, followed by curing at 70 °C during 2 h. After cooling, the PDMS slab is peeled off from the master, and holes are punched to provide fluid access (Stephan et al., 2007).

3.3.2 Fabrication of the microelectrodes

Most of the microfluidic devices employ patterned electrodes positioned in parallel on the bottom wall or on sides of the channel (Kjeang et al., 2008). Electrodes, with varying length and wide, are patterned by coating glass slides with conductive materials such as gold, graphite over an adhesive layer (often chromium or titanium) by standard sputtering techniques (Zebda et al., 2010) or by photolithography and sputtering (Moore et al., 2005; Lim et al., 2007; Togo et al., 2008). The inter-electrode gap varies between 0.2 mm and 1.4 mm (Lee et al., 2007; Togo et al., 2008; Zebda et al., 2010).

3.3.3 Fabrication of the microfluidic devices

The microfluidic device is finally obtained by physically clamping the PDMS slab with the glass substrate that accommodated the electrode pattern. This approach works well with elastomer polymer like PDMS. Alternatively, an irreversible seal may be achieved between both parts by oxygen-plasma treating prior to improve the adhesion (Lim et al., 2007; Lee et al., 2007). Alignment of the flow channel over the microelectrodes is often aided by a microscope. As an example, a device, consisted of a Y-shaped channel with two inlets and two outlets, is presented in Fig. 4. The pressure-driven laminar flow required for injection of fuel and oxidant is typically driven by a syringe pump via polyethylene tubing.

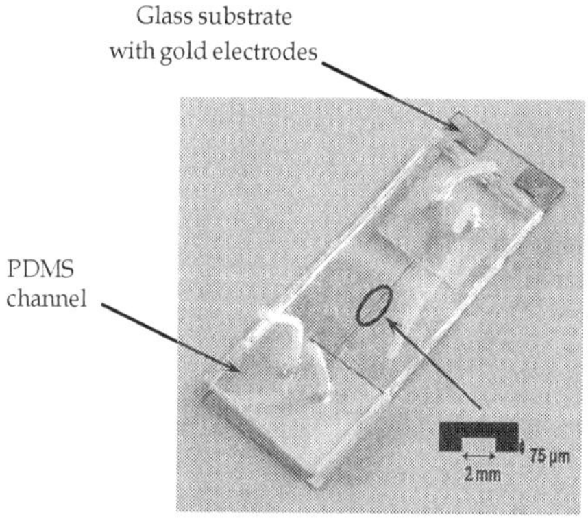

Fig. 4. Schematic microfluidic BFC based on a Y-shaped channel with two inlets and two outlets.

3.4 Performances of microfluidic biofuel cells

This paragraph mainly describes microfluidic BFCs involving mediated monoenzymatic systems, which are capable of only partial oxidation of the fuel. Devices have been

Based on this work, we built a methanol/O_2 microfluidic BFC able to completely oxidize methanol, according to the electron steps described for each electrode in Fig. 10. The enzymes and the mediators were immobilized in poly-L-lysine.
These enzymes are NADH-dependant. The electrochemical connection and regeneration of NADH at the anode is achieved using the enzyme diaphorase and the redox mediator benzylviologen as already described (Palmore et al., 1998). At the anode, a mixture of the three dehydrogenase enzymes was deposited along a gold electrode surface according two configurations (Fig. 11). When enzymes are mixed along the anode, optimal power density is achieved as observed in Fig. 12. The Y-channel device delivers a power density of 70 μW cm^{-2} at a cell voltage 0.25 V.

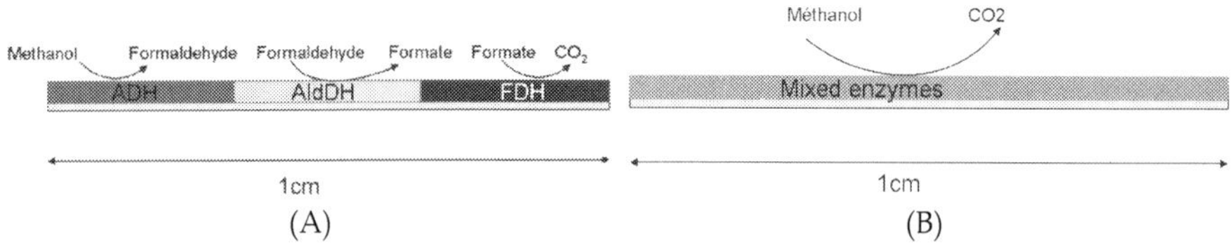

Fig. 11. (A) enzymes are distributed along the anode surface in three distinct, equally sized zones of 10 mm in length, (B) enzymes are mixed randomly along the anode surface of 10 mm in length.

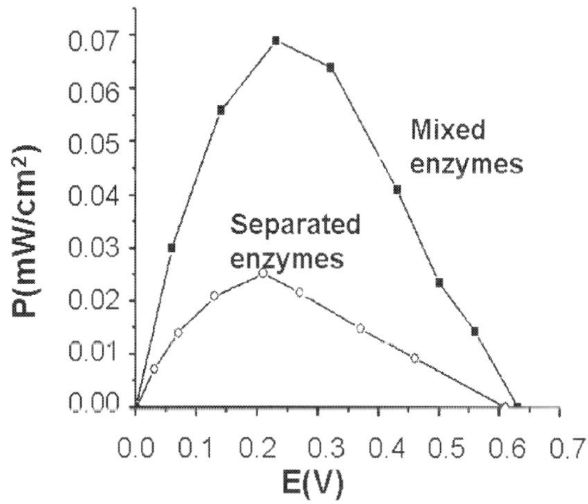

Fig. 12. Power density-voltage plots generated from a microfluidic methanol/O_2 biofuel cell for 2 cases: separated and mixed enzyme patterns at the anode (1:1:1 molecular basis).

4. Conclusion and perspectives

Microfluidic BFCs could be an effective solution for small power sources applications such as biological sensors, implantable medical devices or portable electronics. However significant research efforts must be made for practical applications. Researches must be aimed at identifying most robust and active enzymes, more efficient immobilization environment for enzymes and mediators in microfluidic environment, and at increasing enzyme lifetimes.

To deliver higher power densities, the challenges are in the area of energy density and fuel utilization. Microfluidic BFCs designs should integrate advanced immobilization configurations to improve enzyme performance and high surface area electrodes to enhance rates of convective-diffusive reactant transport. More nanofluidic research and development will be needed to demonstrate the real potential of this form of energy conversion system. Besides, the power output of single microfluidic BFC is inadequate for most practical applications. The enlargement of a single cell by increasing the geometrical electrodes area and the microchannel is constrained by fuel/oxidant crossover and higher ohmic losses. In order to produce adequate power, multiple independent cells could be stacked as in typical fuel cells.

5. Acknowledgment

This work was supported by a CNRS postdoctoral fellowship and by the Project PIE CNRS "Energie" 2010-2013. Nanolyon clean room facilities are acknowledged for the fabrication of the microfluidic devices.

6. References

Addo, P.K., Arechederra, R. L. & Minteer S.D., 2010. Evaluating enzyme cascades for methanol/air biofuel cells based on NAD^+-Dependent Enzymes. *Electroanalysis*, vol. 22, No. 7-8 (April 2010) pp 807-812, ISSN 1040-0397.

Amatore, C.; Da Mota, N.; Della C. & Thouin L. (2007). Theory and experiments of transport at channel microband electrodes under laminar flows. 1. Steady-state regimes at a single electrode, *Analytical Chemistry*, vol. 79, (November 2007), pp. 8502-8510, ISSN 0003-2700.

Barton, S. C. (2005). Oxygen transport in composite mediated cathode. *Electrochimica Acta*, vol. 50, No. 10, (March 2005) pp 2145-2153, ISSN 0013-4686.

Bazylak, A.; Sinton, D. & Djilali, N. (2005). Improved fuel utilization in microfluidic fuel cells: A computational study, *Journal of Power Sources*, vol. 143, No. 1-2, (April 2005), pp 57-66, ISSN 0378-7753.

Bedekar, A.S.; Feng, J.J.; Krishnamoorthy, S.; Lim, K.G.; Palmore, G.T.R. & Sundaram, S. (2008) Oxygen limitation in microfluidic biofuel cells, *Chemical Engineering Communication* 195 pp. 256–266, ISSN 0098-6445.

Bullen, R.A.; Arnot, T.C.; Lakemanc, J.B. & Walsh, F.C. (2006). Biofuel cells and their development. *Biosensors & Bioelectronics*, vol. 2, No. 11, (January 2006), pp. 2015–2045, ISSN 0956-5663.

Calabrese Barton, S.; Gallaway, J. & Atanassov, P. (2004). Enzymatic Biofuel Cells for Implantable and Microscale Devices. *Chemical Reviews*, Vol. 104, No. 10, (october 2004), pp. 4867-4886, ISSN 0009-2665.

Chen, T.; Barton, S.C.; Binyamin, G.; Gao, Z.; Zhang, Y.; Kim, H. & Heller, A. (2001). A Miniature Biofuel Cell, *Journal of the American Chemical Society*, vol. 123, No.35, (august 2001), pp 8630-863, ISSN 0002-7863.

Chen, F.; Chang, M.H. & Lin M.K. (2007). Analysis of membraneless formic acid microfuel cell using a planar microchannel, *Electrochimica acta* Vol. 52, No. 25, (September 2007), pp. 2506-2514, ISSN 0013-4686.

Choban, E.R.; Markoski, L.J.; Wieckowski, A. & Kenis, P.J.A. (2004). Microfluidic fuel cell based on laminar flow, *Journal of Power Sources*, vol. 128, No. 1, (march 2004), pp. 54-60, ISSN 0378-7753.

Choban, E.R., Waszczuk, P. & Kenis P.J.A. (2005). Characterization of limiting factors in laminar flow-based membraneless microfuel cells. *Electrochemical and Solid-State letters*, vol. 8, No. 7, pp. A348-A352, ISSN 1099-0062.

Davis, F. & Higson, S.P.J. (2007). Biofuel cells—Recent advances and applications. *Biosensors & Bioelectronics*, vol. 22, No. 7,(February 2007) pp 1224–1235, ISSN 0956-5663.

Degani Y.& Heller A. (1987). Direct electrical communication between chemically modified enzymes and metal- electrodes. 1. Electron-transfer from glucose oxidase to metal electrode via electron relays, bound covalently to the enzymes, *Journal of Physical Chemistry, vol.* 91, (March 1987), pp. 1285-1289, ISSN 0022-3654.

Duffy D. C.; Cooper McDonald J.; Schueller O. J. A & Whitesides G. M. (1998) Rapid Prototyping of Microfluidic Systems in Poly(dimethylsiloxane), *Analytical Chemistry* 70, pp. 4974-4984, ISSN 0003-2700.

Ferrigno, R.; Stroock, A.D.; Clark, T.D.; Mayer M. & Whitesides M. (2002). Membraneless Vanadium Redox Fuel Cell Using Laminar Flow, *Journal of American Chemical Society*, vol. 124, No. 7, (October 2002), pp. 12930-12931, ISSN 0002-7863.

Fischback, M.B.; Youn, J.K.; Zhao, X.; Wang, P.; Park, H.G.; Chang, H.N.; Kim, J. & Haa, S. (2006). Miniature biofuel cells with improved stability under continuous operation, *Electroanalysis*, vol. 18, No.19-20 (july 2006), pp 2016-2022, ISSN 1040-0397.

Gao, F.; Viry, L.; Maugey, M.; Poulin, P. & Mano, N. (2010). Engineering hybrid nanotube wires for high-power biofuel cells. *Nature Communications*, Vol. 1, No. 1, pp. 1-7, ISSN 2041-1723.

Gellett, W.; Kesmez, M.; Schumacher, J.; Akers N. & Minteer, S.D. (2010). Biofuel Cells for Portable Power. *Electroanalysis* , vol. 22, No. 7-8, (April 2010) pp 727-731, ISSN 1040-0397.

Gervais, T. & Jensen, K.F. (2006) Mass transport and surface reactions in microfluidic systems, *Chemical Engineering Science*, vol. 61, (February 2006), pp. 1102-1121, ISSN 0009-2509.

Hayes, J.R.; Engstrom, A. M. & Friesen, C. (2008). Orthogonal flow membraneless fuel cell. *Journal of Power Sources*, vol. 183, No. 1, (august 2008), pp. 257-259, ISSN 0378-7753.

Heller, A. (2004). Miniature biofuel cells. *Physical Chemistry Chemical Physics*, Vol. 6, No. 2, (January 2004) pp. 209-216, ISSN 1463-9076.

Ismagilov R.F.; Stroock A.D.; Kenis P.J.A.; Whitesides G. & Stone H.A. (2000). Experimental and theoretical scaling laws for transverse diffusive broadening in two-phase laminar flows in microchannels. *Applied Physics Letters*, Vol. 76, (April 2000), pp. 2376-2378, ISSN 0003-6951.

Ivanov, I., Vidaković-Koch, T. & Sundmacher, K. (2010). Recent Advances in Enzymatic Fuel Cells: Experiments and Modeling. *Energies*, Vol. 3, No. 4, (April 2010), pp. 803-846, ISSN 1996-1073.

Jayashree R. S.; Gancs L.; Choban E. R.; Primak A.; Natarajan D.; Markoski L. J. & Kenis P. J. A. (2005) Air-Breathing Laminar Flow-Based Microfluidic Fuel Cell, *Journal of the American Chemical Society*, 127, 48 (December 2005) pp 16758–16759, ISSN 0002-7863.

Kjeang E.; Sinton D. & Harrington D.A. (2006) Strategic enzyme patterning for microfluidic biofuel cells. *Journal of Power Sources* vol.158, No. 1, (July 2006) pp. 1–12, ISSN 0378-7753.

Kjeang E.; Michel R.; Harrington D. A.; Djilali N. & Sinton D. (2008) A Microfluidic Fuel Cell with Flow-Through Porous Electrodes, (2008) *Journal of the American Chemical Society* 130 (March 2008) pp. 4000-4006, ISSN 0002-7863.

Kjeang, E.; Djilali, N. & Sinton, D. (2009). Microfluidic fuel cells: a review, *Journal of Power Sources*, vol. 186, (October 2008), pp 353-369, ISSN 0378-7753.

Lee, J.; Lim, K.G.; Palmore, G.T.R. & Tripathi, A. (2007). Optimization of microfluidic cells using transport principles, *Analytical Chemistry*, vol. 79, No. 19, (October 2007), pp 7301-7307, ISSN 0003-2700.

Lim K.G. & Palmore G. T. R. (2007) Microfluidic biofuel cells: the influence of electrode diffusion layer on performance, *Biosensors & Bioelectronics* 22 (January 2007) pp. 941-947, ISSN 0956-5663.

Luo, T.J.M.; Fei, J.F.; Lim, K.G. & Palmore, G.T.M. (2005). Membraneless fuel cells: an application of microfluidics. In: ACS Symposium Series 890, 2005.

Mano, N.; Mao, F. & Heller, A. (2002). A Miniature Biofuel Cell Operating in A Physiological Buffer. *Journal of the American Chemical Society*, Vol. 124, No. 44, (November 2002) pp. 12962-12963, ISSN 0002-7863.

Mano, N.; Mao, F. & Heller, A. (2003). Characteristics of a miniature compart-mentless glucose-O_2 biofuel cell and its operation in a living plant, *Journal of the American Chemical Society*, vol.125, No.21, (april 2003), pp 6588-6594, ISSN 0002-7863.

Mao, F.; Mano, N. & Heller, A. (2003). Long Tethers Binding Redox Centers to Polymer Backbones Enhance Electron Transport in Enzyme "Wiring" Hydrogels. *Journal of the American Chemical Society*, Vol. 125, No. 16, pp. 4951-4957, ISSN 0002-7863.

Moore, C.M.; Akers, N. L.; Hill, A. D.; Johson, Z. C. & Minteer, S.D. (2004). Improving the environment for immobilized dehydrogenase enzymes by modifying Nafion with tetraalkylammonium bromides. Biomacromolecules, vol. 5, No. 47, (July-August 2004), pp 1241-1247, ISSN 1525-7797.

Moore, C.M.; Minteer, S.D. & Martin, R.S. (2005). Microchip-based Ethanol/Oxygen. Biofuel Cell, *Lab on Chip*, vol. 5, No. 2,pp 218-225, ISSN 1473-0189.

Nick, A.; Akers L.; Moore C. M. & Minteer S. D. (2005) Development of alcohol/O_2 biofuel cells using salt-extracted tetrabutylammonium bromide/Nafion membranes to immobilize dehydrogenase enzymes, *Electrochimica Acta*, Volume 50, Issue 12, (April 2005) pp. 2521-2525, ISSN 0013-4686.

Nishizawa, M.; Togo, M.; Takamura, A. & Abe, T. (2005). Enzyme-based fuel cells for biomedical microdevices, Proceedings of the fifth international workshop on micro and nanotechnology for power generation and energy conversion applications, Takeda Hall, University of Tokyo, November 2005.

Palmore G. T. R.; Bertschy H.; Bergens S. H. & Whitesides G. M. (1998). A methanol/dioxygen biofuel cell that uses NAD+-dependent dehydrogenases as catalysts: application of an electro-enzymatic method to regenerate nicotinamide adenine dinucleotide at low overpotentials . *Journal of Electroanalytical Chemistry*, 443 (February 1998),pp.155–161, ISSN 0022-0728.

Service, R.F. (2002). Shrinking fuel cells promise power in your pocket, *Science*, 296 (May 2002), pp 1222-1224, ISSN 0036-8075.

Sokic-Lazic D.; Arechederra R. L.; Treu B. L. & Minteer S. D. (2010). Oxidation of Biofuels: Fuel Diversity and Effectiveness of Fuel Oxidation through Multiple Enzyme Cascades, *Electroanalysis* 22, No. 7-8, (April 2010) pp. 757 - 764, ISSN 1040-0397.

Stephan K.; Pittet P.; Renaud L.; Kleimann P.; Morin P.; Ouaini N. & Ferrigno R. (2007) Fast prototyping using a dry film photoresist: microfabrication of soft-lithography masters for microfluidic structures, J. Micromech. Microeng. 17 (October 2007) N69–N74, ISSN 0960-1317.

Sun, M.H.; Casquillas, G.V.; Guo, S.S.; Shi, J.; Ji, H.; Ouyang, Q. & Chen, Y. (2007). Characterization of microfluidic fuel cell based on multiple laminar flow. *Microelectronic Engineering*, vol. 84, (May 2007), pp. 1182-1185, ISSN 0167-9317.

Togo M.; Takamura A.; Asai T.; Kaji H. & Nishizawa M. (2007) An enzyme-based microfluidic biofuel cell using vitamin K3-mediated glucose oxidation, *Electrochimica Acta*, Volume 52, Issue 14, (April 2007) pp. 4669-4674, ISSN 0013-4686.

Togo, M.; Takamura, A.; Asai, T.; Kaji, H. & Nishizawa, M. (2008). Structural studies of enzyme-based microfluidic biofuel cells, *Journal of Power Sources*, vol. 178, (December 2007), pp 53-58, ISSN 0378-7753.

Wilson, G.S. (2002). *Bioelectrochemistry*, Encyclopedia of electrochemistry, vol 9, John Willey, 2002, ISBN: 3527304010, Weinheim.

Yahiro A. T.; Lee S. M. & Kimble D. O. (1961). Bioelectrochemistry .I. Enzyme Utilizing Bio-Fuel cell studies, *Biochimica Biophysica Acta*, vol. 88, pp. 375, ISSN 0006-3002.

Yoon S.K.; Fichtl G. W. & Kenis P.J.A. (2006) Active control of the depletion boundary layers in microfluidic electrochemical reactors, *Lab On a Chip* 6 (December 2006) pp. 1516-1524, ISSN 1473-0197.

Zebda, A.; Renaud, L.; Tingry, S; Cretin, M.; Pichot, F.; Ferrigno, R. & Innocent, C. (2009a). Microfluidic Biofuel Cell for Energy Production. *Sensor* letters, vol. 7, No. 5, (October 2009), pp. 824-828, ISSN 1546-198X.

Zebda, A.; Renaud, L.; Cretin, M.; Pichot, F.; Innocent, C.; Ferrigno, R. & Tingry, S. (2009b). Electrochemical performance of a Glucose/Oxygen microfluidic biofuel cell., *Journal of Power Sources*, vol. 193, No. 2, (may 2009), pp 602–606, ISSN0378-7753.

Zebda, A.; Renaud, L.; Cretin, M.; Innocent, C.; Ferrigno, R. & Tingry, S. (2010). Membraneless microchannel glucose biofuel cell with improved electrical performances, *Sensors and Actuators B*, vol. 149, No. 1, (august 2010), pp 44-50, ISSN 0925-4005.

25

Energy Paths due to Blue Tower Process

Kiyoshi Dowaki
Tokyo University of Science
Japan

1. Introduction

In order to solve the environmental problem, decarbonisation has received worldwide attention. In particular, hydrogen energy, which is produced from renewable energy resources of wind power, hydropower, photovoltaic, solar thermal power, geothermal power and biomass, is the most promising. The fuels from biomass feedstock would be electricity and/or thermal energy through typical incineration system (e.g. gas-engine or gas-turbine), or the secondary energy of hydrogen, methanol and/or dimethyl-ether (DME), which are so-called "biomass to liquid" fuels (BTL).

Actually, the biomass energy system is one of the environmentally friendly energy systems which will contribute to the global warming protection. However, there would be somewhat problems at the time when the eco-friendly systems will be promoted to the target area. For instance, it is known well that the resource might cause competition with the biomass as food. Also, even if there are abundant material sources, the products through the system would be costly in comparison to the conventional one due to the characteristics of low energy density.

According to the recent interviews, the cost problem is extremely significant and the market abilities of products are important from the viewpoint of the sustainable business management, too. When we consider the promotion of eco-friendly system, the operational condition on the eco-business, the technological barriers and the environmental contribution effect have to be absolutely estimated before we decide to execute the project or not. That is, we need to create the suitable energy paths and the business model.

In our previous studies, we focused on the biomass gasification system whose scale is not conventional scale in the chemical industrial sector, and we concluded that the comprehensive arguments including the application example through the system were required. This means that the suitable plant scale is necessary due to the low energy density of biomass feedstock. On the promotion of new technology, the public support due to central or local government at the earlier stage, and the good business scheme based on the emission trading or the increased awareness of willingness to pay for eco-products would be required in order to operate the plant operation business, too.

Thus, in this section, we propose the several energy paths on the biomass utilization from the viewpoints of “energy”, “environment” and “economy”. Especially, we consider the biomass energy system due to the Blue-Tower process which will be able to produce electricity and thermal energy by the co-generation unit (ex. gas-engine, gas-turbine or fuel cell), to purify Bio-H_2 through PSA (pressure swing adsorption) unit.

So far, the focus has been on the use of biomass energy systems, such as the biomass gasification process, which have mainly been developed in Europe. Above all, we think that the Blue Tower (BT) process, which is developed by D.M.2 Projekt GmbH, is close to being realized as a commercial plant. Our group of Tokyo University of Science (the Dowaki laboratory) including the Japanese company (Japan Planning Organization Inc.) which purchased the licence of the patent on basic BT process has developed the applied technology through the results of the demo-plant, the lab-scale experiments or the investigations from all viewpoints.
Next, we describe the future projection of eco-friendly fuels supply as follows.
Since the Kyoto protocol has come into effect in 2005, the technology on CO_2 emissions mitigation has developed significantly. In particular, the biomass energy system, that is, the energy system using biomass materials such as wood, organic waste, sewage and similar material, is one of the promising energy systems to abate CO_2 emissions. Although gasification technologies have attracted attention from all over the world, as one of the future technologies, their energy efficiencies are quite different, due to the energy conversion methodologies. Under these circumstances, the energy production cost is likely to still be high, even if related subsidies, environmental regulations and/or policies are put in place to support them.
In the early stages, it was thought that the biomass energy system, through gas-engine cogeneration, would be a mainstream. However, since there is a discrepancy between the demand and the supply, that is, since the produced energy cannot be stored, the energy of electricity and/or heat would be excessive. Consequently, in such a system, CO_2 emissions mitigations could not be realized, and the energy production cost would be expensive.
With this as its background, the Japanese government has tried to promote the biomass energy system as an effort to solve global warming. As an example, the Ministry of Economy, Trade and Industry (METI) predicts the biomass energy use of 39.1 PJ in the future plan of energy demand and supply (METI, 2007). The Ministry of Agriculture, Forestry and Fisheries (MAFF) also predicts biomass energy use of 110.5 to 210.1 PJ in the future energy demand and supply plan of 2010. Especially, MAFF has a project to produce BTL fuel of 5.0×10^8 L in the transportation sector (MAFF, 2006).
On the other hand, the annual amount of hydrogen gas that was consumed in 2005 was 1.5 to 2.0×10^{10} Nm^3 in Japan. Half of this hydrogen energy was produced from fossil fuel (K.K. Gas Review, 2007). For 2015, the many Japanese car makers have effort to promote the fuel cell vehicles (FCVs). Hence, hydrogen energy from renewable energy resources is important from the viewpoint of CO_2 emissions mitigation.
Under the above circumstances, we understood that there was good potentiality to produce Bio-H_2 fuel. However, we have to indicate the successful business model at the earlier stage in order to expand the eco-business using this gasification process. It might be necessary to concentrate the concrete application system through BT process.
Next, we considered the acceptability for the related facilities in agriculture or a forestry field. Because the environmentally friendly system such as a PV system or a fuel cell co-generation system is still not enough to be promoted for those facilities. That is, there would be potential to combine the biomass energy system which is environmentally friendly with the agriculture related facilities. In addition, MAFF contribute to the global warming protection through the carbon-footprint of agricultural products. The ministry has a few subsidy menus on the promotion of the system. Also, on the surplus energy of electricity, there are institutions by which the energy companies are obliged to purchase them with

additional fees. Using the above institutions and/or subsidy menus under the leadership of the Japanese government, we considered the agricultural products harvesting system in which the biomass gasification (BT) process with gas-engine or fuel cell is assumed to be introduced.

In this section, we propose the several energy paths. For the entire system design, we analysed CO_2 emissions and/or energy intensities due to Life Cycle Assessment (LCA) methodology. This assessment follows ISO 14041 guidelines. For each path, we defined the system boundary. The system boundary includes the entire life cycle of each energy input (electricity/thermal energy), including the pre-processing process, the energy conversion process etc. That is, for the concrete energy system, we executed the process design which is based on the simulator using the experimental results, and estimated the energy efficiency, the CO_2 intensity, the energy cost including the business model.

Here is the schematic design of this section in Fig.1. As we mentioned them before, we introduce the biomass energy system of Blue Tower technology and explain about the environmental system analysis in consideration of the eco-business operation.

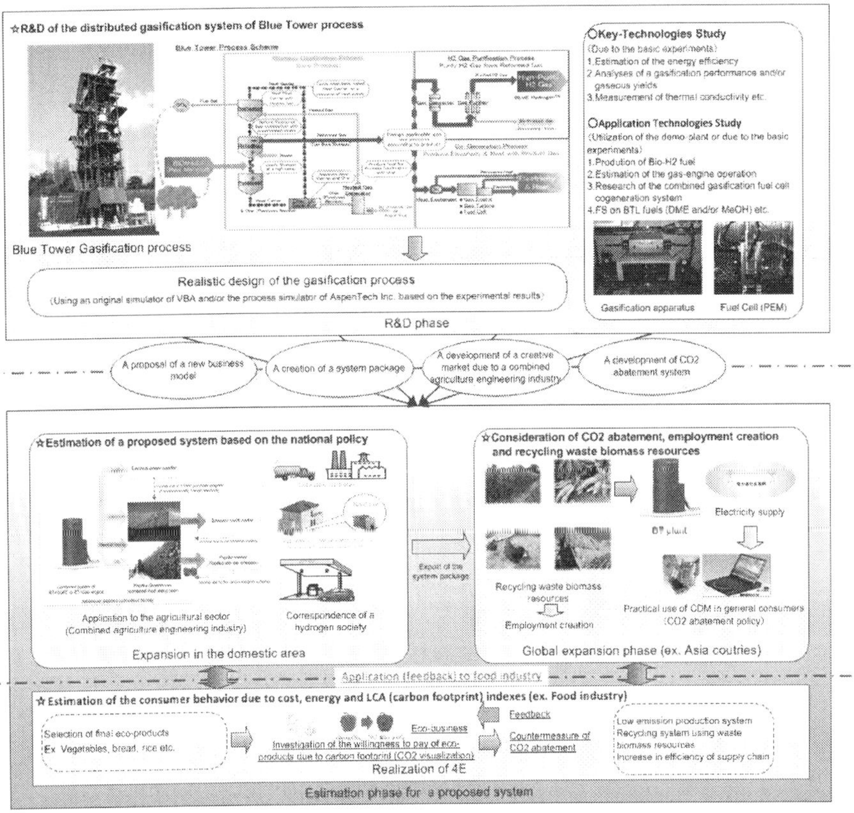

Fig. 1. Outline of biomass energy system through Blue Tower process

2. Blue Tower technology

The characteristics of the BT process are similar to the indirect gasification process with a pyrolyzer. This gasifier has three reactors for the preheating process, the reforming process and the pyrolysis process. The gasifier has the following characteristics; 1) pyrolysis reactions take place in the pyrolyzer, 2) pyrolysis gases are reformed by H_2O (steam), 3) heat required for their reactions is supplied by combustion of off-gas, Tar and Char. Additional heat through a chip boiler might be required in order to accelerate the reactions well. With regard to the gasification performance, the gaseous yield and concentration are dependent upon the kinds of materials, the operating temperature, and the inner pressure (Mühlen et al., 1999, Mayer et al., 2004). Here, in order to evaluate the reliability of each fuel production, we fabricated an apparatus, which included the concept of the BT process (a pyrolyzer and a reformer), and we executed the basic chemical experiments using the biomass samples with the size of 2-3 mm. In the experiments, we measured the syngas components and estimated the equilibrium constants, adjusting the temperature condition and/or steam-carbon ratio (S/C). Also, we ensured the reliability of our simulator which was available for some analyses on the energy cost and/or CO_2 intensity etc. comparing our calculation to the demo-plant data. Assuming that the materials chopped at the size of 20-30 mm are fed into the reactor, H_2 of 54.4 vol.% and CO of 24.4 vol.% were generated at 950°C and Steam/Carbon=1.0 due to our simulator (Dowaki et al., 2007). On the other hand, we executed the studies in order to confirm the absolute proof of the chemical equilibrium reactions, and/or the heat balance in use of the experimental apparatus and/or the demo-plant (1t/d scale) at Izumo, shimane prefecture in Japan. Likewise, the studies on the handling of equipment (the plant operation) have been done at Izumo. In the previous studies, we made the simulator of BT process in order to estimate the operational performance. This simulation program used the parameters estimated by the experimental results of a room condition. Also, the estimation accuracy due to the simulator was analysed. For instance, Kameyama et al. compared the operational result of the demo-plant to that of the simulator. Accordingly, we made sure that the simulated data were corresponding with the practice data to some extent (Kameyama et al., 2010).
Here, we describe the system outline and the performance characteristics of this process which was evaluated through the demonstrated operation with 1t/d plant (see Fig. 2) of the Blue Tower gasification process.

2.1 BT process simulator due to the basic experiments

Based on the basic experiment on the pyrolysis and the reforming, we have developed the simulator by which the gaseous yields and/or the energy efficiency through BT process can be estimated. Here, we compared the practice data through the demo-plant with the result of the simulator.

In general, there would be somewhat deviation between the practice data and the estimated one due to the simulator. That is, it would be extremely significant to identify the deviation from the viewpoint of the reliable plant operation.

The calculation logic of the simulator which we developed is as follows (Dowaki et al. 2007):

1. The reaction temperature in each furnace (pyrolyzer and reformer) and the steam feeding rate are fixed.
2. Based on the gaseous yields in the pyrolysis and/or the reforming reactions, which were analysed by gas chromatograph (GC-8A, Shimadzu), the gaseous components in

each furnace were estimated due to the following two equilibrium reactions. In our experiments, Shincarbon-ST was used as a measure column, and the six kinds of gases which can be measured include H_2, CO, CO_2 CH_4, C_2H_4 and C_2H_6, respectively.

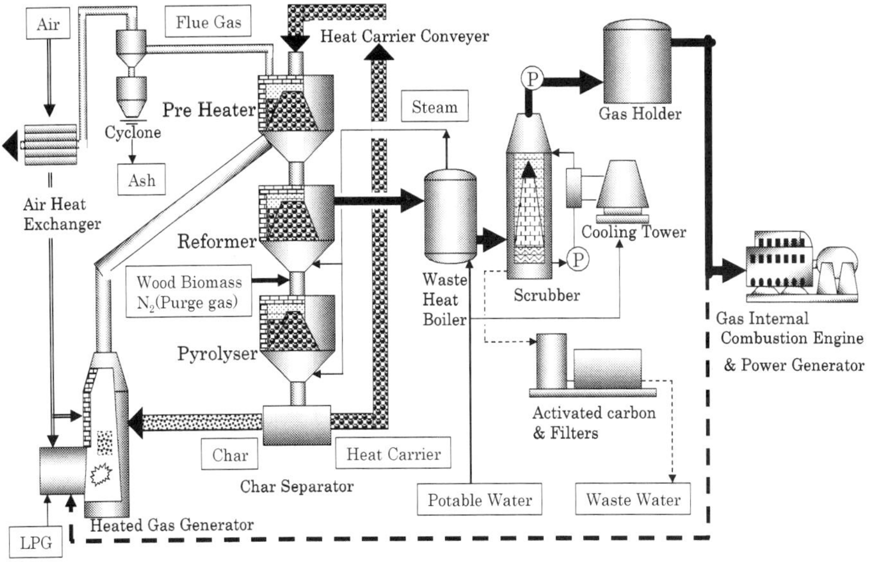

Fig. 2. Schematic design of 1t/d Blue Tower plant

Note that we considered the approach temperature difference between the theoretical one and the reaction one on these equilibrium reactions.
Here, in the gasification process after the pyrolysis reaction, it is usually thought that the following two reactions take place.

$$CH_4 + H_2O \leftrightarrow CO + 3H_2 \tag{1}$$

$$CO + H_2O \leftrightarrow CO_2 + H_2 \tag{2}$$

Using our experimental apparatus, the approach temperature differences between theoretical temperature and the actual temperature was measured. This temperature difference is known as the approach temperature.
In Eqs. (1) and (2), the equilibrium constant of shift reaction and that of methanation reaction are represented as follows:

$$K_s = \exp\left(\frac{-\Delta G^\circ(T-\Delta T)}{RT}\right) = \frac{P_{CO_2}P_{H_2}}{P_{CO}P_{H_2O}} \tag{3}$$

$$K_m = \exp\left(\frac{-\Delta G^\circ(T-\Delta T)}{RT}\right) = \frac{P_{CH_4}P_{H_2O}}{P_{H_2}{}^3 P_{CO}} \tag{4}$$

Where, ΔT, R and P_i are the approach temperature [K], gas constant [J/molK] and the partial pressure of i-component [Pa], respectively.
Next, we compared the measurement data in Izumo plant (1t/d) with the estimated results using the simulator (see Table 1). Here, the tasks of Run 11 and Run 11.2 in which the decomposed reactions are assumed to be completed in the pyrolyzer are described. Both tasks were executed on March, 2009 (Kameyama et al. 2010).

	Deviatoin [-]	Temparature at Refomer [°C]	
		Measured	Estimated
Run 11.1	0.134	820	801
Run 11.2	0.367	750	774

Table 1. Comparison of the measured data and the simulated ones

In this verification, we focused on the molar fractions of H_2, CO, CH_4 and CO_2, and found the average reaction temperature so that the total deviation on molar fraction between the measurement data and the estimated one due to Eqs. (3) and (4) is a minimum. That is, we investigated if the gaseous components based on the temperature which was measured in the plant corresponded to the estimated ones due to the simulator. The reason why we verified the gaseous components using the temperature as a variable is as follows; in the demo-plant, we did not know the temperature profile on the vertical and/or horizontal directions precisely since the sampling point of the temperature in reformer is one position. Thus, assumed that the estimated temperature based on the measured gaseous components would represent the average one of reformer, we made sure that the process simulator would be more suitable.
Next, based on the process simulator which has a precision to some extent, we describe the example due to the biomass feedstock of waste Japanese cedar. With regard to the gasification performance, since gaseous yields and concentrations are dependent upon the kind of materials, the operating temperature, and the inner pressure, they were examined using the gasifier apparatus which has a reformer and a pyrolyzer.
Here, Table 2 shows the ultimate analysis of the waste Japanese cedar.

C*	46.660%	wt.%
H*	5.480%	wt.%
O*	47.351%	wt.%
S*	0.000%	wt.%
N*	0.120%	wt.%
Cl*	0.000%	wt.%
Ash*	0.389%	wt.%
HHV*	18,348	kJ/kg
Moisture Content	20.0	wt.%
Volatile Matter	86.21%	wt.%
Bulk density	0.14	t/m^3

※Dry-Base

Table 2. Ultimate analysis of the waste Japanese cedar

Through the tests, the syngas components and the equilibrium constants were obtained. For instance, Fig.3 illustrates the gaseous yields on the pyrolysis at 550 °C with variation of S/C =0.14 to 0.98, and the reforming reaction at S/C=1.0 with variation of 800 to 950 °C, respectively. Here, a steam carbon ratio is defined as the following equation.

$$S/C\left[mol-H_2O/mol-C\right]=\frac{\text{Added Steam [mol/s]}+\text{Moisture [mol/s]}}{\text{Carbon Content of Material [mol/s]}} \tag{5}$$

Note that the gaseous components are modified at 20% moisture content. Also, the approach temperature for each reaction is shown as Table 3.

Reaction	ΔT	Unit
Pyrolysis	78.3	°C
Reforming	252.0	°C

Table 3. Approach temperature for each reaction (estimated)

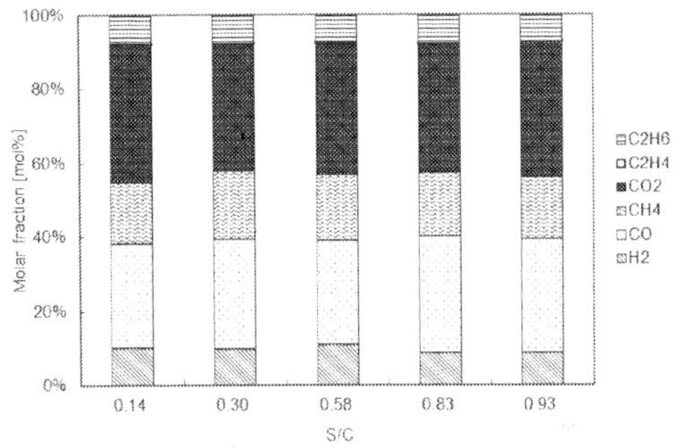

a) Pyrolysis (550 °C, S/C=0.14-0.93)

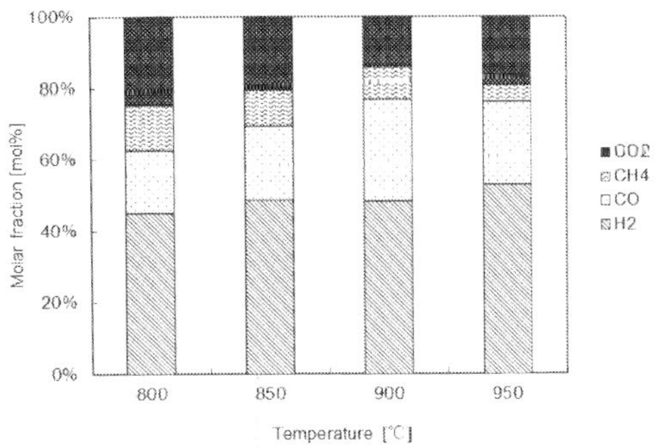

b) Reforming reaction (800-950 °C, S/C=1.0)

Fig. 3. Gaseous yields of pyrolysis (a) and reforming reaction (b).

Based on the above experimental results, we estimated the following material balance:

$$\begin{aligned}&C_{31.078}H_{43.495}O_{23.677}N_{0.069}+31.078H_2O\rightarrow\\&24.052H_2+11.682CO+2.435CH_4+9.442CO_2+23.228H_2O\\&+2.92\times10^{-2}N_2+4.40\times10^{-5}NH_3+C_{2.686}H_{0.482}O_{0.343}N_{0.004}\end{aligned}\tag{6}$$

$C_{2.686}H_{0.482}O_{0.343}N_{0.004}$ is the chemical component of char, and its heating value was 32.0 MJ/kg. In our simulator, the energy performance would be solved so that the input and the output on heat and materials would be balanced.

Next, using ϕ=9.5mm ball, we measured the temperature profiles at the surface of ball and the center of it. In the phase of absorption of heat, the ball was kept at each designed temperature between 200 and 950 °C. At the time, there was difference between the surface temperature and the center one, and the temperature differences were measured. Inversely, in the phase of heat radiation, the ball was heated up to 1,000 °C in the furnace, and it was put in a room temperature. Simultaneously, the temperature differences were measured. Note that these temperature profiles are time series data.

As a result, the thermal conductivities can be obtained. Also, since the thermal circulation time has to be the same as the reacting time on a pyrolysis and a steam reforming reaction, the optimal size of the ball is decided. Thus, the adequate auxiliary power for the circulation of HC would be obtained. Due to this result, we can estimate the suitable residence time in each reactor for the temperature profile which would be led by the simulator. Based on the above concept, we could estimate the syngas through BT process (Dowaki et al., 2008a, Dowaki, 2011a).

2.2 Process design of energy production system through BT process

Next, we introduce the examples of process design through BT process. As we mentioned before, there would be many energy paths through BT process. Here, as the examples, H_2 production and Cogeneration system (CGS) would be concentrated. The purpose of each process design would be due to the energy analysis and/or the environmental one using LCA methodology.

2.2.1 Case study of Bio-H_2 production system

Through a reaction process based on superheated steam, the biomass is converted to the syngas with a high concentration of H_2. In the BT process, pyrolysis gases are reformed with H_2O (steam), and Tar and Char are generated as co-products. Since Tar contents pass through the higher temperature zone, the residual volume would be negligible. Also, due to the recycling of the sensible heat of syngas, the total efficiency of the entire system would be improved.

Here, the process design of Bio-H_2 was executed by the consideration of basic experimental results.

The capability of the biomass gasification plant is 12 t/d, and the annual operation days are 300 day/yr. In the process design, the heat energy generated from the gasifier was assumed to be utilized as the energy for materials dryer. Due to the recycling of thermal energy, the energy of dryer can be reduced at most. For instance, the moisture content can be compensated up to 42 wt.% against the initial moisture content of 50 wt.%. The syngas generated through BT gasifier is transferred to the shift-reaction convertor, and then is fed

into PSA (Pressure Swing Adsorption). In the PSA, the high concentrated H_2 gas was purified to 99.99Vol.% (4N) of H_2 gas.
Here, Tables 4 shows the performance of Bio-H_2 production system. In Tables 4, the cold gas efficiency η_{Cold} is defined as follows:

$$\eta_{Cold} = \frac{Syngas\,[\mathrm{MJ/h}]}{Feedstock[\mathrm{MJ/h}] + Char[\mathrm{MJ/h}] + Offgas[\mathrm{MJ/h}]} \tag{7}$$

Also, the total efficiency η_{Total} of this system is

$$\eta_{Total} = \frac{\text{Bio-H}_2\,[\mathrm{MJ/h}]}{Feedstock[\mathrm{MJ/h}]} \tag{8}$$

BT Process (15 t/d)	Feedstock	635.9 8,415	kg/h MJ/h
	Syngas (For Bio-H_2)	678.5 4,544	Nm^3/h MJ/h
	Cold-Gas Eff. η_{Cold}	62.0%	LHV-%
	Auxiliary Power	247.4	kW
PSA (4N-H_2)	Bio-H_2	303.5 3,275	Nm^3/h MJ/h
	Total Eff. η_{Total}	38.9%	LHV-%

Table 4. Performance of Bio-H_2 production system (estimated)

2.2.2 Case study of Cogeneration system

Next, we explained about the co-generation system by which electricity and thermal energy can be generated. In the case of BT-CGS, due to the heat balance, the reaction energy in the furnace might be shortage. Thus, the additional feedstock would be necessary. In the case of Bio H2 production system, since off-gas through PSA is available, the additional biomass material is not required.
Also, from the viewpoint of the economic condition, the case that the additional one is fed into BT would be much better in comparison to the case without any feedstock. That is, more products (i.e. electricity and/or thermal energy) can be generated. Consequently, the economic condition of BT-CGS operation would be improved by a lot of energy products.
Thus, we consider BT-CGS case in which the additional feedstock is required.
For the operation of gas-engine due to the low calorific heating value of bio-gas which means the syngas of BT gasifier, although there are sometimes problems on the heating value of fuel, we executed the process design using the practice parameters which were analysed by the engine manufacturing maker.
Table 5 shows the performance of Bio-CGS. In Tables 5, the cold gas efficiency η_{Cold}, the net power efficiency η_{Pow}, the heat recovery efficiency η_{Heat} and the net total efficiency η_{Total} are defined as follows:

$$\eta_{Cold} = \frac{Syngas\,[\mathrm{MJ/h}]}{Feedstock[\mathrm{MJ/h}] + Char[\mathrm{MJ/h}] + Add.\ Feedstock[\mathrm{MJ/h}]} \tag{9}$$

$$\eta_{Pow} = \frac{Net\ \mathrm{Power}\ (= \mathrm{Power\text{-}Auxiliary})[\mathrm{MJ/h}]}{Feedstock[\mathrm{MJ/h}] + Add.\ Feedstock[\mathrm{MJ/h}]} \tag{10}$$

$$\eta_{Heat} = \frac{Steam[\mathrm{MJ/h}] + Hot\ water[\mathrm{MJ/h}]}{Feedstock[\mathrm{MJ/h}] + Add.\ Feedstock[\mathrm{MJ/h}]} \tag{11}$$

$$\eta_{Total} = \eta_{Pow} + \eta_{Heat} \tag{12}$$

<table>
<tr><td rowspan="9">¶(9pt)BT Process
(18 t/d)</td><td rowspan="2">Feedstock</td><td>625.0</td><td>kg/h</td></tr>
<tr><td>8,278</td><td>MJ/h</td></tr>
<tr><td rowspan="2">Additional
Feedstock</td><td>139.5</td><td>kg/h</td></tr>
<tr><td>1,846</td><td>MJ/h</td></tr>
<tr><td rowspan="2">Syngas
(For Gas-engine)</td><td>1,021</td><td>Nm³/h</td></tr>
<tr><td>6,922</td><td>MJ/h</td></tr>
<tr><td>Cold-Gas Eff. η_{Cold}</td><td>59.0%</td><td>LHV-%</td></tr>
<tr><td>Auxiliary Power</td><td>111</td><td>kW</td></tr>
<tr><td rowspan="7">Gas-Engine</td><td>Power (Net)</td><td>459</td><td>kW</td></tr>
<tr><td>Steam</td><td>1,344</td><td>MJ/h</td></tr>
<tr><td>Hot water</td><td>1,551</td><td>MJ/h</td></tr>
<tr><td>Power Eff.</td><td>16.3%</td><td>LHV-%</td></tr>
<tr><td>Heat Recovery Eff.</td><td>28.6%</td><td>LHV-%</td></tr>
<tr><td>Total Eff.(Net)</td><td>44.9%</td><td>LHV-%</td></tr>
</table>

Table 5. Performance of BT-CGS (estimated)

3. Concept of the biomass Life Cycle Assessment

So far, the biomass Life Cycle Assessment (LCA) analyses, in which the pre-processing process of chipping, transportation and drying of biomass materials are included, and in which the energy conversion process of a production energy of electricity and/or heat through an integrated gasification combined cycle (IGCC) power system or a co-generation system (CGS) is included, were analysed (Dowaki et al. 2002, Dowaki et al. 2003).

In this section, we describe on the BT-CGS and the production system of Bio-H_2. At the beginning, in this section, we defined the system boundary of the biomass LCA. A target is to estimate a life cycle inventory (CO_2 emissions and/or energy intensities) of the entire system with a biomass gasification system and/or a purification one. That is, we refer to the environmentally friendly system, such as the biomass energy system, considering CO_2 emissions and/or energy intensities from the entire system based on LCA methodology.

In the case of BT-CGS or Bio-H_2, due to the shortage of reaction heat in the furnace or the larger auxiliary power output of PSA, the specific CO_2 emission might be affected. That is, the process design and the energy analysis on basis of the process simulation would be extremely significant.

3.1 System boundary

Following ISO 14041 guidelines, we define the system boundary in the biomass energy system.

The system boundary includes the entire life cycle of Bio-H_2 fuel or electricity and thermal energy products, including the pre-processing process and the energy conversion process. (See Fig. 4). In the pre-processing process, there are sub-processes of chipping, transportation by trucks, and drying. In the energy conversion process, there are sub-processes of the gasification through the BT plant with a purification process or a CGS unit. Also, in our estimation, we focused on "well to tank (WtW)" analysis.

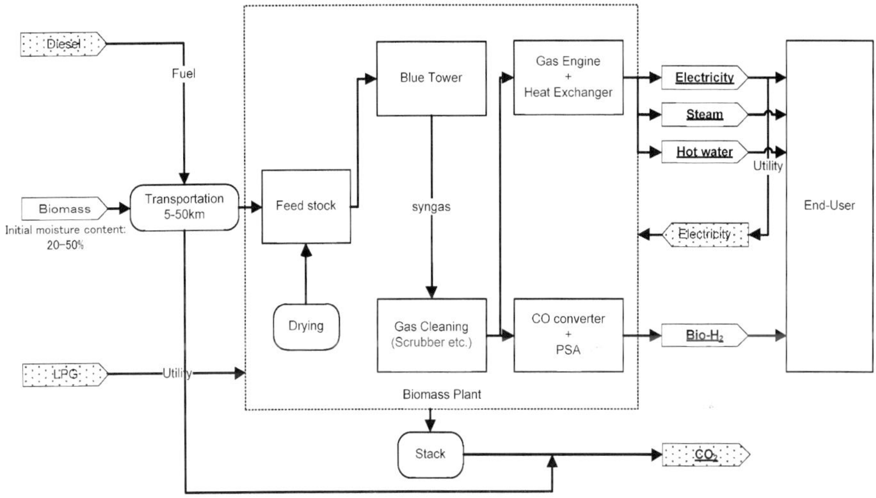

Fig. 4. System boundary of biomass LCA.

3.2 Functional unit

The target product is Bio-H_2 and CGS products of electricity and/or heat energy. Thus, the functional unit is assumed to be the unit per a produced energy. The lower heating values of H_2 and electricity are 10.8 MJ/Nm^3 and 3.6 MJ/kWh, respectively.

3.3 Pre-processing process

In the pre-processing process, there are sub-processes of chipping, transportation, and drying of biomass materials. In particular, within the sub-processes of transportation and drying, we have to consider uncertainties. To date, there are few studies considering these uncertainties. CO_2 emissions and energy intensities in the biomass LCA would be affected by the moisture content of biomass materials, and the transportation distance from the cultivation site, or the site of accumulating waste materials, to the energy plant. Table 6 shows heating values, and that of CO_2 emissions, for each fuel with biomass materials, respectively. Also, CO_2 emissions and energy intensities were estimated using the Monte Carlo simulation in order to consider these uncertainties (Dowaki and Genchi, 2009).

Fuel	CO_2	Note
Feedstock	0.0 g-CO_2/MJ-Fuel	at 20 wt.% (moisture content), Japanese Cedar, HV:13.23 MJ/kg
Diesel	74.4 g-CO_2/MJ-Fuel	Chipping, Transportation, HV: 35.50 MJ/L
Electricity	123.1 g-CO_2/MJ-Fuel	Auxiliary power of the plant (Primary Energy)

Table 6. Data of the specific CO_2 emissions

3.3.1 Sub-processes of chipping, transportation and drying

The energy consumption of chipping, transportation and drying is as follows:

a. **Chipping:** The energy consumption of the chipping process is due to electricity and diesel. The specific units of energy consumption are 13.6 kWh/material-t (122.4 MJ/material-t) and 1.23 L-diesel/material-t (43.7 MJ/material-t), respectively (Hashimoto et al., 2000).
b. **Transportation:** The chopped biomass materials are delivered to the plant by 10 ton diesel trucks. CO_2 emissions and/or energy intensities on a given transportation run would be affected by the weight of biomass materials. That is, the weight of which the materials can be carried is restricted to bulk density. We measured the bulk density (=0.14 t/m^3) in the atmosphere. The bulk density is dependent upon the moisture content. Thus, assuming that the bulk density is at a moisture content of 15 wt.% (ρ_{15}), the bulk density ρ_{MC} at any moisture content (MC wt.%) is

$$\rho_{MC} = \frac{0.85}{1 - MC}\rho_{15} \tag{13}$$

Next, the loading platform of 10t-trucks is to be approximately 24.7 m^3 (Suri-Keikaku Co. Ltd., 2005). Consequently, even a truck with 10 ton's volume cannot always carry that in full weight. Here, CO_2 emissions and/or energy intensities are assumed to be due to the fuel consumption of truck, which is indicated as a function of the loading rate of weight. That is, using the loading rate of λ, the fuel consumption rate of a 10t-truck $f_{FC}(\lambda)$ is

$$f_{FC}(\lambda) = a\lambda + b \tag{14}$$

where, a(=714 g-CO_2/km) and b(=508 g-CO_2/km) are constants on the fuel consumption of the truck (Dowaki et al., 2008b).

The definition of the loading rate of λ is as follows: Assuming that the plant scale is Ps dry-t/d, and that the annual operating time is 300 days, the annual material balance on the feed materials is 300Ps t-dry/yr. Since the throughput per year at MC wt.% is $300Ps/(1 - MC)$, the total number of transportation by 10 t trucks at MC wt.% (N_{mat}) is

$$N_{mat} = \left[\frac{300 Ps/(1-MC)}{24.7 \rho_{MC}} \right] + 1 \quad (15)$$

Where, $[\alpha]$ is represented as the maximum integer, so as not to exceed α. Thus, the average loading rate of a 10 t-truck (λ_{ave}) is

$$\lambda_{ave} = \frac{300 Ps/(1-MC)}{10 N_{mat}} \quad (16)$$

Providing the average loading rate, and multiplying $f_{FC}(\lambda)$ by the transportation distance and the specific CO_2 emissions or the energy consumption of diesel, we can estimate CO_2 emissions or fuel consumption in the transportation sub-process. In this paper, the transportation distance is the range between 5 ($Dist_{\min}$) and 50 km ($Dist_{\max}$), because the wooden materials in Japan are distributed widely. That is, it is assumed that the feed materials are collected within a radius of 50 km.

c. **Drying:** Next, on the sub-process of drying, the energy consumption was estimated under the condition that the moisture content of feed materials would decrease to 20 wt.%. Here, assuming that the initial moisture contents are from 20 ($MC_{\min}$) to 50 wt.% ($MC_{\max}$), the raw materials are dried by a boiler. Also, the auxiliary power of a pump in a boiler is included in the energy consumption of the sub-process. The operational specification of a wood-chip dryer (boiler) is the energy efficiency of 80 %, and the auxiliary power of a pump of 0.195 kWh/t-water (1.75 MJ/t-water). Note that the moisture content of feedstock can be reduced by the hot exhausted gas to some extent.

d. **Monte Carlo simulation on the uncertainties:** As the above, in this paper, we estimated CO_2 emissions and/or energy intensities, considering the uncertainties of the transportation distance and the moisture content. In this paper, the following two uncertainties of the distance and the moisture content were considered by the Monte Carlo simulation.

That is, the uncertainties on the transportation distance ($Dist$ km) and the moisture content (MC wt.%) are represented by uniform random numbers Rnd_i between 0 and 1 in Eqs. (17) and (18). Note that Rnd_1 and Rnd_2 are independent and identically distributed.

$$Dist = Dist_{\min} + Rnd_2 \left(Dist_{\max} - Dist_{\min} \right) \quad (17)$$

$$MC = MC_{\min} + Rnd_1 \left(MC_{\max} - MC_{\min} \right) \quad (18)$$

An iteration count in the simulation was executed up to 10,000. The range within a 95 % significant level was adopted as the uncertain data on the distance and the moisture content, in order to estimate CO_2 emissions. In this case, the gross distributions on CO_2 emissions would be normal distributions.

3.4 CO_2 emissions on CGS and Bio-H_2 fuel

Based on the above data, CO_2 emissions of CGS (electricity and/or thermal energy) and Bio-H_2 fuels are shown in Fig. 5.

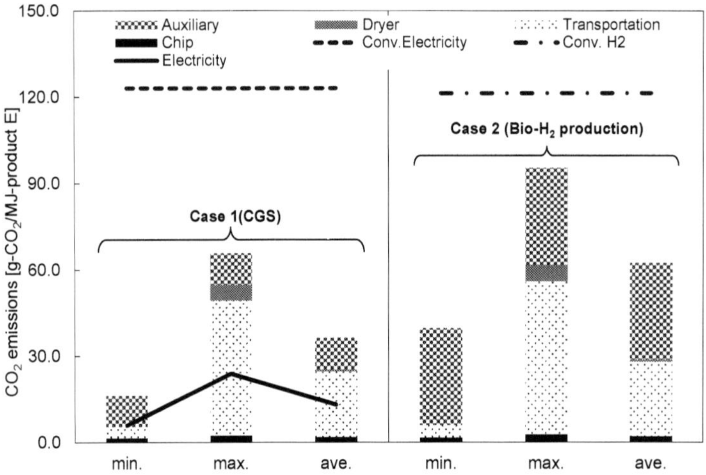

Fig. 5. CO_2 emission in each case (Case 1:CGS, Case 2 Bio-H_2 production).

According to Fig. 5, the entire CO_2 emissions are 16.3-65.7 g-CO_2/MJ of CGS and 39.6-95.3 g-CO_2/MJ of Bio-H_2, respectively. Especially, in the CGS case, the specific CO_2 emissions of electricity are 5.9-23.9 g-CO_2/MJ, and the reduction percentages in comparison to the conventional electricity in Japan are 80.6-95.2%. In the case of Bio-H_2 case, the reduction percentages against the conventional H_2 production (121.3 g-CO_2/MJ, Natural gas origin) are 21.4-67.3%.

CO_2 emissions at the material drying and at the auxiliary power of a purification process of PSA occupy a large portion of the entire CO_2 emission. Especially, the influence due to the compression power of H_2 purification would be significant. In the case of Bio-H_2, the amount of 35.1% to 84.4 % of the total CO_2 emissions would be emitted from the auxiliary power including the power for BT operation. Also, in the case of CGS, that of 16.5% to 66.6 % would be emitted from the auxiliary power origin, even if the PSA operation is not equipped.

The deviations of CO_2 emissions (the maximum value - the minimum one) due to the uncertainties on the moisture content and the transportation distance would be within 49.5 g-CO_2/MJ of CGS and 55.7 g-CO_2/MJ of Bio-H_2, respectively.

That is, the range of collection of biomass feedstock would be extremely significant from the viewpoint of CO_2 emission reduction on basis of LCA methodology.

4. Future application of bio-fuel

As we mentioned before, the renewable energy source, especially, the biomass energy source would be promising for global warming protection. Using the biomass feedstock, there are many fuels which can be converted through the gasification, the fermentation or another process. Here, we concentrated to the biomass gasification process by which electricity and thermal energy or Bio-H_2 fuel are produced. Also, the CO_2 emission due to

LCA methodology, which is estimated in order to understand the impact of Global warming numerically, was estimated. As a next step, we have to create the countermeasure for promotion of our proposed system. However, there is not example in which the relationship between the supply and the demand is argued enough. Based on the sequential and entire system, we have to judge the effects and/or the benefits such as CO_2 emission etc. (See Fig. 1).

Here, as a good example, we introduce the following system. However, that might be difficult to promote our proposed system due to the cost barrier against a conventional system at the present time. The combined system in which the renewable energy such as Bio-H_2 can be available would have a significant meaning in the future utilization for Global warming protection. Simultaneously, we have to create the new business model which would be suitable for the end users.

Now, there is the proposal to install an advanced cell phone (a smart phone) with a PEFC unit so as to get CO_2 benefit. A smart phone is an electronic device used for two-way radio telecommunication over a cellular network of base stations known as cell sites. The sale of mobile phones has been one of the fastest growing markets in the world today. For instance, the cell phone users of Japan were approximately 107 million in 2005 (Infoplease, 2005). At present, around 85% people in America have used cell phone. In addition, new technology of a mobile communication is being developed very quickly. A few years ago, people used their cell phone just for making a call or sending a short mail through a SMS function. However, at the current time, there are a lot of features of a smart phone such as music player, video player, game, chatting, internet browsing and email, etc. These factors should increase energy consumption and increase CO_2 emission.

The current power supply system in a smart phone is dominated by a Li-ion battery, which has some advantage such as wide variety of shapes/sizes without a memory effect. In addition, the rapidly advancing needs for mobile communication are increasing the consumer demand for portable application with even higher power output, longer operation time, smaller size, and lighter weight. A Li-ion and other rechargeable battery system might not be suitable for high power and long time span portable devices due to their lower energy density, shorter operational time, and safety. Li-ion batteries are well established as a power supply for portable devices. Recently, since the power demand has been increasing faster than battery capabilities, the fuel cells might become a promising alternate for niche applications. A fuel cell is an electrochemical device which continuously converts chemical energy into electricity and thermal energy by feeding H_2 fuel and oxygen into it. A fuel cell power supply can be higher energy per a unit mass than conventional batteries. Also, the using of fuel cell system is not harmful to the environment, if compared with a Li-ion battery (Hoogers, 2003). Also, there are the following two types of fuel cell: 1) Polymer Electrolyte Fuel Cell (PEFC) and 2) Direct Methanol Fuel Cell (DMFC), which are operated in low temperature. These two systems are almost same, the difference is only in fuel, that is, the PEFC is operated by H_2 (gas) and DMFC is done by methanol (liquid). Here, we focused on the PEFC into which H_2 fuel is fed. The reason why we concentrate the system is that the fuel for a PEFC can be produced by the renewable resources such as biomass feedstock with a lower CO_2 emission in comparison to the conventional production system. In the area where there is plenty of biomass feedstock (e.g. Indonesia and Malaysia etc.), there is a good potential to install that. A PEFC is applied to replace a Li-ion battery. A comparison of CO_2 emission between a Li-ion battery cell phone and a PEFC cell phone was calculated using Life Cycle Assessment (LCA) methodology, in consideration of the user's behaviour.

4.1 A case study on a smart phone due to LCA methodology

The goal of this study is to compare the CO_2 emission of the conventional Li-ion cell phone and the PEFC cell phone. The functional unit is the specific CO_2 emission per a life cycle (LC) of kg-CO_2/LC. Fig. 6 shows the life cycle stage on the schematic design of system boundary, in which a pre-processing of raw materials, a manufacture, a transportation and distribution, an energy consumption of end users and a disposal process are included. Also, in this study, we referred to the duration time of each operation of cell phone (Dowaki et al., 2010a).

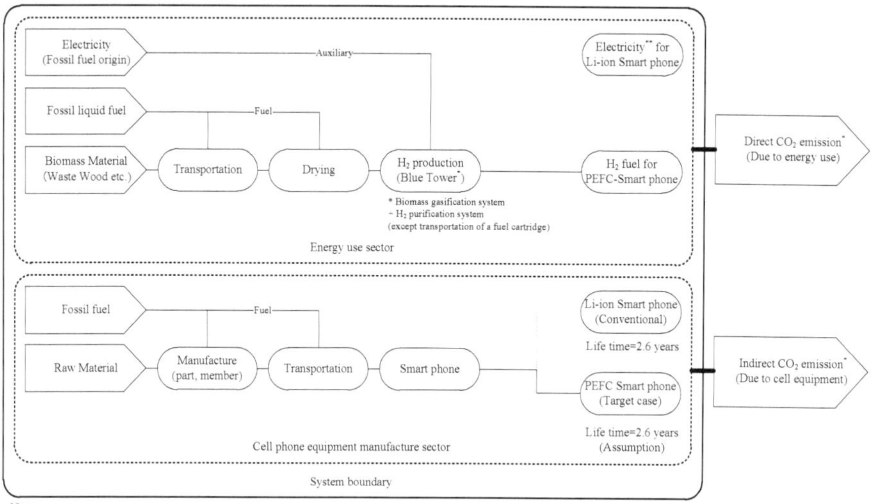

Note:
* Estimated period=2.6 years
** This value is based on the wheel to tank, that is, the input energy for producing the fuel besides raw energy source (primary energy) is considered, too.

Fig. 6. System boundary of a cell phone analysis.

In the system boundary, as we described the prior section, we think about the availability of Bio-H_2 through BT process. For the purpose, we executed the questionnaire on the way to use a smart phone firstly. Also, we executed the performance of a PEM cell which is based on a PEFC unit using the electric power measurement device.

The difference between a Li-ion and a PEFC cell phone is in electrical energy sources. The Li-ion cell phone is supplied by conventional electricity, whereas a PEFC cell phone is done by Bio-H_2 as an energy input. The battery charge due to the conventional electricity emits CO_2 of one of the greenhouse gases. On the other hand, since the Bio-H_2 would be carbon neutral, the CO_2 emission is equivalent to zero in a combustion process. However, the production process of a renewable fuel is accompanied with the conventional energy inputs (i.e. fossil fuels). Thus, it is extremely important to estimate the energy system based on LCA methodology.

4.1.1 A questionnaire for the smart phone users

In order to investigate the way to use a smart phone in each user, we executed the questionnaire between February 17 and February 24, 2011. 200 respondents in Japan participated in this research. Also, target respondents are the users who use a smart phone, with their ages between 15 and 65 years, respectively. In the questionnaire content, the duration time of a talking, a SMS, music (MP3), a game, a web-site (internet), an e-mail checking, and an idle time were estimated for each age category. Fig. 7 shows the result of the duration time of each function. The checking time of internet would be larger in both weekdays and holyday (Dowaki et al., 2011b).

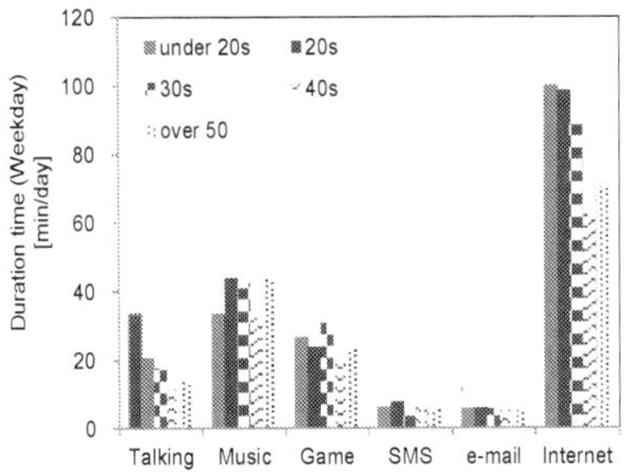

a) Weekdays

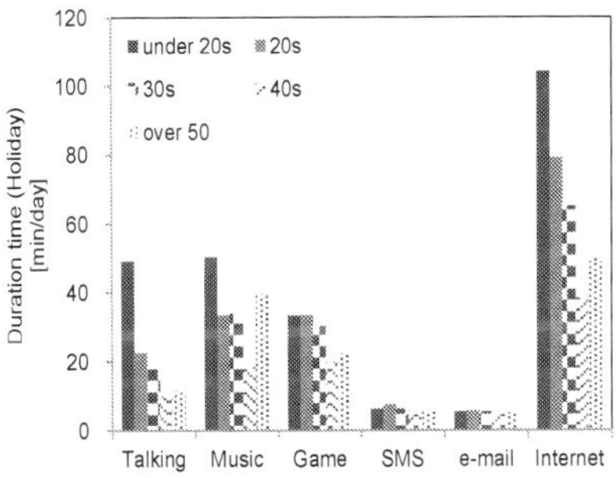

b) Holiday

Fig. 7. Duration time of each function in a smart phone.

4.1.2 Measurement of the performance of a PEFC unit

Based on the duration time, we measured the performance of a PEM-cell which is based on a PEFC unit for a smart phone. That is, using the result of output capability of a PEM-cell and the maximum duration time, we designed the cell area of a PEFC, and estimated the energy consumption for each function.

Here, in order to have a good reliability, 10 times experiments have been done for the following tasks: a talking, a SMS, music (MP3), a game, a web-site (internet) and an e-mail checking, respectively. In our experiments, we used the electric measurement device (AC/DC POWER HiTESTER 3334, HIOKI E.E. Corp.) to measure the voltage and the current, and the power which is obtained by these factors.

Next, for the purpose of estimating the cell performance, we measured the potential of a PEM-cell in varying currents. The apparatus consists of a PEM-cell (Micro Inc.) and a potentiostat (HAB-151, Hokuto Denko Corp.). The size of the cell with low platinum loading electrodes (1.0 Pt mg/cm^2) and Nafion® 115 is 4 cm2×3 cells. Using a potentiostat, 1) an open circuit voltage V_0 [Volt] was measured and 2) the relationship of current density J vs. cell potential V was evaluated between approximately 200 mV/cell and an open circuit potential V_0. Also, H_2 flow in anode was up to 20 ml/min and the concentration of H_2 was 100 vol.% at a constant percentage (see Fig.8). Note that each parameter on the performance of a PEM-cell is decided at the condition which is not rate-limiting. In this case, we adopted the flow rate condition of 20 ml/min. The cathode was stayed at the atmospheric condition. The conditions in both the anode and cathode sides were not saturated by steam (Dowaki et al., 2010a, Dowaki et al., 2011b).

Next, the relationship between a cell's potential and current density, in the low and intermediate current density region of a PEM-cell, has been shown to obey the following Eq. (19) (Kim et al., 1995). Note that the result was shown by the condition of a single PEM- cell.

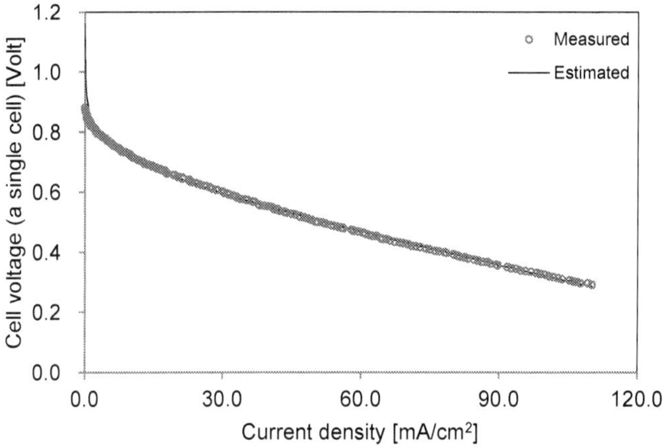

Fig. 8. Relationship between current density and a cell voltage in the single PEM-cell.

$$V = V_0 - b\log J - R_{cell}J - m\exp(nJ) \quad (19)$$

Where, J, b, R_{cell}, m and n are current density [mA/cm^2], Tafel slope [mV/decade], a cell resistance [ohm-cm^2] and constant parameters, respectively.

Based on the experimental result, we analysed each parameter by the approximation formula of Eq. (19). Consequently, the open circuit voltage V_0, the Tafel slope b, a cell resistance R_{cell} and constant parameters of m and n were 1.16 Volt, 54.0 mV/decade, 2.98 ohm-cm^2, 9.08 and -2.53×10^{-3} were obtained. Using these parameters, we designed the PEFC unit of a smart phone as follows: the surface area is 22 cm^2, and the stack number of cell is 3. These conditions would be suitable of a conventional smart phone size and satisfy the maximum output among each function. Also, the stoichiometric ratio is assumed to be 1.00. This means the supplied H_2 would be fully consumed.

Next, due to the questionnaire for smart phone users (see Fig. 7), the energy consumption for each function and the performance of PEFC using a PEM-cell experimental result, we estimated CO_2 emission on basis of LCA methodology. Using Eq. (19), the H_2 flow rate in practice use is able to be calculated. Here, a PEFC would be operated between 1.17 and 2.63 Volt. The specific energy consumption in each function was between 0.39 and 37.7 Nml/min.

4.1.3 Specific CO_2 emission of a smart phone

Based on the above analysed results, we estimated the CO_2 emission of a smart phone use. Here, we considered the indirect and the direct CO_2 emissions. The direct CO_2 emission is equivalent to the fuel consumption origin. On the other hand, the indirect one is mainly on the device of a smart phone. In this study, we focused on HTC Desire X06HT made in Taiwan as a model phone. The indirect CO_2 emission is calculated by Input-Output (IO) table, and this emission referred to the prior result. Also, we estimated the conventional smart phone including Li-ion battery in order to compare to the new one. Assuming that the holding time (life time: LT) when one user has a smart phone until he or she change the new one is 2.6 years, the indirect CO_2 emission of HTC Desire X06HT including Li-ion battery would be 15.32 kg-CO_2/unit. The emission of a smart phone with a PEFC unit would be 15.30 kg-CO_2/unit. Although there are uncertainties on the storage tank of H_2 to some extent, referring to the data of DMFC storage tank which has already developed, we estimated the emission as almost same as the conventional case (Dowaki et al., 2010a).
Next, the direct CO_2 emission is affected by the specific CO_2 emission of each fuel. Here, the CO_2 emissions of conventional electricity, H2 fuel of natural gas origin (on-site) and Bio-H_2 are assumed to be 123.1 g-CO_2/MJ, 121.3 and 39.6 g-CO_2/MJ-H_2, respectively. The CO_2 emission per one life cycle is shown in Fig. 9. Note that the specific emission of Bio-H_2 is a minimum level (see Fig. 5).

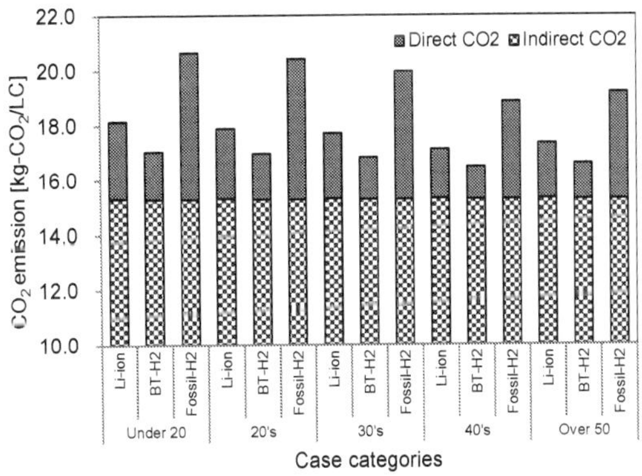

Fig. 9. Life Cycle CO_2 emission for a smart phone use.

According to this result, due to application of a PEFC unit to the smart phone, we would be able to reduce CO_2 emissions of 3.9% to 6.1 % in comparison to the conventional phone. Especially, in the category of younger generation, the CO_2 reduction benefit would be effective.

4.2 A case study on a greenhouse facility due to LCA methodology

Next, we propose the advanced greenhouse system for paprika cultivation with the combined biomass gasification process of BT with SOFC (Solid Oxide Fuel Cell). The BT gasifier which is a biomass gasification process has a characteristic of generating hydrogen gas of high concentration in syngas. Here, we considered the acceptability for the related facilities in agriculture field. Because the environmentally friendly system such as a PV system or a fuel cell co-generation system is still not enough to be promoted for those facilities. That is, there would be potential to combine the biomass energy system which is environmentally friendly with the agriculture related facilities. In addition, MAFF contribute to the global warming protection through the carbon-footprint of agricultural products. The ministry has a few subsidy menus on the promotion of the system. Also, on the surplus energy of electricity and/or thermal energy, there are institutions by which the energy companies are obliged to purchase them with additional fees.

Using the above institutions and/or subsidy menus under the leadership of the Japanese government, we considered the following concrete paprika harvesting system in which the biomass gasification (BT) process with SOFC is assumed to be introduced (Dowaki et al., 2010b).

First, our model site is the paprika harvesting facility in Miyagi of Japan, whose area is 4.6 ha. In our study, through interviews from the owner company, we used the data of not only the energy consumption of electricity and oil, but also the supply of CO_2 gas which is fed into the greenhouse as a growth promoting agent. That is, in the model we proposed, the electricity, the thermal energy and the CO_2 gas which is included in exhausted gas through BT plant are assumed to be available for the greenhouse facility of paprika harvesting. In addition, due to the combination of the advanced power generation such as SOFC, additional benefit of CO_2 emission reduction would be obtained. This may be advantageous from the profit aspect since the surplus electricity would be able to be sold to the commercial energy companies. Also, from the viewpoint of thermal energy use, the combined BT with SOFC units would be advantageous since the exhausted gas with a high temperature (ca.700 °C) is generated. Although the operation of SOFC has been in a developmental stage, we used the published parameters. The initial cost of SOFC unit seems to be costly in comparison to the conventional power system. However, it is said that the commercial stage of SOFC is close. Thus, the initial cost was assumed to be equivalent to the target price as of 2015. The thermal energy for the greenhouse is supplied by the heat pump equipment. This would bring to the benefit of cost and/or CO_2 emission reduction, since there is little waste thermal energy (Dowaki et al., 2011c).

On the other hand, MAFF tries to introduce the carbon-footprint for the agricultural products. It is difficult to estimate the monetary values of CO_2 emissions of agricultural products. For instance, Kikuchi et al. investigated the willingness to pay for CO_2 emission reduction of vegetables (Kikuchi and Itsubo, 2009). They found out that the consumers have a willingness to pay for an additional cost of approximately 5% up against a conventional price. Although this is only a limited effect, there would be a potential to earn income due to the carbon footprint. That is, with regard to income in our system, revenues to the plant owner would include the related subsidy, the processing fee of waste material, the sale of surplus electricity and the paprika sale with low CO_2 emission. The carbon-footprint of agricultural product might be important one of income sources.

In this study, we analysed the CO_2 emission due to LCA methodology.

4.2.1 A LCA for the paprika cultivation

In the LCA concept of this paper, the direct factors and the indirect ones have to be considered. In our definition, fossil fuel energy inputs (primary energy basis) and the electricity of fossil fuel origin are included in the direct factors. Also, chemical fertilizers are included in the indirect ones. Here, note that another greenhouse gases such as N_2O and CH_4 are not taken into consideration.

So far, in the biomass LCA analyses, the pre-processing process of chipping, transportation and drying of biomass materials, and the energy conversion process of a production energy of electricity and/or heat, through an energy system are included. This time, the paprika harvesting process has to be added to the entire life cycle stage. Using the chemical experimental data, the design of BT plant with SOFC units would be extremely significant in the biomass LCA. A target is to estimate a life cycle inventory of the entire system with BT gasifier and SOFC.

Here, we describe on the system boundary in this study. Following ISO 14041 guidelines, we define the system boundary in the biomass energy system (see Fig. 10) (Dowaki et al., 2010b).

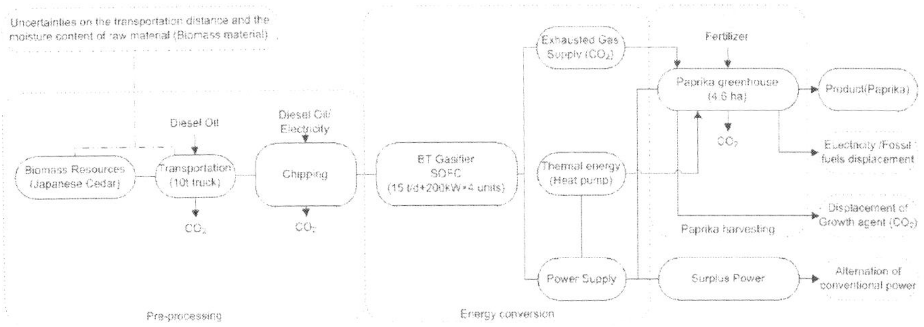

Fig. 10. System boundary of a paprika production system.

The system boundary includes the entire life cycle of each energy input (electricity/thermal energy), including the pre-processing process, the energy conversion process and the paprika harvesting process. In the pre-processing process, there are sub-processes of chipping, transportation by trucks, and drying. In the energy conversion process, there are sub-processes of the gasification through the BT plant (19 t/d) with the four units of SOFC (200 kW/unit) process. In the paprika harvesting process, it is assumed that the exhausted gas of CO_2 is available as a growth agent. Here, the target product is a paprika. Thus, the functional unit is assumed to be the unit per a produced paprika (Dowaki et al., 2011c).

Next, in the pre-processing process, there are sub-processes of chipping, transportation, and drying of biomass materials. In particular, within the sub-processes of transportation and drying, we have to consider uncertainties (see section 3.3.1). To date, there are a few studies considering these uncertainties. CO_2 emissions in the biomass LCA would be affected by the moisture content of biomass materials, and the transportation distance from the cultivation site, or the site of accumulating waste materials, to the energy plant. Hence, it would be extremely significant to consider these factors. Table 7 shows the specific CO_2 emissions, for each fuel with biomass materials, respectively.

Item	CO_2	Note
Feedstock	0.0 g-CO_2/MJ-Fuel	at 20 wt.% (moisture content), Japanese Cedar, HV:13.23 MJ/kg
Diesel	74.4 g-CO_2/MJ-Fuel	Chipping, Transportation, HV: 35.50 MJ/L
Bunker A	76.9 g-CO_2/MJ-Fuel	Paprika production (Boiler)
Kerosene	73.6 g-CO_2/MJ-Fuel	Paprika production (Boiler)
Electricity	123.1 g-CO_2/MJ-Fuel*	Paprika production (Ventilation and lightning)
Fertilizer (N)	5.67 kg-CO_2/kg	Indirect CO_2 emission
Fertilizer (P_2O_5)	0.88 kg-CO_2/kg	Indirect CO_2 emission
Fertilizer (K_2O)	1.85 kg-CO_2/kg	Indirect CO_2 emission

Table 7. Data of the specific CO_2 emissions.

On the energy conversion process, assuming that the 19 t/d BT plant and 4×200 kW SOFC (BT-SOFC system) were installed, we estimated the CO_2 emission in the paprika production system. Here, the operational condition of SOFC unit is assumed to be almost full load operation. Also, the specification of SOFC unit is shown in Table 8.

Unit Scale	[kW]	200
Number of unit	-	4
Operating Temp.	[°C]	900
Current density	[mA/cm²]	612
Stoichiometric ratio	-	1.25
Tafel slope	[mV/dec.]	2.2
Cell Resistance	[ohm]	0.52
Open Circuit Voltage	[mV]	950
DC/AC converter Eff.	[%]	95

Table 8. Specification of SOFC unit.

Due to the specification data in each system, the performance of BT-SOFC system is obtained as Table 9. Also, the thermal energy supply to the facility is assumed to be due to the heat pump (COP: 5.5).

BT Process (19t/d)	Feedstock	781.3	kg/h
		10,338	MJ/h
	Cold-Gas eff.(Eq. (7))	56.2	LHV%
	Auxiliary Power	127.3	kW
SOFC (4×200 kW)	Power eff.vs. syngas	45.5	LHV%
	Power eff. vs. feed	25.0	LHV%
	Net eff. vs. feed	20.6	LHV%
	Net power scale	590	kW

Table 9. Performance of BT-SOFC system.

4.2.2 Paprika cultivation facility

In this study, we investigated the greenhouse facility at Miyagi of Japan where paprika is brought into cultivation. In this facility, the annual product yields are around 200 t/yr. The energy of electricity, kerosene and bunker A for lighting and a heater, and the input of CO_2 gas as a growth agent are consumed. Here, since the energy data of time series was necessary, the boiler fuels of kerosene and/or bunker A were assumed to be in proportion to a difference between the minimum temperature for growing and the atmospheric one. Also, electricity was assumed to be consumed for 12 hours per a day.

Next, the consumption of CO_2 gas as a growth agent would be analysed statistically. In a plant such as paprika, CO_2 is consumed through photosynthesis. That is, this volume would be proportional to the duration of bright sunshine and an intensity of radiation. Fig.11 shows the statistically estimated CO_2 consumption.

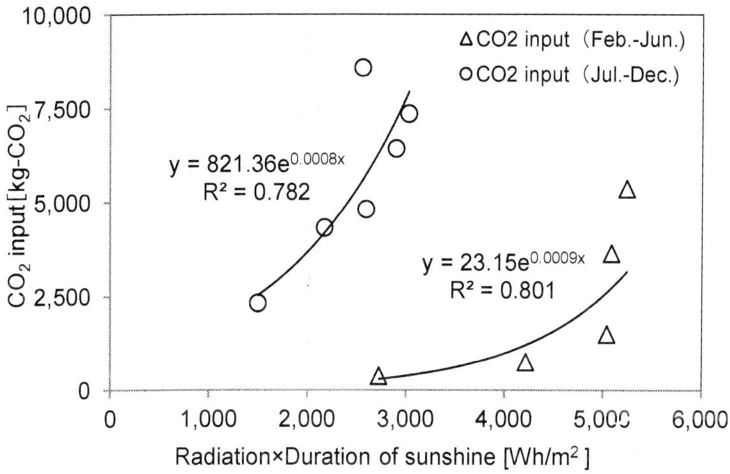

Fig. 11. CO_2 supply volume as a growth agent.

On the other hand, fertilizers of N, P_2O_5 and K_2O only were considered, however another chemical inputs were ignored (Dowaki et al., 2010b).

4.2.3 Specific CO_2 emission of a paprika production

Based on the above results, we estimated the CO_2 emission in conventional case and that of BT-SOFC case (see Fig. 12). In this study, the CO_2 intensities in BT-SOFC case are included on the uncertainties of moisture content and a transportation distance (See section 3.3.1).

In the conventional case, the specific CO_2 emission of 622.6 g-CO_2/paprika was estimated. On the other hand, in the BT-SOFC case, the specific CO_2 emission of 38.1 to 218.4 g-CO_2/paprika was analysed, and CO_2 reduction rate was 64.9% to 93.9%, respectively. Also, since the surplus electricity of 4,137 MWh/yr would be generated through this system, the much CO2 reduction benefit might be obtained due to the alternation with the conventional electricity.

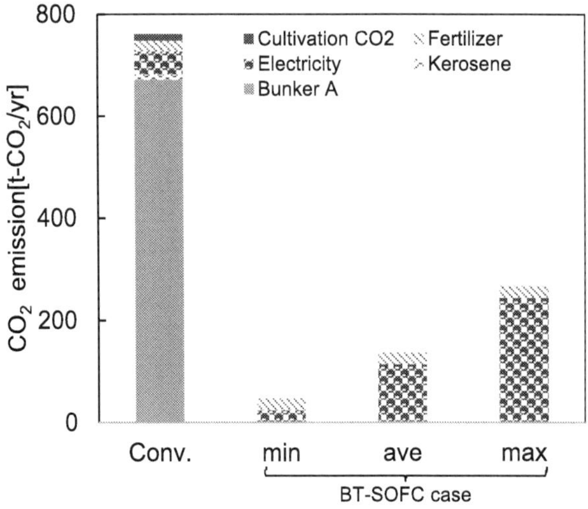

Fig. 12. Specific CO_2 emission of a paprika.

5. Conclusions

As we described before, there is a good potential to install the renewable energy system such as a biomass energy system. In this section, we focused on the Blue Tower gasification process. In the near future, when we consider the promotion of eco-friendly business, we have to realize the sustainable business model which can be operated under a good cost condition and/or a reduction of CO_2 emission. That is, we have to consider not only technological barrier but also the CO_2 abatement effect. In this case, LCA methodology would be reasonable and necessary. Of course, the business scheme would be extremely significant.

In this section, we introduced two cases based on the biomass gasification system of BT process. In both cases, for instance, if we utilize the subsidies due to the central and/or local governments at the initial stage, or the regulation of feed-in tariff is available, the proposed business scheme would become increasingly competitive against the conventional business model. Also, recently, people have a great concern on the carbon-foot print based on LCA methodology. This means that there is potentiality to purchase the product with low-carbon emission. For the future, in order to mitigate GHG gases, we might have to consider the suitable technological system and the effective eco-socio system.

6. References

Ministry of Economy, Trade and Industry, Japan (METI). (2007). Report on Next-Generation Vehicle and Fuel Initiative, 11.07.2007, Available from: www.meti.go.jp/press/20070528001/initiative-torimatome.pdf

The Ministry of Agriculture, Forestry and Fisheries, Japan (MAFF). (2006). 11.07.2007, Available from: http://www.maff.go.jp/biomass/h18_point.pdf

K.K. Gas Review. (2007). Demand and Supply for industrial gas, *2007 Gas Georama in Japan (in Japanse)*. p.65

M .Mayer, H. J. Mühlen and C. Schmid. (2004). Blue Tower hydrogen from biomass, *European Conference on Renewable Energies and Cooperation Exchange,* Vienna, Austria

H. J. Mühlen and C. Schmid. (1999). Staged Reforming-A Thermal Conversion Process Specifically Developed for Biomass, *Proceedings 2nd Olle Lindström Symposium*, pp. 75-82, Stockholm, Sweden

K. Dowaki, T. Ohta, Y. Kasahara, M. Kameyama, K. Sakawaki and S. Mori. (2007) An Economic and Energy Analysis on Bio-hydrogen Fuel using a Gasification Process. *Renewable Energy,* Vol. 32, Issue 1, pp. 80-94

M. Kameyamaa, S. Ogura, H. Kamiuchi and K. Dowaki. (2010). Performance evaluation of 1t/d scale plant through Blue Tower gasification process (in Japanese). *Journal of Japan Society of Energy and Resources,* Vol. 31, No. 2, pp.43-50

K. Dowaki, Y. Kon, and M. Kameyama. (2008a). A Modified Design of CGS Operation or Bio-H_2 Production through Blue Tower (BT) Process. *The 8th International conference on Eco balance,* B-07, Tokyo, Japan

K. Dowaki, M. Nakazono, M. Yamanari, M. Kameyama and H. Kamiuchi. (2011a). An Analysis of Heat and Flow Characteristic of Heat Carrier in BT Process (in Japanese). *Proceedings of the 27th conference on energy, economy, and environment,* pp.271-274, Tokyo, Japan

K. Dowaki, H. Ishitani, R. Matsuhashi and N. Sam. (2002). A comprehensive life cycle analysis of a biomass energy system. *Technology,* Vol. 8, pp. 193–204

K. Dowaki, S. Mori, H. Abe, P. Grierson, M. Adams, N. Sam and P. Nimiago. (2003). A life cycle analysis of a biomass energy system taking sustainable forest management into consideration. *Proceedings of 6th International Conference on Greenhouse Gas Control Technologies,* pp. 1383–1388, Kyoto, Japan

K. Dowaki and Y. Genchi. (2009). Life Cycle Inventory Analysis on Bio-DME and/or Bio-MeOH products through BLUE Tower Process. *International Journal of LCA,* Vol.14, Number 7, pp.611-620

S. Hashimoto, T. Obara and Y. Terashima. (2000). Evaluation of demolition wood waste recycling from environmental aspect (in Japanese), *Journal of environmental systems and engineering VII,* Vol. 643, Issue VII-14, pp.37–48

Suri-Keikaku Co. Ltd. (2005). FY2004 Survey on estimation and investigation of exhaust emission factors and their total amount (in Japanese).

K. Dowaki, T. Eguchi, R. Ohkubo and Y. Genchi (2008b) A life cycle assessment on a fuel production through distributed biomass gasification process (in Japanese). *Transaction of the institute of Electrical Engineers of Japan C,* Vol. 128, Issue 2, pp 168–175

Infoplease. (2005). Cell Phone Usage Worldwide, by Country. 01.04.2011, Available from: http://www.infoplease.com/ipa/A0933605.html

G. Hoogers. (2003). Fuel Cell Technology Handbook, In: *Portable Applications,* G. Hoogers, (Ed.), Capter 9, CRC Press LLC, ISBN 0849308771

K. Dowaki, T. Izawa and K. Sakamoto. (2010a). An analysis of CO_2 emission on the cell phone with PEFC using Bio-H_2 fuel (in Japanese). *Journal of Life Cycle Assessment, Japan,* Vol. 6, Issue 3, pp. 259-269

K. Dowaki, K. Yasumoto, M. Yamanari, Y. Iwai and H. Nishiyama. (2011b). A system analysis on the mobile device in consideration of the characteristics of cell phone users (in Japanese). *Proceedings of The 6th meeting of the institute of Life Cycle Assessment, Japan*, pp.170-171, Sendai, Japan

J. Kim, S. M. Lee and S. Srinivasan. (1995). Modeling of Proton Exchange Membrane Fuel Cell Performance with an Empirical Equation. *Journal of The Electrochemical Society*, Volume 142, Issue 8, pp. 2670-2674

K. Dowaki, M. Yamanari, Y. Takada and J. Kamihirata. (2010b). An indication of a carbon-footprint of Paprika due to the greenhouse facilities with an installation of biomass energy system, *The 9th International conference on Eco balance*, A2-1510, Tokyo, Japan

K. Dowaki, K. Fukumoto, M. Shimoguchi and M. Yamanari. (2011c). Proposals of the agricultural products cultivation system due to Blue Tower Gasification combined-cycle systems to reduce CO_2 emission. *Proceedings of The Life Cycle Management conference (LCM 2011)*, Submitted, Berlin, Germany

K. Kikuchi and N. Itsubo. (2009). Utility of Applying Carbon Footprint at Farmers' Market Aimed at Promotion of Local Production for Local Consumption (in Japanese), *Journal of Life Cycle Assessment, Japan*, Vol. 5, No. 4, 2009, pp.446-455

26

Advances in the Development of Bioethanol: A Review

Giovanni Di Nicola[1], Eleonora Santecchia[1], Giulio Santori[2] and Fabio Polonara[1]
[1]*Dipartimento di Energetica, Università Politecnica delle Marche, Ancona*
[2]*Università degli Studi e-Campus, Via Isimbardi 10, Novedrate (Co)*
Italy

1. Introduction

Henry Ford, father of the modern automobile, constructed his Model T in the early years of the 20th century, when he planned to fuel it with ethanol obtained from cereals. Ford promoted the use of this fuel with such conviction that, by 1938, plants in Kansas were already producing 18 million gallons of ethanol a year (about 54,000 t/year). But interest in ethanol declined after the Second World War because of the enormous availability of natural gas and oil.

At the end of the Seventies, following the first oil crisis, various oil companies began to sell a petrol containing 10% of ethanol, called gasohol, taking advantage of the tax deductions granted on ethanol. Bioethanol did not immediately meet with the success it deserved, however, because it already had competitors on the market, such as methyl tert-butyl ether (MTBE), which was better than ethyl tert-butyl ether (ETBE) in both economic terms and performance. In subsequent years, MTBE proved to be heavily polluting, so it was banned and bioethanol returned to become one of the most attractive prospective solutions for reducing CO_2 emissions.

Another factor that helped to relaunch bioethanol was the growing awareness that we are nearing the so-called tipping point, i.e. the moment commonly indicated as the critical point of no return, when the curve of the demand for oil intersects the declining curve of its availability.

There is an ethical issue, however, that particularly concerns bioethanol, but also affects the other fuels of biological origin. Biofuels are obtained mainly from raw materials such as plants and cereals, that would otherwise be destined for the foodstuffs industry.

To deal with this problem, recent research has been concentrating on an inedible perennial herbaceous plant called Miscanthus giganteus that has a calorific value of approximately 4200 kcal/kg of dry matter. Using lignocellulose materials, municipal solid waste or the wheat wasted each year (around 5%, which would provide about 9.3 Gl of bioethanol) could also overcome the ethical obstacles.

Bioethanol can be used in various forms: added in proportions of 5-10% to the diesel oil in diesel engines; mixed in proportions of 10-85% in petrol for internal combustion engines, or to replace 0-100% of the petrol used in flexible fuel vehicles (FFV). The number of FFV on the roads is constantly increasing: in Brazil their sales now reach 400,000 vehicles/year and

there are approximately 1,500,000 of them (mainly public vehicles) circulating in the USA; in Europe, Sweden has around 15,000 vehicles of this type fueled with E85 (85% ethanol).
Research is also underway on improved engines fueled with bioethanol, and on fuel cells that use the internal reforming of bioethanol to obtain hydrogen.

1.1 Ethyl tertiary butyl ether

Ethyl tertiary butyl ether (ETBE) is a high-octane bioethanol product obtained mainly by making the ethanol react with isobutylene (a byproduct of oil refining) under the effect of heat and various catalysts. It is consequently considered as being partially renewable.
ETBE has technological and functional features that are very like and distinctly better than those of the alcohol it is obtained from. Moreover, it lacks the latter's problems of volatility or miscibility with petrol and it features a high octane number.
Being an ether, it contains oxygen in the molecule, and this enables it to help improve the vehicle's emissions of pollutants. A recent paper (Da Silva et al., 2005) conducted a study on the effects of the anti-detonating properties and Reid vapor pressure (RVP) of petrols mixed with various additives, concluding that adding ETBE improves the mixture's anti-detonating properties and reduces the vapor pressure without interfering with the volatility needed to start a cold engine.
ETBE obtained from bioethanol (also called bioETBE) offers the same benefits as bioethanol, i.e. a lower emission of pollutants, a higher octane number and a reduction in crude oil imports, without the technical and logistic problems posed by the alcoholic nature of bioethanol. BioETBE also contributes to the diffusion of biofuels in the transport sector.

1.2 Diesel and bioethanol mixtures (e-diesel)

The development and increasing use of diesel and bioethanol mixtures in diesel engines has been driven mainly by the European countries needing to comply with the European Union directive 2003/30/CE (which establishes that at least 5.75% of the fuels market must consist of biofuels by the year 2010), as well as the need to dispose of a petrol surplus in the refineries due to the greater demand for diesel vehicles. The drawbacks of the so-called e-diesel mainly concern a reduced viscosity and lubrication issues, a lower cetane number and injection capacity, a greater volatility (which can lead to an increase in the emissions of uncombusted hydrocarbons) and a lesser miscibility (Marek & Evanoff, 2001; Hansen et al., 2005; Lapuerta et al., 2007). In particular, Lapuerta et al. studied different diesel-bioethanol mixtures in different conditions of temperature, water content and additives, developing level maps that give a precise idea of the mixtures' areas of stability and of kinetic separation, that prompt the following conclusions:

- the presence of water in the mixture facilitates the separation of the ethanol phase;
- when its temperature increases, the mixture becomes more stable and the solubility of the ethanol in the diesel also increases;
- the mixture's sensitivity to the effects of water content and additives is higher, the higher the temperature of the mixture;
- mixtures with a bioethanol content up to 10% (v/v) can be used in diesel engines in regions where temperatures in winter rarely drop below 5°C,
- using stability-improving additives can increase the range of ethanol proportions in the mixtures, or the geographical extension of their applicability, enabling any phase separation to be avoided.

1.3 Research projects and bioethanol promotion

To succeed in demonstrating the feasibility of replacing petrol and diesel oil with bioethanol, the European Union developed the BEST project (BioEthanol for Sustainable Transportation) (European Union, 2011), involving six European countries (Sweden, the Netherlands, the United Kingdom, Ireland, Spain and Italy), and also Brazil and China: the global aims of the project are to introduce at least 10,500 FFV and 160 bioethanol-fueled buses, as well as to build 148 service stations, 135 to provide E85 and 13 to provide E95.

The NILE project (New Improvements for Lignocellulosic Ethanol) (Eurec, 2011) focuses instead on proposing the best processes for an economically effective production of bioethanol from lignocellulose biomass, suitable for use in internal combustion engines. The main goal of the NILE project is to reduce the cost of producing bioethanol from this type of raw material so as to make the technology commercially competitive. The NILE project brings together 21 industrial and research organizations from 11 member states, with complementary professional backgrounds and expertise so as to cover the whole cycle of bioethanol production and usage. On a technical level, the problems that remain to be solved concern reducing the cost of the enzymatic hydrolysis process by developing new artificial enzymatic systems, eliminating the current drawbacks intrinsic in converting fermentable sugars into ethanol, and validating the artificial enzymatic systems and yeast strains in a fully-integrated pilot plant.

Finally, there is the European LAMNET research program (Latin America Thematic Network on Bioenergy) (LAMNET, 2011), the main aim of which is to establish a trans-national forum to promote the sustainable use of biomass in Latin America and other emerging countries.

2. Raw materials

One of the great merits of bioethanol consists in the enormous variety of raw materials, and not only plants, from which it can be produced. The production methods vary depending on whether or not the raw material is rich in fiber.

The basic materials for producing biofuels must have certain features, including high carbon and hydrogen concentrations and low concentrations of oxygen, nitrogen and other organic components. The following is a brief description of some of the most important raw materials suitable for use in bioethanol production.

2.1 Alfalfa (medicago sativa)

This is a lucerne of the Fabaceae family that grows in cool subtropical and warm temperate regions. It demands no nitrogen-based fertilizers and its leaves are a precious source of protein in animal fodder. In a recent paper (Dien et al., 2006) it was observed that this plant has a low glucose yield due to a low-efficiency cellulose hydrolysis. The stems contain high concentrations of crude proteins and organic acids.

2.2 Switch grass (panicum virgatum)

This is a perennial herbaceous plant that grows mainly in the United States. Its ethanol yield per hectare is the same as for wheat. It responds to nitrogen fertilizers and can sequester the carbon in the soil. It is a highly versatile plant, capable of adapting easily to lean soils and marginal farmland (Heaton et al., 2004). Like maize, it is a type C_4 plant, i.e. it makes an alternative use of CO_2 fixation (a process forming part of photosynthesis). Most of the genotypes of Panicum virgatum have short underground stems, or rhizomes, that enable them with time to form a grassy carpet. Single hybrids of Panicum virgatum have shown a marked potential for increasing their energy yield (Bouton, 2007), but genetic engineering methods on this plant are still in a developmental stage and for the time being only their tetraploid and octaploid forms are known; we also now know that similar cell types (isotypes) reproduce easily.

2.3 Sweet sorghum (sorghum bicolor L)

The grains obtained from this plant are rich in starch and the stems have a high saccharose content, while the leaves and bagasse have a high lignocellulose content. The plant can be grown in both temperate and tropical countries, and it tolerates drought, flooding and alkalinity. Sorghum is considered an excellent raw material because the methods for growing and transporting it are well established. Ethanol can be obtained from it by exploiting both its starch and its sugar content. Research is currently underway on the use of hybrid or genetically modified species, although those obtained so far are weaker and need to be further refined and tested as concerns energy conversion efficiency (Rooney et al., 2007).

2.4 Cassava (manihoc esculenta)

This tuber is of considerable interest not only for ethanol production but also to produce glucose syrup, and it is available in tropical countries. The ethanol yield from the whole manioc is equivalent to the ethanol produced from cereals using dry milling methods. The only known lies in that the manioc has to be processed 3-4 days after it was harvested. To avoid such lengthy processing times, the manioc is first sliced and then left to dry in the sun. The waste water produced in the process can be treated by means of anaerobic digestion to produce bio gas.

2.5 Spruce (picea abies)

This tree has attracted a great deal of attention as a raw material for ethanol production because it is a lignocellulose material mainly composed of hexose sugars, which are more readily convertible than pentose sugars.

2.6 Willow (salix)

This is a member of the Angiosperm family and is consequently characterized by a hard wood. In this species, a fraction of the xylose units is acetylated. Some of the OH groups of the xylose carbons C_2 and C_3 are replaced by O-acetyl groups. With pretreatment, these groups release acetic acid that, in high enough concentrations, inhibits the yeasts involved in the fermentation process, according to some studies (Sassner et al., 2008a). It was recently demonstrated (Sassner et al., 2008b) that, by pretreating willow with sulfuric acid before the enzymatic hydrolysis process, and then simultaneously performing saccharification and fermentation, they succeeded in obtaining a global ethanol yield of 79%.

2.7 Reed canary grass (phalaris arundinacea)

This is a type C_3 perennial herbaceous plant that grows in the cool season and has an excellent resistance to flooding. Its productivity is strongly influenced by high levels of nitrogen fertilizers, a feature that makes it very useful for the distribution of fertilizer from livestock.

2.8 Sugar cane (saccharum officinarum)

This plant only grows well in tropical and subtropical regions, which is why it is particularly common in Brazil. It has a 12-17% sugar content, 10% of which is glucose and the other 90% is saccharose. Milling can extract 95% of the total sugar content and the juice can subsequently be used to produce sugar or allowed to ferment to produce bioethanol. The bagasse (i.e. the solid residue remaining after milling) can be used as a source of energy and heat.

2.9 Sugar beet (beta vulgaris)

This plant generally grows in the cooler temperate regions, so it is abundant in Europe, North America and Asia. In the ethanol production process, the sugar beet is sliced and, while the juice is used to produce sugar or ethanol, the pulp is dried and used as animal feed or sold for pharmaceutical purposes.

2.10 Cereals

These must be ground to obtain starch, from which bioethanol is subsequently obtained. The cereals containing fewer proteins and more carbohydrates are preferable for distilling purposes because they have a higher bioethanol conversion rate. This means that the nitrogen content in the cereals can be adapted to facilitate starch accumulation instead of proteins synthesis, thereby improving both the energy yield and the quality of the fermentation process (Rosenberg et al., 2001). The principal cereals are:

2.10.1 Wheat

It grows mainly in temperate regions. The wheat treatment process is much the same as for the other cereals and it is best to use high-gravity fermentation to obtain the best performance in the fermentation process.

2.10.2 Barley

The most suitable is the so-called Winter variety, which is often underestimated as a foodstuff, despite the fact that it can tolerate drought and is highly adaptable.

2.10.3 Winter rye (secale cereale L)

This cereal relies heavily on the availability of nitrogen in the soil; it has high contents of both glucan and xylan (40.8% and 22.3% respectively) (Petersson et al., 2007).

2.10.4 Corn stover

This is what remains on the ground after maize has been harvested. This raw material is abundantly available and demands no further investment in biomass, although not all of the corn stover can be removed - 30% of it must be left on the ground to prevent erosion (by facilitating water infiltration and reducing evaporation), and as the main source of soil organic carbon (SOC) in order to preserve the soil's productivity. Corn stover contains polymeric hemicellulose and cellulose, but their biodegradability by glycosidase is strongly inhibited by a small quantity (12-15%) of lignin (Gressel, 2008; Varvel et al., 2007).

2.11 Jerusalem artichoke (helianthus tuberosus)

This plant grows in summer, reaching its maximum height in July and dying in October. The tubers are rich in inulin (a fructose polymer), which can be used to obtain a syrup for use both in the foodstuffs industry and in the production of ethanol. It was demonstrated (Curt et al., 2006) that, towards the end of the season, the potential for bioethanol production of the stems of clones is 38% of that of the tubers.

2.12 Municipal solid waste (MSW)

The most suitable waste for converting into bioethanol is the waste from the fruit and vegetable industries, for instance, cotton fiber, milk whey from cheese-making, the waste products of coffee making, and so on. Generally speaking, such waste contains approximately 45% of cellulose (glucose polymer), which can be simultaneously hydrolyzed and fermented to produce ethanol. SSL (Spent Sulfite Liquor) is a byproduct of bisulfite "pulp" manufacturing that can also be fermented to produce ethanol. Waste varies considerably in content from one area to another, but the majority of the volume generally consists of paper (20-40%), gardening waste (10-20%), plastics, glass, metals and various other materials (Prasad et al., 2007).

2.13 Miscanthus

This is a type C_4 graminaceous perennial that forms rhizomes. Miscanthus x giganteus is generally used to obtain biofuels: this is a sterile tetraploid hybrid obtained from Miscanthus sinensis and Miscanthus sacchariflorus, characterized by a yield that in autumn reaches 30 t ha^{-1} in irrigated soils and 10-25 t ha^{-1} in those without irrigation. The contribution of Miscanthus sacchariflorus to the Miscanthus x giganteus genome lies in its adaptability to warm climates, while Miscanthus sinensis provides the genetic resources needed in the colder regions. It is often used as an ornamental grass or cover crop and it can grow as much as 4 m high. It takes three years to arrive at a stable yield (around 5 years in marginal soils) and in its first year of growth the rhizomes are particularly sensitive to low temperatures, whereas in subsequent years they can even withstand temperatures of around -40°C. The rhizomes remain inactive in winter and begin to grow when the temperatures of the soil reaches 10-12°C. As for the plant's energy value, the dry matter has a net calorific value of approximately 17 MJ/kg. The energy value of 20 t of dry Miscanthus is approximately the same as that of 8 t of coal (Heaton et al., 2004; Sánchez & Cardona, 2008; DEFRA, 2011).

When Panicum virgatum and Miscanthus (Heaton et al., 2004) - both type C_4 plants of considerable interest as energy sources - are compared, Miscanthus produces more biomass per unit than Panicum virgatum (i.e. 12 Mg ha^{-1}). Both plants are perennials and this means a saving because there is no need to replant them. In areas with an abundant rainfall but problems of nitrogen contamination of the water supply, it is better to use Miscanthus as an energy crop, whereas growing Panicum virgatum with adequate nitrogen fertilizing certainly produces a better yield in uncontaminated dry areas.

3. Production processes

Bioethanol production processes vary considerably depending on the raw material involved, but some of the main stages in the process remain the same, even though they take place in different conditions of temperature and pressure, and they sometimes involve different microorganisms. These stages include hydrolysis (achieved chemically and enzymatically), fermentation and distillation.

Hydrolysis is a preliminary treatment that enables sugars to be obtained from the raw materials that are then fermented. In the case of enzymatic hydrolysis, effective pretreatments are needed, however, to increase the susceptibility of lignocellulose materials to the action of the enzymes. The following paragraphs describe the various production methods, distinguishing them according to the type of raw material involved.

3.1 Lignocellulose biomass

The biofuels obtained from wood cellulose and from organic materials in general offer considerable advantages over conventional biofuels. Burning ethanol obtained from cellulose produces 87% lower emissions than burning petrol, while for the ethanol from cereals the figure is no more than 28%. Ethanol obtained from cellulose contains 16 times the energy needed to produce it (Martinez et al., 2008), petrol only 5 times and ethanol from maize only 1.3 times. The problem is a matter of how to disrupt the bonds of this molecule in order to convert it into fermentable sugars.
In fact, this is unquestionably the type of raw material that is the most complicated to process. The starting material may be farming and forest waste, scrap woods, grassy crops grown for energy purposes or even municipal solid waste. Lignocellulose occurs in the walls of vegetable cells and consists of cellulose microfibers contained in the lignin, hemicellulose and pectin. The procedure to obtain ethanol consists first in depolymerizing the carbohydrates into their monomeric sugars, then fermenting the sugars with the aid of appropriate microorganisms. The lignocellulose biomass consists mainly of three basic polymers: cellulose, hemicellulose (such as xylane), lignin and other minor components (essential oils, acids, salts and minerals).

3.1.1 Pretreatments

These are used to modify the structure and dimensions of macroscopic and microscopic raw materials, and also their chemical composition. They have the effect of solubilizing the hemicellulose, reducing the crystallinity, and increasing the available surface area and porosity of the substrate. An effective pretreatment must meet following requirements: - it must increase sugar formation or facilitate the subsequent formation of sugars during hydrolysis, preventing any degradation or the loss of carbohydrates, and avoiding the formation of byproducts capable of inhibiting the subsequent processes of hydrolysis and fermentation, all at a competitive cost (Balat et al., 2008).
Pretreatments are particularly essential before enzymatic hydrolysis and may be of various types, i.e. physical, chemical, biological, steam explosion, and ammonia fiber explosion (AFEX).
Physical pretreatments may or may not be mechanical. The mechanical physical pretreatments include milling and grinding, that not only reduce the substrate, but also increase its surface area to volume ratio, thus making the cellulose easier to convert during hydrolysis. "Ball milling" could also be used to reduce the crystallinity of the cellulose, but this practice is not only very expensive, but also takes a long time (nearly a week) to complete, so it is hardly practicable on an industrial scale. The non-mechanical pretreatments feature a combination of high-power internal and external forces that decompose the lignocellulose.
Chemical pretreatments are used mainly to reduce the crystalline content of the cellulose. Using this type of pretreatment poses plant-related problems, however, since all the structural materials have to be capable of withstanding the severe working conditions imposed by the chemical agents.

The chemical pretreatments most often used are an alkaline treatment to delignify and solubilize the glycan, and an NaOH treatment that dissolves the lignocellulose biomass, destroying its lignin structure. Pretreatment with diluted sulfuric acid is also very important but this poses serious problems if it is associated with diluted acid hydrolysis, because the hydrolyzed end products become scarcely fermentable.

Other chemical pretreatments include: pretreatment with hydrogen peroxide (H_2O_2), which exploits oxidative delignification to separate and solubilize the lignin, and dissolve the lignocellulose matrices, thereby increasing the enzymatic digestibility of the mass; pretreatment with ozone, which degrades the lignin polymers; and pretreatment with liquid hot water (LHW), which is applied mainly to alfalfa. It was demonstrated (Laser et al., 2002) that, in ideal conditions, this method is as effective as diluted acid hydrolysis, without the need to use any acid or create any products of neutralization).

Biological pretreatments involve the use of enzymes, which are already useful in industrial processes on timber waste, in the processing of pulp and scraps. Several microorganisms studied years ago are the enzymes produced by the basidiomycetes Pleurotus ostreatus: these enzymes are homologous proteins characterized by different specifications, depending on which phenols are substituted. Another fungus in the basidiomycetes class that is effective in delignification is the Phanerochaete chrysosporium (Palmieri et al., 1997).

In the steam explosion process, saturated steam is used at very high temperatures and pressures to break up the chemical bonds in the cellulose, hemicellulose and lignin in order to break down the fibers and hydrolyze the biomass. The process consists in delivering steam under high pressure into a sealed chamber containing the lignocellulose material, then reducing the pressure and thus making the steam and matrix expand, and obtaining its explosive decompression through an orifice, which disrupts the cellular structure of the substrate, breaking up the acetyl groups of the hemicellulose. In some cases (e.g. Angiosperm), it is preferable to use acid catalysts, such as H_2SO_2 or SO_2, to make the cellulose-rich components more accessible to the enzymes. SO_2 gas is better able to attack the fibers (Shevchenko et al., 2000), but its use makes it necessary to carefully consider the working conditions in which the steam explosion takes place. In fact, it becomes necessary to find the best compromise between a strong enzymatic hydrolysis (obtainable in very severe conditions) and a good recovery of the components containing hemicellulose, that are in the form of monomeric sugars (which demand much less severe conditions) (Silverstein et al., 2007). That is why a severity indicator has been developed (Overend & Chornet, 1987), which correlates pretreatment temperatures and times, assuming that the pretreatment obeys Arrhenius's equation and has first-order kinetics. The indicator R_0 is:

$$R_o = t \cdot \exp\left[\frac{(T_r - T_b)}{14.75}\right] \tag{1}$$

where t is the duration of the pretreatment (min), T_r is the reaction temperature (°C), T_b is the baseline temperature (100°C) and the constant 14.75 is the conventional activation energy, assuming that the whole conversion is of the first order. If the version with sulfuric acid is being used, then the severity parameter, called M_0 in this case, is slightly modified:

$$M_o = t \cdot C^n \cdot \exp\left(\frac{T_r - T_b}{14.75}\right) \tag{2}$$

where C is the chemical concentration (wt%) and n is an arbitrary constant (Chum et al., 1990).

Ammonia fiber/freeze explosion (AFEX) pretreatment involves the use of liquid ammonia and steam explosion: in this process, the previously-humidified lignocellulose material is placed in a vessel under pressure with liquid NH_3 in proportions of 1-2 kg NH_3/kg of dried biomass. This method is very effective for non-woody materials such as bagasse and newspaper, but less so in the case of "soft" wooden materials. This system does not release any sugars directly, but it does make the polymers (hemicellulose and cellulose) easier for the enzymes to attack. The ammonia can also be replaced with carbon dioxide because the latter is relatively less costly and also because the alcohol waste product contains traces of pollutants that would thus derive only from the lignin.

The most promising pretreatments for farming waste are AFEX and LHW, while pretreatment with steam affords a high output of sugars from both farming waste and forest waste.

3.1.2 The hydrolysis process

Hydrolysis is governed by the law:

$$(C_6H_{10}O_5)_n + nH_2O \rightarrow nC_6H_{12}O_6 \quad (3)$$

and can be mainly of two types: acid (using diluted or concentrated acids) or enzymatic. A lignocellulose biomass is more complicated to hydrolyze than pure cellulose because it contains components that are not glucose-based, such as hemicellulose and lignin.

A lignocellulose biomass undergoing acid hydrolysis mainly produces xylose, while the lignin and cellulose fractions remain unchanged. This is because xylan is more susceptible to hydrolysis in moderately acid conditions because of its amorphous structure, while cellulose demands more severe conditions because of its crystalline nature.

If hydrolysis is implemented using 1% diluted sulfuric acid, the hemicellulose is depolymerized at a lower temperature than the cellulose. This process is usually conducted in two consecutive stages.

One of the most important characteristics of this type of hydrolysis is the rate of the reactions involved, which facilitate the continuity of the process. To speed up the diffusion of the acid, the raw material is mechanically reduced to pieces a few millimeters in size.

Hydrolysis with concentrated acids (10-30%), on the other hand, rapidly and completely converts cellulose into glucose and hemicellulose into xylose, with some degree of degradation. The acids most often used are sulfuric and hydrochloric acid, and hydrogen fluoride.

This type of acid hydrolysis has the great advantage of recovering the sugars very efficiently (approximately 90% of hemicellulose and cellulose are depolymerized into monomeric sugars). From an economic standpoint, this process enables a reduction in production costs by comparison with the diluted acid solution, especially if the acids are retrieved and reconcentrated. The acids and sugars in solution are separated by ion exchange so the acid is reconcentrated by passing it through a series of multiple-effect evaporators. The remaining solid fractions, which are rich in lignin, are collected and can be made into pellets for use as fuel.

So, in short, we can divide concentrated acid hydrolysis into two stages: in the first stage, the concentrated acid (70%) destroys the crystalline structure of the cellulose, breaking up the hydrogen links between the cellulose chains; in the second stage, hydrolysis induces a hydrolytic reaction in the single isolated cellulose chains.

The enzymatic hydrolysis of natural lignocellulose materials is a very slow process, because it is hindered by several structural parameters of the substrate, such as its of cellulose and hemicellulose content, and the surface area and crystallinity of the cellulose. Pretreatments are consequently needed to make the biomass more susceptible to attack by hydrolysis. For the same reason, a cocktail of enzymes has to be used that is capable of breaking the links in the polymeric chains. This cocktail is usually a mixture of various hydrolytic enzymes, including cellulase, xylanase, hemicellulase and mannoxidase. Enzymatic cellulose degradation is a complex process because it takes place in limit conditions between the solid and liquid phases, where the enzymes are the mobile components. Generally speaking, degradation is characterized by a rapid initial phase followed by a slower second phase that can continue until all the substrate has been used up. The reason for this behavior is usually assumed to be because the accessible fraction of cellulose is quick to hydrolyze, followed by the slow activation of the absorbed enzyme molecules.

Chopping up the biomass increases the surface area accessible to the enzymes and reduces the polymerization and crystallinity of the cellulose, thus enabling a smaller quantity of enzymes to be used and the production costs to be contained.

Both bacteria and fungi can produce the cellulase for the hydrolysis of lignocellulose materials. The bacteria may be aerobic or anaerobic, mesophylic or thermophylic. The bacteria most often used are Clostridium, Cellulomonas, Bacillus, Thermomonospora, Ruminococcus, Bacteriodes, Erwinia, Acetovibrio, Microbispora and Streptomyces. The enzymes are usually classified according to their reaction site, so they may be intracellular (or cell-associated) or extracellular. The main function of extracellular enzymes is to convert the substrate into an external medium by taking effect on the cell mass constituents. Conversely, intracellular enzymes need the substrate to spread through the cellular mass before it can be converted.

The most widely accepted mechanism for the enzymatic hydrolysis of cellulose involves the synergic action of the enzymes endoglucanase (or endo-1,4-β-glucanase, EG), exoglucanase (or cellobiohydrolase, CBH), and β-glucosidase. Both EG and CBH are extracellular enzymes, while β-glucosidase is intracellular. EG randomly disrupts the cellulose chains, consequently inducing their strong degradation. It takes effect by hydrolyzing the β-1,4-glucoside bonds, creating new ends in the chains. Exoglucanase breaks up the ends of the chains, thus enabling the release of soluble cellobiose or glucose. BGL hydrolyses the cellobiose into glucose, thus eliminating the inhibitory cellobiose; then BGL completes the process by catalyzing the hydrolysis of cellobiose into glucose. Most cellulase and hemicellulase producers are microorganisms such as the filamentous fungi, e.g. Trichoderma sp., which can be used in their natural form or genetically modified (Trichoderma viride, Trichoderma reesei, Trichoderma longibrachiatum). CBH I and CBH II are the main enzymes of Trichoderma reesei, while EG I and EG II are the dominant endoglucanases.

Enzymatic activity is influenced by various parameters, such as temperature (a 20-30°C increase in temperature leads to a 3- to 5-fold increment in the end products). The crucial issue of temperature lies in the risk of an unwanted denaturation when the temperature is too high (Balat et al., 2008). Enzymatic hydrolysis, with or without the addition of catalysts, has generally proved capable of a high yield of both glucose (>90%) and xylose (>80%).

3.1.3 Fermentation

After hydrolysis, the hydrolyzed products must be fermented by means of microorganisms such as yeasts (Hahn-Hägerdal et al., 2006). Since the hydrolyzed products are composed mainly of glucose, xylose, arabinose and cellobiose, the microorganisms used must be capable of fermenting all of them efficiently for ethanol to be produced on a large scale. The reactions that involve glucose and xylose are respectively:

$$3C_5H_{10}O_5 \rightarrow 5C_2H_5OH + 5CO_2 \tag{4}$$

$$C_6H_{12}O_6 \rightarrow 2C_2H_5OH + 2CO_2 \tag{5}$$

The classic method used in the fermentation of the hydrolyzed biomass is separate hydrolysis and fermentation (SHF), in which the two processes are completed in different units. A commonly used alternative is simultaneous saccharification and fermentation (SSF), in which hydrolysis and fermentation are completed in the same unit. A last option is represented by consolidated bioprocessing (CBP).

When the SHF process is used, the solid fraction of the lignocellulose material undergoes hydrolysis and this process is called saccharification. The liquid fraction, on the other hand, goes first to the reactor for glucose fermentation, then it is distilled to extract bioethanol, leaving behind only the unconverted xylose, which is then fermented in a second reactor and then undergoes a second, final distilling phase.

The main advantage of this process consists in that separating the processes of hydrolysis and fermentation enables optimal working conditions to be adopted in each case. The enzymes are free to work at high temperatures, while the microorganisms can induce fermentation at more moderate temperatures.

Among the disadvantages, in addition to needing two twin reactors, there is the fact that the enzymes for hydrolyzing the cellulose are inhibited end products. The rate of hydrolysis progressively declines due to the accumulation of glucose and cellobiose.

This process has sometimes been used to produce ethanol from a mix of municipal solid waste: in this case, enzyme recycling was improved using micro- and ultra-filtering procedures, thus achieving the hydrolysis of 90% of the cellulose with a net enzyme load of 10 FPU/g of cellulose (where FPU stands for filter paper unit) (Sánchez & Cardona, 2008).

In the SSF procedure, enzymatic hydrolysis and fermentation take place simultaneously. Cellulases and microorganisms take effect in the same process, so the glucose produced by hydrolysis of the cellulose is immediately consumed by the bacterial cells that convert it into ethanol. SSF achieves the highest output of bioethanol at the lowest costs, since the lesser demand for enzymes is lower because the inhibitory effect of the cellobiose and glucose end products is alleviated by fermentation with yeast. This is a discontinuous type of process that uses natural heterogeneous materials containing complex polymers such as lignin, pectin and lignocellulose. The greatest advantages offered by SSF are a faster rate of hydrolysis thanks to the conversion of the sugars that inhibit cellulase activity, a low enzyme demand, a high product yield, the need for less sterile conditions, a shorter process time, and smaller overall reactor dimensions (Sun & Cheng, 2002).

This process also has far from negligible disadvantages, however, the most significant of which consists in the need to complete fermentation and hydrolysis in suboptimal conditions. That is why microorganism selection and preparation is so important for this process. The cocktail of enzymes for hydrolyzing the cellulose must likewise remain stable within a wide range of temperatures and pH. As for the Saccharomyces cerevisiae cultures, the typical working conditions in SSF involve a pH of 4.5 and temperatures of around 310 K. Experiments have recently been conducted with a new variant of this process called simultaneous saccharification and cofermentation (SSCF), in which the five- and six-carbon sugars are fermented simultaneously. In SSCF, hydrolysis continuously releases hexose sugars that increase the rate of glycolysis, so that the pentose sugars can ferment more quickly and produce a higher yield.

In CBP, four biologically-mediated conversions take place in a single process, i.e. the production of glycolytic enzymes (cellulase and hemicellulase), hydrolysis of the carbohydrate component of the pretreated biomass to obtain sugars, fermentation of the six-carbon sugars (mannose, galactose and glucose), and fermentation of the five-carbon sugars (xylose and arabinose).

The main difference between CBP and the other processes consists in that there is no single process focusing on cellulase production. CBP, also known as direct microbial conversion (DMC), requires just one microbial community for both cellulase production and fermentation. The weakness of this approach lies in the difficulty of finding an organism sturdy enough to simultaneously produce cellulase and ethanol with a high yield. Wyman (Wyman, 1994) wrote that many studies on CBP involved the use of the bacterium Clostridium thermocellum for enzyme production, cellulose hydrolysis and glucose fermentation, while Clostridium thermosaccharolyticum enabled the simultaneous conversion of the pentose sugars obtained from hemicellulose hydrolysis into ethanol. Using Clostridium thermocellum in the system also induces a 31% higher conversion of the substrate than when Trichoderma reesei or Saccharomyces cerevisiae are used. Recent studies have focused on cellulase production combined with a high ethanol yield using strains of Escherichia coli, Klebsiella oxytoca and Zymomonas mobilis as well as the yeast Saccharomyces cerevisiae. The expression of cellulase in Klebsiella oxytoca increased the yield from microcrystalline cellulose hydrolysis and enabled an anaerobic growth in the amorphous cellulose. Various cellobiohydrolases have likewise been functionally expressed in the Saccharomyces cerevisiae. Genetic engineering and metabolic studies will enable the development of new stable strains of microorganisms capable of converting the cellulose biomass into bioethanol, leading to improvements in the industrial bioethanol production process (Lynd et al., 2005).

The microorganisms used during the fermentation process must be capable of working efficiently on both monosaccharide and polysaccharide sugars, so they have to be very versatile. The survival of these bacteria is only assured in controlled pH conditions and the majority of the microorganisms cannot tolerate bioethanol concentrations in excess of 10-15% (w/v).

Saccharomyces cerevisiae is one of the microorganisms most often used because it affords a high ethanol yield from hexose sugars, and it can tolerate bioethanol and inhibitory compounds very well. It has the great disadvantage, however, of being unable to assimilate C_6 sugars.

The ethanol-generating bacteria that seem industrially most promising are Escherichia coli, Klebsiella oxytoca and Zymomonas mobilis. Zymomonas, in particular, has demonstrated an aptitude for rapidly and efficiently producing bioethanol from glucose-based raw materials and, by comparison with the other yeasts, it has demonstrated a 5-fold higher yield. The ethanol it produces in the fermentation of the glucose corresponds to a yield that is 97% of the theoretical yield and in concentrations up to 12% (w/v). This bacterium is also

capable of producing bioethanol efficiently from fructose and saccharose (C_5), but not from C_6 sugars.
There are also yeasts that naturally ferment xylose, such as Pichia stipitis, Candida Shehatae and Candida parapsilopis, and they can do so through the action first of xylose reductase (XR), which converts xylose into xylitol, and then of xylitol dehydrogenase (XDH), which converts xylitol into xylulose. Bioethanol fermentation from xylose can also be achieved by recombinant Saccharomyces cerevisiae using the heterologues XR and XDH of Pichia stipitis and xylulose kinase (XK) of Saccharomyces cerevisiae.
Table 1 summarizes the commonly used bacteria and microorganisms (Balat et al. (2008)), highlighting the principal parameters used to assess the performance of the various types of fermentation.

Species	*Characteristics*
Clostridium acetobutilicum	Useful in fermentation of xylose to acetone and Butanol.
Clostridium thermocellum	Capable of converting cellulose directly to ethanol and acetic acid.
Escherichia coli	Native strains ferment xylose to a mixture of Bioethanol.
Klebsiella oxytoca	Native strains rapidly ferment xylose and cellobiose.
Lactobacillus pentoaceticus	Consumes xylose and arabinose.
Latobacillus casei	Ferments lactose very well; particularly useful for bioconversion of whey.
Lactobacillus xylosus	Uses cellobiose if nutrients are supplied: uses D-glucose, D-xylose, and L-arabinose.
Lactobacillus pentosus	Homolactic fermentation. Some strains produce lactic acid from sulfite waste liquors.
Lactobacillus plantarum	Consumes cellobiose more rapidly than glucose, xylose, or arabinose.
Zymomonas mobilis	Normally ferments glucose and fructose.

Table 1. Commonly used bacteria and microorganisms (Balat et al. (2008)).

Fermentation can occur in various ways, i.e. discontinuously, continuously, with cells immobilized, and batch-fed (Chandel et al., 2007).
A problem encountered in enzymatic hydrolysis consists in the formation of inhibitors. The activity of the enzymes is strongly influenced by certain levels of cellobiose, glucose or products such as furfural and organic acids deriving from pretreatments.
Inhibitors form in relation to the conditions in which enzymatic hydrolysis takes place. Conditions can be selected that should provide maximum solubilisation and recovery of the hemicellulose component (low severity), optimum enzymatic hydrolysis of the water

insoluble cellulosic component (high severity), and a compromise between the two conditions (medium severity). The combined severity (CS) links the severity factor (R0) to the ambient pH, and this index expresses the intensity of the previously-described factors. Its value is expressed as:

$$CS = \log R_0 - \text{pH} \qquad (6)$$

When the CS increases beyond the value that generates the highest concentrations of mannose and glucose, the cellulose and hemicellulose break down and there is a drop in the concentration of fermentable sugars that coincides with the formation of furfural and hydroxy methyl furfural (HMF), which subsequently degrade into levulinic and formic acids. To achieve both the maximum fermentability and a high yield of fermentable sugars, the *CS* should be around 3 (Palmqvist & Hahn-Hägerdal, 2000)).

Inhibitors can come from various sources, e.g. equipment, carbohydrate degradation, lignin decomposition, wood extracts and their decomposition. They can be classified according to their structure as organic, acid, furanes and phenolic compounds. The fermentation inhibitors in particular include the furane derivatives, such as furfural and 5-hydroxy-methyl-furfural (5-HMF), the aliphatic acids, such as acetic acid, formic acid and levulinic acid, and the phenolic compounds. The furane derivatives can further react to form certain polymeric materials. The formation of inhibitory compounds makes it necessary to introduce changes in the production process, such as process water recirculation. It was demonstrated (Palmqvist et al., 1996), for instance, that unconcentrated hydrolyzed products have a moderately inhibitory action, while five-fold concentrations of nonvolatile components almost completely inhibit the fermentation of ethanol by Saccharomyces cerevisiae.

The formation of inhibitors and consequently of toxic compounds is a problem that has a negative fallout on the rate of both enzymatic hydrolysis and fermentation. The toxic compounds can form during steam explosion pretreatments and also during hydrolysis in the presence of low acid concentrations, and they are mainly the products of lignin degradation.

Four main groups of inhibitors have been identified in hydrolyzed lignocellulose products these are acetic acid from the hemicellulose fraction, products of lignin degradation, products of sugar degradation, and extracts that have been solubilized during the pretreatment.

The fermentation inhibitors, on the other hand, can be divided into various groups, depending on their origin:

- substances released during pre-hydrolysis and hydrolysis: acetic acid and extracts including terpenes, alcohols and aromatic compounds (e.g. tannins);
- inhibitors produced as a byproduct of pre-hydrolysis and hydrolysis, due to sugar degradation (furfural, 5-HMF);
- products of lignin degradation, including sizable groups of aromatic and polyaromatic compounds with a great variety of constituents (cinnamaldehyde, p-hydroxybenzaldehyde, syringaldehyde);
- products of the fermentation process, such as ethanol, acetic acid, glycerol and lactic acid,
- metals released by equipment and additives, e.g. nickel, iron, copper and chrome.

The compounds revealing the greatest inhibitory potential are acetic acid and the products of lignin degradation (Larsson et al., 1999).

A detoxification procedure can be used to improve the sugars' fermentability. Detox methods may be physical, chemical or biological, and they are impossible to compare directly with one another because the degree to which they can neutralize the inhibitors varies. The different microorganisms suitable for this purpose can tolerate the inhibitors to varying degrees. The choice of the most suitable method consequently depends on the raw materials involved and the composition of the hydrolyzed products. Figure 1 shows a flowchart of ethanol production from lignocellulose raw materials.

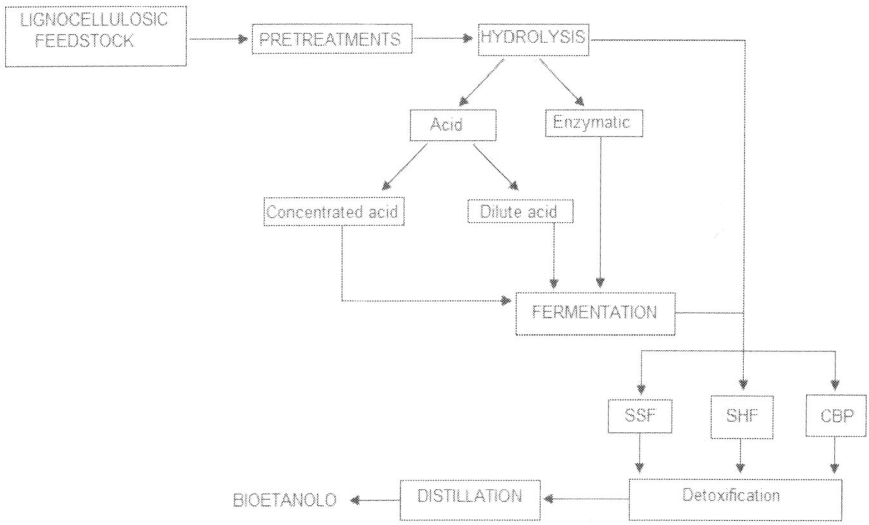

Fig. 1. Flowchart of ethanol production from lignocellulose raw materials

3.2 Raw materials containing starch

Starch is a biopolymer defined as a homopolymer. The constituent monomers are molecules of glucose held together by bonds between the oxygen atom of one molecule and the carbon atoms of adjacent molecules. These particular bonds are described as glycosidic and may be type α or type β, depending on the stereoisomery of the anomeric carbon in the molecule. Starch in plants occurs in the form of granules containing two main constituents in variable proportions, depending on the resource, i.e. amylose (16-30%) and amylopectin (65-85%). These are both type α glucose polymers. Amylose is a glucose polymer held together by α-1,4 bonds in linear chains, while amylopectin is a highly-branched glucose polymer with type α-1,6 bonds. Inside the cell, the starch is in the form of granules located in the amyloplasts. The granules contain both amorphous and crystalline regions, in proportions of approximately 30-70, respectively.

The starch for ethanol production comes mainly from cereals, wheat or corm being at the top of the list in North America and Europe, and tubers such as manioc in the tropical regions. In order to produce bioethanol from starch, its carbohydrate chains have to be broken down to obtain glucose syrup, which can then be converted into bioethanol with the aid of yeasts.

3.2.1 Starch hydrolysis

Various microorganisms are capable of hydrolyzing starch, though a preliminary process called gelatinization is needed to ensure an efficient hydrolysis. During this preliminary process, the starch granules swell, particularly rupturing the hydrogen bonds in the crystalline regions. The long glucose chains comprising the starch must be converted into fermentable sugars by means of a process called the "hydrolysis technique", during which the starch reacts with the water normally used to break down the starch into its fermentable sugars. There are numerous microorganisms capable of hydrolyzing starch, but those involved in the starch degradation process are generally amylase, α-amylase, β-amylase and isoamylase. The most important for the purposes of the SSF process are certainly the first two. α-amylase is an endo-amylase that randomly attacks the α-1,4 bonds, rapidly reducing the starch molecule's dimensions and consequently also its viscosity, i.e. it liquefies the starch. α-amylase can be obtained by means of heat-resistant bacteria such as Bacillus licheniformis, or by means of new strains of Escherichia coli or Bacillus subtilis, used on the starch suspensions during the first hydrolysis stage. For amylase to succeed in attacking these suspensions, they must be brought up to high temperatures (90-110°C) to rupture the starch cell nuclei. The products of this preliminary hydrolysis phase, called liquefaction, is a solution containing dextrins and a small amount of glucose.

$$\text{Starch} \xrightarrow[\text{(liquefaction)}]{\alpha\text{-amylase}} \text{Dextrins} \xrightarrow[\text{(saccharification)}]{\text{glycoamylase}} \text{Glucose} \quad (7)$$

At this point, the liquefied starch undergoes saccharification at low temperatures (60-70°C), induced by the action of glycoamylase generally obtained from Aspergillus or Rhizopus species. This enzyme is an exo-amylase capable of producing units of glucose from amylose and amylopectin chains.

The factors that influence starch hydrolysis include the substrate, enzyme activity and the reaction conditions (temperature, pH and other parameters) (Prasad et al., 2007). The microorganisms take effect more easily on gelatinized starch, but this process demands large amounts of energy so on an industrial level there has been a tendency to focus on using microorganisms capable of growing on ungelatinized starch. Various studies on this issue have considered certain species of fungi for producing enzymes capable of degrading starch in its natural state (Soccol et al., 1994). Liquefaction is followed by a saccharification stage under the effect of glycoamylase.

3.2.2 Milling

The milling phase enables the starch to be extracted from the biomass and it is very important for the purposes of analyzing the bioethanol production process as a whole because it strongly influences not only the subsequent stages but also the co-products obtained at the end of intermediate stages, which also vary according to the specific raw material entering the process (wheat, barley, corn, oats). The two main options are wet milling and dry milling.

Wet milling is the standard procedure generally used in the starch-based foodstuffs industry. Though this procedure demands more energy and more economic resources, and it delivers a smaller quantity of ethanol, it is still preferred at industrial level because its capacity to separate the grain into its components enables a purer form of starch to be obtained, along with more valuable byproducts. Wet milling can be used to obtain not only ethanol, but also products such as corn oil, gluten-based foods and flour, and corn steep liquor (CSL).

Dry milling means there is no need to pre-treat the raw material, which simply has to be ground before going through the other processing stages (hydrolysis, fermentation, distillation), which are identical to those following the wet milling process. Because dry milling does not break down the cereals into their various components, the unfermentable residue leaving the process that extracts the ethanol from the fermentation broth is rich in proteins, fibers, fats and sugars.

3.2.3 From hydrolysis to bioethanol

After the preparatory stage, the glucose solution can be fermented in ethanol. The temperature of the glucose is lowered to around 35°C, and then the yeast (usually Saccharomyces cerevisiae) is added and the anaerobic fermentation process begins, which converts the glucose into ethanol and carbon dioxide.

$$\text{Glucose} \xrightarrow[\text{Alcoholic fermentation}]{\text{Saccharomyces cerevisia}} 2\ \text{EtOH} + 2\ CO_2 \tag{8}$$

As a rule, the preferred method is to conduct the saccharification and fermentation steps during the same stage of the production process. Fermentation can be completed in two stages (Verma et al., 2000) using starch treated with α-amylase and glycoamylase.

Fermentation may be continuous or discontinuous, it makes no difference. When the fermentation broth reaches an ethanol content of around 8-10% v/v (beyond which the yeast can no longer survive), the ternary mixture is distilled by adding benzene or cyclohexanone, or using molecular sieves. After distillation, the ethanol is 95% pure.

In 2006, a research group (Robertson et al., 2006) experimented with the so-called "cold hydrolysis" of starch, concluding that the potential use of this method relies on the discovery and characterization of more efficient enzymes and the development of processes with a high level of integration, such as simultaneous liquefaction, saccharification and fermentation, along with other factors. Figure 2 shows the flow chart for bioethanol production from materials containing starch.

3.3 Raw materials containing saccharose

For the purposes of bioethanol production, the most important raw materials containing saccharose are unquestionably sugar cane and sugar beet. Two thirds of the world's sugar production derives from cane, the other third from beet.

3.3.1 Sugar cane (saccharum officinarum)

Sugar cane contains approximately 12-17% of total sugars, 90% of which are saccharose and 10% are glucose. Milling can extract approximately 95% of the sugar, leaving behind the solid residue. This cane residue goes by the name of bagasse. Sugar cane is washed in order to undergo a primary "crushing" process before milling. The cane juice obtained undergoes a clarification process in which the pH is balanced and cachaça is obtained, which can be sold as animal feed or as a component in mixtures. Fermentation is usually done with the aid of a yeast, Saccharomyces cerevisiae, which is separated in a continuous phase by centrifugation and reused in the fermenter. The fermentation process differs slightly, depending on whether all the juice is used to obtain bioethanol or whether part of it is drawn off to obtain sugar: in the former case, the juice is heated up to 110°C (to reduce the risk of bacterial contamination), then decanted and fermented; in the latter, the crystals formed by concentration are centrifuged, leaving a viscous syrup called molasses.

The extract leaving the fermenter must then be distilled to extract the hydrated ethanol (an azeotropic solution containing 95.5% v/v of ethanol and 4.5% v/v of water), which is dehydrated using molecular sieves or azeotropic distillation (i.e. with cyclohexanone or benzene) to obtain a higher-grade, anhydrous ethanol. In addition to ethanol, there is also an aqueous solution leaving the distillation process, that is called residue.

Molasses obtained from sugar cane are the most important raw material for the purposes of bioethanol production. In recent years, however, there have been rising prices and restrictions on the availability of molasses, which have strongly influenced the production of bioethanol (Quintero et al., 2008). Figure 3 shows the flow chart for bioethanol, energy and sugar production from sugar cane.

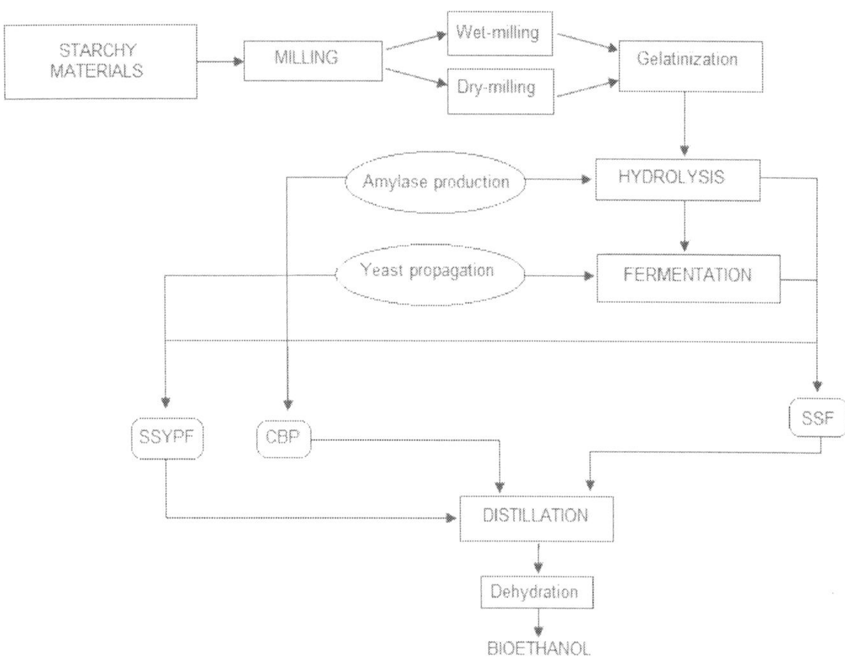

Fig. 2. Flow chart for bioethanol production from materials containing starch

3.3.2 Sugar beet (beta vulgaris)

Like sugar cane, sugar beet can also be used to obtain bioethanol by fermenting and distilling its juice. The beet is first cut into thin slices, then placed in contact with a medium (water or juice extracted from a previous process) and brought up to a temperature of about 70-80°C. In the case of sugar beet, temperature is a fundamental extraction parameter because it must be high enough to rupture the proteins in the cell walls containing the sugars, which has the effect of allowing the sugars to spread through the medium. Once this process has been completed, the sugar beet pulp is dried and sold as animal feed or to the pharmaceutical industry for use in the production of citric acid and its esters. The beet juice proceeds instead through the stages that convert it into bioethanol. At plants where sugar and bioethanol are both produced together, the juice can either be used directly or it can be concentrated in evaporators and stored for several months. Both the fresh and the concentrated sugar juice can be used in production processes involving cold crystallization and fermentation. The fermentation process relies on the use of yeasts (preferably Saccharomyces cerevisiae) or bacteria such as Zymomonas mobilis (Linde et al., 1998), which is only used in the case of a discontinuous fermentation. The great interest focusing on the bacteria is due to their capacity to convert the glucose into ethanol more efficiently than yeasts succeed in doing. Figure 4 shows the flow chart for the production of bioethanol and byproducts from sugar beet.

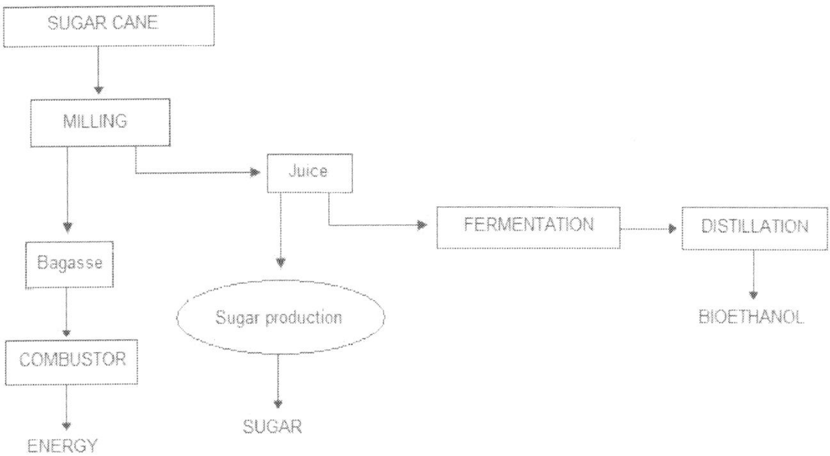

Fig. 3. Flow chart for bioethanol, energy and sugar production from sugar cane

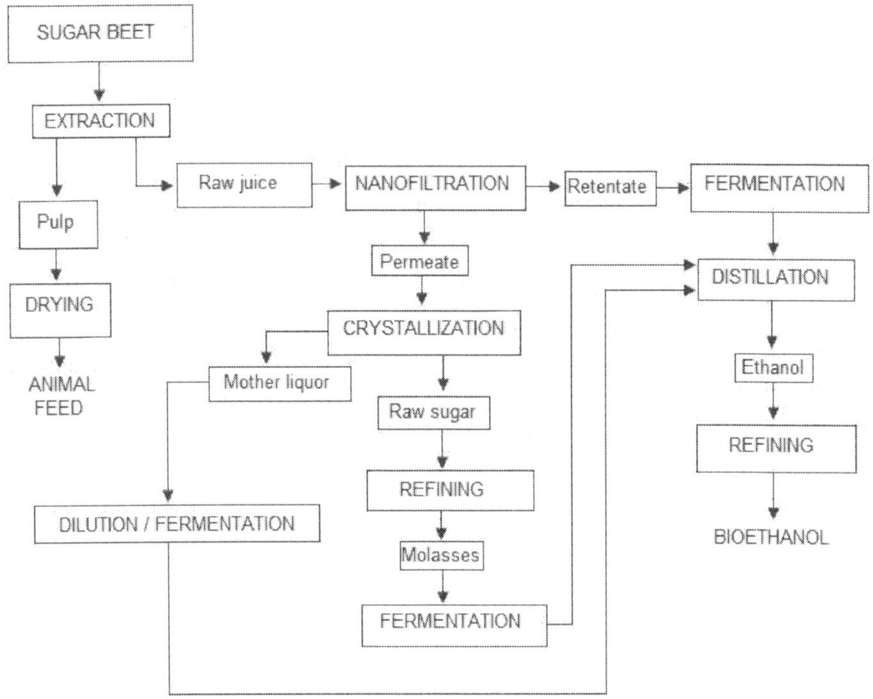

Fig. 4. Flow chart for the production of bioethanol and byproducts from sugar beet

3.4 Comparing the various raw materials

The choice of the raw materials to use to produce ethanol depends largely on local climatic conditions. North America and Europe, for instance, have based their ethanol production on materials containing starch, because of their particular farming and ecological conditions, which make it unfeasible to sugar cane adequately, although this plant offers a higher ethanol yield. In these areas, the most often grown energy crops are cereals. Using these raw materials poses some energetic sustainability limits (Patzek et al., 2005; Pimentel, 2003). The yield per ton of raw material is higher for sugar beet molasses than for cereals, so although growing sugar beet is less productive in quantitative terms than growing cereals, the annual ethanol yield from beet is higher than from cereals. The importance of analyzing the geographical position of crops helps us to see that growing the same type of cereal in tropical regions would produce a distinctly lower yield than could be achieved from the same plant grown in more suitable areas (Espinal et al., 2005). The lignocellulose materials represent the future as concerns raw materials for ethanol, because of their excellent energy value, great availability, low cost and high bioethanol yield.

These materials cannot be used to produce food, but they provide important secondary products such as methanol, syngas, hydrogen and electricity. The choice of which lignocellulose material also depends on the nature of the waste products in a given country (Kim & Dale, 2004). Cereals that are discarded during the distribution process can be destined to ethanol production, together with farming waste and sugar cane bagasse. The drawback of these raw materials consists in the complexity of the phenomena involved in converting the biomass into ethanol. Various studies have been conducted on the process of bioethanol production starting from various raw materials, including lignocellulose materials, cereals (McAloon et al., 2000; Cardona et al., 2005), and sugar cane (Quintero et al., 2008).

3.5 Converting syngas into ethanol

Bioethanol can also be obtained by means of chemical processes (Sánchez & Cardona, 2008; Demirbas, 2005), which may or may not demand the presence of microorganisms in the fermentation stage. Gasification of a biomass to obtain syngas ($CO + H_2$), followed by the catalytic conversion of the syngas, has the potential for producing ethanol in large quantities. The catalysts most often used and studies are those based on rhodium (Rh) (Holy & Carey, 1985; Yu-Hua et al., 1987; Gronchi et al.; 1994).
The geometrical structure of the active site seems to be:

$$\left(Rh_x^{0}Rh_y^{+}\right)-O-M^{n+} \tag{9}$$

where part of the Rh occurs as Rh^+ and the promoter ion (M^{n+}) is in close contact with these Rh species. The carbon monoxide is then hydrogenated to form an absorbed species $-CH_x-$ that is then inserted in the absorbed CO. Hydrogenation of these absorbed species leads to the formation of ethanol (Subramani & Gangwal, 2008).
Another mechanism considered valid for ethanol formation involves the use of acetate (acetaldehyde formation followed by reduction) and is known, in the cases of Rh-based catalysts, to be promoted by manganese (Luo et al., 2001).
In this case, ethanol is formed by direct hydrogenation of tilt-absorbed CO molecules, followed by CH_2 insertion on the surface of the CH_2-O species to form an absorbed intermediate species. Ethanol is produced by hydrogenation of the intermediate species of CH_2-O. Acetaldehyde is formed by the insertion of CO on the surface of the CH_3-Rh species, followed by hydrogenation. The catalyst's performance can be improved by modifying its composition and preparing the ideal conditions for the reaction (Subramani & Gangwal, 2008). Manganese (Lin et al., 1995), Samarium and Vanadium (Luo et al., 2001) can also be used as promoter ions in processes involving Rh.

4. Environmental Issues

The greenhouse gases (GHGs) are gases occurring in the Earth's atmosphere that absorb in the infrared field (carbon dioxide, ozone, methane, nitrogen oxides, carbon monoxide and so on). This feature enables them to trap the heat of the sun reflected back from the Earth's surface.
The GHG that occurs in the largest quantities is carbon dioxide, and that is why it attracts so much attention. In fact, the carbon cycle is a delicate balance between carbon accumulation, release and recycling that enables vegetable and animal species to survive. Problems linked to CO_2 began to emerge at the start of the industrial era: the ever-increasing use of fossil fuels as a source of energy meant that the carbon dioxide trapped for centuries in the fossils was being put back into the atmosphere, with no correspondingly reinforced recycling mechanism, which relies on chlorophyllic photosynthesis).

In addition to reducing carbon dioxide emissions, bioethanol can be seen as a no-impact fuel because the amount of CO_2 released into the atmosphere is compensated by the amount of CO_2 converted into oxygen by the plants grown to produce the bioethanol (Ferrel & Glassner, 1997).

4.1 Carbon sequestering

In the analysis of the environmental impact of bioethanol (and other biofuels too), some of the key factors concern the impact of the increasing quantities of dedicated crops on soil carbon levels and subsequent photosynthesis: these changes will also influence the atmospheric concentrations of GHG such as CO_2 and CH_4.
The main problem concerns the fact that, when a system in equilibrium experiences persistent changes, it can take decades before a new equilibrium with a constant carbon level is reached. Taking the current situation in Europe as concerns wheat and sugar beet crops, there is an estimated depletion of approximately 0.84 t of C or 3.1 t CO_2 equivalent ha^{-1} $years^{-1}$ from the ground. If no crops were grown on the soil, this depletion would be even greater, i.e. 6.5 t of C each year for sugar beet and 4.9 t of C for wheat. Apart from the effects on ground carbon levels, there are also signs of other adverse effects indirectly linked to crops grown for energy purposes, such as the increase in the amount of C in the atmospheric levels of GHG. Irrigation with good-quality water also exacerbates carbon sequestering: the water used for irrigation contains dissolved calcium and carbon dioxide (in the form of HCO_3^-); Ca and HCO_3^- react together, giving rise to the precipitation of $CaCO_3$ and the consequent release of CO_2 into the atmosphere. In the typical dry conditions of the USA, further reactions take place and irrigation is responsible for the transfer of CO_2 from the ground into the atmosphere (Rees et al., 2005). An important type of crop that can be used to reduce soil carbon sequestering is defined as "zero tillage", which means that it can be grown year after year without disturbing the soil. Seed crops (such as wheat) may be zero tillage, but not root crops (such as Panicum virgatum). Zero tillage has variable effects, and in some cases carbon sequestering in the soil may even increase, but this phenomenon can be completely overturned by a one-off application of conventional tillage. If only the carbon in the soil is considered, zero tillage leads in the long term to less global warming than growing conventional crops in damp climates, but in areas with dry climates, there is no certainty of any such beneficial effect (Six et al., 2004). Using straw from cereals can increase the carbon levels in the soil. Such residue is useful in maintaining soil carbon levels (Blair et al., 1998; Blair and Crocker, 2000) because it has a low rate of breakdown, so it is important for the residue to go back into the ground in order to keep the farming system sustainable. Since removing the residue from the ground has other negative effects too, such as an increased soil erosion and a lesser availability of macro- and micronutrients, some have suggested in the United States (Lal, 2005) that it would be advisable to remove only 20-40% of the residue for the purposes of bioethanol production, whereas it was claimed (Sheehan et al., 2004) that if up to 70% of the residue were removed to produce bioethanol, the carbon levels would initially decline and then remain stable for about 90 years. Increasing the land used to grow energy crops would have a substantial impact on the concentrations of carbon-containing gases in the atmosphere. If areas covered with forest were converted into arable land, the carbon sequestering would go from values of around 50-145 $t{\cdot}ha^{-1}$ to approximately 50-200 $t{\cdot}ha^{-1}$, assuming a 60-year rotation (Reijinders & Huijbregts, 2007).

4.2 Emissions

Mixing bioethanol with petrol, even in modest proportions, increases the octane number of the fuel and reduces the percentage of aromatic and carcinogenic compounds, and emissions of NO_x, smoke, CO, SO_x and volatile organic compounds (VOC). But there is also an increase in the emissions of formaldehyde and acetaldehyde. On the other hand, modern bioethanol production systems have an energy ratio (or net usable energy) of around 2 to 7, depending on the crops and processes used. The composition of petrols can influence the emissions of organic compounds: those containing aromatic hydrocarbons such as benzene, toluene, xylene and olefins produce relatively high concentrations of reactive hydrocarbons, while petrols formulated using oxygenated compounds (such as those mixed with bioethanol) may contain lower quantities of aromatic compounds.

The problem of petrols with high concentrations of aromatic compounds lies in their marked tendency to emit uncombusted hydrocarbons, which are difficult for catalytic converters to oxidize as well as being precursors of photochemical contamination. All oxygenated fuels have the potential for reducing the emissions of carbon monoxide (CO) and uncombusted hydrocarbons, which are also "photochemically" less reactive than the hydrocarbons of normal petrols. Because ethanol acts as an oxygenating agent on the exhaust gases of an internal combustion engine fitted with a three-way catalytic converter, adding ethanol to petrol (Poulopoulos et al., 2001) leads to an effective 10% reduction in the emission of CO, as well as a general reduction in aromatic hydrocarbon emissions. Using four-stroke engines, with four cylinders and electronic injection, fueled with various ethanol and petrol mixtures (Al-Hasan, 2003) reduced the CO emissions by about 46.5%. The anti-detonating features of petrols are very important and depending essentially on their chemical composition.

Life cycle analysis taking the "well to wheel" approach showed that the GHG emissions from bioethanol obtained from sugar beet are around 40-60% lower than the emissions from petrols obtained from fossil fuels (Reijinders & Huijbregts, 2007). Mixing bioethanol with diesel oil improves the fuel's combustion (Lapuerta et al., 2008) and reduces the size of the particles in the exhaust without increasing their quantity. Using an E10 mixture reduces the total hydrocarbon emissions because of ethanol's greater heat of vaporization.

CO emissions increase if moderate amounts of ethanol are added to diesel oil, while they diminish as the proportion of ethanol increases (Li et al., 2005). Conversely, NO_x emissions decrease with a low or moderate quantity of ethanol, but increase if more ethanol is added. The total hydrocarbons (THC) also increase with different proportions of ethanol and different speeds.

5. Conclusions

Although bioethanol is a valid alternative to fossil fuels and has a low environmental impact, its use is nonetheless posing problems relating to the use of raw materials such as cereals, which are fundamental to the food industry.

Increasing the farmland used to grow energy crops for the production of biofuels means competing with food crops. Many studies have attempted to assess the need for farmland for crops for producing ethanol. The yield in bioethanol per hectare naturally depends on the crops used, but reference can be made to the mean productivity in Europe (weighted according to the type of crop), which is currently estimated at around 2790 liters/hectare (based on a mean yield in seeds of 7 tons/hectare and 400 liters/ton).

Although bioethanol can be produced successfully in temperate climates too, the tropical climates are better able to ensure a high productivity. In Brazil, sugar cane is used to produce approximately 6200 liters/hectare (an estimate based on a crop yield of 69 tons/hectare and 90 liters/ton). The productivity of bioethanol from sugar cane is high in India too, with a yield of approximately 5300 liters/hectare. If bioethanol from sugar cane becomes a commodity used worldwide, then South America, India, Southeast Asia and Africa could become major exporters.

Research is focusing on alternatives, concentrating on innovative raw materials such as Miscanthus Giganteus, an inedible plant with a very high calorific value (approximately 4200 Kcal/kg of dry matter), or filamentous fungi such as Trichoderma reesei, which can break down the bonds of complex lignocellulose molecules.

This article summarizes the main raw materials that can be used to produce bioethanol, from the traditional to the more innovative, and the principal production processes involved. It also analyses the issues relating to emissions and carbon sequestering.

6. References

Al-Hasan, M. (2003). Effect of ethanol unleaded gasoline blends on engine performance and exhaust emission. *Energy Conversion Management,* Vol.44, No.9, pp.1547-1561.

Balat, M.; Balat, H. & Öz, C. (2008). Progress in bioethanol processing. *Progress Energy Combustion Science,* Vol.34, No.5, pp.551-573.

Blair, N.; Chapman, L.; Whitbread, A.M.; Ball-Coelho, B.; Larsen, P. & Tiessen, H. (1998). Soil carbon changes resulting from sugarcane trash management at two locations in Queensland, Austria, and in N-E Brazil. *Australian Journal of Soil Research,* Vol.36, pp.873-881.

Blair, N. & Crocker, G.J. (2000). Crop rotation effects on soil carbon and physical fertility of two Australian soils. Australian Journal of Soil research, Vol.38, No.1, pp.71-84.

Bouton, J.H. (2007). Molecular breeding of switchgrass for use as a biofuel crop. *Current Opinion in Genetics & Development,* Vol.17, No.6, pp.1-6.

Cardona, C.A.; Sánchez, O.J.; Montoya, M.I. & Quintero, J.A. (2005). Analysis of fuel ethanol production processes using lignocellulosic biomass and starch as feedstocks. *Proceeding of the Seventh World Congress of Chemical Engineering,* Glasgow, Scotland, UK, 2005.

Chandel, A.K.; Chan, E.S.; C, Rydravaram, R.; Narasu M,L.; Rao, L.V. & Ravindra, P. (2007). Economics and environmental impact of bioethanol production technologies: an appraisal. *Biotechnology and Molecular Biology Review,* Vol.2, No.1, pp.14-32.

Chum, H.L.; Johnson, D.K.; Balck, S.K. & Overend, R.P. (1990). Pretreatment-catalyst effects and combined severity parameter. *Applied Biochemistry and Biotechnology,* Vol.24-25, No.1, pp.1-14.

Curt, M.D.; Aguado, P.; Sanz, M; Sánchez, G. & Fernández, J. (2006). Clone precocity and the use of Helianthus tuberosus L. stems for bioethanol. *Industrial Crops and Products,* Vol.24, No.3, pp.314-320.

Da Silva, R.; Cataluña, R.; Menezes, E.W.; Samios, D. & Piatnicki, C.M.S. (2005). Effect of additives on the antiknock properties and Reid vapor pressure of gasoline. *Fuel,* 2005; Vol.84, No.7-8, pp.951-959.

DEFRA, *Planting and growing Mischanthus. Policy for Energy Crops Scheme.* London: DEFRA Publications; 31 January 2001, Available from: http://www.defra.gov.uk/

Demirbas, A. (2005). Bioethanol from cellulosic materials: a renewable motor fuel from biomass. *Energy Sources*, Vol.27, pp.327-337.

Dien, B.; Jung, H.J.G.; Vogel, K.P.; Casler, M.D.; Lamb, J.F.S.; Iten, L.; Mitchell, R.B. & Sarath, G. (2006). Chemical composition and response to dilute acid pretreatment and enzymatic saccharification, *Biomass and Bioenergy*, Vol.30, No.10, pp.880-891.

Espinal, C.F.; Martínez, H.J. & Acevedo, X. (2005). The chain of cereals balanced feed, aviculture, and pig farming in Colombia. An overview of its structure and dynamics 1991-2005. *Working report n. 87, 2005.* Observatorio Agrocadenas Colombia, Ministry of Agriculture and Rural Development.

Eurec Agency, *New improvements for lingo-cellulosic ethanol*, 31 January 2011, Available from: www.nile-bioethanol.org

European Union, *Fuel the future with bioethanol. A realistic large scale alternative to fossil fuels*, 31 January 2001, Available from: www.best-europe.org

Ferrel, J. & Glassner, D. (1997). *Bioethanol-the climate-cool fuel.* NREL, produced for the U.S. Department of Energy (DOE) 1997, DOE/GO-10097-515.

Gressel, J. (2008). Transgenics are imperative for biofuel crops. *Plant Science*; Vol.174, No.3, pp.246-263.

Gronchi, P.; Tempesti, E. & Mazzocchia, C. (1994). Metal dispersion dependent selectivities for syngas conversion to ethanol on V_2O_3 supported rhodium. *Applied Catalyst A: General*, Vol.120, No.1, pp.115-126.

Hahn-Hägerdal, B.; Galbe, M.; Gorwa-Grauslund, M.F.; Lidèn, G. & Zacchi, G. (2006). Bio-ethanol - the fuel of tomorrow from the residues of today. *Trends in Biotechnology*, Vol.24, No.12, pp.549-556.

Hansen, A.C.; Lyne, P.W.L. & Zhang, Q. Ethanol-diesel blends: a step towards a biobased fuel for diesel engines. (2005) *ASAE paper 01-6048.*

Heaton, H.; Voigt, T. & Long, S.P. (2004). A quantitative review comparing the yields of two candidate C4 perennial biomass crops in relation to nitrogen, temperature and water. *Biomass and Bioenergy*, Vol.27, No.1, pp.21-30.

Holy, N.L. & Carey, T.F. (1985). Ethanol and n-propanol from syngas. *Applied Catalysts*, Vol.19, No.2, pp.219-223.

Kim, S. & Dale, B.E. (2004). Global potential bioethanol production from wasted crops and crop residues. *Biomass and Bioenergy*, Vol.26. No.4, pp.361-375.

Lal, R. (2005). World crop residues production and implications of its use as a biofuel. *Environment International*, Vol.31, No.4, pp.575-584.

Lamnet, *Latin America Thematic Network on Bioenergy*, 31 January 2001, Available from: http://www.bioenergy-lamnet.org/

Lapuerta, M; Armas, O & Garcìa-Contreras, R. (2007). Stability of diesel-bioethanol blends for use in diesel engines. *Fuel*, Vol.86, No.10-11, pp.1351-1357.

Lapuerta, M.; Armas, O. & Herreros, J.M. (2008). Emissions from a diesel-bioethanol blend in an automotive diesel engine. *Fuel*, Vol.87, No.1, pp.25-31.

Larsson, S.; Palmqvist, E.; Hahn-Hägerdal, B.; Tengborg, C.; Stenberg, K.; Zacchi, G. & Nilvebrant, N.O. (1999). The generation of fermentation inhibitors during dilute acid hydrolysis of softwood. *Enzyme Microbial Technology*, Vol.24, No.3-4, pp.151-159.

Laser, M.; Schulman, D.; Allen, S.G.; Lichwa, J.; Anta, Jr M.J. & Lynd, L.R. (2002). A comparison of liquid hot water and steam pretreatments of sugar cane bagasse for bioconversion to ethanol. *Bioresource Technology*, Vol.81, No.1, pp.33-44.

Li, D.; Zhen, H.; Xingcai, L.; Wu-gao, Z. & Jian-guang, Y. (2005). Physico-chemical properties of ethanol-diesel blend fuel and its effect on performance and emissions of diesel engines. *Renewable energy*, Vol.30, No.6, pp.967-976.

Lin, P.Z.; Liang, D.B.; Lou, H.Y.; Xu, C.H.; Zhou, H.W.; Huang, S.Y. & Lin, L.W. (1995). Synthesis of C_{2+}-oxygenated compounds directly from syngas. *Applied Catalyst A: General*, Vol.131, No.2, pp.207-214.

Linde, R.; Bekers, M.; Ventina, E.; Vina, I.; Kaminska, H.; Upite, D.; Scherbaka, R. & Danilevich, A. (1998). Ethanol and Fructose from Sugar Beets. Biomass for Energy and Industry, *Proceeding of the C.A.R.M.E.N. International Conference*, pp.464-467, Würzburg, Germany, June 8-11, 1998.

Luo, H.Y.; Zhang, W.; Zhou, H.W.; Huang, S.Y.; Lin, P.Z.; Ding, Y.J. & Lin, L.W. (2001). A study of Rh-Sm-V/SiO_2 catalysts for the preparation of C_2-oxygenates from syngas. *Applied Catalyst A: General*, Vol.214, No.2, pp.161-166.

Lynd, L.R.; van Zyl, W.H.; McBride, J.E. & Laser, M. (2005). Consolidated bioprocessing of cellulosic biomass: an update. *Current Opinion in Biotechnology*, Vol.16, No.5, pp.577-583.

Marek, N.J. & Evanoff J. The use of ethanol-blended diesel fuel in unmodified compression ignition engines: an interim case study. *Air and Waste Management Association (AWMA) Annual Conference* ,Orlando, Florida, USA, June 2001.

Martinez, D.; Berka, R.M.; Henrissat, B.; Saloheimo, M.; Arvas, M.; Baker, S.E., Chapman, J.; Chertkov, O.; Coutinho, P.M.; Cullen, D.; Danchin, E.G., Grigoriev, I.V.; Harris, P.; Jackson, M.; Kubicek, C.P.; Han, C.S.; Ho, I.; Larrondo, L.F.; de Leon, A.L.; Magnuson, J.K.; Merino, S.; Misra, M.; Nelson, B.; Putnam, N.; Robbertse, B.; Salamov, A.A.; Schmoll, M. Terry A, Thayer N, Westerholm-Parvinen A, Schoch CL, Yao J, Barbote R, Nelson MA, Detter C, Bruce D, Kuske CR, Xie G.; Richardson, P.; Rokhsar, D.S.; Lucas, S.M.; Rubin, E.M.; Dunn-Coleman, N.; Ward, M. & Brettin, T.S. (2008). Genome sequencing and analysis of the biomass-degrading fungus Trichoderma reesei (syn. Hypocrea jecorina). *Nature Biotechnology*, Vol.26, No.5, pp.553-60.

McAloon, A.; Taylor, F.; Yee, W.; Ibsen, K. & Wooley, R. (2000). Determining the cost of producing ethanol from corn starch and lignocellulosic feedstocks. *Technical report NREL/TP-580-28893*. Golden, Co (USA), National Renewable Energy Laboratory 2000.

Overend, R.P. & Chornet, E. (1987). Fractionation of lignocellulosics by steam-acqueous pretreatments. *Philosophical Transactions for the Royal Society of London. Series A, Mathematical and Physical Sciences*,Vol.321, No.1561, pp.523-536.

Palmieri, G.; Giardina, P. & Sannia, G. (1997). A novel white laccase from Pleurotus ostreatus. *The journal of biological chemistry*, Vol.272, No.50, pp.31301-31307.

Palmqvist, E.; Hahn-Hägerdal, B.; Galbe, M. & Zacchi, G. (1996). The effect of water-soluble inhibitors from steam-pretreated willow on enzymatic hydrolysis and ethanol fermentation. *Enzyme and Microbial Technology*, Vol.19, No.6, pp.470-476.

Palmqvist, E. & Hahn-Hägerdal, B. (2000). Fermentation of lignocellulosic hydrolysates. II: inhibitors and mechanisms of inhibition. *Bioresource Technology*, Vol.74, No.1, pp.25-33.

Patzek, T.W.; Anti, S.M.; Campos, R. Ha, K.W.; Lee, J.; Li, B.; Padnick, J. & Yee, S.A. (2005). Ethanol from corn: Clean renewable fuel for the future, or drain on our resources and pockets? *Environment, Development, and Sustainability*, Vol.7, No.3, pp.319-336.

Prasad, S.; Singh, A. & Joshi, H.C. (2007). Ethanol as an alternative fuel from agricultural, industrial and urban residues. Resources. *Conservation and Recycling*, Vol.50, No.1, pp.1-39.

Petersson, A.; Thomsen, M.H.; Hauggaard-Nielsen H. & Thomsen, A.B. (2007). Potential bioethanol and biogas production using lignocellulosic biomass from winter rye, oilseed rape and faba bean. *Biomass and Bioenergy*, Vol.31, No.11-12, pp.812-819.

Pimentel, D. (2003). Ethanol fuels: energy balance, economics, and environmental impacts are negative. *Natural Resources Research* , Vol.12, No.2, pp.127-134.

Poulopoulos, S.G.; Samaras, D.P. & Philippopoulos, C.J. (2001). Regulated and unregulated from an internal combustion engine operating on ethanol-containing fuels. *Atmosphere Environment*, Vol.35, No.26, pp.4399-4406.

Quintero, J.A.; Montoya, M.I.; Sánchez, O.J.; Giraldo, O.H. & Cardona, C.A. (2008). Fuel ethanol production from sugarcane and corn: comparative analysis for a Colombian case. *Energy*, Vol.33, No.3, pp.385-399.

Rees, R.M.; Bingham, I.J.; Baddeley, J.A. & Watson, C.A. (2005). The role of plants and land management in sequestering carbon temperate arable and grassland systems. *Geoderma*, Vol.128, No.1-2, pp.130-154.

Reijinders, L. & Huijbregts, M.A.J. (2007). Life cycle greenhouse gas emissions, fossil fuel demand and solar energy conversion efficiency in European bioethanol production for automotive purposes. *Journal of Cleaner Production*, Vol. 15, No.18, pp.1806-1812.

Robertson, G.H.; Wong, D.W.S.; Lee, C.C.; Wagschal, K.; Smith, M.R. & Orts, W.J. (2006). Native or raw starch digestion: a key step in energy efficient biorefining of grain. *Journal of Agricultural and Food Chemistry*, Vol.54, No.2, pp.353-365.

Six, J.; Ogle, S.M.; Breidt, F.J.; Conant, R.T.; Mosiers, A.R. & Paustian, K. (2004). The potential to mitigate global warming with no-tillage management is only realized when practiced in the long term. *Global Change Biology*, Vol.10, pp.155-160.

Rooney, L.W.; Blumenthal, J.; Bean, B. & Mullet, J.E. (2007). Designing sorghum as a dedicated bioenergy feedstock. *Biofuels, Bioproducts and Biorefining*, Vol.1, No.2, pp.147-157.

Rosenberg, A.; Kaul, H.P.; Senn, T. & Aufhammer, W. (2001). Improving the energy balance of bioethanol production from winter cereals: the effect of crop production intensity. *Applied Energy*, Vol.68, No.1, pp.51-67.

Sánchez, Ó. & Cardona, C.A. (2008). Trends in biotechnological production of fuel ethanol from different feedstocks. *Bioresource Technology*, Vol.99, No.13, pp.5270-5295.

Sassner, P.; Galbe, M. & Zacchi, G. (2008a). Techno-economic evaluation for bioethanol production from three different lignocellulosic materials. *Biomass and Bioenergy*, Vol.32, No.5, pp.422-430.

Sassner, P.; Mårtensson, C.G.; Galbe, M. & Zacchi, G. (2008b). Steam pretreatment of H_2SO_4-impregnated Salix for the production of bioethanol. *Bioresource Technology*, Vol.99, No.1, pp.137-145.

Sheehan, J.; Aden, A.; Paustian, K.; Killian, K.; Walsh, M. & Nelson, R. (2004). Energy and environmental aspects of using corn stover for fuel ethanol. *Journal of Industrial Ecology*, Vol.7, No.3-4, pp.117-146.

Shevchenko, S.M.; Chang, K., Robinson, J. & Saddler, J.N. Optimization of monosaccharide recovery by post-hydrolysis of the water-soluble hemicellulose component after steam explosion of softwood chips. *Bioresource Technology*, Vol.72, No.3, pp.207-211.

Silverstein, R.A.; Chen, Y.; Sharma-Shivappa, R.R.; Boyette, M.D. & Osborne, J. (2007). A Comparison of chemical pretreatment methods for improving saccharification of cotton stalks. *Bioresource Technology*, Vol.98, No.16, pp.3000-3011.

Soccol, C.; Marin, B.; Raimbault, M. & Lebeault, J.M. (1994). Breeding and growth of Rhizopus in raw cassava by solid state fermentation. *Applied Microbiology and Biotechnology*, Vol.41, No.3, pp.330-336.

Subramani, V. & Gangwal, S.K. (2008). A Review of Recent Literature to Search for an Efficient Catalytic Process for the Conversion of Syngas to Ethanol. *Energy & Fuels*, Vol.22, No.2, pp.814-839.

Sun, Y. & Cheng, J. (2002). Hydrolysis of lignocellulosic materials for ethanol production: a review. *Bioresource Technology*, Vol.83, No.1, pp.1-11.

Varvel, G.E.; Vogel, K.P.; Mitchell, R.B. & Kimble, J.M. (2007) Comparison of corn and switchgrass on marginal soils for bioenergy. *Biomass and Bioenergy* , Vol.32, No.1, pp.18-21.

Verma, G.; Nigam, P.; Singh, D. & Chaudhary, K. (2000). Bioconversion of starch to ethanol ion a single-step by co-culture of amylolytic yeasts and Saccharomyces cerevisiae 21. *Bioresource Technology*, Vol.72, No.3, pp.261-266.

Wyman, C.E. (1994). Ethanol from lignocellulosic biomass: technology, economics, and opportunities. *Bioresource Technology*, Vol.50, No.1, pp.3-16.

Yu-Hua, D.; De-An, C. & Khi-Rui, T. (1987). Promoter action of rare earth oxides in rhodium/silica catalysts for the conversion of syngas to ethanol. *Applied Catalyst A*, Vol.35, No.1, pp.77-92.

27

Effect of Fried Dishes Assortment on Chosen Properties of Used Plant Oils as Raw Materials for Production of Diesel Fuel Substitute

Marek Szmigielski, Barbara Maniak,
Wiesław Piekarski and Grzegorz Zając
University of Life Sciences in Lublin
Poland

1. Introduction

Utilization of post-frying plant oils which are waste product of operation of, serving fried products, gastronomical points, for many years has been growing and complex problem of technological, ecological and economical nature. It must be noted that methods of solving this problem were subject of numerous research [Alcantara 2000; Buczek and Chwiałkowski 2005; Dzieniszewski 2007; Leung and Guo 2006]. Conception of utilization of post-frying plant oils as components for production of substitute of diesel fuel seems to be promising. However, it is necessary to investigate in detail properties of such oils, so that elaborated technologies of their utilization are optimal. Answer to question concerning influence of assortment of fried products on quality of post-frying oil, and its usefulness, when aspect of differences in utilization of particular batches of such oil, obtained after frying various food products, seems to be the most significant issue.

Most commonly used method of frying food in gastronomical points is deep frying. During this type of frying, processed food is submerged in frying medium and contacts oil or fat with most of its external surface. The main role of frying medium is keeping processed food in proper position to source of heat and transferring proper amount of heat energy into a fried product [Drozdowski, 2007; Ledóchowska and Hazuka, 2006]. Frying fat, which is a frying medium, and products subjected to culinary processing form a specific system in which partial penetration of these two compounds and two-way transfer of energy and weight take place. As a result of frying, product loses significant amount of water and, depending on its composition, some of its compounds e.g. food dyes, taste and flavour compounds and partially, transferred to frying fat, lipids. They are replaced with some amount of frying fat, which content in fried food, according to approximate data, may vary significantly and reach even 40% [Ledóchowska and Hazuka, 2006].

Water present in processed products and released during submersion frying has got diverse and multi directional influence on changes occurring in oil, among which is, causing partial increase of acid number (AN) of oil, fat hydrolysis. Moreover, transport of heat emitted with released water vapours favours decrease of temperature of fried food and partly inhibits oxidation transformations of fat by displacing oxygen in it [Ledóchowska and Hazuka, 2006].

Oxygen dissolved in frying fat together with water vapour are also significant factors of so called thermooxidative transformations, which have not been fully explained yet. As a result of these transformations numerous substances, having complex and not fully determined structure, are formed. They are precursors of secondary transformations, products of which can be usually classified in one of two categories: volatile compounds (hydrocarbons, fatty acids and carboxylic compounds) and non-volatile (monomers, dimers, polymers and also some aldehydes and ketones, as well as fatty acids characterizing with changed melting point) [Drozdowski 2007, Paul and Mittal 1996, Blumenthal 1991, Choe and Min 2006, Clark and Serbia 1991, Hoffman 2004, Ledóchowska and Hazuka 2006].
Gastronomical fryers are usually containers having fairly high capacity, in which, next to the surface layer, which is environment determining properties of processed product, some volume of oil deposited near a bottom of a fryer can be distinguished. Bottom zone of a fryer, adjacent usually to the source of heat emission and having relatively low content of oxygen and water vapour, favours free radical or polymerization transformations of unsaturated fatty acids occurring in frying fat. The most common result of these transformations are numerous, having complex structure, non-polar thermal polymers. Macroscopic result of this type of reactions are increase of viscosity and darkening as well as increase of melting point of frying medium, what results in change of its state of aggregation. Products of these transformations are main components of dark brown deposits found on walls of a fryer, which can be a reason for many problems related to utilization of such oil [Hoffman 2004].
It should be noted that direction and intensity of frying fat transformations depends on numerous factors accompanying this process during frying of food products. In literature [Ledóchowska and Hazuka 2006] at least few groups of such factors are named. As the basic ones, conditions of carrying out the process (its duration, temperature and periodicity) and degree of unsaturation of fatty acids in triglycerides of fat, are mentioned. Among all factors affecting properties of frying medium many other, accompanying frying process, like oxygen availability and amount and composition of compounds released from food (e.g. pro and antioxidants and presence of water), play a significant role [Ledóchowska and Hazuk'a 2006].

2. Assessment of usability of post-frying edible oils as a raw material for production of diesel fuel substitute

2.1 Materials and methods

2.1.1 Preparation of samples for Investigation

In this research, comparison of influence of fried dishes assortment (potato chips and breadcrumbs coated fish fingers) on physicochemical properties and quality of post-frying plant oils to be utilized as raw materials for production of, used as a substitute of diesel fuel, fatty acids methyl esters, was conducted. Main focus of the research was on evaluation of effect of fried dishes assortment on quality of obtained post-frying oils (rapeseed, sunflower and soybean) with regard to their utilization as a substrate for production of engine biofuel.
In model conditions of laboratory investigation, usability of post-frying waste oils as raw materials for production of fatty acids methyl esters was evaluated. Three most commonly used edible oils (rapeseed, sunflower and soybean) were used as material for this research. From total amount of each of raw oils, sample for laboratory analyses was taken. It was marked as "0" and was used as a reference sample. Remaining amount of each of oils was

divided into three batches and poured into separate containers. Batch no. 1 was prepared by means of cyclic, five-time heating without fried product. Particular cycle within this batch comprised of heating whole amount of oil to temperature of approximately 180°C, and than maintaining it in such temperature for 10 min. Next, oil was left to cool down in room temperature and than a sample, to be used for laboratory analyses, was taken. The sample was marked as "heating I - without fried product". After 24 hours all described above actions were repeated yielding sample marked as "heating II - without fried product". Whole process of heating, cooling and sampling was repeated, yielding samples marked as "heating without fried product" bearing following, respective to number of cycle, labels: III, IV and V.

Preparation of oil from batch no. 2 was differed from previously presented in only one way. After heating it to 180°C, in each of three investigated oils, potato chips, prepared of purchased raw potatoes and cut to the size and shape of frozen potato chips found in trade, were fried.

After frying and separating chips, oil was cooled down to room temperature and than samples for research were taken. They were marked as "heating I - process of chips frying". Repeating whole process enabled obtaining samples marked following, respective to number of cycle, labels: III, IV and V.

Third part of oil (batch no. 3) was heated same way as batch no. 2 but in this case purchased breadcrumbs coated fish fingers were the fried product. After frying and separating breadcrumbs coated fish fingers, oil was cooled down to room temperature and than samples for research were taken. They were marked as "heating I - process of breadcrumbs coated fish fingers frying". Repeating whole process enabled obtaining samples marked following, respective to number of cycle, labels: III, IV and V.

2.2 Laboratory test

Oil samples obtained in conformity with chosen methodology were subjected to laboratory tests, which comprised of following analyses: determination of peroxide number (PN), acid number (AN) and composition of fatty acids. Determination of peroxide number (PN) in conformity with [ISO 3960] was based on titration of iodine released from potassium iodide by peroxides present in the sample, calculated per their weight unit. Results of analyses were expressed in millimoles of oxygen per weight unit of the sample.

Determination of acid number (AN) in conformity with [PN-ISO 660] was conducted by means of titration and evaluation of acidity of a sample, and expressed in numeric form in millilitres of 0,1M solution of sodium hydroxide, calculated per weight unit of analysed oil.

Determination of fatty acids composition was conducted by means of method based on utilization of gas chromatography [Krełowska - Kułas, 1993]. Sample of fat was subjected to alkaline hydrolysis in anhydrous environment with utilization of methanol solution of sodium hydroxide. As a result of this reaction, fatty acids of investigated oil were transformed into a mixture of sodium soaps, which than were subjected of reaction of esterification with anhydrous solution of hydrogen chloride in methanol, yielding mixture of fatty acids methyl esters.

Obtained methyl esters were separated in a chromatographic column and than their participation in a sum of fatty acids was determined [Krełowska - Kułas, 1993]. Chromatographic separation was conducted by means of gas chromatograph with nitrogen as carrier gas, packed column (2,5 m with stationary phase PEGA - polyethylene glycol adipate on carrier GAZ-ChROM-Q) and flame ionization detector.

2.3 Engine tests

Samples of soybean oil which remained after laboratory samples had been taken from each of three batches, differentiated by type of initial preparation (frying potato chips, frying breadcrumbs coated fish fingers and heating without fried product), were separately subjected to esterification with methanol. Fatty acids methyl esters were obtained by method analogous to the one used in investigation of fatty acids composition by means of gas chromatography. Fuel obtained this way was used in engine tests including main engine work parameters. Four mixtures were prepared, each containing 90% diesel fuel and 10% addition of fatty acids methyl esters obtained in research and marked as:

a. M1 - esters obtained from purchased fresh soybean oil,
b. M2 - esters obtained from soybean oil subjected to five-time cyclic heating, without addition of fried product,
c. M3 - esters obtained from soybean oil, previously used for five-time cyclic frying of potato chips,
d. M4 - esters obtained from soybean oil, previously used for five-time cyclic frying of breadcrumbs coated fish fingers.

Results of internal combustion engine running on diesel fuel (DF) were used as reference for determination of work parameters of engine powered with fuel blends. Above mentioned fuel mixtures, were used for powering 2CA90 diesel engine installed on dynamometric stand for purpose of conducting measurements of its energetic work parameters. Test bed comprised of following devices:

- internal combustion diesel engine 2CA90;
- dynamometric stand composed of eddy-current brake AMX210 and control-measurement system AMX201, AMX 211;
- fuel consumption measuring system;
- system measuring engine parameters: exhaust gasses temperature - tsp, engine oil temperature - tol, oil pressure -pol;
- system measuring state of environment: temperature of environment - tot, atmospheric pressure - pa, and air humidity - φ.

Measurements for each of investigated fuels were conducted and obtained results of energetic parameters were elaborated. Data yielded by measurements was used to draw external characteristics of the engine for rotational speed ranging from minimal to nominal. Carried out research included kinematic and dynamic parameters of the engine: torque - Mo, rotational speed - n, time in which set amount of investigated fuel was used - τ. Amount of fuel used for purpose of this characteristic was 50 g. Methodology of measurements and methods of measurements and results reduction of power and torque, were in conformity with norms: PN-88/S-02005, BN-79/1374-03.

3. Result of investigation

Raw, purchased plant oils characterised with typical properties, fulfilling requirements of recommended in Poland norm [PN – A - 86908] with regard to peroxide number (PN) and acid number (AN) (fig. 2 and 3).

Heating edible oils in conditions corresponding to frying potato chips, breadcrumbs coated fish fingers and heating without a product lead to significant changes of investigated oils properties. It caused mainly distinct changes of acid number (AN) and peroxide number (PN).

Differences in properties of oils subjected to cyclic heating without fried product and in which potato chips or breadcrumbs coated fish fingers were fried may result from course of temperature changes for various investigated batches (Fig. 1).

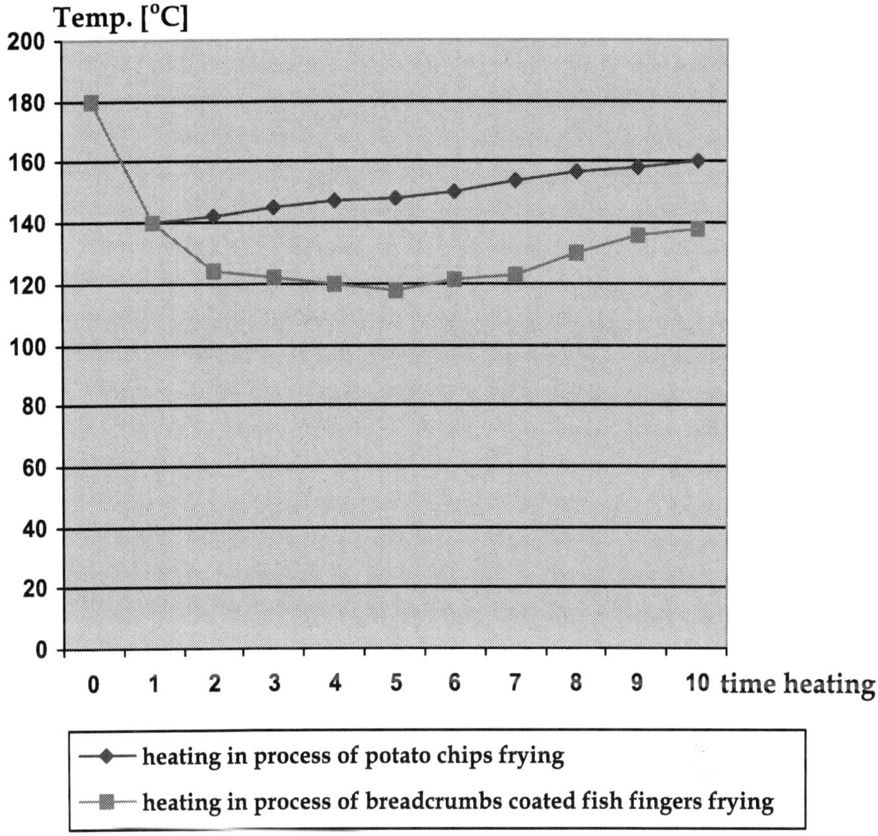

Fig. 1. Course of rapeseed oil temperature changes in relation to time of potato chips and breadcrumbs coated fish fingers frying (presented data based on authors own research [Szmigielski et al. 2009])

The highest temperature for each of investigated oils and in each of five heating cycles was observed in case of samples heated without fried products, in which temperature remained at 180°C. Changes of temperature of oil heated in the process of frying potato chips or breadcrumbs coated fish fingers had dynamic course, reaching the lowest value in approximately beginning of fifth minute. However, value of this minimum was depended on weight of fried product but main factor was fried product to frying medium weight ratio (fig. 1).

Conducted research show that heating plant oils caused noticeable increase of peroxide number (PN) value, when compared to samples not subjected to thermal processing. It must be noted that diverse course and intensity of these changes were observed in case of samples heated without product, samples heated in process of potato chips and breadcrumbs coated fish fingers frying (Fig. 2).

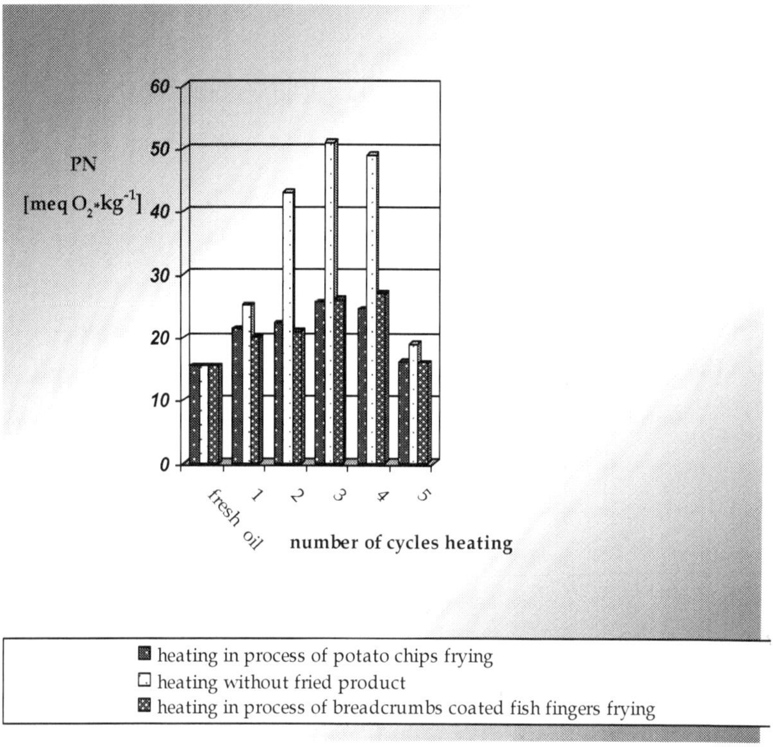

Fig. 2. Peroxide number of rapeseed oil subjected to cyclic heating [mMO kg-1]/ data for oil heated in process of potato chips frying and heated without addition of product according to Szmigielski et al. 2008/

Typical course of peroxide number changes in relation to number of frying cycle was presented in Fig. 2. In case of each of five heating cycles, highest value of peroxide number in rapeseed and soybean oils was observed in samples heated without a product [Szmigielski et al. 2008]. It was characteristic, that in these samples peroxide number value increased fast until third or fourth cycle, after which decrease of its value was noted (Fig. 2). Most probable cause of such course of peroxide number changes, in relation to heating cycles, is formation of oxidation products, which partially evaporate from the environment of reaction in form of volatile products. An exception to the rule were analyses conducted for samples of soybean oil (firs and second cycle of heating), in which temporarily highest value was observed in samples heated in process of breadcrumbs coated fish fingers frying [Szmigielski et al. 2011]. Most probably it results from influence of fat present in fried product on a final result of determination.

Samples of oil heated in the process of potato chips frying, characterised with lower values of peroxide number, for each of five heating cycles, when compared to samples heated without the fried product. Stabilizing effect of potato chips, caused by sorption of oxidation products on their surface or partial absorption of frying fat, is most commonly mentioned probable cause of such course of PN changes in these samples [Maniak et al. 2009, Szmigielski et al. 2008, 2009, 2011]. It should be noted (Fig. 1) that PN level in samples of rapeseed and soybean oil heated in process of potato chips frying [Szmigielski et al. 2008], had similar course, stabilizing respectively at approx. 2 mMo/100g and approx. 1,5 mMo/100g [Szmigielski et al. 2011] (with exemption of null samples and first cycle of soybean oil heating). Results of investigation of soybean and rapeseed oil samples heated without fried product differed significantly - reaching almost two times higher value of peroxide number (PN) than respective samples heated in process of potato chips frying. As opposed to this research, heating sunflower oil in process of potato chips frying caused only slight decrease of its peroxide number (PN) when compared to samples heated without fried product [Maniak et al. 2009].

Typical course of acid number (AN) changes of heated oil samples in relation to number of frying cycles was presented in Fig. 3. Acid number of heated oil samples was higher than in raw oil, however, heating in process of potato chips frying caused stabilization of acid number value (AN) at similar level (0,02 mgKOH/g) regardless of number of oil heating cycles, while heating without the product caused systematic increase of AN. Very similar course of acid number changes of investigated post-frying oils was also observed in analogous research on rapeseed oil [Szmigielski et al. 2008] and sunflower oil samples [Maniak et al. 2009]. It is believed, that the most probable cause of observed changes of acid number of these samples is sorption of oxidation products on surface of, subjected to culinary processing, potato chips or partial absorption of oil surrounding the product into its deeper, more distant from surface of investigated raw product layers.

Acid number (AN) of plant oils (rapeseed and soybean) heated in the process of frying breadcrumbs coated fish fingers was increasing systematically. It should be noted that AN for first two cycles of heating remained at level similar or lower than AN determined in respective samples heated in the process of potato chips frying. However, starting from the third heating cycle AN exceeded this value and was systematically increasing with each of heating cycles, reaching values lower than in respective samples of soybean oil heated without fried product (fig. 3). It is believed that two opposing processes were the most probable cause of above described course of changes of acid number (AN) in samples of oils heated in process of breadcrumbs coated fish fingers frying. Increase of AN value should probably be explained with oxidation of fatty acids and hydrolytic effect of water vapour, released from product as a result of frying, while reduction of its level occurred as an effect of sorption of oxidation products on surface of fried product [Szmigielski et al. 2009; 2011].

Five-time cyclic heating of plant oils caused significant changes in composition of fatty acids, which can be simply characterised as significant decrease of fatty acids content. It concerns mainly unsaturated fatty acids, and significant increase of oxidation products content, what can be easily observed on example of soybean oil (fig. 4-6). Similar course of fatty acids composition changes of investigated post-frying oils was also observed in research of, subjected to cyclic heating, samples of rapeseed oil [Szmigielski et al. 2008;2009] and sunflower oil [Maniak et al. 2009]. Five-time cyclic frying of breadcrumbs coated fish fingers or potato chips caused partial stabilization of fatty acids composition, what can be noted in case of two, dominating in soybean, fatty acids i.e. oleic and linolic. Their content in typical raw soybean oil often exceeds 75% (fig. 3-5), [Staat and Vallet 1994, Tys et al. 2003].

Heating this oil only slightly changed proportion of oleic to linolic acid, for in raw oil, on one particle of oleic acid approx. two particles of linolic acid are found. After process of heating, this rate is approx. 1,5 - from 1,4 for sample heated without fried product to 1,50 for sample heated in the process of frying potato chips, and up to 1,63 when sample of oil heated in the process of frying breadcrumbs coated fish fingers is investigated.

Similar effect, when ratio of unsaturated fatty acids (oleic and linolic) is taken into consideration, was also observed in research of sunflower oil used as frying fat in cyclic frying of potato chips.

Fresh sunflower oil usually contains over 80% of these fatty acids, while, statistically on one particle of oleic acid 2,47 particles of linolic acids are found. After five-time cyclic heating in process of potato chips frying this proportion remains unchanged, while it changes only in case of oil heated without fried product [Maniak et al. 2009].

In fresh rapeseed oil, proportion of linolic acid to oleic acid is 1 : 2,72. Five-time cyclic heating in process of potato chips frying caused significant change of this proportion to 1 : 2,37, while, for example, effect of disturbance of this fatty acids ratio occurring during similar cycle of heating without fried product reached 1:3,77 [Szmigielski et al. 2008]. The same processes of heating caused also slight changes of saturated fatty acids ratio. In fresh soybean oil, on one particle of stearic acid 2,66 particles of palmitic acid are found, while after five cycles of heating this ratio was from 1 : 2,1 in oil heated without product (Fig. 4), 1 : 2,31 in oil subjected to heating in process of potato chips frying (Fig. 6) to 1 : 2,38 in oil subjected to heating in process of breadcrumbs coated fish fingers frying (Fig. 5).

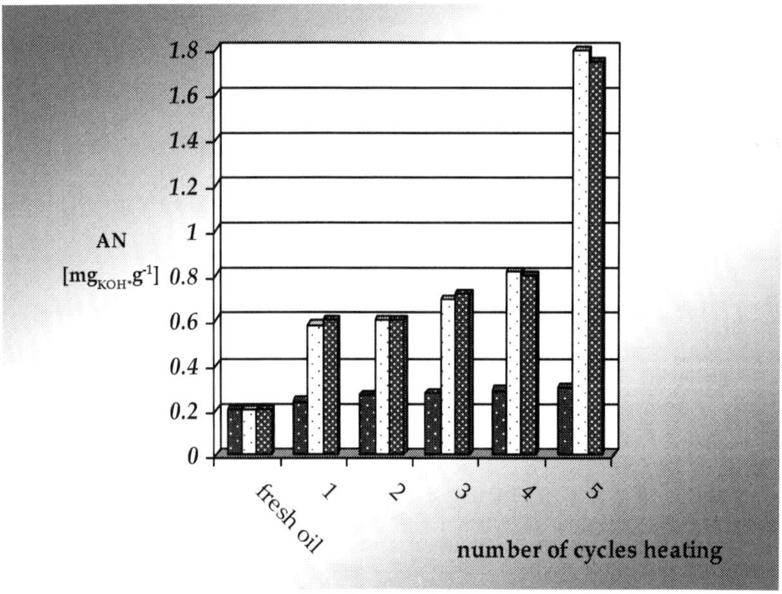

Fig. 3. Acid number of rapeseed oil subjected to cyclic heating [mgKOH g^{-1}]/ data for oil heated in process of potato chips frying and heated without addition of product according to Szmigielski et al. 2008/

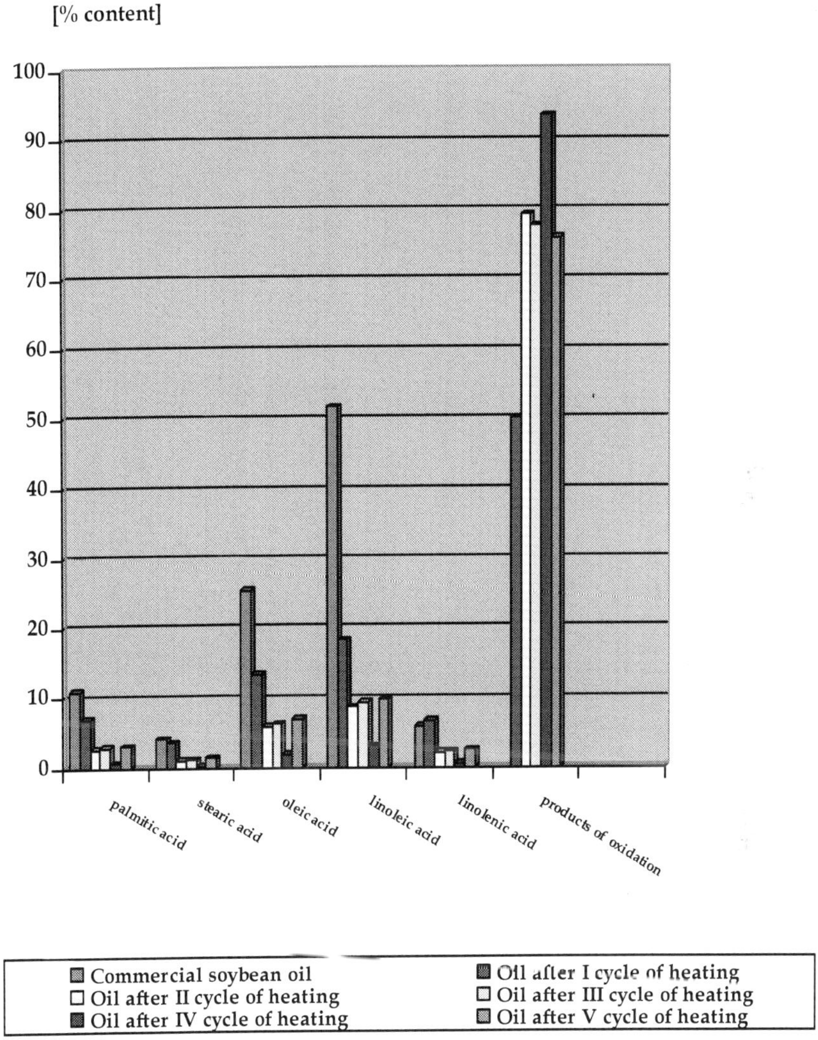

Fig. 4. The composition of fatty acids of soybean oil treated five-time cyclic heating, heating without fried product /presented data based on authors own research [Szmigielski et al. 2011]/

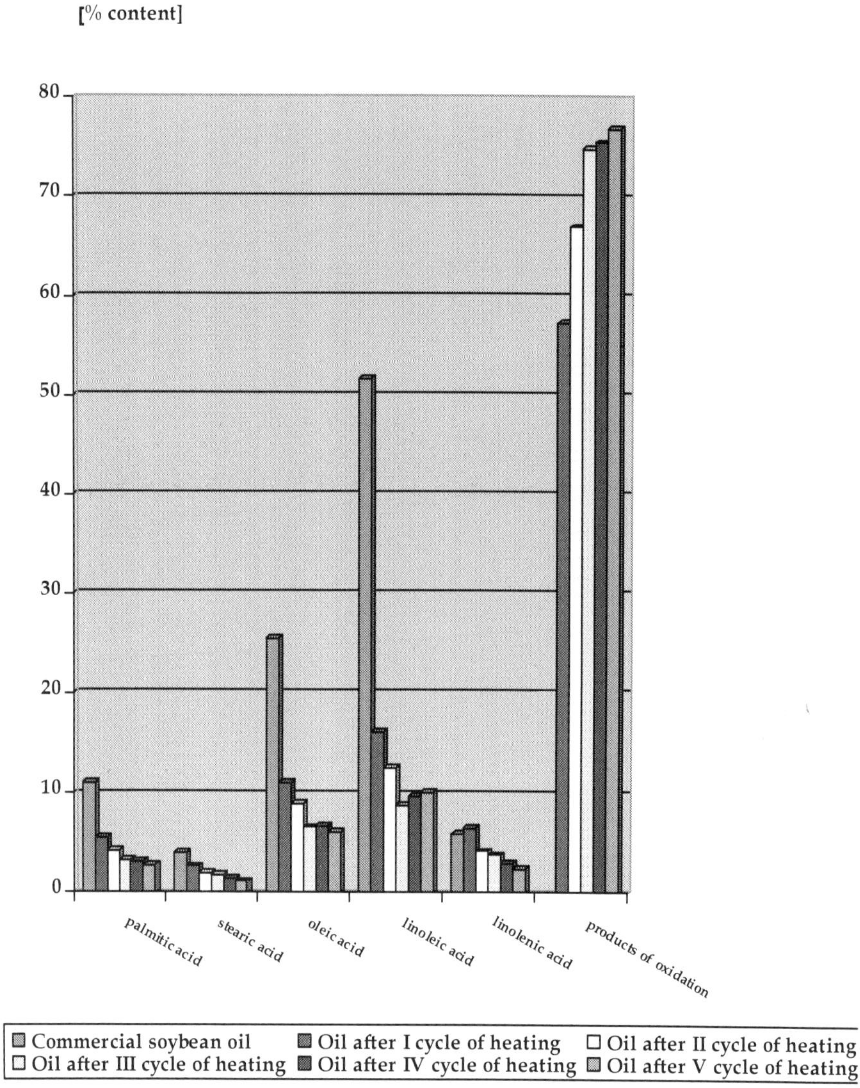

Fig. 5. The composition of fatty acids of soybean oil treated five-time cyclic heating, heating in process of breadcrumbs coated fish fingers frying /presented data based on authors own research [Szmigielski et al. 2011]/

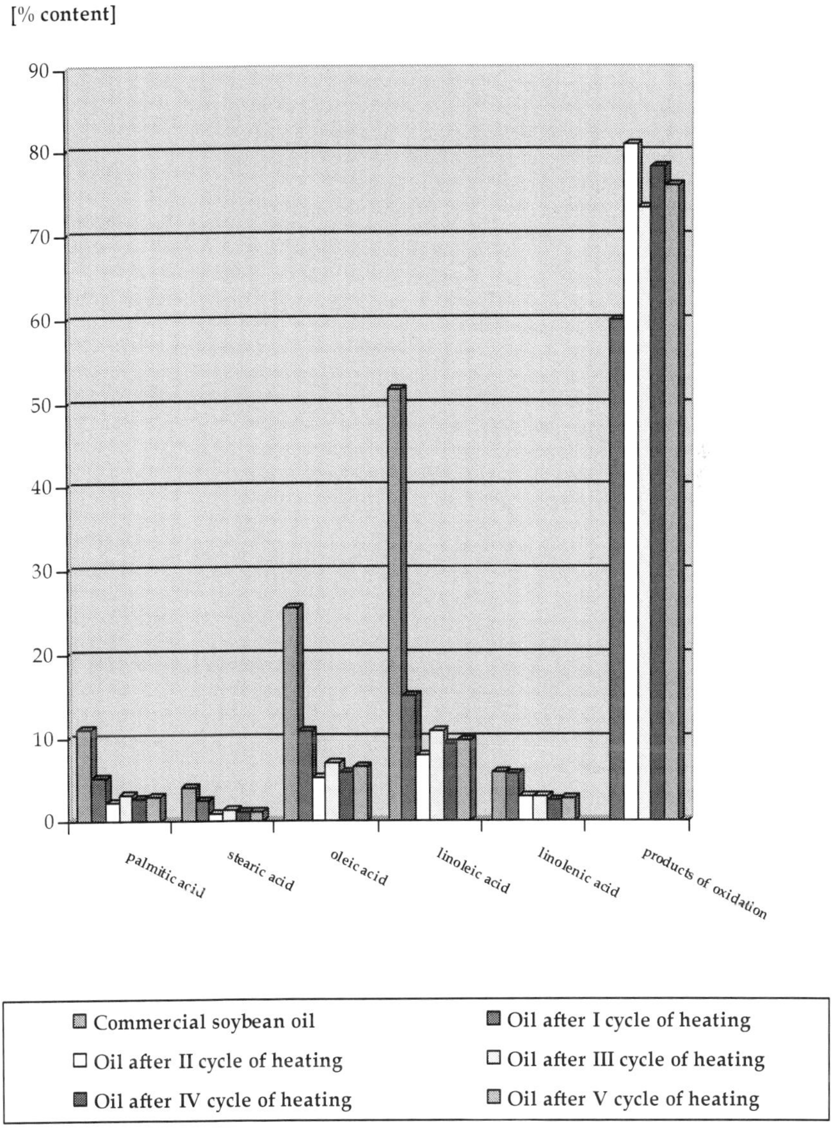

Fig. 6. The composition of fatty acids of soybean oil treated five-time cyclic heating heating in process of potato chips frying /presented data based on authors own research [Szmigielski et al. 2011]/

Similar, slight fluctuations of stearic and palmitic acids ratio were noted after five-time cyclic heating of rapeseed oil, and ranged from 1 : 2,97 in fresh oil, to 1 : 3,03 after heating in process of cyclic potato chips frying and 1 : 2,94 after cyclic heating without fried product [Szmigielski et al. 2008].

It should be noted that similar cyclic heating of sunflower oil did not cause change of ratio of two main saturated fatty acids present in investigated oil e.g. stearic acid and palmitic acid. The ratio was 1 : 1,69. Both in fresh sunflower oil and in oil after five-time cyclic heating without fried product or heated in process of potato chips frying, this ratio did not change [Maniak et al. 2009].

Presented in graphs 7-10 research data, obtained during engine tests, in which mixtures of diesel fuel containing 10% of fatty acids methyl esters were utilised, indicate on similar character of changes of investigated parameters of 2CA90 engine powered with methyl esters obtained from purchased raw soybean oil and post-frying oils obtained after five-time cyclic heating without fried product as well as five-time cyclic frying of potato chips or breadcrumbs coated fish fingers. Mixtures containing 10% addition of esters have similar influence on changes of power and torque of investigated engine in relation to its rotational speed (Fig. 7 and 8).

Curves of specific and hourly fuel consumption for investigated fuel mixtures, containing 10% addition of fatty acids methyl esters, characterised with higher values of energetic parameters, when compared to diesel fuel, for each of five investigated rotational speeds. It should be noted that they characterize with identical nature and high similarity of their course, what suggests insignificance of differences between them. Analogous results of research were obtained by Szmigielski et al. [2009], who, in similar conditions, investigated rapeseed oil samples.

4. Conclusion

1. Model, cyclic heating of plant oils, and especially three first cycles, contribute to significant changes in composition of their fatty acids. Significant changes of peroxide number (PN) and acid number (AN) of investigated oils were noted. Content of unsaturated fatty acids decreases, while increase of oxidation products is observed.
2. Heating plant oils in process of frying products like breadcrumbs coated fish fingers or potato chips affects stabilization of amount of peroxide products present in post-frying oil, what leads to decrease of peroxide number (PN) of such oil in comparison to process of heating without fried products.
3. Acid number (AN) of post-frying oils obtained after frying potato chips stabilized, while frying breadcrumbs coated fish fingers and heating oil without fried product contributed to gradual increase of AN.
4. Frying breadcrumbs coated fish fingers and potato chips favours stabilisation of proportion of fatty acids in investigated post-frying oils, and the proportion is similar to one noted in case of purchased raw oils.
5. Change of properties of post-frying plant oils occurring during stage of chemical conversion to fatty acids methyl esters, contributes to unification of properties of biofuels prepared on base of various batches of post-frying oils and favours utilization of post-frying oils which proved as suitable for production of biofuel as fresh vegetable oils.
6. Unidentified oxidation products undergo similar transformations in the process of fatty acids methyl esters formation, and are not a significant obstacle in correct operation of diesel engines powered with such biofuel.

7. Results of research confirmed usability of post-frying plant oils as a raw material for production of diesel fuel biocomponent.
8. It is currently necessary to elaborate efficient ways of recovery of post-frying fats from points of small gastronomy, and technology of their purification and utilization as components of fuel for diesel engines.

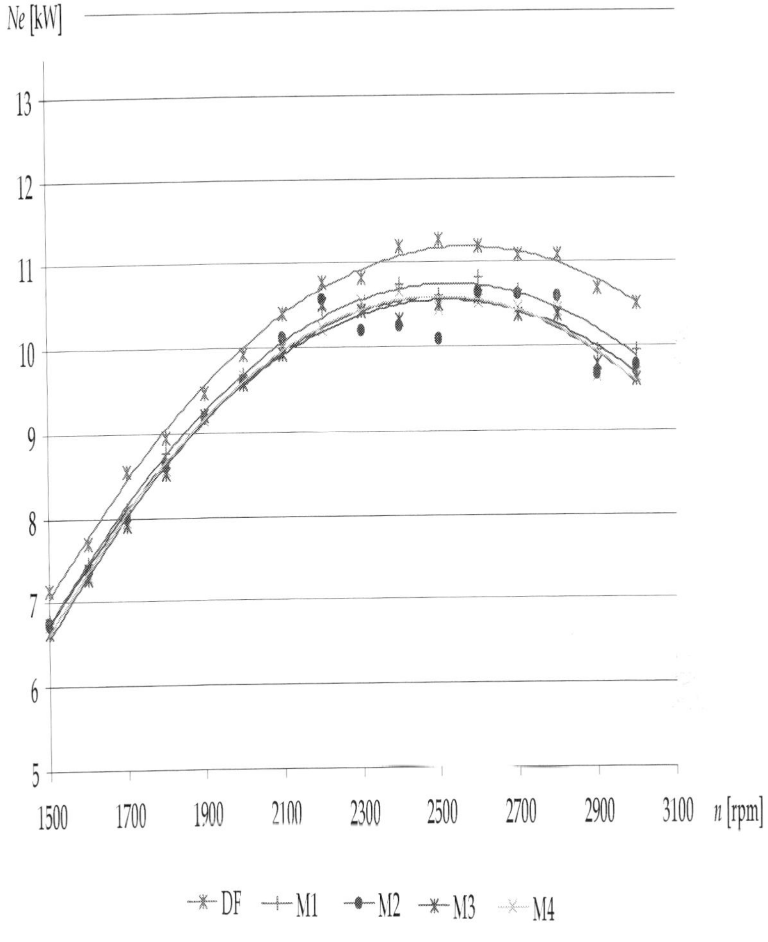

Fig. 7. Changes of the course of engine power 2CA90 powered by diesel fuel (ON) and mixtures containing 90% diesel fuel and 10% methyl esters of fatty acids and diesel fuel: M1 - esters obtained from purchased fresh soybean oil, M2 - esters obtained from soybean oil without addition of fried product, M3 - esters obtained from soybean oil frying of potato chips, M4 - esters obtained from soybean oil frying of breadcrumbs coated fish fingers. /presented data based on authors own research / [Szmigielski et al. 2011]/

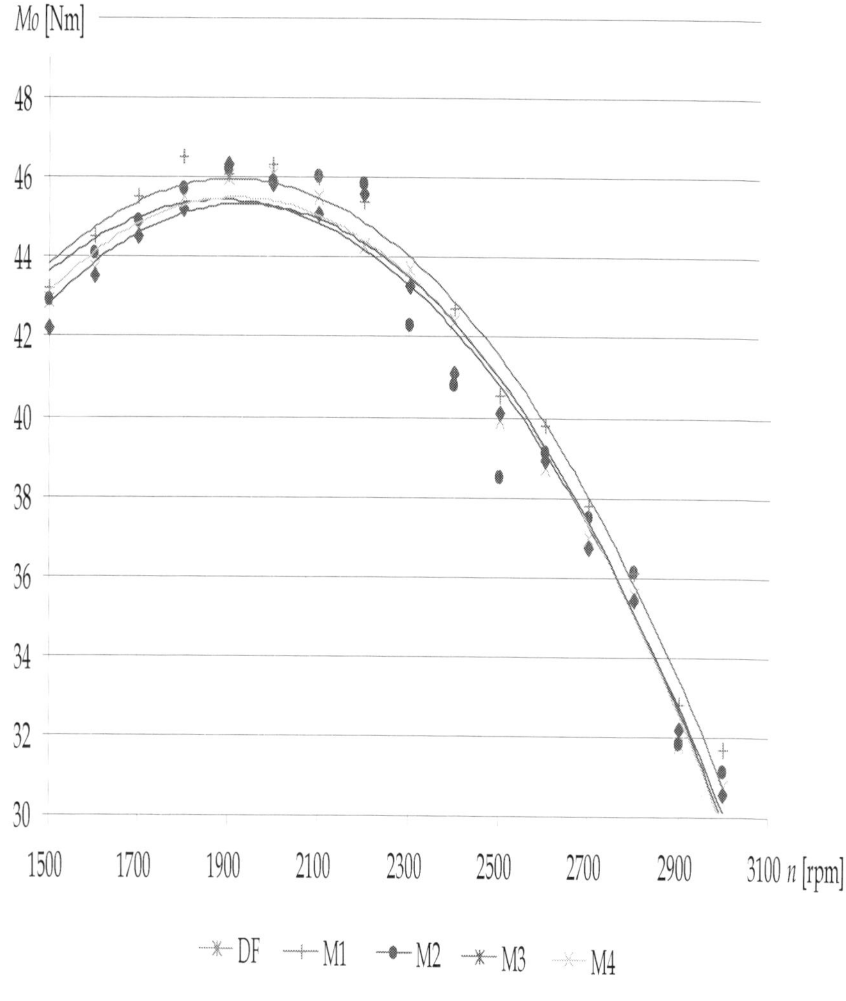

Fig. 8. Changes of the course of engine torque 2CA90 powered by diesel fuel (ON) and mixtures containing 90% diesel fuel and 10% methyl esters of fatty acids and diesel fuel: M1 - esters obtained from purchased fresh soybean oil, M2 - esters obtained from soybean oil without addition of fried product, M3 - esters obtained from soybean oil frying of potato chips, M4 - esters obtained from soybean oil frying of breadcrumbs coated fish fingers. /presented data based on authors own research / [Szmigielski et al. 2011]/

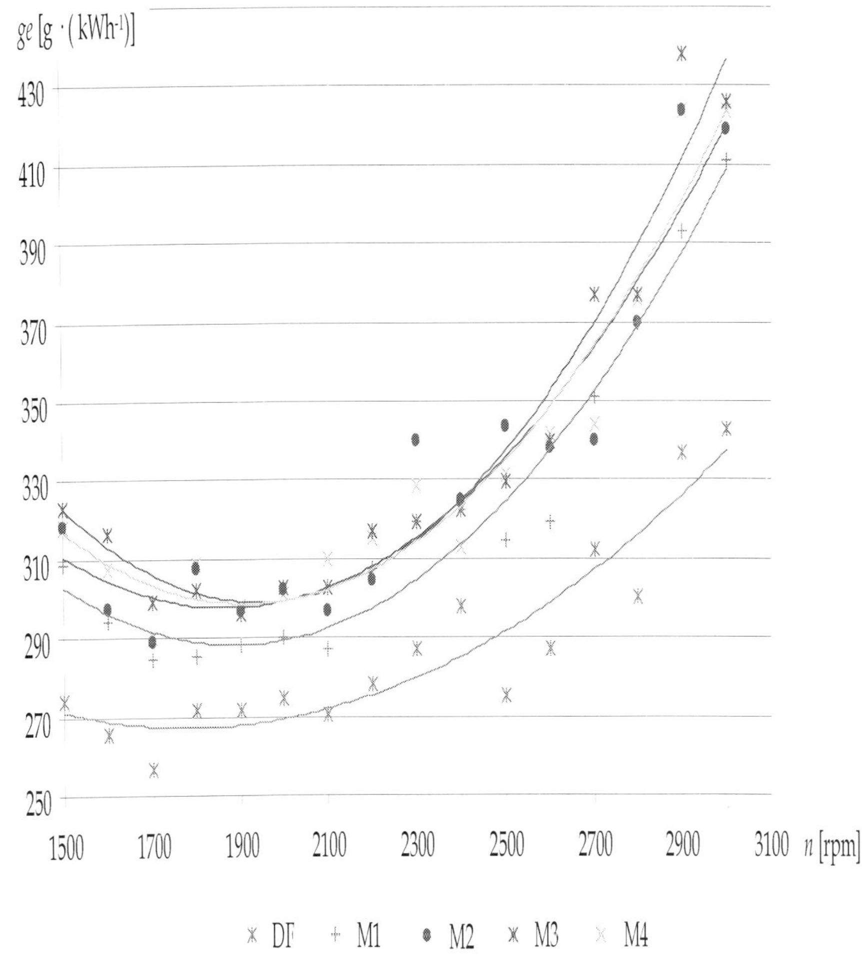

Fig. 9. Changes of the course of unitary fuel consumption engine 2CA90 powered by diesel fuel (ON) and mixtures containing 90% diesel fuel and 10% methyl esters of fatty acids and diesel fuel: M1 - esters obtained from purchased fresh soybean oil, M2 - esters obtained from soybean oil without addition of fried product, M3 - esters obtained from soybean oil frying of potato chips, M4 - esters obtained from soybean oil frying of breadcrumbs coated fish fingers. / presented data based on authors own research / [Szmigielski et al. 22011]/

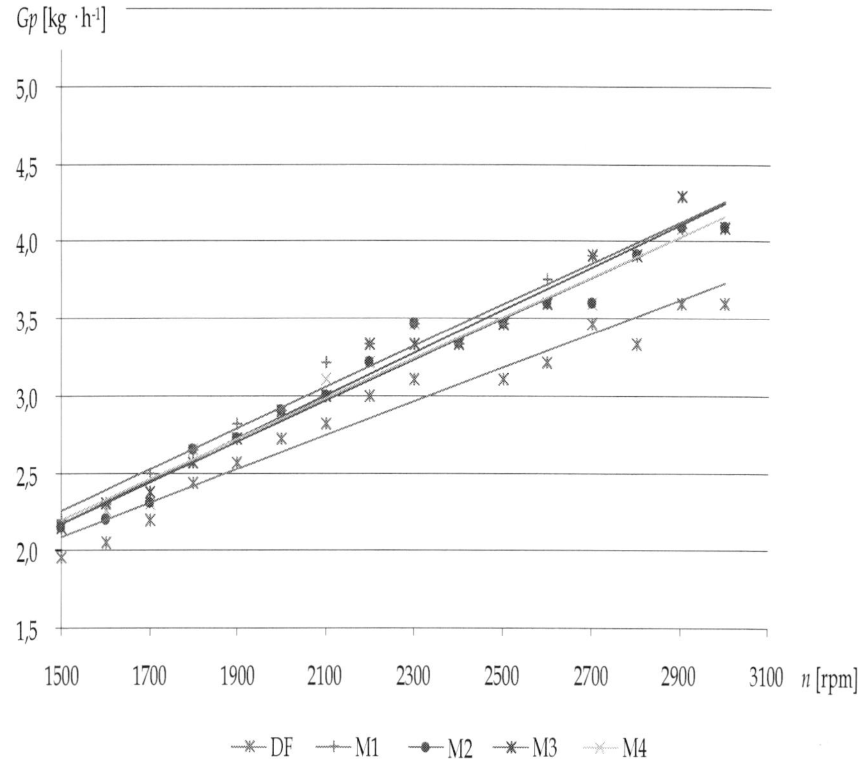

Fig. 10. Changes of the course of hourly fuel consumption engine 2CA90 powered by diesel fuel (ON) and mixtures containing 90% diesel fuel and 10% methyl esters of fatty acids and diesel fuel: M1 - esters obtained from purchased fresh soybean oil, M2 - esters obtained from soybean oil without addition of fried product, M3 - esters obtained from soybean oil frying of potato chips, M4 - esters obtained from soybean oil frying of breadcrumbs coated fish fingers. /presented data based on authors own research /[Szmigielski et al. 2011]/

5. References

Alcantara R., Amores J., Canoira L., Fidalgo E., Franco M.J., & Navarro A. (2000). Catalytic production of biodiesel from soy-bean oil, used frying oil and tallow. *Biomass and Bioenergy* 18, 515-527, ISSN.0961-9534

Blumenthal, M. M. (1991). A new look at the chemistry and physics of deep-fat frying. *Food Technology,* 45, 68-94, ISSN 0015-6639.

Buczek B.& Chwiałkowski W. (2005). Wpływ modyfikacji węgla aktywnego na jego zdolność do oczyszczania zużytego oleju smażalniczego. *Żywność, Nauka, Technologia, Jakość.* 4(45) Supl. 85-99, ISSN 1425-6959.

Dzieniszewski G. (2007). Wybrane problemy zasilania silnika diesla zużytym olejem roślinnym. *Inżynieria Rolnicza* 9, 49-55, ISSN 1429-7264.

Choe E. & Min D. B. (2006). Mechanisms and factors for edible oil oxidation. *Comprehensive Reviews in Food Science and Food Safety*, 5, 169-186, ISSN 1541-4337.

Clark W L & Serbia G. (1991). Safety aspects of frying fats and oils. *Food Technology*, 45, 84-94, ISSN 0015-6639.

Drozdowski B. (2007). Reakcje chemiczne kwasów tłuszczowych i acylogliceroli. In: *Chemia żywności T.* 2. 112-140. Praca zbiorowa pod red. Z.E. Sikorskiego WNT ISBN 978-83-204-375-6 Warszawa.

Hoffman M. (2004) Niebezpieczne przemiany. *Przegląd Gastronomiczny*. 10/2004, 14-16, ISSN 0033-2119

Krełowska-Kułas M. (1993) Metody ilościowego oznaczania i badania tłuszczów In: *Badanie jakości produktów spożywczych* 63-90. PWE, ISBN 83-208-0902-9 Warszawa..

Ledóchowska E. & Hazuka Z. (2006) Przemiany termooksydatywne wybranych olejów oliwkowych i oleju rzepakowego zachodzące w czasie ogrzewania i smażenia. *Tłuszcze Jadalne* 41, 193-196, ISSN 0371-9227.

Leung D.Y.C. & Guo Y. (2006). Transesterification of neat and used frying oil: Optimization for biodiesel production. *Fuel Processing Technology* 87, 883-890, ISSN 0378-3820.

Maniak B., Szmigielski M., Piekarski W. & Markowska A. (2009). Physicochemical changes of post frying sunflower oil. *International Agrophysics* 23, 243-248, ISSN 0236-8722.

Paul S. & Mittal G. S. (1996). Dynamics of fat/oil degradation during frying based on optical properties. *Journal of Food Engineering*, 30, 389-403, ISSN 0260-8774.

Polish Norm PN-A- 86908, Oleje i tłuszcze roślinne oraz zwierzęce - Rafinowane oleje roślinne, PKN 2000.

Polish Norm PN-ISO 660, Oleje i tłuszcze roślinne oraz zwierzęce - Oznaczanie liczby kwasowej i kwasowości, PKN 1998.

Polish Norm PN-ISO 3960, Oleje i tłuszcze roślinne oraz zwierzęce - Oznaczanie liczby nadtlenkowej, PKN 1996.

Polish Norm PN-S-02005, Silniki samochodowe -- Badania stanowiskowe --Wyznaczanie podstawowych parametrów pracy, PKN 1988.

Polish Norm BN-1374-03, Silniki samochodowe -- Badania stanowiskowe - Wykonywanie charakterystyk, PKN 1979.

Staat F. & Vallet E. (1994). Vegetable oil. methyl ester as a diesel substite. *Chemistry and Industry*, 7, 863-865, ISSN 0009-3068.

Szmigielski M., Maniak B. & Piekarski W. (2008). Evaluation of chosen quality parameters of used frying rape oil as fuel biocomponent. *International Agrophysics* 22, 361-364, ISSN 0236-8722

Szmigielski M., Maniak B., Piekarski W. & Zając G. (2009): Wybrane cechy posmażalniczego oleju rzepakowego jako surowca do produkcji komponentów paliw silnikowych w aspekcie zróżnicowanego asortymentu dań smażonych. In: *Biopaliwa i smary ekologiczne*. 41-51 pod red. naukową W. Piekarskiego, ISBN 83-89503-75-1, Warszawa.

Szmigielski M., Piekarski W., Maniak B., Zając G., Masłowski A.& Biernacka B. (2011) Influence of fried meals assortment on selected postfrying soya-bean oil features as components to produce diesel biocomponents *International Agrophysics* (in review), ISSN 0236-8722.

Tys J, Piekarski W, Jackowska I, Kaczor A, Zając G & Starobrat P. (2003). Technologiczne i ekonomiczne uwarunkowania produkcji biopaliw z rzepaku. *Acta Agrophysica, Rozprawy i Monografie* 99, 1-162, ISSN 1234-4125.

28

Recent Development of Miniatured Enzymatic Biofuel Cells

Yin Song, Varun Penmasta and Chunlei Wang
Florida International University
USA

1. Introduction

The global energy demands have increased significantly every year and current reliance on fossil fuels is unsustainable due to finite supplies from environment. In addition, the products from using fossil fuels cause pollution and global warming. Fuel cells offer an alternative solution to this issue. A fuel cell is an electrochemical cell that converts chemical energy from a fuel to electrical energy. In a fuel cell, an oxidation reaction occurs at the anode generating electrons that transfer to the cathode through the external circuit and a reduction reaction occurs at the cathode. Conventional fuel cells, for example, can be operated by using hydrogen or methanol (MeOH) as fuels to produce energy, releasing water and carbon dioxide as by-products. However, hydrogen is gaseous which gives rise to safety issues in storage and transport. Besides, many of the alternative fuels that can be used for fuel cells still rely on petroleum products. Therefore, it is well recognized that alternative sources of renewable energy are urgently required. Numerous efforts have been made to develop different power sources alternatives that are capable of performing in physiological conditions for prolonged lifetime without recharging. More recently, miniaturized medical implants such as pacemaker, defibrillator, insulin pumps, sensor-transmitter systems for animals and plants, nano-robots for drug delivery and health monitoring systems gain increasing attention which led to an upsurge in research and development in micropower source, especially, biofuel cells (Ramanavicius, 2005; Liu & Dong, 2007; Zhu et al., 2007; Moehlenbrock & Minteer, 2008 and Wang et al., 2009). Biofuel cell is a particular kind of fuel cell, which converts biochemical energy to electrical energy by using biocatalysts (Palmore & Whitesides, 1994). The two major types of biofuel cells are microbial fuel cells and enzymatic biofuel cells. Microbial fuel cells employ living cells such as microorganism as the catalyst to convert chemical energy into electricity while enzymatic biofuel cells use enzymes to catalyze the redox reaction of the fuels. In this chapter, we will first introduce both kinds of biofuel cells along with the type of catalysts used, electron transfer mechanism, electrode materials and cell performance. Then we will briefly review recent progress in miniaturized biofuel cells, which offer possibilities for implantable devices within the human body. Carbon-microelectromechanical system (C-MEMS) based miniaturized enzymatic biofuel cells are also highlighted in the chapter.

1.1 History

The earliest discovery between biology and electrical energy was demonstrated by Galvani in 1791 showing the frog leg twitching from an electric current (Galvani, 1791). The first fuel cell, which involved electrolysis of water, was discovered by Grove in 1839. An electrical stimulation can induce a biological reaction and vice-versa a biological process can also generate electricity. The first half-cell using microorganism (E.coli) was demonstrated by Potter at University of Durham (Potter, 1910). Further development of half-cell by Cohen from University of Cambridge led to one of the major types of biofuel cells, i.e., microbial fuel cells. Cohen applied a number of microbial half-cells connected in series, which generated over 35 volts (Cohen, 1931). In the late 1950s and early 1960s, the interest in development of biofuel cells received a boost by the USA space program, which led to the application of microbial fuel cells as an advanced technology for waste disposal treatment in space flights. Also, in the late 1960s, a biofuel cell using cell-free enzyme systems was discovered aiming to permanently power medical implants by utilizing specific body fluids as fuel (Yahiro et al., 1964). The concept of using microorganism as a biocatalyst in microbial fuel cells was widely applied since the 1970s (Suzuki, 1976 and Roller et al., 1984) and in the 1980s it was found out that power output could be greatly improved by using electron mediators (Vega & Fernandez, 1987; Habermann & Pommer, 1991 and Allen & Bennetto, 1993). However, the toxicity and instability of mediators limited the cell performance. A breakthrough was made when some microorganisms were found to transfer electrons directly to the electrode which led to the mediator-less microbial fuel cells first used in wastewater treatment and electricity generation (Kim et al., 1999; Chaudhuri & Lovley, 2003). These microorganisms are stable and yield a high Coulombic efficiency which facilitates the direct electron transfer (Scholz & Schroder, 2003). Shewanella putrefaciens (Kim et al., 2002), Geobacteraceae sulferreducens (Bond & Lovley, 2003), Geobacter metallireducens (Min et al., 2005) and Rhodoferax ferrireducens (Chaudhuri & Lovley, 2003) are all bioelectrochemically active microbes and can transfer electrons directly through the membrane. On the other hand, since the first enzymatic biofuel cell was reported in 1964, noticeable developments have been made in the terms of the power density, cell lifetime, operational stability (Bockris & Srinivasan, 1969; Govil & Saran, 1982 and Palmore & Whitesides, 1994). However, the output potential generated from enzymatic biofuel cells was still far beyond the demand of commercial application. Therefore, instead of considering enzymatic biofuel cells as a conventional power source, most of the researches on enzymatic biofuel cells have been aimed toward special applications such as implantable medical devices (Katz & Willner, 2003; Barton et al., 2004 and Heller, 2004). In the past ten years, cell performances on both types of biofuel cells have been improved significantly and we will discuss the detailed development in the following sessions.

1.2 Microbial fuel cells

A microbial fuel cell (MFC) converts chemical energy to electrical energy by the catalytic processes of microorganisms. Microorganisms in the MFC oxidize organic substrates and generate both electrons and protons on the anode. Electrons transfer from the anode to the cathode through an external circuit and simultaneously the protons migrate to the cathode and reduce the oxygen with the electrons available at the cathode surface. Various kinds of microorganisms are reported in association with electrodes in MFC systems. For example, brevibacillus sp. PTH1 is one of the most abundant microorganisms in a MFC system. Pure cultures used for generating current in a MFC include firmicutes, acidobacteria, proteobacteria and yeast strains Saccharomyces cerevisiae and hansenula anomala (Allen & Bennetto, 1993). These microorganisms interact with fuels through a variety of direct and indirect processes to generate energy. Microbial biofuel cells have major advantage of thorough oxidation of the fuels due to the use of microorganism as catalyst system and they can be typically operated for long lifetimes. Besides, a MFC has no intermediated processes thus it is a very efficient energy producing process. In addition, as a fuel cell, a MFC does not need charging during operation (Willner et al., 1996; Katz et al., 2003 and Calabrese et al., 2004). However, the bottlenecks of MFC still remain. Power generation of a MFC is affected by many factors including microbe type, fuel biomass type and concentration, ionic strength, pH, temperature, and reactor configuration.

The principle cell performance of MFCs lies in the electron transfer from microbial cells to the anode electrode. The direct electron transfer from the microorganism to electrodes is hindered by overpotential due to transfer resistance. The overpotential lowers the potential of a MFC and significantly affects the cell efficiency. In this case, the practical output potential is less than ideal because the electron transfer efficiency from the substrate to the anode varies from microbe to microbe. Microorganism species do not readily release electrons and hence the redox mediators are needed. A desirable mediator should have a whole range of properties: Firstly, its potential should be different from the microorganism potential to facilitate electron transfer. Secondly, it should have a high diffusion coefficient in the solution. Lastly, it is suitable for repeatable redox cycles in order to remain active in the electrolyte. Widely used Dye mediators such as neutral red (NR), methylene blue (MB), thionine (Th), meldola's blue (MelB) and 2-hydroxy-1,4-naphthoquinone (HNQ) can facilitate electron transfer for microorganism such as Proteus, Enterobacter, Bacillus, Pseudomonas and Escherichia coli. In the electron transfer process, these mediators are reduced by interacting with electron generated within the cell then these mediators in reduced form diffuse out of the cell to the anode surface where they are electrocatalytically oxidised. The oxidised mediator is then capable to repeat this redox cycle.

Better performing electrodes can improve the cell performance of a MFC because different anode materials can result in different activation of a polarization loss, which is attributed to an activation energy that must be overcome by the reactants. Carbon or graphite based materials are widely used as electrodes due to their large surface area, high conductivity, biocompatibility and chemical stability according to Table 1. Also, platinum and gold are popular as electrode system although they are expensive. Compared with carbon based electrode materials, platinum and gold electrodes are superior in the performance of the cells based on the Table 1. Besides, they have a higher catalytic kinetics towards oxygen compared to carbon based materials and hence the MFCs with Pt based cathodes yielded higher power densities than those with carbon based cathodes (Moon et al., 2009).

Electrode modification is another way to improve MFC performance of cells. (Park & Zerkeis, 2003) reported an increase of 100-folds in current compared to the previous results by using (neutral red) NR-woven graphite and Mn^{4+}-graphite anode instead of the woven graphite anode alone. Four times higher current was reported in 2004 using the combination of Mn^{4+}-graphite anode and Fe^{3+}-graphite cathode (Niessen et al., 2004). NR and Mn^{4+} doping ions serve as mediators in their MFC systems and also catalyze the cathodic reactions to facilitate electricity generations. Electrodes modifications including adsorption of AQDS or 1,4-naphthoquinone (NQ) and incorporation with Mn^{2+}, Ni^{2+}, Fe_3O_4 increased the cell performance of MFCs in their long-term operations (Lowy et al., 2006). In addition, the fluorinated polyanilines, poly (2-fluoroaniline) and poly (2, 3, 5, 6-tetrafluoroaniline) outperformed polyaniline were applied for electrode modification (Niessen et al., 2006). These conductive polymers also serve as mediators due to their structural similarities to conventional redox mediators.

Fuel	Organism	Electrode (cm^2)	Electron transfer	OCV (V)	Current density ($\mu A\ cm^{-2}$)	Reference
Sugar/ferricyanide	Suspended Proteus vulgaris/anaerobic	RVC anode (30.4), platinum cathode (16)	MET	0.52	5.26	Kim et al. (2000)
Glucose/ferricyanide	Suspended E. coli/ anaerobic	Woven graphite	MET	0.85	5.3	Park & Zeikus (2000)
H_2/O_2	Desulphovibrio vulgaris,	Carbon felt mat (5.1)	MET	1.17	176	Tsujimura et al. (2001)
Lactate/O_2	Suspended Shewanella putrefaciens and E. coli , anaerobic	Graphite felt (56)	DET	0.5	0.02	Kim et al. (2002)
Marine sediment con-stituents/ seawater constituents	Mixed natural bacteria	Drilled graphite discs	DET & MET	0.75	3.2	Tender et al. (2002)
Lactate/O_2	Suspended Shewanella putrefaciens/ anaerobic	Mn^{4+} graphite plate (80) anode, Fe^{3+} modified graphite plate cathode (50)	MET	0.6	0.94	Park & Zeikus (2002)
Glucose/ferricyanide	Rhodoferax ferrireducens/ anaerobic	Graphite rod (0.65)	DET	N/A	N/A	Chaudhuri & Lovley (2003)
Glucose /O_2	Suspended E. coli/ anaerobic	NR-woven graphite (80) or Mn^{4+}graphite plate anode (80), woven graphite or Fe^{3+} graphite plate (80) cathode	MET	N/A	N/A	Park & Zeikus (2003)
Glucose /O_2	Mixed culture	Graphite plate electrodes (50)	MET and DET	N/A	36	Rabaey et al. (2003)
Glucose/ ferricyanide	Suspended E. coli/ anaerobic	Woven graphite cloth	DET	0.895	120	Schroder et al. (2003)
Glucose/ ferricyanide	Clostridium butyricum	Woven graphite cloth	MET	0.759	200	Niessen et al. (2004)

Fuel	Organism	Electrode (cm^2)	Electron transfer	OCV (V)	Current density ($\mu A\ cm^{-2}$)	Reference
Glucose/ ferricyanide	Mixed culture	Graphite plate electrodes (50)	MET & DET	N/A	231	Rabaey et al. (2004)
Glucose/ ferricyanide	Shewanella oneidensis DSP-10	Graphite Felt (610)	DET	N/A	110	Ringeisen et al. (2006)
Glucose $/O_2$	Saccharomyces cevevisiae	Gold (0.51)	MET	N/A	15	Chiao et al. (2006)
Glucose $/O_2$	Mixed bacterial culture	Carbon cloth (7)	MET	N/A	90	Fan et al. (2007)
Glucose/ ferricyanide	Geobacter sulfurrenducens	Pt (7.8)	DET	N/A	688	Richter et al. (2008)
Glucose $/O_2$	Shewanella oneidensis MR-1	Pt (1.2)	MET	N/A	302	Siu & Chiao (2008)
Glucose/ ferricyanide	Shewanella oneidensis MR-1	Pt (0.49)	MET	N/A	370	Hou et al. (2009)
Glucose/ ferricyanide	Shewanella oneidensis MR-1	Gold (0.15)	MET	N/A	130	Qian et al. (2009)

Table 1. Summary of microbial biofuel cells

Proton exchange membrane (PEM) can also significantly affect a MFC system's internal resistance and concentration polarization loss because the internal resistance of MFC decreases with the increase in the PEM surface area (Oh & Logan, 2006). Nafion (DuPont, Wilmington, Delaware) is the most popular proton exchange membrane material due to its highly selective permeability of protons (Min et al., 2005). Compared with the performance of MFC using a PEM or a salt bridge, the power density using the salt bridge MFC was 2.2 mW/m^2 that was an order of magnitude lower than that attained using Nafion. However, side effect is unavoidable with the use of PEM. For example, the concentration of cation species such as Na^+, K^+, NH^{4+}, Ca^{2+}, Mg^{2+} is much higher than that of proton so that transportation of cation species dominates. In this case, Nafion used in the MFCs is not an efficient proton specific membrane but actually a cation specific membrane (Rozendal et al., 2006). Subsequent studies have implied that anion-exchange or bipolar membranes has better properties than cation exchange membranes regarding to cell performance (Zhang et al., 2009).

Two promising applications of MFCs in the future are wastewater treatment and electricity generation (Feng et al., 2008 and Katuri & Scott, 2010). Although some noticeable development has been made in the MFC research, there are still a lot of challenges to be overcome for large-scale applications. The primary challenge is how to improve the cell performance in terms of power density and energy efficiency. In addition, catalytic effect of bioelectrodes needs to be further enhanced to solve the problems caused by enzyme activity loss and other degradation processes. Moreover, the lifetime of the MFC must be significantly improved.

1.3 Enzymatic biofuel cells

Enzymatic biofuel cells (EBFCs) utilize redox enzymes such as glucose oxidase (GOx), laccase as the catalysts that can facilitate the electron generation between substrates and electrode surface, hence generating the output potential. There are two types of electron transfer mechanisms which are direct electron transfer (DET) and mediator electron transfer (MET). In DET based EBFCs, the substrate is enzymatically oxidized at the anode, producing protons and electrons which directly transfer from enzyme moleculars to anode surface. At the cathode, the oxygen reacts with electrons and protons, generating water. However, DET between an enzyme and electrode has only been reported with several enzymes such as cytochrome c, laccase, hydrogenase, and several peroxidases (Schuhmann, 2002 and Freire et al., 2003). Some enzymes have nonconductive protein shell so that the electron transfer is inefficient. To overcome this barrier, MET was used to enhance the transportation of electrons. The selection and mechanism of MET in EBFCs are quite similar to those of MFCs that are discussed before. Similarly, there are still some challenges in using MET in EBFCs, such as poor diffusion of mediators and non-continuous supply. Therefore, modification of bioelectrodes to realize DET based EBFCs attracted most attention. In EBFCs system, power density and lifetime are two important factors which determine the cell performance in the application of EBFCs. Significant improvements have been made during the last decade (Katz et al., 2003; Calabrese et al., 2004; Zhu et al., 2007; Moehlenbrock & Minteer, 2008; Wang et al., 2009; Lee et al. 2010 and Saleh et al. 2011). Noticeably, these advancements have been mostly achieved by modification of electrode with better performance, improving enzyme immobilization methods as well as optimizing the cell configuration. Recent development of enzymatic biofuel cells is shown in Table 2.

The performance of electrodes for EBFCs mainly depends on: electron transfer kinetics, mass transport, stability, and reproducibility. The electrode is mostly made of gold, platinum or carbon (Katz & Willner 2003). Besides these conventional materials, biocompatible conducting polymers are widely used because they can facilitate electron transfer and co-immobilize the enzymes at the same time (Schuhmann & Muenchen, 1992; Haccoun et al., 2006 and Nagel et al. 2007). In order to maximize the cell performance, mesoporous materials have been applied in many studies because of their high surface areas thus high power density could be achieved. Moreover, many attempts using nanostructures such as nanoparticles, nanofibers, and nanocomposites as electrode materials have also been made to fabricate electrodes for EBFCs. The large surface area by using these nanostructures leads to high enzyme loading and enables to improve the power density of the cells. Recently, one of the most significant advances in EBFCs is electrode modification by employing carbon nanotubes. (Wang et al.2009; Lee et al. 2010; Tanne et al. 2010 and Saleh et al. 2011.) Several research activities have addressed the application of single wall carbon nanotube hybrid system. The oriented assembly of short SWNT normal to electrode surfaces was accomplished by the covalent attachment of the CNT to the electrode surface. It was reported that surface assembled GOx is in good electric contact with electrode due to the application of SWNT, which acted as conductive nanoneedles that electrically wire the enzyme active site to the transducer surface. Other studies have been reported on improving electrochemical and electrocatalytic behavior and fast electron transfer kinetics of CNTs. Improved enzyme activity was observed in comparison to similar enzyme-containing composites without using SWNTs. It was discussed that the application of SWNTs, which

Fuel	Enzyme	Electrode	Electron transfer	OCV (V)	Current density (μA cm^{-2})	Reference
Glucose/ O_2	GOx/laccase	Carbon fiber electrodes	MET	0.8	64	Chen et al. (2001)
Glucose/ O_2	GOx/BOx	Carbon fiber electrodes	MET	0.84	432	Mano et al. (2002)
Glucose/ O_2	GDH/BOx	Glassy carbon disc electrodes	MET	0.44	58	Tsujimura et al. (2002)
Glucose/ O_2	GOx/COx	Gold electrodes coated with Cu	MET	0.12	4.3	Katz & Willner (2003)
Glucose/ O_2	GOx/BOx	Carbon fiber electrodes	MET	0.68	50	Kim et al. (2003)
Glucose/ O_2	GOx/BOx	Carbon fiber electrodes	MET	0.8	440	Mano et al. (2003)
Glucose/ O_2	GOx/BOx	Carbon fiber electrodes	MET	0.63	244	Mano & Heller (2003)
Glucose/ O_2	GOx/laccase	Carbon fiber electrodes	MET	1.0	350	Heller (2004)
EtOH to CH_3 CHO to CH_3COOH	ADH, ADH +AldDH, formaldehyde dehydroge-nase + FDH	Carbon coated with poly(methylene)	MET	0.62	1160	Akers et al. (2005)
Glucose/ O_2	PLL-VK3 / PDMS	Pt	MET	0.55	130	Togo et al. (2007)
Ethanol/H_2O_2	QH-ADH/AOx	Pt	DET	0.24	30	Ramanavicius et al. (2008)
Glucose/ O_2	GDH/PDMS	Pt	DET	0.80	11000	Sakai et al. (2009)
Glucose/ O_2	GOx/laccase	Silicon/ SWNTs	DET	N/A	30	Wang et al. (2009)
Glucose/ O_2	GOx/laccase	Au/SWNTs	DET	0.46	960	Lee et al. (2010)
Glucose/ O_2	PQQ-GDH/BOD	Au/MWNTs substrates	DET	0.60	200	Tanne et al. (2010)
Glucose/O_2	GDH/NB	Glass carbon/ SWNTs	DET	0.35	100	Saleh et al. (2011)

Table 2. Summary of enzymatic biofuel cells.

possesses a high specific surface area, may effectively adsorb enzyme molecules and retain the enzyme within the polymer matrix, whereas other forms of enzyme-composites may suffer from enzyme loss when they were placed in contact with aqueous solutions. Although recent advancement in modification of electrodes appears to be promising due to the improvement of cell performance obtained, biocompatibility and nanotoxicity need to be further studied and addressed.

Successful immobilization of the enzymes on the electrode surface is considered as another critical factor that affects cell performance. The immobilization of enzyme can be achieved physically or chemically. There are two major types of physical methods, physical absorption and entrapment. The first one is to absorb the enzymes onto conductive particles such as carbon black or graphite powders. For example, hydrogenase and laccase were immobilized by using physical absorption on carbon black particles to construct composite electrodes and the EBFCs could continuously work for 30 days. Another physical immobilization method is based on polymeric matrices entrapment, which usually shows more stabilized enzyme immobilization (Mano et al., 2002; Mano et al., 2003, Heller, 2004 & Soukharev et al., 2004). For example, Soukharev utilized redox polymers to fabricate enzymatic biofuel cells system. The electrodes were built by casting the enzyme-polymer mixed solution onto 7 μm diameter, 2 cm length carbon fibers. It showed that the glucose-oxygen biofuel cell was capable of generating a power density up to 0.35 mW/cm^2 at 0.88 V (Soukharev et al., 2004). Compared with the physical immobilization which is unstable during the operation, the chemical immobilization methods with the efficient covalent bonding of enzymes and mediators are more reliable. Katz et al. reported a biofuel cell using co-immobilized enzyme-cofactor-mediator composites on metal electrodes to functionalize the electrode surface with a monolayer then integrate with enzymes via bioaffinity (Willner et al., 1998; Katz et al., 2001 and Katz et al., 2003). Another example is that a redox monolayer was covalently grafted with pyrroloquinoline quinone (PQQ) to Au-electrode. Then GOx-FAD electrode was assembled with PQQ as mediators (Willner et al., 1996). Other widely used materials to functionalize electrode surface have also been reported, such as nitrospiropyran (Blonder et al., 1998), rotaxane (Katz et al., 2004), C-60 (Patolsky et al., 1998) and Au nanoparticles (Xiao et al., 2003).

Rapid development on EBFCs has been achieved in the past decade with the arised demands for reliable power supplies for implantable medical device. It has shown particular advantages over conventional batteries because of the specific biocatalysts and the possibility of miniaturization. However, there are still challenges for further development of long term stability of the enzymatic bioelectrodes and efficient electron transfer between enzymes and electrode surfaces. Recent efforts have been given to protein engineering, reliable immobilization method and novel cell configuration.

2. Miniature biofuel cells

Miniature power systems using biocatalysts have received increased attention associated with demand for micro-scale power supplies for implantable medical devices. Development of miniature biofuel cells offers a great opportunity to serve as long-term power sources in implantable device where frequent replacement of battery is not practical. The ability of biocatalyst in converting available indigenous fuels into electrical energy makes miniature biofuel cells attractive to enable long-term and self-sustained power system. The success of medical implants is akin with the effective miniaturization of power sources. This can be achieved by miniaturization of different functional components such as electrodes, power supply, and signal processing units. Some of the effective techniques for miniaturization involve fabricating microfluidic systems using photolithography, etching, polymer molding, and metal deposition (Kim et al., 2008). For example, Siu and Chiao (Siu & Chiao, 2008) applied photolithography and polymer molding to fabricate polydimethylsiloxane (PDMS) electrodes. It was also used by Hou et al. (Hou et al., 2009) to fabricate gold electrode arrays for the microbe screening. Besides polymer molding, etching can also be used to transfer micro-patterns onto device-building substrate. Chiao (Chiao et al., 2006) applied wet etching to construct silicon-based chambers containing serpentine channels. Additionally, C-MEMS microfabrication technique for 3D microstrustures, involving the pyrolysis of patterned photoresist has been developed which can be used as microelectrodes for miniature biofuel cells (Wang & Madou, 2006). With current microfabrication processes, the miniature biofuel cells offer unique advantages such as large surface area to volume ratio, short distance between the electrode, fast response time and low Reynolds number. In the following section, we will discuss the developments of both miniature MFCs and EBFCs.The experimental demonstration of miniature biofuel cells, along with the discussion of the key challenges and opportunities for realizing the practical potential of miniaturized biofuel cells for medical implants will be discussed.

2.1 Miniature microbial biofuel cells and its state of the art

One of the early efforts on miniature microbial biofuel cells reported a surface power output of 0.023 mW/m^2 and current density of 150 mA/m^2 based on the 10μm diameter circular anodic electrode (Chiao et al., 2006). The miniature microbial biofuel cells were limited by relatively low volumetric power density and coulombic efficiency due to their high internal resistances. Compared with macro scale microbial biofuel cells using the same microbes (Ringeisen et al., 2006), miniature microbial biofuel cells generated similar volumetric current density but significantly lower volumetric power density, which is insufficient for the anticipated applications. It was pointed out that the internal resistances of miniature microbial biofuel cells were around 40 fold higher than that in the macro scale microbial biofuel cells. The ohmic loss was higher in the micropillar devices with the same catholyte and anolyte; however, they generated higher volumetric power density (32 A/m^3) than the serpentine-channel devices (0.5 A/m^3) (Ringeisen et al., 2006). The high surface area-to-volume ratio and good microbe adaptivity of the micropillar electrodes decreased the anode resistance and resulted in higher volumetric power output. Carbon based anodes are known for high surface area-to-volume ratio and easy adaptation of microorganism and they are widely used in macro scale microbial biofuel cells. The recent investigations using carbon nanotubes (CNT) as electrodes (Qiao et al., 2007 and Timur et al., 2007) provide promising solutions for constructing carbon-based anodes in miniature microbial biofuel cells. The CNT based electrodes showed great improvement in the electricity generation and biocompatibility. Its maximum power density was 42 mW/m^2 using E.coli as the microbial catalyst.

In the pursuit to improve the miniature microbial biofuel cell performance, different strategies were employed such as increasing the anode surface area, improving coupling of microorganism to anode surface, developing electrochemically active microbes and decreasing proton diffusion resistance. In summary, the enhancement strategies resulted in enhanced mass transport, improved reaction kinetics, and reduced ohmic resistance. Based on these developments, the ability to generate sufficient current and power from miniature devices was realized, thus breaking the conventional concept that small scale microbial biofuel cells would perform unsatisfactorily due to limited amount of substrates and microorganism. Since the development of the first miniature microbial biofuel cells in 2006, the volumetric power density and coulombic efficiency have been increased over 5 times. Although the output potential from the miniature MFCs is still insufficient for powering conventional equipment, they are promising options for on-chip power sources, especially for medical implants, which only require several millivolts to operate. Given the evidence that volumetric current density of the miniature MFC was achieved to be 2400 mA/m^3 and required power from the cell was therefore 960 mW/m^3, which is sufficient for existing devices (Wang & Lu, 2008). However, higher current density can result in excessive ohmic heating and electrolysis during the operation. Therefore, study in optimizing current density, overall output voltage and stability of the miniature MFCs as well as electrode design and device configuration for implantation rejection, microbe leakage, and analysis of the composition and distribution of internal resistances is necessary before further implementation in practical applications.

2.2 Miniature enzymatic biofuel cells and its state of the art

The first micro-sized enzymatic biofuel cells reported in 2001 (Chen et al., 2001). A glucose/O_2 biofuel cell consisted of two 7 μm diameter, 2 cm long electrocatalyst-coated carbon fibers operating at ambient temperature in a pH 5 aqueous solution. The areas of the anode and the cathode of the cell were about 60 times smaller than those of the smallest reported fuel cell and 180 times smaller than those of the previously reported smallest biofuel cell. The power density of the cell is 64 $\mu W/cm^2$ at 23 °C and 137 $\mu W/cm^2$ at 37 °C, and its power output is 280 nW at 23 °C and 600 nW at 37 °C. The results revealed that the miniature enzymatic biofuel cells could generate sufficient power for small power-consuming CMOS circuit. Later, a miniature enzymatic biofuel cell with the same carbon fibers operating in a physiological buffer was reported (Mano et al., 2002). In a week operation the cell generated 0.9 J of electrical energy while passing 1.7 C charge. Based on this result, Mano developed a miniature compartment-less glucose/O_2 biofuel cell operating in a living plant. Implantation of the fibers in the grape leads to an operating biofuel cell producing 2.4 μW at 0.52 V, which is adequate for operation of low-voltage CMOS/SIMOX integrated circuits. The performance of the miniature enzymatic biofuel cell was upgraded to 0.78 V operating at 37 °C in ph 5 buffer later on (Mano et al., 2003). In 2004, a miniature single-compartment glucose/O_2 biofuel cell made with the novel cathode operated optimally at 0.88 V, the highest operating voltage for a compartmentless miniature fuel cell (Soukharev et al., 2004). The enzyme was formed by "wiring" laccase to carbon through an electron conducting redox hydrogel, its redox functions tethered through long and flexible spacers to its cross-linked and hydrated polymer, which led to the apparently increased electron diffusion coefficient. The latest report on miniature glucose/O_2 biofuel cells demonstrated a new kind of carbon fiber microelectrodes modified with single-wall carbon nanotubes (CNTs) (Li et al., 2008). The power density of this assembled miniature compartment-less glucose/O_2 BFC reaches 581 Wcm^{-1} at 0.40 V. When the cell operated continuously with an external loading of 1 M resistance, it lost 25% of its initial power in the first 24 h and the power output dropped by 50% after a 48 h continuous work. Although from the practical application point of view, the performance and the stability of the current emzymatic biofuel cells remain to be improved, the miniature feature and the compartment-less property as well as the tissue-implantable biocapability of enzymatic biofuel cell essentially enable the future studies on in vivo evaluation of the cell performance and stability in real implantable systems.

In an effort to miniaturize the EBFCs, we have developed a versatile technique based on C-MEMS process for the miniaturization of electrodes. Our research focuses on the fabrication of 3D microelectrodes for miniature enzymatic biofuel cells. First, the functionalization methods for EBFCs enzyme immobilization were studied. Then we apply finite element approach to simulate the miniature EBFCs to attain the design rule such as electrode aspect ratio, configuration as well as orientation of the chip. Building an EBFC based on the design rule we obtained is on-going.

3. C-MEMS 3D architecture electrodes

The surface area of biofuel cells determines its amperage, meaning that cell power is directly proportional to the electrode surface area. A conventional 2D power system is typically a parallel arrangement of a planar cathode and an anode separated by a solid or liquid electrolyte. More recently, carbon-microelectromechanical (C-MEMS) fabrication technology has offered the flexibility to fabricate complex carbon-based EBFCs, with 3D dense microelectrode arrays. C-MEMS, describes a manufacturing technique in which carbon microstructures are fabricated by baking UV sensitive polymers at high temperatures in an inert environment. It has been demonstrated that 3D high-aspect-ratio carbon structures can be made from carbonizing (pyrolysis) patterned NANO™ SU-8 negative photoresist (Wang et al., 2005). The four steps involved in converting organic polymer to pyrolytic carbon is shown as schematic in Fig. 1. Positive photoresist (AZ4620, AZ1518) as well as negative photoresist (SU-8) can be converted to carbon by pyrolysis depending on the application. Electrodes based on 3D microstructures are expected to offer higher surface area and significant advantages in comparison to thin-film devices for powering MEMS and miniaturized electronic devices.

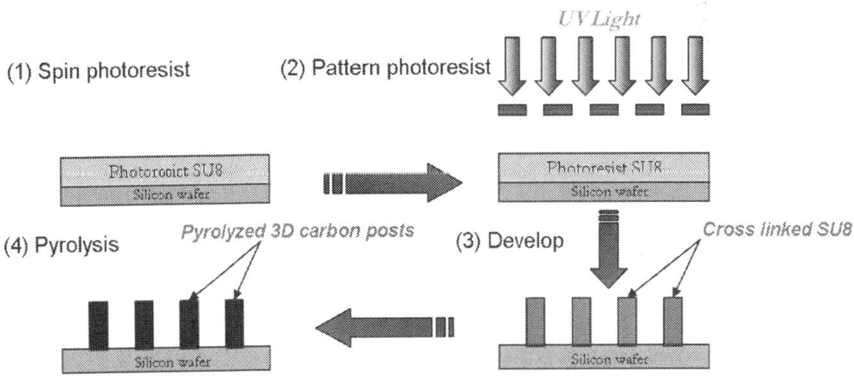

Fig. 1. Schematic showing the typical C-MEMS process.

3.1 C-MEMS microelectrodes

Fig. 2. shows the various carbon architectures possible using C-MEMS technique. The versatility in the technique gives us the opportunity to integrate nanofeatures such as suspended carbon nanowires, carbon nanofibers with microelectrodes (Wang et al., 2005; Malladi et al., 2006).

Fig. 2. SEM images of different carbon architectures possible by C-MEMS technique.

Besides, in order to increase the surface area, we reported a modified C-MEMS process using a block copolymer F127 as porogen capable of producing porous carbon microelectrodes (Penmatsa et, al., 2010). These results indicated that porous carbon thin film electrodes derived from 10% F127 mixed in SU-8 had an A_{eff} 185% compared to the conventional photoresist derived carbon electrode. This fabrication approach can be employed to produce reproducible high aspect ratio carbon microelectrodes with different shapes for various electrochemical devices.

Although the 3D structures compared to 2D planar electrodes or thin films have advantages, such as an increase in the surface area and power density for same foot print area, there are yet certain important issues which need to be addressed in order to use these structures effectively. Anandan and Godino have studied the mass transport phenomenon in micro and nano-electrodes by finite element analysis approach. They suggest that in order to accommodate the specific analyte species in terms of reaction kinetics and mass transport, it is necessary to optimize the geometry of nanopillars (their diameter, spacing and height), to reap the true benefit of using micro-nanostructured electrodes for enhancing the performance of biosensors. They reveal that the glucose immediately react with the top portions of the nanopillars due to higher reaction rate of enzymes and hence the bottom portion of the pillars lack the diffusion of glucose, which may not be favorable to improve the performance of EBFC. Jeffrey suggests that in contrast to the 2D electrodes, in which uniform current density is naturally obtained over the surface of the cathodes and anodes, the current density in the 3D microelectrode array suffer from a non-uniform primary current distribution. These non-uniform currents result in utilization of the electrode materials and are thus associated with lower cell efficiencies, reduced electrodes stability due to non-uniform stresses, and non-uniform heat dissipation. Therefore, it is essential to optimize the geometries of electrodes to homogenize the current density distribution around microelectrodes surfaces.

In the later section, we introduced finite element method based simulative approach to understand the effect of 3D design rule and spatial distributions of the microelectrodes in the arrays with respect to the mass transport of glucose, enzymatic reaction rate and open circuit output potential.

4. Surface functionalization of C-MEMS

There are several factors regulating the lifetime of biofuel cells, which has always been a concern for their practical application. In most cases, the stability of enzyme determines the lifetime of biofuel cells. Immobilization of enzyme through covalent bonding on solid surface has attracted great attention for applications in catalytic processes. Therefore, in our research, covalent attachment of enzyme on supports was studied to promote rigidification of enzyme structure of the immobilized enzyme.

We have studied three different types of covalent surface functionalization for enzyme immobilization in EBFCs - (1) Direct amination; (2) Diazonium grafting; (3) Diamine grafting. In all these methods, the functional groups are realized on the pyrolyzed carbon surface.

4.1 Direct amination on pyrolyzed carbon

Direct amination was conducted by functionalizing its surface with ultraviolet (UV) irradiation under ammonia gas (Yang et, al., 2009). Quantified amino groups on the carbon surface were estimated by X-ray photoelectron spectroscopy in Fig. 3. It is found out that the amino

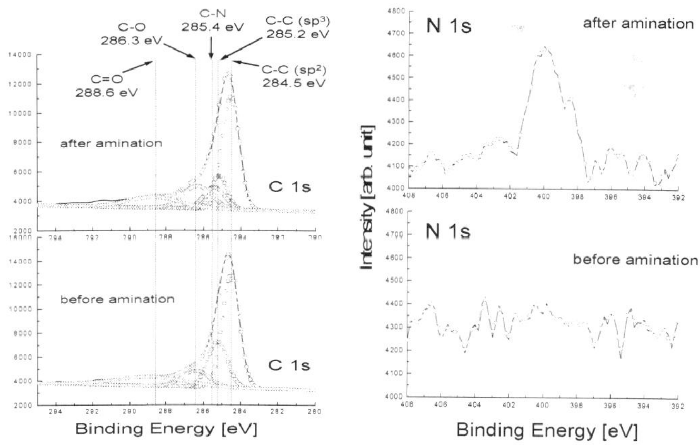

Fig. 3. X-ray photoelectron spectra of pyrolyzed carbon surface. (a) High-resolution scan of the C 1s peak and (b) high-resolution scan of the N 1s peak are compared before and after direct amination.

groups exist at surface by amination processes due to the high density of carbon. Ammonia gas forms as C-NH_2 on carbon substrate because C-H bonds react easily with ammonia gas and undergo photochemical reaction on exposure to UV irradiation although steric limitations will limit the amine group coverage on the surface. The results showed that the amino groups were successfully formed on pyrolyzed carbon surface by direct amination.

4.2 Surface functionalization of carbon surface by diazonium grafting

Diazonium grafting is a promising alternative to conventional electrode functionalization method. In this approach the electrochemical reduction of diazonium forms an aryl centered radical. The resulting aryl radical can then form a covalent bond with conducting and semiconducting surfaces. The CV results shown in Fig. 4. indicate the first cycle of electro-reduction process from NO_2 to NH_2 at different diazonium concentrations. The reduction of NO_2 to NH_2 occurs on the first negative-going sweep in a range of potential from -1.0 V to -1.1 V forming a clear irreversible anodic peak. From the results, it was confirmed that the amino groups are successfully grafted on the carbon surface for further immobilization.

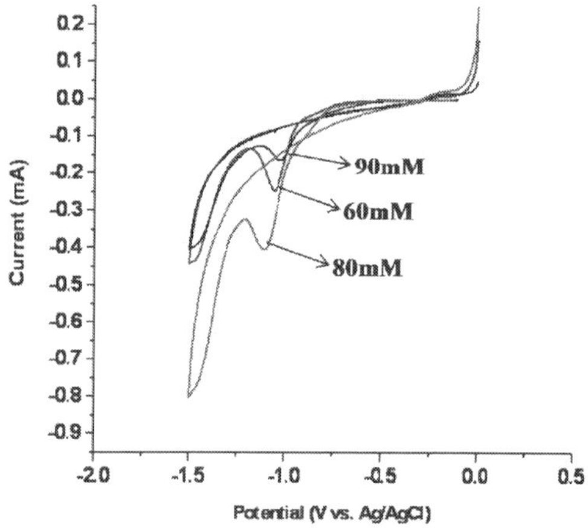

Fig. 4. CV curves showing the electroreduction from NO_2 to NH_2 at different concentrations of diazonium. The irreversible peak indicates the electroreduction process.

4.3 Functionalization of carbon surface by diamine grafting

Ethylene diamine grafting on the carbon surface was conducted using the electrochemical reduction. As the potential sweeps to negative direction during the first cycle, the oxidation of amino group in ethylene diamine is reduced on pyrolyzed carbon surface at potentials between -0.8 V and -1.4 V, leading to a clear irreversible anodic peak. It also shows that any peaks by oxidation and reduction do not exist after first reduction sweep even though we notice that the anodic current of electrode is decreased. This result indicates that the ethylene diamine was grafted on the surface of pyrolyzed carbon by applying potential and pyrolyzed surface is successfully functionalized by the electrochemical method.

We have introduced three types of functionalization on carbon surface in this session. In the future work, we will immobilize different biomolecules based on these functionalization methods for EBFCs device. Work on building a prototype EBFC consisting of glucose oxidase immobilized anode and a laccase immobilized cathode using C-MEMS based interdigitated electrode arrays is underway.

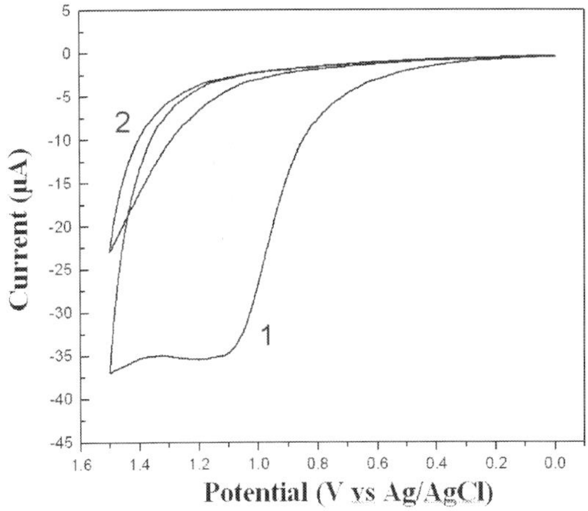

Fig. 5. Cyclic voltammograms showing the first and second cycle confirming the surface functionalization completed in the first irreversible cycle.

5. Simulation of C-MEMS based EBFCs

5.1 Finite element approach for optimization of electrodes design

For our simulation approach, we used commercially available COMSOL 3.5 software multiphysics software, which solves partial differential equations (PDEs) by finite element technique. In the model we assume that 3D carbon microelectrode arrays were uniformly immobilized with glucose oxidase and laccase on anode and cathode respectively with out the use of any mediators. The proposed implantable membraneless EBFC is assumed to be placed inside a blood artery of the human body thus utilizes the glucose extracted from blood as a fuel. In principle, glucose oxidase reacts with glucose and produces gluconolactone and hydrogen peroxide. This hydrogen peroxide oxidizes on the anode to generate electron and hydrogen ions. The hydrogen ions travel from electrolyte to cathode, while electrons flow through an external load and generate electricity. On cathode, dissolved oxygen is reduced via laccase enzyme and by combining with electrons and hydrogen ions forms water.

We applied Michaelis-Menten theory in our 2D model to analyze phenomenon between enzyme kinetics on the electrode surface and glucose diffusion and thus optimize the electrode microarray design rule according to the enzyme reaction rate In order to determine the output potential in developing biofuel cell, we also incorporated Nernst equation. The numerical simulations have been performed with various electrodes heights and well widths (distance between any two electrodes) to obtain the relation between design rule and EBFCs performance. Various 2D models are investigated for same foot print length (600 μm) of SiO_2, with fixed electrode diameter of 30 μm and fixed enzyme layer thickness of 10 μm. The height of electrodes is chosen as 60 μm, 120 μm and 240 μm for different well widths (WW-distance between any two electrodes) of 10μm, 20 μm, 40 μm, 60 μm, 80 μm, 100 μm, 120 μm, 140 μm, 160 μm, 180 μm and 200 μm.

The quantification of reaction rates of enzymes on anode and cathode is showed in Fig. 6.

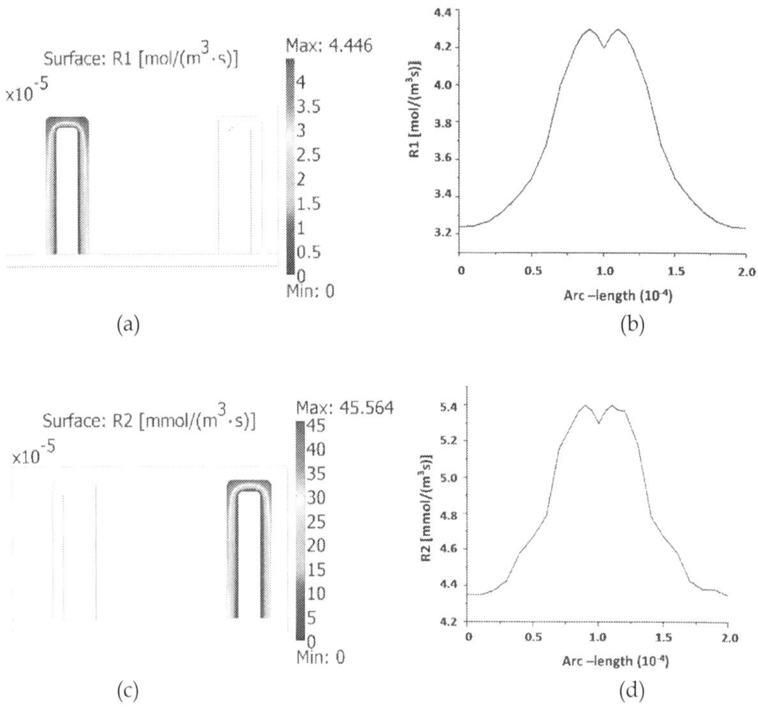

Fig. 6. (a) Subdomain plot of anode reaction rate (R1); (b) reaction rates from the whole surface of anode. (c) Subdomain plot of cathode reaction rate (R2); (d) reaction rates from the whole surface of cathode.

From the results, we observe that the reaction rate decreased from the top to the bottom along the surface of both electrodes due to the lack of diffusion of the substrate as we go towards the bottom; also the outer surfaces of the electrodes have the larger reaction rate in the enzyme layer. The reaction rate along the surface of both electrodes is plotted in Fig. 6. The reaction rate is increased from the bottom to the top along the electrode surface and reached the maximum at edge of the top due to the edge effect. The maximum reaction rates of GOx enzymes vs. different well widths is shown in Fig. 7. for three different heights of electrodes: 60 μm, 120 μm and 240 μm. with 10 μm, 20 μm, 40 μm, 60 μm, 80 μm, 100 μm and 120 μm well widths, respectively. In the case of 60 μm height of electrodes, the maximum reaction rate is obtained when the well width is about 30 μm. For the height of 120 μm and

240 μm, reaction rate reached the highest at the well width of 60 μm and 120 μm respectively. From all these three sets of models both in anode and cathode, we can conclude that the reaction rates of one pair of electrodes reach the maximum when the well width is half as the height of electrodes.

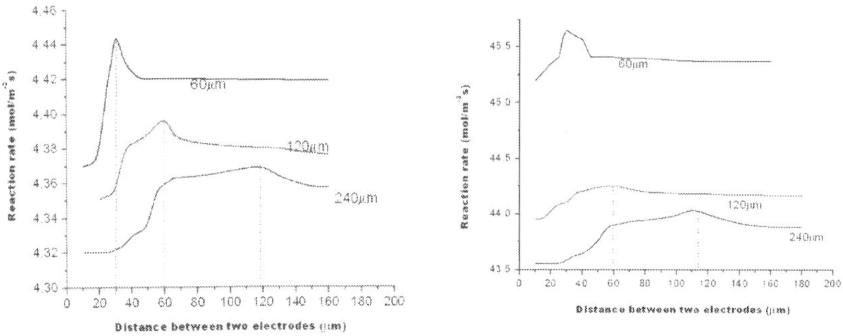

Fig. 7. (a) Anode reaction rate curves vs. well width at different ratio of electrode dimensions; (b) Cathode reaction rate curves vs. well width at different ratio of electrode dimensions.

The open circuit output potential also has been simulated for the same heights and well widths of electrodes by applying the Nernst equation. The current collectors are assumed at the bottom of the electrodes and hence these potentials are calculated from the bottom. Fig. 8. shows the open circuit output potential vs. well width of electrodes at different height of electrodes. From the results of simulation, we could find out an empirical relationship between electrodes height and well width to achieve optimized output potential is when height of electrodes is twice than that of well width which is in agreement to the results we obtain for the diffusion of the substrate.

5.2 Finite element approach for optimization of orientation of microelectrodes chip for enzymatic biofuel cells

Until now, majority of the research was focused on in-vitro experiments by mimicking physiological conditions. The additional complex problems may arise when a BFC chip is placed inside a blood artery. The first is with implantation process itself, which involves a surgery for the insertion of a BFC, and other necessary electronics components. The second is the stability of this chip inside an artery and how/where this chip can be fixed such that it can survive against the blood flow. Third problem is the clotting of the blood. The goal is to put this EBFC chip in such a way that it does not obstruct the flow of blood and lead to substantial pressure drop inside an artery. The fixation of this chip with the blood artery also should not harm the blood vessel walls (Parikh et al., 2010).

In order to improve mass transport around microelectrodes by optimizing the positioning of an EBFC chip, we have adopted the finite element analysis approach to look into the stability of an EBFC inside a blood artery. On the initial stage, we have analyzed only two orientations: horizontal position (HP) and vertical position (VP). The stability of the chip in these positions, diffusion and convectional fluxes around microelectrodes has been finely investigated. We have proposed a novel chip design, with holes in between all electrodes on the substrate, which can drastically improve the diffusion in between microelectrodes.

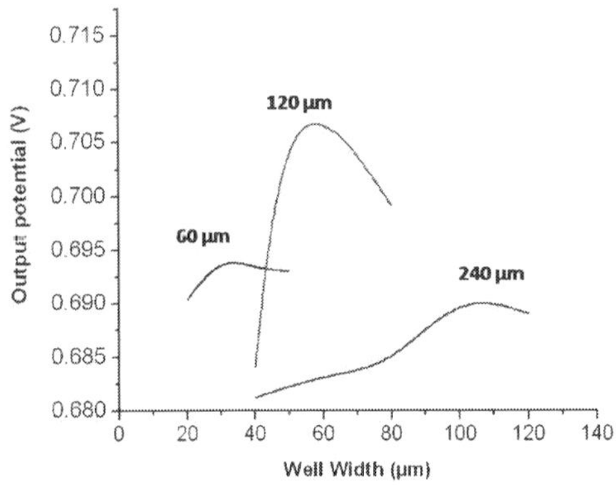

Fig. 8. Output potential vs. well width for different ratio of electrode dimensions.

The diffusion between the microelectrodes has shown in Fig. 9, where Fig. 9a and b shows the simulation profiles for diffusive flux along with the streamlines around microelectrodes in HP and VP, respectively. In HP, it is observed that the diffusive flux is less near the central electrodes and increases when going towards outer electrodes. However, the diffusive flux is almost same on top of all electrodes in VP. It is observed that in both the positions, the diffusive flux is following laminar pattern. The diffusive flux from bottom of an electrode to top of an electrode is investigated in HP and VP as shown in Fig. 9c and d, respectively. The flux is not uniform from the central to outer electrodes. The electrodes located at the circumference of a chip are having more flux compared to those located in the centre of the chip. The variation of the diffusive flux distribution around inner to outer electrodes is high in HP. The flux is not constant at every instance, but it is oscillating as shown in inset figures. The diffusive flux profiles in these figures are considered at the time, when the flux reaches its maximum value. This is also evident from Fig. 9e and f, the flux is higher exactly at the top of electrodes while lesser in the vicinity between any two electrodes. In comparison of HP and VP, the diffusive flux is 8 orders larger in case of VP than in HP.

Total flux is the combination of a diffusive flux and a convective flux. Fig. 10 depicts the total flux data for (a) HP and (b) VP of a chip. In HP, flux is negligible up to almost 275 µm height of electrodes and then increasing at the top. Total flux is highest at the top of outer most electrodes and then reducing to the central electrodes. In case of VP, the flux is almost uniform on top of all electrodes, with negligible value in between electrodes up to 200 µm height and then gradually increasing to about 2000-3500 mmol m^{-2} s^{-1} at the top of all electrodes.

Based on the results, the new design with the holes in between all microelectrodes has been inspected precisely and compared with the prototype design. The diffusive flux (Fig. 11a, c, e) and convective flux (Fig. 11b, d, f) profiles for the new design are compared with diffusive flux and convective flux profiles of the prototype model, respectively. The streamlines present the lines of motion of glucose at a particular instance.

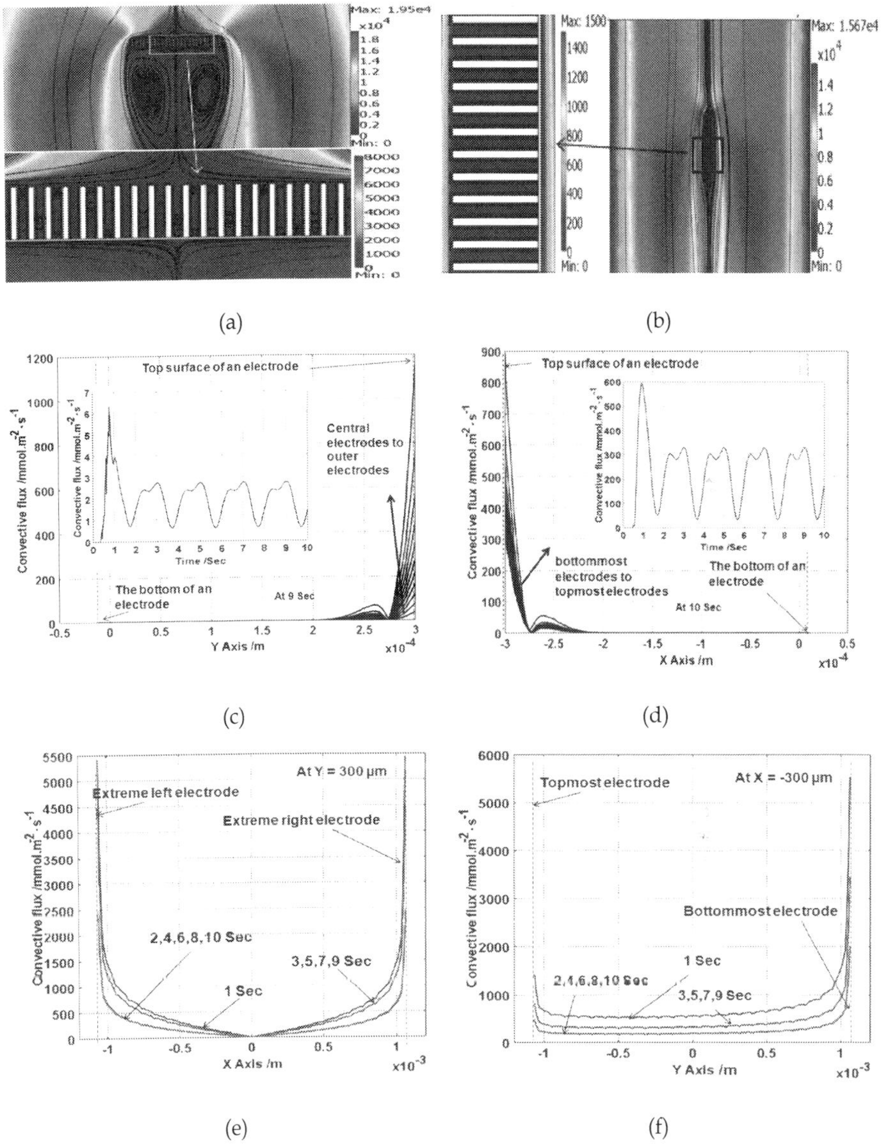

Fig. 9. Surface plot with streamlines for convective flux of glucose around microelectrodes for a) HP, b) VP, convective flux in between all 24 electrodes from bottom of electrodes to up to 300 μm height is shown for c) HP and d) VP, convective flux at top of all the electrodes from leftmost to right most electrodes for 0 - 10 secs in e) HP and f) VP.

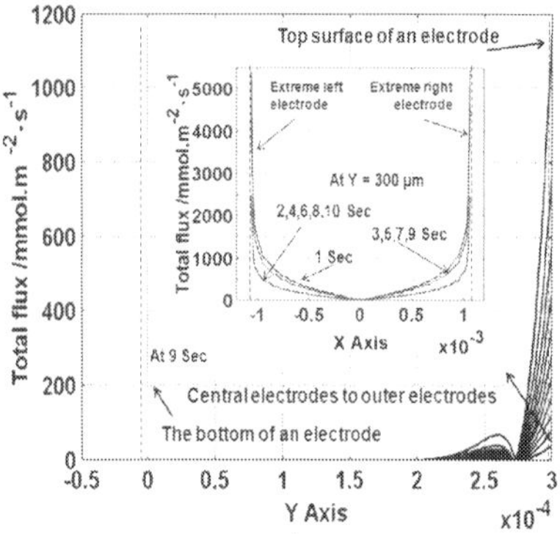

a)

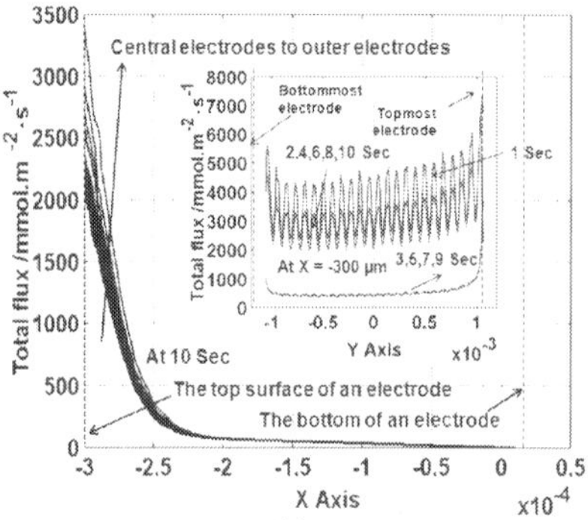

b)

Fig. 10. Total fluxes in between micro-electrodes for a) HP and b) VP. Insets provide the total flux on top of all electrodes.

From Fig. 11 it is inferred that the total flux (combined diffusive and convective flux) has been improved between all microelectrodes in terms of values and their uniformity for the chip with the holes. This enhanced mass transport around microelectrodes is significantly important for an EBFC performance. This proposed design could also be advantageous to prevent blood clotting. Human blood is mainly consisted of red blood cells and white blood cells. The sizes of all these cells such as red blood cells (6 μm), lymphocyte (7–8 μm), neutrophil (10–12 μm), eosinophil (10–12 μm), basophil (12–15 μm), and monocytes (14-17 μm) are mostly smaller than 20 μm, the size of the holes provided in the chip. So these cells can pass through the holes in between microelectrodes without blocking the way in between micro-electrodes. These holes can be made bigger depending on the requirement. The improved convection in between microelectrodes may also be forceful enough to eliminate the bubble formation. However, the biomechanical process and hemodynamic process are more complex than convection and diffusion, especially on the micro-scale level. Cell growth and clotting phenomenon are related to many aspects, such as: biocompatibility, bending of blood artery, platelet and protein components. More detailed research needs to be done with biologists in order to obtain more sufficient and helpful information and further reach the applicable level of the EBFCs.

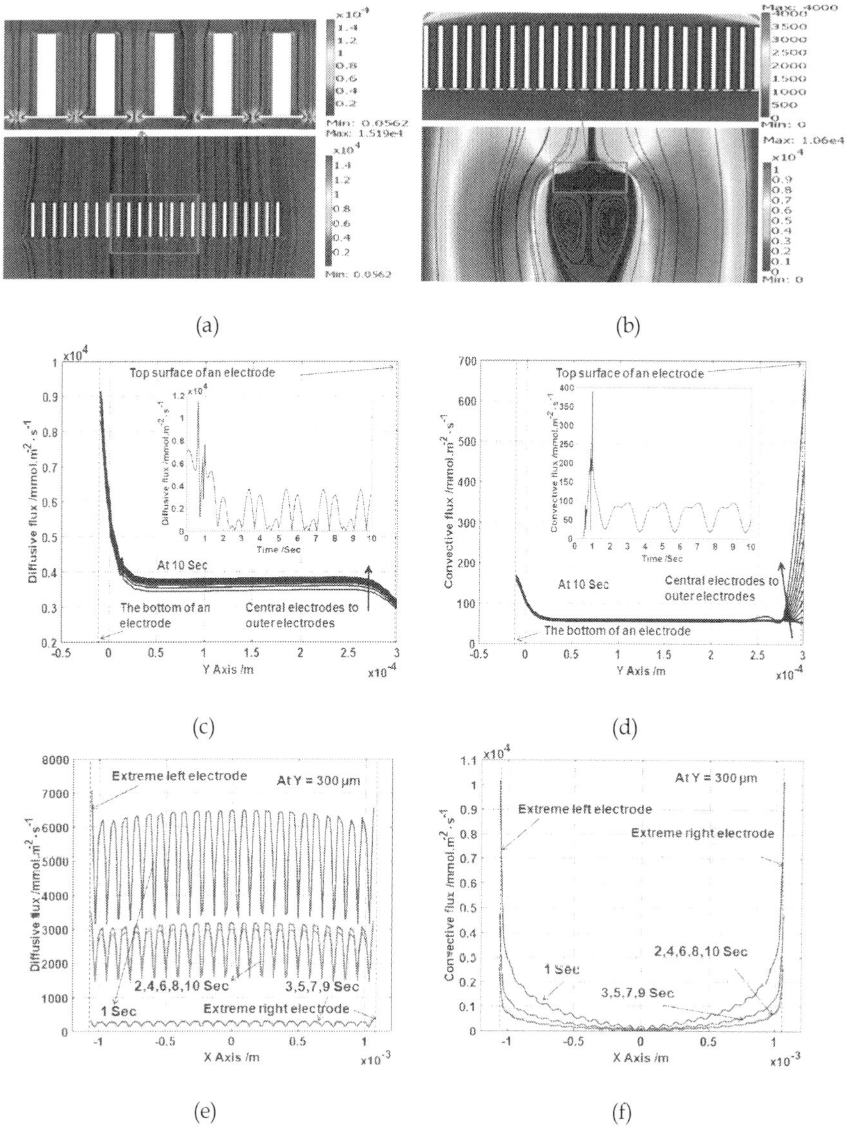

Fig. 11. Surface plot with streamlines for (a) diffusive flux and (b) convective flux of glucose around microelectrodes; (c) diffusive flux and (d) convective flux in between all 24 electrodes from bottom of electrodes to up to 300 µm height; (e) diffusive and (f) convective flux at top of all the electrodes from leftmost to right most electrodes for 0 - 10 secs.

6. Conclusion

In this chapter, we have introduced the two major kinds of biofuel cells-microbial fuel cells and enzymatic biofuel cells. Significant development on both biofuel cells has been achieved in the past decade. With the demands for reliable power supplies for medical devices for implantable applications, great effort has been made to make the miniaturized biofuel cells. The past experiment results revealed that the enzymatic miniature biofuel cells could generate sufficient power for slower and less power-consuming CMOS circuit. In addition, we have also presented simulation results showing that the theoretical power output generated from C-MEMS enzymatic biofuel cells can satisfy the current implantable medical devices. However, there are some challenges for further advancements in miniaturized biofuel cells. The most significant issues include long term stability and non-sufficient power output. Successful development of biofuel cell technology requires the joint efforts from different disciplines: biology to understand biomolecules, chemistry to gain knowledge on electron transfer mechanisms; material science to develop novel materials with high biocompatibility and chemical engineering to design and establish the system.

7. Acknowledgements

This project is supported by national Science Foundation (CBET# 0709085).

8. References

Aelterman, P.; Versichele, M.; Marzorati, M.; Boon, N. & Verstraete, W. (2010). Loading rate and external resistance control the electricity generation of microbial fuel cells with different three-dimensional anodes, *Bioresource Technology* 99 (18), pp. 8895-8902.

Akers, N.L.; Moore, C.M. & Minteer, S.D. (2005). Development of alcohol/O_2 biofuel cells using salt-extracted tetrabutylammonium bromide/Nafion membranes to immobilize dehydrogenase enzymes, *Electrochim. Acta* 50 (12), pp. 2521-2525.

Allen, R.M. & Bennetto, H.P. (1993). Microbial Fuel Cells: Electricity Production from Carbohydrates. *Appl. Biochem. Biotechnol.*, 39/40, pp. 27-40.

Barton, S.C.; Gallaway, J. & Atanassov, P. (2004). Enzymatic biofuel cells for implantable and microscale devices, *Chem Rev* 104, pp. 4867-4886.

Blonder, R.; Willner, I. & Bueckmann, AF. (1998). Reconstitution of apo-glucose oxidase on nitrospiropyran and FAD mixed monolayers on gold electrodes: photostimulation of bioelectrocatalytic features of the biocatalyst. *J Am Chem Soc* 120, pp. 9335- 41.

Bockris, JOM & Srinivasan, S. (1969). Fuel cells: their electrochemistry. *NewYork; McGraw-Hill.*

Bond, DR.& Lovley (2003). Electricity production by Geobacter sulferreducens attached to electrodes. *Appl. Environ. Microbiol.* 69, pp. 1548-1555.

Calabrese Barton, S.; Gallaway, J. & Atanassov, P. (2004). Enzymatic Biofuel Cells for Implantable and Microscale Devices.*Chem. Rev.* 104 (10), pp. 4867-4886.

Chaudhuri, S.K. & D.R. Lovley. (2003). Electricity generation by direct oxidation of glucose in mediatorless microbial fuel cells ,*Nat. Biotechnol.* 21, (10), pp. 1229–1232.

Chen, T.; Barton, S.C.; Binyamin, G.; Gao, Z.Q.; Zhang, Y.C.; Kim, H.H. & Heller, A. (2001). *J. Am. Chem. Soc.* 123 (35), pp. 8630–8631.

Chiao, M., Lam, K.B. & Lin, L.W. (2006). Micromachined microbial and photosynthetic fuel cells, *Journal of Micromechanics and Microengineering* 16 (12), pp. 2547–2553.

Cohen, B. *J. (1931). Bacteriol.* 21, pp. 18.

Crittenden, S.R.; Sund, C.J. & Sumner, J.J. (2006). Mediating electron transfer from bacteria to a gold electrode via a self-assembled monolayer, *Langmuir* 22 (23), pp. 9473–9476.

Debabov, V.G. (2008). Electricity from microorganisms, *Microbiology* 77 (2), pp. 123–131.

Feng, Y.; Wang, X.; Logan, B.E. & Lee, H. (2008). *Appl. Microbiol. Technol.* 78, pp. 873–880.

Fan, Y.Z., Hu, H.Q. & Liu, H. (2007). Enhanced coulombic efficiency and power density of air-cathode microbial fuel cells with an improved cell configuration, *Journal of Power Sources* 171 (2), pp. 348–354.

Freire, R.S.; Pessoa, C.A.; Mello, L.D.; & Kubota, L.T. (2003). Direct electron transfer: an approach for electrochemical biosensors with higher selectivity and sensitivity, *J Braz Chem Soc* 14, pp. 230–243.

Galvani, L. (1791). *De bononiensi scientiarum et artium instituto atque academia Comentarrii* 7, pp. 363–418.

Ghangrekar, M.M. & Shinde, V.B. (2007). Performance of membrane-less microbial fuel cell treating wastewater and effect of electrode distance and area on electricity production, *Bioresource Technology* 98 (15), pp. 2879–2885.

Gorge, G; Kirstein, M & Erbel, R. (2001). Microgenerators for Energy Autarkic Pacemakers and Defibrillators: Fact or Fiction. *Herz* 26, pp. 64–68.

Govil, G & Saran, A. (1982). Biochemical fuel cells. *J Indian Chem. Soc.* (1982), 59: p. 1226– 8.

Grove, W.(1839). *Philos. Mag. Ser.* 3 (14), pp. 127.

Heller, A. (2004). Miniature biofuel cells. Phys. Chem. Chem. Phys. 6, pp. 209–216.

Katz, E.; Shipway, A.; & Willner, I. (2003). In: *Handbook of Fuel Cells – Fundamentals, Technology and Applications* Chichester, W. Vielstich, H.A. Gasteiger and A. Lamm, England, John Wiley & Sons Ltd., pp. 355–381.

Habermann, W. & Pommer, E.H. (1991). Biological Fuel Cells with Sulphide Storage Capacity. *Appl. Microbiol. Biotechnol.*, 35, pp. 128-133.

Haccoun, J.; Piro, B.; Noel, V. & Pham, M.C. (2006). *Bioelectrochemistry* 68, pp. 218–226.

Heller, A. (2006). Potentially implantable miniature batteries. *Anal. Bioanal. Chem.* 385, pp. 469–473.

Hou H.J., Li L., Cho Y., de Figueiredo P. & Han, A. (2009). Microfabricated microbial fuel cell arrays reveal electrochemically active microbes, *Plos One* 4 (8), pp. e6570.

Ieropoulos, I. Winfield, J. & Greenman, J. (2010). Effects of flow-rate, inoculum and time on the internal resistance of microbial fuel cells, *Bioresource Technology* 101 (10), pp. 3520–3525.

Katuri, K.P. & Scott, K. (2010). *Biotechnol. Bioeng.* 107, pp. 52–58.

Katz, E.; Buckmann, A.F. & Willner, I. (2001). *J. Am. Chem. Soc.* 123 (43), pp. 10752–10753.

Katz, E.; Shipway, A.N. & Willner, I. (2003). Biochemical fuel cells. In: *Handbook of Fuel Cells – Fundamentals, Technology and Applications,* W. Vielstich, H.A. Gasteiger and A. Lamm, *Fundamentals and Survey of Systems* vol. 1, John Wiley & Sons, Ltd., Hoboken, NJ, pp. 355–381.

Katz, E. & Willner, I. (2003). Biofuel cells based on monolayer-functionalized biocatalytic electrodes. In: K.E. Geckeler, Editor, *Advanced macromolecular and supramolecular materials and processes,* Kluwer Academic/Plenum Publishers, New York, pp. 175-196.

Katz, E.; Sheeney-Haj-Ichia, L. & Willner, I. (2004). Electrical contacting of glucose oxidase in a redox-active rotaxane configuration. *Angew Chem Int Ed* 43, pp. 3292–300.

Kim, B.H.; Kim, H.J.; Hyun, M.S. & Park, D.H. (1999). Direct Electrode Reaction of Fe (III)-reducing Bacterium, *Shewanella putrefaciens*. J. Microbiol. Biotechnol. 9, pp. 127-131.

Kim, H.J. H.S. Park, M.S. Hyun, I.S. Chang, M. Kim & B.H. Kim. (2002). A mediator-less microbial fuel cell using a metal reducing bacterium, Shewanella putrefaciens *Enzyme Microb. Technol.* 30 (2), pp. 145–152.

Kim, N. Y. Choi, S. Jung & S. Kim. (2000). Development of microbial fuel cells using Proteus vulgaris, *Korean Chem. Soc.* 21 (1), pp. 44–48.

Kim, H.H.; Mano, N.; Zhang, X.C. & Heller, A. (2003). *J. Electrochem. Soc.*150 (2), pp. 209-213.

Kim, P.; Kwon, K.W.; Park, M.C.; Lee, S.H.; Kim S.M. & K.Y. Suh. (2008). Soft lithography for microfluidics: a review, Biochip Journal 2 (1), pp. 1–11.

Lee, J.; Shin, H. Y.; Kang, S.W.; Park, C. & Kim, S.W. (2010). Use of bioelectrode containing DNA-wrapped single-walled carbon nanotubes for enzyme-based biofuel cell, *Journal of Power Sources* 195, pp. 750–755.

Li, X.; Zhou, H.; Yu, P.; Su, L.; Ohsaka, T. & Mao, L. (2008). A Miniature glucose/O_2 biofuel cell with single-walled carbon nanotubes-modified carbon fiber microelectrodes as the substrate, *Electrochemistry Communications*. 10 (6), pp. 851-854

Liu, Y. & Dong, S.J. (2007). A biofuel cell harvesting energy from glucose-air and fruit juice-air. *Biosens. Bioelectron. 23*, pp. 593-597.

Logan, B.E. (2008). Microbial Fuel Cells, John Wiley & Sons, Inc.

Lowy, D.; Tender, L.; Zeikus, J.; Park, D. & Lovley, D. (2006). Harvesting energy from the marine sediment-water interface II - Kinetic activity of anode materials. *Biosensors & Bioelectronics* 21, pp. 2058-2063

Lovley, D.R. (2008). The microbe electric: conversion of organic matter to electricity, *Current Opinion in Biotechnology* 19, pp. 564–571.

Malladi, K.; Wang, C. & Madou, M. (2006). Microfabrication of Suspended C-MEMS Structures by EB Writer and Pyrolysis. *Carbon*. 44(13), pp. 2602-07.

Mano, N.; Mao, F. & Heller, A., (2002). *J. Am. Chem. Soc.* 124 (44), pp. 12962–12963.

Mano, N.; Mao, F & Heller, A. (2003). Characteristics of a Miniature Compartment-less Glucose/O2 Biofuel Cell and Its Operation in a Living Plant. *J. Am. Chem. Soc.* 125, pp. 6588–6594.

Mano, N. & Heller, A (2003). *J. Electrochem. Soc.* 150 (8), pp. A1136–A1138.

Min, B; Cheng, S & Logan, BE. (2005) Electricity generation using membraneand salt bridge microbial fuel cells. *Water Res*, 39, pp. 1675–1686.

Moehlenbrock, M.J. & Minteer, S.D. (2008). Extended lifetime biofuel cells. *Chem. Soc. Rev.*, *37*, pp. 1188-1196.

Moon, H.; Komlos, J. & Jaffe, P. (2009) Biogenic U(IV) oxidation by dissolved oxygen and nitrate in sediment after prolonged U(VI)/Fe(III)/$SO_4{}^{2-}$ reduction. *J Contam Hydrol* 105,pp. 18–27.

Nagel, B.; Warsinke, A. & Katterle, M. (2007). *Langmuir* 23, pp. 6807–6811.

Niessen, J.; Schroder, U. & Scholz, F. (2004). Exploiting complex carbohydrates for microbial electricity generation - A bacterial fuel cell operating on starch, *Electrochem. Commun.* 6 (9), pp. 955–958.

Niessen, J.; Harnisch, F.; Rosenbaum, M.; Schroder, U. & Scholz, F. (2006). Heat treated soil as convenient and versatile source of bacterial communities for microbial electricity generation. *Electrochem Commun* 8, pp. 869–73.

Oh, S.E.; Min, B. & Logan, B.E. (2004). Cathode performance as a factor in electricity generation in microbial fuel cells. *Environ. Sci. Technol.* 38, pp. 4900–4904.

Palmore, G & Whitesides, G.M. (1994). Microbial and enzymatic biofuel cells. In: *Enzymatic Conversion of Biomass for Fuels Production,* E. Himmel, Editor,vol. 566, American Chemical Society, pp. 271–290.

Parikh, Y.; Yang, J. H. & Wang, C. (2010). Optimizing the mass transport phenomenon around micro-electrodes of an enzymatic biofuel cell inside a blood artery via finite element analysis method, *J Power Sources.* 195 (15), pp. 4685-4694.

Park, D.H. & Zeikus, J.G. (2000). Electricity Generation in Microbial Fuel Cells Using Neutral Red as an Electronophore, *Appl. Environ. Microbiol.* 66 (4), pp. 1292–1297.

Park, D.H. & Zeikus, J.G. (2002). Impact of electrode composition on electricity generation in a single-compartment fuel cell using Shewanella putrefaciens, *Appl. Microbiol. Biotechnol.* 59 (1), pp. 58–61.

Park, D.H. & Zeikus, J.G. (2003). Improved fuel cell and electrode designs for producing electricity from microbial degradation, *Biotechnol. Bioeng.* 81 (3), pp. 348–355.

Patolsky, F.; Tao, G.; Katz, E. & Willner, I. (1998). C60-Mediated bioelectrocatalyzed oxidation of glucose with glucose oxidase. *J Electroanal Chem* 454, pp. 9–13.

Penmatsa, V.; Yang, J. H.; Yu, Y. & Wang, C. (2010). Fabrication of porous carbon micropillars using a block copolymer as porogen, *Carbon.* 48 (14), pp. 4109–4115.

Potter, M.C. (1910). *Proceedings of the Royal Society B, vol. 84*, p. 260.

Qian, F.; Baum, M.; Gu, Q. & Morse, D.E. (2009). A 1.5 μL microbial fuel cell for on-chip bioelectricity generation, *Lab on a Chip* 9 (21), pp. 3076–3081.

Qiao, Y.; Li, C.M.; Bao, S.J & Bao, Q.L. (2007). Carbon nanotube/polyaniline composite as anode material for microbial fuel cells, *Journal of Power Sources* 170 (1), pp. 79–84.

Rabaey, C.; K. Lissens.; G. Siciliano S.D. & Verstraete, W. (2003). A microbial fuel cell capable of converting glucose to electricity at high rate and efficiency, *Biotechnol. Lett.* 25, pp. 1531–1535.

Rabaey, K.; Boon, N.; Siciliano, S.D.; Verhaege, M. & Verstraete, W. (2004). Biofuel Cells Select for Microbial Consortia That Self-Mediate Electron Transfer, *Appl. Environ. Microbiol.* 70 (9), pp. 5373–5382.

Ramanavicius, A.; Kausaite, A. & Ramanaviciene, A. (2005) Biofuel cell based on direct bioelectrocatalysis. *Biosens. Bioelectron., 20*, pp. 1962-1967.

Ramanavicius, A.; Kausaite, A. & Ramanaviciene, A. (2008). Enzymatic biofuel cell based on anode and cathode powered by ethanol. *Biosens. Bioelectron., 24*, pp. 761-766.

Richter, H.; McCarthy, K.; Nevin, K.P.; Johnson, J.P.; Rotello, V.M. & Lovley, D.R. (2008). Electricity generation by *Geobacter sulfurreducens* attached to gold electrodes, *Langmuir* 24 (8), pp. 4376–4379.

Ringeisen, B.R.; Henderson, E.; Wu, P.K.; Pietron, J.; Ray, R.; Little, B.; Biffinger, J.C. & Jones-Meehan, J.M. (2006). High power density from a miniature microbial fuel cell using Shewanella oneidensis DSP10, *Environmental Science & Technology* 40 (8), pp. 2629–2634.

Roller, S.B.; Bennetto, H.P.; Delancy, G.M.; Mason, J.R.; Stirling, J.L. & Thurston, C.F. (1984). Electron-transfer coupling in microbial fuel cells: 1. comparison of redox-mediator reduction rates and respiratory rates of bacteria. *J. Chem. Technol. Biotechnol.* 34B, pp. 3–12.

Roundy, S. (2005). On the Effectiveness of Vibration-based Energy Harvesting.*Intell. Mater. Syst. Struct.* 16, pp. 809–823.

Rozendal, R. A.; Hamelers, H. V. M. & Buisman, C. J. N. (2006) Effects of membrane cation transport on pH and microbial fuel cell performance. Environ. Sci. Technol. 40(17), pp. 5206–5211.

Saleh, F. S.; Mao, L. & Ohsaka, T. (2011).Development of a dehydrogenase-based glucose anode using a molecular assembly composed of nile blue and functionalized SWCNTs and its applications to a glucose sensor and glucose/O_2 biofuel cell, *Sensors and Actuators B* 152, pp. 130-135.

Tanne, C.; Göbel, G. & Lisdat F. (2010). Development of a (PQQ)-GDH-anode based on MWCNT-modified gold and its application in a glucose/O_2-biofuel cell, *Biosensors and Bioelectronics* 26, pp. 530-535.

Tender, L.M.; Reimers, C.E.; Stecher, H.A.; Holmes, D.E.; Bond, D.R.; Lowy, D.A.; Pilobello, K.; Fertig, S.J. & Lovley, D.R. (2002). Harnessing microbially generated power on the seafloor, *Nat. Biotechnol.* 20 (8), pp. 821–825.

Timur, S.; Anik, U.; Odaci, D. & Gorton, L. (2007). Development of a microbial biosensor based on carbon nanotube (CNT) modified electrodes, *Electrochemistry Communications* 9 (7), pp. 1810–1815.

Togo, M.; Takamura, A.; Asai, T.; Kaji, H. & Nishizawa, M. (2007). An enzyme-based microfluidic biofuel cell using vitamin K-3-mediated glucose oxidation. *Electrochim. Acta, 52*, pp. 4669-4674.

Tsujimura, S., Kano, K., Ikeda, T., 2002. *Electrochemistry* 70 (12), pp. 940–942.

Sakai, H.; Nakagawa, T.; Tokita, Y.; Hatazawa, T.; Ikeda, T.; Tsujimura, S. & Kano, K. (2009). A high-power glucose/oxygen biofuel cell operating under quiescent conditions, *Energy Environ.Sci.* 2, pp. 133-138.

Schroder, U.; Niessen J. & Scholz, F. (2003). A Generation of Microbial Fuel Cells with Current Outputs Boosted by More Than One Order of Magnitude, *Angew. Chem. Int. Ed.* 42 (25), pp. 2880–2883.

Schuhmann, W. (2002). Amperometric enzyme biosensors based on optimized electron-transfer pathways and non-manual immobilization procedures, *Rev Mol Biotechnol* 82, pp. 425–441.

Schuhmann, W. & Muenchen, T.U. (1992). *DECHEMA Monograph.* 126 (1992), pp. 237–253.

Siu, C.P.B. & Chiao, M. (2008). A microfabricated PDMS microbial fuel cell, *Journal of Microelectromechanical Systems* 17 (6), pp. 1329–1341.

Soukharev, V.; Mano, N. & Heller, A. (2004). *J. Am. Chem. Soc.* 126 (27), pp. 8368–8369

Suzuki, S.(1976). *Hosp. Hyg. Gesundheitswesen Desinfect.*, pp. 159.

Tsujimura, S.; M. Fujita, H. Tatsumi, K. Kano & T. Ikeda. (2001). Bioelectrocatalysis-based dihydrogen/dioxygen fuel cell operating at physiological pH, *Phys. Chem. Chem. Phys.* 3 (7), pp. 1331–1335.

Vega, C.A. & Fernandez, I. (1987). Mediating Effect of Ferric Chelate Compounds in Microbial Fuel Cells with *Lactobacillus Plantarum, Streptococcus lactis and Erwinia dissolvens*. Bioelectrochem. Bioenerg., 17, pp. 217-222.

Wang, C.; Jia, G.; Taherabadi, L. H. & Madou, M. J. (2005). "A Novel Method for the Fabrication of High Aspect Ratio C-MEMS Structures," *J. of Microelectromechal Sys.* 14, pp. 348

Wang, C. & Madou, M. J. (2006). Sensors: C-MEMS Based Microbattery Arrays for Miniature Sensors. *2006 NSF Design, Service, and Manufacturing Grantees and Research Conference.*

Wang, S.C.; Yang, F.; Silva, M.; Zarow, A.; Wang, Y.B. & Iqbal, Z. (2009). Membrane-less andmediator-free enzymatic biofuel cell using carbon nanotube/porous silicon electrodes. *Electrochem. Commun., 11*, pp. 34-37.

Watkins, C; Shen, B & Venkatasubramanian, R. (2005) Low-grade-heat energy harvesting using superlattice thermoelectrics for applications in implantable medical devices and sensors. *Proceedings of the 24th International Conference on Thermoelectronics (ICT 2005)* Clemson, USA, pp. 265–267.

Willner, I.; HelegShabtai, V; Blonder, R; Katz, E. & Tao, G.L. (1996). Electrical Wiring of Glucose Oxidase by Reconstitution of FAD-Modified Monolayers Assembled onto Au-Electrodes. *J. Am. Chem. Soc.* 118 (42), pp. 10321–10322.

Willner, I.; Arad, G. & Katz, E. (1998). A biofuel cell based on pyrroloquinoline quinone and microperoxidase-11 monolayer-functionalized electrodes. *Bioelectrochem. Bioenerget.* 44 (2), pp. 209–214.

Yahiro, A.T.; Lee, S.M. & Kimble, D.O. (1964) Bioelectrochemistry: I. Enzyme utilizing bio-fuel cell studies. *Biochim. Biophys. Acta.* 88 (1964), pp. 375–383.

Yang, J.; Penmatsa, V.; Tajima, S.; Kawarada, H. & Wang, C. (2009). Direct amination on 3-dimensional pyrolyzed carbon micropattern surface for DNA detection, *Materials Letters.* 12, pp. 2680-2683.

Zhang, X.; Cheng, S.; Wang, X.; Huang, X. & Logan, B.E. (2009). Environ. Sci. Technol. 43, pp. 8456–8461.

Zhu, M.W.; Gautam, A.; Nazor, J.; Momeu, C.; Prodanovic R. & Schwaneberg, U. (2007). Directed evolution of glucose oxidase from Aspergillus Niger for ferrocenemethanol mediated electron transfer. *Biotechnol. J.*, *2*, pp. 241-248.

29

Biorefining Lignocellulosic Biomass via the Feedstock Impregnation Rapid and Sequential Steam Treatment

Jean-Michel Lavoie[1,2], Romain Beauchet[1], Véronique Berberi[1,2] and Michel Chornet[1,2]
[1]*Industrial Research Chair on Cellulosic Ethanol Department of chemical and biotechnological engineering, Université de Sherbrooke Québec*
[2]*CRB Innovations Sherbooke, Québec*
Canada

1. Introduction

The first generation of biofuels, made out of starch (ethanol) or triacyglycerol (biodiesel) uses expensive homogeneous feedstocks (sugar cane, corn, wheat and edible oils) coupled with relatively inexpensive technologies known and practiced for years at an industrial level. First generation biofuels have had a bad press: high water and energy consumption (very significant is the energy used in the production of the fertilizers needed by agriculture) and the fuel versus food controversy. Increased use of biofuels requires alternative sources of biomass that lower water and energy consumption and do not compete with food supplies. Lignocellulosic biomass, either from forestry or agriculture offers such potential. Cellulose, the most abundant carbohydrate on the planet, is a fraction of the complex lignocellulosic matrix along with other macromolecules, lignin, hemicelluloses, and extractives. The cellulose macromolecule is composed of glucose units linked together via β-1,4-glycosidic bonds (or acetal bonds) creating long chains that combine together to form fibrils and eventually fibres. The polar hydroxyl groups are oriented one toward the other so that interaction with a polar medium (as a solvent) is fairly difficult making cellulose water resistant. The natural macromolecule is usually present in nature in two forms: crystalline and amorphous. A typical fibril will have zones that are crystalline separated by zones that are amorphous. Whilst the crystalline form is difficult to disassemble with hydrolyzing agents, the amorphous phase has a certain level of disorder that makes relatively easy the penetration and action of hydrolysing agents, either enzymes or ionic species. The other macromolecules found in the lignocellulosic matrix are also of interest. Lignin is a macromolecule composed of phenylpropane units bond together via, predominantly, ether bonds although C-C between moieties are also significant. Lignin, has low oxygen content and thus a high energetic value. Hemicelluloses are, as cellulose, macromolecules composed of carbohydrates. Upon hydrolysis, the C6 fraction of these carbohydrates can effectively be converted to ethanol via fermentation using classical yeast strands (Gírio, 2009). Studies have shown that fermentation of all the glucidic part of the hemicelluloses, both C6 and C5 sugars, was feasible using nontraditional microorganisms (Agbogbo & Coward-Kelly, 2008; Casey et al., 2009; Chu & Lee, 2007). It is also well known that hydrolysis of hemicelluloses produces / liberates organic acids that could inhibit fermentation of the carbohydrate.

Cellulosics being an alternative for ethanol production, there is still an important aspect that has to be considered: the cost of the feedstock. Lignocellulosics for sugar production and subsequent fermentation can be considered to belong to three categories which interlink biomass cost, quality and transformability: homogeneous, quasi-homogeneous and non-homogeneous. The first category comprises structural and furniture wood and chips for pulp production which requires a single species (or a mixtures of comparable species). Such homogeneous biomass has also a rather homogenous chemical composition and it is used for high end products with well established markets. Homogeneous biomass is expensive, above $US 100 / tonne (anhydrous basis equivalent) in the NorthEast of America (2011 basis, prices range is a courtesy of CRB Innovations). Besides cost, such biomass also has a large market, in structural wood and in the pulp and paper industry. Quasi homogeneous biomass is usually composed of a mixture of species and to a certain extent, of tissues. This category embraces the residual lignocellulosic biomass produced during forest or agricultural operations. Cost range for this biomass will vary, FOB conversion plant in 2011, NorthEast of America, from $US 60 to 80 per tonne, dry basis, mostly related to transportation costs in a radius of a maximum of 100 km from harvesting operations (prices are a courtesy of CRB Innovations). Contrary to the homogeneous feedstock which has a wide and diversified market (yet very competitive), the quasi-homogeneous feedstock, since such biomass is of lesser homogeneity and often includes a higher quantity of ashes, is less coveted. Therefore, this biomass could be the main feedstock of the upcoming cellulosic ethanol market since the competition is actually low, the feedstock does not compete with food supply and the biomass is readily available close to major cities in the world. The last category of biomass is non-homogeneous. It is of lesser quality than the previous categories and usually costs close to 0 USD (it may even come with a tipping fee in some cases). The low cost is of course attractive but the conversion process will have to use such biomass as a whole complex mixture converting it to a more homogeneous intermediary. Although we acknowledge the availability and the potential of each type of biomass, this chapter will be focused on the residual lignocellulosic material generated from established forest and agricultural industries (quasi-homogeneous biomass) as well as on plantation biomass.

Plantation biomass can be identified as "energy crops" which are grown, ideally on marginal lands, with two objectives; sequestering carbon and bioconverting it into carbon-based structured macromolecules that could be used for the production of bioenergy. In North America, some cultures that have gained attention during the last 10 years, amongst many: willows, poplars, miscanthus, switchgrass, panic, reed canary grass, etc. Depending on the targeted market, these energy crops could be oriented towards high yields of cellulose (if the ethanol market is the main target) or high yield of lignin and less ashes (if the combustion market is targeted).

Biomass cost and composition are the main concerns of a cellulosic ethanol plant. A technological issue is how to convert the carbonhydrates to low cost monomeric sugars in high yields. Cellulose has been separated from plants for decades by the pulp and paper industry, the latter having developed industrial scale facilities that converted large quantity of lignocellulosic biomass to pulp and paper. However, the established processes to isolate the cellulose use large quantities of water putting a stress on water supplies. Furthermore, the pulp and paper does not actually use the hemicellulose and lignin other than for CHP production.. Research around the world have been focusing in the past decades toward processes that recover and use most of the carbon present in biomass to create true biomass refineries from which multiproducts would be obtained. This requires a careful consideration of which biomass to use to achieve valuable multiproducts and which biomass to use to provide heat and power.

The key technological challenge for the production of cellulosic ethanol is depolymerizing the cellulose to obtain high yields of glucose. As mentioned earlier, cellulose is a compact macromolecule, particularly its crystalline fractions, and it requires specific enzymes or chemicals to allow hydrolysis of the β-1,4-glycosidic bonds. Accessibility of enzymes and chemical hydrolytic agents is a function of the three-dimensional ultrastructure of cellulose. Therefore, before going forward with production of cellulosic ethanol, the composition of the original feedstock and the ultrastructure of its isolated cellulosic fraction has to be known in order to adapt the hydrolytic processes to such ultrastructure.

This chapter focuses on three aspects that should be closely related to the production of second generation ethanol. In a first section, the composition of different substrate will be review as for their cellulosic, hemicellulosic and lignin contents. These data are essential for adaptation of the downstream process of a biorefinery. The second section of this chapter will be aimed at reviewing the steam treatments from our experience with the Feedstock Impregnation Rapid and Sequential Steam Treatment (FIRSST) process developed through two and a half decades of effort within our extended team (fundamentals at the academic level; engineering and technology via the spin-off company, CRB). Finally, the third section of this chapter will be an overview of the chemical treatments for cellulose hydrolysis compared with the CRB decrystallyzation and depolymerization process whose fundamental basis was developed by our team at the Université de Sherbrooke.

2. Biomass

Lignocellulosic biomass is a readily available feedstock that can be purchased yearly from forest and agricultural operations. Forest residues comprise unusable trunk sections, limbs and tops. Typical composition of these residues is similar to that of common wood chips shown in Table 1. We define, as agricultural residues, the non-edible part of the plant which is let on the field after the harvest and the latter are usually composed of straw and stalk. It also comprises the parts of the cultivated plants that are thrown out after industrial processes. A specific example of such biomass includes but is not limited to corn cobs. Agricultural residues are the most probable feedstock that will be the original source for the production of ethanol from lignocellulsic materials due to their availability, their quantity and their proximity to the existing grain to ethanol platform. Non conventional plantation crops (i.e. 'energy crops') are also to be considered as feed for biorefineries. Most of these crops have not reach industrial scale production (in North America) but an increasing amount of information has been published during the last few years about their chemical composition. Pricing for this biomass has been evaluated to 100-120$ per (dry basis; prices courtesy of CRB Innovations) metric ton but it tends to decrease because of a reduced use of fertilizers and the utilisation of *marginal lands* instead of high value agricultural land. The characteristics which make energy crops, especially perennial grasses, attractive for ethanol production, are the high amount of cellulose and hemicellulose as well as, under certain restriction, the favorable environmental impact.

Lignocellulosic biomass is composed of cellulose, hemicelluloses, lignin, extractives and ashes. The quantities of each fraction are detailed below for a large range of lignocellulosic materials including agricultural residues, energy crops and forest residues which are divided into leafy hardwoods and coniferous families (Table 1). Cellulose is the principal constituent of lignocellulosic plants representing 30-50 wt% of its composition. It is a polymer composed of D-glucose. Contrarily to cellulose, hemicelluloses are a heterogeneous polymer principally composed of pentoses (β-D-xylose, α-L-arabinose), hexoses (β-D-mannose, β-D-glucose, α-D-galactose) and/or uronic acids (α-D-glucuronic, α-D-4-O-methylgalacturonic, α-D-galacturonic acids). Among them, xylans and glucomannans are the most common compounds. Hemicelluloses represent 15-35 wt% of the plant. We can define lignin as a relatively hydrophobic amorphous polymer which consists of phenylpropane units. This macromolecule occurs primarily between the fibre cells, acting as a cementing material and giving the wood its rigidity and its impact resistance. It is always associated with hemicelluloses through carbon-carbon and ether linkages (Xu et al., 2008) and can be classified following 2 major classes (Gibbs and Thimann, 1958): (1) the guaiacyl which includes most of the lignins of softwoods (gymnosperms), (2) the guaiacyl–syringyl which comprises the lignins of hardwoods (angiosperms) and the lignins of grasses (non woody or herbaceous crops) (angiosperms). Extractives are composed of resins, fats and fatty acids, phenolics, phytosterols, terpenes, salts and minerals. This fraction is not used for the production of ethanol, and, for obvious reasons, neither are the ashes. The latter is defined as the residue remaining after total combustion. It is composed of elements such as silicon, aluminum, calcium, magnesium, potassium, and sodium. Typically, the amount of each fraction can also differ within a single biological species (following the environment: soil composition, water supply and weather patterns) and also during the growth of the plant making the quantification of sugar present in the holocellulose (sum of hemicelluloses and cellulose) difficult to specify.

The valorization of the lignocellulosic materials and more particularly of the carbohydrates (composing the holocellulose) into ethanol is made possible through their fermentation. Each plant has different composition (Table 1) but, as detailed before, contains the same major compounds. All the biomasses show comparable characteristics (Table 2) with the following order by quantity: Glucan>Xylan>Mannan-Galactan-Arabinan, except for the coniferous forest residues which show a high amount of mannan, which leads to a shift between the glucan and the xylan. Table 2 also shows an average of the 6 carbons sugars (C6) which can be fermented by most common yeast (including but not limited to S.cerevisiae) to give ethanol. Agricultural residues, energy crops and leafy forest residues present high averages of 41.5, 46.6 ans; 48.2 wt% respectively. Furthermore, coniferous forest residues could potentially produce more ethanol as they possess a very high amount of C_6 (56 %) sugars which can be explained by a high amount of mannose (about 10 % more than the other species).

North America produced 46% of the world biofuels in 2008 (IEA, 2009)and the R &D efforts on second generation biofuels have been widely orientated toward the production of ethanol. From the results in Table 2, it is possible to estimate the ethanol production directly from C_6 sugars using S. cerevisiae. Production of ethanol from C_5 sugar was not taken into account in our study as these sugars require the use of special microorganisms. C_5 sugars, although hard to ferment to ethanol could be converted into other value added products as ethyl levulinate (considered as part of the extended P-fuel pool) by successive dehydration, reduction and ethanolysis. Table 3 shows a comparison between the actual possible production of ethanol (not operational) using the forest residues, the agricultural residues and the unexploited forest biomass available in Quebec, Canada and North America versus the operational ethanol production from energy crop (first generation of biofuel) and the consumption of gasoline. Only 25 % of the forest and agricultural residues have been taken into account for ethanol production estimate as the rest of the biomass is already dedicated for others purposes. In the case of the unexploited forest, our study presents a result based on the forest zone which can be used without causing damage on biodiversity (Ministère des Resources Naturelles et de la Faune MNRF, Quebec, 2009).

	Cellulose (wt%)	Hemicellulose (wt%)	Lignin (wt%)	Extractives (wt%)	Ashes (wt%)
Agricol residues					
Rice straw[a]	41.2	19.5	21.9	-	-
Wheat straw [a]	39.7	36.5	17.3	-	-
Rye straw[b]	37.9	36.9	17.6	-	-
Tritical straw[v]	34	31.7	17	12.4	
Flax straw[c]	53.8	17.1	23.3	-	3.6
Sun flower stalk [a]	37.6	29.3	10.3	-	-
Sorghum stalk [a]	41.5	24.4	15.6	-	-
Cotton stalk [a]	58.5	14.4	21.5	-	-
Corn stover[d]	38	26	19	-	6
Barley straw[e]	42	28	-	-	11
Vine shoots [a]	41.1	26	20.4	-	-
Olive prunings[a]	35.7	25.8	19.7	-	-
Energy crops					
Switchgrass	37	28	16.4	15	3.7
Miscanthus	40	18	25	-	-
Big bluestern [f,g,h]	37	28	18		6
Little bluestern[h]	35	31	-	-	7
Prairie cordgrass[i]	41	33	-	-	6
Indian grass[f,h]	39	29	-	-	8
Intermediate wheatgrass[g]	35	29	-		6
Reed canarygrass[j,k]	24	36	-	-	8
Smooth bromegrass [j,k]	32	36	-	-	6
Tymothyb [l,m]	28	30	-	-	6
Tall fescue [n]	25	25	14	-	11
Sundan grass [j,k]	33	27	-	-	8
Jatropha stem[l]	37.1	30.6	22.3	-	-
Jatropha seed cake[o]	13.5	26.8	12.4	-	-
Cannabis sativa[v]	43	26	14.5	16	3.2
Salix viminalis[v]	30.2	33.5	29.2	8.8	-
Forest residues					
Leafy					
Soft maple	41	35	24	-	-
Red oak[p]	35.5	18.8	29	-	-
European oak[q]	38	29	25	4.4	0.3
White oak[r]	44	24	24	5.4	1
Chesnut oak [r]	41	30	22	6.6	0.4
Post oak [r]	38	30	26	5.8	0.5

	Cellulose (wt%)	Hemicellulose (wt%)	Lignin (wt%)	Extractives (wt%)	Ashes (wt%)
White birch [r]	45	33	18	5	0.3
Yellow birch	40	39	21		
Quaking aspen [r]	49	29	19	6	0.4
White elm	49	27	24	-	-
Beech	42	36	22	-	-
Basswood	-	-	-	-	-
Poplarwood	41.4	23.7	24.5	-	-
Eucalyptus [a]	52.8	27.7	20	-	-
Coniferous					
Pinus pinaster [a]	55.9	13.7	26.2	-	-
Pinus radiate chips[c]	53	15.8	23.7	-	-
Cedar[s]	43.5	20.3	-	-	-
Eastern Hemlock	42	26	33	-	-
Eastern white cedar	44	25	31	-	-
White spruce	44	29	27	-	-
Jack pine	41	30	29	-	-
Tamarack	43	28	29	-	-
Spruce[t]	54.1	21.4	24.4	-	-
Loblolly pine[u]	43.6	21.2	26.8	3.2	0.4
Balsam-fir	44	27	29	-	-

[a](Rodríguez et al., 2010); [b](Sun et al., 2000); [c](Schafer & Bray, 1929); [d](Lee et al., 2007); [e](Mani et al., 2008); [f](Lee & Owens, 2008); [g](Owens et al., 2006); [h](Jefferson et al., 2004); [i](Boe & Lee, 2006); [j](Jung et al., 1997); [k](Jurgens, 1997), [l](Alvo et al., 1996), [m](Claessens et al., 2004); [n](Department of energy, 2006); [o](Liang etal., 2010); [p](Mazlan et al., 1999); [q](Bednar & Fengel, 1974); [r](Pettersen, 1984); [s](Yamashita et al., 2010); [t](Yildiz et al., 2006); [u](FrederickJr et al., 2008); [v]Mesured in our laboratory

Table 1. Chemical composition of various lignocellulosic materials

In the case of forest residues, we can assume that for 1 m^3 of roundwood exploited, 0.6 m^3 of residual biomass is left behind (Smeets & Faaij, 2007). In the province of Quebec, forest residues have been estimated to 6.9 millions of tons per year (Goyette & Boucher, 2009). Thus, the production of ethanol from glucose fermentation can be determinated assuming that the average of this sugar in such materials is about 52.4% (calculated from the Table 2) and that the maximum yield is equal to 0.51g of ethanol per g of glucose. Thus, 584 millions liters of ethanol could be produced in Quebec. To put such a value in perspective, consumption of refined petroleum in Quebec reached 9 billion liters in 2007 (Ministères de Ressources Naturelles et de la Faune, MNRF, Quebec, 2009). The production of ethanol from forest residues is sufficient to reach the objective fixed by the government (5 vol% in gasoline in 2012) since it represents 6.5 vol%. The North American consumption of gasoline for transport was estimated in 2008 at 518 739 millions liters with respectively 479 243 millions liters for the United States of America and 39 496 millions liters for Canada (IEA energy statistic, 2010). More ethanol can be produced by using agricultural residues, energy crops and unexploited forest zone. In Quebec, the latter represents 14,100,000 m^3 or 5.6 millions tons assuming an average density of 400kg/m^3. Thus, 1 638 millions liters of ethanol can be produced per year. Quebec will be able to replace 24.7 vol% of its gasoline by ethanol just by using exploited forest zone and forest residues, thus using only residual biomass. The comparison between the gasoline consumption and the possible production of ethanol from residues shows undeniably the importance of such source of raw material. Furthermore, the production of ethanol coming from the latter could rise above the production of ethanol coming from the first generation of biofuel. As an example, in North America, the nameplate production of ethanol from grain represents 53 949 millions of liters per year while the exploitation of residues could give more than 60 000 millions of liters per year.

	Glucan (wt%)	Xylan (wt%)	Mannan (wt%)	Galactan (wt%)	Arabinan (wt%)	Lignin (wt%)
Agricol residues						
Corn stover[a]	39	14.8	0.3	0.8	3.2	13.1
Rice straw[a]	41	14.8	1.8	0.4	4.5	9.9
Rice hulls[a]	36.1	14	3	0.1	2.6	19.4
Wheat straw[a]	36.6	19.2	0.8	2.4	2.4	14.5
Triticale straw[j]	43.5	17.7	-	-	2.3	17
Sugar cane	41.3	21.8	0.3	0.5	1.8	-
C_6 Average	*39.5*		*1.2*	*0.8*		
Energy crops						
switchgrass	35.2	21.7	0.2	0.9	2.8	27.4
Miscanthus[b]	44	21	-	-	-	-
Hemp[j]	51	14.3	1.5	0.7	1.3	14.5
Sweet sorghum[c]	44.6	25.3	-	-	-	18
Bagasse fiber	38.1	13	8	-	2	20
C_6 Average	*42.6*		*3.2*	*0.8*		
Forest residues						
Leafy						
Populus tristis[a]	40	13	8	-	2	20
Oak	45.2	20.3	4.2	-	-	-
Red Maple[i]	46.0	19.0	2.4	0.6	0.5	24
Aspen[d]	45.9	16.7	1.2	0	0	23
Salix[e]	41.4	15.0	3.2	2.3	1.2	26.4
Yellow poplar[f]	42.1	15.1	2.4	1	0.5	23.3
Eucalyptus[f]	48.1	10.4	1.3	0.7	0.3	26.9
C_6 Average	*44.1*		*3.2*	*0.9*		
Coniferous						
Spruce[g]	43.2	5.7	11.5	2.7	1.4	28.3
Lodgepole pine[h]	42.5	5.5	11.6	2.1	1.6	27.9
Ponderosa pine[h]	41.7	6.3	10.8	4.7	1.8	26.9
Douglas-fir[i]	44	2.8	11.0	4.7	2.7	32
Loblolly pine[i]	45	6.8	11.0	2.3	1.7	28
Red pine[i]	42	9.3	7.4	1.8	2.4	29
C_6 Average	*43.1*		*10.6*	*3*		

a(Lee, 1997); b(Sørensen et al., 2008); c(Ballesteros et al., 2004); d(Wang et al. 2008); e(Sassner et al., 2008); f(Zhu & Pan, 2010); g(Zhu et al., 2009); h(Youngblood et al., 2009); i(Pettersen, 1984); jMesured in our laboratory

Table 2. Details of carbohydrates and lignin amounts present in various lignocellulosic materials

	Possible Production of ethanol (Millions of liters per year) from			Production of ethanol operational from energy crop (Millions of liters per year)	Gasoline consumption (Millions of liters per year)
	Agricultural residues (25%)	Forest residues (25%)	Unexploited Forest		
Quebec	-	584	1638	155(a)	9000(d)
Canada	5097(e)	3353	-	1821(a)	39496(f)
North America	41483(e)	20322(c)	-	53949(b)	518739(f)

(Canadian Renewalable Fuels Association, 2010), (b) (Renewable Fuels Association, 2010) (c) Estimated from the production of roundwood; (d) (Natural Ressources Canada, 2007) (e) Estimated from FAOSTAT (FAOSTAT, 2010) thanks to the coeffient of residues proposed by D. Bellerini (Bellerini, 2006); (f) IEA energy statistic (IEA, 2011)

Table 3. Comparison between the actual possible production of ethanol from C_6 sugars contained in the lignocellulosic biomass, the operational ethanol production from energy crop and the gasoline consumption

In this estimation of the possible production of ethanol, marginal lands have not been taken into account. The potential of surplus land for the cultivation of energy crops like willows, poplars, miscanthus, switchgrass, panic, reed canary grass (second generation of biofuel) considerably depends on the regions. Numerous constraints exist for the implementation of new energy crops, making the estimation of biofuel production very approximative. The first one is food competition. In fact, in major countries, populations are growing, consequently increasing the food demand therefore reducing surplus land, which overall limits the additionnal production of energy crops. Among the others constraints we can mention, the water shortages, the implementation of indigeneous species (could be a probleme for biodiversity), the type of plant, improvement of agricultural system (allowing the cultivation of other land), etc. Depending on the scenario, disparate results are obtained; bioenergy could reach between 39EJ to 204 EJ in 2050, furthermore, Smeets et al. 2007 show that the biggest energy apport will be done by dedicated energy crops with 20-174 EJ of biomass against 6-11 EJ of agricultural residues and 6 EJ of forestry residues.

More than replacing gasoline coming from fossil ressources, the employement of biofuels in well defined conditions can contribute to reduce the Greenhouse Gas (GHG) emissions. The GHG balance varies significantly following the choice of biomass, the technology employed throughout the full « fuel cycle » from biomass production to final fuel consumption, the caracteristic of the land and climate, the crop management, etc. Thus, the choice of biomass is essential. As for ethanol production, the potential of lignocellulosic biomass to reduce GHG is comprised between 60 and 120 % and it is comparable to the high diminution of GHG observed with sugar cane (90%). In comparison, production of ethanol from wheat grain brings a lesser gain of 20 to 50% (IEA, 2004). The reduction, especially for lignocellulosic biomass, is due to the compostion of plant itself, to fertilizer loading and to the efficiency of vehicles. The high reduction of carbon dioxide emissions in the case of lignocellulosic biomass (cellulose to ethanol) essentially comes from the use of the other part of the plant (mainly lignin) as a source of energy for the process. However the previous estimations do not take into account the modifications affecting the lands. In fact, the GHG balance of the second-generation biofuels are closely related to the land use change (LUC) and the indirect use change (ILUC) which could in certain case conduct to a negative GHG balance. When a prior land-use like forest is replaced by culture for biofuel production, a direct land-use change occurs which can change the carbon stock of that land. This aspect has been widely studied and factors of changing balance can be found in the literature. However, the changes on GHG are induced by ILUC (takes place when land use change implies the displacement of the previous activity on another land). Thus, the replacement of sparsely vegetated and certain grass land by energy crops could generate a positive effect on GHG and in the mean time participate to the stockage of carbon in soil. Contrarily, the estimation made by Farrel and O'hare (2008) on the ILUC GHG emission shows that if the actual crops of soybean are used for the production of ethanol, the result could lead to the expansion of soybean for food into forests and will conduct to more emission than the use of fossil ressources (6 times more). These cultivations can also have several positive or negative impacts on soil, water and biodiversity. ILUC are normally less important if residues are used as feedstock since there is no need for additional land to be cultivated.

The second-biofuel generation, and in particular in the case of lignocellulosic ethanol production, should start with a sustainable development of agricultural and forestry residues which are at that time very interesting in terms of productivity (as it was shown on the Table 3) and environment. Even if some species contains much more C_6 sugars like coniferous and notably loblolly Pine, the use of a wide range of biomass genotypes is advised and needed. All the species presented show an interest for ethanol production. In fact, the use of just the best species would be catastrophic for biodiversity. In general, lignocellulosic biomass is a promising source of fuel as it is shown on the figure 1. As an example, in about ten years the production of biofuel via the lignocellulosic biomass in the US could almost reach the same production as from the other sources of biofuel.

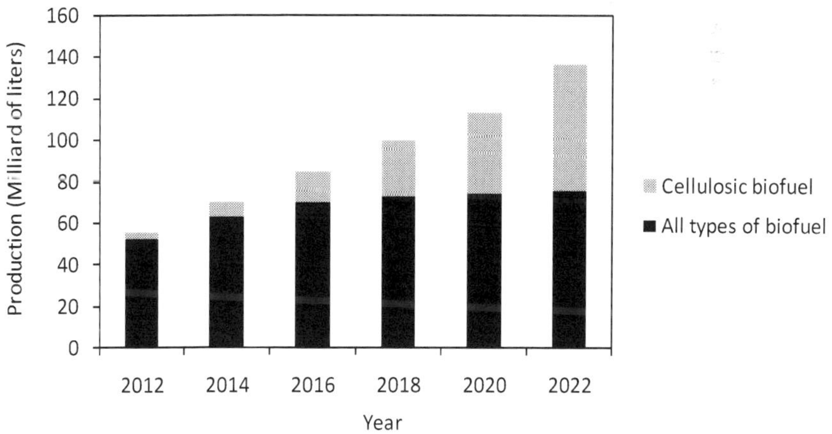

Fig. 1. Biofuel Mandate from lignocellulosic materials in the United States Renewable Fuels Standard

3. Fractionation

3.1 Basics informations about steam treatments

The first technological challenge restraining the commercialisation of the cellulosic biomass industry is the fractionation of the lignocellulosic biomass to isolate the cellulose macromolecule. Similar processes have been applied by the pulp and paper industry for decades but there is a new objective now, the complete utilisation of the carbon-based structures of the biomass as well as a reduction of the water consumption, hence, the concept of biorefinery. Although the traditional pulping processes are being remodelled to fit with the biorefinery approach, other processes are also investigated among which are the organosolv process and steam treatments. There is a significant variety of different steam treatments which all rely on the same concept: biomass is first saturated with a solvent (usually water) with or without the utilisation of a catalyst (acid or basic depending on the targeted macromolecule). The mixture is then "cooked" by addtion of steam in a pressure-resistant vessel for a certain period after which a valve is open, and the vapor phase exits the vessel through a nozzle entraining the solids. The exiting vapor reaches very high velocities as a function of the geometry of the nozzle thus reaching a sonic velocity. A "explosion" takes place while induced by the sonic field. The water saturating the biomass in its pores rapidly expands to vapour causing cell changes which vary from simple fibrilar disaggregation to fragmentation. During cooking, water, at high pressure and temperature, has a high dissociation constant leading to the occurence of a larger quantity of hydronium ions directly formed in the saturated pores of the biomass. Hemicelluloses, which are highly ramified and relatively easy to hydrolyse (in comparison to cellulose), will be affected by the increasing concentration of ions in the solution. The reaction of water and biomass is of course a major concern when considering steam treatments and it was found that in the absence of a catalyst, the two most important factors that were related to fractionation of the biomass and the hydrolysis of hemicelluloses were temperature and cooking period. The relationship between both parameters has been related to a mathematical equation called the "severity factor". This equation, reported first by Overend and Chornet (1987) has been from this point forward a significant contribution for the homogeneization of the steam processes.

$$R_0 = \int_0^{t[\text{min}]} \exp\left(\frac{T\left[^{\circ}\text{C}\right]-100}{14.75}\right) * dt \tag{1}$$

$$S_0 = \log R_0 \tag{2}$$

Where S_0 is the severity fractor, T is the temperature (expressed in degree celcius) and the overall equation relates on the integral of the temperature curve between the start and the end of the cooking period, including the preliminary heating leading to operating temperature. Severity factors were also related to the relative hydrolysis of the hemicelluloses macromolecule and (Overend and Chornet., 1987) showed that at a severity factor of 4, hemicelluloses were completely hydrolysed and there was starting to be an impact on the cellulosic fiber. This concept, can serve as a guide for other substrates although it was shown to vary from one feedstock to another, mostly because of the varying nature and amounts of hemicelluloses found in the biomass (Lavoie et al., 2010a, 2010b, 2010c). Impact of the calculated severity factor has also been reported for other feedstocks as residual cotton and recycled paper (Shen et al., 2008), aspen wood (Li et al., 2005), douglas fir (Wu et al., 1999), from rice husk and straws (Gerardi et al., 1999), from yellow poplar, from peanut hulls and from sugar cane (Glasser et al., 1998). In most of the cases reported previously, the ideal severity factor for the isolation of cellulose and hydrolysis of hemicelluloses was found to be between a severity factor of 3 and 4. A non-catalytic steam process of biomass usually leads to a brown lignocellulosic fibre as depicted in Figure 2 below:

Fig. 2. Lignocellulosic matrix obtained after a non-catalytic steam treatment of different biomasses with a severity factor between 3 and 4

Although hydrolysis of hemicelluloses can be performed only using the natural dissociative potential of water, many researches have investigated the effects of including a catalyst on the overall outcome of the process. Both acid and basic catalyst has been considered and each will have the tendency to target one type of macromolecule more than the other. Whilst acids will have a more pronounced effect on cellulose, bases will have a more significant effect on lignin. Among the acids that were used for catalysis of steam explosion reaction, sulphuric acid is one of the most common but also one of the less expensive at an industrial level. Utilisation of the latter has been reported repetitively in literature (Lawford et al., 2003; Emmel et al., 2003; Ballesteros et al., 2001). Directly comparable to sulphuric acid, sulphur dioxide was also widely used as a catalyst for steam explosion. The latter will interact with biomass and react with water to produce in turn sulphuric acid. The main difference might be that utilisation of SO_2 would allow a more homogeneous distribution of the acid catalyst in the biomass since the diffusion of the gas should be higher than the sulphuric acid molecule. Such a treatment has been effectively applied on lodgepole pine (Ewanick et al., 2007), poplar (Lu et al., 2009), aspen (De Bari et al., 2007) and eucalyptus (Ramos et al., 1999). The acid catalyst will have a direct effect on the hydrolytic potential of the mixture increasing the natural hydrolytic potential of water considerabily. As for the previously mentioned severity factor, researchers have tried to translate this phenomenon into an equation. Abatzoglou et al. (1992) were able to introduce the concentration of acid into the calculation of the severity factor as depicted below:

$$R_{0H} = \exp\left(\frac{X - X_{ref}}{\lambda X_{ref}}\right)\exp\left\{\frac{T - T_{ref}}{\omega'}\right\}t_R \tag{3}$$

Where X and X_{ref} is the acid loading (g of acid/g of dry biomass) and reference (acid loading, g of acid/g of dry biomass) respectively, λ is a parameter expressing the acid catalyst role in conversion of the system, ω' is parameter expressing the temperature role in conversion of the catalysed reaction system and t_R is the reaction time. A couple of years later, Montane and co-workers (Montane et al., 1998) developed a new version of the equation which included slight modifications over the equation proposed by Abatzoglou et al. The equation is depicted below:

$$R_0 = \exp\left(\frac{1}{\omega_0}\left(1-\frac{T_{ref}}{T}\right)\right)\frac{t^{\gamma}}{\gamma} \tag{4}$$

In this equation the ω_0 parameter express the energetic of the process respect to a reference reaction temperature, T and t remains the temperature and time whilst γ defines the shape of the distribution of activation energies. The research also showed that it was possible, for a specific species, to estimate the whole conversion of the process using a single equation:

$$(1-f) = \exp\left(-1.06x10^{10}\exp\left(\frac{-9733}{T}\right)C^{0.674}\frac{t^{0.608}}{0.608}\right) \tag{5}$$

Where f is the conversion parameter and C is the catalyst concentration. Equation 5 has been developed by Montane et al. using birch as substrate for the steam explosion process. Utilisation of an acid catalyst, for similar temperatures and times, should allow a more complete hydrolysis of the hemicelluloses but should also attack the cellulose molecule which is overall sensitive to the occurence of protons. It has been mentioned and it is still widely studied that the interactions between the hydronium and the cellose macromolecule may lead to a more efficient hydrolysis to glucose when used as a pretreatment for an enzymatic treatment (Dererie et al., 2011; Khunrong et al., 2011; Zhang et al., 2009). In most of the previously mentioned situations, utilisation of the catalyst leaded to increased value for conversion following the enzymatic hydrolysis, although in some specific cases, even if the conversion to glucose was increased, the fermentation was strongly inhibited by the production of furfural-derived compounds. Dehydration of xylose to furfural is depicted in Figure 3.

Dehydration of carbohydrates is strongly induced by acid catalysts at temperature higher than 150 °C with a classical inorganic catalyst although lower operating conditions were reported for the utilisation of ionic liquids (Tao et al., 2011). Five-carbon carbohydrates will dehydrate to furfural whilst dehydration of C_6 sugars will lead to 5-hydroxymethylfurfural (5-HMF) (Zhang et al., 2010). The latter will usually be less concentrated in a steam explosion process since it will require an isomerisation of the aldohexose sugars to a ketohexose form. Furthermore, under acid catalyst, 5-HMF has been reported to undego spontaneous hydrolysis to levulinic acid and formic acid which are both fermentation inhibitors. The minimal concentration at which furfural starts to inhibit fermentation has been reported to be at 2-3 g/L (Palmqvist et al., 1999) whilst as for 5-HMF, it has been reported that the concentration that causes 50% inhibition of fermentation was of 8 g/L.

Base-catalysed steam explosion may also be a potential pathway since the occurence of hydroxide ions, as in the case of kraft pulping, would lead to the hydrolysis of the hemicelluloses as well as the lignin whilst allowing the isolation of cellulose. Utilisation of NaOH as a catalyst for steam explosion has been reported in literature (Zhuang et al., 1997;

Li et al., 2005), although less frequently in comparison to the acid-catalysed reaction. Another process called ammonium fiber explosion is also slightly comparable to a base-catalysed steam explosion since it will allow defibration of the feedstock in a first time, then the interaction of ammonia with water can be directly related to an hydroxide ion catalyst although part of the hydrolysis process could be related to the ammonia itself although it is highly soluble in water and it interacts in an classical acido-basic reaction to produce ammonium hydroxide. This concept was efficiently tested on rice straws (Vlasenko et al., 1997) and the process itself has been patented by Dale et al. (2008).

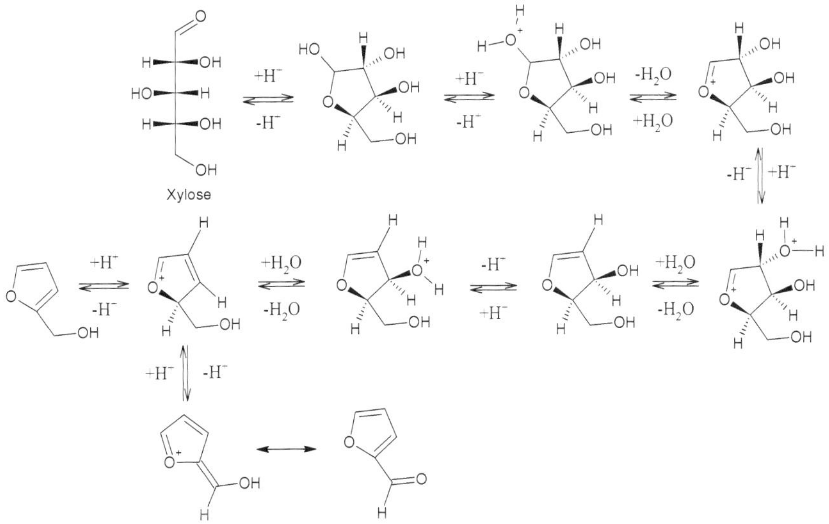

Fig. 3. Example of dehydration of a carbohydrate to furfural under acid catalyst (in this case xylose)

3.2 The Feedstock Impregnation Rapid and Sequential Steam Treatment (FIRSST)

Since a non-catalytic and an acid steam treatment allowed targeting the carbohydrate-based macromolecules from the biomass and the based-catalysed reaction allowed partial depolymerisation and solubilisation of lignin, our group has developed the two step FIRSST (Feedstock Impregnation Rapid and Sequential Steam Treatment) process. The biomass is first reduced in size to a range 3 to 6 cm long. It then follows the process flow diagram depicted below (Figure 4).

Biomass is first extracted with water, solvent and/or a mixture of both to extract the secondary metabolites. Two reasons justify the preliminary extraction, first some of the compounds could have a bioactive potential thus leading to applications in cosmetics and pharmaceuticals. Secondly, the extractives could act as inhibitors for fermentation and depending what is the targeted application for the broth obtained after the first steamexplosion, it might be beneficial to remove such compounds. After extraction, biomass is rinced with a minimal amount of water to remove traces of the residual solvent or to ensure maximal removal of extracts. Typically, at the bench scale level, a 5/1 massic ratio of water/biomass is used at this point. Biomass is then impregnated with water to ensure

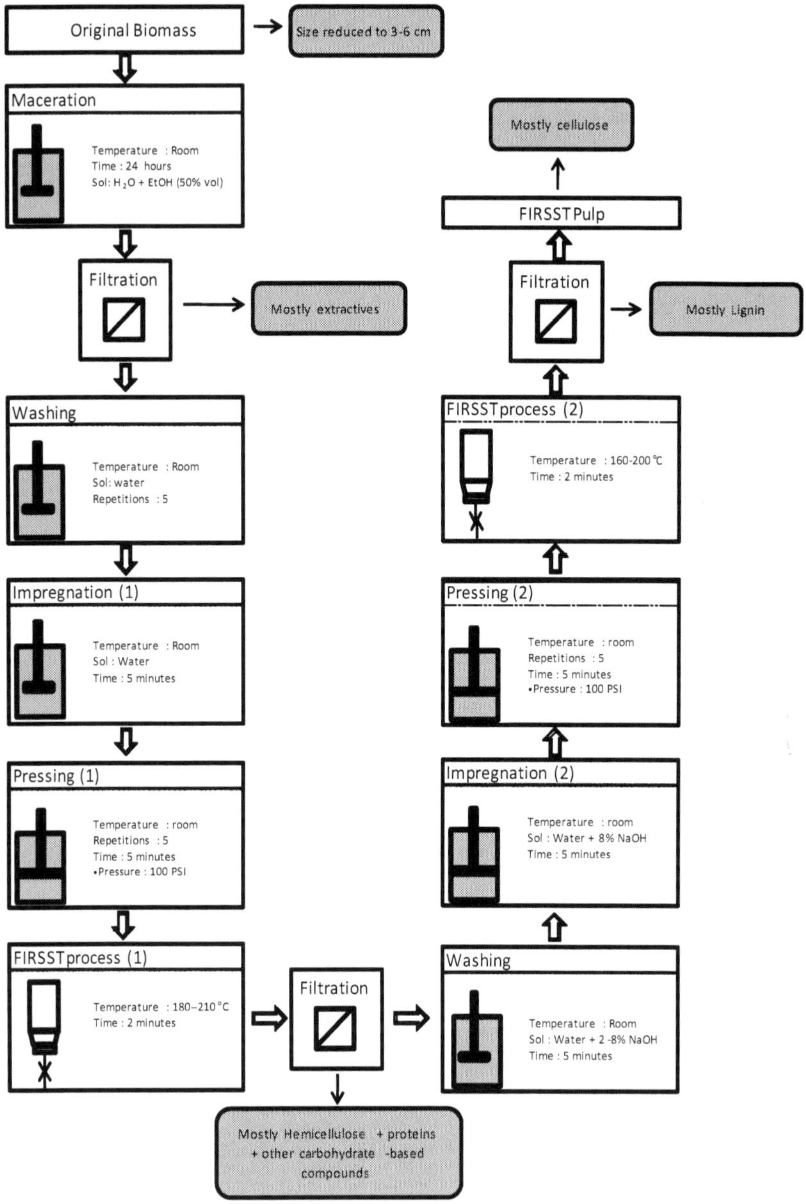

Fig. 4. Process flow diagram for the two step FIRSST process allowing isolation of the cellulose fibre as well as lignin, hemicelluloses and extractives with minimal purification required.

maximal penetration of the aqueous medium in the biomass' pores. Impregnation could be performed with or without pressure, either positive or negative. Saturation with water is one of the key elements for performing an efficient steam process and has to be monitored carefully. Impregnation could be performed by letting the biomass soak in water for a time period (typically up to 24h) allowing the water molecules to fill the small pores via capillarity. Whilst both positive and negative pressure might be used, utilisation of a positive pressure to ensure water penetration is by far the most efficiently scalable approach. After impregnation, excess water has to be removed, a pressure of 100 psi is sufficient both for the pressurized impregnation process as well as the following excess water removal. Once excess water is removed, the biomass is transferred in the FIRSST reactor where it is cooked for 2-4 minutes whilst monitoring the severity factor of the whole process. In a two step FIRSST process, one must ensure that the severity factor of the first process is not excessively high or the following delignification process, although efficient, will lead to excessive conversion and a lignin-carbohydrates broth after the second process.

Once the first FIRSST process is completed, the biomass is once more impregnated but this time with an aqueous diluted NaOH solution (typically 1-10%). Impregnation as well as removal of excessive solution was performed at 100 psi and room temperature. The residual lignocellulosic matrix is then cooked at temperature comparable to the first process although typically in a 10 °C inferior temperature range. Once the second cooking period is completed, biomass is rinced with a 10/1 ratio water/fibre to ensure removal of the remaining sodium ions. Such process has been tested on different feedstock including energy crops (Lavoie et al., 2010a), residual forest biomass (Lavoie et al., 2010b) and different agricultural residues (Lavoie et al., 2011). When using based-catalysed steam process, the catalyst itself, as in the case of acid hydrolysis, becomes an important factor of the reaction. Increasing the concentration of the basic catalyst showed to have a direct impact on lignin removal and using the same conditions (time and temperature), it was shown that an increasing alkali concentration in the mixture allowed the production of a whiter fibre. The texture of the fibres are strongly affected by the severity of the steam processes, example of the different textures of fibres produced according to the two-step FIRSST process used on triticale straws are depicted below (Figure 5).

At this point, the two-step FIRSST process has allowed the isolation, in high yields, of all the fractions of lignocellulosic biomass whilst producing pulp with good mechanical properties. Example of mechanical properties obtained from FIRSST pulps are depicted in the Table 4 for *Salix viminalis* (Lavoie et al., 2010a), for a mixture of softwood (Lavoie et al., 2010b) and for *Cannabis sativa* (Lavoie et al. unpublished results). Opportunity to produce quality pulp is yet another advantage of this technique which would, from the same feedstock, allow the production of many derived products including ethanol and cellulose

For the first step of the steam explosion and as mentionned earlier, no catalyst is used since the hemicelluloses and/or protein found in the lignocellulosic matrix were shown to be directly affected by water at the operating conditions. The first step of the FIRSST process is usually performed between temperatures of 180-230 °C depending on the nature of the biomass used as a feedstock. As an example, residual forest biomass was shown to require more severe conditions in comparison to residual agricultural biomass as triticale or hemp. Cooking period is usually ranging from 2-4 minutes, but it most of the cases investigated by our team, a 2 minute cooking period was shown sufficient. The uncatalyzed steam explosion process can be related to the severity factor that is calculated from the cooking temperature and time, needless to underline the fact that similar severity could be obtained by increasing the cooking period whilst decreasing the temperature. The 2 minutes cooking time usually excludes the heating period where biomass is heated to the operating conditions, nevertheless, this heating period is taken into consideration when calculating the severity factor. The heating period for the FIRSST process varies from 10-30 seconds and the temperature of biomass is monitored with thermocouples strategically located in the steam explosion reactor and recorded on an acquisition system allowing control and downstream calculations with regards to the conditions of operation.

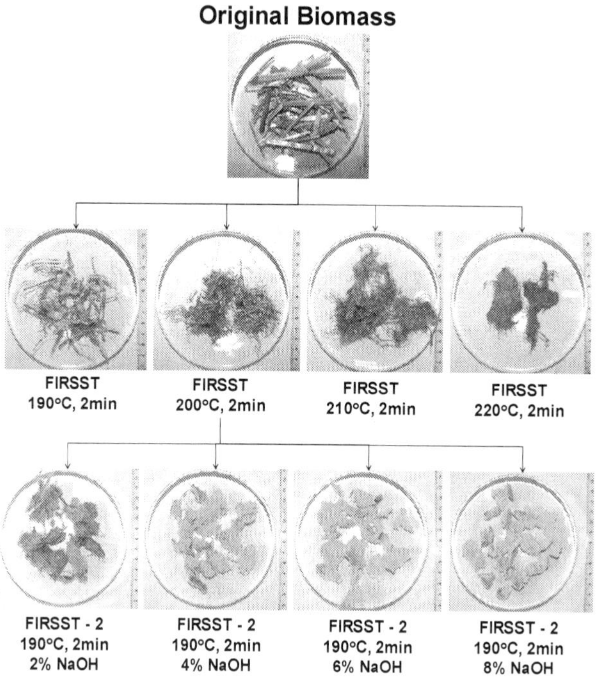

Fig. 5. Lignocellulosic and cellulosic fibres obtaineds with the two step FIRSST process applied to triticale straws

Another important factor that has to be taken into consideration when applying the 2-steps FIRSST process concerns the production of high quality pulp which may be tricky because of the severity of the reaction. Therefore, even if the second steam process is often based catalyzed, the severity of the treatment will lead to an indirect attack on cellulose which will reduce the fibre length if not suitably controlled. If ethanol production is intented, then the quality of the fibres produced after the FIRSST process is of lesser interest and both FIRSST treatments can be relatively severe. When dealing with quality cellulose fibres production, one must think about reducing the strength of the first FIRSST treatment in order to cope with the severity of the second treatment. So far the approach that has been developed in order to cope with such problem was to evaluate the composition of the lignocellulosic matrix at different severities for a first uncatalyzed FIRSST process. Optimal severity for the first process would lead to a drastic decrease of the hemicelluloses content whilst alterating minimally the long cellulose fibres (which are determined by quantification of the hemi, holo and α-cellulose). During the first FIRSST process, the holocellulose concentration will progressively decrease whilst α-cellulose will remain constant. We identify the maximal severity at the point where α-cellulose starts to significantly decrease. An example of such is depicted in Table 5 below:

Standard test	Willow[1]	Softwood[2]	Hemp
ATPPC C.1	454	664	407
ATPPC C.12	0.16	5.58	n.d.
ATPPC C.5U			
+ 14		47.8	
+ 28		20.0	
+ 48	n.d.	11.8	n.d.
+ 100		9.5	
+ 200		4.0	
- 200		7.0	
ATPPC B.4P	0.41	2.08	n.d.
ATPPC D.3	60.1	59.6	59.2
ATPPC E.1	33.3	24.8	29.4
ATPPC E.5			
L*	71.45 ± 0.01	67.26	68.90
a*	2.18 ± 0.01	4.41	2.74
b*	12.42 ± 0.01	18.31	13.79
ATPPC E.2	99.5	98.5	99.0
ATPPC D.4	1.79	2.24	2.90
ATPPC D.9	3.14	8.04	11.8
ATPPC D.8	1.56	3.05	<0.34
ATPPC D.34 Lenght of rupture	4.24	5.03	2.29
ATPPC D.34 TEA	19.6	37.8	8.8

[1]From Lavoie et al. 2010a
[2]From Lavoie et al. 2010b

Table 4. Mechanical properties of FIRSST for different types of feedstock.

Conditions	Severity	Conversion	Lignin	Holocellulose	α-cellulose*
-	Eqn [1,2]	%wt	%wt ± SD	%wt ± SD	%wt ± SD
210 °C, 2 min.	3.64	20.03	28 ± 2	51 ± 2	36.2 ± 0.7
220 °C, 2 min.	3.95	22.00	29.8 ± 0.5	47 ± 5	37 ± 1
220 °C, 4 min.	4.20	23.02	32 ± 3	44 ± 6	34.6 ± 0.4
230 °C, 2 min.	4.27	24.94	31 ± 8	43 ± 3	33 ± 1

Percentages are expressed in terms of bone dry biomass. Deviations from 100% closure are due to ash content.
** Comprised in the holocellulose*

Table 5. Composition of the lignocellulosic fibers following the first steam treatment under various conditions for softwood (From Lavoie et al. 2010b)

For the specific case depicted in Table 5, we can assume the severity factor varying from 3.64 to 3.95 only had a minor impact on the alpha cellulose and therefore we can assume that the treatment targeted the hemicelluloses principally. The quantity of removed hemicelluloses can be assimilated to the conversion also presented in the same table which shows an increasing mass content in the lignin broth produced from the FIRSST process. In a situation where quality pulp would be intended, although the best conditions for removing all the hemicelluloses would be at a severity factor of approximatively 4.00, it would be more strategic to use a lower severity first process (210°C, 2 minutes as an example) to ensure that the second steam treatment will not affect to much the pulp quality downstream.
The second process is also tricky since ideally and economically, lignin should be recuperated at >80%. Increasing the severity of the treatment will lead to very low lignin content but will also affect the cellulose fibres. To a certain extent, concentration of lignin inferior to 1% can be obtained using steam treatment. The downside will be that the higher severity will also affect the cellulose fibres and overall, the conversion will be higher and the quantity of pulp recovered downstream will also be lower with regards to the original quantity of biomass used for the process. An example of this is depicted in Table 6 below:

Conditions	**Yield***	**Lignin**	**Holocellulose**
-	%wt	*%wt ± SD*	*%wt ± SD*
190 °C, 2 min.	48.52	24.8 ± *0.2*	75.3 ± *0.2*
200 °C, 2 min.	41.35	14.5 ± *0.1*	85.6 ± *0.1*
210 °C, 2 min.	37.60	6.9 ± *0.1*	93.2 ± *0.1*
220 °C, 2 min.	37.23	3.2 ± *0.4*	96.9 ± *0.4*

* *Yield is expressed in terms of %wt of dry biomass and can therefore be considered as the rate of pulping for the overall process.*

Table 6. Pulp yield and composition (lignin and holocellulose) following delignification of the filtered solids (76.98 wt% of the original biomass, dry basis) obtained from the first steam treatment performed at 220 °C during 4 minutes from softwood biomass(From Lavoie et al. 2010b)

In light of the results depicted in Table 5, it is clear that increasing severity will also affect the overall conversion which will lead, at higher severity to a lower production of pulp. Other less severe technique could be used for removing the residual lignin and it should be considered to accentuate the pulping yields and to reduce to chemical alterations that the process may have induce to the cellulose fibres. On the other hand, if hydrolysis of glucose is intended, it was previously reported that severe steam explosion process could have a beneficial effect on enzyme hydrolysis and therefore, reaching higher severity may be beneficial to the whole process even if part of the cellulose is solubilized in the lignin broth.
Typical energy consumption for each of the steam process is approximatively 7% of the net biomass calorific value which represents approximatively 1.4 GJ per tonne of biomass process. A two step process could easily lead to a 15 % which now justifies investigations towards a one step process. Using triticale, our group compared the two- and one- step steam process. Since lignin has to be partially hydrolysed and removed, the one step steam process will require utilisation of a base-catalyst.

Althought the process was show efficient for the isolation of cellulose from triticale straws (unpublished results), the downside of such process is that it will generate a broth containing both hemicelluloses and lignin. Although lignin can be precipitated by reducing significantly the pH and mildly heating, it will nevertheless produce an ion-rich solution that will be overall hard to ferment without prior purification. Depending on the targeted downstream processes for hemicelluloses and lignine, the one-step or the two-step process may be more suitable and the choice of one or another should be influenced by the economic of the added-vapue products that will be generated from this biorefinery.

4. Hydrolysis

The cellulose macromolecule is composed of glucose units that are linked via a β 1-4 acetal bond. Despite the fact that the glucose units composing it are highly polar and completely soluble in water, cellulose itself is rather hydrophobic. The reason explaining such a phenomenon is because the polar functional groups of the cellulose macromolecules in the fibrils are oriented facing one another in such a fashion that render penetration by a polar solvent as water fairly hard to achieve. Furthermore, cellulose is composed of highly structured crystalline form as well as a less structured and fairly more ramified amorphous phase. The amorphous phase is more accessible to hydrolysing agents as acids or enzymes and usually bares the sites where original hydrolysis is going to take place. Although it is fairly uncommon for solvent to slide between the cellulose macromolecules, some solvent have been shown to be able to accomplish such task. The latter are usually highly polar and often have an ionic part.

The key to chemical hydrolysis of cellulose is to break the hydrogen bonding between the cellulose macromolecules in the fibrils. To do so, a compound or a mixture of compound that can efficiently rupture the hydrogen bonding is used. Furthermore, the compound has to be made in a specific shape capable of penetration in the tight hydrogen bonded ultrastructure. The compounds that have shown such potential all have in common an "arrow head" structure the ionic part of the molecule being the tip of the arrow. Once the ion is inserted between the cellulosic fibril layers, much effort has to be made in order to avoid reformation of the previous hydrogen bonding. Therfore, at this point, the utilisation of a larger quantity of solvent is of the essence. A classical example of this theory is the ASTM D1106-56 method which allows the quantification of lignin in a lignocellulosic matrix. The process relies on the acidification of the fibers with a strong sulphuric acid mixture. The mixture is kept at low temperature, allowing the ionic sulphuric acid to progress through the cellulosic matrix after which water is added to the mixture prior to hydrolysis. In this specific case, once the cellulose molecules have been spaced one from the other, water is added in large quantity to avoid reformation of hydrogen bonding between the cellulose macromolecules, thus isolating them one from the other. The solution is still acidic enough to allow hydrolysis and since the process requires energy to activate the bond, heating the solution close to ebullition of water allows the hydrolysis of cellulose to glucose. This process is highly effective and allows the conversion of cellulose up to a 100% into glucose. The downside of this process is directly related to the low cellulose concentration in the solution and the high residual ionic composition which makes this classic ASTM method fairly hard to scale up into an industrial fully functional process.

Sulphuric acid, although probably the more inexpensive of the potential reactants for cellulose hydrolysis is not the only one that has been reported so far. Formic acid, the smallest organic acid, has also been reported to be efficient for the swelling and overall hydrolysis of water. A few researchers have reported the utilisation of formic acid in relation with cotton (Sun et al. 2008). Cotton is a good example of a cellulose rich medium and it was reported that formic acid was shown effective for the hydrolysis of these fibres. Neverheless, formic acid was never efficient at a 100% concentration and for a suitable hydrolysis, it had to be diluted somewhat with water. This organic acid, as sulphuric acid, absolutely requires water as a medium for propagation of the protons. In a pure formic acid medium, the auto-hydrolysis process will generate far less free hydronium ions which will overall reduce the catalytic potential on the mixture. On the other hand, adding too much water will lead to dilution of the acid which will not be able to performed efficiently as a swelling agent in cellulose, since, as mentioned earlier, water is not strong enough to break the actual bonds between the cellulose macromolecules. Other reports have mentioned that they could increase efficiency of a formic acid hydrolysis by adding a specific amount of sulphuric acid in the mixture. In this specific case, sulphuric acid is not used as a swelling agent (small concentration has been reported to be efficient) but only as a proton provider, accentuating the actual contribution made by formic acid. And since sulphuric acid is significantly more acidic than formic acid, there will be available protons in the mixture, allowing hydrolysis of cellulose without however justifying the addition of water to the mixture. Another catalyst that has been reported often for the hydrolysis of cellulose using formic acid as a solvent is chlorhydric acid (Sun et al., 2010). Although it was shown effective for cellulose hydrolysis, such a mixture also has major drawbacks, first of which is the high concentration of formic acid which is a fermentation inhibitor. Production of glucose according to this process would therefore require a very extensive purification process to ensure minimal occurrence of formic acid in the mixture. Another problem that may arise from a formic-acid process would be the removal of sulphuric acid. Althought it may not be too much of a problem at low concentration, removing formic acid would concentrate the sulphuric acid in the mixture leading to a strong acidic catalyst at the end of the distillation process. Even in small amount, the inorganic acid would more likely interact with glucose at the boiling point of formic acid and induce dehydration of the glucose molecule leading yet to another major problem via the occurrence of another major inhibitor, 5-hydroxymethylfurfural (5-HMF).

Another compound that was shown to be effective to separate the cellulose macromolecules is cupryethylene diamine, an organometallic compound currently used to evaluate the polymerisation degree of the cellulose fibres. This compound also demonstrates the "arrow head" structure and the major interaction between this molecule and the cellulose fibrils is made via the amine groups (see Figure 6). The target of cupryethylene diamine is to solubilise cellulose in order to evaluate the polymerisation level of the fibres, therefore, it does not allow hydrolysis of cellulose. Nonetheless, insertion of an acid catalyst may also lead to an efficient hydrolysis although the acid would most probably interact with the strongly basic amine compounds.

In recent literature, other compounds have also been reported to interact favourably with cellulose and most of these compounds are related to the ionic liquids group. Such liquids have gained attention around the 90's and are currently involved in many scientific studies and specially in the field of high efficiency electrochemical cells. There also have been some reports that ionic liquids could have an impact on cellulose, allowing dissolution of the latter. A recent patent by Braun et al. (2010) reported that they were able to produce cellulose beads by dissolving a cellulose medium in a solution containing an imidazolium compound (see Figure 7 below).

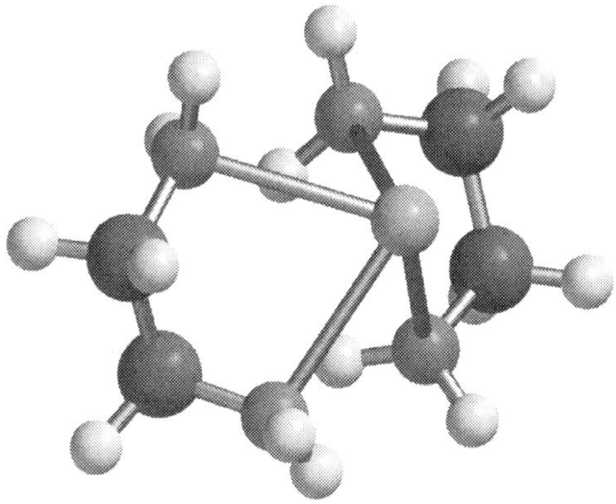

Fig. 6. Cupryethylene diamine molecule in aqueous solution (copper is depicted in orange, nitrogen in blue and carbon in black), the model was elaborated with Spartan'10 software

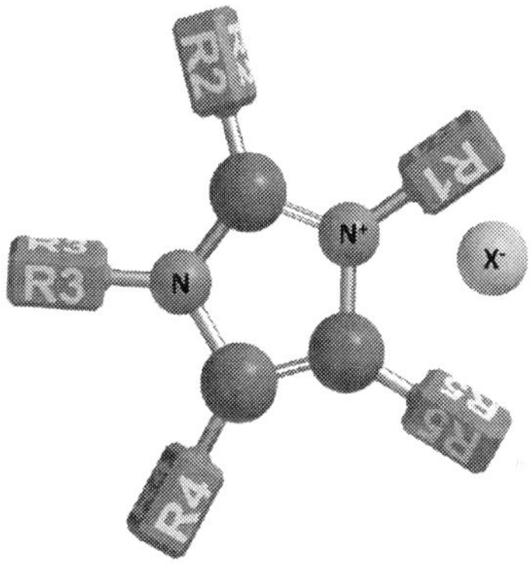

Fig. 7. General structure of an imidazolium compound, in which R1 and R3 are each an identical C2-20 org. radical, R2, R4 and R5 are each an H atom, X- is an anion. The model was elaborated with Spartan'10 software. A few reports also mentioned the utilisation of 1-alkyl-3-methylimidazolium chloride for the suitable dispersion of the cellulose macromolecule in solution (Ignatyev et al. 2010 and Vitz et al. 2010).

Our group has developed a process "the CRB-UdeS process" by which cellulose is hydrolysed to glucose (Lavoie et al., 2010a, 2010b; Chornet et al., 2008) being scaled to a demonstration level by CRB. The production of glucose out of lignocellulosics is overall a tricky process and one should keep in mind that before attacking the acetal bonding of the cellulose, it is primordial that the targeted bonds are exposed to the catalyst. The process has been divided in three steps: swelling, hydrolysis and purification. Swelling of cellulose implies the insertion of a suitable molecule, that breaks the strong hydrogen bonding among cellulose macromolecules in the fibrils. A stable gel is formed. Our group has investigated the different parameters which favour an efficient swelling and the most important factors to be taken into account are temperature and time of swelling. The concentration of the swelling agent is of crucial importance if cellulose macromoilecules in the fibrils are to be separated one from the others. Figure 8 below depicts the influence of the concentration of sulphuric acid as well as the acid/cellulose ratio in relationship with total hydrolysis of cellulose.

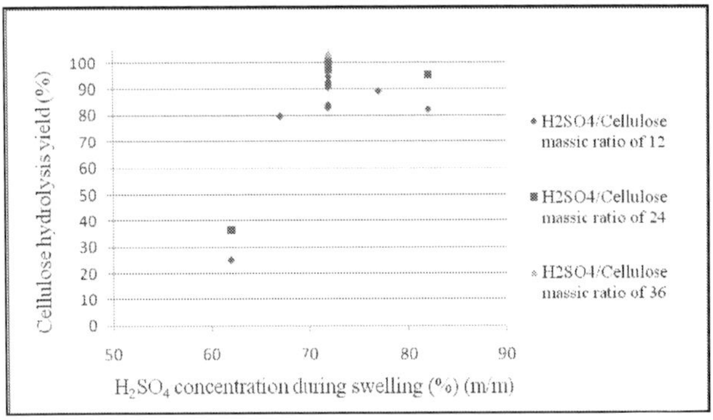

Fig. 8. Acid quantity and concentration during swelling vs tritical cellulose hydrolysis yield (for 10min hydrolysis at 121◦C in autoclave with H_2SO_4 concentration of 208-267 g/L and with H+/OH- molar ratio of 2.3 during NaOH 18-23 % addition)

Results shows that the there is an optimal concentration of acid leading to an optimized swelling which facilitates the hydrolysis of cellulose with a yield close to 100% at around 72-74 wt% acid in solution. A 72 %wt of sulphuric acid correspond to a 2.0 molar ratio of water/sulphuric acid which is stoichiometrically sufficient to completely ionise sulphuric acid which leads back to the arrowhead structure. The swelling period is also fairly important for such process, getting optimal after half an hour. Results (depicted in Figure 9 below) shows that increasing the time of swelling does not lead to significant increases in terms of rate of hydrolysis downstream. The temperature at which the cellulose is swelled can be varied from 20-40 ◦C after which degradation of the cellulose is observed (Belanger, 2005).

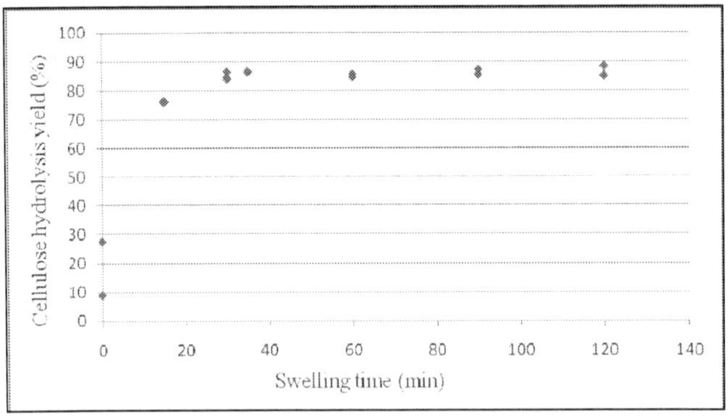

Fig. 9. Time of swelling vs tritical cellulose hydrolysis yield (for 72% H₂SO4 during swelling at 30°C, 10 min hydrolysis at 121°C in autoclave, H_2SO_4/Cellulose massic ratio of 36, H+/OH- molar ratio of 2.5 during NaOH 20 % addition)

Once the hydrogen bonding between the cellulose macromolecules is replaced by hydrogen bonding to the aqueous solvent, the next step leading to the production of glucose is hydrolysis. The latter is performed under a decreased free acid concentration. The effect of the pH on the hydrolysis process is depicted in Figure 10 below.

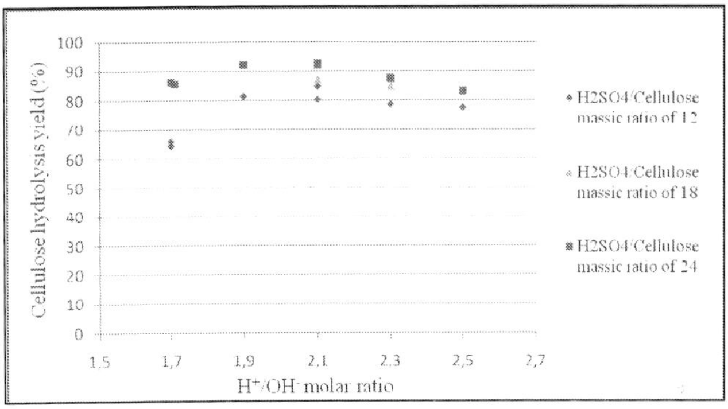

Fig. 10. H+/OH- molar ratio vs tritical cellulose hydrolysis yield (for 72% H2SO4 during the 2h at 30°C swelling, with 10 min hydrolysis at 121°C in autoclave, with NaOH 40% addition)

The previously mentioned results show that a minimal amount of hydronium ions are required in order to perform hydrolysis to an optimal value. The concentration of available protons is of course an important factor but temperature and time are also parameters that have to be assessed in order to control optimially the production of cellulosic glucose. Hydrolysis of the acetal bonding requires activation. Figure 11 below shows the effect of temperature on the hydrolysis process.

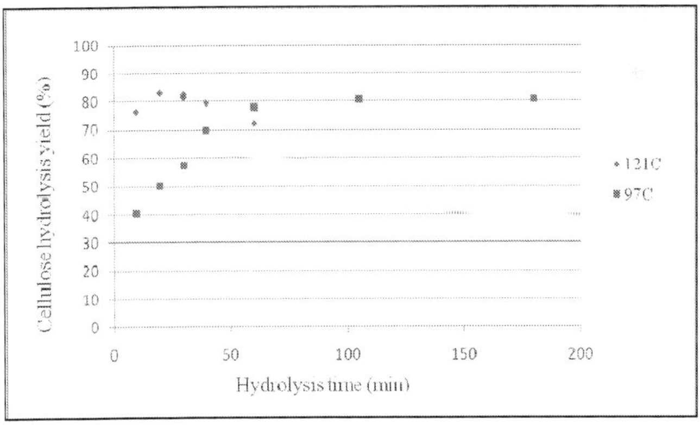

Fig. 11. Hydrolysis time and temperature vs tritical cellulose hydrolysis yield (for 72% H_2SO_4 during the 2h at 30°C swelling, with H_2SO_4/Cellulose massic ratio of 12, with H+/OH- molar ratio of 2.0 and partial neutralization. All the results at 121°C include 20 min more above 97°C due to the preheat and the cool down of temperature in the autoclave.

Figure 11 shows that for both temperatures, a maximum hydrolysis yield is reached but the required time for achieve such hydrolysis is different which is perfectly in occordance to the kinetics of the hydrolysis reaction. As for steam explosion, hydrolysis of cellulose could as well be depicted in terms of a severity factor but no mention of such mathematical relationship has been reported so far. Finally, time of hydrolysis is of course an important factor when considering hydrolysis of cellulose. Once cellulose has undergone a sufficient swelling, it is completely exposed to acid attacks on the acetal bonds of the glucose units. Results depicted in Figure 12 below clearly shows that the hydrolysis is fast and efficient at 121 °C.

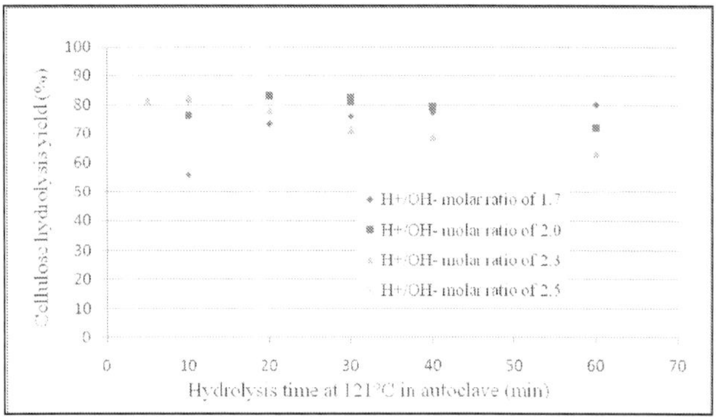

Fig. 12. Hydrolysis time and H+/OH- molar ratio vs tritical cellulose hydrolysis yield (for 72% H_2SO_4 during the 2h at 30°C swelling, with 121°C hydrolysis in autoclave, with H_2SO_4/Cellulose mass ratio of 12, and partial neutralization with NaOH 32.8%)

5. Conclusions

The production of cellulosic ethanol via the fractionation process requires an understanding of the availability of the cellulose in different lignocellulosic substrates. The cellulose content of a biomass can lead to the estimation of ethanol productivity via glucose fermentation. In order to convert biomass to glucose, two important steps have to be controlled and optimized, fractionation of the biomass to yield cellulose and hydrolysis of the latter. Aqueous/Steam treatents are an efficient alternative to isolate cellulose. They relie on the utilisation of water with or without catalyst to saturate the biomass and induce, via steam addition, the increase in temperature that results in, the hydrolysis of bonds between constitutive fractions liberating hemicelluloses and lignin. The impact of steam processes on biomass can be correlated by a severity factor which allows estimating the quantity of hemicelluloses and to certain extent cellulose that is hydrolyzed during the process. Utilization of a base catalyst can allow hydrolysis of lignin and production of a relatively clean cellulosic fibre. The two step FIRSST process first hydrolyzes the hemicelluloses whilst lignin is separated from the cellulose in a second based-catalyzed treatment. This process thus allows the fractionation of biomass while producing a high quality pulp and a variety of smaller fibers called fines which can be hydrolyzed via penetration of strong ion solution, (concentrated sulphuric acid) which swells the crystalline cellulose. The"swelling" was shown to be more efficient when using ionic compounds that show an "arrowhead" configuration. Such a pattern is representative of a few compounds including sulphuric acid. The step for hydrolyzing cellulose is the acidic attack on the acetal bonds. The latter can be performed by an hydronium ion and accessibility to the acetal bonding of cellulose is directly related to the efficiency of the previous swelling. It was shown that once the targeted bonds are exposed, the swelling is of the order of minutes (and can be performed at different temperatures near the boiling point of water. It is an alternative to enzyme hydrolysis.

6. Acknowledgment

We would like to acknowledge our industrial partners; Enerkem, Greenfield Ethanol, CRB Innovations and the Ministry of Natural Resources of Quebec for financial support of the Industrial Chair in Cellulosic Ethanol. We would also like to thank the ABIN network for partial financial support of this project.

7. References

Abatzoglou, N.; Chornet, E.; Belkacemi, K. & Overend, R. P. (1992). Phenomenological kinetics of complex systems: the development of a generalized severity parameter and its application to lignocellulosics fractionation. *Chemical Engineering Science*, Vol. 47, No. 5, (April 1992), pp. 1109-22, ISSN 0009-2509

Agbogbo, F.K. & Coward-Kelly, G. (2008). Cellulosic ethanol production using the naturally occurring xylose-fermenting yeast, Pichia stipitis. *Biotechnology Lett*ers, Vol. 30, No. 9,(May 2006), pp.1515–1524, 0141-5492

Alvo, P.; Savoice, P.; Tremblay, D.; Emond, J.P. & Turcotte., G. (1996). A system approach for evaluation of ethanol production based on forages. *Bioresource Technology*, Vol. 56, No. 1, (April 1996), pp. 61–68, ISSN 0960-8524

Canadian Renewalable Fuels Association CRFA. (2010). In : *Industry information*, December 2010, http://www.greenfuels.org/en/industry-information/plants.aspx

Ballesteros, I.; Oliva, J. M.; Saez, F. & Ballesteros, M. (2001) Ethanol production from lignocellulosic by products of olive oil extraction. *Applied Biochemistry and Biotechnology*, Vol. 91-93, No. 1-9, (2001), 91-93, pp. 237-252, ISSN 0273-2289

Ballesteros, M.; Oliva, J.M.; Negro, M.J.; Manzanares, P. & Ballesteros, I. (2004). Ethanol from lignocellulosic materials by a simultaneous saccharification and fermentation process (SFS) with Kluyveromyces marxianus CECT 10875. *Process Biochemistry*, Vol. 39, No. 12, (October 2004), pp. 1843–1848, ISSN 1359-5113

Bednar, H. & Fengel, D. (1974) Physikalische, chemische und strukturelle Eigenschaften von rezentem und subfossilem Eichenholz. *Holz-Roh Werskt*, Vol. 32, No. 3, pp. 99-107

Belanger, k. (2005). Controlled depolymerisation and decrystallization of cellulose-rich substrates into glucose. Research master thesis in chemical engineering. Sherbrooke, Canada, August 2005

Bellerini, D. (2006). *Les Biocarburants, états des lieux, perspectives et enjeux du développement*, Technip, IFP publications, ISBN 9782710808695, Paris, France

Berberi, V.; Turcotte, F.; Lantagne, G.; Chornet, M. & Lavoie, J.-M. (2011). High Yields of Sugars via the Non-Enzymatic Hydrolysis of Cellulose. Energy and sustainability III, Alicante, Espagne, April 2011

Boe, A. & D.K. Lee. (2006). Genetic variation for biomass production in prairie cordgrass and switchgrass. *Crop Science*, Vol. 47, No. 3, pp. 929-934, ISSN 0011-183x

Boulanger, F.; Bourke, P. & Dunsky. P. (2009). Imaginons le Quebec sans pétrole, In : les rendez-vous de l'energie, February 2011, Available from : <http://www.rdvenergie.qc.ca/wp-content/uploads/2010/09/Cahier_de_reference_RDV/ Energie1.pdf>

Braun, M.; Guentherberg, N.; Lutz, M.; Magin, A.; Siemer, M.; Swaminathan, V.N.; Linner, B.; Ruslim, F. & Fernadez Ramierz, G.A. (2010). Process for producing cellulose beads from solutions of cellulose in ionic liquid. US Patent Application 20100331222, December 30, 2010

Buranov, A.U. & Mazza, G. Lignin in straw of herbaceous crops. *Industrial Crops and Products*, Vol. 28, No. 3, (November 2008), pp. 237-259, ISSN 0926-6690

Casey, E.; Sedlak, M.; Ho, N.W.Y. & Mosier N.S. (2010). Effect of acetic acid and pH on the cofermentation of glucose and xylose to ethanol by a geneticallyengineered strain of Saccharomyces cerevisiae. *FEMS Yeast Res*, Vol 10, (April 2010), pp. 385–393, ISSN 1574-6968

Chornet, E.; Chornet, M. & Lavoie, J.-M. (2008) Conversion of cellulosic biomass to sugar. US Provisional Patent Application #80685-2 filed Oct 8, 2008

Chu, B. & Lee, C.H., (2007). Genetic improvement of Saccharomyces cerevisiae for xylose fermentation. *Biotechnology Advances* Vol. 25, No. 5, (September 2007), pp. 425–41, ISSN 1873-1899

Claessens, A.; Michaul, R.; Belanger, G. & Mather., D.E. (2004). Characteristics of timothy genotypes divergently selected for fiber traits. *Crop Science*, Vol. 44, No.1, pp. 81–88, ISSN 0011-183X.

Dale, B.E. & Bals, B. (2008). Separation of proteins from grasses integrated with ammonia fiber explosion (AFEX) pretreatment and cellulose hydrolysis. PCT Int. Appl. (2008), WO 2008020901 A2 20080221

De Bari, I.; Nanna, F. & Braccio, G. (2007). SO_2-Catalyzed Steam Fractionation of Aspen Chips for Bioethanol Production: Optimization of the Catalyst Impregnation. *Industrial & Engineering Chemistry Research*, Vol. 46, No. 23, (October 2007) , pp. 7711-7720, ISSN 0888-5885

Department of Energy (US), DOE. (2006). Biomass feedstock composition and property database, In: *Biomass Program*. Available from: <http://www.eere.energy.gov/biomass/progs/search1.cgi>

Dererie, D. Y.; Trobro, S.; Momeni, M. H.; Hansson, H.; Blomqvist, J.; Passoth, V.; Schnuerer, A.; Sandgren, M. & Stahlberg, J. (2011). Improved bio-energy yields via sequential ethanol fermentation and biogas digestion of steam exploded oat straw. *Bioresource Technology*, Vol. 102, No. 6, (March 2011), pp. 4449-4455, ISSN 0960-8524

Emmel, A.; Mathias, A.L.; Wypych, F. & Ramos, L. P. (2003). Fractionation of Eucalyptus grandis chips by dilute acid-catalysed steam explosion. *Bioresource Technology*, Vol. 86, No. 2 (January 2003), pp. 105-115, ISSN 0960-8524

Ewanick, S.M.; Bura, R. & Saddler, J.N. (2007). Acid-catalyzed steam pretreatment of lodgepole pine and subsequent enzymatic hydrolysis and fermentation to ethanol. Biotechnology and Bioengineering, Vol. 98, No. 4, (November 2007), pp. 737-746, ISSN 1097-0290

Food and Agriculture Organization of the United Nations FAOSTAT. (2010). In : *Production*, December 2010, http://faostat.fao.org/site/339/default.aspx

Farrell, A. E. & O'Hare, M. (2008). Greenhouse gas (GHG) emissions from indirect land use change (LUC). Memorandum for the California Air Resources Board. Available from : <http://www.arb.ca.gov/fuels/lcfs/011608ucb_luc.pdf>

FrederickJr, W.J.; Lien, S.J.; Courchene, C.E.; DeMartini, N.A.; Ragauskas A.J. & Iisa K. (2008). Co production of ethanol and cellulose fiber from Southern Pine: A technical and economic assessment. *Biomass and Bioenergy*, Vol. 32, No. 12, (december 2008), pp.1293–1302, ISSN 0961-9534

Fitzpatrick, S.W. (2010) Production of formic acid from hydrolysis of carbohydrate-containing compounds such as paper pulp and wood feed stocks. PCT Int. Appl. (2010), WO 2010104722 A1 20100916

Gerardi, V.; Minelli, F. & Viggiano, D. (1998). Steam treated rice industry residues as an alternative feedstock for the wood based particleboard industry in Italy. *Biomass and Bioenergy*, Vo. 14, No. 3, (March 1998), pp. 295-299, ISSN 0961-9534

Gibbs, R.D. (1958). The Physiology of Forest Trees, Thimann, pp. 269, John Wiley and Sons, New York, USA

Gírio, F.M.; Fonseca, C.; Carvalheiro, F.; Duarte, L.C.; Marques, S. & Bogel-Łukasik, R. (2009). Hemicelluloses for fuel ethanol: A review. *Bioresource Technology*, Vol. 101, No. 13, (July 2010), pp. 4775-4800, ISSN 0960-8524

Glasser, W. G. & Wright, R. S. (1998). Steam-assisted biomass fractionation. II. Fractionation behavior of various biomass resources. Biomass and Bioenergy, Vol. 14, No. 3, (March 1998), pp. 219-235, ISSN 0961-9534.

Goyette, J. & Boucher, S. (2008). Communication from the engineering firm roche, Etablissement d'une chaufferie central à la biomasse forestière: lignes directrices dans un context Québécois. (November 2008)

Ibrahim, M., & Glasser, W.G. (1999). Steam-assisted biomass fractionation. Part III: a quantitative evaluation of the ``clean fractionation'' concept . *Bioresource Technology*, Vol. 70, No. 2, (October 1999), pp. 181-192, ISSN 0960-8524

Ignatyev, I.A.; Mertens, P. G. N.; Van Doorslaer, C.; Binnemans, K. & de Vos, D.E. (2010). Cellulose conversion into alkylglycosides in the ionic liquid 1-butyl-3-methylimidazolium chloride. *Green Chemistry*, Vol. 12, No. 10, pp. 1790-1795, ISSN 1463-9270

International Energy Agency IEA. (2009). *Medium Term Oil Market Report*, OECD/IEA, Paris.

International Energy Agency IEA. (2004). *Biofuel for transport an international perspective*. Chirat, 9264015124, France.

International Energy Agency IEA. (2011). In : *IEA energy statistic*, January 2011, Available from: http://www.iea.org/stats/index.asp

Jefferson, P.G.; McCaughey, W.P.; May, K., Woosaree, J. & McFarlane, L. (2004). Potential utilization of native prairie grasses from western Canada as ethanol feedstock. *Canadian Journal of Plant Science*, Vol. 84, (May 2004), pp. 1067–1075, ISSN 0008-4220

Jung, H.G.; D.R. Mertens & Payne, A.J. (1997). Correlation of acid detergent lignin and klason lignin with digestibility of forage dry matter and neutral detergent fiber. *Journal of Dairy Science*, Vol. 80, No. 8, (August 1997), pp. 1622–1628, ISSN 1525-3198

Jurgens, M.H. (1997). Animal Feeding and Nutrition, 8th ed. Kendall/Hunt Publishing Company. Dubuque, Iowa.

Khunrong, T.; Punsuvon, V.; Vaithanomsat, P. & Pomchaitaward, C. (2011). Production of ethanol from pulp obtained by steam explosion pretreatment of oil palm trunk. *Energy Sources, Part A: Recovery, Utilization, and Environmental Effects*, Vol. 33, No. 3, (January 2006) pp. 221-228, ISSN 1556-7230

Lavoie, J.-M.; Capek, E.; Gauvin, H. & Chornet, E. (2010a). Production of pulp from Salix viminalis energy crops using the FIRSST process. *Bioresource Technology*, Vol. 101, No. 13, (July 2010), pp. 4940-4946, ISSN 0960-8524

Lavoie, J.-M.; Capek, E.; Gauvin, H. & Chornet, E., (2010b). Production of quality pulp from mixed softwood chips as one of the added value product using the FIRSST process in a general biorefinery concept. *Industrial & Ingineering Chemistry*, Vol. 49, pp. 2503-2509, ISSN 1520-5045

Lavoie, J.-M., Capek, E., Chornet, E., (2010c). Evaluation of the co-product pulp from Salix viminalis energy crops. *Biomass & bioenergy*, Vol. 34, No. 9, (September 2010), pp. 1342-1347, ISSN 0961-9534

Lawford, H. G. & Rousseau, J. D. (2003). Cellulosic fuel ethanol. Alternative fermentation process designs with wild-type and recombinant *Zymomonas mobilis*. *Applied Biochemistry and Biotechnology*, Vol. 106, No. 1-3, (March 2003), pp. 105-108, pp. 457-469, ISSN ***1559-0291.***

Lee, J. (1997). Biological conversion of lignocellulosic biomass to ethanol. *Journal of Biotechnology*, Vol. 56, No.1, (July 1997), pp. 1–24, ISSN 0168-1656

Lee, D.K. & V.N. Owens. (2005). Biomass production potential of native warm-season grass monocultures and mixtures. Agronomy abstracts. ASA

Lee, D.K., Owens, V.N., Boe, A., Jeranyama, P. (2007). Composition of Herbaceous Biomass Feedstocks, February 2011, Available from: <http://ncsungrant1.sdstate.org/uploads/publications/SGINC1-07.pdf>

Li, J.; Henriksson, G. & Gellerstedt, G. (2005). Carbohydrate reactions during high-temperature steam treatment of aspen wood. *Applied Biochemistry and Biotechnology*, Vol. 125, No. 3, (June 2005), pp. 175-188, ISSN ***1559-0291***

Liang, Y.; Siddaramu, T.; Yesuf, J. & Sarkany, N. (2010). Fermentable sugar release from Jatropha seed cakes following lime pretreatment and enzymatic hydrolysis. *Bioresource Technology*, Vol. 101, No. 16, (April 2010), pp.6417–6424, ISSN 0960-8524

Lu, Y.; Warner, R.; Sedlak, M.; Ho, N. & Mosier, N. S. (2009). Comparison of glucose/xylose cofermentation of poplar hydrolysates processed by different pretreatment technologies. *Biotechnology Progress*, Vol. 25, No. 2, (March 2009), pp. 349-356, ISSN 1520-6033

Mani, S.; Tabil, L.G. & Sokhansanj. S. (2006). Effects of compressive force, particle size and moisture content on mechanical properties of biomass pellets from grasses. *Biomass and Bioenergy*, Vol. 30, No. 7, (July 2006), pp. 648–654, ISSN 0961-9534

Mazlan I. & W.G. Glasser. (1999) Steam-assisted biomass fractionation. Part III: a quantitative evaluation of the ``clean fractionation'' concept. *Bioresource Technology*, Vol. 70 , No. 2, (November 1999), pp. 181-182, ISSN 0960-8524

Ministères de Ressources Naturelles et de la Faune MRNF. (2009). Chiffres clés du Quebec Forestier, December 2010, Available from : <www.mrnf.gouv.qc.ca>

Montané, D.; Overend R.P. & Chornet, E., (1998). Kinetic Models for Non-Homogeneous Complex systems with a Time-Dependent Rate Constant. *Canadian Journal of Chemical Engineering*, Vol. 76, No. 1, (February 1998), pp. 58-68, ISSN 1939-019X

Natural Ressources Canada. (2007). Canadian Petroleum Product Market, December 2010, Available from : <http://www.nrcan-rncan.gc.ca/eneene/sources/petpet/reprap/2007-11/supoff-eng.php>

Overend, R. P. & Chornet, E. (1987). Phil. Trans. R. S. Lond. A321, 523–536

Owens, V.N.; Lee, D.K. & Boe, A. (2006). Manure and harvest timing effects on biomass and seed yield of three perennial grasses in South Dakota. The World Congress on Industrial Biotechnology and Bioprocessing. Biotechnology Industry Organization.

Palmqvist, E. & Grage, H. (1999). Main and interaction effects of acetic acid, furfural, and p-hydroxybenzoic acid on growth and ethanol productivity of yeasts. *Biotechnology and Bioenergy*, Vo. 63, No. 1 (April 1999) pp. 46–55, 1097-0290

Pettersen, R.C. (1984). In: *'The Chemistry of Solid Wood'*, R. Rowell, pp. (57-127), American Chemical Society, ISBN 9780841207967, Washington D.C.

Pienkos, P. T. & Zhang, M., (2009). Role of pretreatment and conditioning processes on toxicity of lignocellulosic biomass hydrolysates. *Cellulose*, Vol. 16, No. 4, (August 2009), pp. 743-762, ISSN 1572-882X

Ramos, L. P.; Mathias, A.L.; Silva, F.T.; Cotrim, A.R.; Ferraz, A.L. & Chen, C.-L, (1999). Characterization of Residual Lignin after SO2-Catalyzed Steam Explosion and Enzymatic Hydrolysis of *Eucalyptus viminalis* Wood Chips. *Agricultural and Food Chemistry*, Vol. 47, No. 6, (May 1999), pp. 2295-2302, ISSN 1520-5118

Renewable Fuels Association. 2010. Ethanol facilities capacity by state, December 2010, Available from: <http://www.neo.ne.gov/statshtml/121.htm>

Rodríguez, A.; Sánchez, R.; Requejo, A. & Ferrer, A. (2010). Feasibility of rice straw as a raw material for the production of soda cellulose pulp. *Journal of Cleaner Production*, Vol. 18, No. 10-11, (March 2010), pp. 1084-1091, ISSN 0959-6526

Sassner, P.; Martensson, C.G.; Galbe, M. & Zacchi, G. (2008). Steam pretreatment of H_2SO_4-impregnated Salix for the production of bioethanol. *Bioresource Technology*, Vo. 99, No. 1, (January 2008), pp. 137–145, ISSN 0960-8524

Schafer E.R. & Bray M.W. (1929). Pulping Flax Straw. VI. Properties of Flax straw cellulose and its value in the cellulose industries. Industrial Engineering Chemistry, Vol. 21, No. 3, (March 1929), pp. 278–280, ISSN **0888-5885**

Shen, J., Agblevor, F. A. (2008). Ethanol production from cotton gin waste by simultaneous saccharification and fermentation. Beltwide Cotton Conferences (2008), pp. 569-572, Nashville, USA

Sørensen, A.; Teller, P.J.; Hilstrøm, T. & Ahring, B.K. (2008). Hydrolysis of Miscanthus for bioethanol production using dilute acid presoaking combined with wet explosion pre-treatment and enzymatic treatment. *Bioresource Technology*, Vol. 99, No. 14, (September 2008), pp. 6602–6607, ISSN 0960-8524

Sun, R.C.; Fang, J.M. & Tomkinson, J. (2000). Delignification of rye straw using hydrogen peroxide. *Industrial Crops and Products*, Vol. 12, No.2, pp. 71–83, ISSN 0926-6690

Sun, Y.; Lin, L.; Pang, C.; Deng, H.; Peng, H.; Li, J.; He, B. & Liu, S. (2007). Hydrolysis of Cotton Fiber Cellulose in Formic Acid. *Energy & Fuels*, Vol. 21, No. 4, (June 2007), pp. 2386-2389, ISSN 0887-0624

Sun, Y.; Zhang, B.; Lin, L. & Liu, S. (2010). Clean conversion to fermentable glucose from wheat straw. Journal of Biobased Materials and Bioenergy, Vo. 4, No. 1, (March 2010), pp. 27-34, ISSN 1556-6579

Smeets, E. & Faaij, A. (2007). Bioenergy Potentials from Forestry in 2050. *Climatic Change*, Vol. 81, (November 2006), pp. 353-390, ISSN 0165-0009

Tao, F.; Song, H. & Chou, L. (2011). Efficient process for the conversion of xylose to furfural with acidic ionic liquid. *Canadian Journal of Chemistry*, Vol. 89, No. 1, (January 2011), pp. 83-87, ISSN 1480-3291

Vitz, J.; Yevlampieva, N.P.; Rjumtsev, E. & Schubert, U.S. (2010). Cellulose molecular properties in 1-alkyl-3-methylimidazolium-based ionic liquid mixtures with pyridine. Carbohydrate Polymers, Vol. 82, No. 4, (July 2010), pp. 1046-1053, ISSN 0144-8617

Vlasenko, E. Y.; Ding, H.; Labavitch, J. M. & Shoemaker, S. P. (1997). Enzymic hydrolysis of pretreated rice straw. *Bioresource Technology*, Vol. 59, No. 2-3, pp. 109-119, ISSN 0960-8524

Wang, G.S.; Pan, X.J.; Zhu, J.Y. & Gleisner, R.L. (2008). Sulfite pretreatment for biorefining biomass. US Patent Application. Filed by Wisconsin Alumni Research Foundation (WARF) 61/047024. Provisional application 4/22/2008. Utility Conversion, 20090298149, April 2009

Wu, M.M.; Chang, K.; Gregg, D. J.; Boussaid, A.; Beatson, R. P. & Saddler, J. N. (1999). Optimization of steam explosion to enhance hemicellulose recovery and enzymatic hydrolysis of cellulose in softwoods. *Applied Biochemistry and Biotechnology*, Vo. 77-79, No. 1-3, (March 1999), pp. 47-54, ISSN ***1559-0291***

Xu, F.; Sun, J.; Sun, R.; Fowler, P. & Baird, M.S. (2006). Comparative study of organosolv lignins from wheat straw. *Industrial Crops and Products*, Vol. 23, No. 2, (March 2006), pp. 180–193, ISSN 0926-6690

Xu, F.; Sun, R.C.; Zhai, M.Z.; Sun, J.X.; Jiang, J.X. & Zhao, G.J. (2008). Comparative study of three lignin fractions isolated from mild ball-milled Tamarix austromogoliac and Caragana sepium. *Journal of Applied Polymers Science*, Vol. 108, No. 2, (April 2008), pp. 1158–1168, ISSN 0021-8995

Yamashita, Y.; Sasaki, C. & Nakamura, Y. (2010). Effective enzyme saccharification and ethanol production from Japanese cedar using various pretreatment methods. *Journal of Bioscience and Bioengineering*, Vol. 110, No. 1, (July 2010), pp. 79–86, ISSN 1389-1723

Yildiz, S. & Gezerb, E.D. (2006). Mechanical and chemical behavior of spruce wood modified by heat. *Building and Environment*, Vol. 41, No. 12, (December 2006), pp. 1762–1766, ISSN 0360-1323

Youngblood, A.; Zhu, J.Y. & Scott, C.T. (2009). Ethanol Production from Woody Biomass: Silvicultural Opportunities for Suppressed Western Conifers. 2009 National Silviculture Workshop, June 2009, Boise, Idaho

Zhang, J., Chu, D.; Huang, J.; Yu, Z.; Dai, G. & Bao, J., (2009). Simultaneous saccharification and ethanol fermentation at high corn stover solids loading in a helical stirring bioreactor. *Biotechnology and Bioengineering*, Vol. 105, No. 4, (March 2010), pp. 718-728, ISSN 1097-0290

Zhang, W. & Woelk, K. (2010). NMR Investigations into hydrothermal biomass-to-fuel (BTF) conversion using glucose as model substrates for cellulosic biomass. 45th Midwest Regional Meeting of the American Chemical Society, Wichita, KS, United States, October 27-30, MWRM-417

Zhu, J.Y.; Pan, X.J.; Wang, G.S. & Gleisner, R., (2009). Sulfite pretreatment for robust enzymatic saccharification of spruce and red pine. *Bioresource Technology*, Vol. 100, No. 8, (April 2009), pp. 2411–2418, ISSN 0960-8524

Zhu, J.Y. & Pan X.J. (2010). Woody biomass pretreatment for cellulosic ethanol production: Technology and energy consumption evaluation. *Bioresource Technology*, Vol. 101, No. 13, (July 2010), pp. 4992–5002, ISSN 0960-8524

Zhuang, Q. & Vidal, P.F. (1997). Hemicelluloses solubilization from Populus tremuloides via steam explosion and characterization of the water-soluble fraction. II. Alkali-catalytic process. *Cellulose Chemistry and Technology* (1997), Vol. 31, No. 1-2, pp. 37-49, ISSN 0576-9787

30

Biomethanol Production from Forage Grasses, Trees, and Crop Residues

Hitoshi Nakagawa et al.*
Biomass Research and Development Center
National Agriculture and Food Research Organization (NARO)
Japan

1. Introduction

About 12 billion tons of fossil fuels (oil equivalent) are consumed in the world in 2007 (OECD 2010) and these fuels influence the production of acid rain, photochemical smog, and the increase of atmospheric carbon dioxide (CO_2). Researchers warn that the rise in the earth's temperature resulting from increasing atmospheric concentrations of CO_2 is likely to be at least 1°C and perhaps as much as 4°C if the CO_2 concentration doubles from pre-industrial levels during the 21st century (Brown *et al.* 2000). A second global problem is the likely depletion of fossil fuels in several decades even though new oil resources are being discovered. To address these issues, we need to identify alternative fuel resources.

Stabilizing the earth's climate depends on reducing carbon emissions by shifting from fossil fuels to the direct or indirect use of solar energy. Among the latter, utilization of biofuel is most beneficial because; 1) the solar energy that produces biomass is the final sustainable energy resource; 2) it reduces atmospheric CO_2 through photosynthesis and carbon sequestration; 3) even though combustion produces CO_2, it does not increase total global CO_2; 4) liquid fuels, especially bioethanol and biomethanol, provide petroleum fuel alternatives for various engines and machines; 5) it can be managed to eliminate output of soot and SO_x; and 6) in terms of storage, it ranks second to petroleum and is far easier to store than batteries, natural gas and hydrogen.

Utilization of biomass to date has been very limited and has primarily included burning wood and the production of bioethanol from sugarcane in Brazil or maize in the USA. The necessary raw materials for bioethanol production by fermentation are obtained from crop plants with high sugar or high starch content. Since these crops are primary sources of human nutrition, we cannot use them indiscriminately for biofuel production when the

---

* Masayasu Sakai[2], Toshirou Harada[3], Toshimitsu Ichinose[4], Keiji Takeno[4], Shinji Matsumoto[4], Makoto Kobayashi[5], Keigo Matsumoto[4] and Kenichi Yakushido[6]
[2]*Nagasaki Institute of Applied Science*
[3]*Forestry and Forest Production Research Institute*
[4]*Nagasaki Research and Development Center, Mitsubishi Heavy Industries ltd.*
[5]*National Institute of Livestock and Grassland Science, NARO*
[6]*National Agricultural Research Center, NARO*
Japan

demand for food keeps increasing as global population increases. Although fermentation of lignocellulosic materials, such as wood of poplar (*Populus* spp.) (Wyman *et al.* 2009), switchgrass (*Panicum virgatum*) (Keshwani and Cheng 2009) and Miscanthus (*Miscanthus* spp.) (Sørensen *et al.* 2008), straw of rice (*Oryza sativa*) (Binod 2010), old trunks of oil palm (*Elaeis guineensis*) (Kosugi *et al.* 2010) are being attempted by improving pre-treatment of the materials, yeast and enzymes, establishment of the technology with low cost and high ethanol yield will be required. Recently, a new method of gasification by partial oxidation and production of biomethanol from carbohydrate resources has been developed (Sakai 2001). This process enables any source of biomass to be used as a raw material for biomethanol production. We report on the estimated gas mixture and methanol yield using this new technology for biofuel production from gasification of diverse biomass resources, such as wood, forages, and crop residues etc. Data obtained from test plant operation is also provided.

2. Gasification technology and the test plants

The idea and technology of gasification systems that generate soot and tar is not new. Our methods of gasification technology through partial oxidation and implementation of a new high calorie gasification technology, has been developed focusing on the perfect gasification at 900-1,000°C without the production of soot and tar. The result of these technologies is the production of a superior mixture of biogases for producing liquid biofuels through thermo-chemical reaction with Zn/Cu-based catalyst or electricity through generator. The first test plant, named "Norin Green No. 1 (the "Norin" means Ministry of Agriculture, Forestry and Fisheries in Japanese; later renamed as "Norin Biomass No. 1")" was completed on April 18, 2002 and second plant with a new high calorie gasification technology, named "Norin Biomass No. 3" was completed in March in 2004.

2.1 Gasification technology of partial oxidation

Figure 1 shows the concept of our new method of gasification by partial oxidation. This production of biomethanol from carbohydrate (Sakai 2001) has been given the term "C1

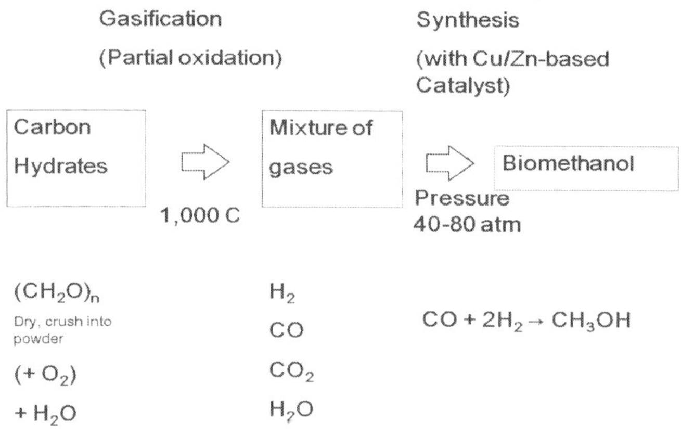

Fig. 1. Principle of methanol synthesis by gasification method (the C1 chemical transformation technology)

chemical transformation technology". In this process, the biomass feedstock must be dried and crushed into powder (ca. 1mm in diameter). When the crushed materials are gasified at 900-1000°C with gasifying agent (steam and oxygen), all carbohydrates are transformed to hydrogen (H_2), carbon monoxide (CO), carbon dioxide (CO_2) and vapor (H_2O). The mixture of gases is readily utilized for generating electricity. The mixture of gases is transformed by thermo-chemical reaction to biomethanol under pressure (40-80 atm) with Cu/Zn-based catalyst, too. That is,

$$CO + 2H_2 \rightleftarrows CH_3OH + Q \text{ (Radiation of heat)}$$

$$CO_2 + 3H_2 \rightleftarrows CH_3OH + H_2O + Q \text{ (Radiation of heat)}$$

All the ash contained in the materials is collected in the process (Fig. 2). This process enables any source of biomass to be used as a raw material for biomethanol production.

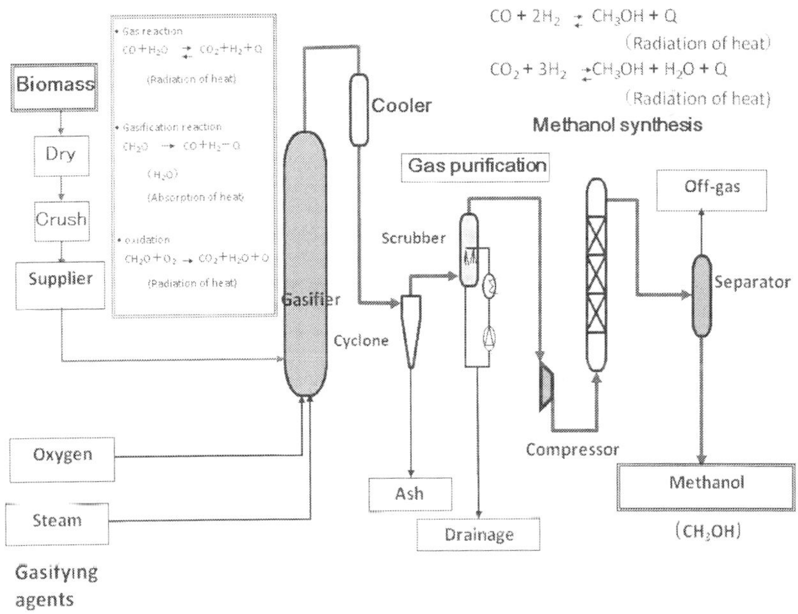

Fig. 2. Gasification and biomethanol synthesis system (Nakagawa *et al.* 2007)

2.1.1 Materials and methods

Twenty materials were tested: 1) sawdust (wood of Japanese cedar (*Cryptomeria japonica*), without bark, was isolated by passing through a 2 mm mesh sieve); 2) rice bran (*Oryza sativa*: cv. Koshihikari); 3) rice straw (cv. Yumehitachi: only the inflorescences are harvested in September and the plants were left in the field until cutting in December); 4) rice husks (cv. Koshihikari: rice was threshed in October and kept in plastic bags following typical post-harvest practices); 5) sorghum heads (*Sorghum bicolor*: var. Chugoku Kou 34 (medium maturing hybrid line between a male sterile grain sorghum line and sudangrass; with mature seeds at ripened stage); 6) leaf and stem of sorghum (*ibid.*; at ripened stage, cut to a length of 30 cm and dried in a dryer for 7 days at 70°C); 7) total plant of sorghum (*ibid.*); 8) sorghum (cv. Kazetachi; extremely late maturing dwarf type; before flowering); 9) sorghum (cv. Ultra sorgo; late maturing tall type; heading stage); 10) sorghum (cv. Green A; medium maturing hybrid between sudangrass and grain sorghum; heading stage); 11) sorghum (cv. Big Sugar: late maturing tall sweet sorghum: milk-ripe stage); 12) guineagrass (*Panicum maximum* cv. Natsukaze ; heading stage); 13) rye (*Secale cereale* cv. Haru-ichiban; heading stage); 14) Japanese lawngrass (*Zoysia japonica* cv. Asamoe; before flowering); 15) *Erianthus* sp. Line NS-1; heading stage; 16) bark of Japanese cedar; 17) chipped Japanese larch (*Larix leptolepis*); 18) bamboo (*Phyllostachys pubescens*); 19) salix (*Salix sachalinensis* and *S. pet-susu*); 20) cut waste wood: sawn wood and demolition waste (raw material for particle board).

Characteristics important for gasification were evaluated for the above materials: 1) Water content and ash were measured following drying at 107 ± 10°C for 1 hour; then followed by combustion at 825 ± 10°C for 1 hour; 2) Percent carbon (C), hydrogen (H), oxygen (O), nitrogen (N), total sulfur (T-S), and total chloride (T-Cl): C and H weights were estimated by CO_2 and H_2O weight after combustion at 1,000 ± 10°C by adding oxygen. The estimate of O was calculated by the equation, O = 100 - (C + H + T-S + T-Cl); estimates of N were determined by the amount of ammonia produced by oxidation with sulfuric acid to generate ammonium sulfate. Following distillation, total sulfur was estimated by SO_2 following combustion at 1,350°C with oxygen. Total chloride was estimated by the water soluble remains following combustion with reagent and absorption of the gas; 3) The higher heating values were measured by the rise in temperature in water from all the heat generated through combustion. The lower heating value was estimated by the calculation (the higher heating value - (9×h+w)×5.9) [h: hydrogen content (%); w: water content (%)]; 4) Chemical composition (molecular) of the biomass was calculated based on molecular weight of the elements; 5) Size distribution of the various biomass types was measured (diameter, density of materials [g/ml]); 6) Gas yield and generated heat gas were estimated by the process calculation on the basis of chemical composition and the heating value. Heat yield or cold gas efficiency was calculated by (total heating value of synthesized gases)/(total heating value of supplied biomass); 7) The weight and calories generated as methanol, given a production gasifier capacity of 100 tons dry biomass/day, were estimated by the process calculation. These data were obtained in different years.

2.1.2 Results and discussion

Water and ash content for some materials evaluated are shown in Fig. 3. The materials were prepared in various ways. Water contents ranged from 3.4% (wood waste) to 13.1% (bark). Water content of sorghum was low (4.6%) because this material was dried in a mechanical drier. The other materials were not mechanically dried and the water content averaged ca. 10%. Although individual elements are not reported, the ash content of wood materials, such as sawdust, bark, chip, and bamboo was very low, 0.3% for sawdust, 1.8% for bark, and 2.2% for wood waste. Although the ash content of rice straw and husks was very high (22.6% and 14.6%), probably due to the high Si content of rice plants, the ash content of rice bran was much lower (8.1%). The ash content of sorghum plant was 5.8%.

The percent by weight of some elements in the raw materials are shown in Fig. 4 and Table 1. Carbon content was high in wood materials and averaged 48.3% for wood waste and 51.8% for bark. Rice bran carbon content was 48.3% and sorghum carbon content was ca.

45%. Carbon content of rice straw and husks were lower at 36.9 and 40.0%, respectively. Four sorghum cultivars with different plant types exhibited a narrow range of carbon content (45.5 - 46.1%). Carbon content of the sorghum heads (with seeds), is higher than leaf and stem of sorghum (with lignin) by 2.3%. Rye, Japanese lawngrass and Erianthus exhibited slightly higher carbon content and guineagrass was at the lower end of the range.

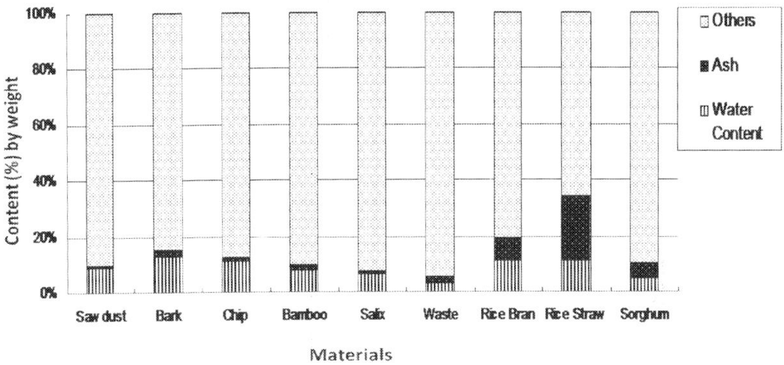

The numbers of materials are same as those in Materials and Methods. Saw dust (1); Bark (16); Chip (17); Bamboo (18); Salix (19); Waste (20); Rice Bran (2); Rice straw (3); sorghum (7).

Fig. 3. Content of water and ash in materials (Nakagawa *et al.* 2007).

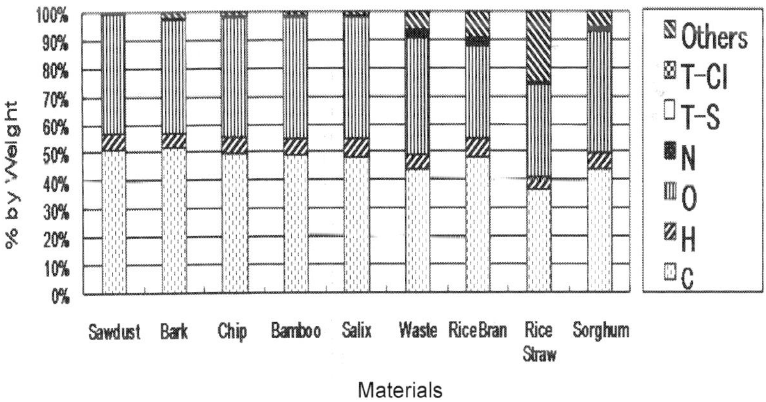

The numbers of materials are same as those in Materials and Methods. C: carbon; H: hydrogen; O: oxygen; N: nitrogen; T-S: total sulfur; T-Cl: total chloride; Saw dust (1); Bark (16); Chip (17); Bamboo (18); Salix (19); Waste (20); Rice Bran (2); Rice straw (3); sorghum (7)

Fig. 4. Content of some elements in materials without water (% by weight) (Nakagawa *et al.* 2007).

Hydrogen content ranged from 4.7 to 7.0% for rice straw and rice bran, respectively. Although rice bran had the highest hydrogen content, the others were only marginally different and the range of wood materials was narrow (from 5.6 to 5.9% for bark and salix, respectively). Oxygen content ranged between 32.5% and 43.9% for rice straw and salix, respectively with wood materials and sorghum in the higher range. Nitrogen content was between 0.12% (sawdust) and 2.44% (rice bran), with wood materials exhibiting low values except for wood waste (1.92%). Nitrogen contents of sorghum cultivars ranged from 0.80 to 1.30 % and sorghum heads exhibited 1.68%. The sulfur content was very low in all of the materials and ranged between 0.02% (sawdust) and 0.30% (Japanese lawngrass). Chlorine content ranged from 0.01% (sawdust) to 1.31% (rye). These data demonstrates that these materials are much cleaner than coal and other fossil fuels and, we expect chemical properties of harvested tropical grasses to be similar to the grasses used in this report.

Biomass Materials	C	H	O	N	T-Cl	T-S	Ash
Sawdust (1)	51.1	5.9	42.5	0.12	0.01	0.02	0.3
Rice bran (2)	48.3	7.0	33.0	2.44	0.05	0.21	8.1
Rice straw (3)	36.9	4.7	32.5	0.30	0.08	0.06	22.6
Rice husk (4)	40.0	5.2	37.3	0.76	0.41	0.22	14.6
Sorghum Head(5)	46.7	6.1	40.7	1.68	0.08	0.14	4.3
Leaf and Stem of Sorghum (6)	44.4	5.8	42.9	0.45	0.23	0.15	5.8
Sorghum 'Kazetachi' (8)	45.7	5.8	39.5	1.30	0.78	0.08	6.2
Sorghum 'Ultra sorgo' (9)	45.5	5.7	41.6	0.80	0.79	0.03	5.4
Sorghum 'Green A' (10)	46.1	5.7	40.6	1.20	0.57	0.04	5.5
Sorghum 'Big sugar' (11)	45.9	5.7	41.2	1.00	0.50	0.05	5.4
Guineagrass (12)	42.8	5.4	37.9	1.50	0.89	0.11	10.4
Rye 'Haruichiban' (13)	45.7	5.8	39.2	1.40	1.21	0.07	6.2
Japanese lawngrass 'Asamoe' (14)	46.4	6.1	37.9	2.15	0.43	0.30	6.4
Erianthus sp. Line 'NS-1' (15)	47.1	6.1	42.3	0.80	0	0.22	3.5

C: carbon, H: hydrogen, O: oxygen, N: nitrogen, T-Cl: total chloride, T-S: Total sulfur

Table 1. Content of some elements in dry matter (% by weight). (The numbers of materials are same as those in Materials and Methods)

The higher and lower heating values of materials are shown in Fig. 5 and Table 2. Among the materials tested, the higher heating values of wood materials were high and ranged between 4,570 kcal/kg (sawdust: 19.13 MJ/kg) and 4,320 kcal/kg (bark: 18.08 MJ/kg). Rice bran was also high (4,520 kcal/kg: 18.92 MJ/kg), although rice straw and husks were at the low end, 3,080 kcal/kg (12.89 MJ/kg) and 3,390 kcal/kg (14.19 MJ/kg), respectively. The higher heating value of total sorghum plant of Chugoku Kou 34 was intermediate among the materials evaluated and 3,940 kcal/kg. Sorghum cultivars exhibited mostly similar higher heating value of 17.4 MJ/kg.

Molecular ratios of C, H and O in various materials are shown in Table 3. Most of the materials had similar ratios for $C_nH_2O_m$ (n between 1.28 and 1.54, and m between 0.87 and 0.93) except for rice bran which contains considerable quantities of lipid resulting in an n =

1.15 and m = 0.59. This ratio is important since it will affect the condition of gasification when oxygen and steam are added as gasifying agents.

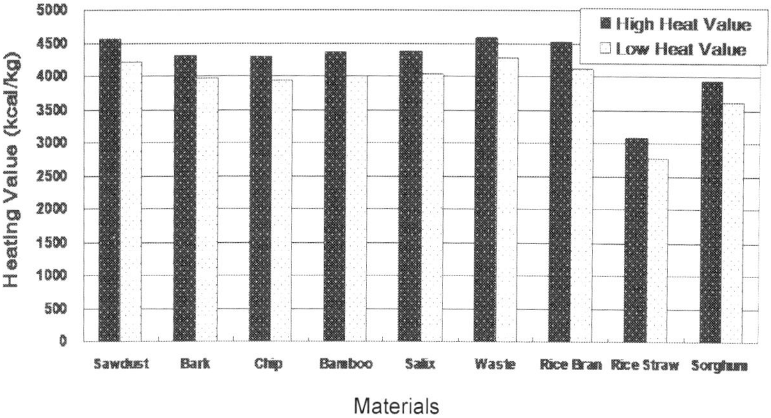

The numbers of materials are same as those in Materials and Methods. Sawdust (1) ; Bark (16); Chip (17); Bamboo (18); Salix (19); Waste (20); Rice Bran (2); Rice straw (3); sorghum (7)

Fig. 5. Higher and lower heating value of materials (Nakagawa *et al.* 2007).

Biomass Materials	HHV (MJ/kg)	LHV (MJ/kg)
Sawdust (1)	19.13	17.66
Bark of Japanese Cedar (16)	18.08	16.65
Waste Wood (20)	19.08	17.91
Rice Bran (2)	18.92	17.25
Rice Straw (3)	12.89	11.64
Rice Husk (4)	14.19	12.89
Sorghum Head(5)	17.41	15.99
Leaf and Stem of Sorghum (6)	16.49	15.15
Sorghum 'Kazetachi' (8)	17.04	15.56
Sorghum 'Ultra sorgo' (9)	17.50	16.12
Sorghum 'Green A' (10)	17.41	16.03
Sorghum 'Big sugar' (11)	17.45	16.11
Guineagrass (12)	15.82	14.48
Rye 'Haruichiban' (13)	17.58	15.57
Japanese Lawngrass cv. 'Asamoe' (14)	18.59	17.17
Erianthus sp. Line 'NS-1' (15)	18.56	17.16

Table 2. Higher and lower heating value of Materials. (The numbers of materials are same as those in Materials and Methods)

Material	C (n)	H	O (m)
Sawdust (1)	1.44	2	0.90
Bark (16)	1.54	2	0.90
Chips (17)	1.39	2	0.88
Bamboo (18)	1.42	2	0.93
Salix (19)	1.38	2	0.93
Waste (20)	1.42	2	0.90
Rice Bran (2)	1.15	2	0.59
Rice Straw (3)	1.31	2	0.87
Sorghum (7)	1.28	2	0.93

Table 3. Molecular ratios of C, H, O ($C_nH_2O_m$) in the materials (The numbers of materials are same as those in Materials and Methods)

Estimated volume percent for each gas in the gas mixtures produced from various materials using the gasification by partial oxidation process are shown in Fig. 6 and Table 4. In the mixture of produced gases, contents of hydrogen (H_2) and carbon monoxide (CO) are the most important compounds for methanol production. Although the variation across values is small, H_2 percentage and CO percentage are high in wood materials, ranging from 46.8% for bark, 47.9% for wood waste, 47.3% for salix, and 47.7% for sawdust, respectively, for H_2, and 18.3% for bark, 18.2% for wood waste, 17.9% for salix, and 18.7% for sawdust, respectively, for CO. The H_2 percentage of rice straw and husks was the same (44.7%) and CO percentage was 17.1% and 17.3%, respectively. H_2 and CO values of sorghum were intermediate among these materials tested, and those of sorghum varieties have narrow range such as 46.3-47.0% for H_2 and 17.7-17.8% for CO. Japanese lawngrass and Erianthus exhibited a slightly elevated CO percentage than sorghum and approximated that of the wood materials.

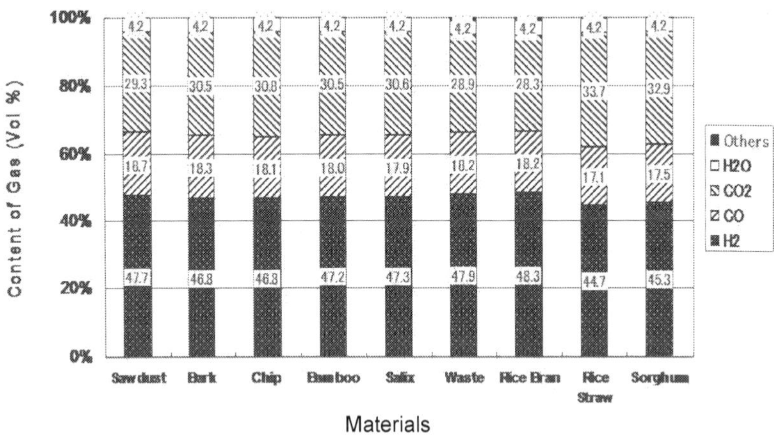

The numbers of materials are same as those in Materials and Methods Sawdust (1) ; Bark (16); Chip (17); Bamboo (18); Salix (19); Waste (20); Rice Bran (2); Rice straw (3); sorghum (7)

Fig. 6. Volume % of H_2, CO, CO_2, H_2O and other gasses produced from some materials (before the entrance to methanol synthetic system) (Nakagawa *et al.* 2007) .

Biomass Materials	H_2 (vol%)	CO (vol%)	CO_2 (vol%)	H_2O (vol%)	HHV (MJ/kg -wet)
Sawdust (1)	47.7	18.7	29.3	4.2	9.54
Rice Bran (2)	48.3	18.2	28.3	4.2	9.67
Rice Straw (3)	44.7	17.1	33.7	4.2	8.25
Rice Husk (4)	44.7	17.3	33.4	4.2	8.29
Sorghum Head (5)	46.1	17.7	31.3	4.2	8.83
Leaf and Stem of Sorghum (6)	45.3	17.5	32.9	4.2	8.50
Sorghum 'Kazetachi' (8)	46.3	17.7	31.1	4.2	8.92
Sorghum 'Ultra sorgo' (9)	47.0	17.7	30.7	4.2	9.13
Sorghum 'Green A' (10)	46.7	17.8	30.8	4.2	9.00
Sorghum 'Big sugar' (11)	46.6	17.7	31.1	4.2	8.96
Guineagrass (12)	46.2	17.7	31.1	4.2	8.92
Rye 'Haruichiban' (13)	46.9	17.8	30.5	4.2	9.13
Japanese Lawngrass 'Asamoe' (14)	47.6	18.0	29.3	4.2	9.00
Erianthus sp. Line 'NS-1' (15)	46.6	18.6	30.3	4.2	8.96

HHV: higher heating value.

Table 4. Composition and higher heating value of product gas derived from biomass gasification for methanol synthesis. (The numbers of materials are same as those in Materials and Methods)

The estimated methanol yield by weight and by heating value for each material tested, calculated from the contents of the gas mixtures produced by gasification, are shown in Fig. 7 and Table 5. The values are correlated to carbon content and heat emission. Wood materials exhibited high methanol yield by weight and ranged from 53.0% (salix) to 55.8% (sawdust) based on dry biomass weight. This means that 530 kg and 558 kg of methanol will be produced through the gasification of 1 ton of dry salix and sawdust, respectively. Rice bran also exhibited a high methanol yield potential (ca. 55%) but rice straw and rice husks had considerably lower potentials, 35.8% and 39.4%, respectively. Sorghum varieties with different plant types exhibited similar potentials (ca. 47 - 49%). Interestingly, methanol yield potential of sorghum heads (with starch of the grain) exhibited 48.6% and higher than that of leaf and stem of sorghum (with higher amount of lignin) at the ripening stage (44.1%) but the difference is only 4.5%. This means that there is little difference in methanol yield between starch and lignin. This suggests that there may be little need to utilize food starch resources and competitive methanol yields can be generated through the utilization of crude lignocellulosic materials. This indicates that significant levels of methanol can be produced from previously cast-off residues and byproducts of agriculture and forest industries. Japanese lawngrass and Erianthus exhibited higher methanol yield potentials than sorghum cultivars. Although estimated methanol yield by weight differed among materials, the estimated heat yield of 54 - 59% by heating value was rather constant in the different materials. Heat yield of the various materials tested, regardless of their heating values, was high and demonstrate the efficiency of this technology.

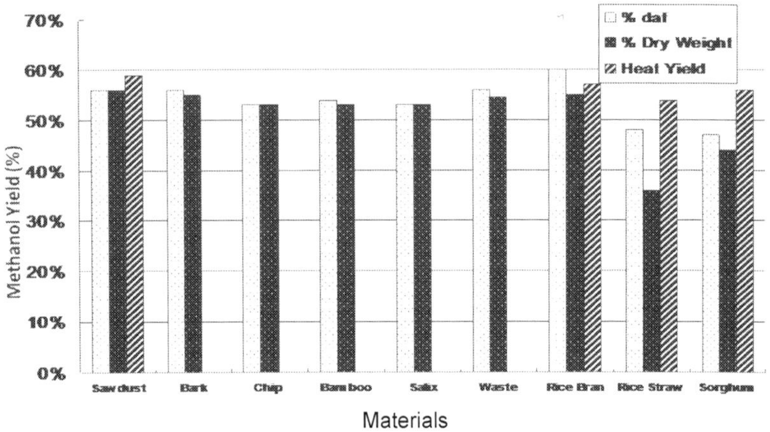

The numbers of materials are same as those in Materials and Methods.Sawdust (1) ; Bark (16); Chip (17); Bamboo (18); Salix (19); Waste (20); Rice Bran (2); Rice straw (3); sorghum (7) daf: percentage of methanol weight to dry biomass weight without ash

Fig. 7. Estimated methanol yield (weight %) and heat yield of various biomass materials (Nakagawa *et al.* 2007).

Biomass Materials	methanol yield	
	daf	dry
Sawdust (1)	56.0	55.8
Rice Bran (2)	60.0	54.5
Rice Straw (3)	48.0	35.8
Rice Husk (4)	47.0	39.4
Sorghum Head (5)	51.0	48.6
Leaf and Stem of Sorghum (6)	47.0	44.1
Sorghum 'Kazetachi' (8)	51.1	47.7
Sorghum 'Ultra sorgo' (9)	51.3	48.4
Sorghum 'Green A' (10)	52.0	49.0
Sorghum 'Big sugar' (11)	51.0	48.1
Guineagrass (12)	51.0	45.2
Rye 'Haruichiban' (13)	55.9	52.2
Japanese Lawngrass 'Asamoe' (14)	55.1	51.4
Erianthus sp. Line 'NS-1' (15)	53.4	51.5

daf: based on dry ash free biomass weight, dry: based on total dry biomass weight with ash

Table 5. Estimated methanol yield (% by weight) (The numbers of materials are same as those in Materials and Methods).

For perfect gasification of any biomass material, it is necessary to convert the materials into ca. 0.1-0.9 mm in diameter powder through micro-crushing. The physical characteristics of the raw materials and the handling procedures needed to prepare these raw materials for biomethanol production are shown in Table 6. As rice bran is very fine, there was no need for any prior preparation. Although the diameter of sawdust is ca. 0.8 mm, we can utilize the powder smaller than ca. 1 mm directly for the gasification. Though the rice straw dried in the field was long in length, it required only micro-crushing without extra-drying in a dryer. Sorghum was harvested at the ripened stage with sickles, cut to a length of 30 cm and dried in a dryer. This procedure made this material very hard to process and both rough-crushing (1.0-3.0 mm) and micro-crushing were needed to prepare sorghum for gasification. Usually, a mechanical harvester is used to cut sorghum plants into lengths of less than 10 cm. This latter harvest method will require much less subsequent preparation than was needed in this study.

Material	Size (mm)		Density	Handling
	Diameter	Length	(g/ml)	Characteristics
Sawdust (1)	0.78	-	0.07	No micro-crushing needed
Rice Bran (2)	0.31	-	0.31	No micro-crushing needed
Rice Straw (3)	3.0-4.0	400	-	Micro-crushing needed
Rice Husk (4)	2.05	-	0.11	Micro-crushing needed
Sorghum head (5)	15.0	250	-	Rough- and micro-crushing needed
Leaf and Stem of Sorghum (6)	20 (width) 0.5 (thickness)	300	-	Rough- and micro-crushing needed
Sorghum (7)	7.9	50	0.07	Rough- and micro-crushing needed

Table 6. Size and handling characteristics of various materials (The numbers of materials are same as those in Materials and Methods) (Nakagawa *et al.*, 2000) .

2.1.3 "Norin Green No. 1" test plant

The test plant, named "Norin Green No. 1" (Fig. 8), was utilized to obtain data for heat yield and methanol yield through the gasification and biomethanol synthesis system shown in Fig. 2. The test plant comprises a supplier of crushed biomass, a gasifier for gasification, and an apparatus for gas purification and methanol synthesis by the use of a Cu/Zn-based catalyst. The practical methanol yield of crushed waste wood (ca. 1 mm in diameter) produced by pin-type-mill was also measured by operating both "Norin Green No. 1" test plant at Nagasaki Research and Development Center, Mitsubishi Heavy Industries ltd. with a gasifier capacity of 240 kg dry biomass/day and another test plant at Kawagoe Power Station of Chubu Electric Power Co., Inc. with a gasifier capacity of 2 t dry biomass/day (Matsumoto *et al.* 2008).

Table 7 indicates the heat yield and methanol yield of the two test plants (the test plant gasifier can process 240 kg/day (Norin Green No. 1 test plant) or 2t/day (a test plant constructed by Chubu Electric Power Co., Inc) of dry biomass) by operating these plants, when crushed waste wood is utilized as a raw material, and the estimated capacity of a commercial scale plant (a gasifier capable of processing 100 t/day of dry biomass). The commercial scale plant would be large enough (larger than 100 t/day) to maintain critical temperature (900 to 1,000°C) within the gasifier by adding the raw materials into the gasifier without the addition of supplemental heat. Our data indicate that the estimated heat yield of

methanol production by commercial scale plants is 54 - 59 % (Fig. 7). However, the real heat yield of a commercial scale plant after reducing the energy needed for crushing of the biomass (1.0 - 5.0 % of the quantity of heat; biomass feedstock with 2-3 mm in diameter is available), operation of the plant (5 - 10 %), and heat loss from the surface of the gasifier (1-2 %), estimated by simulation using the test plant data will be ca. 40 %.
The cold gas efficiency, that represents a percentage of the total heating value of synthesized gases by gasification divided by the total heating value of supplied biomass of the test plant, varies from 65 to 70% and methanol yield varied from 9 to 13% in the "Norin Green No. 1" test plant. Heat yield and methanol yield of another test plant capable of processing 2 t/day constructed by Chubu Electric Power Co., Inc. with the support of New Energy and Industrial Technology Development Organization (NEDO) Project, has, however, achieved ca. 20% of methanol yield by weight by its operation (Ishii *et al.* 2005; Matsumoto *et al.* 2008; Ogi *et al.* 2008).

Item	Test Plant		Practical Plant
(Gasifier Size: Dry biomass to be processed)	(240kg/day)	(2t/day)	(100t/day)
Heat Yield (Heating Value %)	60-70%	65%	70-75%
Methanol Yield (by weight)	9-13%	20%	40-50%

Table 7. Ability of test plants and estimated practical plant.

2.1.4 Conclusion

The above data indicate that the most important advantage of this technology is that it can utilize any form of biomass feedstock for H_2 and CO generation. The mixture of gases can be utilized as direct burning both for heat and for electricity generation. High yields of methanol will be efficiently produced from the mixture of gases by using a Cu/Zn-based catalyst. The disadvantage of the system is with the size of the processing facility. The larger the plant, the higher the efficiency. The biomethanol yield from a 100 t/day gasifier would be more than twice that of the 2 t/day gasifier from the same raw materials. Although it is feasible to construct a biomethanol plant of this size, it may be very difficult to collect and provide the required 100 dry matter tons of biomass each day for the operation; with the possible exception of large sugarcane mills and palm oil industry in Southeast Asian countries. In addition, required permits and a license of boiler operation to operate such a large-scale gasifier in Japan, would add additional costs.
Prof. Sakai, of the Nagasaki Institute of Applied Science, one of the authors of this report, has developed another type of plant in the university, named "Norin Biomass No. 3" test plant (Fig. 9) by using another gasification technology, called as "high-calorie gasification reaction", that is introduced in the following sentence.

2.2 "Norin Biomass No. 3" test plant

The "Norin Biomass No. 3" Test Plant (Fig. 9, 10), which represents a "Suspension/external heating type" with a boiler (1-2 Dry t/day) was newly developed for improving the disadvantages associated with the "Norin Green No. 1" test plant through the introduction of a new type of gasification called "high-calorie gasification reaction".

Fig. 8. "Norin Green No. 1", a test plant of gasification and biomethanol production, located in Nagasaki Research & Development Center, Mitsubishi Heavy Industries ltd, Nagasaki, Japan.

Fig. 9. "Norin Biomass No. 3" test plant located in Sakai Kougyo, ltd., Isahaya City, Nagasaki, Japan.

2.2.1 Essential of the high-calorie gasification reaction

In the high-calorie gasification method, finely crushed biomass of 1-3 mm in diameter is subjected to the gasification reaction together with steam in an atmosphere of 800-1000°C within the reaction tube. At this time, the reaction tube is heated using high-temperature combustion gas that is separately combusted using additional biomass. The introduced biomass raw materials leave only ash, and all organic content is gasified, resulting in a clean, high-calorie gas (ca. 12MJ/Nm^3) composed of H_2, CO, CH_4, etc. The basic principle of the technology is illustrated in Fig. 10.

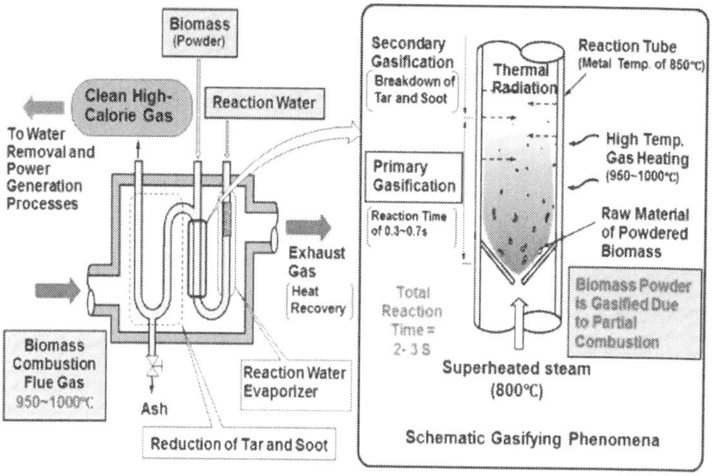

Fig. 10. Suspension/external heating type high-calorie gasification.

The gas composition varies with gasification reaction conditions such as reaction temperature, residence time (reaction time), and the [steam]/[biomass carbon] mode ratio, but an example is represented by the following reaction formula.

(Endothermic reaction)

$C_{1.3}H_2O_{0.9} + 0.4\ H_2O \rightarrow 0.8H_2 + 0.7CO + 0.3CH_4 + 0.3CO_2 + 165.9kJ/mol$

(Biomass) (Steam) (Resulting gas fuel) (Absorbed heat)

In this process, the total biomass material reacts with steam and is converted to an [H_2, CO, CH_4, CO_2] gas mixture. The application of external heat is required due to the fact that the gasification reaction is endothermic. However, the potential heat stored in the gas mixture generated in the reaction is greater than that contained in the raw biomass material, such that the cold gas efficiency surpasses 100%. In the formula shown above, a figure of ca. 115% is obtained by solving for cold gas efficiency (Ec). On the other hand, the externally supplied heat used in the reaction, is not considered in the calculation of Ec, and the total gasification efficiency is ca. 85% when this external heat is taken into account.

While the previous biomass gasification technology of "Norin Green No. 1" test plant mentioned above uses the partial oxidation technology, this high-calorie gasification technology enables the production of a high-calorie gas fuel that was not possible with the conventional method due to the formation of an exhaust gas. The principle is illustrated in Fig. 11.

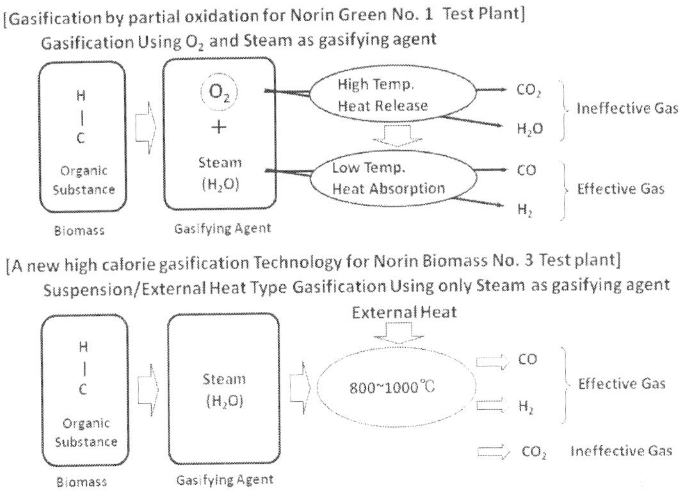

Fig. 11. Principle of a new high calorie gasification technology compared with gasification by partial oxidation technology.

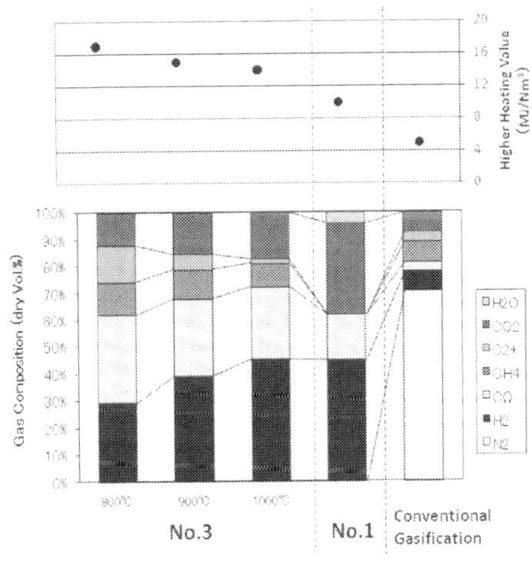

Fig. 12. Composition (bar chart) and higher heating value (line chart) of high calorie gasification gas (No. 3), partial oxidation (No. 1) and conventional gas generated by partial oxidation with air as gasifying agent.
No.1: "Norin Green No. 1": gasification by partial oxidation using O_2 and H_2O as gasifying agents.
No. 3: "Norin Biomass No. 3": high calorie gasification using only H_2O as a gasifying agent at 800, 900, and 1000°C
Conventional gasification: gasification using air as a gasifying agent.

2.2.2 Basic high-calorie gasification reaction

According to the results obtained by the operation, it has been confirmed that the output gas mixture possesses the properties indicated in Fig. 12. As shown in the figure, high-calorie gas featuring 15-18MJ/Nm³, that could not be achieved by gasification by "Norin Green No. 1" test plant through partial oxidation (ca. 10 MJ/Nm³; using O_2 and H_2O as gasifying agents) or conventional gasification using air as a gasifying agent (ca. 5 MJ/Nm³), can be produced when the reaction temperature is 800-900°C, H_2O/C mole ratio is lower than 5.0, and the reaction time is ca. 2 seconds. In addition, this gas mixture contains over 20% hydrogen (H_2). This value is higher than the threshold value of 10% for applicability in terms of ignition and combustion rate for gas engines and micro gas turbines, which indicate that the gas mixture is a high-quality gas fuel. Besides, given that the compositional ratio of H_2 to CO is higher, the threshold combustion temperature is 90°C higher than that of methane. Figure 13 explains a comparison of theoretical combustion temperatures of this gas mixture and various fuels, such as methane, gasoline, propane, methanol and ethanol.

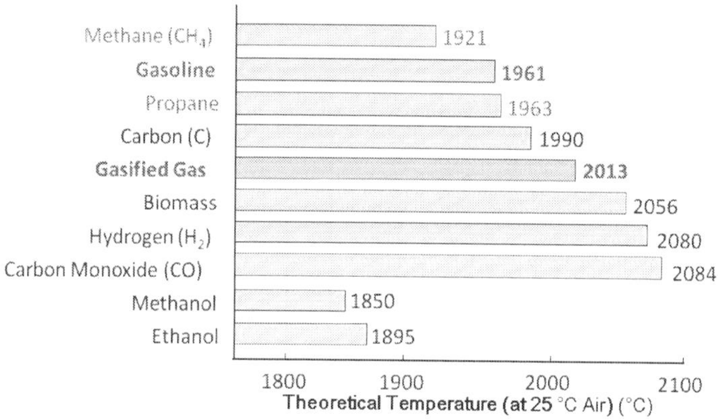

Fig. 13. Comparison of theoretical combustion temperature for gasified gas generated by Norin Biomass No. 3 Test Plant and various fuels.

Biomass means any form of lignocellulosic materials.

3. Conclusion

As the gas mixture generated with "Norin Green No. 1" test plant and high-calorie gas produced with "Norin Biomass No. 3" test plant using the high calorie gasification technology is temporarily stored in a cold gas state, it can be used in a manner similar to natural or city gas, with widespread applications.

Obviously, since "Norin Biomass No. 3" plant, which efficiently converts biomass into high calorie gas mixture with a small system, can be easily used as a fuel for gas engines and micro gas turbines, it can also be used for small-scale power generation and co-generation. Accordingly, high-efficiency and small-scale power generation can be achieved.

The potential applications of the gas mixture generated by gasification through high-calorie gas production are as follows;

1. Co-generation in buildings, hospitals, industrial parks, factories, etc.

2. Commercial power (targeted efficiency of 25-35%, with at least 1 million kWh/year)
3. Peak cut (reduction of contracted power) and emergency use for large-scale factories
4. Gas fuel for industrial parks (e.g. ceramics and porcelain)
5. Supplementary fuel for incinerator (dioxin countermeasure for industrial waste processing)
6. Fuel for boilers of greenhouse agriculture
7. Fuel for food processing industries by the use of residues produced in the process.
8. Synthesis of biomethanol for BDF production, for batteries of direct methanol fuel cell (DMFT), and a liquid fuel mixed with gasoline for flexible fuel vehicles (FFV).

Three "Norin biomass No. 3" plants processing 4-6 dry t/day of biomass feedstock are under construction by private companies and local government with the 50% financial support from the Government.

This study demonstrates that the gasification of readily available biomass materials both by partial oxidation technology and by high calorie gasification technology could be optimized for generation of gas mixtures primarily composed of H_2, CO and producing methanol yields ranging theoretically from ca. 40 to 60% by dry weight. A test plant utilizing gasification through partial oxidation with 2t/day gasifier can achieve a methanol yield of ca. 20% from the biomass raw material (by weight). This creates an opportunity to utilize a wide range of high yielding with low sugar and starch content such as Erianthus and Miscanthus. Non-palatable lignocellulosic byproducts such as sawdust and crop residues such as straw and husks of rice from various industries would also have suitable application. Sawdust, rice bran, refuse of sugarcane mills (bagasse etc.) and rice husks are particularly attractive and provide a ready-to-use biofuel resource. It is anticipated that the cultivation and utilization of biomass crops will be attractive as carbon neutral biomass feedstocks for biofuel production in the future.

The potentially positive economic impact of biomethanol production on Japanese farming and social systems from planting grasses and trees in unutilized land is immense (Nakagawa 2001; Harada 2001). Reduced CO_2 emissions, recycling of abandoned upland and paddy field and woodland in mountainous areas, and recycling of wastes of agricultural products would all be possible by promoting biofuel production system based on the gasification technologies. This technology is particularly attractive since biomethanol can be produced from a wide range of biomass raw materials.

4. Acknowledgements

Authors would like to express their sincere thanks to Dr. Bryan Kindiger, USDA-ARS, Grazinglands Research Laboratory for his critical reading of the manuscript.

This researches were supported by a grant from Ministry of Agriculture, Forestry and Fisheries of Japan, named "Development of sustainable ecosystem for primary industries towards the 21st century" (2000-2002), "Bio-recycle of wastes from agriculture, forestry, and fisheries" (2003-2005), and "Rural Biomass Research Project, BEC (Biomass Ethanol Conversion)" (2006-2010).

5. References

Binod, P., Sindhu, R., Singhania, R. R., Vikram, S., Devi, L., Nagalakshimi, S., Kurien, N., Sukumaran, R. K., Pandey, A. (2010) Bioethanol production from rice straw : an overview, *Bioresource Technology*, 101, 4767-4774.

Brown, L. R. et al. (2000). State of the world 2000, W. W. Norton & Company Ltd., New York, pp. 276.

Harada, Toshirou (2001). Utilization of wooden biomass resources as energy. *Farming Japan*, 35-2, 34-39.

Ishii, H., Takeno, K., Ichinose, T. (2005) Development of integrated system of biomass gasification using partial oxidizing process and liquid fuel synthesis with catalyst, *J. Jpn. Inst. Energy* 84: 420-425 (in Japanese with English abstract).

Keshwani, D. R. and Cheng, J. J. (2009) Switchgrass for bioethanol and other value-added applications: a review, *Bioresource Technology*, 100, 1515-1523.

Kosugi, A., Tanaka, R., Magara, K., Murata, Y., Arai, T., Sulaiman, O., Hashim, R., Abdul, H. Z. A., Azri, Y. M. K., Mohd, Y. M. N., Ibrahim, W. A., Mori, Y. (2010) Ethanol and lactic acid production using sap squeezed from old oil palm trunks felled for replanting, *J. Biosci. Bioeng.*, 110-3, 322-325.

Matsumoto, S. Takeno, K., Ichinose, T., Ishii, H., Nishimura, K. (2008) Analysis of a two ton/day test plant with an integrated biomass gasification and liquid fuel synthesis system, *Proceedings of 15th European Biomass Conference and Exhibition from Research to Market Development*, Berlin, Germany, 7-11 May, 2008.

Nakagawa, H. (2001). Development and cultivation of forage grasses and crops for clean bio-methanol production to keep global environment. *Farming Japan*, 35-2, 22-31.

Nakagawa, H., Harada, T., Ichinose, T., Takeno, K., Matsumoto, S., Kobayashi, M., Sakai, M. (2007) Biomethanol production and CO_2 emission reduction from forage grasses, trees, and crop residues, *JARQ*, 41(2), 173-180.

Nakagawa, H.; Sakai, M.; Harada, T.; Kitamura, Y.; Taniwaki, K. & Hashimoto, A (2000). Biomethanol production and CO_2 emission reduction from various forms of biomass. *Proceedings of the fourth International conference on ecobalance, Ecomaterials Forum, Society for Non-Traditional Technology*, pp. 405-408, Tokyo, Japan, October 31-November 2, 2000.

OECD (2010). OECD Factbook 2010, economic, environmental and social statistics. OECD, Paris, France, OECD Factbook 2010: Economic, Environmental and Social Statistics - ISBN 92-64-08356-1 - © OECD 2010.

Ogi, T. and Nakanishi, M. (2008) Biomass-derived liquid fuel via gasification), *Journal of JSES* 34 (5), 13-20 (in Japanese).

Sakai, M. (2001) Technological innovation of production method of liquefied biomass fuels: thermochemical production method of methanol fuels from any type of biomass. *Farming Japan*, 35-2, 10-15.

Sørensen, A., Teller, P. J., Hilstrøm, T., Ahring B. K. (2008) Hydrolysis of Miscanthus for bioethanol production using dilute acid presoaking combined with wet explosion pre-treatment and enzymatic treatment, *Bioresource Technology*, 99, 6602-6607.

Wyman, C., Dale, B. E., Elander, R. T., Holtzapple, M., Landisch, M. R., Lee, Y. Y., Mitchinson, C., Saddleer, J. N. (2009) Comparative sugar recovery and fermentation data following pretreatment of poplar wood by leading technologies, *Biotechnol. Prog.*, 25-2, 333 339.